Introduction to General and Organic Chemistry

Chapters 1-18

11th Edition

Bettelheim | Brown | Campbell | Farrell | Torres

Australia • Brazil • Japan • Korea • Mexico • Singapore • Spain • United Kingdom • United States

Introduction to General and Organic Chemistry, Chapters 1-18, 11th Edition

Introduction to General, Organic and Biochemistry, 11th Edition
Frederick A. Bettelheim | William H. Brown | Mary K. Campbell | Shawn O. Farrell | Omar Torres

ALL RIGHTS RESERVED. No part of this work covered by the copyright herein may be reproduced or distributed in any form or by any means, except as permitted by U.S. copyright law, without the prior written permission of the copyright owner.

> For product information and technology assistance, contact us at
> **Cengage Learning Customer & Sales Support, 1-800-354-9706**
> For permission to use material from this text or product,
> submit all requests online at **cengage.com/permissions**
> Further permissions questions can be emailed to
> **permissionrequest@cengage.com**

This book contains select works from existing Cengage Learning resources and was produced by Cengage Learning Custom Solutions for collegiate use. As such, those adopting and/or contributing to this work are responsible for editorial content accuracy, continuity and completeness.

Compilation © 2016 Cengage Learning

ISBN: 978-1-337-04669-5

Cengage Learning
20 Channel Center Street
Boston, MA 02210
USA

Cengage Learning is a leading provider of customized learning solutions with office locations around the globe, including Singapore, the United Kingdom, Australia, Mexico, Brazil, and Japan. Locate your local office at:
www.international.cengage.com/region.

Cengage Learning products are represented in Canada by Nelson Education, Ltd.

For your lifelong learning solutions, visit **www.cengage.com/custom.**

Visit our corporate website at **www.cengage.com.**

To Carolyn, with whom life is a joy. —WB

To my family and friends – thank you for all your support.

I couldn't have done it without you. — MC

To my lovely wife, Courtney – between textbook revisions,

a full-time job, and school, I have been little more than

a ghost around the house, hiding in my study writing.

Courtney held the family together, taking care of our children

and our home while maintaining her own writing schedule.

None of this would have been possible without her love,

support, and tireless effort. —SF

To my loving family and friends who have supported me through

this journey: Mom, Dad, Lisa, Abuela, René, Ryan,

and Dianne. I could not have made it without your urging

and support. I am truly blessed to have each of

you in my life. — OT

Contents in Brief

General Chemistry

Chapter 1 Matter, Energy, and Measurement   1
Chapter 2 Atoms   27
Chapter 3 Chemical Bonds   58
Chapter 4 Chemical Reactions   91
Chapter 5 Gases, Liquids, and Solids   117
Chapter 6 Solutions and Colloids   147
Chapter 7 Reaction Rates and Chemical Equilibrium   175
Chapter 8 Acids and Bases   200
Chapter 9 Nuclear Chemistry   233

Organic Chemistry

Chapter 10 Organic Chemistry   260
Chapter 11 Alkanes   273
Chapter 12 Alkenes and Alkynes   298
Chapter 13 Benzene and Its Derivatives   322
Chapter 14 Alcohols, Ethers, and Thiols   338
Chapter 15 Chirality: The Handedness of Molecules   358
Chapter 16 Amines   376
Chapter 17 Aldehydes and Ketones   389
Chapter 18 Carboxylic Acids   404

Contents

Chapter 1 Matter, Energy, and Measurement 1

- 1-1 Why Do We Call Chemistry the Study of Matter? 1
- 1-2 What Is the Scientific Method? 3
- 1-3 How Do Scientists Report Numbers? 5
  - *How To...* **Determine the Number of Significant Figures in a Number** 6
- 1-4 How Do We Make Measurements? 7
- 1-5 What Is a Handy Way to Convert from One Unit to Another? 12
  - *How To...* **Do Unit Conversions by the Factor-Label Method** 13
- 1-6 What Are the States of Matter? 17
- 1-7 What Are Density and Specific Gravity? 18
- 1-8 How Do We Describe the Various Forms of Energy? 20
- 1-9 How Do We Describe Heat and the Ways in Which It Is Transferred? 21
- Summary of Key Questions 26
- Problems 26

Chemical Connections

- 1A Drug Dosage and Body Mass 11
- 1B Hypothermia and Hyperthermia 22
- 1C Cold Compresses, Waterbeds, and Lakes 23

Chapter 2 Atoms 27

- 2-1 What Is Matter Made Of? 27
- 2-2 How Do We Classify Matter? 28
- 2-3 What Are the Postulates of Dalton's Atomic Theory? 31
- 2-4 What Are Atoms Made Of? 34
- 2-5 What Is the Periodic Table? 39
- 2-6 How Are the Electrons in an Atom Arranged? 45
- 2-7 How Are Electron Configuration and Position in the Periodic Table Related? 52
- 2-8 What Is a Periodic Property? 53
- Summary of Key Questions 56
- Problems 57a

Chemical Connections

- 2A Elements Necessary for Human Life 29
- 2B Abundance of Elements Present in the Human Body and in the Earth's Crust 33
- 2C Strontium-90 43
- 2D The Use of Metals as Historical Landmarks 44

Chapter 3 Chemical Bonds 58

- 3-1 What Do We Need to Know Before We Begin? 58
- 3-2 What Is the Octet Rule? 58
- 3-3 How Do We Name Anions and Cations? 61
- 3-4 What Are the Two Major Types of Chemical Bonds? 63
- 3-5 What Is an Ionic Bond? 65
- 3-6 How Do We Name Ionic Compounds? 67
- 3-7 What Is a Covalent Bond? 70
  - *How To...* **Draw Lewis Structures** 73
- 3-8 How Do We Name Binary Covalent Compounds? 77
- 3-9 What Is Resonance? 78
  - *How To...* **Draw Curved Arrows and Push Electrons** 80
- 3-10 How Do We Predict Bond Angles in Covalent Molecules? 82
- 3-11 How Do We Determine If a Molecule Is Polar? 87
- Summary of Key Questions 89
- Problems 90

Chemical Connections

- 3A Coral Chemistry and Broken Bones 63
- 3B Ionic Compounds in Medicine 70
- 3C Nitric Oxide: Air Pollutant and Biological Messenger 79

Chapter 4 Chemical Reactions 91

- 4-1 What Is a Chemical Reaction? 91
- 4-2 How Do We Balance Chemical Equations? 92
  - *How To...* **Balance a Chemical Equation** 92
- 4-3 How Can We Predict If Ions in Aqueous Solution Will React with Each Other? 96

4-4 What Are Oxidation and Reduction? 99
4-5 What Are Formula Weights and Molecular Weights? 104
4-6 What Is a Mole and How Do We Use It to Calculate Mass Relationships? 105
4-7 How Do We Calculate Mass Relationships in Chemical Reactions? 109
4-8 What Is Heat of Reaction? 115

Summary of Key Questions 116
Problems 116a

Chemical Connections
4A Solubility and Tooth Decay 99
4B Voltaic Cells 102
4C Artificial Pacemakers and Redox 103

Chapter 5 Gases, Liquids, and Solids 117

5-1 What Are the Three States of Matter? 117
5-2 What Is Gas Pressure and How Do We Measure It? 118
5-3 What Are the Laws That Govern the Behavior of Gases? 119
5-4 What Are Avogadro's Law and the Ideal Gas Law? 123
5-5 What Is Dalton's Law of Partial Pressures? 125
5-6 What Is the Kinetic Molecular Theory? 127
5-7 What Types of Intermolecular Attractive Forces Exist Between Molecules? 128
5-8 How Do We Describe the Behavior of Liquids at the Molecular Level? 132
5-9 What Are the Characteristics of the Various Types of Solids? 139
5-10 What Is a Phase Change and What Energies Are Involved? 141

Summary of Key Questions 146
Problems 146a

Chemical Connections
5A Breathing and Boyle's Law 120
5B Hyperbaric Medicine 126
5C Blood Pressure Measurement 134
5D The Densities of Ice and Water 137
5E Supercritical Carbon Dioxide 145

Chapter 6 Solutions and Colloids 147

6-1 What Do We Need to Know as We Begin? 147
6-2 What Are the Most Common Types of Solutions? 148
6-3 What Are the Distinguishing Characteristics of Solutions? 148
6-4 What Factors Affect Solubility? 150
6-5 What Are the Most Common Units for Concentration? 153
6-6 Why Is Water Such a Good Solvent? 159
6-7 What Are Colloids? 164
6-8 What Is a Colligative Property? 166

Summary of Key Questions 173
Problems 174

Chemical Connections
6A Acid Rain 149
6B The Bends 152
6C Electrolyte Solutions in Body and Intravenous Fluids 160
6D Hydrates and Air Pollution: The Decay of Buildings and Monuments 163
6E Emulsions and Emulsifying Agents 166
6F Reverse Osmosis and Desalinization 171
6G Hemodialysis 173

Chapter 7 Reaction Rates and Chemical Equilibrium 175

7-1 How Do We Measure Reaction Rates? 175
7-2 Why Do Some Molecular Collisions Result in Reaction Whereas Others Do Not? 177
7-3 What Is the Relationship Between Activation Energy and Reaction Rate? 179
7-4 How Can We Change the Rate of a Chemical Reaction? 181
7-5 What Does It Mean to Say That a Reaction Has Reached Equilibrium? 185
7-6 What Is an Equilibrium Constant and How Do We Use It? 188

How To... Interpret the Value of the Equilibrium Constant, K 191

7-7 What Is Le Chatelier's Principle? 193

Summary of Key Questions 199
Problems 199a

Chemical Connections
7A Why High Fever Is Dangerous 184
7B The Effects of Lowering Body Temperature 186
7C Timed-Release Medication 187
7D Sunglasses and Le Chatelier's Principle 196
7E The Haber Process 198

Chapter 8 Acids and Bases 200

8-1 What Are Acids and Bases? 200
8-2 How Do We Define the Strength of Acids and Bases? 202
8-3 What Are Conjugate Acid–Base Pairs? 204

How To... Name Common Acids 206

8-4 How Can We Tell the Position of Equilibrium in an Acid–Base Reaction? 207
8-5 How Do We Use Acid Ionization Constants? 209

How To... Use Logs and Antilogs 210

Contents

- 8-6 What Are the Properties of Acids and Bases? 212
- 8-7 What Are the Acidic and Basic Properties of Pure Water? 215
- 8-8 What Are pH and pOH? 217
- 8-9 How Do We Use Titrations to Calculate Concentration? 220
- 8-10 What Are Buffers? 222
- 8-11 How Do We Calculate the pH of a Buffer? 226
- 8-12 What Are TRIS, HEPES, and These Buffers with the Strange Names? 228

  Summary of Key Questions 231
  Problems 232

  ### Chemical Connections
  - 8A Some Important Acids and Bases 203
  - 8B Drugstore Antacids 216
  - 8C Respiratory and Metabolic Acidosis 229
  - 8D Alkalosis and the Sprinter's Trick 231

Chapter 9 Nuclear Chemistry 203

- 9-1 How Was Radioactivity Discovered? 233
- 9-2 What Is Radioactivity? 234
- 9-3 What Happens When a Nucleus Emits Radioactivity? 235

  How To... Balance a Nuclear Equation 237

- 9-4 What Is Nuclear Half-Life? 240
- 9-5 How Do We Detect and Measure Nuclear Radiation? 243
- 9-6 How Is Radiation Dosimetry Related to Human Health? 246
- 9-7 What Is Nuclear Medicine? 249
- 9-8 What Is Nuclear Fusion? 254
- 9-9 What Is Nuclear Fission and How Is It Related to Atomic Energy? 256

  Summary of Key Questions 258
  Summary of Key Reactions 259
  Problems 259a

  ### Chemical Connections
  - 9A Radioactive Dating 242
  - 9B The Indoor Radon Problem 248
  - 9C How Radiation Damages Tissues: Free Radicals 250
  - 9D Magnetic Resonance Imaging 252
  - 9E Radioactive Fallout from Nuclear Accidents 258

Chapter 10 Organic Chemistry 260

- 10-1 What Is Organic Chemistry? 260
- 10-2 Where Do We Obtain Organic Compounds? 262
- 10-3 How Do We Write Structural Formulas of Organic Compounds? 264
- 10-4 What Is a Functional Group? 266

  Summary of Key Questions 272
  Problems 272

  ### Chemical Connections
  - 10A Taxol: A Story of Search and Discovery 263

Chapter 11 Alkanes 273

- 11-1 What Are Alkanes? 273
- 11-2 How Do We Write Structural Formulas of Alkanes? 274
- 11-3 What Are Constitutional Isomers? 275
- 11-4 How Do We Name Alkanes? 278
- 11-5 Where Do We Obtain Alkanes? 282
- 11-6 What Are Cycloalkanes? 282
- 11-7 What Are the Shapes of Alkanes and Cycloalkanes? 284

  How To... Draw Alternative Chair Conformations of Cyclohexane 286

- 11-8 What Is *Cis-Trans* Isomerism in Cycloalkanes? 288
- 11-9 What Are the Physical Properties of Alkanes and Cycloalkanes? 290
- 11-10 What Are the Characteristic Reactions of Alkanes? 293
- 11-11 What Are Some Important Haloalkanes? 295

  Summary of Key Questions 296
  Summary of Key Reactions 297
  Problems 297a

  ### Chemical Connections
  - 11A The Poisonous Puffer Fish 287
  - 11B Octane Rating: What Those Numbers at the Pump Mean 293
  - 11C The Environmental Impact of Freons 295

Chapter 12 Alkenes and Alkynes 298

- 12-1 What Are Alkenes and Alkynes? 298
- 12-2 What Are the Structures of Alkenes and Alkynes? 299
- 12-3 How Do We Name Alkenes and Alkynes? 301
- 12-4 What Are the Physical Properties of Alkenes and Alkynes? 308
- 12-5 What Are the Characteristic Reactions of Alkenes? 308
- 12-6 What Are the Important Polymerization Reactions of Ethylene and Substituted Ethylenes? 317

  Summary of Key Questions 320
  Summary of Key Reactions 321
  Problems 321

  ### Chemical Connections
  - 12A Ethylene: A Plant Growth Regulator 299
  - 12B *Cis* Double Bonds in Unsaturated Fatty Acids 300
  - 12C The Case of the Iowa and New York Strains of the European Corn Borer 305
  - 12D *Cis-Trans* Isomerism in Vision 307
  - 12E Recycling Plastics 318

Chapter 13 Benzene and Its Derivatives 322

- 13-1 What Is the Structure of Benzene? 322
- 13-2 How Do We Name Aromatic Compounds? 324
- 13-3 What Are the Characteristic Reactions of Benzene and Its Derivatives? 328
- 13-4 What Are Phenols? 330

  Summary of Key Questions 337
  Summary of Key Reactions 337
  Problems 337

  ### Chemical Connections

  - 13A DDT: A Boon and a Curse 327
  - 13B Carcinogenic Polynuclear Aromatic Hydrocarbons and Smoking 328
  - 13C Iodide Ion and Goiter 329
  - 13D The Nitro Group in Explosives 330
  - 13E FD & C No. 6 (a.k.a. Sunset Yellow) 334
  - 13F Capsaicin, for Those Who Like It Hot 335

Chapter 14 Alcohols, Ethers, and Thiols 338

- 14-1 What Are the Structures, Names, and Physical Properties of Alcohols? 339
- 14-2 What Are the Characteristic Reactions of Alcohols? 342
- 14-3 What Are the Structures, Names, and Physical Properties of Ethers? 349
- 14-4 What Are the Structures, Names, and Physical Properties of Thiols? 353
- 14-5 What Are the Most Commercially Important Alcohols? 355

  Summary of Key Questions 356
  Summary of Key Reactions 357
  Problems 357

  ### Chemical Connections

  - 14A The Importance of Hydrogen Bonding in Drug-Receptor Interactions 343
  - 14B Nitroglycerin: An Explosive and a Drug 344
  - 14C Breath-Alcohol Screening 349
  - 14D Ethylene Oxide: A Chemical Sterilant 351
  - 14E Ethers and Anesthesia 352

Chapter 15 Chirality: The Handedness of Molecules 358

- 15-1 What Is Enantiomerism? 358
  - *How To... Draw Enantiomers* 362
- 15-2 How Do We Specify the Configuration of a Stereocenter? 365
- 15-3 How Many Stereoisomers Are Possible for Molecules with Two or More Stereocenters? 368
- 15-4 What Is Optical Activity, and How Is Chirality Detected in the Laboratory? 372
- 15-5 What Is the Significance of Chirality in the Biological World? 374

  Summary of Key Questions 375
  Problems 375a

  ### Chemical Connections

  - 15A Chiral Drugs 372

Chapter 16 Amines 376

- 16-1 What Are Amines? 376
- 16-2 How Do We Name Amines? 379
- 16-3 What Are the Physical Properties of Amines? 381
- 16-4 How Do We Describe the Basicity of Amines? 382
- 16-5 What Are the Characteristic Reactions of Amines? 384

  Summary of Key Questions 388
  Summary of Key Reactions 388
  Problems 388

  ### Chemical Connections

  - 16A Amphetamines (Pep Pills) 377
  - 16B Alkaloids 378
  - 16C Tranquilizers 382
  - 16D The Solubility of Drugs in Body Fluids 385
  - 16E Epinephrine: A Prototype for the Development of New Bronchodilators 387

Chapter 17 Aldehydes and Ketones 389

- 17-1 What Are Aldehydes and Ketones? 389
- 17-2 How Do We Name Aldehydes and Ketones? 390
- 17-3 What Are the Physical Properties of Aldehydes and Ketones? 393
- 17-4 What Are the Characteristic Reactions of Aldehydes and Ketones? 394
- 17-5 What Is Keto-Enol Tautomerism? 401

  Summary of Key Questions 402
  Summary of Key Reactions 402
  Problems 403

  ### Chemical Connections

  - 17A From Moldy Clover to a Blood Thinner 393

Chapter 18 Carboxylic Acids 404

- 18-1 What Are Carboxylic Acids? 404
- 18-2 How Do We Name Carboxylic Acids? 404
- 18-3 What Are the Physical Properties of Carboxylic Acids? 407
- 18-4 What Are Soaps and Detergents? 408
- 18-5 What Are the Characteristic Reactions of Carboxylic Acids? 414

  Summary of Key Questions 421
  Summary of Key Reactions 421
  Problems 422

Chemical Connections

18A *Trans* Fatty Acids: What Are They and How Do You Avoid Them? 410
18B Esters as Flavoring Agents 418
18C Ketone Bodies and Diabetes 421

Appendix I Exponential Notation A1
Appendix II Significant Figures A5

Answers to In-Text and Odd-Numbered End-of-Chapter Problems A8
Glossary G1

Preface

The cure for boredom is curiosity
There is no cure for curiosity.
—DOROTHY PARKER

Perceiving order in nature is a deep-seated human need. It is our primary aim to convey the relationship among facts and thereby present a totality of the scientific edifice built over the centuries. In this process, we marvel at the unity of laws that govern everything in ever-exploding dimensions: from photons to protons, from hydrogen to water, from carbon to DNA, from genome to intelligence, from our planet to the galaxy and to the known Universe. Unity in all diversity.

As we prepare the eleventh edition of our textbook, we cannot help but be struck by the changes that have taken place in the last 40 years. From the slogan of the '70s, "Better living through chemistry," to today's saying, "Life by chemistry," one can sense the change in the focus. Chemistry helps to provide not just the amenities of a good life, but it is at the core of our conception of and preoccupation with life itself. This shift in emphasis demands that our textbook, designed primarily for the education of future practitioners of health sciences, should attempt to provide both the basics as well as a scope of the horizon within which chemistry touches our lives.

The increasing use of our textbook made this new edition possible, and we wish to thank our colleagues who adopted the previous editions for their courses. Testimony from colleagues and students indicates that we managed to convey our enthusiasm for the subject to students, who find this book to be a great help in studying difficult concepts.

Therefore, in the new edition we strive further to present an easily readable and understandable text along with more application problems related to health sciences. At the same time, we emphasize the inclusion of new relevant concepts and examples in this fast-growing discipline, especially in the biochemistry chapters. We maintain an integrated view of chemistry. From the very beginning of the book, we include organic compounds and biochemical substances to illustrate basic principles. This progression ascends from the simple to the complex. We urge our colleagues to advance to the chapters of biochemistry as fast as possible, because there lies most of the material that is relevant to the future professions of our students.

Dealing with such a giant field in one course, and possibly the only course in which our students get an exposure to chemistry, makes the selection of the material an overarching enterprise. We are aware that even though we tried to keep the book to a manageable size and proportion, we included more topics than could be covered in one course. Our aim was to provide enough material from which the instructor can select the topics he or she deems important. The wealth of problems, both drill and challenging, provide students with numerous ways to test their knowledge from a variety of angles.

Audience and Unified Approach

This book is intended for non-chemistry majors, mainly those entering health sciences and related fields, such as nursing, medical technology, physical therapy, and nutrition. In its entirety, it can be used for a one-year (two-semester or three-quarter) course in chemistry, or parts of the book can be used in a one-term chemistry course.

We assume that the students using this book have little or no background in chemistry. Therefore, we introduce the basic concepts slowly at the beginning and increase the tempo and the level of sophistication as we go on. We progress from the basic tenets of general chemistry to organic and then to biochemistry. Throughout, we integrate the parts by keeping a unified view of chemistry. For example, we frequently use organic and biological substances to illustrate general principles.

While teaching the chemistry of the human body is our ultimate goal, we try to show that each subsection of chemistry is important in its own right, besides being necessary for understanding future topics.

Chemical Connections (Medical and Other Applications of Chemical Principles)

The Chemical Connections boxes contain applications of the principles discussed in the text. Comments from users of earlier editions indicate that these boxes have been especially well received, and provide a much-requested relevance to the text. For example, in Chapter 1, students can see how cold compresses relate to waterbeds and to lake temperatures (Chemical Connections 1C). New up-to-date topics include coverage of omega-3 fatty acids and heart disease (Chemical Connections 21F), and the search for treatments for cystic fibrosis (Chemical Connections 26F).

The inclusion of Chemical Connections allows for a considerable degree of flexibility. If an instructor wants to assign only the main text, the Chemical Connections do not interrupt continuity, and the essential material will be covered. However, because they enhance the core material, most instructors will probably wish to assign at least some of the Chemical Connections. In our experience, students are eager to read the relevant Chemical Connections, without assignments, and they do with discrimination. From such a large number of boxes, an instructor can select those that best fit the particular needs of the course. So that students can test their knowledge, we provide problems at the end of each chapter for all of the Chemical Connections; these problems are now identified within the boxes.

Metabolism: Color Code

The biological functions of chemical compounds are explained in each of the biochemistry chapters and in many of the organic chapters. Emphasis is placed on chemistry rather than physiology. Positive feedback about the organization of the metabolism chapters has encouraged us to maintain the order (Chapters 26–28).

First, we introduce the common metabolic pathway through which all food is utilized (the citric acid cycle and oxidative phosphorylation), and only after that do we discuss the specific pathways leading to the common pathway. We find this a useful pedagogic device, and it enables us to sum the caloric values of each type of food because its utilization through the common pathway has already been learned. Finally, we separate the catabolic pathways from the anabolic pathways by treating them in different chapters, emphasizing the different ways the body breaks down and builds up different molecules.

The topic of metabolism is a difficult one for most students, and we have tried to explain it as clearly as possible. We enhance the clarity of presentation by the use of a color code for the most important biological compounds. Each type of compound is screened in a specific color, which remains the same throughout the three chapters. These colors are as follows:

In figures showing metabolic pathways, we display the numbers of the various steps in yellow. In addition to this main use of a color code, other figures in various parts of the book are color coded so that the same color is used for the same entity throughout. For example, in all figures that show enzyme–substrate interactions, enzymes are always shown in blue and substrates in orange.

Features

- **Problem-Solving Strategies** The in-text examples include a description of the strategy used to arrive at a solution. This will help students organize the information in order to solve the problem.
- **Visual Impact** We have introduced illustrations with heightened pedagogical impact. Some of these show the microscopic and macroscopic aspects of a topic under discussion, such as Figures 6-4 (Henry's Law) and 6-11 (electrolytic conductance). The Chemical Connections essays have been enhanced further with more photos to illustrate each topic.
- **Key Questions** We use a Key Questions framework to emphasize key chemical concepts. This focused approach guides students through each chapter by using section head questions.
- **[UPDATED] Chemical Connections** Over 150 essays describe applications of chemical concepts presented in the text, linking the chemistry to their real uses. Many new application boxes on diverse topics were added.
- **Summary of Key Reactions** In each organic chemistry chapter (10–19) there is an annotated summary of all the new reactions introduced. Keyed to sections in which they are introduced, there is also an example of each reaction.
- **Chapter Summaries** Summaries reflect the Key Questions framework. At the end of each chapter, the Key Questions are restated and the summary paragraphs that follow are designed to highlight the concepts associated with the questions.
- **Looking Ahead Problems** At the end of most chapters, the challenge problems are designed to show the application of principles in the chapter to material in the following chapters.
- **Tying-It-Together and Challenge Problems** At the end of most chapters, these problems build on past material to test students' knowledge of these concepts. In the Challenge Problems, associated chapter references are given.

- **How To Boxes** These boxes emphasize the skills students need to master the material. They include topics such as, "How to Determine the Number of Significant Figures in a Number" (Chapter 1) and "How to Draw Enantiomers" (Chapter 15).
- **Molecular Models** Ball-and-stick models, space-filling models, and electron-density maps are used throughout the text as appropriate aids for visualizing molecular properties and interactions.
- **Margin Definitions** Many terms are also defined in the margin to help students learn terminology. By skimming the chapter for these definitions, students will have a quick summary of its contents.
- **Answers to all in-text and odd-numbered end-of-chapter problems** Answers to selected problems are provided at the end of the book. Detailed worked-out solutions to these same problems are provided in the Student Solutions Manual.
- **Glossary** The glossary at the back of the book gives a definition of each new term along with the number of the section in which the term is introduced.

Organization and Updates

General Chemistry (Chapters 1–9)

- **Chapter 1, Matter, Energy, and Measurement,** serves as a general introduction to the text and introduces the pedagogical elements that are new to this edition, with an emphasis on solving conversion problems related to a clinical setting. Six new problems were added.
- In **Chapter 2, Atoms,** we introduce four of the five ways used to represent molecules throughout the text: we show water as a molecular formula, a structural formula, a ball-and-stick model, and a space-filling model. Twelve new problems were added.
- **Chapter 3, Chemical Bonds,** begins with a discussion of ionic compounds, followed by a discussion of molecular compounds. Twenty-one new problems were added.
- **Chapter 4, Chemical Reactions,** introduces the various intricacies in writing and balancing chemical reactions before stoichiometry is introduced. This chapter includes the How To box, "How to Balance a Chemical Equation," which illustrates a step-by-step method for balancing an equation. Twenty-three new challenge problems were added.
- In **Chapter 5, Gases, Liquids, and Solids,** we present intermolecular forces of attraction in order of increasing energy, namely London dispersion forces, dipole–dipole interactions, and hydrogen bonding. Fifteen new challenge problems were added.
- **Chapter 6, Solutions and Colloids,** opens with a listing of the most common types of solutions, followed by a discussion of the factors that affect solubility and the most common units for concentration, and closes with an enhanced discussion of colligative properties. Seven new challenge problems were added.
- **Chapter 7, Reaction Rates and Chemical Equilibrium,** shows how these two important topics are related to one another. A How To box shows how to interpret the value of the equilibrium constant, K. In addition, six new problems were added.
- **Chapter 8, Acids and Bases,** introduces the use of curved arrows to show the flow of electrons in organic reactions. Specifically, we use them

here to show the flow of electrons in proton-transfer reactions. The major theme in this chapter is the discussion of acid–base buffers and the Henderson–Hasselbalch equation. Information was added on solving problems using the activity series, along with six new problems.

- **Chapter 9, Nuclear Chemistry,** highlights nuclear applications in medicine. A new Chemical Connections box on magnetic resonance imaging (MRI) was added, along with two new problems.

Organic Chemistry (Chapters 10–19)

- **Chapter 10, Organic Chemistry,** is an introduction to the characteristics of organic compounds and to the most important organic functional groups. The list of common organic functional groups has been expanded in this edition to include amides, and a schematic of a tripeptide is presented to illustrate the importance of amide bonds in the structure of polypeptides and proteins.
- In **Chapter 11, Alkanes,** we introduce the concept of a line-angle formula, which we will continue to use throughout the organic chapters. These are easier to draw than the usual condensed structural formulas and are easier to visualize. The discussion on the conformation of alkanes has been reduced and instead concentrates on the conformations of cycloalkanes. Chemical Connections Box 11C, "The Environmental Impact of Freons," has been extended to include some possible replacements for refrigerant gases and their ozone-depleting potential.
- In **Chapter 12, Alkenes and Alkynes,** we introduce a new, simple way of looking at reaction mechanisms: add a proton, take a proton away, break a bond, and make a bond. The purpose of this introduction to reaction mechanisms is to demonstrate to students that chemists are interested not only in what happens in a chemical reaction, but also in how it happens. We refined the discussion of these reaction mechanisms in this edition and added a new problem to the end-of-chapter exercise about a compound once used as a flame retardant in polystyrene-foam building insulation and why its use is now prohibited.
- **Chapter 13, Benzene and Its Derivatives,** includes a discussion of phenols and antioxidants. A short history of chemistry was added for this edition that discusses structures proposed in the 19th century as alternatives to those proposed by Kekulé. We lengthened the discussion on the reactions of phenols to include the oxidation of phenols to quinones, the use of the hydroquinone-to-quinone interconversion in black-and-white photography, the role of Coenzyme Q (ubiquinone) in the respiratory chain as a carrier of electrons, and the structure of vitamin K and its role in blood clotting. A new problem at the end of the chapter challenges students to propose additional alternative structures for C_6H_6 consistent with the tetravalence of carbon.
- **Chapter 14, Alcohols, Ethers, and Thiols,** discusses the structures, names, and properties of alcohols first, and then gives a similar treatment to ethers, and finally thiols. A new Chemical Connections box, "The Importance of Hydrogen Bonding in Drug-Receptor Interactions," was added, and several new puzzle problems dealing with the interconversion of alcohol and alkenes were added to the end-of-chapter exercises.
- In **Chapter 15, Chirality: The Handedness of Molecules,** the concept of a stereocenter and enantiomerism is slowly introduced, using 2-butanol as a prototype. We then treat molecules with two or more stereocenters and show how to predict the number of stereoisomers possible for a particular molecule. We also explain R,S convention for assigning absolute

configuration to a tetrahedral stereocenter. The discussion on the structure and stereochemistry of Tamiflu has been expanded, and a problem showing the different odors of the enantiomers of carvone was added.

- In **Chapter 16, Amines,** we trace the development of new asthma medications from epinephrine, which can be viewed as a historical precursor to albuterol (Proventil).
- **Chapter 17, Aldehydes and Ketones,** has a discussion of $NaBH_4$ as a carbonyl-reducing agent with emphasis on its use as a hydride-transfer reagent. We then make the parallel to NADH as a carbonyl-reducing agent and hydride-transfer agent. In this edition, the discussion on the mechanism of the acid-catalyzed formation of acetals has been broadened.
- **Chapter 18, Carboxylic Acids,** focuses on the chemistry and physical properties of carboxylic acids. There is a brief discussion of *trans* fatty acids, omega-3 fatty acids, and the significance of their presence in our diets. The discussion on carboxylic acids has been expanded to include molecules that contain an aldehyde or ketone group in addition to a carboxyl group.
- **Chapter 19, Carboxylic Anhydrides, Esters, and Amides,** describes the chemistry of these three important functional groups with emphasis on their acid-catalyzed and base-promoted hydrolysis and reactions with amines and alcohols. A short presentation was inserted on the structure and nomenclature of lactones and how to recognize them when they are embedded in macromolecules, and two new end-of-chapter problems have been added: Problem 19-48, which illustrates how an insect utilizes a plant-derived chemical as a raw material from which to synthesize a compound that impacts its species survival, and Problem 19-49, which describes polyester polymers that are biodegradable by microbial enzymes by composting.

Biochemistry (Chapters 20–31)

- **Chapter 20, Carbohydrates,** begins with the structure and nomenclature of monosaccharides, including their oxidation, reduction, and the formation of glycosides, then concludes with a discussion of the structure of disaccharides, polysaccharides, and acidic polysaccharides. The descriptions of these structures, especially glucose stereochemistry, have been clarified in this edition. Eight new end-of-chapter problems were added.
- **Chapter 21, Lipids,** covers the most important features of lipid biochemistry, including membrane structure and the structures and functions of steroids. In this edition, we have stressed the need for students to recall material from earlier chapters, especially structure and reactions of carboxylic acids. The chapter also has an increased emphasis on membrane transport and an update on possible classification of *trans* fatty acids as food additives. The Chemical Connections "Anabolic Steroids" has been updated to reflect new and continuing incidents in professional sports. One new end-of-chapter exercise has been added.
- **Chapter 22, Proteins,** covers the many facets of protein structure and function. It gives an overview of how proteins are organized, beginning with the nature of individual amino acids and how this organization leads to their many functions. This supplies the student with the basics needed to lead into the sections on enzymes and metabolism. Points causing difficulty for students in the last edition, mostly pertaining to the roles of amino acids in proteins and bonding in transition-metal complexes, have been clarified. Eight new end-of-chapter problems were added.

- **Chapter 23, Enzymes,** covers the important topic of enzyme catalysis and regulation. This discussion has been modified for a stronger correlation with pathways to be discussed in Chapter 28. Specific medical applications of enzyme inhibition are included, as well as an introduction to the fascinating topic of transition-state analogs and their use as potent inhibitors. One of these medical applications is enhanced in an updated section on the use of abzymes in treatment of AIDS. A new Chemical Connections box discussing enzyme mechanisms using chymotrypsin was included as an example, and a new figure was added to illustrate the binding of effectors to allosteric enzymes. Nine new end-of-chapter problems were added.

- In **Chapter 24, Chemical Communications,** we see the biochemistry of hormones and neurotransmitters. This chapter has been reorganized for better flow in introducing the different ways of classifying neurotransmitters. The health-related implications of how these substances act in the body is the main focus of this chapter. Along with a new Chemical Connections box focusing on Alzheimer's disease and diabetes, a new section on the fight against depression was added.

- **Chapter 25, Nucleotides, Nucleic Acids, and Heredity,** introduces DNA and the processes encompassing its replication and repair. How nucleotides are linked together and the flow of genetic information due to the unique properties of these molecules is emphasized. The sections on the types of RNA have been expanded again as our knowledge increases daily about these important nucleic acids. This edition introduces three of the newest RNA types to be discovered: long non-Coding RNA, Piwi-associated RNA, and circular RNA. The uniqueness of an individual's DNA is described with a Chemical Connections box that introduces DNA fingerprinting and how forensic science relies on DNA for positive identification.

- **Chapter 26, Gene Expression and Protein Synthesis,** shows how the information contained in the DNA blueprint of a cell is used to produce RNA and, eventually, protein. The focus is on how organisms control the expression of genes through transcription and translation. A new section was added on epigenetics, one of the hottest topics in the field. Two new Chemical Connections boxes were added. The first explores how protein synthesis is related to creating memories. The second expands on the importance of epigenetics by looking at how it affects disease states.

- **Chapter 27, Bioenergetics,** is an introduction to metabolism that focuses strongly on the central pathways, namely the citric acid cycle, electron transport, and oxidative phosphorylation. A new Chemical Connections box on role of ATP in cell signaling was added, along with four new end-of-chapter problems.

- In **Chapter 28, Specific Catabolic Pathways,** we address the details of carbohydrate, lipid, and protein breakdown, concentrating on energy yield. A new Chemical Connections box on manipulating carbohydrate metabolism to treat obesity was added, and four new end-of-chapter exercises were also included.

- **Chapter 29, Biosynthetic Pathways,** starts with a general consideration of anabolism and proceeds to carbohydrate biosynthesis in both plants and animals. Lipid biosynthesis is linked to the production of membranes, and the chapter concludes with an account of amino-acid biosynthesis. A new Chemical Connections box on statin drugs as inhibitors of cholesterol biosynthesis was inserted, and two new end-of-chapter exercises were also included.

- In **Chapter 30, Nutrition,** we take a biochemical approach to understanding nutrition concepts. Along the way, we look at a revised version of the Food Guide Pyramid and debunk some of the myths about carbohydrates and fats. A new Chemical Connections box on the causes of obesity was added.

- **Chapter 31, Immunochemistry,** covers the basics of our immune system and how we protect ourselves from foreign invading organisms. Considerable time is spent on the acquired immunity system. No chapter on immunology would be complete without a description of the Human Immunodeficiency Virus. The chapter includes a new section on immunization and a new Chemical Connections box, "Immunology and Oncology." The chapter has also been shortened and streamlined to make some of the very technical material simpler to digest.

- **Chapter 32, Body Fluids**
  To access this online-only chapter, search for ISBN 978-1-285-86975-9 at www.cengagebrain.com and visit this book's companion website.

Supporting Materials

Please visit http://www.cengage.com/chemistry/bettelheim/gob11E for information about the student and instructor resources for this text.

Acknowledgments

The publication of a book such as this requires the efforts of many more people than merely the authors. We would like to thank the following professors who offered many valuable suggestions for this new edition:

We are especially grateful for David Shinn, United States Merchant Marine Academy, and Jordan Fantini, Denison University, who read page proofs with eyes for accuracy.

We give special thanks to Alyssa White, Content Developer, who has been a rock of support through the entire revision process. We appreciate her constant encouragement as we worked to meet deadlines; she has also been a valuable resource person. We appreciate the help of our other colleagues at Cengage Learning: Teresa Trego—Senior Content Project Manager and Maureen Rosener—Senior Product Manager. We would also like to give special thanks to Matt Rosenquist, our Production Editor at Graphic World, Inc.

We so appreciate the time and expertise of our reviewers who have read our manuscript and given us helpful comments. They include:

Reviewers of the 11th Edition:

Jennifer Barber, *Atlanta Metropolitan State College*

Ling Chen, *Borough of Manhattan Community College*

Kyle Craig, *Walla Walla University*

Sidnee-Marie Dunn, *South Puget Sound Community College*

Timothy Marshall, *Pima Community College*

Lynda Peebles, *Texas Woman's University*

Rill Reuter, *Winona State University*

Susan Sawyer, *Kellogg Community College*

Theresa Thewes, *Edinboro University of Pennsylvania*

Reviewers of the 10th Edition:

Julian Davis, *University of the Incarnate Word*
Robert Keil, *Moorpark College*
Margaret Kimble, *Indiana University–Purdue University Fort Wayne*
Bette Kruez, *University of Michigan, Dearborn*
Timothy Marshall, *Pima Community College*
Donald Mitchell, *Delaware Technical and Community College*
Paul Root, *Henry Ford Community College*
Ahmed Sheikh, *West Virginia University*
Steven Socol, *McHenry County College*
Susan Thomas, *University of Texas–San Antonio*
Holly Thompson, *University of Montana*
Janice Webster, *Ivy Tech Community College*

Reviewers of the 9th Edition:

Allison J. Dobson, *Georgia Southern University*
Sara M. Hein, *Winona State University*
Peter Jurs, *Pennsylvania State University*
Delores B. Lamb, *Greenville Technical College*
James W. Long, *University of Oregon*
Richard L. Nafshun, *Oregon State University*
David Reinhold, *Western Michigan University*
Paul Sampson, *Kent State University*
Garon C. Smith, *University of Montana*
Steven M. Socol, *McHenry County College*

About the Cover

Honeycomb consists of a repeating structural pattern of hexagons, and just as hexagons are the repeating structural unit of honeycomb, the hexagonal-six-membered rings of cyclohexane and benzene are repeating structural units of the organic and bioorganic molecules that make up the molecules of the natural world around us. Honey, which fills each hexagonal unit of honeycomb, derives its sweetness from the carbohydrates glucose and fructose ($C_6H_{12}O_6$). Glucose and fructose have the same molecular formula and both can exist in nature in the form of six-membered-rings as seen on the cover, which shows a six-membered ring form of fructose. Chemically, honey is not only a nutritional powerhouse, but it also has medicinal uses due to its antibacterial and healing properties. Another major component of honeycomb is beeswax, or triacontyl palmitate, $CH_3(CH_2)_{14}COO(CH_2)_{29}CH_3$, an ester of palmitic acid. Beeswax is also an important chemical substance because it is used in the production of foods, cosmetics, and pharmaceuticals.

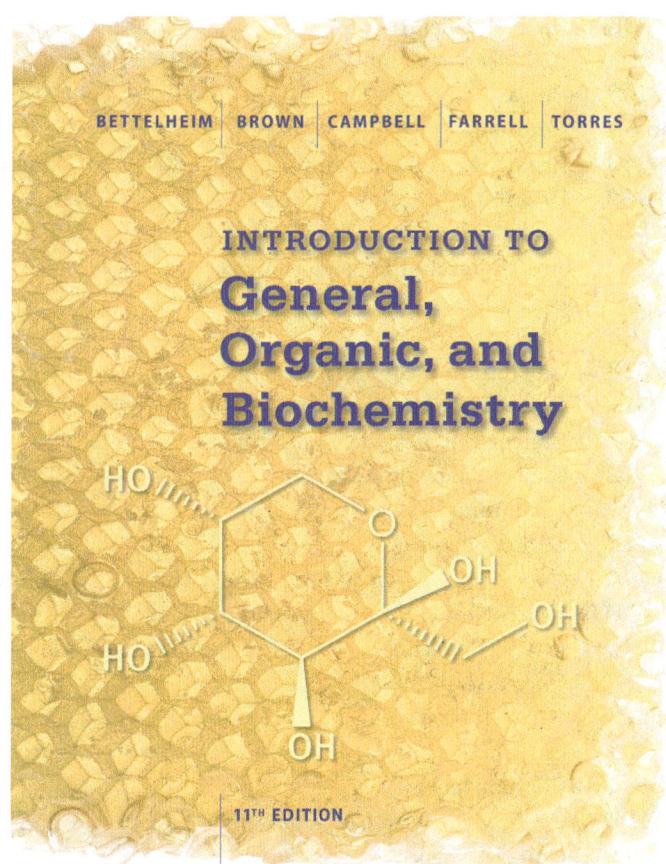

Health-Related Topics

Key

ChemConn = Chemical Connections Box number
Sect. = Section number
Prob. = Problem number

A, B, AB, and O Blood Types	ChemConn 20D
Abundance of Elements in the Human Body and in the Earth's Crust	ChemConn 2B
Acetaminophen (Tylenol)	Probs. 1-45, 1-112
Acid Rain	ChemConn 6A
Acidic Polysaccharides	Sect. 20-6
Acidosis	ChemConn 8C
Acquired Immunity	Sect. 31-1
Adjuvants	Sect. 31-6D
Advanced Glycation End Products	ChemConn 22B
AGE and Aging	ChemConn 22B
AIDS	Sect. 31-8
Alkaloids	ChemConn 16B
Alkalosis and the Sprinter's Trick	ChemConn 8D
Alzheimer's Disease	ChemConn 24C
Amoxicillin	Prob. 15-32
Amphetamines	ChemConn 16A, Prob. 24-50
Anabolic Steroids	ChemConn 21C
Androstenedione	Prob. 3-126
Antacids	ChemConn 8B
Antibiotics	ChemConn 31B
Antibodies and Cancer Therapy	ChemConn 31A
Antidepressants	Probs. 24-79, 24-80, Sect. 24-5F
Antigens	Sect. 31-1C
Antihistamines	Sect. 24-5G
Anti-inflammatory Drugs	ChemConn 21E
Anticoagulants	ChemConn 17A
Antioxidants	Sect. 13-4D
Artificial Pacemakers and Redox	ChemConn 4C
Ascorbic Acid (Vitamin C)	ChemConn 20B
Aspartame	Prob. 19-13, ChemConn 22A
Aspirin and Other NSAIDs	ChemConn 19C
Asthma	Sect. 21-12
Atherosclerosis: Levels of LDL and HDL	Sect. 21-9E
Atomic Energy	Sect. 9-9
Atropine	Prob. 16-54
Attention Deficit Disorder (ADD)	ChemConn 24D
Autoimmune Diseases	Sect. 31-7C
Automobile Air Bags	Prob. 5-54
Autoxidation	Sect. 13-4D
Azithromycin	Prob. 18-54
B Cells	Sect. 31-2C
Barbiturates	ChemConn 19E
Basal Caloric Requirement	Sect. 30-2
Base Excision Repair (BER) of DNA	Sect. 25-7
Bends, the	ChemConn 6B
BHT as an Antioxidant in Foods	Sect. 13-4D
Bile Salts	Sect. 21-11
Blood Alcohol Screening	ChemConn 14C
Blood Buffers	Sect. 8-10D
Blood pH	ChemConn 8C
Blood Pressure Measurement	Sect. 32-8, Probs. 32-36–32-38, ChemConn 5C
Bone, Artificial	ChemConn 3A
Bone Density and Solubility Equilibrium	Sect. 7-6
Botox	ChemConn 24B
Botulism	ChemConn 24B
Breath Alcohol Screening	ChemConn 14C
Breathing and Boyle's Law	ChemConn 5A
Bronchodilators and Asthma	ChemConn 16E
Brown Fat and Hibernation	ChemConn 27A
Buffers	Sect. 8-10
Caffeine	Prob. 4-103
Calcium as a Signaling Agent	Sect. 24-3C
Calorie Counting	Sect. 30-2
Cancer Cell Growth and Metabolic State	Sect. 27-1
Cancer Therapy Antibodies	ChemConn 31A
Cannabinoid Receptors	ChemConn 30F
Captopril and ACE Inhibitors	Prob. 15-31
Capsaicin, for Those Who Like It Hot	ChemConn 13F
Carcinogenic Polynuclear Aromatics and Smoking	ChemConn 13B
Carcinogens	Sect. 26-7
β-Carotene	Prob. 12-72
Cataracts	ChemConn 22B
Cephalosporins	ChemConn 19B
Cetylpyridinium Chloride	Prob. 18-25
Chiral Drugs	ChemConn 15A
Chirality in the Biological World	Sect. 15-5
Chlorine Dioxide	Probs. 3-102, 4-59
Cholera	Sect. 24-5E

Cholesterol	Sect. 21-9F
Cis-Trans Isomerism in	
European Corn Borer	ChemConn 12C
Cocaine	ChemConn 16B
Cocaine Addiction	ChemConn 23G
Cold Compresses	ChemConn 1C
Chondroitin Sulfate	Prob. 20-79
Coniine	ChemConn 16B
Connection between	
Carbohydrates and Obesity	ChemConn 20E
Coral Chemistry and Broken Bones	ChemConn 3A
COX-2 Inhibitor Drugs	ChemConn 21G
Creatine: Performance Enhancement	ChemConn 30D
Crenation	Sect. 6-8C
Cystic Fibrosis	ChemConn 26F
Cytokines	Sect. 31-2A
DDT	ChemConn 13A
DEET	Prob. 19-43
Depression	ChemConn 24G, 30G
Diabetes	ChemConn 24F
Dialysis	Sect. 6-8D
Dietary Protein	Prob. 22-96
Dietary Reference Intake	Sect. 30-1
Dieting and Weight Loss	Prob. 20-77, Sect. 30-2
2,4-Dinitrophenol as an Uncoupling	
Agent	ChemConn 27A
DNA Fingerprinting	ChemConn 25C
Drug Dosage and Body Mass	Prob. 1-100, ChemConn 1A
Eicosapentenoic Acid	ChemConn 21F
Emulsions and Emulsifying Agents	ChemConn 6E
Enantiomers of Ibuprofen	Sect. 15-2
Environmental Impact of Freons	ChemConn 11C
Enzyme Regulation	ChemConn 23E, Sect. 23-6A
Enzymes in Medical Diagnosis	Sect. 23-7
Enzymes in Therapy	Sect. 23-7
Ephedrine	Probs. 3-107, 15-24
Epigenetics and DNA	Sect. 26-10
Epigenetics, Cancer, and Aging	ChemConn 26G
Epinephrine	ChemConn 16E
Erythromycin A	Prob. 14-65
Essential Amino Acids	Sect. 30-5, ChemConn 29D
Esters, as Flavoring Agents	ChemConn 18B
Ethers and Anesthesia	ChemConn 14E
Ethylene Oxide, as a Chemical Sterilant	ChemConn 14D
Ethylene, a Plant Growth Regulator	ChemConn 12A
Fatty Acids	ChemConn 12B, Sect. 18-4A
Fever, as a Protective Mechanism	ChemConn 7A
Fever, Dangers of	ChemConn 7A
Fluid Mosaic, Model of Membranes	Sect. 21-5
Fluoride Ion in Dental Decay	ChemConn 4A
Fluoxetine (Prozac)	Probs. 24-49, 24-81
Fish in Diet – Health Benefits	ChemConn 21F
Food for Performance	ChemConn 30E
Food Guide Pyramid	ChemConn 30A
Free Radicals	ChemConn 9C, Sect. 13-4D
Galactosemia	ChemConn 20A
Gallstones	Sect. 21-9A
Gene Patents	ChemConn 25A
Gene Therapy	Sect. 26-9
Genetic Code	Sect. 26-4
Gout	Sect. 6-6B
G-protein/cAMP Cascade	Sect. 24-5C
Haber Process	ChemConn 7E
Heart Enzymes	Sect. 23-7
Heavy Metal Poisoning	Sect. 22-13
Heme Products in Bruises	Sect. 28-10
Hemodialysis	ChemConn 6G
Hemoglobin	ChemConn 22G
Hemolysis	Sect. 6-8C
Heparin	Sect. 20-6B, Prob. 20-68
High Fever	ChemConn 7A
HIV Protease Inhibitors	ChemConn 23D
Hormones	Sect. 24-2
How Soap Cleans	Sect. 18-4C
Human Insulin	ChemConn 22C
Hyaluronic Acid	Sect. 20-6A
Hyperbaric Medicine	ChemConn 5B
Hypoglycemic Awareness	ChemConn 22C
Hypothermia and Hyperthermia	ChemConn 1B
Immune System	Sect. 31-1
Immunization	Sect. 31-6
Immunoglobin	Sect. 31-4
Immunology and Oncology	ChemConn 31C
Immunosuppressant FK-506	Probs. 15-44, 17-55
Indoor Radon Problem	ChemConn 9B
Influenza Vaccine	ChemConn 31E
Innate Immunity	Sect. 31-1A
Insulin	Sect. 22-8
Insulin, Structure	Sect. 22-8
Insulin, Use	ChemConn 22C
Iodide Ion and Goiter	ChemConn 13C
Ionic Compounds in Medicine	ChemConn 3B
Ionizing Radiation	Sect. 9-5
Iron and Mineral Requirements	ChemConn 30C
Ketoacidosis in Diabetes	ChemConn 28C
Ketone Bodies	ChemConn 18C, Sect. 28-7
Lactate Accumulation	ChemConn 28A
Laetrile	Prob. 20-71
Laser In Situ Keratomileusis (LASIK)	ChemConn 22H
Laser Surgery and Protein Denaturation	ChemConn 22H
Librium	ChemConn 16C
Lipid Storage Diseases	ChemConn 21B
Local Anesthetics for Dentistry	Prob. 19-44
Lowering Body Temperature	ChemConn 7B

Lunesta	Prob. 15-46
Lycopene	Prob. 12-71
Mad Cow Disease	ChemConn 22E
Magnetic Resonance Imaging	ChemConn 9D
Medical Uses of Inhibitors	ChemConn 23D
Memory and Protein Synthesis	ChemConn 26B
Menstrual Cycle	Sect. 21-10B
Metformin (Glucophage)	Prob. 16-28
Methadone	ChemConn 16D
Methamphetamine	ChemConn 16A
Methylparaben	Prob. 18-47
Milk of Magnesia	ChemConn 8A
Monoclonal Antibodies	ChemConn 31A
Morphine and Enkephalins	Sect. 24-6
Morphine and Morphine Analogs	Prob. 16-59
MSG and Headaches	ChemConn 24E
Muscle Relaxants	ChemConn 23A
Mutagens	Sect. 26-7
Mutations and Biochemical Evolution	ChemConn 26C
Naproxen	Sects. 15-2, 15-5B
Neurotransmitters	Sect. 22-5, Sect. 24-1
Nicotine	ChemConn 16B
Nitric Oxide	ChemConn 3C, 24E
Nitrous Oxide ("Laughing Gas")	Prob. 3-78
Nitroglycerin, an Explosive and a Drug	ChemConn 14B
NMDA Receptors	Sect. 24-4B
NO Release and Impotence	ChemConn 24E
Nuclear Medicine	Sect. 9-7
Nutritional Causes of Depression	ChemConn 30G
Nutritional Daily Values	Sect. 30-1
Obesity	ChemConn 28B, 30C, Sect. 30-2
Omega-3 Fatty Acids	ChemConn 21F
Oncogenes and Cancer	ChemConn 26E
Oral Contraception	ChemConn 21D
Organic Food and Health	ChemConn 30F
Oseltamivir (Tamiflu)	Prob. 15-45
Osmotic Pressure	Sect. 6-8C
Parkinson's Disease	ChemConn 24D, Sect. 22-5
Paroxetine	Prob. 15-35
Paternity Testing	ChemConn 25C
Penicillins	ChemConn 19B
Peramivir for Influenza A (H1N1) and Swine Flu	Prob. 19-46
Phenols as Antioxidants	Sect. 13-4D
Phencyclidine (PCP)	Sect. 24-4B
Phenylketonuria (PKU)	ChemConn 28D
Photorefractive Keratectomy (PRK)	ChemConn 22H
Photosynthesis	ChemConn 29A
Poison Ivy	Sect. 13-4A
Poisonous Puffer Fish	ChemConn 11A
Polynuclear Aromatic Hydrocarbons (PAHs)	Sect. 13-2D, ChemConn 13B
Positron Emission Tomography (PET)	Sect. 9-7A
Prion Disease	ChemConn 22E
Propofol	Sect. 20-3D
Proteomics	ChemConn 22F
Pyrethrins	ChemConn 19A
Quaternary Structure and Allosteric Proteins	ChemConn 22G
Radiation	Sect. 9-2
Radiation Dosimetry	Sect. 9-6
Radiation Sickness	Sect. 9-6
Radioactive Dating	ChemConn 9A
Radioactive Fallout from Nuclear Accidents	ChemConn 9E
Radioactive Isotopes, in Nuclear Imaging	Sect. 9-7A
Radioactive Isotopes, in Medical Therapy	Sect. 9-7B
Recommended Daily Allowances (RDA)	Sect. 30-1
Relative Sweetness	Sect. 20-4D
Reverse Osmosis and Desalinization	ChemConn 6F
Roentgens, Rads, and Rems	Sect. 9-6
SCID	Sect. 26-9
SCUBA Diving	Prob. 5-111
Sickle Cell Anemia	ChemConn 22D
Significance of Chirality in the Biological World	Sect. 15-5
Side Effects of COX Inhibitors	Sect. 21-12
Solubility of Drugs in Body Fluids	ChemConn 16D
Solubility and Tooth Decay	ChemConn 4A
Sports Drinks	Sect. 21-11, ChemConn 30D
Statins	ChemConn 29C
Stem Cells	Sect. 31-2C
Stitches That Dissolve	ChemConn 19F
Strontium-90	ChemConn 2C
Sunglasses and Le Chatelier's Principle	ChemConn 7D
Sunscreens and Sunblocks	ChemConn 19D
Swine Flu	ChemConn 31D
Synapses and Sleep	ChemConn 24A
Synthetic Food Dyes	ChemConn 13E
Synthetic Genome	ChemConn 25E
T Cells	Sect. 31-2C
Tamoxifen and Breast Cancer	Prob. 16-32
Taxol	ChemConn 10A
Tears	ChemConn 32E, Probs. 32-47–32-50
Telomeres and Immortality	ChemConn 25B
Terpin Hydrate	Prob. 12-52
Testing for Glucose	ChemConn 20C
Tetrodotoxin	ChemConn 11A
Timed-Released Medication	ChemConn 7C
Tranquilizers	ChemConn 16C

Trans Fatty Acids	Sect. 21-3B, ChemConn 18A	Viruses	Sect. 26-1
Transport Across Cell Membranes	Sect. 21-5B	Vitamin A and Vision	ChemConn 12D
Tumor Suppressors	ChemConn 26E	Vitamin Excess	Sect. 30-6
Urinometer	Sect. 1-7B	Vitamins and Depression	ChemConn 30G
Vaccines	Sect. 31-6	Vitamins and Minerals	Sect. 30-6
Valium	ChemConn 16C	Vitamins, in Diet	Sect. 30-6
Viagra	ChemConn 23D, 24E	Xeroderma Pigmentosum	Sect 25-7
Viagra and Blood Vessel Dilation	ChemConn 24E	Zebrafish Synapses	ChemConn 24A

Matter, Energy, and Measurement

A woman climbing a frozen waterfall.

KEY QUESTIONS

- 1-1 Why Do We Call Chemistry the Study of Matter?
- 1-2 What Is the Scientific Method?
- 1-3 How Do Scientists Report Numbers?
- *How To . . .* Determine the Number of Significant Figures in a Number
- 1-4 How Do We Make Measurements?
- 1-5 What Is a Handy Way to Convert from One Unit to Another?
- *How To . . .* Do Unit Conversions by the Factor-Label Method
- 1-6 What Are the States of Matter?
- 1-7 What Are Density and Specific Gravity?
- 1-8 How Do We Describe the Various Forms of Energy?
- 1-9 How Do We Describe Heat and the Ways in Which It Is Transferred?

1-1 Why Do We Call Chemistry the Study of Matter?

The world around us is made of chemicals. Our food, our clothing, the buildings in which we live are all made of chemicals. Our bodies are made of chemicals too. To understand the human body, its diseases, and its cures, we must know all we can about those chemicals. There was a time—only a few hundred years ago—when physicians were powerless to treat many diseases. Cancer, tuberculosis, smallpox, typhus, plague, and many other sicknesses struck people seemingly at random. Doctors, who had no idea what caused any of these diseases, could do little or nothing about them. Doctors treated them with magic or by such measures as bleeding, laxatives, hot plasters, and pills made from powdered stag-horn, saffron, or gold. None of these treatments were effective, and the doctors, because they came into direct contact with highly contagious diseases, died at a much higher rate than the general public.

Medicine has made great strides since those times. We live much longer, and many once-feared diseases have been essentially eliminated or are curable. Smallpox has been eradicated, and polio, typhus, bubonic plague, diphtheria, and other diseases that once killed millions no longer pose a serious problem, at least not in developed countries.

How has this medical progress come about? The answer is that diseases could not be cured until they were understood, and this understanding has emerged through greater knowledge of how the body functions. It is

▶ Medical practice over time.
(a) A woman being bled by a leech on her left forearm; a bottle of leeches is on the table. From a 1639 woodcut.
(b) Modern surgery in a well-equipped operating room.

(a)

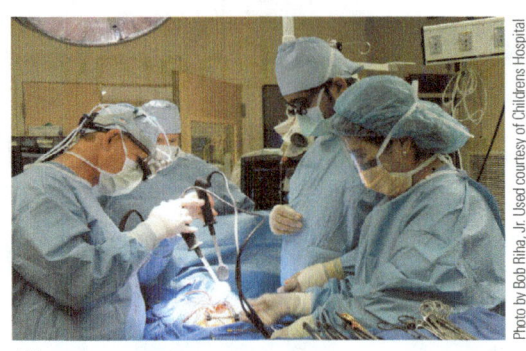
(b)

progress in our understanding of the principles of biology, chemistry, and physics that has led to these advances in medicine. Because so much of modern medicine depends on chemistry, it is essential that students who intend to enter the health professions have some understanding of basic chemistry. This book has been written to help you achieve that goal. Even if you choose a different profession, you will find that the chemistry you learn in this course will greatly enrich your life. ◀

The universe consists of matter, energy, and empty space. **Matter** is anything that has mass and takes up space. **Chemistry** is the science that deals with matter: the structure and properties of matter and the transformations from one form of matter to another. We will discuss energy in Section 1-8.

It has long been known that matter can change, or be made to change, from one form to another. In a **chemical change,** more commonly called a **chemical reaction,** some substances are used up (disappear) and others are formed to take their place. An example is the burning of a mixture of hydrocarbons, usually called "bottled gas." In this mixture of hydrocarbons, the main component is propane. When this chemical change takes place, propane and oxygen from the air are converted to carbon dioxide and water. Figure 1-1 shows another chemical change.

(a)

(b)

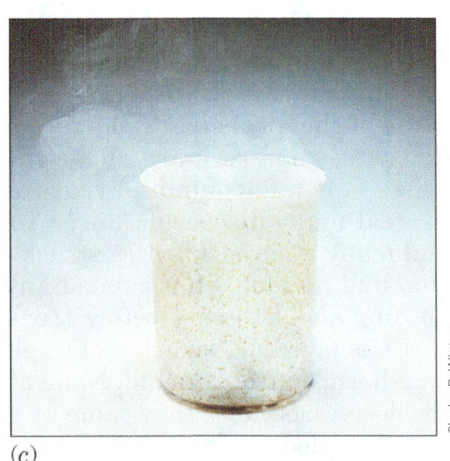
(c)

FIGURE 1-1 A chemical reaction. (a) Bromine, an orange-brown liquid, and aluminum metal. (b) These two substances react so vigorously that the aluminum becomes molten and glows white hot at the bottom of the beaker. The yellow vapor consists of vaporized bromine and some of the product of the reaction, white aluminum bromide. (c) Once the reaction is complete, the beaker is coated with aluminum bromide and the products of its reaction with atmospheric moisture. (*Note:* This reaction is dangerous! Under no circumstances should it be done except under properly supervised conditions.)

Matter also undergoes other kinds of changes, called **physical changes.** These changes differ from chemical reactions in that the identities of the substances do not change. Most physical changes involve changes of state—for example, the melting of solids and the boiling of liquids. Water remains water whether it is in the liquid state or in the form of ice or steam. The conversion from one state to another is a physical—not a chemical—change. Another important type of physical change involves making or separating mixtures. Dissolving sugar in water is a physical change.

When we talk about the **chemical properties** of a substance, we mean the chemical reactions that it undergoes. **Physical properties** are all properties that do not involve chemical reactions. For example, density, color, melting point, and physical state (liquid, solid, gas) are all physical properties.

1-2 What Is the Scientific Method?

Scientists learn by using a tool called the **scientific method.** The heart of the scientific method is the testing of theories. It was not always so, however. Before about 1600, philosophers often believed statements just because they sounded right. For example, the great philosopher Aristotle (384–322 BCE) believed that if you took the gold out of a mine it would grow back. He believed this idea because it fit with a more general picture that he had about the workings of nature. In ancient times, most thinkers behaved in this way. If a statement sounded right, they believed it without testing it.

About 1600 CE, the scientific method came into use. Let us look at an example to see how the scientific method operates. The Greek physician Galen (200–130 BCE) recognized that the blood on the left side of the heart somehow gets to the right side. This is a fact. A **fact** is a statement based on direct experience. It is a consistent and reproducible observation. Having observed this fact, Galen then proposed a hypothesis to explain it. A **hypothesis** is a statement that is proposed, without actual proof, to explain the facts and their relationship. Because Galen could not actually see how the blood got from the left side to the right side of the heart, he came up with the hypothesis that tiny holes must be present in the muscular wall that separates the two halves. ◂

Up to this point, a modern scientist and an ancient philosopher would behave the same way. Each would offer a hypothesis to explain the facts. From this point on, however, their methods would differ. To Galen, his explanation sounded right and that was enough to make him believe it, even though he couldn't see any holes. His hypothesis was, in fact, believed by virtually all physicians for more than 1000 years. When we use the scientific method, however, we do not believe a hypothesis just because it sounds right. We test it, using the most rigorous testing we can imagine. ◂

William Harvey (1578–1657) tested Galen's hypothesis by dissecting human and animal hearts and blood vessels. He discovered that one-way valves separate the upper chambers of the heart from the lower chambers. He also discovered that the heart is a pump that, by contracting and expanding, pushes the blood out. Harvey's teacher, Fabricius (1537–1619), had previously observed that one-way valves exist in the veins, so that blood in the veins can travel only toward the heart and not the other way.

▶ Galen did not do experiments to test his hypothesis.

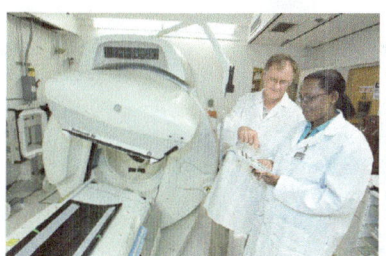

▶ Using a PET scanner is an example of how modern scientists collect information to confirm a diagnosis and test a hypothesis.

Hypothesis A statement that is proposed, without actual proof, to explain a set of facts and their relationship

Harvey put these facts together to come up with a new hypothesis: blood is pumped by the heart and circulates throughout the body. This was a better hypothesis than Galen's because it fit the facts more closely. Even so, it was still a hypothesis and, according to the scientific method, had to be tested further. One important test took place in 1661, four years after Harvey died. Harvey had predicted that because there had to be a way for the blood to get from the arteries to the veins, tiny blood vessels must connect them. In 1661, the Italian anatomist Malpighi (1628–1694), using the newly invented microscope, found these tiny vessels, which are now called capillaries.

Malpighi's discovery supported the blood circulation hypothesis by fulfilling Harvey's prediction. When a hypothesis passes enough tests, we have more confidence in it and call it a theory. A **theory** is the formulation of an apparent relationship among certain observed phenomena, which has been verified to some extent. In this sense, a theory is the same as a hypothesis except that we have a stronger belief in it because more evidence supports it. No matter how much confidence we have in a theory, however, if we discover new facts that conflict with it or if it does not pass newly devised tests, the theory must be altered or rejected. In the history of science, many firmly established theories have eventually been thrown out because they could not pass new tests.

One of the most important ways to test a hypothesis is by a controlled experiment. It is not enough to say that making a change causes an effect, we must also see that the lack of that change does not produce the observed effect. If, for example, a researcher proposes that adding a vitamin mixture to the diet of children improves growth, the first question is whether children in a control group who do not receive the vitamin mixture do not grow as quickly. Comparison of an experiment with a control is essential to the scientific method.

The scientific method is thus very simple. We don't accept a hypothesis or a theory just because it sounds right. We devise tests, and only if the hypothesis or theory passes the tests do we accept it. The enormous progress made since 1600 in chemistry, biology, and the other sciences is a testimony to the value of the scientific method.

You may get the impression from the preceding discussion that science progresses in one direction: facts first, hypothesis second, theory last. Real life is not so simple, however. Hypotheses and theories call the attention of scientists to discover new facts. An example of this scenario is the discovery of the element germanium. In 1871, Mendeleev's Periodic Table—a graphic description of elements organized by properties—predicted the existence of a new element whose properties would be similar to those of silicon. Mendeleev called this element eka-silicon. In 1886, it was discovered in Germany (hence the name), and its properties were truly similar to those predicted by theory.

On the other hand, many scientific discoveries result from serendipity, or chance observation. An example of serendipity occurred in 1926, when James Sumner of Cornell University left an enzyme preparation of jack bean urease in a refrigerator over the weekend. Upon his return, he found that his solution contained crystals that turned out to be a protein. This chance discovery led to the hypothesis that all enzymes are proteins. Of course, serendipity is not enough to move science forward. Scientists must have the creativity and insight to recognize the significance of their observations. Sumner fought for more than 15 years for his hypothesis to gain acceptance because people believed that only small molecules can form crystals. Eventually his view won out, and he was awarded a Nobel Prize in chemistry in 1946.

Theory The formulation of an apparent relationship among certain observed phenomena, which has been verified. A theory explains many interrelated facts and can be used to make predictions about natural phenomena. Examples are Newton's theory of gravitation and the kinetic molecular theory of gases, which we will encounter in Section 6-6. This type of theory is also subject to testing and will be discarded or modified if it is contradicted by new facts.

1-3 How Do Scientists Report Numbers?

Scientists often have to deal with numbers that are very large or very small. For example, an ordinary copper penny (dating from before 1982, when pennies in the United States were still made of copper) contains approximately

$$29{,}500{,}000{,}000{,}000{,}000{,}000{,}000 \text{ atoms of copper}$$

and a single copper atom weighs

$$0.00000000000000000000000023 \text{ pound}$$

which is equal to

$$0.00000000000000000000000104 \text{ gram}$$

Many years ago, an easy way to handle such large and small numbers was devised. This method, which is called **exponential notation,** is based on powers of 10. In exponential notation, the number of copper atoms in a penny is written

$$2.95 \times 10^{22}$$

and the weight of a single copper atom is written

$$2.3 \times 10^{-25} \text{ pound}$$

which is equal to

$$1.04 \times 10^{-22} \text{ gram}$$

The origin of this shorthand form can be seen in the following examples:

$$100 = 1 \times 10 \times 10 = 1 \times 10^2$$

$$1000 = 1 \times 10 \times 10 \times 10 = 1 \times 10^3$$

What we have just said in the form of an equation is "100 is a one with two zeros after the one, and 1000 is a one with three zeros after the one." We can also write

$$1/100 = 1/10 \times 1/10 = 1 \times 10^{-2}$$

$$1/1000 = 1/10 \times 1/10 \times 1/10 = 1 \times 10^{-3}$$

where negative exponents denote numbers less than 1. The exponent in a very large or very small number lets us keep track of the number of zeros. That number can become unwieldy with very large or very small quantities, and it is easy to lose track of a zero. Exponential notation helps us deal with this possible source of determinant error. ◄

When it comes to measurements, not all the numbers you can generate in your calculator or computer are of equal importance. Only the number of digits that are known with certainty are significant. Suppose you measured the weight of an object as 3.4 g on a balance that reads to the nearest 0.1 g. You can report the weight as 3.4 g but not as 3.40 or 3.400 g because you do not know the added zeros with certainty. This becomes even more important when you use a calculator. For example, you might measure a cube with a ruler and find that each side is 2.9 cm. If you are asked to calculate the volume, you multiply $2.9 \text{ cm} \times 2.9 \text{ cm} \times 2.9 \text{ cm}$. The calculator will then give you an answer that is 24.389 cm^3. However, your initial measurements were good to only one decimal place, so your final answer cannot be good to three decimal places. As a scientist, it is important to report data that have the correct number of **significant figures.** A detailed account of using significant figures is presented in Appendix II. The following How To box describes the way to determine the number of significant figures in a number. You will find boxes like this at places in the text where detailed explanations of concepts are useful.

▶ Photos showing different orders of magnitude.

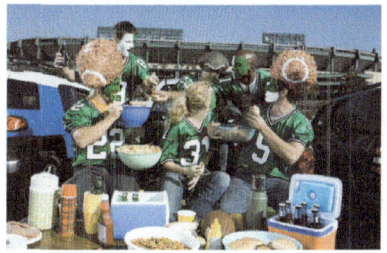

1. Group picnic in stadium parking lot (~10 meters)

2. Football field (~100 meters)

3. Vicinity of stadium (~1000 meters).

HOW TO...

Determine the Number of Significant Figures in a Number

1. **Nonzero digits are always significant.**
   For example, 233.1 m has four significant figures; 2.3 g has two significant figures.
2. **Zeros at the beginning of a number are never significant.**
   For example, 0.0055 L has two significant figures; 0.3456 g has four significant figures.
3. **Zeros between nonzero digits are always significant.**
   For example, 2.045 kcal has four significant figures; 8.0506 g has five significant figures.
4. **Zeros at the end of a number that contains a decimal point are always significant.**
   For example, 3.00 L has three significant figures; 0.0450 mm has three significant figures.
5. **Zeros at the end of a number that contains no decimal point may or may not be significant.**

We cannot tell whether they are significant without knowing something about the number. This is the ambiguous case. If you know that a certain small business made a profit of $36,000 last year, you can be sure that the 3 and 6 are significant, but what about the rest? The profit might have been $36,126 or $35,786.53, or maybe even exactly $36,000. We just don't know because it is customary to round off such numbers. On the other hand, if the profit were reported as $36,000.00, then all seven digits would be significant.

In science, to get around the ambiguous case, we use exponential notation. Suppose a measurement comes out to be 2500 g. If we made the measurement, then we know whether the two zeros are significant, but we need to tell others. If these digits are *not* significant, we write our number as 2.5×10^3. If one zero is significant, we write 2.50×10^3. If both zeros are significant, we write 2.500×10^3. Because we now have a decimal point, all the digits shown are significant. We are going to use decimal points throughout this text to indicate the number of significant figures.

EXAMPLE 1-1 Exponential Notation and Significant Figures

Multiply:
(a) $(4.73 \times 10^5)(1.37 \times 10^2)$ (b) $(2.7 \times 10^{-4})(5.9 \times 10^8)$

Divide:
(c) $\dfrac{7.08 \times 10^{-8}}{300.}$ (d) $\dfrac{5.8 \times 10^{-6}}{6.6 \times 10^{-8}}$ (e) $\dfrac{7.05 \times 10^{-3}}{4.51 \times 10^5}$

Strategy and Solution

The way to do calculations of this sort is to use a button on scientific calculators that automatically uses exponential notation. The button is usually labeled "E." (On some calculators, it is labeled "EE." In some cases, it is accessed by using the second function key.)

(a) Enter 4.73E5, press the multiplication key, enter 1.37E2, and press the "=" key. The answer is 6.48×10^7. The calculator will display this number as 6.48E7. This answer makes sense. We add exponents when we multiply, and the sum of these two exponents is correct (5 + 2 = 7). We also multiply the numbers, 4.73 × 1.37. This is approximately 4 × 1.5 = 6, so 6.48 is also reasonable.

(b) Here we have to deal with a negative exponent, so we use the "+/−" key. Enter 2.7E+/−4, press the multiplication key, enter 5.9E8, and press the "=" key. The calculator will display the answer as 1.593E5. To have the correct number of significant figures, we should report our answer as 1.6E5. This answer makes sense because 2.7 is a little less than 3 and 5.9 is a little less than 6, so we predict a number slightly less than 18; also, the algebraic sum of the exponents (−4 + 8) is equal to 4. This gives 16×10^4. In exponential notation, we normally prefer to report numbers between 1 and 10, so we rewrite our answer as 1.6×10^5. We made the first number 10 times smaller, so we increased the exponent by 1 to reflect that change.

(c) Enter 7.08E+/−8, press the division key, enter 300., and press the "=" key. The answer is 2.36×10^{-10}. The calculator will display this number as 2.36E − 10. We subtract exponents when we divide, and we can also write 300. as 3.00×10^2.

(d) Enter 5.8E+/−6, press the division key, enter 6.6E+/−8, and press the "=" key. The calculator will display the answer as 87.878787878788. We report this answer as 88 to get the right number of significant figures. This answer makes sense. When we divide 5.8 by 6.6, we get a number slightly less than 1. When we subtract the exponents algebraically (−6 − [−8]), we get 2. This means that the answer is slightly less than 1×10^2, or slightly less than 100.

(e) Enter 7.05E+/−3, press the division key, enter 4.51E5, and press the "=" key. The calculator displays the answer as 1.5632E−8, which, to the correct number of significant figures, is 1.56×10^{-8}. The algebraic subtraction of exponents is −3 − 5 = −8.

Problem 1-1

Multiply:
(a) $(6.49 \times 10^7)(7.22 \times 10^{-3})$
(b) $(3.4 \times 10^{-5})(8.2 \times 10^{-11})$

Divide:
(a) $\dfrac{6.02 \times 10^{23}}{3.10 \times 10^5}$
(b) $\dfrac{3.14}{2.30 \times 10^{-5}}$

1-4 How Do We Make Measurements?

In our daily lives, we are constantly making measurements. We measure ingredients for recipes, driving distances, gallons of gasoline, weights of fruits and vegetables, and the timing of TV programs. Doctors and nurses measure pulse rates, blood pressures, temperatures, and drug dosages. Chemistry, like other sciences, is based on measurements.

A measurement consists of two parts: a number and a unit. A number without a unit is usually meaningless. If you were told that a person's weight is 57, the information would be of very little use. Is it 57 pounds, which would indicate that the person is very likely a child or a midget, or 57 kilograms, which is the weight of an average woman or a small man? Or is it perhaps some other unit? Because so many units exist, a number by itself is not enough; the unit must also be stated.

In the United States, most measurements are made with the English system of units: pounds, miles, gallons, and so on. In most other parts of the world,

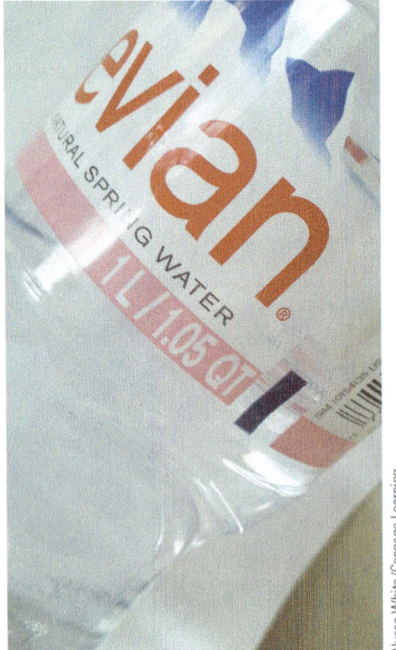

▶ The label on this bottle of water shows the metric size (one liter) and the equivalent in quarts.

Metric system A system of units of measurement in which the divisions to subunits are made by a power of 10

Table 1-1 Base Units in the Metric System

Length	meter (m)
Volume	liter (L)
Mass	gram (g)
Time	second (s)
Temperature	Kelvin (K)
Energy	joule (J)
Amount of substance	mole (mol)

FIGURE 1-2 Road sign in California showing metric equivalents of mileage.

however, few people could tell you what a pound or an inch is. Most countries use the **metric system,** a system that originated in France about 1800 and that has since spread throughout the world. Even in the United States, metric measurements are slowly being introduced (Figure 1-2). For example, many soft drinks and most alcoholic beverages now come in metric sizes. Scientists in the United States have been using metric units all along. ◀

Around 1960, international scientific organizations adopted another system, called the **International System of Units** (abbreviated **SI**). The SI is based on the metric system and uses some of the metric units. The main difference is that the SI is more restrictive: It discourages the use of certain metric units and favors others. Although the SI has advantages over the older metric system, it also has significant disadvantages. For this reason, U.S. chemists have been very slow to adopt it. At this time, approximately 40 years after its introduction, not many U.S. chemists use the entire SI, although some of its preferred units are gaining ground.

In this book, we will use the metric system (Table 1-1). Occasionally we will mention the preferred SI unit.

A. Length

The key to the metric system (and the SI) is that there is one base unit for each kind of measurement and that other units are related to the base unit by powers of 10. As an example, let us look at measurements of length. In the English system, we have the inch, the foot, the yard, and the mile (not to mention such older units as the league, furlong, ell, and rod). If you want to convert one unit to another unit, you must memorize or look up these conversion factors:

$$5280 \text{ feet} = 1 \text{ mile}$$
$$1760 \text{ yards} = 1 \text{ mile}$$
$$3 \text{ feet} = 1 \text{ yard}$$
$$12 \text{ inches} = 1 \text{ foot}$$

All this is unnecessary in the metric system (and the SI). In both systems the base unit of length is the **meter (m).** To convert to larger or smaller units, we do not use arbitrary numbers like 12, 3, and 1760, but only 10, 100, 1/100, 1/10, or other powers of 10. This means that *to convert from one metric or SI unit to another, we only have to move the decimal point.* Furthermore, the other units are named by putting prefixes in front of "meter," and *these prefixes are the same throughout the metric system and the SI.*

Table 1-2 The Most Common Metric Prefixes

Prefix	Symbol	Value
giga	G	10^9 = 1,000,000,000 (one billion)
mega	M	10^6 = 1,000,000 (one million)
kilo	k	10^3 = 1000 (one thousand)
deci	d	10^{-1} = 0.1 (one-tenth)
centi	c	10^{-2} = 0.01 (one-hundredth)
milli	m	10^{-3} = 0.001 (one-thousandth)
micro	μ	10^{-6} = 0.000001 (one-millionth)
nano	n	10^{-9} = 0.000000001 (one-billionth)
pico	p	10^{-12} = 0.000000000001 (one-trillionth)

Conversion factors are defined numbers. We can use them to have as many significant figures as needed without limit.

Table 1-2 lists the most important of these prefixes. If we put some of these prefixes in front of "meter," we have

1 kilometer (km) = 1000 meters (m)

1 centimeter (cm) = 0.01 meter

1 nanometer (nm) = 10^{-9} meter

For people who have grown up using English units, it is helpful to have some idea of the size of metric units. Table 1-3 shows some conversion factors.

Table 1-3 Some Conversion Factors Between the English and Metric Systems

Length	Mass	Volume
1 in. = 2.54 cm	1 oz = 28.35 g	1 qt = 0.946 L
1 m = 39.37 in.	1 lb = 453.6 g	1 gal = 3.785 L
1 mile = 1.609 km	1 kg = 2.205 lb	1 L = 33.81 fl oz
	1 g = 15.43 grains	1 fl oz = 29.57 mL
		1 L = 1.057 qt

Some of these conversions are difficult enough that you will probably not remember them and must, therefore, look them up when you need them. Some are easier. For example, a meter is about the same as a yard. A kilogram is a little over two pounds. There are almost four liters in a gallon. These conversions may be important to you someday. For example, if you rent a car in Europe, the price of gas listed on the sign at the gas station will be in Euros per liter. When you realize that you are spending two dollars per liter and you know that there are almost four liters to a gallon, you will realize why so many people take the bus or a train instead.

B. Volume

Volume is space. The volume of a liquid, solid, or gas is the space occupied by that substance. The base unit of volume in the metric system is the **liter (L)**. This unit is a little larger than a quart (Table 1-3). The only other common metric unit for volume is the milliliter (mL), which is equal to 10^{-3} L. ◀

$$1 \text{ mL} = 0.001 \text{ L (or } 1 \times 10^{-3} \text{ L)}$$

$$1000 \text{ mL (or } 1 \times 10^3 \text{ mL)} = 1 \text{ L}$$

▶ Hypodermic syringe. Note that the volumes are indicated in milliliters.

One milliliter is exactly equal to one cubic centimeter (cc or cm³):

$$1 \text{ mL} = 1 \text{ cc}$$

Thus, there are 1000 (1×10^3) cc in 1 L.

C. Mass

Mass is the quantity of matter in an object. The base unit of mass in the metric system is the **gram (g).** As always in the metric system, larger and smaller units are indicated by prefixes. The ones in common use are

$$1 \text{ kilogram (kg)} = 1000 \text{ g}$$

$$1 \text{ milligram (mg)} = 0.001 \text{ g}$$

The gram is a small unit; there are 453.6 g in one pound (Table 1-3).

We use a device called a balance to measure mass. Figure 1-3 shows two types of laboratory balances.

There is a fundamental difference between mass and weight. Mass is independent of location. The mass of a stone, for example, is the same whether we measure it at sea level, on top of a mountain, or in the depths of a mine. In contrast, weight is not independent of location. **Weight** is the force a mass experiences under the pull of gravity. This point was dramatically demonstrated when the astronauts walked on the surface of the Moon. The Moon, being a smaller body than the Earth, exerts a weaker gravitational pull. Consequently, even though the astronauts wore space suits and equipment that would be heavy on Earth, they felt lighter on the Moon and could execute great leaps and bounces during their walks.

Although mass and weight are different concepts, they are related to each other by the force of gravity. We frequently use the words interchangeably because we weigh objects by comparing their masses to standard reference masses (weights) on a balance, and the gravitational pull is the same on the unknown object and on the standard masses. Because the force of gravity is essentially constant, mass is always directly proportional to weight.

FIGURE 1-3 Two laboratory balances for measuring mass.

CHEMICAL CONNECTIONS 1A

Drug Dosage and Body Mass

In many cases, drug dosages are prescribed on the basis of body mass. For example, the recommended dosage of a drug may be 3 mg of drug for each kilogram of body weight. In this case, a 50 kg (110 lb) woman would receive 150 mg and an 82 kg (180 lb) man would get 246 mg. This adjustment is especially important for children, because a dose suitable for an adult will generally be too much for a child, who has much less body mass. For this reason, manufacturers package and sell smaller doses of certain drugs, such as aspirin, for children.

Drug dosage may also vary with age. Occasionally, when an elderly patient has an impaired kidney or liver function, the clearance of a drug from the body is delayed, and the drug may stay in the body longer than is normal. This persistence can cause dizziness, vertigo, and migraine-like headaches, resulting in falls and broken bones. Such delayed clearance must be monitored and the drug dosage adjusted accordingly.

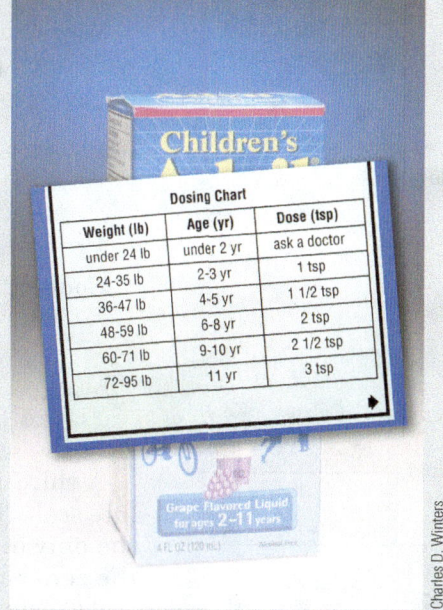

This package of Advil has a chart showing the proper doses for children of a given weight.

Test your knowledge with Problems 1-69 and 1-70.

D. Time

Time is the one quantity for which the units are the same in all systems: English, metric, and SI. The base unit is the second(s):

$$60 \text{ s} = 1 \text{ min}$$
$$60 \text{ min} = 1 \text{ h}$$

E. Temperature

Most people in the United States are familiar with the Fahrenheit scale of temperature. The metric system uses the centigrade, or **Celsius,** scale. In this scale, the boiling point of water is set at 100°C and the freezing point at 0°C. We can convert from one scale to the other by using the following formulas:

$$°F = \frac{9}{5} °C + 32$$

$$°C = \frac{5}{9} (°F - 32)$$

The 32 in these equations is a defined number and is, therefore, treated as if it had an infinite number of zeros following the decimal point. (See Appendix II.)

> **EXAMPLE 1-2** Temperature Conversion
>
> Normal body temperature is 98.6°F. Convert this temperature to Celsius.
>
> **Strategy**
>
> We use the conversion formula that takes into account the fact that the freezing point of water, 0°C is equal to 32°F.
>
> **Solution**
>
> $$°C = \frac{5}{9}(98.6 - 32) = \frac{5}{9}(66.6) = 37.0°C$$
>
> **Problem 1-2**
>
> Convert:
> (a) 64.0°C to Fahrenheit
> (b) 47°F to Celsius

FIGURE 1-4 Three temperature scales.

Figure 1-4 shows the relationship between the Fahrenheit and Celsius scales.

A third temperature scale is the **Kelvin (K)** scale, also called the absolute scale. The size of a Kelvin degree is the same as that of a Celsius degree; the only difference is the zero point. The temperature −273°C is taken as the zero point on the Kelvin scale. This makes conversions between Kelvin and Celsius very easy. To go from Celsius to Kelvin, just *add* 273; to go from Kelvin to Celsius, *subtract* 273:

$$K = °C + 273$$

$$°C = K - 273$$

Figure 1-4 also shows the relationship between the Kelvin and Celsius scales. Note that we don't use the degree symbol in the Kelvin scale: 100°C equals 373 K, not 373°K.

Why was −273°C chosen as the zero point on the Kelvin scale? The reason is that *−273°C, or 0 K, is the lowest possible temperature*. Because of this, 0 K is called **absolute zero.** Temperature reflects how fast molecules move. The more slowly they move, the colder it gets. At absolute zero, molecules stop moving altogether. Therefore, the temperature cannot get any lower. For some purposes, it is convenient to have a scale that begins at the lowest possible temperature; the Kelvin scale fulfills this need. The Kelvin is the SI unit.

It is very important to have a "gut feeling" about the relative sizes of the units in the metric system. Often, while doing calculations, the only thing that might offer a clue that you have made an error is your understanding of the sizes of the units. For example, if you are calculating the amount of a chemical that is dissolved in water and you come up with an answer of 254 kg/mL, does your answer make sense? If you have no intuitive feeling about the size of a kilogram or a milliliter, you will not know. If you realize that a milliliter is about the volume of a thimble and that a standard bag of sugar might weigh 2 kg, then you will realize that there is no way to pack 254 kg into a thimble of water, and you will know that you made a mistake.

1-5 What Is a Handy Way to Convert from One Unit to Another?

Factor-label method A procedure in which equations are set up so that all the unwanted units cancel and only the desired units remain

We frequently need to convert a measurement from one unit to another. The best and most foolproof way to do this is the **factor-label method,** a procedure in which the equations are set up so that all unwanted units cancel and only the desired units remain. In this method, we follow the rule that *when multiplying numbers, we also multiply units and when dividing numbers, we also divide units.*

For conversions between one unit and another, it is always possible to set up two fractions, called **conversion factors.** Suppose we wish to convert the weight of an object from 381 grams to pounds. We are converting the units, but we are not changing the object itself. We want a ratio that reflects the change in units. In Table 1-3, we see that there are 453.6 grams in 1 pound. That is, the amount of matter in 453.6 grams is the same as the amount in 1 pound. In that sense, it is a one-to-one ratio, even though the units are not numerically the same. The conversion factors between grams and pounds therefore are

Conversion factors The ratio of two different units

$$\frac{1 \text{ lb}}{453.6 \text{ g}} \quad \text{and} \quad \frac{453.6 \text{ g}}{1 \text{ lb}}$$

To convert 381 grams to pounds, we must multiply by the proper conversion factor—but which one? Let us try both and see what happens.

First, let us multiply by 1 lb/453.6 g:

$$381 \text{ g} \times \frac{1 \text{ lb}}{453.6 \text{ g}} = 0.840 \text{ lb}$$

Following the procedure of multiplying and dividing units when we multiply and divide numbers, we find that dividing grams by grams cancels out the grams. We are left with pounds, which is the answer we want. Thus, 1 lb/453.6 g is the correct conversion factor because it converts grams to pounds.

Suppose we had done it the other way, multiplying by 453.6 g/1 lb:

$$381 \text{ g} \times \frac{453.6 \text{ g}}{1 \text{ lb}} = 173,000 \frac{\text{g}^2}{\text{lb}} \left(1.73 \times 10^5 \frac{\text{g}^2}{\text{lb}}\right)$$

When we multiply grams by grams, we get g^2 (grams squared). Dividing by pounds gives g^2/lb. This is not the unit we want, so we used the incorrect conversion factor.

HOW TO...

Do Unit Conversions by the Factor-Label Method

One of the most useful ways of approaching conversions is to ask three questions:

- What information am I given? This is the starting point.
- What do I want to know? This is the answer that you want to find.
- What is the connection between the first two? This is the conversion factor. Of course, more than one conversion factor may be needed for some problems.

Let's look at how to apply these principles to a conversion from pounds to kilograms. Suppose we want to know the weight in kilograms of a woman who weighs 125 lb. We see in Table 1-3 that there are 2.205 lb in 1 kg. Note that we are starting out with pounds and we want an answer in kilograms.

$$125 \text{ lb} \times \frac{1 \text{ kg}}{2.205 \text{ lb}} = 56.7 \text{ kg}$$

- The mass in pounds is the starting point. We were given that information.

- We wanted to know the mass in kilograms. That was the desired answer, and we found the number of kilograms.
- The connection between the two is the conversion factor in which the unit of the desired answer is in the numerator of the fraction, rather than the denominator. It is not simply a mechanical procedure to set up the equation so that units cancel; it is a first step to understanding the underlying reasoning behind the factor-label method. If you set up the equation to give the desired unit as the answer, you have made the connection properly.

If you apply this kind of reasoning, you can always pick the right conversion factor. Given the choice between

$$\frac{2.205 \text{ lb}}{1 \text{ kg}} \quad \text{and} \quad \frac{1 \text{ kg}}{2.205 \text{ lb}}$$

you know that the second conversion factor will give an answer in kilograms, so you use it. When you check the answer, you see that it is reasonable. You expect a number that is about one half of 125, which is 62.5. The actual answer, 56.7, is close to that value. The number of pounds and the number of kilograms are not the same, but they represent the same mass. That fact makes the use of conversion factors logically valid; the factor-label method uses the connection to obtain a numerical answer.

The advantage of the factor-label method is that it lets us know when we have made an incorrect calculation. *If the units of the answer are not the ones we are looking for, the calculation must be wrong.* Incidentally, this principle works not only in unit conversions but in all problems where we make calculations using measured numbers. Keeping track of units is a sure-fire way of doing conversions. It is impossible to overemphasize the importance of this way of checking on calculations.

The factor-label method gives the correct mathematical solution for a problem. However, it is a mechanical technique and does not require you to think through the problem. Thus, it may not provide a deeper understanding. For this reason and also to check your work (because it is easy to make mistakes in arithmetic—for example, by punching the wrong numbers into a calculator), you should always ask yourself if the answer you have obtained is reasonable. For example, the question might ask the mass of a single oxygen atom. If your answer comes out 8.5×10^6 g, it is not reasonable. A single atom cannot weigh more than you do! In such a case, you have obviously made a mistake and should take another look to see where you went wrong. Of course, everyone makes mistakes at times, but if you check, you can at least determine whether your answer is reasonable. If it is not, you will immediately know that you have made a mistake and can then correct it.

Checking whether an answer is reasonable gives you a deeper understanding of the problem because it forces you to think through the relationship between the question and the answer. The concepts and the mathematical relationships in these problems go hand in hand. Mastery of the mathematical skills makes the concepts clearer, and insight into the concepts suggests ways to approach the mathematics. We will now give a few examples of unit conversions and then test the answers to see whether they are reasonable. To save space, we will practice this technique mostly in this chapter, but you should use a similar approach in all later chapters.

In unit conversion problems, you should always check two things. First, the numeric factor by which you multiply tells you whether the answer will be larger or smaller than the number being converted. Second, the factor tells you how much greater or smaller your answer should be when compared to your starting number. For example, if 100 kg is converted to pounds and there are 2.205 lb in 1 kg, then an answer of about 200 is reasonable—but an answer of 0.2 or 2000 (2.00×10^3) is not.

Interactive Example 1-3 — Unit Conversion: Volume

The label on a container of olive oil says 1.844 gal. How many milliliters does the container hold?

Strategy
Here we use two conversion factors, rather than a single one. We still need to keep track of units.

Solution
Table 1-3 shows no factor for converting gallons to milliliters, but it does show that 1 gal = 3.785 L. Because we know that 1000 mL = 1 L, we can solve this problem by multiplying by two conversion factors, making certain that all units cancel except milliliters:

$$1.844 \text{ gal} \times \frac{3.785 \text{ L}}{1 \text{ gal}} \times \frac{1000 \text{ mL}}{1 \text{ L}} = 6980. \text{ mL}$$

Is this answer reasonable? The conversion factor in Table 1-3 tells us that there are more liters in a given volume than gallons. How much more? Approximately four times more. We also know that any volume in milliliters is 1000 times larger than the same volume in liters. Thus, we expect that the volume expressed in milliliters will be 4×1000, or 4000 times more than the volume given in gallons. The estimated volume in milliliters will be approximately 1.8×4000, or 7000 mL. But we also expect that the actual answer should be somewhat less than the estimated figure because we overestimated the conversion factor (4 rather than 3.785). Thus, the answer, 6980. mL, is quite reasonable. Note that the answer is given to four significant figures. The decimal point after the zero makes that point clear. We do not need a period after 1000 in the defined conversion factor; that is an exact number.

Problem 1-3
Calculate the number of kilometers in 8.55 miles. Check your answer to see whether it is reasonable.

Interactive Example 1-4 — Unit Conversion: Multiple Units

The maximum speed limit on many roads in the United States is 65 mi/h. How many meters per second (m/s) is this speed?

Strategy
We use four conversion factors in succession. It is more important than ever to keep track of units.

Solution

Here, we have essentially a double conversion problem: We must convert miles to meters and hours to seconds. We use as many conversion factors as necessary, always making sure that we use them in such a way that the proper units cancel:

$$65 \frac{\cancel{\text{mi}}}{\cancel{\text{h}}} \times \frac{1.609 \, \cancel{\text{km}}}{1 \, \cancel{\text{mi}}} \times \frac{1000 \, \text{m}}{1 \, \cancel{\text{km}}} \times \frac{1 \, \cancel{\text{h}}}{60 \, \cancel{\text{min}}} \times \frac{1 \, \cancel{\text{min}}}{60 \, \text{s}} = 29 \, \frac{\text{m}}{\text{s}}$$

Is this answer reasonable? To estimate the 65 mi/h speed in meters per second, we must first establish the relationship between miles and meters. As there are approximately 1.5 km in 1 mi, there must be approximately 1500 times more meters. We also know that in one hour, there are 60 × 60 = 3600 seconds. The ratio of meters to seconds will be approximately 1500/3600, which is about one half. Therefore, we estimate that the speed in meters per second will be about one half of that in miles per hour, or 32 m/s. Once again, the actual answer, 29 m/s, is not far from the estimate of 32 m/s, so the answer is reasonable.

Problem 1-4

Convert the speed of sound, 332 m/s to mi/h. Check your answer to see whether it is reasonable.

EXAMPLE 1-5 — Unit Conversion: Multiple Units and Health Care

A physician recommends adding 100. mg of morphine to 500. cc of IV fluid and administering it at a rate of 20. cc/h to alleviate a patient's pain. Determine how many grams per second (g/s) the patient is receiving.

Strategy

Here, we use four conversion factors, rather than a single one. It is important to keep track of the desired units in the calculation and set up the other units in a way that allows us to cancel them out. It is also important to note that certain conversion factors do not need to be looked up in a table. Instead, they can be found in the problem (100. mg = 500. cc and 20. cc = 1 h).

Solution

We must convert the numerator from milligrams to grams and the denominator from cubic centimeters to seconds using the provided information. Because we know that 1000 mg = 1 g, 60 min = 1 h, and 60 s = 1 min, we can solve this problem by multiplying these conversion factors and making sure the units provided in the problem cancel out:

$$\frac{100. \, \cancel{\text{mg}}}{500. \, \cancel{\text{cc}}} \times \frac{20. \, \cancel{\text{cc}}}{1 \, \cancel{\text{h}}} \times \frac{1 \, \text{g}}{1000 \, \cancel{\text{mg}}} \times \frac{1 \, \cancel{\text{h}}}{60 \, \cancel{\text{min}}} \times \frac{1 \, \cancel{\text{min}}}{60 \, \text{s}} = 1.1 \times 10^{-6} \, \frac{\text{g}}{\text{s}}$$

Is this answer reasonable? Because this problem involves manipulating various conversion units, one way to estimate the final answer is to examine the ratio of the numerator to denominator. In this case, we know from our setup that the answer has to be less than one since it is obtained by dividing by a larger quantity than itself.

As shown in these examples, when canceling units, we do not cancel the numbers. The numbers are multiplied and divided in the ordinary way.

Problem 1-5

An intensive care patient is receiving an antibiotic IV at the rate of 50. mL/h. The IV solution contains 1.5 g of the antibiotic in 1000. mL. Calculate the mg/min of the drip. Check your answer to see if it is reasonable.

1-6 What Are the States of Matter?

Matter can exist in three states: gas, liquid, and solid. **Gases** have no definite shape or volume. They expand to fill whatever container they are put into. On the other hand, they are highly compressible and can be forced into small containers. Liquids also have no definite shape, but they do have a definite volume that remains the same when they are poured from one container to another. **Liquids** are only slightly compressible. **Solids** have definite shapes and definite volumes. They are essentially incompressible.

Whether a substance is a gas, a liquid, or a solid depends on its temperature and pressure. On a cold winter day, a puddle of liquid water turns to ice; it becomes a solid. If we heat water in an open pot at sea level, the liquid boils at 100°C; it becomes a gas—we call it steam. If we heated the same pot of water on the top of Mount Everest, it would boil at about 70°C due to the reduced atmospheric pressure. Most substances can exist in the three states: they are gases at high temperature, liquids at a lower temperature, and solids when their temperature becomes low enough. Figure 1-5 shows a single substance in the three different states.

The chemical identity of a substance does not change when it is converted from one state to another. Water is still water whether it is in the form of ice, steam, or liquid water. We discuss the three states of matter, and the changes between one state and another, at greater length in Chapter 5.

(a)

(b)

(c)

FIGURE 1-5 The three states of matter for bromine: (a) bromine as a solid, (b) bromine as a liquid, and (c) bromine as a gas.

1-7 What Are Density and Specific Gravity?

A. Density

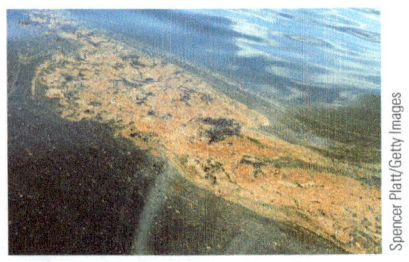

▶ The *Deepwater Horizon* oil spill (also referred to as the BP oil spill) in April 2010 flowed for three months in the Gulf of Mexico, releasing about 200 million barrels of crude oil. It is the largest accidental marine oil spill in the history of the petroleum industry. The spill continues to cause extensive damage to marine and wildlife habitats, as well as the Gulf's fishing and tourism industries.

One of the many pollution problems that the world faces is the spillage of petroleum into the oceans from oil tankers or from offshore drilling. When oil spills into the ocean, it floats on top of the water. ◀ The oil doesn't sink because it is not soluble in water and because water has a higher density than oil. When two liquids are mixed (assuming that one does not dissolve in the other), the one of lower density floats on top (Figure 1-6).

The **density** of any substance is defined as its *mass per unit volume*. Not only do all liquids have a density, but so do all solids and gases. Density is calculated by dividing the mass of a substance by its volume:

$$d = \frac{m}{V} \quad d = \text{density}, \ m = \text{mass}, \ V = \text{volume}$$

EXAMPLE 1-6 Density Calculations

If 73.2 mL of a liquid has a mass of 61.5 g, what is its density in g/mL?

Strategy

We use the formula for density and substitute the values we are given for mass and volume.

Solution

$$d = \frac{m}{V} = \frac{61.5 \text{ g}}{73.2 \text{ mL}} = 0.840 \ \frac{\text{g}}{\text{mL}}$$

Problem 1-6

The density of titanium is 4.54 g/mL. What is the mass, in grams, of 17.3 mL of titanium? Check your answer to see whether it is reasonable.

EXAMPLE 1-7 Using Density to Find Volume

The density of iron is 7.86 g/cm³. What is the volume in milliliters of an irregularly shaped piece of iron that has a mass of 524 g?

Strategy

We are given density and mass. The volume is the unknown quantity in the equation. We substitute the known quantities in the formula for density and solve for volume.

Solution

Here, we are given the mass and the density. In this type of problem, it is useful to derive a conversion factor from the density. Since 1 cm³ is exactly 1 mL, we know that the density is 7.86 g/mL. This means that 1 mL of iron has a mass of 7.86 g. From this, we can get two conversion factors:

$$\frac{1 \text{ mL}}{7.86 \text{ g}} \quad \text{and} \quad \frac{7.86 \text{ g}}{1 \text{ mL}}$$

FIGURE 1-6 Two separatory funnels containing water and another liquid. The density of carbon tetrachloride is 1.589 g/mL, that of water is 1.00 g/mL, and that of diethyl ether is 0.713 g/mL. In each case, the liquid with the lower density is on top.

As usual, we multiply the mass by whichever conversion factor results in the cancellation of all but the correct unit:

$$524 \text{ g} \times \frac{1 \text{ mL}}{7.86 \text{ g}} = 66.7 \text{ mL}$$

Is this answer reasonable? The density of 7.86 g/mL tells us that the volume in milliliters of any piece of iron is always less than its mass in grams. How much less? Approximately eight times less. Thus, we expect the volume to be approximately 500/8 = 63 mL. As the actual answer is 66.7 mL, it is reasonable.

Problem 1-7

An unknown substance has a mass of 56.8 g and occupies a volume of 23.4 mL. What is its density in g/mL? Check your answer to see whether it is reasonable.

The density of any liquid or solid is a physical property that is constant, which means that it always has the same value at a given temperature. We use physical properties to help identify a substance. For example, the density of chloroform (a liquid formerly used as an inhalation anesthetic) is 1.483 g/mL at 20°C. If we want to find out if an unknown liquid is chloroform, one thing we might do is measure its density at 20°C. If the density is, say, 1.355 g/mL, we know the liquid isn't chloroform. If the density is 1.483 g/mL, we cannot be sure the liquid is chloroform, because other liquids might also have this density, but we can then measure other physical properties (the boiling point, for example). If all the physical properties we measure match those of chloroform, we can be reasonably sure the liquid is chloroform.

We have said that the density of a pure liquid or solid is a constant at a given temperature. Density does change when the temperature changes. Almost always, density decreases with increasing temperature. This is true because mass does not change when a substance is heated, but volume almost always increases because atoms and molecules tend to get farther apart as the temperature increases. Since $d = m/V$, if m stays the same and V gets larger, d must get smaller.

The most common liquid, water, provides a partial exception to this rule. As the temperature increases from 4°C to 100°C, the density of water does decrease, but from 0°C to 4°C, the density increases. That is, water has its maximum density at 4°C. This anomaly and its consequences are due to the unique structure of water and will be discussed in Chemical Connections 5D.

B. Specific Gravity

Because density is equal to mass divided by volume, it always has units, most commonly g/mL or g/cc or (g/L for gases). **Specific gravity** is numerically the same as density, but it has no units (it is dimensionless). The reason why there are no units is because specific gravity is defined as a comparison of the density of a substance with the density of water, which is taken as a standard. For example, the density of copper at 20°C is 8.92 g/mL. The density of water at the same temperature is 1.00 g/mL. Therefore, copper is 8.92 times as dense as water, and its specific gravity at 20°C is 8.92. Because water is taken as the standard and because the density of water is 1.00 g/mL at 20°C, the specific gravity of any substance is always numerically equal to its density, provided that the density is measured in g/mL or g/cc.

FIGURE 1-7 Urinometer.

Specific gravity is often measured by a hydrometer. This simple device consists of a weighted glass bulb that is inserted into a liquid and allowed to float. The stem of the hydrometer has calibration marks, and the specific gravity is read where the meniscus (the curved surface of the liquid) hits the marking. The specific gravity of the acid in your car battery and that of a urine sample in a clinical laboratory are measured by hydrometers. A hydrometer measuring a urine sample is also called a urinometer (Figure 1-7). Normal urine can vary in specific gravity from about 1.010 to 1.030. Patients with diabetes mellitus have an abnormally high specific gravity of their urine samples, while those with some forms of kidney disease have an abnormally low specific gravity.

EXAMPLE 1-8 Specific Gravity

The density of ethanol at 20°C is 0.789 g/mL. What is its specific gravity?

Strategy

We use the definition of specific gravity.

Solution

$$\text{Specific gravity} = \frac{0.789 \text{ g/mL}}{1.00 \text{ g/mL}} = 0.789$$

Problem 1-8

The specific gravity of a urine sample at 20°C is 1.016. What is its density, in g/mL?

1-8 How Do We Describe the Various Forms of Energy?

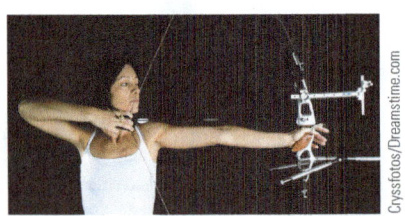

▶ Potential energy is stored in this drawn bow and becomes kinetic energy in the arrow when released.

Energy is defined as the capacity to do work. It can be described as being either kinetic energy or potential energy.

Kinetic energy (KE) is the energy of motion. Any object that is moving possesses kinetic energy. We can calculate how much kinetic energy by the formula $KE = \frac{1}{2}mv^2$, where m is the mass of the object and v is its velocity. This means that kinetic energy increases (1) when an object moves faster and (2) when a heavier object is moving. When a truck and a bicycle are moving at the same velocity, the truck has more kinetic energy.

Potential energy is stored energy. ◀ The potential energy possessed by an object arises from its capacity to move or to cause motion. For example, body weight in the up position on a seesaw contains potential energy—it is capable of doing work. If given a slight push, it will move down. The potential energy of the body in the up position is converted to kinetic energy as the body moves down on the seesaw. Work is done by gravity in the process. Figure 1-8 shows another way in which potential energy is converted to kinetic energy.

An important principle in nature is that things have a tendency to seek their lowest possible potential energy. We all know that water always flows downhill and not uphill.

Several forms of energy exist. The most important are (1) mechanical energy, light, heat, and electrical energy, which are examples of kinetic energy possessed by all moving objects, whether elephants or molecules or electrons, and (2) chemical energy and nuclear energy, which are examples of potential energy or stored energy. In chemistry, the more common form of potential energy is chemical energy—the energy stored within chemical substances

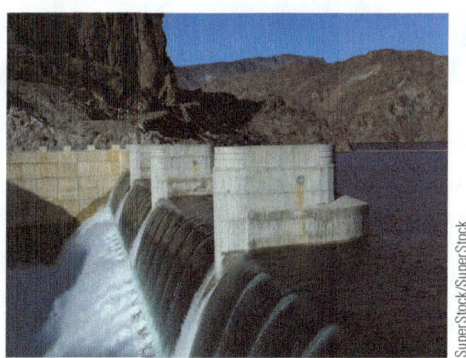

FIGURE 1-8 The water held back by the dam possesses potential energy, which is converted to kinetic energy when the water is released.

and given off when they take part in a chemical reaction. For example, a log possesses chemical energy. When the log is ignited in a fireplace, the chemical energy (potential) of the wood is turned into energy in the form of heat and light. Specifically, the potential energy has been transformed into thermal energy (heat makes molecules move faster) and the radiant energy of light.

The various forms of energy can be converted from one to another. In fact, we make such conversions all the time. A power plant operates either on the chemical energy derived from burning fuel or on nuclear energy. This energy is converted to heat, which is converted to the electricity that is sent over transmission wires into houses and factories. There, the electricity is converted to light, heat (in an electrical heater, for example), or mechanical energy (in the motors of refrigerators, vacuum cleaners, and other devices). ◄

Although one form of energy can be converted to another, the *total amount of energy in any system does not change. Energy can be neither created nor destroyed.* This statement is called the **law of conservation of energy.***

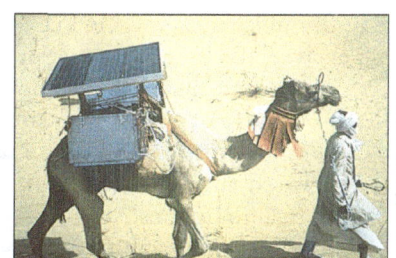

▶ An example of energy conversion. Light energy from the sun is converted to electrical energy by solar cells. The electricity runs a refrigerator on the back of the camel, keeping the vaccines cool so that they can be delivered to remote locations.

1-9 How Do We Describe Heat and the Ways in Which It Is Transferred?

A. Heat and Temperature

One form of energy that is particularly important in chemistry is **heat.** This is the form of energy that most frequently accompanies chemical reactions. Heat is not the same as temperature, however. Heat is a form of energy, but temperature is not.

The difference between heat and temperature can be seen in the following example. If we have two beakers, one containing 100 mL of water and the other containing 1 L of water at the same temperature, the heat content of the water in the larger beaker is ten times that of the water in the smaller beaker, even though the temperature is the same in both. If you were to dip your hand accidentally into a liter of boiling water, you would be much more severely burned than if only one drop fell on your hand. Even though the water is at the same temperature in both cases, the liter of boiling water has much more heat.

As we saw in Section 1-4, temperature is measured in degrees. Heat can be measured in various units, the most common of which is the **calorie,** which is defined as the amount of heat necessary to raise the temperature

---

*This statement is not completely true. As discussed in Sections 9-8 and 9-9, it is possible to convert matter to energy, and vice versa. Therefore, a more correct statement would be *matter-energy can be neither created nor destroyed*. However, the law of conservation of energy is valid for most purposes and is highly useful.

CHEMICAL CONNECTIONS 1B

Hypothermia and Hyperthermia

The human body cannot tolerate temperatures that are too low. A person outside in very cold weather (say, −20°F [−29°C]) who is not protected by heavy clothing will eventually freeze to death because the body loses heat. Normal body temperature is 37°C. When the outside temperature is lower than that, heat flows out of the body. When the air temperature is moderate (10°C to 25°C), this poses no problem and is, in fact, necessary because the body produces more heat than it needs and must lose some. At extremely low temperatures, however, too much heat is lost and body temperature drops, a condition called **hypothermia.** A drop in body temperature of 1 or 2°C causes shivering, which is the body's attempt to increase its temperature by the heat generated through muscular action. An even greater drop results in unconsciousness and eventually death.

The opposite condition is **hyperthermia.** It can be caused either by high outside temperatures or by the body itself when an individual develops a high fever. A sustained body temperature as high as 41.7°C (107°F) is usually fatal.

The label on this sleeping bag indicates the temperature range in which it can be used safely.

Test your knowledge with Problems 1-71 and 1-72.

---

of 1 g of liquid water by 1°C. This is a small unit, and chemists more often use the kilocalorie (kcal):

$$1 \text{ kcal} = 1000 \text{ cal}$$

Nutritionists use the word "Calorie" (with a capital "C") to mean the same thing as "kilocalorie"; that is, 1 Cal = 1000 cal = 1 kcal. The calorie is not part of the SI. The official SI unit for heat is the **joule (J),** which is about one-fourth of a calorie:

$$1 \text{ cal} = 4.184 \text{ J}$$

B. Specific Heat

As we noted, it takes 1 cal to raise the temperature of 1 g of liquid water by 1°C. **Specific heat (SH)** is the amount of heat necessary to raise the temperature of 1 g of any substance by 1°C. Each substance has its own specific heat, which is a physical property of that substance, like density or melting point. Table 1-4

Table 1-4 Specific Heats for Some Common Substances

Substance	Specific Heat (cal/g · °C)	Substance	Specific Heat (cal/g · °C)
Water	1.00	Wood (typical)	0.42
Ice	0.48	Glass (typical)	0.22
Steam	0.48	Rock (typical)	0.20
Iron	0.11	Ethanol	0.59
Aluminum	0.22	Methanol	0.61
Copper	0.092	Ether	0.56
Lead	0.031	Carbon tetrachloride	0.21

CHEMICAL CONNECTIONS 1C

Cold Compresses, Waterbeds, and Lakes

The high specific heat of water is useful in cold compresses and makes them last a long time. For example, consider two patients with cold compresses: one compress made by soaking a towel in water and the other made by soaking a towel in ethanol. Both are at 0°C. Each gram of water in the water compress requires 25 cal to make the temperature of the compress rise to 25°C (after which it must be changed). Because the specific heat of ethanol is 0.59 cal/g · °C (see Table 1-4), each gram of ethanol requires only 15 cal to reach 25°C. If the two patients give off heat at the same rate, the ethanol compress is less effective because it will reach 25°C a good deal sooner than the water compress and will need to be changed sooner.

The high specific heat of water also means that it takes a great deal of heat to increase its temperature. That is why it takes a long time to get a pot of water to boil. Anyone who has a waterbed (300 gallons) knows that it takes days for the heater to bring the bed up to the desired temperature. It is particularly annoying when an overnight guest tries to adjust the temperature of your waterbed because the guest will probably have left before the change is noticed, but then you will have to set it back to your favorite temperature. This same effect in reverse explains why the outside temperature can be below zero (°C) for weeks before a lake will freeze. Large bodies of water do not change temperature very quickly.

The ice on this lake will take days, or even weeks, to melt in the spring.

Test your knowledge with Problem 1-73.

---

lists specific heats for a few common substances. For example, the specific heat of iron is 0.11 cal/g · °C. Therefore, if we had 1 g of iron at 20°C, it would require only 0.11 cal to increase the temperature to 21°C. Under the same conditions, aluminum would require twice as much heat. Thus, cooking in an aluminum pan of the same weight as an iron pan would require more heat than cooking in the iron pan. Note from Table 1-4 that ice and steam do not have the same specific heat as liquid water.

It is easy to make calculations involving specific heats. The equation is

Amount of heat = specific heat × mass × change in temperature

Amount of heat = SH × m × ΔT

where ΔT is the change in temperature.

We can also write this equation as

$$\text{Amount of heat} = \text{SH} \times m \times (T_2 - T_1)$$

where T_2 is the final temperature and T_1 is the initial temperature in °C.

EXAMPLE 1-9 Specific Heat

How many calories are required to heat 352 g of water from 23°C to 95°C?

Strategy

We use the equation for the amount of heat and substitute the values given for the mass of water and the temperature change. We have already seen the value for the specific heat of water.

Solution

$$\text{Amount of heat} = SH \times m \times \Delta T$$

$$\text{Amount of heat} = SH \times m \times (T_2 - T_1)$$

$$= \frac{1.00 \text{ cal}}{g \cdot °C} \times 352 \text{ g} \times (95 - 23)°C$$

$$= 2.5 \times 10^4 \text{ cal}$$

Is this answer reasonable? Each gram of water requires one calorie to raise its temperature by one degree. We have approximately 350 g of water. To raise its temperature by one degree would therefore require approximately 350 calories. But we are raising the temperature not by one degree but by approximately 70 degrees (from 23 to 95). Thus, the total number of calories will be approximately $70 \times 350 = 24,500$ cal, which is close to the calculated answer. (Even though we were asked for the answer in calories, we should note that it will be more convenient to convert the answer to 25 kcal. We are going to see that conversion from time to time.)

Problem 1-9

How many calories are required to heat 731 g of water from 8°C to 74°C? Check your answer to see whether it is reasonable.

EXAMPLE 1-10 Specific Heat and Temperature Change

If we add 450. cal of heat to 37 g of ethanol at 20.°C, what is the final temperature?

Strategy

The equation we have has a term for temperature change. We use the information we are given to calculate that change. We then use the value we are given for the initial temperature and the change to find the final temperature.

Solution

The specific heat of ethanol is 0.59 cal/g · °C (see Table 1-4).

$$\text{Amount of heat} = SH \times m \times \Delta T$$

$$\text{Amount of heat} = SH \times m \times (T_2 - T_1)$$

$$450. \text{ cal} = 0.59 \text{ cal/g} \cdot °C \times 37 \text{ g} \times (T_2 - T_1)$$

We can show the units in fraction form by rewriting this equation.

$$450. \text{ cal} = 0.59 \frac{\text{cal}}{g \cdot °C} \times 37 \text{ g} \times (T_2 - T_1)$$

$$(T_2 - T_1) = \frac{\text{amount of heat}}{SH \times m}$$

$$(T_2 - T_1) = \frac{450. \text{ cal}}{\left[\frac{0.59 \text{ cal} \times 37 \text{ g}}{g \cdot °C}\right]} = \frac{21}{1/°C} = 21°C$$

(Note that we have the reciprocal of temperature in the denominator, which gives us temperature in the numerator. The answer has units of degrees Celsius). Because the starting temperature is 20°C, the final temperature is 41°C.

Is this answer reasonable? The specific heat of ethanol is 0.59 cal/g · °C. This value is close to 0.5, meaning that about half a calorie will raise the temperature of 1 g by 1°C. However, 37 g of ethanol need approximately 40 times as many calories for a rise, and $40 \times \frac{1}{2} = 20$ calories. We are adding 450. calories, which is about 20 times as much. Thus, we expect the temperature to rise by about 20°C, from 20°C to 40°C. The actual answer, 41°C, is quite reasonable.

Problem 1-10
A 100 g piece of iron at 25°C is heated by adding 230. cal. What will be the final temperature? Check your answer to see whether it is reasonable.

EXAMPLE 1-11 Calculating Specific Heat

We heat 50.0 g of an unknown substance by adding 205 cal, and its temperature rises by 7.0°C. What is its specific heat? Using Table 1-4, identify the substance.

Strategy
We solve the equation for specific heat by substituting the values for mass, amount of heat, and temperature change. We compare the number we obtain with the values in Table 1-4 to identify the substance.

Solution

$$SH = \frac{\text{Amount of heat}}{m \times (\Delta T)}$$

$$SH = \frac{\text{Amount of heat}}{m \times (T_2 - T_1)}$$

$$SH = \frac{205 \text{ cal}}{50.0 \text{ g} \times 7.0°C} = 0.59 \text{ cal/g} \cdot °C$$

The substance in Table 1-4 having a specific heat of 0.59 cal/g · °C is ethanol.

Is this answer reasonable? If we had water instead of an unknown substance with SH = 1 cal/g · °C, raising the temperature of 50.0 g by 7.0°C would require $50 \times 7.0 = 350$ cal. But we added only approximately 200 cal. Therefore, the SH of the unknown substance must be less than 1.0. How much less? Approximately $200/350 = 0.6$. The actual answer, 0.59 cal/g · °C, is quite reasonable.

Problem 1-11
It required 88.2 cal to heat 13.4 g of an unknown substance from 23°C to 176°C. What is the specific heat of the unknown substance? Check your answer to see whether it is reasonable.

Summary of Key Questions

Section 1-1 Why Do We Call Chemistry the Study of Matter?
- **Chemistry** is the science that deals with the structure of matter and the changes it can undergo. In a **chemical change,** or **chemical reaction,** substances are used up and others are formed.
- Chemistry is also the study of energy changes during chemical reactions. In **physical changes,** substances do not change their identity.

Section 1-2 What Is the Scientific Method?
- The **scientific method** is a tool used in science and medicine. The heart of the scientific method is the testing of **hypotheses** and **theories** by collecting facts.

Section 1-3 How Do Scientists Report Numbers?
- Because we frequently use very large or very small numbers, we use powers of 10 to express these numbers more conveniently, a method called **exponential notation.**
- With exponential notation, we no longer have to keep track of so many zeros, and we have the added convenience of being able to see which digits convey information (**significant figures**) and which merely indicate the position of the decimal point.

Section 1-4 How Do We Make Measurements?
- In chemistry, we use the **metric system** for measurements.
- The base units are the meter for length, the liter for volume, the gram for mass, the second for time, and the joule for heat. Other units are indicated by prefixes that represent powers of 10. Temperature is measured in degrees Celsius or in Kelvins.

Section 1-5 What Is a Handy Way to Convert from One Unit to Another?
- Conversions from one unit to another are best done by the **factor-label method,** in which units are multiplied and divided to yield the units requested in the answer.

Section 1-6 What Are the States of Matter?
- There are three states of matter: **solid, liquid,** and **gas.**

Section 1-7 What Are Density and Specific Gravity?
- **Density** is mass per unit volume. **Specific gravity** is density relative to water and thus has no units. Density usually decreases with increasing temperature.

Section 1-8 How Do We Describe the Various Forms of Energy?
- **Kinetic energy** is energy of motion; **potential energy** is stored energy. Energy can be neither created nor destroyed, but it can be converted from one form to another.
- Examples of kinetic energy are mechanical energy, light, heat, and electrical energy. Examples of potential energy are chemical energy and nuclear energy.

Section 1-9 How Do We Describe Heat and the Ways in Which It Is Transferred?
- **Heat** is a form of energy and is measured in calories. A calorie is the amount of heat necessary to raise the temperature of 1 g of liquid water by 1°C.
- Every substance has a **specific heat,** which is a physical constant. The specific heat is the number of calories required to raise the temperature of 1 g of a substance by 1°C.

Problems

Orange-numbered problems are applied.

Section 1-1 Why Do We Call Chemistry the Study of Matter?
1-12 The life expectancy of a citizen in the United States is 76 years. Eighty years ago it was 56 years. In your opinion, what was the major contributor to this spectacular increase in life expectancy? Explain your answer.

1-13 Define the following terms:
(a) Matter   (b) Chemistry

Section 1-2 What Is the Scientific Method?
1-14 In Table 1-4, you find four metals (iron, aluminum, copper, and lead) and three organic compounds (ethanol, methanol, and ether). What kind of hypothesis would you suggest about the specific heats of these chemicals?

1-15 In a newspaper, you read that Dr. X claimed that he has found a new remedy to cure diabetes. The remedy is an extract of carrots. How would you classify this claim: (a) fact, (b) theory, (c) hypothesis, or (d) hoax? Explain your choice of answer.

1-16 Classify each of the following as a chemical or physical change:
(a) Burning gasoline
(b) Making ice cubes
(c) Boiling oil
(d) Melting lead
(e) Rusting iron
(f) Making ammonia from nitrogen and hydrogen
(g) Digesting food

Section 1-3 How Do Scientists Report Numbers?

Exponential Notation

1-17 Write in exponential notation:
(a) 0.351 (b) 602.1 (c) 0.000128 (d) 628122

1-18 Write out in full:
(a) 4.03×10^5 (b) 3.2×10^3
(c) 7.13×10^{-5} (d) 5.55×10^{-10}

1-19 Multiply:
(a) $(2.16 \times 10^5)(3.08 \times 10^{12})$
(b) $(1.6 \times 10^{-8})(7.2 \times 10^8)$
(c) $(5.87 \times 10^{10})(6.6 \times 10^{-27})$
(d) $(5.2 \times 10^{-9})(6.8 \times 10^{-15})$

1-20 Divide:
(a) $\dfrac{6.02 \times 10^{23}}{2.87 \times 10^{10}}$ (b) $\dfrac{3.14}{2.93 \times 10^{-4}}$
(c) $\dfrac{5.86 \times 10^{-9}}{2.00 \times 10^3}$ (d) $\dfrac{7.8 \times 10^{-12}}{9.3 \times 10^{-14}}$
(e) $\dfrac{6.83 \times 10^{-12}}{5.02 \times 10^{14}}$

1-21 Add:
(a) $(7.9 \times 10^4) + (5.2 \times 10^4)$
(b) $(8.73 \times 10^4) + (6.7 \times 10^3)$
(c) $(3.63 \times 10^{-4}) + (4.776 \times 10^{-3})$

1-22 Subtract:
(a) $(8.50 \times 10^3) - (7.61 \times 10^2)$
(b) $(9.120 \times 10^{-2}) - (3.12 \times 10^{-3})$
(c) $(1.3045 \times 10^2) - (2.3 \times 10^{-1})$

1-23 Solve:
$$\dfrac{(3.14 \times 10^3) \times (7.80 \times 10^5)}{(5.50 \times 10^2)}$$

1-24 Solve:
$$\dfrac{(9.52 \times 10^4) \times (2.77 \times 10^{-5})}{(1.39 \times 10^7) \times (5.83 \times 10^2)}$$

Significant Figures

1-25 How many significant figures are in the following?
(a) 0.012   (b) 0.10203
(c) 36.042   (d) 8401.0
(e) 32100   (f) 0.0402
(g) 0.000012

1-26 How many significant figures are in the following?
(a) 5.71×10^{13} (b) 4.4×10^5
(c) 3×10^{-6} (d) 4.000×10^{-11}
(e) 5.5550×10^{-3}

1-27 Round off to two significant figures:
(a) 91.621   (b) 7.329
(c) 0.677   (d) 0.003249
(e) 5.88

1-28 Multiply these numbers, using the correct number of significant figures in your answer:
(a) 3630.15×6.8
(b) 512×0.0081
(c) $5.79 \times 1.85825 \times 1.4381$

1-29 Divide these numbers, using the correct number of significant figures in your answer:
(a) $\dfrac{3.185}{2.08}$ (b) $\dfrac{6.5}{3.0012}$ (c) $\dfrac{0.0035}{7.348}$

1-30 Add these groups of measured numbers using the correct number of significant figures in your answer:
(a) $37.4083 + 5.404 + 10916.3 + 3.94 + 0.0006$
(b) $84 + 8.215 + 0.01 + 151.7$
(c) $51.51 + 100.27 + 16.878 + 3.6817$

Section 1-4 How Do We Make Measurements?

1-31 In the SI system, the second is the base unit of time. We talk about atomic events that occur in picoseconds (10^{-12} s) or even in femtoseconds (10^{-15} s). But we don't talk about megaseconds or kiloseconds; the old standards of minutes, hours, and days prevail. How many minutes and hours are 20. kiloseconds?

1-32 How many grams are in the following?
(a) 1 kg   (b) 1 mg

1-33 Estimate without actually calculating which one is the shorter distance:
(a) 20 mm or 0.3 m
(b) 1 in. or 30 mm
(c) 2000 m or 1 mi

1-34 For each of these, tell which figure is closest to the correct answer:
(a) A baseball bat has a length of 100 mm or 100 cm or 100 m
(b) A glass of milk holds 23 cc or 230 mL or 23 L
(c) A man weighs 75 mg or 75 g or 75 kg
(d) A tablespoon contains 15 mL or 150 mL or 1.5 L
(e) A paper clip weighs 50 mg or 50 g or 50 kg
(f) Your hand has a width of 100 mm or 100 cm or 100 m
(g) An audiocassette weighs 40 mg or 40 g or 40 kg

1-35 You are taken for a helicopter ride in Hawaii from Kona (sea level) to the top of the volcano Mauna Kea. Which property of your body would change during the helicopter ride?
(a) height   (b) weight   (c) volume   (d) mass

1-36 Convert to Celsius and to Kelvin:
(a) 320°F (b) 212°F (c) 0°F (d) −250°F

1-37 Convert to Fahrenheit and to Kelvin:
(a) 25°C (b) 40°C (c) 250°C (d) −273°C

Section 1-5 What Is a Handy Way to Convert from One Unit to Another?

1-38 Make the following conversions (conversion factors are given in Table 1-3):
(a) 42.6 kg to lb (b) 1.62 lb to g
(c) 34 in. to cm (d) 37.2 km to mi
(e) 2.73 gal to L (f) 62 g to oz
(g) 33.61 qt to L (h) 43.7 L to gal
(i) 1.1 mi to km (j) 34.9 mL to fl oz

1-39 Make the following metric conversions:
(a) 96.4 mL to L (b) 275 mm to cm
(c) 45.7 kg to g (d) 475 cm to m
(e) 21.64 cc to mL (f) 3.29 L to cc
(g) 0.044 L to mL (h) 711 g to kg
(i) 63.7 mL to cc (j) 0.073 kg to mg
(k) 83.4 m to mm (l) 361 mg to g

1-40 There are 2 bottles of cough syrup available on the shelf at the pharmacy. One contains 9.5 oz and the other has 300. cc. Which one has the larger volume?

1-41 A humidifier located at a nursing station holds 4.00 gallons of water. How many fluid ounces of water will completely fill the reservoir?

1-42 You drive in Canada where the distances are marked in kilometers. The sign says you are 80 km from Ottawa. You are traveling at a speed of 75 mi/h. Would you reach Ottawa within one hour, after one hour, or later than that?

1-43 The speed limit in some European cities is 80 km/h. How many miles per hour is this?

1-44 Your car gets 25.00 miles on a gallon of gas. What would be your car's fuel efficiency in km/L?

1-45 Children's Chewable Tylenol contains 80. mg of acetaminophen per tablet. If the recommended dosage is 10. mg/kg, how many tablets are needed for a 70.-lb child?

1-46 A patient weighs 186 lbs. She must receive an IV medication based on body weight. The order reads, "Give 2.0 mg per kilogram." The label reads "10. mg per cc." How many mL of medication would you give?

1-47 The doctor orders administration of a drug at 120. mg per 1000. mL at 400. mL/24 h. How many mg of drug will the patient receive every 8.0 hours?

1-48 The recommended pediatric dosage of Velosef is 20. mg/kg/day. What is the daily dose in mg for a child weighing 36 pounds? If the stock vial of Velosef is labeled 208 mg/mL, how many mL would be given in a daily dose?

1-49 A critical care physician prescribes an IV of heparin to be administered at a rate of 1100 units per hour. The IV contains 26,000 units of heparin per liter. Determine the rate of the IV in cc/h.

1-50 If an IV is mixed so that each 150 mL contains 500. mg of the drug lidocaine, how many minutes will it take for 750 mg of lidocaine to be administered if the rate is set at 5 mL/min?

1-51 A nurse practitioner orders isotonic sodium lactate 50. mL/kg body mass to be administered intravenously for a 139-lb patient with severe acidosis. The rate of flow is 150 gtts/min, and the IV administration set delivers 20. gtts/mL, where the unit "gtts" stands for drops of liquid. What is the running time in minutes?

1-52 An order for a patient reads "Give 40. mg of pantoprazole IV and 5 g of $MgSO_4$ IV." The pantoprazole should be administered at a concentration of 0.4 mg/mL and the $MgSO_4$ should be administered at a concentration of 0.02 g/mL in separate IV infusion bags. What is the total fluid volume the patient has received from both IV infusions?

Section 1-6 What Are the States of Matter?

1-53 Which states of matter have a definite volume?

1-54 Will most substances be solids, liquids, or gases at low temperatures?

1-55 Does the chemical nature of a substance change when it melts from a solid to a liquid?

Section 1-7 What Are Density and Specific Gravity?

1-56 The volume of a rock weighing 1.075 kg is 334.5 mL. What is the density of the rock in g/mL? Express it to three significant figures.

1-57 The density of manganese is 7.21 g/mL, that of calcium chloride is 2.15 g/mL, and that of sodium acetate is 1.528 g/mL. You place these three solids in a liquid, in which they are not soluble. The liquid has a density of 2.15 g/mL. Which will sink to the bottom, which will stay on the top, and which will float in the middle of the liquid?

1-58 The density of titanium is 4.54 g/mL. What is the volume, in milliliters, of 163 g of titanium?

1-59 An injection of 4 mg of Valium has been prescribed for a patient suffering from muscle spasms. A sample of Valium labeled 5 mg/mL is on hand. How many mL should be injected?

1-60 The density of methanol at 20°C is 0.791 g/mL. What is the mass, in grams, of a 280 mL sample?

1-61 The density of dichloromethane, a liquid insoluble in water, is 1.33 g/cc. If dichloromethane and water are placed in a separatory funnel, which will be the upper layer?

1-62 A sample of 10.00 g of oxygen has a volume of 6702 mL. The same weight of carbon dioxide occupies 5058 mL.
(a) What is the density of each gas in g/L?
(b) Carbon dioxide is used as a fire extinguisher to cut off the fire's supply of oxygen. Do the densities of these two gases explain the fire-extinguishing ability of carbon dioxide?

1-63 Crystals of a material are suspended in the middle of a cup of water at 2°C. This means that the densities of the crystal and of the water are the same. How might you enable the crystals to rise to the surface of the water so that you can harvest them?

Section 1-8 How Do We Describe the Various Forms of Energy?

1-64 On many country roads, you see telephones powered by a solar panel. What principle is at work in these devices?

1-65 While you drive your car, your battery is being charged. How would you describe this process in terms of kinetic and potential energy?

Section 1-9 How Do We Describe Heat and the Ways in Which It Is Transferred?

1-66 How many calories are required to heat the following (specific heats are given in Table 1-4)?
(a) 52.7 g of aluminum from 100°C to 285°C
(b) 93.6 g of methanol from −35°C to 55°C
(c) 3.4 kg of lead from −33°C to 730°C
(d) 71.4 g of ice from −77°C to −5°C

1-67 If 168 g of an unknown liquid requires 2750 cal of heat to raise its temperature from 26°C to 74°C, what is the specific heat of the liquid?

1-68 The specific heat of steam is 0.48 cal/g · °C. How many kilocalories are needed to raise the temperature of 10.5 kg of steam from 120°C to 150°C?

Chemical Connections

1-69 (Chemical Connections 1A) If the recommended dose of a drug is 445 mg for a 180-lb man, what would be a suitable dose for a 135-lb man?

1-70 (Chemical Connections 1A) The average lethal dose of heroin is 1.52 mg/kg of body weight. Estimate how many grams of heroin would be lethal for a 200-lb man.

1-71 (Chemical Connections 1B) How does the body react to hypothermia?

1-72 (Chemical Connections 1B) Low temperatures often cause people to shiver. What is the function of this involuntary body action?

1-73 (Chemical Connections 1C) Which would make a more efficient cold compress, ethanol or methanol? (Refer to Table 1-4.)

Additional Problems

1-74 The meter is a measure of length. Tell what each of the following units measures:
(a) cm^3 (b) mL (c) kg (d) cal
(e) g/cc (f) joule (g) °C (h) cm/s

1-75 A brain weighing 1.0 lb occupies a volume of 620 mL. What is the specific gravity of the brain?

1-76 If the density of air is 1.25×10^{-3} g/cc, what is the mass in kilograms of the air in a room that is 5.3 m long, 4.2 m wide, and 2.0 m high?

1-77 Classify these as kinetic or potential energy:
(a) Water held by a dam
(b) A speeding train
(c) A book on its edge before falling
(d) A falling book
(e) Electric current in a lightbulb

1-78 The kinetic energy possessed by an object with a mass of 1 g moving with a velocity of 1 cm/s is called 1 erg. What is the kinetic energy, in ergs, of an athlete with a mass of 127 lb running at a velocity of 14.7 mi/h?

1-79 A European car advertises an efficiency of 22 km/L, while an American car claims an economy of 30 mi/gal. Which car is more efficient?

1-80 In Potsdam, New York, you can buy gas for US$3.93/gal. In Montreal, Canada, you pay US$1.22/L. (Currency conversions are outside the scope of this text, so you are not asked to do them here.) Which is the better buy? Is your calculation reasonable?

1-81 Shivering is the body's response to increase the body temperature. What kind of energy is generated by shivering?

1-82 When the astronauts walked on the Moon, they could make giant leaps in spite of their heavy gear.
(a) Why were their weights on the Moon so small?
(b) Were their masses different on the Moon than on the Earth?

1-83 Which of the following is the largest mass and which is the smallest?
(a) 41 g (b) 3×10^3 mg
(c) 8.2×10^6 μg (d) 4.1310×10^{-8} kg

1-84 Which quantity is bigger in each of the following pairs?
(a) 1 gigaton : 10. megaton
(b) 10. micrometer : 1 millimeter
(c) 10. centigram : 200. milligram

1-85 In Japan, high-speed "bullet trains" move with an average speed of 220. km/h. If Dallas and Los Angeles were connected by such a train, how long would it take to travel nonstop between these cities (a distance of 1490. miles)?

1-86 The specific heats of some elements at 25°C are as follows: aluminum = 0.215 cal/g · °C; carbon (graphite) = 0.170 cal/g · °C; iron = 0.107 cal/g · °C; mercury = 0.0331 cal/g · °C.
(a) Which element would require the smallest amount of heat to raise the temperature of 100 g of the element by 10°C?
(b) If the same amount of heat needed to raise the temperature of 1 g of aluminum by 25°C were applied to 1 g of mercury, by how many degrees would its temperature be raised?
(c) If a certain amount of heat is used to raise the temperature of 1.6 g of iron by 10°C, the temperature of 1 g of which element would also be raised by 10°C, using the same amount of heat?

1-87 Water that contains deuterium rather than ordinary hydrogen (see Section 2-4D) is called heavy water. The specific heat of heavy water at 25°C is 4.217 J/g · °C. Which requires more energy to raise the temperature of 10.0 g by 10°C, water or heavy water?

1-88 One quart of milk costs 80 cents and one liter costs 86 cents. Which is the better buy?

1-89 Consider butter, density 0.860 g/mL, and sand, density 2.28 g/mL.
(a) If 1.00 mL of butter is thoroughly mixed with 1.00 mL of sand, what is the density of the mixture?
(b) What would be the density of the mixture if 1.00 g of the same butter were mixed with 1.00 g of the same sand?

1-90 Which speed is the fastest?
(a) 70 mi/h (b) 140 km/h
(c) 4.5 km/s (d) 48 mi/min

1-91 In calculating the specific heat of a substance, the following data are used: mass = 92.15 g; heat = 3.200 kcal; rise in temperature = 45°C. How many significant figures should you report in calculating the specific heat?

1-92 A solar cell generates 500. kJ of energy per hour. To keep a refrigerator at 4°C, one needs 250. kcal/h. Can the solar cell supply sufficient energy per hour to maintain the temperature of the refrigerator?

1-93 The specific heat of urea is 1.339 J/g · °C. If one adds 60.0 J of heat to 10.0 g of urea at 20°C, what would be the final temperature?

1-94 You are waiting in line in a coffee shop. As you look at the selections, you see that the decaffeinated coffee is labeled "chemical-free." Comment on this label in light of the material in Section 1-1.

1-95 You receive an order for 60. mg of meperidine (Demerol) for your postsurgical patient. The injection syringe is prepackaged with 75 mg/mL. How many mL will you administer?

1-96 You are on vacation in Europe. You have bought a loaf of bread for a picnic lunch, and you want to buy some cheese to go with it. Do you buy 200. mg, 200. g, or 200. kg?

1-97 You have just left Tucson, Arizona, on I-19 to go on a trip to Mexico. Distances on this highway are shown in kilometers. A sign says that the border crossing is 95 km away. You estimate that is about 150 miles to go. When you get to the border, you find that you have traveled less than 60 miles. What went wrong in your calculation?

1-98 The antifreeze-coolant compound used in cars does not have the same density as water. Would a hydrometer be useful for measuring the amount of antifreeze in the cooling system?

1-99 In photosynthesis, light energy from the sun is used to produce sugars. How does this process represent a conversion of energy from one form to another?

1-100 What is the difference between aspirin tablets that contain 81 mg of aspirin and tablets that contain 325 mg?

1-101 In Canada, a sign indicates that the current temperature is 30°C. Are you most likely to be wearing a down parka and wool slacks, jeans and a long-sleeved shirt, or shorts and a T-shirt? What is the reason for your answer?

1-102 In very cold weather, ice fishing enthusiasts build small structures on the ice, drill holes, and put their fishing lines through the holes in the ice. How can fish survive under these conditions?

1-103 Most solids have a higher density than the corresponding liquid. Ice is less dense than water, with corresponding expansion on freezing. How can this property be used to disrupt living cells by cycles of freezing and thawing?

1-104 A scientist claims to have found a treatment for ear infections in children. All the patients given this treatment showed improvement within three days. What comments do you have on this report?

Special Categories

Three special categories of problems—Tying It Together, Looking Ahead, and Challenge Problems—will appear from time to time at the ends of chapters. Not every chapter will have these problems, but they will appear to make specific points.

Tying It Together

1-105 Heats of reaction are frequently measured by monitoring the change in temperature of a water bath in which the reaction mixture is immersed. A water bath used for this purpose contains 2.000 L of water. In the course of the reaction, the temperature of the water rose 4.85°C. How many calories were liberated by the reaction? (You will need to use what you know about unit conversions and apply that information to what you know about energy and heat.)

1-106 You have samples of urea (a solid at room temperature) and pure ethanol (a liquid at room temperature). Which technique or techniques would you use to measure the amount of each substance?

Looking Ahead

1-107 You have a sample of material used in folk medicine. Suggest the approach you would use to determine whether this material contains an effective substance for treating disease. If you do find a new and effective substance, can you think of a way to determine the amount present in your sample? (Pharmaceutical companies have used this approach to produce many common medications.)

1-108 Many substances that are involved in chemical reactions in the human body (and in all organisms) contain carbon, hydrogen, oxygen, and nitrogen arranged in specific patterns. Would you expect new medications to have features in common with these substances, or would you expect them to be drastically different? What are the reasons for your answer?

Challenge Problems

1-109 If 2 kg of a given reactant is consumed in the reaction described in Problem 1-105, how many calories are liberated for each kilogram?

1-110 A patient is to receive 1 Liter of IV fluid over 12 hours. The drop factor for the tubing is 15 gtts/mL. What should the flow rate be in gtts/min?

1-111 In the hospital, your doctor orders 100. mg of medication per hour. The label on the IV bag reads 5.0 g/1000. mL.

(a) How many mL should infuse each hour?

(b) The IV administration set delivers 15 gtts/mL, where the unit gtts denotes drops of liquid as explained in Problem 1-51. The current drip rate is set to 10. gtts/min. Is this correct? If not, what is the correct drip rate?

1-112 A febrile, pediatric patient weighs 42 pounds. You need to administer acetaminophen (Tylenol) 15 mg/kg.

(a) How many mg will you administer?

(b) The acetaminophen (Tylenol) packages come in liquid form 160 mg/5.0 mL. How many mL will you administer to your 42 pound patient?

Atoms

2

Image of atoms by SEM (scanning electron microscope).

KEY QUESTIONS

2-1 What Is Matter Made Of?
2-2 How Do We Classify Matter?
2-3 What Are the Postulates of Dalton's Atomic Theory?
2-4 What Are Atoms Made Of?
2-5 What Is the Periodic Table?
2-6 How Are the Electrons in an Atom Arranged?
2-7 How Are Electron Configuration and Position in the Periodic Table Related?
2-8 What Is a Periodic Property?

2-1 What Is Matter Made Of?

This question was discussed for thousands of years, long before humans had any reasonable way of getting an answer. In ancient Greece, two schools of thought tried to answer this question. One group, led by a scholar named Democritus (about 460–370 BCE), believed that all matter is made of very small particles—much too small to see. Democritus called these particles atoms (Greek *atomos*, meaning "not to cut"). Some of his followers developed the idea that there were different kinds of atoms, with different properties, and that the properties of the atoms caused ordinary matter to have the properties we all know.

Not all ancient thinkers, however, accepted this idea. A second group, led by Zeno of Elea (born about 450 BCE), did not believe in atoms at all. They insisted that matter is infinitely divisible. If you took any object, such as a piece of wood or a crystal of table salt, you could cut it or otherwise divide it into two parts, divide each of these parts into two more parts, and continue the process forever. According to Zeno and his followers, you would never reach a particle of matter that could no longer be divided.

Today we know that Democritus was right and Zeno was wrong. Atoms are the basic units of matter. Of course, there is a great difference in the way we now look at this question. Today our ideas are based on evidence. Democritus had no evidence to prove that matter cannot be divided an infinite number of times, just as Zeno had no evidence to support his claim that matter can be divided infinitely. Both claims were based not on evidence, but on visionary belief: one in unity, the other in diversity. In Section 2-3 we will discuss the evidence for the existence of atoms, but first we need to look at the diverse forms of matter.

2-2 How Do We Classify Matter?

Matter can be divided into two classes: pure substances and mixtures. Each class is then subdivided as shown in Figure 2-1.

A. Elements

An **element** is a substance (for example, carbon, hydrogen, and iron) that consists of identical atoms. At this time, 118 elements are known. Of these, 98 occur in nature; chemists and physicists have made the others in the laboratory. A list of the known elements appears on the inside front cover of this book, along with their symbols, which consist of one or two letters. Many symbols correspond directly to the name in English (for example, C for carbon, H for hydrogen, and Li for lithium), but a few are derived from the Latin or German names. Others are named for people who played significant roles in the development of science—in particular, atomic science (see Problem 2-12). Still other elements are named for geographic locations (see Problem 2-13).

B. Compounds

A **compound** is a pure substance made up of two or more elements in a fixed ratio by mass. The properties of a compound are different from those of a mixture of its constituent elements. For example, water is a compound made up of hydrogen and oxygen. The properties of water bear no resemblance to the properties of hydrogen and oxygen. At room temperature, their densities are known to be 1.00 g/mL, 0.084 g/L, and 1.33 g/L, respectively. There are an estimated 20 million known compounds, only a few of which we will introduce in this book.

A compound is characterized by its formula. The formula gives us the ratios of the compound's constituent elements and identifies each element by its atomic symbol. For example, in table salt (or NaCl, which consists of

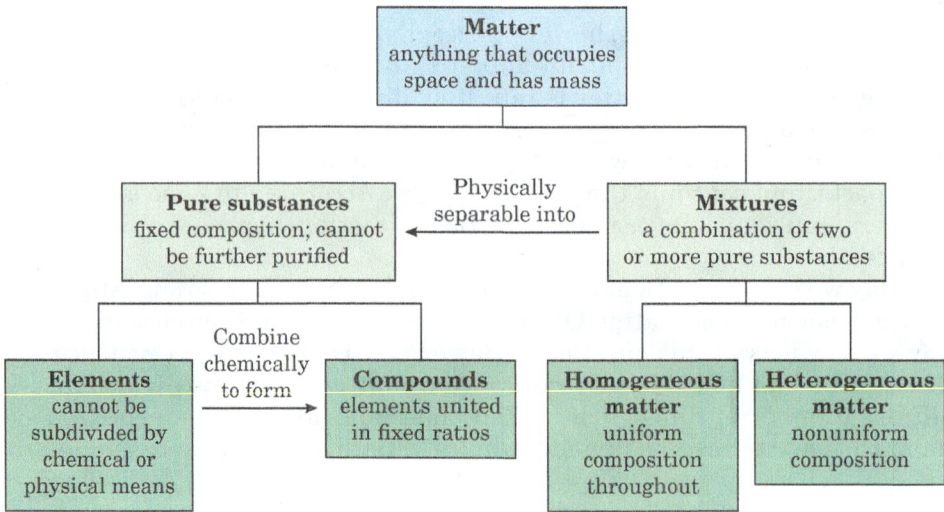

FIGURE 2-1 Classification of matter. Matter is divided into pure substances and mixtures. A pure substance may be either an element or a compound. A mixture may be either homogeneous or heterogeneous.

CHEMICAL CONNECTIONS 2A

Elements Necessary for Human Life

To the best of our knowledge, 20 of the 118 known elements are necessary for human life. The six most important of these—carbon, hydrogen, nitrogen, oxygen, phosphorus, and sulfur—are the subjects of organic chemistry and biochemistry (Chapters 10–31). Carbon, hydrogen, nitrogen, and oxygen are the big four in the human body. Seven other elements are also quite important, and our bodies use at least nine additional ones (trace elements) in very small quantities. Table 2A lists these 20 major elements and their functions in the human body. Many of these elements are more fully discussed later in the book. For the average daily requirements of these elements, their sources in foods, and symptoms of their deficiencies, see Chapter 30.

Table 2A Elements and Their Functions in the Human Body

Element	Function	Element	Function
The Big Four		**The Trace Elements**	
Carbon (C)	The subject of Chapters 10–19 (organic chemistry) and 20–31 (biochemistry)	Chromium (Cr)	Increases effectiveness of insulin
Hydrogen (H)		Cobalt (Co)	Part of vitamin B_{12}
Nitrogen (N)		Copper (Cu)	Strengthens bones; assists in enzyme activity
Oxygen (O)			
		Fluorine (F)	Reduces the incidence of dental cavities
The Next Seven			
Calcium (Ca)	Strengthens bones and teeth; aids blood clotting	Iodine (I)	An essential part of thyroid hormones
		Iron (Fe)	An essential part of some proteins, such as hemoglobin, myoglobin, cytochromes, and FeS proteins
Chlorine (Cl)	Necessary for normal growth and development		
Magnesium (Mg)	Helps nerve and muscle action; present in bones	Manganese (Mn)	Present in bone-forming enzymes; aids in fat and carbohydrate metabolism
Phosphorus (P)	Present as phosphates in bone, in nucleic acids (DNA and RNA), and involved in energy storage and transfer	Molybdenum (Mo)	Helps regulate electrical balance in body fluids
		Zinc (Zn)	Necessary for the action of certain enzymes
Potassium (K)	Helps regulate electrical balance in body fluids; essential for nerve conduction		
Sulfur (S)	An essential component of proteins		
Sodium (Na)	Helps regulate electrical balance in body fluids		

Test your knowledge with Problem 2-69.

sodium and chlorine), the ratio of sodium atoms to chlorine atoms is 1:1. Given that Na is the symbol for sodium and Cl is the symbol for chlorine, the formula of table salt is NaCl. In water, the combining ratio is two hydrogen atoms to one oxygen atom. The symbol for hydrogen is H, that for oxygen is O, and the formula of water is H_2O. The subscripts following the atomic symbols indicate the ratio of the combining elements. The number 1 in these ratios is omitted from the subscript. It is understood that NaCl means a ratio of 1:1 and that H_2O represents a ratio of 2:1. You will learn more about the nature of combining elements in a compound and their names and formulas in Chapter 3.

Figure 2-2 shows four representations for a water molecule. We will have more to say about molecular models as we move through this book.

FIGURE 2-2 Four representations of a water molecule.

| Molecular formula | Structural formula | Ball-and-stick model | Space-filling model |

> **EXAMPLE 2-1** Formula of a Compound
>
> (a) In the compound magnesium fluoride, magnesium (atomic symbol Mg) and fluorine (atomic symbol F) combine in a ratio of 1:2. What is the formula of magnesium fluoride?
> (b) The formula of perchloric acid is $HClO_4$. What are the combining ratios of the elements in perchloric acid?
>
> ### Strategy
>
> The formula gives the atomic symbol of each element combined in the compound, and subscripts give the ratio of its constituent elements.
>
> ### Solution
>
> (a) The formula is MgF_2. We do not write a subscript of 1 after Mg.
> (b) Both H and Cl have no subscripts, which means that hydrogen and chlorine have a combining ratio of 1:1. The subscript on oxygen is 4. Therefore, the combining ratios in $HClO_4$ are 1:1:4.
>
> ### Problem 2-1
>
> Write the formulas of compounds in which the combining ratios are as follows:
> (a) Sodium: chlorine: oxygen, 1:1:3
> (b) Aluminum (atomic symbol Al): fluorine (atomic symbol F), 1:3

C. Mixtures

A **mixture** is a combination of two or more pure substances. Most of the matter we encounter in our daily lives (including our own bodies) consists of mixtures rather than pure substances. For example, blood, butter, gasoline, soap, the metal in a ring, the air we breathe, and the Earth we walk on are all mixtures of pure substances. An important difference between a compound and a mixture is that the ratios by mass of the elements in a compound are fixed, whereas in a mixture, the pure substances can be present in any mass ratio.

For some mixtures—blood, for example (Figure 2-3)—the texture of the mixture is even throughout. However, if you examine blood under magnification, you can see that it is composed of different substances. This nonuniform observation is known as a heterogeneous mixture, where at least two components can be observed.

Other mixtures are homogeneous throughout, and no amount of magnification will reveal the presence of different substances. The air we breathe, for example, is a mixture of gases, primarily nitrogen (78%) and oxygen (21%). A metal alloy such as brass, which consists of copper and zinc, is another example of a homogeneous mixture.

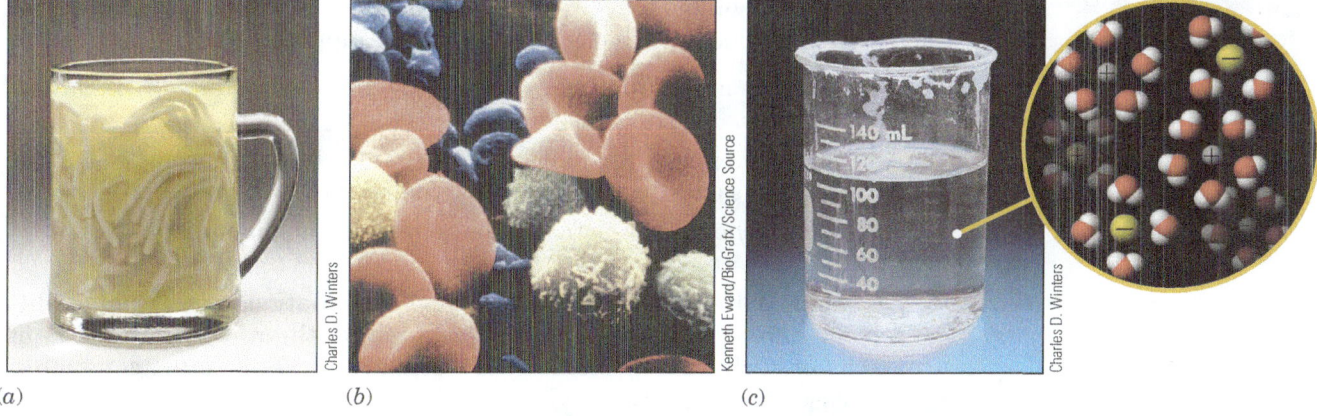

FIGURE 2-3 Mixtures. (a) A cup of noodle soup is a heterogeneous mixture. (b) A sample of blood may look homogeneous, but examination with an optical microscope shows that it is, in fact, a heterogeneous mixture of liquid and suspended particles (blood cells). (c) A homogeneous solution of salt, NaCl, in water. The models show that the salt solution contains Na$^+$ and Cl$^-$ ions as separate particles in water, with each ion being surrounded by a sphere of six or more water molecules. The particles in this solution cannot be seen with an optical microscope because they are too small.

(a) Iron and sulfur can be separated by stirring with a magnet.

(b) The first time that the magnet is removed, much of the iron is removed with it.

(c) The sulfur still looks dirty because a small quantity of iron remains.

Repeated stirrings eventually leave a bright yellow sample of sulfur that cannot be purified further by this technique.

FIGURE 2-4 Separating a mixture of iron and sulfur. (a) The iron–sulfur mixture is stirred with a magnet, which attracts the iron filings. (b) Much of the iron is removed after the first stirring. (c) Stirring continues until no more iron filings can be removed.

An important characteristic of a mixture is that it consists of two or more pure substances, each having different physical properties. If we know the physical properties of the individual substances, we can use appropriate physical means to separate the mixture into its component parts. Figure 2-4 shows one example of how a mixture can be separated.

2-3 What Are the Postulates of Dalton's Atomic Theory?

In 1808, the English chemist John Dalton (1766–1844) put forth a model of matter that underlies modern scientific atomic theory. The major difference between Dalton's theory and that of Democritus (Section 2-1) is that Dalton

Atoms The smallest particles of an element that retain the chemical properties of the element; the interaction among atoms accounts for the properties of matter

based his theory on evidence rather than on a belief. First, let us state his theory. We will then see what kind of evidence supported it.

1. All matter is made up of very tiny, indivisible particles, which Dalton called **atoms.**
2. All atoms of a given element have the same chemical properties. Conversely, atoms of different elements have different chemical properties.
3. In ordinary chemical reactions, no atom of any element disappears or is changed into an atom of another element.
4. Compounds are formed by the chemical combination of two or more different kinds of atoms. In a given compound, the relative numbers of atoms of each kind of element are constant and are most commonly expressed as integers.
5. A **molecule** is a tightly bound combination of two or more atoms that acts as a single unit.

A. Evidence for Dalton's Atomic Theory

The Law of Conservation of Mass

The great French chemist Antoine Laurent Lavoisier (1743–1794) discovered the **law of conservation of mass,** which states that matter can neither be created nor destroyed. In other words, there is no detectable change in mass in an ordinary chemical reaction. Lavoisier proved this law by conducting many experiments in which he showed that the total mass of matter at the end of the experiment was exactly the same as that at the beginning. Dalton's theory explained this fact in the following way: If all matter consists of indestructible atoms (postulate 1) and if no atoms of any element disappear or are changed into an atom of a different element (postulate 3), then any chemical reaction simply changes the attachments between atoms but does not destroy the atoms themselves. Thus, mass is conserved in a chemical reaction.

In the following illustration, a carbon monoxide molecule reacts with a lead oxide molecule to give a carbon dioxide molecule and a lead atom. All of the original atoms are still present at the end; they have merely changed partners. Thus, the total mass after this chemical change is the same as the mass that existed before the reaction took place.

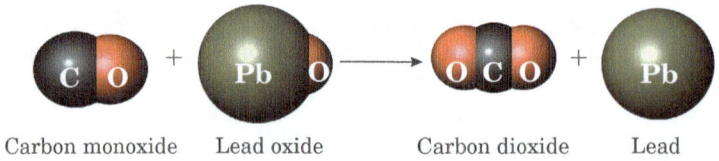

Carbon monoxide   Lead oxide   Carbon dioxide   Lead

The Law of Constant Composition

Another French chemist, Joseph Proust (1754–1826), demonstrated the **law of constant composition,** which states that any compound is always made up of elements in the same proportion by mass. For example, if you decompose water, you will always get 8.0 g of oxygen for each 1.0 g of hydrogen. The mass ratio of oxygen to hydrogen in pure water is always 8.0 to 1.0, whether the water comes from the Atlantic Ocean or the Missouri River or is collected as rain, squeezed out of a watermelon, or distilled from urine.

This fact was also evidence for Dalton's theory. If a water molecule consists of one atom of oxygen and two atoms of hydrogen and if an oxygen atom has a mass 16 times that of a hydrogen atom, then the mass ratio of these two elements in water must always be 8.0 to 1.0. The two elements can never be found in water in any other mass ratio.

CHEMICAL CONNECTIONS 2B

Abundance of Elements Present in the Human Body and in the Earth's Crust

Table 2B shows the abundance of the elements present in the human body. As you can see, oxygen is the most abundant element by mass, followed by carbon, hydrogen, and nitrogen. If we go by number of atoms, however, hydrogen is even more abundant in the human body than oxygen.

The table also shows the abundance of elements in the Earth's crust. Although 98 elements are found in the Earth's crust (we know very little about the interior of the Earth because we have not been able to penetrate into it very far), they are not present in anything close to equal amounts. In the Earth's crust as well as the human body, the most abundant element by mass is oxygen. But there the similarity ends. Silicon, aluminum, and iron, which are the second, third, and fourth most abundant elements in the Earth's crust, respectively, are not major elements in the body, whereas carbon, the second most abundant element by mass in the human body, is present to the extent of only 0.08 percent in the Earth's crust.

Table 2B The Relative Abundance of Elements Present in the Human Body and in the Earth's Crust, Including the Atmosphere and Oceans

Element	Percentage in Human Body By Number of Atoms	Percentage in Human Body By Mass	Percentage in Earth's Crust by Mass
H	63.0	10.0	0.9
O	25.4	64.8	49.3
C	9.4	18.0	0.08
N	1.4	3.1	0.03
Ca	0.31	1.8	3.4
P	0.22	1.4	0.12
K	0.06	0.4	2.4
S	0.05	0.3	0.06
Cl	0.03	0.2	0.2
Na	0.03	0.1	2.7
Mg	0.01	0.04	1.9
Si	—	—	25.8
Al	—	—	7.6
Fe	—	—	4.7
Others	0.01	—	—

Test your knowledge with Problem 2-70.

Now consider the compound hydrogen peroxide, which has the formula H_2O_2. If you decompose hydrogen peroxide, you will always get 16.0 g of oxygen for each 1.0 g of hydrogen. Once more, this takes into account an oxygen atom has a mass 16 times that of a hydrogen atom. The mass ratio of these two elements in hydrogen peroxide must always be 16.0 to 1.0, resulting in a different ratio of the elements in hydrogen peroxide (H_2O_2) compared to water (H_2O).

Thus, if the atomic ratio of the elements in a compound is fixed (postulate 4), then their proportions by mass must also be fixed.

B. Monatomic, Diatomic, and Polyatomic Elements

Some elements—for example, helium and neon—consist of single atoms that are not connected to each other—that is, they are **monatomic elements.** In contrast, oxygen, in its most common form, contains two atoms in each molecule, connected to each other by a chemical bond. We write the formula for an oxygen molecule as O_2, with the subscript showing the number of atoms in the molecule. Six other elements also occur as diatomic molecules (that is, they contain two atoms of the same element per molecule): hydrogen (H_2), nitrogen (N_2), fluorine (F_2), chlorine (Cl_2), bromine (Br_2), and iodine (I_2). It is important to understand that under normal conditions, free atoms of O, H, N, F, Cl, Br, and I do not exist. Rather, these seven elements occur only as **diatomic elements** (Figure 2-5).

Some elements have even more atoms in each molecule. Ozone, O_3, has three oxygen atoms in each molecule. In one form of phosphorus, P_4, each

FIGURE 2-5 Some diatomic, triatomic, and polyatomic elements. Hydrogen, nitrogen, oxygen, and chlorine are diatomic elements. Ozone, O_3, is a triatomic element. One form of sulfur, S_8, is a polyatomic element.

molecule has four atoms. One form of sulfur, S_8, has eight atoms per molecule. Some elements have molecules that are much larger. For example, diamond has millions of carbon atoms all bonded together in a gigantic cluster. Diamond and S_8 are referred to as **polyatomic elements.**

2-4 What Are Atoms Made Of?

A. Three Subatomic Particles

Today, we know that matter is more complex than Dalton believed. A wealth of experimental evidence obtained over the last 100 years or so has convinced us that atoms are not indivisible, but rather, consist of even smaller particles called subatomic particles. Three subatomic particles make up all atoms: protons, electrons, and neutrons. There are many other subatomic particles, but we will not deal with them in this book. Table 2-1 shows the charge, mass, and location of these particles in an atom.

Table 2-1 Properties and Location within Atoms of Protons, Neutrons, and Electrons

Subatomic Particle	Charge	Mass (g)	Mass (amu)	Mass (amu); Rounded to One Significant Figure	Location in an Atom
Proton	+1	1.6726×10^{-24}	1.0073	1	In the nucleus
Electron	−1	9.1094×10^{-28}	5.4858×10^{-4}	0.0005	Outside the nucleus
Neutron	0	1.6749×10^{-24}	1.0087	1	In the nucleus

Proton A subatomic particle with a charge of +1 and a mass of approximately 1 amu; it is found in a nucleus

Atomic mass unit (amu) A unit of the scale of relative masses of atoms: 1 amu = 1.6605×10^{-24} g; by definition, 1 amu is 1/12 the mass of a carbon atom containing 6 protons and 6 neutrons

Electron A subatomic particle with a charge of −1 and a mass of approximately 0.0005 amu; it is found in the space surrounding a nucleus

A **proton** has a positive charge. By convention we say that the magnitude of the charge is +1. Thus, one proton has a charge of +1, two protons have a charge of +2, and so forth. The mass of a proton is 1.6726×10^{-24} g, but this number is so small that it is more convenient to use another unit, called the **atomic mass unit (amu),** to describe its mass.

$$1 \text{ amu} = 1.6605 \times 10^{-24} \text{ g}$$

Thus, a proton has a mass of 1.0073 amu. For most purposes in this book, it is sufficient to round this number to one significant figure, and therefore, we say that the mass of a proton is 1 amu.

An **electron** has a charge of −1, equal in magnitude to the charge on a proton, but opposite in sign. The mass of an electron is approximately 5.4858×10^{-4} amu or 1/1837 that of the proton. It takes approximately 1837 electrons to equal the mass of one proton.

Like charges repel, and unlike charges attract. Two protons repel each other, just as two electrons also repel each other. A proton and an electron, however, attract each other.

A **neutron** has no charge. Therefore, neutrons neither attract nor repel each other or any other particle. The mass of a neutron is slightly greater than that of a proton: 1.6749×10^{-24} g or 1.0087 amu. Again, for our purposes, we round this number to 1 amu.

These three particles make up atoms, but where are they found? Protons and neutrons are found in a tight cluster in the center of an atom (Figure 2-6), which is called the **nucleus.** We will discuss the nucleus in greater detail in Chapter 9. Electrons are found as a diffuse cloud outside the nucleus.

B. Mass Number

Each atom has a fixed number of protons, electrons, and neutrons. One way to describe an atom is by its **mass number** (A), which is the sum of the number of protons and neutrons in its nucleus. Note that an atom also contains electrons, but because the mass of an electron is so small compared to that of protons and neutrons (Table 2-1), electrons are not counted in determining mass number.

$$\text{Mass number (A)} = \text{the number of protons + neutrons in the nucleus of an atom}$$

For example, an atom with 5 protons, 5 electrons, and 6 neutrons has a mass number of 11.

Neutron A subatomic particle with a mass of approximately 1 amu and a charge of zero; it is found in the nucleus

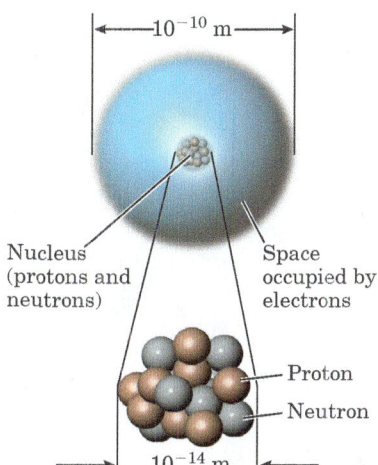

FIGURE 2-6 Relative sizes of the atomic nucleus and an atom (not to scale). The diameter of the region occupied by the electrons is approximately 10,000 times the diameter of the nucleus.

Interactive Example 2-2   **Mass Number**

What is the mass number of an atom containing:
(a) 58 protons, 58 electrons, and 78 neutrons?
(b) 17 protons, 17 electrons, and 20 neutrons?

Strategy

The mass number of an atom is the sum of the number of protons and neutrons in its nucleus.

Solution

(a) The mass number is 58 + 78 = 136.
(b) The mass number is 17 + 20 = 37.

Problem 2-2

What is the mass number of an atom containing:
(a) 15 protons, 15 electrons, and 16 neutrons?
(b) 86 protons, 86 electrons, and 136 neutrons?

C. Atomic Number

The **atomic number** (Z) of an element is the number of protons in its nucleus.

$$\text{Atomic number (Z)} = \text{number of protons in the nucleus of an atom}$$

Note that in a neutral atom, the number of electrons is equal to the number of protons. Atomic numbers for all the known elements are given in the atomic weight table on the inside front cover. They are also given in the Periodic Table on the inside front cover.

At the present time, 118 elements are known. These elements have atomic numbers from 1 to 118. The smallest atomic number belongs to the element hydrogen, which has only one proton and the largest (so far), to the as-yet-unnamed heaviest known element, which contains 118 protons.

If you know the atomic number and the mass number of an element, you can properly identify it. For example, the element with 6 protons, 6 electrons, and 6 neutrons has an atomic number of 6 and a mass number of 12. The element with atomic number 6 is carbon, C. Because its mass number is 12, we call this atomic nucleus carbon-12. Alternatively, we can write the symbol for this atomic nucleus as $^{12}_{6}C$. In this symbol, the mass number of the element is always written in the upper-left corner (as a superscript) of the symbol of the element and the atomic number in the lower-left corner (as a subscript). See the two tables on the inside front cover of the textbook for more information.

Mass number (number of protons + neutrons) $^{12}_{6}C$ ← Symbol of the element
Atomic number (number of protons)

EXAMPLE 2-3 Atomic Number

Name the elements given in Example 2-2 and write the symbols for their atomic nuclei.

Strategy

Determine the atomic number (the number of protons in the nucleus) and then locate the element in the Periodic Table on the inside front cover.

Solution

(a) This element has 58 protons. We find in the Periodic Table that the element with atomic number 58 is cerium, and its symbol is Ce. An atom of this element has 58 protons and 78 neutrons, and therefore, its mass number is 136. We call it cerium-136. Its symbol is $^{136}_{58}Ce$.

(b) This atom has 17 protons, making it a chlorine (Cl) atom. Because its mass number is 37, we call it chlorine-37. Its symbol is $^{37}_{17}Cl$.

Problem 2-3

Name the elements given in Problem 2-2. Write the symbols of their atomic nuclei.

EXAMPLE 2-4 Atomic Nuclei

Several elements have an equal number of protons and neutrons in their nuclei. Among these are oxygen, nitrogen and, neon. What are the atomic numbers of these elements? How many protons and neutrons does an atom of each have? Write the name and the symbol of each of these atomic nuclei.

Strategy

Look at the Periodic Table to determine the atomic number of each element. Mass number is the number of protons plus the number of neutrons.

Solution

Atomic numbers for these elements are found in the list of elements on the inside back cover. This table shows that oxygen (O) has atomic number 8, nitrogen (N) has atomic number 7, and neon (Ne) has atomic number 10.

This means that oxygen has 8 protons and 8 neutrons. Its name is oxygen-16, and its symbol is $^{16}_{8}O$. Nitrogen has 7 protons and 7 neutrons, its name is nitrogen-14, and its symbol is $^{14}_{7}N$. Neon has 10 protons and 10 neutrons, its name is neon-20, and its symbol is $^{20}_{10}Ne$.

Problem 2-4

(a) What are the atomic numbers of mercury (Hg) and lead (Pb)?
(b) How many protons does an atom of each have?
(c) If both Hg and Pb have 120 neutrons in their nuclei, what is the mass number of each?
(d) Write the name and the symbol of each.

D. Isotopes

Although we can say that an atom of carbon always has 6 protons and 6 electrons, we cannot say that an atom of carbon must have any particular number of neutrons. Some of the carbon atoms found in nature have 6 neutrons; the mass number of these atoms is 12, they are written as carbon-12, and their symbol is $^{12}_{6}C$. Other carbon atoms have 6 protons and 7 neutrons and, therefore, a mass number of 13; they are written as carbon-13, and their symbol is $^{13}_{6}C$. Still other carbon atoms have 6 protons and 8 neutrons; they are written as carbon-14 or $^{14}_{6}C$. Atoms with the same number of protons but different numbers of neutrons are called **isotopes**. All isotopes of carbon contain 6 protons and 6 electrons (or they wouldn't be carbon atoms). Each isotope, however, contains a different number of neutrons and, therefore, has a different mass number.

The properties of isotopes of the same element are almost identical, and for most purposes, we regard them as identical. They differ, however, in radioactive properties, which we discuss in Chapter 9. The fact that isotopes exist means that the second statement of Dalton's atomic theory (Section 2-3) is not correct.

EXAMPLE 2-5   Isotopes

How many neutrons are in each isotope of oxygen? Write the symbol of each isotope.
(a) Oxygen-16   (b) Oxygen-17   (c) Oxygen-18

Strategy

Each oxygen atom has 8 protons. The difference between the mass number and the number of protons gives the number of neutrons.

Solution

(a) Oxygen-16 has $16 - 8 = 8$ neutrons. Its symbol is $^{16}_{8}O$.
(b) Oxygen-17 has $17 - 8 = 9$ neutrons. Its symbol is $^{17}_{8}O$.
(c) Oxygen-18 has $18 - 8 = 10$ neutrons. Its symbol is $^{18}_{8}O$.

Problem 2-5

Two iodine isotopes are used in medical treatments: iodine-125 and iodine-131. How many neutrons are in each isotope? Write the symbol for each isotope.

Most elements are found on Earth as mixtures of isotopes, in a more or less constant ratio. For example, all naturally occurring samples of the element chlorine contain 75.77% chlorine-35 (18 neutrons) and 24.23% chlorine-37 (20 neutrons). Silicon exists in nature in a fixed ratio of three isotopes, with 14, 15, and 16 neutrons, respectively. For some elements, the ratio of isotopes may vary slightly from place to place, but for most purposes, we can ignore these slight variations. The atomic masses and isotopic abundances are determined using an instrument called a mass spectrometer.

E. Atomic Weight

Atomic weight The weighted average of the masses of the naturally occurring isotopes of the element. The units of atomic weight are atomic mass units (amu).

The **atomic weight** of an element given in the Periodic Table is a weighted average of the masses (in amu) of its isotopes found on the Earth. As an example of the calculation of atomic weight, let us examine chlorine. As we have just seen, two isotopes of chlorine exist in nature, chlorine-35 and chlorine-37. The mass of a chlorine-35 atom is 34.97 amu, and the mass of a chlorine-37 atom is 36.97 amu. Note that the atomic weight of each chlorine isotope (its mass in amu) is very close to its mass number (the number of protons and neutrons in its nucleus). This statement holds true for the isotopes of chlorine and those of all elements, because protons and neutrons have a mass of approximately (but not exactly) 1 amu.

The atomic weight of chlorine is a weighted average of the masses of the two naturally occurring chlorine isotopes:

$$\left(\frac{75.77}{100} \times 34.97 \text{ amu}\right) + \left(\frac{24.23}{100} \times 36.97 \text{ amu}\right) = 35.45 \text{ amu}$$

(Chlorine-35)   (Chlorine-37)

| 17 |
| Cl |
| 35.4527 |

Atomic weight in the Periodic Table is given to four decimal places using more precise data than given here

Some elements—for example, gold, fluorine, and aluminum—occur naturally as only one isotope. The atomic weights of these elements are close to whole numbers (gold, 196.97 amu; fluorine, 18.998 amu; aluminum, 26.98 amu). A table of atomic weights is found facing the inside front cover of this book.

EXAMPLE 2-6 Atomic Weight

The natural abundances of the three stable isotopes of magnesium are 78.99% magnesium-24 (23.98504 amu), 10.00% magnesium-25 (24.9858 amu), and 11.01% magnesium-26 (25.9829 amu). Calculate the atomic weight of magnesium and compare your value with that given in the Periodic Table.

Strategy

To calculate the weighted average of the masses of the isotopes, multiply each atomic mass by its abundance and then add.

Solution

$$\left(\frac{78.99}{100} \times 23.99 \text{ amu}\right) + \left(\frac{10.00}{100} \times 24.99 \text{ amu}\right) + \left(\frac{11.01}{100} \times 25.98 \text{ amu}\right) =$$

(Magnesium-24)   (Magnesium-25)   (Magnesium-26)

$$18.95 \quad + \quad 2.499 \quad + \quad 2.860 \quad = 24.31 \text{ amu}$$

The atomic weight of magnesium given in the Periodic Table to four decimal places is 24.3050.

Problem 2-6

The atomic weight of lithium is 6.941 amu. Lithium has only two naturally occurring isotopes: lithium-6 and lithium-7. Estimate which isotope of lithium is in greater natural abundance.

F. The Mass and Size of an Atom

A typical heavy atom (although not the heaviest) is lead-208, a lead atom with 82 protons, 82 electrons, and $208 - 82 = 126$ neutrons. It has a mass of 3.5×10^{-22} g. You would need 1.3×10^{24} atoms (a very large number) of lead-208 to make 1 lb. of lead. There are approximately 7 billion people on Earth right now. If you divided 1 lb. of these atoms among all the people on Earth, each person would get about 2.2×10^{14} atoms.

An atom of lead-208 has a diameter of about 3.1×10^{-10} m. If you could line them up with the atoms just touching, it would take 82 million lead atoms to make a line 1 inch long. Despite their tiny size, we can actually see atoms, in certain cases, by using a special instrument called a scanning tunneling microscope (Figure 2-7).

Virtually all of the mass of an atom is concentrated in its nucleus (because the nucleus contains the protons and neutrons). The nucleus of a lead-208 atom, for example, has a diameter of about 1.6×10^{-14} m. When you compare this with the diameter of a lead-208 atom, which is about 3.1×10^{-10} m, you see that the nucleus occupies only a tiny fraction of the total volume of the atom. If the nucleus of a lead-208 atom were the size of a baseball, then the entire atom would be much larger than a baseball stadium. In fact, it would be a sphere about one mile in diameter. Because a nucleus has such a relatively large mass concentrated in such a relatively small volume, a nucleus has a very high density. The density of a lead-208 nucleus, for example, is 1.6×10^{14} g/cm^3. Nothing in our daily life has a density anywhere near as high. If a paper clip had this density, it would weigh about 10 million (10^7) tons.

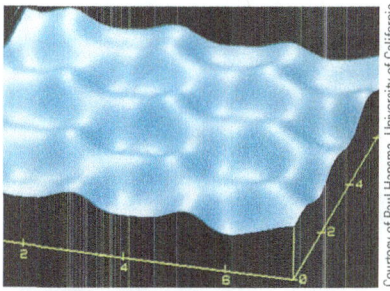

FIGURE 2-7 The surface of graphite is revealed with a scanning tunneling microscope. The contours represent the arrangement of individual carbon atoms on a crystal surface.

2-5 What Is the Periodic Table?

A. Origin of the Periodic Table

In the 1860s, the Russian scientist Dmitri Mendeleyev (1834–1907), then professor of chemistry at the University of St. Petersburg, produced one of the first Periodic Tables, the form of which we still use today. Mendeleyev started by arranging the known elements in order of increasing atomic weight beginning with hydrogen. He soon discovered that when the elements are arranged in the order of increasing atomic weight, certain sets of properties recur periodically. Mendeleyev then arranged those elements with recurring properties into **periods** (horizontal rows) by starting a new row each time he came to an element with properties similar to hydrogen. In this way, he discovered that lithium, sodium, potassium, and so forth, each start new rows. All are metallic solids at room temperature, all form ions with a charge of +1 (Li$^+$, Na$^+$, K$^+$, and so on), and all react with water to form metal hydroxides (LiOH, NaOH, KOH, and so on). Mendeleyev also discovered that elements in other vertical columns (families) have similar properties. ◀

For example, the elements fluorine (atomic number 9), chlorine (17), bromine (35), and iodine (53) all fall in the same column of the table. These elements, which are called halogens, are all colored substances, with the color deepening as we go down the table (Figure 2-8). The symbol "X" is

▶ Dmitri Mendeleyev.

Periods The elements in a horizontal row of the Periodic Table

FIGURE 2-8 Four halogens. Fluorine and chlorine are gases, bromine is a liquid, and iodine is a solid.

commonly used to represent a halogen. They all form compounds with sodium that have the general formula NaX (for example, NaCl and NaBr), but not NaX_2, Na_2X, Na_3X, or anything else. Only the elements in this column share this property.

At this point, we must say a word about the numbering of the columns (**families** or groups) of the Periodic Table. Mendeleyev gave them numerals and added the letter A for some columns and B for others. This numbering pattern remains in common use in the United States today. In 1985, an alternative pattern was recommended by the International Union of Pure and Applied Chemistry (IUPAC). In this system, the groups are numbered 1 to 18, without added letters, beginning on the left. Thus, in Mendeleyev's numbering system, the halogens are in Group 7A; in the new international numbering system, they are in Group 17. Although this book uses the Mendeleyev numbering system, both patterns are shown on the Periodic Table on the inside front cover. The A group elements (Groups 1A and 2A on the left side of the table and Groups 3A through 8A at the right) are known collectively as **main-group elements.**

Families The elements in the vertical columns of the Periodic Table

Main-group elements The elements in the A groups (Groups 1A, 2A, and 3A–8A) of the Periodic Table

The elements in the B columns (Groups 3 to 12 in the new numbering system) are called **transition elements.** Notice that elements 58 to 71 and 90 to 103 are not included in the main body of the table but rather are shown separately at the bottom. These sets of elements, called **inner transition elements,** actually belong in the main body of the Periodic Table, between columns 3 and 4 (between La and Hf and Ac and Rf). As is customary, we put them outside the main body solely to make a more compact presentation. If you like, you may mentally take a pair of scissors, cut through the heavy line between columns 3B and 4B, move them apart, and insert the inner transition elements. You will now have a table with 32 columns.

B. Classification of the Elements

Metals Elements that are solid at room temperature (except for mercury, which is a liquid), shiny, conduct electricity, are ductile and malleable, and form alloys; in their reactions, metals tend to give up electrons

There are three classes of elements: metals, nonmetals, and metalloids. The majority of elements are **metals**—only 24 are not. Metals are solids at room temperature (except for mercury, which is a liquid), shiny, conductors of electricity, ductile (they can be drawn into wires), and malleable (they can be hammered and rolled into sheets). In their reactions, metals tend to give up electrons. They also form alloys, which are solutions of one or more metals dissolved in another metal. Brass, for example, is an alloy of copper and zinc. Bronze is an alloy of copper and tin, and pewter is an alloy of tin,

FIGURE 2-9 Classification of the elements.

antimony, and lead. In their chemical reactions, metals tend to give up electrons (Section 3-2). Figure 2-9 shows a form of the Periodic Table in which the elements are classified by type.

Nonmetals are the second class of elements. With the exception of hydrogen, the 18 nonmetals appear to the right side of the Periodic Table. Excluding graphite, which is one form of carbon, nonmetals do not conduct electricity. At room temperature, nonmetals such as phosphorus and iodine are solids. Bromine is a liquid, and the elements of Group 8A (the noble gases)—helium through radon—are gases. In their chemical reactions, nonmetals tend to accept electrons (Section 3-2). Virtually all of the compounds we will encounter in our study of organic and biochemistry are built from just six nonmetals: H, C, N, O, P, and S.

Six elements are classified **metalloids**: boron, silicon, germanium, arsenic, antimony, and tellurium.

B	Si	Ge	As	Sb	Te
Boron	Silicon	Germanium	Arsenic	Antimony	Tellurium

These elements have some properties of metals and some of nonmetals. For example, some metalloids are shiny like metals, but do not conduct electricity. One of these metalloids, silicon, is a semiconductor—that is, it does not conduct electricity under certain applied voltages, but becomes a conductor at higher applied voltages. This semiconductor property of silicon makes it a vital element for Silicon Valley–based companies and the entire electronics industry (Figure 2-10).

C. Examples of Periodicity in the Periodic Table

Not only do the elements in any particular column (group or family) of the Periodic Table share similar properties, but the properties also vary in some fairly regular ways as we go up or down a column (family). For instance, Table 2-2 shows that the melting and boiling points of the **halogens** regularly increase as we go down a column.

Another example involves the Group 1A elements, also called the **alkali metals.** All alkali metals are soft enough to be cut with a knife, and their softness increases going down the column. They have relatively low melting and boiling points, which decrease going down the columns (Table 2-3). ◂

Nonmetals Elements that do not have the characteristic properties of a metal and, in their reactions, tend to accept electrons; eighteen elements are classified as nonmetals

Metalloids Elements that display some of the properties of metals and some of the properties of nonmetals; six elements are classified as metalloids

Halogens The elements in Group 7A of the Periodic Table

Alkali metals The elements, except hydrogen, in Group 1A of the Periodic Table

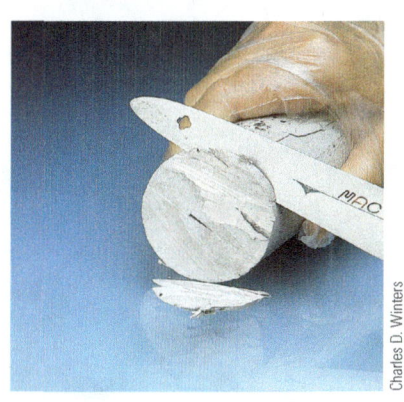

▸ Sodium metal can be cut with a knife.

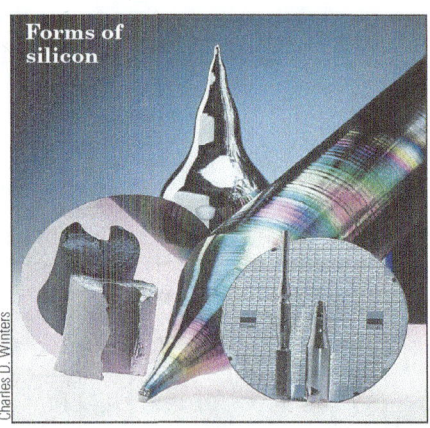

(a) Metals (b) Nonmetals (c) Metalloids

FIGURE 2-10 Representative elements. (a) Magnesium, aluminum, and copper are metals. All can be drawn into wires and conduct electricity. (b) Only 18 or so elements are classified as nonmetals. Shown here are liquid bromine and solid iodine. (c) Only six elements are generally classified as metalloids. This photograph is of solid silicon in various forms, including a wafer on which electronic circuits are printed.

Table 2-2 Melting and Boiling Points of the Halogens (Group 7A Elements)

Element	Melting point (°C)	Boiling point (°C)
Fluorine	−220	−188
Chlorine	−101	−35
Bromine	−7	59
Iodine	114	184
Astatine	302	337

Table 2-3 Melting and Boiling Points of the Alkali Metals (Group 1A Elements)

Element	Melting point (°C)	Boiling point (°C)
Lithium	180	1342
Sodium	98	883
Potassium	63	760
Rubidium	39	686
Cesium	28	669

CHEMICAL CONNECTIONS 2C

Strontium-90

Elements in the same column of the Periodic Table show similar properties. One important example is the similarity of strontium (Sr) and calcium (strontium is just below calcium in Group 2A). Calcium is an important element for humans because our bones and teeth consist largely of calcium compounds. We need some of this mineral in our diet every day, and we get it mostly from milk, cheese, and other dairy products.

One of the products released by test nuclear explosions in the 1950s and 1960s was the isotope strontium-90. This isotope is radioactive, with a half-life of 28.1 years. (Half-life is discussed in Section 9-4.) Strontium-90 was present in the fallout from aboveground nuclear test explosions. It was carried all over the Earth by winds and slowly settled to the ground, where it was eaten by cows and other animals. Strontium-90 got into milk and eventually into human bodies as well. If it were not so similar to calcium, our bodies would eliminate it within a few days. Because it is similar, however, some of the strontium-90 became deposited in bones and teeth (especially in children), subjecting all of us to a small amount of radioactivity for long periods of time.

In 1958, pathologist Walter Bauer helped start the St. Louis Baby Tooth Survey to study the effects of nuclear fallout on children. The study helped establish an early 1960s ban on aboveground A-bomb testing and led to similar surveys across the United States and the rest of the world. By 1970, the team had collected 300,000 shed primary teeth, which they discovered had absorbed nuclear waste from the milk of cows that were fed contaminated grass.

A 1963 treaty between the United States and the former Soviet Union banned aboveground nuclear testing. Although a few other countries still conduct occasional aboveground tests, there is reason to hope that such testing will be completely halted in the future.

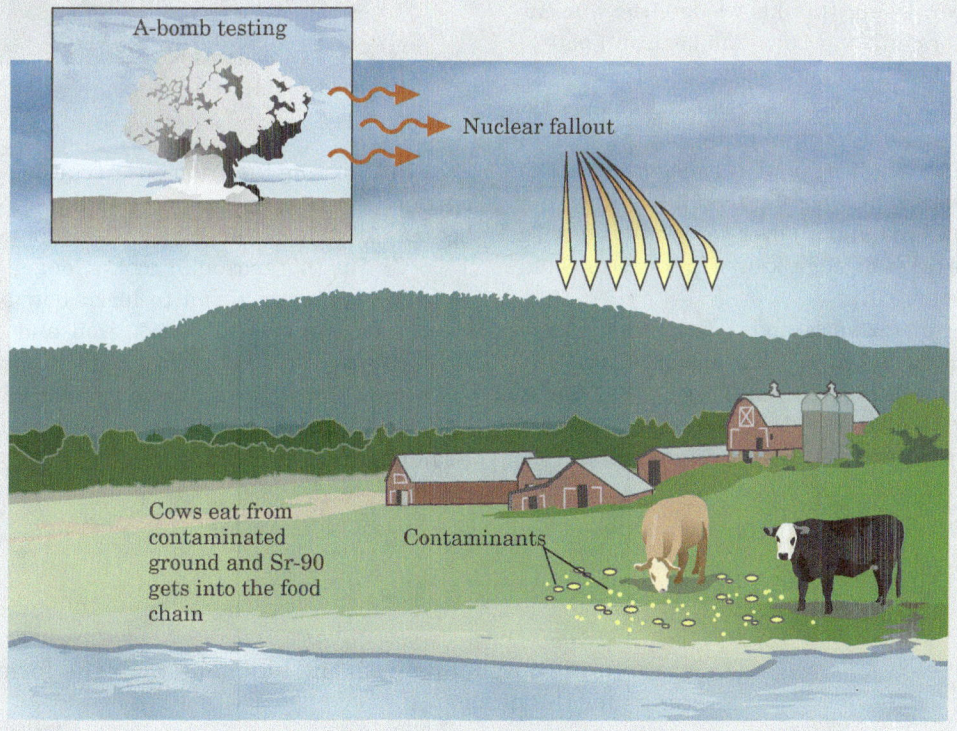

Test your knowledge with Problem 2-71.

All alkali metals react with water to form hydrogen gas, H_2, and a metal hydroxide with the formula MOH, where "M" stands for the alkali metal. The violence of their reaction with water increases in going down the column.

$$2Na + 2H_2O \longrightarrow 2NaOH + H_2$$

Sodium   Water   Sodium hydroxide   Hydrogen gas

CHEMICAL CONNECTIONS 2D

The Use of Metals as Historical Landmarks

The malleability of metals played an important role in the development of human society. In the Stone Age, tools were made from stone, which has no malleability. Then, about 11,000 BCE, it was discovered that the pure copper found on the surface of the Earth could be hammered into sheets, which made it suitable for use in vessels, utensils, and religious and artistic objects. This period became known as the Copper Age. Pure copper on the surface of the Earth, however, is scarce. Around 5000 BCE, humans found that copper could be obtained by putting malachite, $Cu_2CO_3(OH)_2$, a green copper-containing stone, into a fire. Malachite yielded pure copper at the relatively low temperature of 200°C.

Copper is a soft metal made of layers of large copper crystals. It can easily be drawn into wires because the layers of crystals can slip past one another. When hammered, the large crystals break into smaller ones with rough edges and the layers can no longer slide past one another. Therefore, hammered copper sheets are harder than drawn copper. Using this knowledge, the ancient profession of coppersmith was born, and beautiful plates, pots, and ornaments were produced.

Around 4000 BCE, it was discovered that an even greater hardness could be achieved by mixing molten copper with tin. The resulting alloy is called bronze. The Bronze Age was born somewhere in the Middle East and quickly spread to China and all over the world. Because hammered bronze takes an edge, knives and swords could be manufactured using it.

An even harder metal was soon to come. The first raw iron was found in meteorites. (The ancient Sumerian name of iron is "metal from heaven.") Around 2500 BCE, it was discovered that iron could be recovered from its ore by smelting, the process of recovering a metal from

Bronze Age artifact.

its ore by heating the ore. Thus began the Iron Age. More advanced technology was needed for smelting iron ores because iron melts only at a high temperature (about 1500°C). For this reason, it took a longer time to perfect the smelting process and to learn how to manufacture steel, which is about 90–95% iron and 5–10% carbon. Steel objects appeared first in India around 100 BCE.

Modern anthropologists and historians look back at ancient cultures and use the discovery of a new metal as a landmark for that age.

Test your knowledge with Problems 2-72 and 2-73.

They also form compounds with the halogens with the formula MX, where "X" stands for the halogen.

$$2Na + Cl_2 \longrightarrow 2NaCl$$

Sodium Chlorine Sodium chloride

The elements in Group 8A, often called the **noble gases,** provide yet another example of how the properties of elements change gradually within a column. Group 8A elements are gases under normal temperature and pressure, and they form either no compounds or very few compounds. Notice how close the melting and boiling points of the elements in this series are to one another (Table 2-4).

Table 2-4 Melting and Boiling Points of the Noble Gases (Group 8A Elements)

Element	Melting point (°C)	Boiling point (°C)
Helium	−272	−269
Neon	−249	−246
Argon	−189	−186
Krypton	−157	−152
Xenon	−112	−107
Radon	−71	−62

The Periodic Table is so useful that it hangs in nearly every chemistry classroom and chemical laboratory throughout the world. What makes it so useful is that it correlates a vast amount of data about the elements and their compounds and allows us to make many predictions about both chemical and physical properties. For example, if you were told that the boiling point of germane (GeH_4) is −88°C and that of methane (CH_4) is −164°C, could you predict the boiling point of silane (SiH_4)? The position of silicon in the table, between germanium and carbon, might lead you to a prediction of about −125°C. The actual boiling point of silane is −112°C, not far from this prediction.

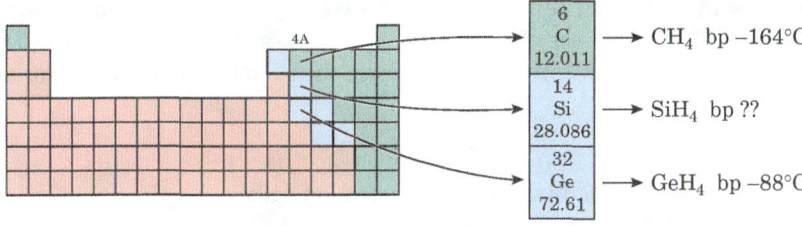

2-6 How Are the Electrons in an Atom Arranged?

We have seen that the protons and neutrons of an atom are concentrated in the atom's very small nucleus and that the electrons of an atom are located in the considerably larger space outside the nucleus. We can now ask how the electrons of an atom are arranged in this extranuclear space. Are they arranged randomly like seeds in a watermelon, or are they organized into layers like the layers of an onion?

Let us begin with hydrogen because it has only one electron and is the simplest atom. Before we do so, however, it is necessary to describe a discovery made in 1913 by the Danish physicist Niels Bohr (1885–1962). At the time, it was known that an electron is always moving around the nucleus and so possesses kinetic energy. Bohr discovered that only certain values are possible for this energy. This was a very surprising discovery. If you were told that you could drive your car at 23.4 mi/h or 28.9 mi/h or 34.2 mi/h, but never at any speed in between these values, you wouldn't believe it. Yet, that is

FIGURE 2-11 An energy stairway. A ramp, foreground (not quantized), and stair steps, background (quantized).

Ground state The electron configuration of the lowest energy state of an atom

Principal energy levels The energy levels containing orbitals of the same number (1, 2, 3, 4, and so forth)

Shells All orbitals of a principal energy level of an atom

Subshells All of the orbitals of an atom having the same principal energy level and the same letter designation (either s, p, d, or f)

Orbitals The regions of space around a nucleus that can hold a maximum of two electrons

just what Bohr discovered about electrons in atoms. The lowest possible energy level is the **ground state.**

If an electron is to have more energy than it has in the ground state, only certain values are allowed; values in between are not permitted. Bohr was unable to explain why these energy levels of electrons exist in atoms, but the accumulated evidence forced him to the conclusion that they do. We say that the energy of electrons in atoms is quantized. We can liken quantization to walking up a ramp compared with walking up a flight of stairs (Figure 2-11). You can put your foot on any stair step, but you cannot stand any place between two steps. You can stand only on steps.

A. Electrons Are Distributed in Shells, Subshells, and Orbitals

One conclusion reached by Bohr is that electrons in atoms do not move freely in the space around the nucleus, but rather remain confined to specific regions of space called **principal energy levels,** or more simply, **shells.** These shells are numbered 1, 2, 3, and 4, and so on, from the inside out. Table 2-5 gives the number of electrons that each of the first four shells can hold.

Table 2-5 Distribution of Electrons in Shells

Shell	Number of Electrons Shell Can Hold	Relative Energies of Electrons in Each Shell
4	32	Higher
3	18	↑
2	8	
1	2	Lower

Electrons in the first shell are closest to the positively charged nucleus and are held most strongly by it; these electrons are said to be the lowest in energy (hardest to remove). Electrons in higher-numbered shells are farther from the nucleus and are held less strongly to it; these electrons are said to be higher in energy (easier to remove).

Shells are divided into **subshells** designated by the letters s, p, d, and f. Within these subshells, electrons are grouped in **orbitals.** An orbital is a region of space and can hold two electrons (Table 2-6). The first shell contains a single s orbital and can hold two electrons. The second shell contains one s orbital and three p orbitals. All p orbitals come in sets of three and can

Table 2-6 Distribution of Orbitals within Shells

Shell	Orbitals Contained in Each Shell	Maximum Number of Electrons Shell Can Hold
4	One 4s, three 4p, five 4d, and seven 4f orbitals	2 + 6 + 10 + 14 = 32
3	One 3s, three 3p, and five 3d orbitals	2 + 6 + 10 = 18
2	One 2s and three 2p orbitals	2 + 6 = 8
1	One 1s orbital	2

hold six electrons. The third shell contains one s orbital, three p orbitals, and five d orbitals. All d orbitals come in sets of five and can hold ten electrons. The fourth shell also contains a set of f orbitals. All f orbitals come in sets of seven and can hold 14 electrons.

B. Orbitals Have Definite Shapes and Orientations in Space

All s orbitals have the shape of a sphere with the nucleus at the center of the sphere. Figure 2-12 shows the shapes of the 1s and 2s orbitals. Of the s orbitals, the 1s is the smallest sphere, the 2s is a larger sphere, and the 3s (not shown) is a still larger sphere. Figure 2-12 also shows the three-dimensional shapes of the three 2p orbitals. Each 2p orbital has the shape of a dumbbell with the nucleus at the midpoint of the dumbbell. The three 2p orbitals are at right angles to each other, with one orbital on the x-axis, the second on the y-axis, and the third on the z-axis. The shapes of 3p orbitals are similar, but larger.

Because the vast majority of organic compounds and biomolecules consist of the elements H, C, N, O, P, and S, which use only 1s, 2s, 2p, 3s, and 3p orbitals for bonding, we will concentrate on just these and other elements of the first, second, and third periods of the Periodic Table.

C. Electron Configurations of Atoms Are Governed by Three Rules

The **electron configuration** of an atom is a description of the orbitals that its electrons occupy. The orbitals available to all atoms are the same—namely, 1s, 2s, 2p, 3s, 3p, and so on. In the ground state of an atom, only

Electron configuration A description of the orbitals of an atom or ion occupied by electrons

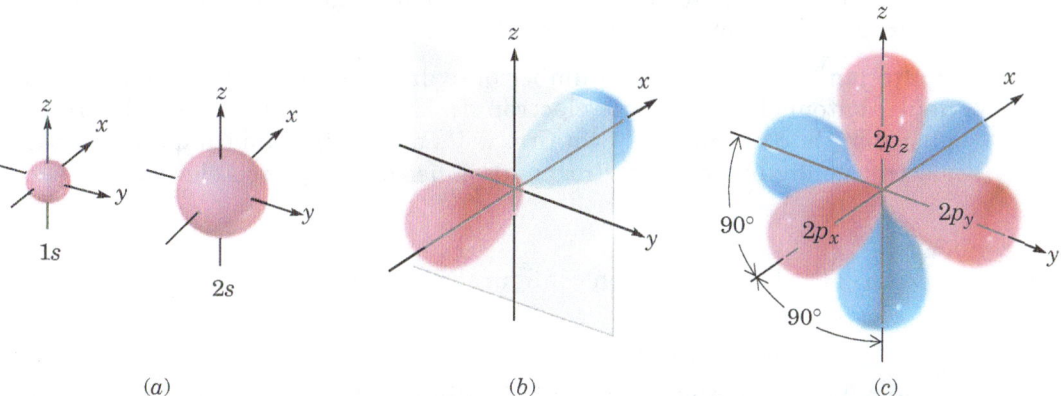

FIGURE 2-12 The 1s, 2s, and 2p orbitals. (a) A 1s orbital has the shape of a sphere, with the nucleus at the center of the sphere. A 2s orbital is a larger sphere than a 1s orbital, and a 3s orbital (not shown) is larger still. (b) A 2p orbital has the shape of a dumbbell, with the nucleus at the midpoint of the dumbbell. (c) Each 2p orbital is perpendicular to the other two. The 3p orbitals are similar in shape but larger. To make it easier for you to see the two lobes of each 2p orbital, one lobe is colored red and the other is colored blue.

48 Chapter 2 Atoms

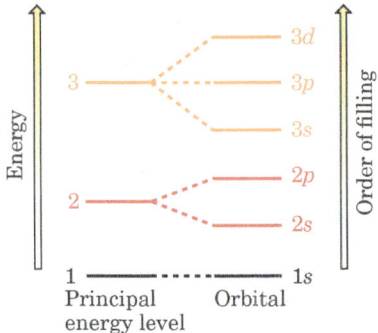

FIGURE 2-13 Energy levels for orbitals through the third shell.

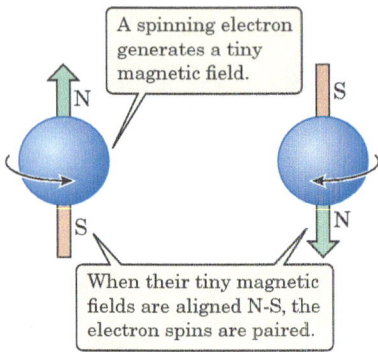

FIGURE 2-14 The pairing of electron spins.

the lowest-energy orbitals are occupied; all other orbitals are empty. We determine the ground-state electron configuration of an atom using the following rules:

Rule 1: Orbitals fill in the order of increasing energy from lowest to highest.

Example: In this book, we are concerned primarily with elements of the first, second, and third periods of the Periodic Table. Orbitals in these elements fill in the order 1s, 2s, 2p, 3s, 3p, and 3d. Figure 2-13 shows the order of filling through the third period.

Rule 2: Each orbital can hold up to two electrons with spins paired.

Example: With four electrons, the 1s and 2s orbitals are filled and we write them as $1s^2 2s^2$. With an additional six electrons, the three 2p orbitals are filled and we write them either in the expanded form of $2p_x^2\ 2p_y^2\ 2p_z^2$, or in the condensed form $2p^6$. Spin pairing means that the electrons spin in opposite directions (Figure 2-14).

Rule 3: When there is a set of orbitals of equal energy, each orbital becomes half-filled before any of them becomes completely filled.

Example: After the 1s and 2s orbitals are filled, a fifth electron is put into the $2p_x$ orbital, a sixth into the $2p_y$ orbital, and a seventh into the $2p_z$ orbital. Electrons are placed into orbitals of equal energy using the same spin direction before pairing begins. An arrow pointing up (↑) represents an electron with a positive spin, and an arrow pointing down (↓) represents an electron with a negative spin. Table 2-7 illustrates how to fill orbitals of equal energy using spin direction. Therefore, only after each 2p orbital has one electron is a second added to any 2p orbital.

D. Showing Electron Configurations: Orbital Box Diagrams

To illustrate how these rules are used, let us write the ground-state electron configurations for several of the elements in periods 1, 2, and 3. In the following **orbital box diagrams,** we use a box to represent an orbital, an arrow with its head up to represent a single electron, and a pair of arrows with heads in opposite directions to represent two electrons with paired spins. In addition, we show both expanded and condensed electron configurations. Table 2-7 gives the complete condensed ground-state electron configurations for elements 1 through 18.

Hydrogen (H) The atomic number of hydrogen is 1, which means that its neutral atoms have a single electron. In the ground state, this electron is placed in the 1s orbital. Shown first is its orbital box diagram and then its electron configuration. A hydrogen atom has one unpaired electron.

Helium (He) The atomic number of helium is 2, which means that its neutral atoms have two electrons. In the ground state, both electrons are placed in the 1s orbital with paired spins, which fill the 1s orbital. All electrons in helium are paired.

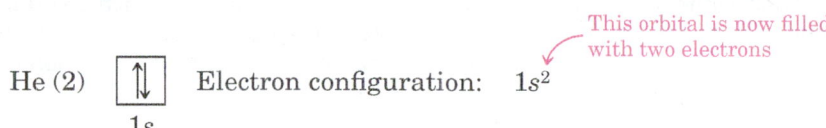

Table 2-7 Ground-State Electron Configurations of the First 18 Elements

	Orbital Box Diagram								Electron Configuration (Condensed)	Noble Gas Notation
	1s	2s	2p$_x$	2p$_y$	2p$_z$	3s	3p$_x$	3p$_y$	3p$_z$		
H (1)	↑									$1s^1$	
He (2)	↑↓									$1s^2$	
Li (3)	↑↓	↑								$1s^2\,2s^1$	[He] $2s^1$
Be (4)	↑↓	↑↓								$1s^2\,2s^2$	[He] $2s^2$
B (5)	↑↓	↑↓	↑							$1s^2\,2s^2\,2p^1$	[He] $2s^2\,2p^1$
C (6)	↑↓	↑↓	↑	↑						$1s^2\,2s^2\,2p^2$	[He] $2s^2\,2p^2$
N (7)	↑↓	↑↓	↑	↑	↑					$1s^2\,2s^2\,2p^3$	[He] $2s^2\,2p^3$
O (8)	↑↓	↑↓	↑↓	↑	↑					$1s^2\,2s^2\,2p^4$	[He] $2s^2\,2p^4$
F (9)	↑↓	↑↓	↑↓	↑↓	↑					$1s^2\,2s^2\,2p^5$	[He] $2s^2\,2p^5$
Ne (10)	↑↓	↑↓	↑↓	↑↓	↑↓					$1s^2\,2s^2\,2p^6$	[He] $2s^2\,2p^6$
Na (11)	↑↓	↑↓	↑↓	↑↓	↑↓	↑				$1s^2\,2s^2\,2p^6\,3s^1$	[Ne] $3s^1$
Mg (12)	↑↓	↑↓	↑↓	↑↓	↑↓	↑↓				$1s^2\,2s^2\,2p^6\,3s^2$	[Ne] $3s^2$
Al (13)	↑↓	↑↓	↑↓	↑↓	↑↓	↑↓	↑			$1s^2\,2s^2\,2p^6\,3s^2\,3p^1$	[Ne] $3s^2\,3p^1$
Si (14)	↑↓	↑↓	↑↓	↑↓	↑↓	↑↓	↑	↑		$1s^2\,2s^2\,2p^6\,3s^2\,3p^2$	[Ne] $3s^2\,3p^2$
P (15)	↑↓	↑↓	↑↓	↑↓	↑↓	↑↓	↑	↑	↑	$1s^2\,2s^2\,2p^6\,3s^2\,3p^3$	[Ne] $3s^2\,3p^3$
S (16)	↑↓	↑↓	↑↓	↑↓	↑↓	↑↓	↑↓	↑	↑	$1s^2\,2s^2\,2p^6\,3s^2\,3p^4$	[Ne] $3s^2\,3p^4$
Cl (17)	↑↓	↑↓	↑↓	↑↓	↑↓	↑↓	↑↓	↑↓	↑	$1s^2\,2s^2\,2p^6\,3s^2\,3p^5$	[Ne] $3s^2\,3p^5$
Ar (18)	↑↓	↑↓	↑↓	↑↓	↑↓	↑↓	↑↓	↑↓	↑↓	$1s^2\,2s^2\,2p^6\,3s^2\,3p^6$	[Ne] $3s^2\,3p^6$

Lithium (Li) Lithium has atomic number 3, which means that its neutral atoms have three electrons. In the ground state, two electrons are placed in the 1s orbital with paired spins and the third electron is placed in the 2s orbital. A lithium atom has one unpaired electron.

Li (3)　↑↓　↑　Electron configuration:　$1s^2 2s^1$
　　　　1s　2s

Li has one unpaired electron

Carbon (C) Carbon, atomic number 6, has six electrons in its neutral atoms. Two electrons are placed in the 1s orbital with paired spins and two are placed in the 2s orbital with paired spins. The fifth and sixth electrons are placed one each in the $2p_x$ and $2p_y$ orbitals. The ground state of a carbon atom has two unpaired electrons.

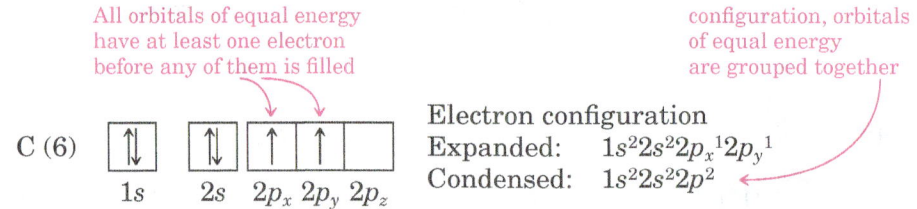

Oxygen (O) Oxygen, atomic number 8, has eight electrons in its neutral atoms. The first four electrons fill the $1s$ and $2s$ orbitals. The next three electrons are placed in the $2p_x$, $2p_y$, and $2p_z$ orbitals so that each $2p$ orbital has one electron. The remaining electron now fills the $2p_x$ orbital. The ground state of an oxygen atom has two unpaired electrons.

Neon (Ne) Neon, atomic number 10, has ten electrons in its neutral atoms, which completely fill all orbitals of the first and second shells. The ground state of a neon atom has no unpaired electrons.

Sodium (Na) Sodium, atomic number 11, has 11 electrons in its neutral atoms. The first 10 fill the $1s$, $2s$, and $2p$ orbitals. The 11th electron is placed in the $3s$ orbital. The ground state of a sodium atom has one unpaired electron.

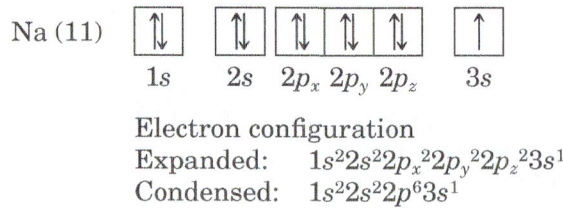

Electron configuration
Expanded: $1s^22s^22p_x^22p_y^22p_z^23s^1$
Condensed: $1s^22s^22p^63s^1$

Phosphorus (P) Phosphorus, atomic number 15, has 15 electrons in its neutral atoms. The first 12 fill the $1s$, $2s$, $2p$, and $3s$ orbitals. Electrons 13, 14, and 15 are placed one each in the $3p_x$, $3p_y$, and $3p_z$ orbitals. The ground state of a phosphorus atom has three unpaired electrons.

Electron configuration
Expanded: $1s^22s^22p_x^22p_y^22p_z^23s^23p_x^13p_y^13p_z^1$
Condensed: $1s^22s^22p^63s^23p^3$

E. Showing Electron Configurations: Noble Gas Notations

An alternate way of writing ground-state electron configurations uses the symbol of the noble gas immediately preceding the particular atom to indicate the electron configuration of all filled shells. The first shell of lithium, for example, is abbreviated [He] and the single electron in its 2s shell is indicated by $2s^1$. Thus, the electron configuration of a lithium atom is $[He]2s^1$ (right column of Table 2-7).

F. Showing Electron Configurations: Lewis Dot Structures

When discussing the physical and chemical properties of an element, chemists often focus on the outermost electron shell because the electrons in this shell are involved in the formation of chemical bonds (Chapter 3) and in chemical reactions (Chapter 4). Outer-shell electrons are called **valence electrons,** and the energy level in which they are found is called the **valence shell.** Carbon, for example, with a ground-state electron configuration of $1s^2 2s^2 2p^2$, has four valence (outer-shell) electrons.

To show the outermost electrons of an atom, we commonly use a representation called a **Lewis dot structure,** named after the American chemist Gilbert N. Lewis (1875–1946), who devised this notation. A Lewis structure shows the symbol of the element surrounded by a number of dots equal to the number of electrons in the outer (valence) shell of an atom of that element. In a Lewis structure, the atomic symbol represents the nucleus and all filled inner shells. Table 2-8 shows Lewis structures for the first 18 elements of the Periodic Table.

Valence electrons The electrons in the outermost occupied (valence) shell of an atom

Valence shell The outermost occupied shell of an atom

Lewis dot structure The symbol of the element surrounded by a number of dots equal to the number of electrons in the valence shell of an atom of that element

Table 2-8 Lewis Dot Structures for Elements 1–18 of the Periodic Table

1A	2A	3A	4A	5A	6A	7A	8A
H·							He:
Li·	Be:	B:	·C:	·N:	:O:	:F:	:Ne:
Na·	Mg:	Al:	·Si:	·P:	:S:	:Cl:	:Ar:

Each dot represents one valence electron.

The noble gases helium and neon have filled valence shells. The valence shell of helium is filled with two electrons ($1s^2$); that of neon is filled with eight electrons ($2s^2 2p^6$). Neon and argon have in common an electron configuration in which the s and p orbitals of their valence shells are filled with eight electrons. The valence shells of all other elements shown in Table 2-8 contain fewer than eight electrons.

At this point, let us compare the Lewis structures given in Table 2-8 with the ground-state electron configurations given in Table 2-7. The Lewis structure of boron (B), for example, is shown in Table 2-8 with three valence electrons; these are the paired 2s electrons and the single $2p_x$ electron shown in Table 2-7. The Lewis structure of carbon (C) is shown in Table 2-8 with four valence electrons; these are the two paired 2s electrons and the unpaired $2p_x$ and $2p_y$ electrons shown in Table 2-7.

> **EXAMPLE 2-7** **Electron Configuration**
>
> The Lewis dot structure for nitrogen shows five valence electrons. Write the expanded electron configuration for nitrogen and show to which orbitals its five valence electrons are assigned.
>
> ### Strategy
>
> Locate nitrogen in the Periodic Table and determine its atomic number. In an electrically neutral atom, the number of negatively charged extranuclear electrons is the same as the number of positively charged protons in its nucleus. The order of filling of orbitals is $1s\ 2s\ 2p_x\ 2p_y\ 2p_z\ 3s$, etc.
>
> ### Solution
>
> Nitrogen, atomic number 7, has the following ground-state electron configuration:
>
> $$1s^2\ 2s^2\ 2p_x^1\ 2p_y^1\ 2p_z^1$$
>
> The five valence electrons of the Lewis dot structure are the two paired electrons in the $2s$ orbital and the three unpaired electrons in the $2p_x$, $2p_y$, and $2p_z$ orbitals.
>
> ### Problem 2-7
>
> Write the Lewis dot structure for the element that has the following ground-state electron configuration. What is the name of this element?
>
> $$1s^2\ 2s^2\ 2p_x^2\ 2p_y^2\ 2p_z^2\ 3s^2\ 3p_x^1$$

2-7 How Are Electron Configuration and Position in the Periodic Table Related?

When Mendeleyev published his first Periodic Table in 1869, he could not explain why it worked—that is, why elements with similar properties became aligned in the same column. Indeed, no one had a good explanation for this phenomenon. It was not until the discovery of electron configurations that chemists finally understood why the Periodic Table works. The answer, they discovered, is very simple: Elements in the same column have the same ground-state electron configuration in their outer valence shells. Figure 2-15 shows the relationship between shells (principal energy levels) and orbitals being filled.

All main-group elements (those in columns A) have in common the fact that either their s or p orbitals are being filled. Notice that the $1s$ shell is filled with two electrons; there are only two elements in the first period. The $2s$ and $2p$ orbitals are filled with eight electrons; there are eight elements in period 2. Similarly, the $3s$ and $3p$ orbitals are filled with eight electrons; there are eight elements in period 3.

To create the elements of period 4, one $4s$, three $4p$, and five $3d$ orbitals are available. These orbitals can hold a total of 18 electrons; there are 18 elements in period 4. Similarly, there are 18 elements in period 5. Inner transition elements are created by filling f orbitals, which come in sets of seven and can hold a total of 14 electrons; there are 14 inner transition elements in the lanthanide series and 14 in the actinide series.

To see the similarities in electron configurations within the Periodic Table, let us look at the elements in column 1A. We already know the

FIGURE 2-15 Electron configuration and the Periodic Table.

configurations for lithium, sodium, and potassium (Table 2-7). To this list we can add rubidium and cesium. All elements in column 1A have one electron in their valence shell (Table 2-9).

All Group 1A elements are metals, with the exception of hydrogen, which is a nonmetal. The properties of elements largely depend on the electron configuration of their outer valence shell. As a consequence, it is not surprising that Group 1A elements, all of which have similar outer-shell configurations, are metals (except for hydrogen) and have such similar physical and chemical properties.

2-8 What Is a Periodic Property?

As we have now seen, the Periodic Table originally was constructed on the basis of trends (periodicity) in physical and chemical properties. With an understanding of electron configurations, chemists realized that the periodicity in chemical properties could be explained in terms of the periodicity in ground-state electron configuration. As we noted in the opening of Section 2-7, the Periodic Table works because "elements in the same column have similar ground-state electron configurations in their outer shells." Thus, chemists could now explain why certain chemical and physical properties of elements changed in predictable ways in going down a column or going across

Table 2-9 Noble Gas Notation and Lewis Dot Structures for the Alkali Metals (Group 1A Elements)

Noble Gas Notation	Lewis Dot Structure
$[He]2s^1$	Li•
$[Ne]3s^1$	Na•
$[Ar]4s^1$	K•
$[Kr]5s^1$	Rb•
$[Xe]6s^1$	Cs•

a row of the Periodic Table. In this section, we will examine the periodicity of one physical property (atomic size) and one chemical property (ionization energy) to illustrate how periodicity is related to position in the Periodic Table.

A. Atomic Size

The size of an atom is determined by the size of its outermost occupied orbital. The size of a sodium atom, for example, is the size of its singly occupied $3s$ orbital. The size of a chlorine atom is determined by the size of its three $3p$ orbitals ($3s^2 3p^5$). The simplest way to determine the size of an atom is to determine the distance between bonded nuclei in a sample of the element. A chlorine molecule, for example, has an internuclear bond distance of 198 pm (pm = picometer; 1 pm = 10^{-12} meter). The radius of a chlorine atom is thus 99 pm, which is one-half of the distance between two bonded chlorine nuclei in Cl_2.

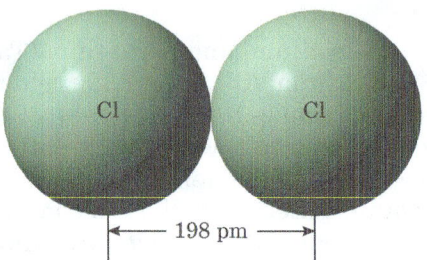

Similarly, the distance between bonded carbon nuclei in diamond is 154 pm, and so the radius of a carbon atom is 77 pm.

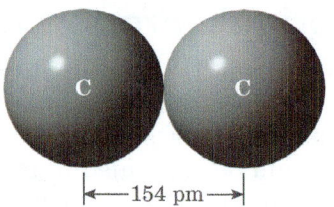

From measurements such as these, we can assemble a set of atomic radii (Figure 2-16).

From the information in this figure, we can see that for main group elements, (1) atomic radii increase going down a group and (2) decrease going from left to right across a period. Let us examine the correlation between each of these trends and ground-state electron configuration.

1. The size of an atom is determined by the size of its outermost electrons. In going down a column, the outermost electrons are assigned to higher and higher principal energy levels. The electrons of lower principal energy levels (those lying below the valence shell) must occupy some space, so the outer-shell electrons must be farther and farther from the nucleus, which rationalizes the increase in size in going down a column.

2. For elements in the same period, the principal energy level remains the same (for example, the valence electrons of all second-period elements occupy the second principal energy level). But in going from one element to the next across a period, one more proton is added to the nucleus, thus increasing the nuclear charge by one unit for each step from left to right. The result is that the nucleus exerts an increasingly stronger pull on the valence electrons and atomic radius decreases.

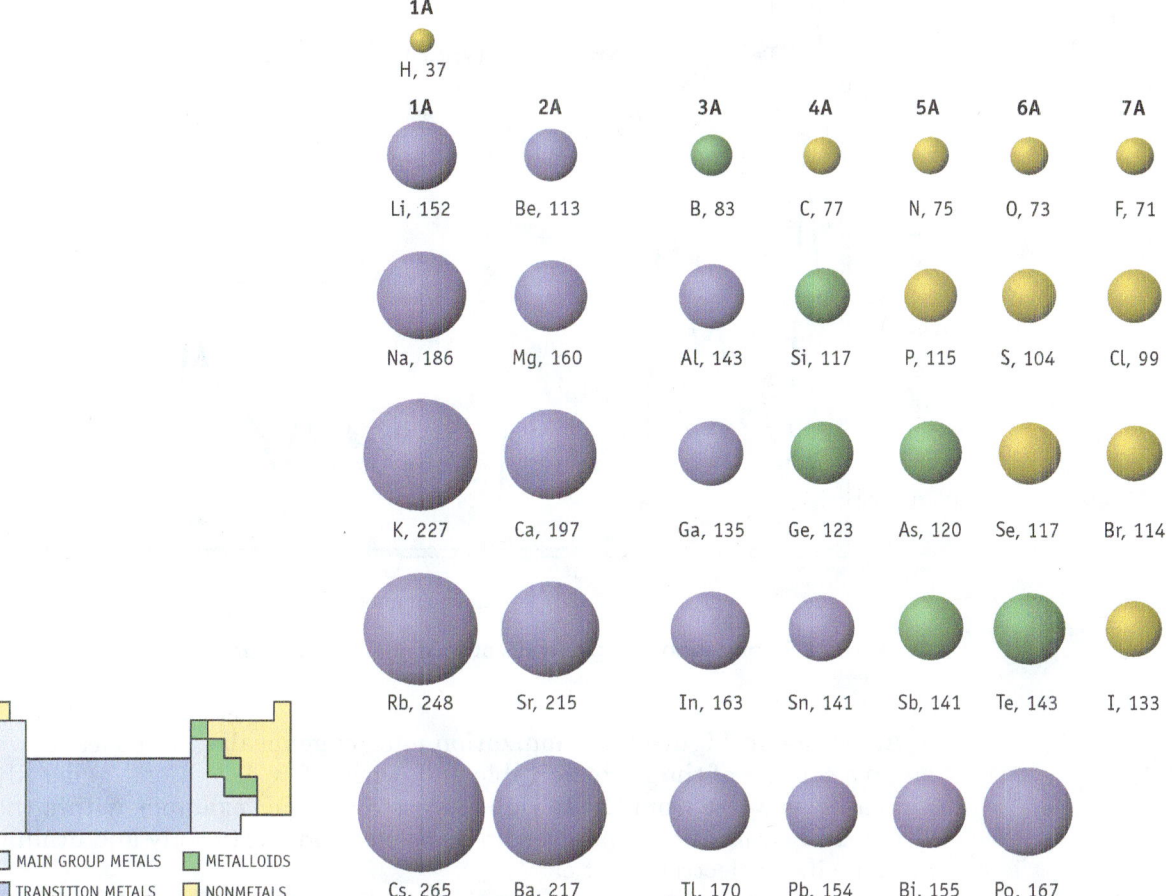

FIGURE 2-16 Atomic radii of main-group elements (in picometers, 1 pm = 10^{-12} m).

B. Ionization energy

Atoms are electrically neutral—the number of electrons outside the nucleus of an atom is equal to the number of protons inside the nucleus. Atoms do not normally lose or gain protons or neutrons, but they can lose or gain electrons. When a lithium atom, for example, loses one electron, it becomes a lithium ion. A lithium atom has three protons in its nucleus and three electrons outside the nucleus. When a lithium atom loses one of these electrons, it still has three protons in its nucleus (and, therefore, is still lithium), but now it has only two electrons outside the nucleus. The two remaining electrons cancel the charge of two of the protons, but there is no third electron to cancel the charge of the third proton. Therefore, a lithium ion has a charge of +1 and we write it as Li^+. The ionization energy for a lithium ion in the gas phase is 0.52 kJ/mol.

Ion An atom with an unequal number of protons and electrons

$$\underset{\text{Lithium atom}}{Li} + \underset{\text{Ionization energy}}{energy} \longrightarrow \underset{\text{Lithium ion}}{Li^+} + \underset{\text{Electron}}{e^-}$$

Ionization energy is a measure of how difficult it is to remove the most loosely held electron from an atom in the gaseous state. The more difficult it is to remove the electron, the higher the ionization energy. Ionization energies are always positive because energy must be supplied to overcome the attractive force between the electron and the positively charged nucleus. Figure 2-17 shows the ionization energies for the atoms of main-group elements 1 through 37 (hydrogen through rubidium).

Ionization energy The energy required to remove the most loosely held electron from an atom in the gas phase

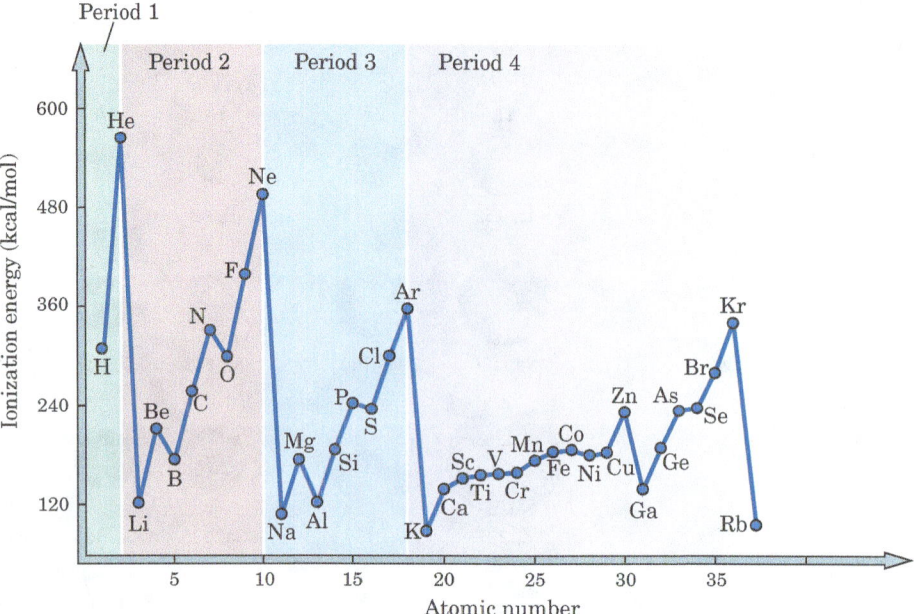

FIGURE 2-17 Ionization energy versus atomic number for elements 1–37.

As we see in Figure 2-17, ionization energy generally increases as we go up a column of the Periodic Table and, with a few exceptions, generally increases as we go from left to right across a row. For example, within the Group 1A metals, rubidium gives up its 5s electron most easily and lithium gives up its 2s electron least easily.

We explain this trend by saying that the 5s electron of rubidium is farther from the positively charged nucleus than is the 4s electron in potassium, which in turn is farther from the positively charged nucleus than is the 3s electron of sodium, and so forth. Furthermore, the 5s electron of rubidium is more "shielded" by inner-shell electrons from the attractive force of the positive nucleus than is the 4s electron of potassium, and so forth. The greater the shielding, the lower the ionization energy. Thus, going down a column of the Periodic Table, the shielding of an atom's outermost electrons increases and the element's ionization energies decrease.

We explain the increase in ionization energy across a row by the fact that the valence electrons across a row are in the same shell (principal energy level). Because the number of protons in the nucleus increases regularly across a row, the valence electrons experience an increasingly stronger pull by the nucleus, which makes them more difficult to remove. Thus, ionization energy increases from left to right across a row of the Periodic Table.

Summary of Key Questions

Section 2-1 What Is Matter Made Of?
- The Greek philosopher Democritus (circa 460–370 BCE) was the first person to propose an atomic theory of matter. He stated that all matter is made of very tiny particles, which he called atoms.

Section 2-2 How Do We Classify Matter?
- We classify matter as **elements, compounds,** or **mixtures.**

Section 2-3 What Are the Postulates of Dalton's Atomic Theory?
- (1) All matter is made up of atoms; (2) all atoms of a given element are identical, and the atoms of any one element are different from those of any other element; (3) compounds are formed by the chemical combination of atoms; and (4) a molecule is a cluster of two or more atoms that acts as a single unit.

- Dalton's theory is based on the **law of conservation of mass** (matter can be neither created nor destroyed) and the **law of constant composition** (any compound is always made up of elements in the same proportion by mass).

Section 2-4 What Are Atoms Made Of?

- Atoms consist of protons and neutrons found inside the nucleus and electrons located outside it. An **electron** has a mass of approximately 0.0005 amu and a charge of -1. A **proton** has a mass of approximately 1 amu and a charge of $+1$. A **neutron** has a mass of approximately 1 amu and no charge.
- The **mass number** of an atom is the sum of the number of its protons and neutrons.
- The **atomic number** of an element is the number of protons in the nucleus of an atom of that element.
- **Isotopes** are atoms with the same atomic number but different mass numbers; that is, they have the same number of protons but different numbers of neutrons in their nuclei.
- The **atomic weight** of an element is a weighted average of the masses (in amu) of its isotopes as they occur in nature.
- Atoms are very tiny, with a very small mass, almost all of which is concentrated in the nucleus. The nucleus is tiny, with an extremely high density.

Section 2-5 What Is the Periodic Table?

- The **Periodic Table** is an arrangement of elements with similar chemical properties into columns; the properties gradually change as we move down a column.
- **Metals** are solids (except for mercury, which is a liquid), shiny, conductors of electricity, ductile, malleable, and form alloys, which are solutions of one or more metals dissolved in another metal. In their chemical reactions, metals tend to give up electrons.
- With the exception of hydrogen, the **nonmetals** appear on the right side of the Periodic Table. With the exception of graphite, they do not conduct electricity. In their chemical reactions, nonmetals tend to accept electrons.
- Six elements are classified as **metalloids:** boron, silicon, germanium, arsenic, antimony, and tellurium. These elements have some properties of metals and some properties of nonmetals.

Section 2-6 How Are the Electrons in an Atom Arranged?

- Electrons in atoms exist in **principal energy levels** or **shells.**
- All principal energy levels except the first are divided into **subshells** designated by the letters s, p, d, and f. Within each subshell, electrons are grouped into **orbitals.** An orbital is a region of space that can hold two electrons with paired spins. All s orbitals are spherical and can hold two electrons. All p orbitals come in sets of three, and each is shaped like a dumbbell, with the nucleus at the center of the dumbbell. A set of three p orbitals can hold six electrons. A set of five d orbitals can hold ten electrons, and a set of seven f orbitals can hold fourteen electrons.
- Electrons are arranged in orbitals according to the following rules.
  (1) Orbitals fill in order of increasing energy; (2) each orbital can hold a maximum of two electrons with paired spins; (3) when filling orbitals of equivalent energy, each orbital adds one electron before any orbital adds a second electron.
- The electron configuration of an atom may be shown by an orbital notation, an orbital box diagram, or a noble gas notation.
- Electrons in the outermost or **valence shell** of an atom are called **valence electrons.** In a **Lewis dot structure** of an atom, the symbol of the element is surrounded by a number of dots equal to the number of its valence electrons.

Section 2-7 How Are Electron Configuration and Position in the Periodic Table Related?

- The Periodic Table works because elements in the same column have the same outer-shell electron configuration.

Section 2-8 What Is a Periodic Property?

- **Ionization energy** is the energy necessary to remove the most loosely held electron from an atom in the gas phase to form an **ion.** Ionization energy increases from bottom to top within a column of the Periodic Table because the valence shell of the atom becomes closer to the positively charged nucleus. It increases from left to right within a row because the positive charge on the nucleus increases in this direction.
- The **size of an atom (atomic radius)** is determined by the size of its outermost occupied orbital. Atomic size is a periodic property. For main-group elements, atomic size increases going down a group and decreases going from left to right across a period. In going down a column, the outermost electrons are assigned to higher and higher principal energy levels. For elements in the same period, the principal energy level remains the same from one element to the next, but the nuclear charge increases by one unit (by one proton). As a result of this increase across a period, the nucleus exerts a stronger pull on the valence electrons and atomic size decreases.

Problems

Orange-numbered problems are applied.

Section 2-1 What Are Atoms Made Of?

2-8 In what way(s) was Democritus's atomic theory similar to that of Dalton's atomic theory?

Section 2-2 How Do We Classify Matter?

2-9 Answer true or false.
   (a) Matter is divided into elements and pure substances.
   (b) Matter is anything that has mass and volume (occupies space).
   (c) A mixture is composed of two or more pure substances.
   (d) An element is a pure substance.
   (e) A heterogeneous mixture can be separated into pure substances, but a homogeneous mixture cannot.
   (f) A compound consists of elements combined in a fixed ratio.
   (g) A compound is a pure substance.
   (h) All matter has mass.
   (i) All of the 118 known elements occur naturally on Earth.
   (j) The first six elements in the Periodic Table are the most important for human life.
   (k) The combining ratio of a compound tells you how many atoms of each element are combined in the compound.
   (l) The combining ratio of 1:2 in the compound CO_2 tells you that this compound is formed by the combination of one gram of carbon with two grams of oxygen.

2-10 Classify each of the following as an element, a compound, or a mixture:
   (a) Oxygen   (b) Table salt
   (c) Sea water (d) Wine
   (e) Air   (f) Silver
   (g) Diamond   (h) A pebble
   (i) Gasoline (j) Milk
   (k) Carbon dioxide (l) Bronze

2-11 Name these elements (try not to look at a Periodic Table):
   (a) O   (b) Pb (c) Ca (d) Na
   (e) C   (f) Ti (g) S   (h) Fe
   (i) H   (j) K   (k) Ag (l) Au

2-12 The elements game, Part I. Name and give the symbol of the element that is named for each person.
   (a) Niels Bohr (1885–1962), Nobel Prize for physics in 1922
   (b) Pierre and Marie Curie, Nobel Prize for chemistry in 1903
   (c) Albert Einstein (1879–1955), Nobel Prize for physics in 1921
   (d) Enrico Fermi (1901–1954), Nobel Prize for physics in 1938
   (e) Ernest Lawrence (1901–1958), Nobel Prize for physics in 1939
   (f) Lise Meitner (1868–1968), codiscoverer of nuclear fission
   (g) Dmitri Mendeleyev (1834–1907), first person to formulate a workable Periodic Table
   (h) Alfred Nobel (1833–1896), discoverer of dynamite
   (i) Ernest Rutherford (1871–1937), Nobel Prize for chemistry in 1908
   (j) Glen Seaborg (1912–1999), Nobel Prize for chemistry in 1951

2-13 The elements game, Part II. Name and give the symbol of the element that is named for each geographic location.
   (a) The Americas
   (b) Berkeley, California
   (c) The state and University of California
   (d) Dubna, location in Russia of the Joint Institute of Nuclear Research
   (e) Europe
   (f) France
   (g) Gallia, the Latin name for ancient France
   (h) Germany
   (i) Hafnia, the Latin name for ancient Copenhagen
   (j) Hesse, a German state
   (k) Holmia, the Latin name for ancient Stockholm
   (l) Lutetia, the Latin name for ancient Paris
   (m) Magnesia, a district in Thessaly
   (n) Poland, the native country of Marie Curie
   (o) Rhenus, the Latin name for the river Rhine
   (p) Ruthenia, the Latin name for ancient Russia
   (q) Scandia, the Latin name for ancient Scandinavia
   (r) Strontian, a town in Scotland
   (s) Ytterby, a village in Sweden (three elements)
   (t) Thule, the earliest name for Scandinavia

2-14 The elements game, Part III. Give the names and symbols for the two elements named for planets. Note that the element plutonium was named for Pluto, which is no longer classified as a planet.

2-15 Write the formulas of compounds in which the combining ratios are as follows:
   (a) Potassium: oxygen, 2:1
   (b) Sodium: phosphorus: oxygen, 3:1:4
   (c) Lithium: nitrogen: oxygen, 1:1:3

2-16 Write the formulas of compounds in which the combining ratios are as follows:
(a) Sodium: hydrogen: carbon: oxygen, 1:1:1:3
(b) Carbon: hydrogen: oxygen, 2:6:1
(c) Potassium: manganese: oxygen, 1:1:4

Section 2-3 What Are Postulates of Dalton's Atomic Theory?

2-17 How does Dalton's atomic theory explain:
(a) the law of conservation of mass?
(b) the law of constant composition?

2-18 When 2.16 g of mercuric oxide is heated, it decomposes to yield 2.00 g of mercury and 0.16 g of oxygen. Which law is supported by this experiment?

2-19 The compound carbon monoxide contains 42.9% carbon and 57.1% oxygen. The compound carbon dioxide contains 27.3% carbon and 72.7% oxygen. Does this disprove Proust's law of constant composition?

2-20 Calculate the percentage of hydrogen and oxygen in water, H_2O, and hydrogen peroxide, H_2O_2.

Section 2-4 What Are Atoms Made Of?

2-21 Answer true or false.
(a) A proton and an electron have the same mass but opposite charges.
(b) The mass of an electron is considerably smaller than that of a neutron.
(c) An atomic mass unit (amu) is a unit of mass.
(d) One amu is equal to 1 gram.
(e) The protons and neutrons of an atom are found in the nucleus.
(f) The electrons of an atom are found in the space surrounding the nucleus.
(g) All atoms of the same element have the same number of protons.
(h) All atoms of the same element have the same number of electrons.
(i) Electrons and protons repel each other.
(j) The size of an atom is approximately the size of its nucleus.
(k) The mass number of an atom is the sum of the numbers of protons and neutrons in the nucleus of that atom.
(l) For most atoms, their mass number is the same as their atomic number.
(m) The three isotopes of hydrogen (hydrogen-1, hydrogen-2, and hydrogen-3) differ only in the number of neutrons in the nucleus.
(n) Hydrogen-1 has one neutron in its nucleus, hydrogen-2 has two neutrons in its nucleus, and hydrogen-3 has three neutrons.
(o) All isotopes of an element have the same number of electrons.
(p) Most elements found on Earth are mixtures of isotopes.
(q) The atomic weight of an element given in the Periodic Table is the weighted average of the masses of its isotopes found on Earth.
(r) The atomic weights of most elements are whole numbers.
(s) Most of the mass of an atom is found in its nucleus.
(t) The density of a nucleus is its mass number expressed in grams.

2-22 Where in an atom are these subatomic particles located?
(a) Protons   (b) Electrons   (c) Neutrons

2-23 It has been said, "The number of protons determines the identity of the element." Do you agree or disagree with this statement? Explain.

2-24 What is the mass number of an atom with:
(a) 22 protons, 22 electrons, and 26 neutrons?
(b) 76 protons, 76 electrons, and 114 neutrons?
(c) 34 protons, 34 electrons, and 45 neutrons?
(d) 94 protons, 94 electrons, and 150 neutrons?

2-25 Name and give the symbol for each element in Problem 2-24.

2-26 Given these mass numbers and number of neutrons, what is the name and symbol of each element?
(a) Mass number 45; 24 neutrons
(b) Mass number 48; 26 neutrons
(c) Mass number 107; 60 neutrons
(d) Mass number 246; 156 neutrons
(e) Mass number 36; 18 neutrons

2-27 If each atom in Problem 2-26 acquired two more neutrons, what element would each then be?

2-28 How many neutrons are in:
(a) a carbon atom of mass number 13?
(b) a germanium atom of mass number 73?
(c) an osmium atom of mass number 188?
(d) a platinum atom of mass number 195?

2-29 How many protons and how many neutrons does each of these isotopes of radon contain?
(a) Rn-210   (b) Rn-218   (c) Rn-222

2-30 How many neutrons and protons are in each isotope?
(a) ^{22}Ne   (b) ^{104}Pd
(c) ^{35}Cl   (d) Tellurium-128
(e) Lithium-7   (f) Uranium-238

2-31 Tin-118 is one of the isotopes of tin. Name the isotopes of tin that contain two, three, and six more neutrons than tin-118.

2-32 What is the difference between atomic number and mass number?

2-33 Define:
(a) Ion   (b) Isotope

2-34 There are only two naturally occurring isotopes of antimony: ^{121}Sb (120.90 amu) and ^{123}Sb (122.90 amu). The atomic weight of antimony given in the Periodic Table is 121.75. Which of the two isotopes has the greater natural abundance?

2-35 The two most abundant naturally occurring isotopes of carbon are carbon-12 (98.90%, 12.000 amu) and carbon-13 (1.10%, 13.003 amu). From these abundances, calculate the atomic weight of carbon and compare your calculated value with that given in the Periodic Table.

2-36 Another isotope of carbon, carbon-14, occurs in nature but in such small amounts relative to carbon-12 and carbon-13 that it does not contribute to the atomic weight of carbon as recorded in the Periodic Table. Carbon-14 is invaluable in the science of radiocarbon dating (see Chemical Connections 9A). Give the number of protons, neutrons, and electrons in an atom of carbon-14.

2-37 The isotope carbon-11 does not occur in nature but has been made in the laboratory. This isotope is used in a medical imaging technique called positron emission tomography (PET, see Section 9-7A). Give the number of protons, neutrons, and electrons in an atom of carbon-11.

2-38 Other isotopes used in PET imaging are fluorine-18, nitrogen-13, and oxygen-15. None of these isotopes occurs in nature; all must be produced in the laboratory. Give the number of protons, neutrons, and electrons in an atom of each of these artificial isotopes.

2-39 Americium-241 is used in household smoke detectors. This element has 11 known isotopes, none of which occurs in nature, but must be made in the laboratory. Give the number of protons, neutrons, and electrons in an atom of americium-241.

2-40 In dating geological samples, scientists compare the ratio of rubidium-87 to strontium-87. Give the number of protons, neutrons, and electrons in an atom of each element.

Section 2-5 What Is the Periodic Table?

2-41 Answer true or false.
  (a) Mendeleyev discovered that when elements are arranged in order of increasing atomic weight, certain sets of properties recur periodically.
  (b) Main-group elements are those in the columns 3A to 8A of the Periodic Table.
  (c) Nonmetals are found at the top of the Periodic Table, metalloids in the middle, and metals at the bottom.
  (d) Among the 118 known elements, there are approximately equal numbers of metals and nonmetals.
  (e) A horizontal row in the Periodic Table is called a group.
  (f) The Group 1A elements are called the "alkali metals."
  (g) The alkali metals react with water to give hydrogen gas and a metal hydroxide, MOH, where "M" is the metal.
  (h) The halogens are Group 7A elements.
  (i) The boiling points of noble gases (Group 8A elements) increase going from top to bottom of the column.

2-42 How many metals, metalloids, and nonmetals are there in the third period of the Periodic Table?

2-43 Which group(s) of the Periodic Table contain(s):
  (a) Only metals?   (b) Only metalloids?
  (c) Only nonmetals?

2-44 Which period(s) in the Periodic Table contain(s) more nonmetals than metals? Which contain(s) more metals than nonmetals?

2-45 Group the following elements according to similar properties (look at the Periodic Table): As, I, Ne, F, Mg, K, Ca, Ba, Li, He, N, P.

2-46 Which are transition elements?
  (a) Pd   (b) K   (c) Co
  (d) Ce   (e) Br   (f) Cr

2-47 Which element in each pair is more metallic?
  (a) Silicon or aluminum   (b) Arsenic or phosphorus
  (c) Gallium or germanium   (d) Gallium or aluminum

2-48 Classify these elements as metals, nonmetals, or metalloids:
  (a) Argon   (b) Boron   (c) Lead
  (d) Arsenic   (e) Potassium (f) Silicon
  (g) Iodine   (h) Antimony   (i) Vanadium
  (j) Sulfur   (k) Nitrogen

Section 2-6 How Are the Electrons in an Atom Arranged?

2-49 Answer true or false.
  (a) To say that "energy is quantized" means that only certain energy values are allowed.
  (b) Bohr discovered that the energy of an electron in an atom is quantized.
  (c) Electrons in atoms are confined to regions of space called "principal energy levels."
  (d) Each principal energy level can hold a maximum of two electrons.
  (e) An electron in a 1s orbital is held closer to the nucleus than an electron in a 2s orbital.
  (f) An electron in a 2s orbital is harder to remove from an atom than an electron in a 1s orbital.
  (g) An s orbital has the shape of a sphere, with the nucleus at the center of the sphere.
  (h) Each 2p orbital has the shape of a dumbbell, with the nucleus at the midpoint of the dumbbell.
  (i) The three 2p orbitals in an atom are aligned parallel to each other.
  (j) An orbital is a region of space that can hold two electrons.
  (k) The second shell contains one s orbital and three p orbitals.
  (l) In the ground-state electron configuration of an atom, only the lowest-energy orbitals are occupied.
  (m) A spinning electron behaves as a tiny bar magnet, with a North Pole and a South Pole.
  (n) An orbital can hold a maximum of two electrons with their spins paired.
  (o) Paired electron spins means that the two electrons are aligned with their spins North Pole to North Pole and South Pole to South Pole.
  (p) An orbital box diagram puts all of the electrons of an atom in one box with their spins aligned.

(q) An orbital box diagram of a carbon atom shows two unpaired electrons.

(r) A Lewis dot structure shows only the electrons in the valence shell of an atom of the element.

(s) A characteristic of Group 1A elements is that each has one unpaired electron in its outermost occupied (valence) shell.

(t) A characteristic of Group 6A elements is that each has six unpaired electrons in its valence shell.

2-50 How many periods of the Periodic Table have two elements? How many have eight elements? How many have 18 elements? How many have 32 elements?

2-51 What is the correlation between the group number of the main-group elements (those in the A columns of the Mendeleyev system) and the number of valence electrons in an element in the group?

2-52 Given your answer to Problem 2-51, write the Lewis dot structure for each of the following elements using no information other than the number of the group in the Periodic Table to which the element belongs.
(a) Carbon (4A) (b) Silicon (4A)
(c) Oxygen (6A) (d) Sulfur (6A)
(e) Aluminum (3A) (f) Bromine (7A)

2-53 Write the condensed ground-state electron configuration for each of the following elements. The element's atomic number is given in parentheses.
(a) Li (3) (b) Ne (10) (c) Be (4)
(d) C (6) (e) Mg (12)

2-54 Write the Lewis dot structure for each element in Problem 2-53.

2-55 Write the condensed ground-state electron configuration for each of the following elements. The element's atomic number is given in parentheses.
(a) He (2) (b) Na (11) (c) Cl (17)
(d) P (15) (e) H (1)

2-56 Write the Lewis dot structure for each element in Problem 2-55.

2-57 What are the similarities and differences in the electron configurations of:
(a) Na and Cs? (b) O and Te? (c) C and Ge?

2-58 Silicon, atomic number 14, is in Group 4A. How many orbitals are occupied by the valence electrons of Si in its ground state?

2-59 You are presented with a Lewis dot structure of element X as X:. To which two groups in the Periodic Table might this element belong?

2-60 The electron configurations for the elements with atomic numbers higher than 36 follow the same rules as given in the text for the first 36 elements. In fact, you can arrive at the correct order of filling of orbitals from Figure 2-15 by starting with H and reading the orbitals from left to right across the first row, then the second row, and so on. Write the condensed ground-state electron configuration for:
(a) Rb (b) Sr (c) Br

Section 2-7 How Are Electron Configuration and Position in the Periodic Table Related?

2-61 Answer true or false.
(a) Elements in the same column of the Periodic Table have the same outer-shell electron configuration.
(b) All Group 1A elements have one electron in their valence shell.
(c) All Group 6A elements have six electrons in their valence shell.
(d) All Group 8A elements have eight electrons in their valence shell.
(e) Period 1 of the Periodic Table has one element, period 2 has two elements, period 3 has three elements, and so forth.
(f) Period 2 results from filling the $2s$ and $2p$ orbitals, and therefore, there are eight elements in period 2.
(g) Period 3 results from filling the $3s$, $3p$, and $3d$ orbitals, and therefore, there are nine elements in period 3.
(h) The main-group elements are s block and p block elements.

2-62 Why do the elements in column 1A of the Periodic Table (the alkali metals) have similar but not identical properties?

Section 2-8 What Is a Periodic Property?

2-63 Answer true or false.
(a) Ionization energy is the energy required to remove the most loosely held electron from an atom in the gas phase.
(b) When an atom loses an electron, it becomes a positively charged ion.
(c) Ionization energy is a periodic property because ground-state electron configuration is a periodic property.
(d) Ionization energy generally increases going from left to right across a period of the Periodic Table.
(e) Ionization energy generally increases in going from top to bottom within a column in the Periodic Table.
(f) The sign of an ionization energy is always positive; the process is always endothermic.

2-64 Consider the elements B, C, and N. Using only the Periodic Table, predict which of these three elements has:
(a) the largest atomic radius.
(b) the smallest atomic radius.
(c) the largest ionization energy.
(d) the smallest ionization energy.

2-65 Account for the following observations.
(a) The atomic radius of an anion is always larger than that of the atom from which it is derived. Examples: Cl 99 pm and Cl$^-$ 181 pm; O 73 pm and O^{2-} 140 pm.
(b) The atomic radius of a cation is always smaller than that of the atom from which it is derived. Examples: Li 152 pm and Li$^+$ 76 pm; Na 156 pm and Na$^+$ 98 pm.

2-66 Using only the Periodic Table, arrange the elements in each set in order of increasing ionization energy:
(a) Li, Na, K (b) C, N, Ne
(c) O, C, F (d) Br, Cl, F

2-67 Account for the fact that the first ionization energy of oxygen is less than that of nitrogen.

2-68 Every atom except hydrogen has a series of ionization energies (IE) because they have more than one electron that can be removed. Following are the first three ionization energies for magnesium:

$$Mg(g) \longrightarrow Mg^+(g) + e^-(g) \quad IE_1 = 738 \text{ kJ/mol}$$
$$Mg^+(g) \longrightarrow Mg^{2+}(g) + e^-(g) \quad IE_2 = 1450 \text{ kJ/mol}$$
$$Mg^{2+}(g) \longrightarrow Mg^{3+}(g) + e^-(g) \quad IE_3 = 7734 \text{ kJ/mol}$$

(a) Write the ground-state electron configuration for Mg, Mg^+, Mg^{2+}, and Mg^{3+}.
(b) Account for the large increase in ionization energy for the removal of the third electron compared with the ionization energies for removal of the first and second electrons.

Chemical Connections

2-69 (Chemical Connections 2A) Why does the body need sulfur, calcium, and iron?

2-70 (Chemical Connections 2B) Which are the two most abundant elements, by weight, in:
(a) the Earth's crust? (b) the human body?

2-71 (Chemical Connections 2C) Why is strontium-90 more dangerous to humans than most other radioactive isotopes that were present in the Chernobyl fallout?

2-72 (Chemical Connections 2D) Bronze is an alloy of which two metals?

2-73 (Chemical Connections 2D) Copper is a soft metal. How can it be made harder?

Additional Problems

2-74 Give the designations of all subshells in the:
(a) 1 shell (b) 2 shell
(c) 3 shell (d) 4 shell

2-75 Tell whether metals or nonmetals are more likely to have each of the following characteristics:
(a) Conduct electricity and heat
(b) Accept electrons
(c) Be malleable
(d) Be a gas at room temperature
(e) Be a transition element
(f) Lose electrons

2-76 Explain why:
(a) atomic radius decreases going across a period in the Periodic Table.
(b) energy is required to remove an electron from an atom.

2-77 Name and give the symbol of the element with the given characteristic.
(a) Largest atomic radius in Group 2A.
(b) Smallest atomic radius in Group 2A.
(c) Largest atomic radius in the second period.
(d) Smallest atomic radius in the second period.
(e) Largest ionization energy in Group 7A.
(f) Lowest ionization energy in Group 7A.

2-78 What is the outer-shell electron configuration of the elements in:
(a) Group 3A? (b) Group 7A?
(c) Group 5A?

2-79 Determine the number of protons, electrons, and neutrons present in:
(a) ^{32}P (b) ^{98}Mo (c) ^{44}Ca
(d) ^{3}H (e) ^{158}Gd (f) ^{212}Bi

2-80 What percentage of the mass of each element do neutrons contribute?
(a) Carbon-12 (b) Calcium-40
(c) Iron-55 (d) Bromine-79
(e) Platinum-195 (f) Uranium-238

2-81 Do isotopes of the heavy elements (for example, those from atomic number 37 to 53) contain more, the same, or fewer neutrons than protons?

2-82 What is the symbol for each of the following elements? (Try not to look at a Periodic Table.)
(a) Phosphorus (b) Potassium
(c) Sodium (d) Nitrogen
(e) Bromine (f) Silver
(g) Calcium (h) Carbon
(i) Tin (j) Zinc

2-83 The natural abundance of boron isotopes is as follows: 19.9% boron-10 (10.013 amu) and 80.1% boron-11 (11.009 amu). Calculate the atomic weight of boron (watch the significant figures) and compare your calculated value with that given in the Periodic Table.

2-84 How many electrons are in the outer shell of each of the following elements?
(a) Si (b) Br
(c) P (d) K
(e) He (f) Ca
(g) Kr (h) Pb
(i) Se (j) O

2-85 The mass of a proton is 1.67×10^{-24} g. The mass of a grain of salt is 1.0×10^{-2} g. How many protons would it take to have the same mass as a grain of salt?

2-86 (a) What are the charges of an electron, a proton, and a neutron?
(b) What are the masses (in amu, to one significant figure) of an electron, a proton, and a neutron?

2-87 What is the name of this element, and how many protons and neutrons does this isotope have in its nucleus: $^{131}_{54}X$?

2-88 Based on the data presented in Figure 2-16, which atom would have the highest ionization energy: I, Cs, Sn, or Xe?

2-89 Assume that a new element has been discovered with atomic number 117. Its chemical properties should be similar to those of astatine (At). Predict whether the new element's ionization energy

will be greater than, the same as, or smaller than that of:
(a) At (b) Ra

2-90 Explain why the sizes of atoms change when proceeding across a period of the Periodic Table.

2-91 These are the first two ionization energy for lithium:

$Li(g) \longrightarrow Li^+(g) + e^-(g)$
Ionization energy = 523 kJ/mol

$Li^+(g) \longrightarrow Li^{2+}(g) + e^-(g)$
Ionization energy = 7298 kJ/mol

(a) Explain the large increase in ionization energy that occurs for the removal of the second electron.

(b) The radius of Li^+ is 78 pm (1 pm = 10^{-12} m) while that of a lithium atom, Li, is 152 pm. Explain why the radius of Li^+ is so much smaller than the radius of Li.

2-92 Which has the largest radius: O^{2-}, F^- or F? Explain your reasoning.

2-93 Arrange the following elements in order of increasing size: Al, B, C, and Na. Try doing it without looking at Figure 2-16 and then check yourself by looking at the figure.

2-94 Using your knowledge of trends in element sizes in going across a period of the Periodic Table, explain why the density of the elements increases from potassium through vanadium. (Recall from Section 1-7 that specific gravity is numerically the same as density but has no units.)

Element	Specific Gravity
K	0.862
Ca	1.55
Se	2.99
Ti	4.54
V	6.11

2-95 Name the elements in Group 3A. What does the group designation tell you about the electron configuration of these elements?

2-96 Using the orbital box diagrams and the noble gas notation, write the electron configuration for each atom and ion.
(a) Ti (b) Ti^{2+} (c) Ti^{4+}

2-97 Explain why the Ca^{3+} ion is not found in chemical compounds.

2-98 Explain how the ionization energy of atoms changes when proceeding down a group of the Periodic Table and explain why this change occurs.

2-99 A 7.12 g sample of magnesium is heated with 1.80 g of bromine. All the bromine is used up, and 2.07 g of magnesium bromide is produced. What mass of magnesium remains unreacted?

2-100 A 0.100 g sample of magnesium, when combined with oxygen, yields 0.166 g of magnesium oxide. What masses of magnesium and oxygen must be combined to make exactly 2.00 g of magnesium oxide?

2-101 Complete the following table:

Symbol	Atomic number	Atomic weight	Mass number	# of protons	# of neutrons	# of electrons
H					0	
Li					4	3
Al						
	26		58			
			78			
					17	20
			16			

2-102 An element consists of 90.51% of an isotope with a mass of 19.992 amu, 0.27% of an isotope with a mass of 20.994 amu, and 9.22% of an isotope with a mass of 21.990 amu. Calculate the average atomic mass and identify the element.

2-103 The element silver has two naturally occurring isotopes: ^{109}Ag and ^{107}Ag with a mass of 106.905 amu. Silver consists of 51.82% ^{107}Ag and has an average atomic mass of 107.868 amu. Calculate the mass of ^{109}Ag.

2-104 The average atomic weight of lithium is 6.941 amu. The two naturally occurring isotopes of lithium have the following masses: ^{6}Li, 6.01512 amu; ^{7}Li, 7.01600 amu. Calculate the percent abundance of ^{6}Li and ^{7}Li in naturally occurring lithium.

Looking Ahead

2-105 Suppose that you face a problem similar to Mendeleyev: You must predict the properties of an element not yet discovered. What will element 118 be like if and when enough of it is made for chemists to study its physical and chemical properties?

2-106 Compare the neutron to proton ratio for the heavier and lighter elements. Does the value of this ratio generally increase, decrease, or remain the same as atomic number increases?

3 Chemical Bonds

KEY QUESTIONS

- 3-1 What Do We Need to Know Before We Begin?
- 3-2 What Is the Octet Rule?
- 3-3 How Do We Name Anions and Cations?
- 3-4 What Are the Two Major Types of Chemical Bonds?
- 3-5 What Is an Ionic Bond?
- 3-6 How Do We Name Ionic Compounds?
- 3-7 What Is a Covalent Bond?
- *How To . . .* Draw Lewis Structures
- 3-8 How Do We Name Binary Covalent Compounds?
- 3-9 What Is Resonance?
- *How To . . .* Draw Curved Arrows and Push Electrons
- 3-10 How Do We Predict Bond Angles in Covalent Molecules?
- 3-11 How Do We Determine If a Molecule Is Polar?

Sodium chloride crystal

3-1 What Do We Need to Know Before We Begin?

In Chapter 2, we stated that compounds are tightly bound groups of atoms. In this chapter, we will see that the atoms in compounds are held together by powerful forces of attraction called chemical bonds. There are two main types: ionic bonds and covalent bonds. We begin by examining ionic bonds. To talk about ionic bonds, however, we must first discuss why atoms form the ions they do.

3-2 What Is the Octet Rule?

In 1916, Gilbert N. Lewis (Section 2-6) devised a beautifully simple model that unified many of the observations about chemical bonding and chemical reactions. He pointed out that the lack of chemical reactivity of the noble gases (Group 8A) indicates a high degree of stability of their electron configurations: helium with a filled valence shell of two electrons ($1s^2$), neon with a filled valence shell of eight electrons ($2s^2 2p^6$), argon with a valence shell of eight electrons ($3s^2 3p^6$), and so forth.

The tendency of atoms to react in ways that achieve an outer shell of eight valence electrons is particularly common among Group 1A–7A elements and is given the special name of the **octet rule**. An atom with almost eight valence electrons tends to gain the needed electrons to have eight electrons in its valence shell and an electron configuration like that of the noble gas nearest to it in atomic number. In gaining electrons, the atom becomes a negatively charged ion called an **anion**. An atom with only one or two valence electrons tends to lose the number of electrons required to have an electron configuration like the noble gas nearest it in atomic number. In losing electrons, the atom becomes a positively charged ion called a **cation**. It is important to note that when an atom gains or loses electrons to achieve an outer shell of eight valence electrons, it still retains its same atomic number (Z) as noted in Chapter 2. That is, the resulting ion, which achieves the same electron configuration as a noble gas, is still uniquely characterized by its original atomic number, which identifies the number of protons in the nucleus. For example, although both Cl⁻ (Example 3-1) and Ar have the same number of 18 electrons, Cl⁻ has 17 protons in its nucleus while Ar has 18 protons in its nucleus, thus distinguishing between these two species. Therefore, when an ion forms, the number of protons (and neutrons) in the nucleus remains unchanged; only the number of electrons in the valence shell changes.

Noble Gas	Electron Configuration
He	$1s^2$
Ne	$[He]2s^2\,2p^6$
Ar	$[Ne]3s^2\,3p^6$
Kr	$[Ar]4s^2\,4p^6\,3d^{10}$
Xe	$[Kr]5s^2\,5p^6\,4d^{10}$

Octet rule When undergoing a chemical reaction, atoms of Group 1A–7A elements tend to gain, lose, or share sufficient electrons to achieve an electron configuration having eight valence electrons

Anion An ion with a negative electric charge

Cation An ion with a positive electric charge

EXAMPLE 3-1 The Octet Rule

Show how the following chemical changes obey the octet rule:

(a) A sodium atom loses an electron to form a sodium ion, Na⁺.

$$\text{Na} \longrightarrow \text{Na}^+ + e^-$$
A sodium atom   A sodium ion   An electron

(b) A chorine atom gains an electron to form a chloride ion, Cl⁻.

$$\text{Cl} + e^- \longrightarrow \text{Cl}^-$$
A chlorine atom   An electron   A chloride ion

Strategy
To see how each chemical change follows the octet rule, first write the condensed ground-state electron configuration (Section 2-6C) of the atom involved in the chemical change and of the ion it forms and then compare them.

Solution
(a) The condensed ground-state electron configurations for Na and Na⁺ are:

Na (11 electrons): $1s^2 2s^2 2p^6 3s^1$

Na⁺ (10 electrons): $1s^2 2s^2 2p^6$

A Na atom has one electron ($3s^1$) in its valence shell. The loss of this one valence electron changes the Na atom to a Na ion, Na⁺, which has a complete octet of electrons in its valence shell ($2s^2 2p^6$) and the same electron configuration as Ne, the noble gas nearest to it in atomic number. We can write this chemical change using Lewis dot structures (Section 2-6F):

$$\text{Na·} \longrightarrow \text{Na}^+ + e^-$$
A sodium atom   A sodium ion   An electron

(a) Sodium chloride

(b) Sodium

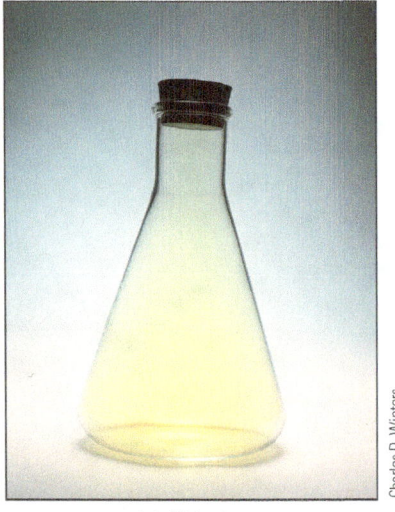

(c) Chlorine

▶ (a) The chemical compound sodium chloride (table salt) is composed of the elements (b) sodium and (c) chlorine in chemical combination. Salt is very different from the elements that constitute it.

(b) The condensed ground-state electron configurations for Cl and Cl⁻ are:

$$\text{Cl (17 electrons): } 1s^2 2s^2 2p^6 3s^2 3p^5$$

$$\text{Cl}^- \text{ (18 electrons): } 1s^2 2s^2 2p^6 3s^2 3p^6$$

A Cl atom has seven electrons in its valence shell ($3s^2 3p^5$). The gain of one electron changes the Cl atom to a Cl ion, Cl⁻, which has a complete octet of electrons in its valence shell ($3s^2 3p^6$) and the same electron configuration as Ar, the noble gas nearest to it in atomic number. We can write this chemical change using Lewis dot structures:

$$:\!\ddot{\text{Cl}}\!\cdot \; + \; e^- \; \longrightarrow \; :\!\ddot{\text{Cl}}\!:^-$$

A chlorine atom   An electron   A chloride ion

Problem 3-1

Show how the following chemical changes obey the octet rule:
(a) A magnesium atom forms a magnesium ion, Mg^{2+}.
(b) A sulfur atom forms a sulfide ion, S^{2-}.

The octet rule gives us a good way to understand why Group 1A–7A elements form the ions that they do. It is not perfect, however, for two reasons:

1. Ions of period 1A and 2A elements with charges greater than +2 are unstable. Boron, for example, has three valence electrons. If it lost these three electrons, it would become B^{3+} and have a complete outer shell like that of helium. It seems, however, that this is far too large a charge for an ion of this second-period element; consequently, this ion is not found in stable ionic compounds. By the same reasoning, carbon does not lose its four valence electrons to become C^{4+}, nor does it gain four valence electrons to become C^{4-}. Either of these changes would place too great a charge on this period 2 element.

2. The octet rule does not apply to Group 1B–7B elements (the transition elements), most of which form ions with two or more different positive charges. Copper, for example, can lose one valence electron to form Cu^+; alternatively, it can lose two valence electrons to form Cu^{2+}.

It is important to understand that there are enormous differences between the properties of an atom and those of its ion(s). Atoms and their ions are completely different chemical species and have completely different chemical and physical properties. Consider, for example, sodium and chlorine. Sodium, a soft metal made of sodium atoms, reacts violently with water. Chlorine atoms are very unstable and even more reactive than sodium atoms. Both sodium and chlorine are poisonous. NaCl, common table salt, is made up of sodium ions and chloride ions. These two ions are quite stable and unreactive. Neither sodium ions nor chloride ions react with water at all. ◀

Because atoms and their ions are different chemical species, we must be careful to distinguish one from the other. Consider the drug commonly known as "lithium," which is used to treat bipolar disorder (also known as manic depression). The element lithium, like sodium, is a soft metal that reacts violently with water. The drug used to treat bipolar disorder is not composed of lithium atoms, Li, but rather lithium ions, Li$^+$, usually administered in the form of lithium carbonate, Li$_2$CO$_3$. Another example comes from the fluoridation of drinking water and of toothpastes and dental gels. The element fluorine, F$_2$, is an extremely poisonous and corrosive gas: it is not what is used for fluoridation. Instead, this process uses fluoride ions, F$^-$, in the form of sodium fluoride, NaF, a compound that is unreactive and nonpoisonous in the concentrations used.

3-3 How Do We Name Anions and Cations?

We form names for anions and cations using a system developed by the International Union of Pure and Applied Chemistry. We will refer to these names as "systematic" names. Many ions have "common" names that were in use long before chemists undertook the effort to name them systematically. In this and the following chapters, we will make every effort to use systematic names for ions, but where a long-standing common name remains in use, we will give it as well.

A. Naming Monatomic Cations

A monatomic (containing only one atom) cation forms when a metal loses one or more valence electrons. Elements of Groups 1A, 2A, and 3A form only one type of cation. For ions of these metals, the name of the cation is the name of the metal followed by the word "ion" (Table 3-1). There is no need to specify the charge on these cations, because only one charge is possible. For example, Na$^+$ is sodium ion and Ca^{2+} is calcium ion. The charges for the species listed in Table 3-1 are based on the number of electrons each atom must lose in order to achieve an outer shell of eight valence electrons, the octet rule. For example, the magnesium atom will readily lose 2 electrons in order to obtain the same number of 10 electrons as neon.

Table 3-1 Names of Cations from Some Metals That Form Only One Positive Ion

Group 1A		Group 2A		Group 3A	
Ion	Name	Ion	Name	Ion	Name
H$^+$	Hydrogen ion	Mg^{2+}	Magnesium ion	Al^{3+}	Aluminum ion
Li$^+$	Lithium ion	Ca^{2+}	Calcium ion		
Na$^+$	Sodium ion	Sr^{2+}	Strontium ion		
K$^+$	Potassium ion	Ba^{2+}	Barium ion		

Most transition and inner transition elements form more than one type of cation, and therefore, the name of the cation must show its charge. To show the charge in a systematic name, we write a Roman numeral (enclosed in parentheses), immediately following (with no space) the name of the metal (Table 3-2). For example, Cu$^+$ is copper(I) ion and Cu^{2+} is copper(II) ion. Note that even though silver is a transition metal, it forms only Ag$^+$; therefore, there is no need to use a Roman numeral to show this ion's charge.

In the older common system for naming metal cations with two different charges, the suffix -*ous* is used to show the smaller charge and -*ic* is used to show the larger charge (Table 3-2). These suffixes are often added to the stem part of the Latin name for the element.

▶ Copper(I) oxide and copper(II) oxide. The different copper ion charges result in different colors.

Table 3-2 Names of Cations from Four Metals That Form Two Different Positive Ions

Ion	Systematic Name	Common Name	Origin of the Symbol of the Element or the Common Name of the Ion
Cu^+	Copper(I) ion	Cuprous ion	*Cupr-* from *cuprum*, the Latin name for copper
Cu^{2+}	Copper(II) ion	Cupric ion	
Fe^{2+}	Iron(II) ion	Ferrous ion	*Ferr-* from *ferrum*, the Latin name for iron
Fe^{3+}	Iron(III) ion	Ferric ion	
Hg^{2+}	Mercury(II) ion	Mercuric ion	*Hg* from *hydrargyrum*, the Latin name for mercury
Sn^{2+}	Tin(II) ion	Stannous ion	*Sn* from *stannum*, the Latin name for tin
Sn^{4+}	Tin(IV) ion	Stannic ion	

Table 3-3 Names of the Most Common Monatomic Anions

Anion	Stem Name	Anion Name
H^-	*hydr*	Hydride
F^-	*fluor*	Fluoride
Cl^-	*chlor*	Chloride
Br^-	*brom*	Bromide
I^-	*iod*	Iodide
O^{2-}	*ox*	Oxide
S^{2-}	*sulf*	Sulfide

B. Naming Monatomic Anions

A monatomic anion is named by adding *-ide* to the stem part of the name. Table 3-3 gives the names of the monatomic anions we deal with most often. The charges for the species listed in Table 3-3 are based on the number of electrons needed to be gained by each atom in order to achieve an outer shell of eight valence electrons, the octet rule.

C. Naming Polyatomic Ions

A **polyatomic ion** contains more than one atom. Examples are the hydroxide ion, OH^-, and the phosphate ion, PO_4^{3-}. We will not be concerned with how these ions are formed, only that they exist and are present in the materials around us. Table 3-4 lists several important polyatomic ions.

The preferred system for naming polyatomic ions that differ in the number of hydrogen atoms is to use the prefixes di-, tri-, and so forth, to show the presence of more than one hydrogen. For example, HPO_4^{2-} is the hydrogen phosphate ion and $H_2PO_4^-$ is the dihydrogen phosphate ion. Because

Table 3-4 Names of Common Polyatomic Ions (Common names, where still widely used, are given in parentheses.)

Polyatomic Ion	Name	Polyatomic Ion	Name
NH_4^+	Ammonium	HCO_3^-	Hydrogen carbonate (bicarbonate)
OH^-	Hydroxide	SO_3^{2-}	Sulfite
NO_2^-	Nitrite	HSO_3^-	Hydrogen sulfite (bisulfite)
NO_3^-	Nitrate	SO_4^{2-}	Sulfate
CH_3COO^- or $C_2H_3O_2^-$	Acetate	HSO_4^-	Hydrogen sulfate (bisulfate)
ClO_4^-	Perchlorate		
CN^-	Cyanide	PO_4^{3-}	Phosphate
MnO_4^-	Permanganate	HPO_4^{2-}	Hydrogen phosphate
CrO_4^{2-}	Chromate	$H_2PO_4^-$	Dihydrogen phosphate
$Cr_2O_7^{2-}$	Dichromate		

CHEMICAL CONNECTIONS 3A

Coral Chemistry and Broken Bones

Bone is a highly structured matrix consisting of both inorganic and organic materials. The inorganic material is chiefly hydroxyapatite, $Ca_5(PO_4)_3OH$, which makes up about 70% of bone by dry weight. By comparison, the enamel of teeth consists almost entirely of hydroxyapatite. Chief among the organic components of bone are collagen fibers (proteins, see Chapter 22), which thread their way through the inorganic matrix, providing extra strength and allowing bone to flex under stress. Also weaving through the hydroxyapatite-collagen framework are blood vessels that supply nutrients.

A problem faced by orthopedic surgeons is how to repair bone damage. For a minor fracture, usually a few weeks in a cast suffices for the normal process of bone growth to repair the damaged area. For severe fractures, especially those involving bone loss, a bone graft may be needed. An alternative to a bone graft is an implant of synthetic bone material. One such material, called Pro Osteon®, is derived by heating coral (calcium carbonate) with ammonium hydrogen phosphate to form a hydroxyapatite similar to that of bone. Throughout the heating process, the porous structure of the coral, which resembles that of bone, is retained.

The surgeon can shape a piece of this material to match the bone void, implant it, stabilize the area by

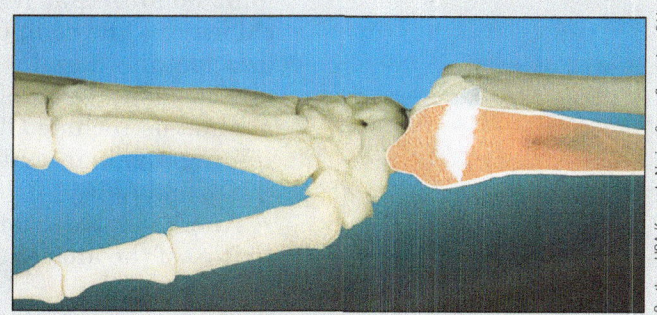

A wrist fracture repaired with bone cement (white area).

inserting metal plates and/or screws, and let new bone tissue grow into the pores of the implant.

In an alternative process, a dry mixture of calcium dihydrogen phosphate monohydrate, $Ca(H_2PO_4)_2 \cdot H_2O$; calcium phosphate, $Ca_3(PO_4)_2$; and calcium carbonate, $CaCO_3$, is prepared. Just before the surgical implant occurs, these chemicals are mixed with a solution of sodium phosphate to form a paste that is then injected into the bony area to be repaired. In this way, the fractured bony area is held in the desired position by the synthetic material while the natural process of bone rebuilding replaces the implant with living bone tissue.

Test your knowledge with Problems 3-92 and 3-96.

---

several hydrogen-containing polyatomic anions have common names that are still widely used, you should memorize them as well. In these common names, the prefix *bi-* is used to show the presence of one hydrogen.

3-4 What Are the Two Major Types of Chemical Bonds?

A. Ionic and Covalent Bonds

According to the Lewis model of chemical bonding, atoms bond together in such a way that each atom participating in a bond acquires a valence-shell electron configuration the same as that of the noble gas nearest to it in atomic number. Atoms acquire completed valence shells in two ways:

1. An atom may lose or gain enough electrons to acquire a filled valence shell, becoming an ion as it does so (Section 3-2). An **ionic bond** results from the force of electrostatic attraction between a cation and an anion.

2. An atom may share electrons with one or more other atoms to acquire a filled valence shell. A **covalent bond** results from the force of attraction between two atoms that share one or more pairs of electrons. A molecule or polyatomic ion is formed.

Ionic bond A chemical bond resulting from the attraction between positive and negative ions

Covalent bond A chemical bond resulting from the sharing of electrons between two atoms

We can now ask how to determine whether two atoms in a compound are bonded by an ionic bond or a covalent bond. One way to do so is to consider the relative positions of the two atoms in the Periodic Table. Ionic bonds usually form between a metal and a nonmetal. An example of an ionic bond is that formed between the metal sodium and the nonmetal chlorine in the compound sodium chloride, Na^+Cl^-. When two nonmetals or a metalloid and a nonmetal combine, the bond between them is usually covalent. Examples of compounds containing covalent bonds between nonmetals include Cl_2, H_2O, CH_4, and NH_3. Examples of compounds containing covalent bonds between a metalloid and a nonmetal include BF_3, $SiCl_4$, and AsH_3.

Another way to determine the bond type is to compare the electronegativities of the atoms involved, which is the subject of the next subsection.

B. Electronegativity and Chemical Bonds

Electronegativity is a measure of an atom's attraction for the electrons it shares in a chemical bond with another atom. The most widely used scale of electronegativities (Table 3-5) was devised in the 1930s by Linus Pauling. On the Pauling scale, fluorine, the most electronegative element, is assigned an electronegativity of 4.0 and all other elements are assigned values relative to fluorine.

As you study the electronegativity values in Table 3-5, note that they generally increase from left to right across a row of the Periodic Table and from bottom to top within a column. Values increase from left to right because of the increasing positive charge on the nucleus, which leads to a stronger attraction for electrons in the valence shell. Values increase going up a column because the decreasing distance of the valence electrons from the nucleus leads to a stronger attraction between the nucleus and its valence electrons.

You might compare these trends in electronegativity with the trends in ionization energy (Section 2-8B). Each illustrates the periodic nature of elements within the Periodic Table. Ionization energy measures the amount of energy necessary to remove an electron from an atom. Electronegativity measures how tightly an atom holds the electrons that it shares with another atom. Notice that both electronegativity and ionization energy generally increase from left to right across a row of the Periodic Table from columns 1A to 7A. In addition, both electronegativity and ionization energy increase going up a column.

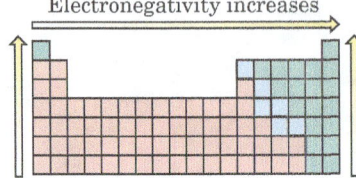

Electronegativity increases

Table 3-5 Electronegativity Values of the Elements (Pauling Scale)

1A	2A	3B	4B	5B	6B	7B	8B			1B	2B	3A	4A	5A	6A	7A
H 2.1																
Li 1.0	Be 1.5											B 2.0	C 2.5	N 3.0	O 3.5	F 4.0
Na 0.9	Mg 1.2											Al 1.5	Si 1.8	P 2.1	S 2.5	Cl 3.0
K 0.8	Ca 1.0	Sc 1.3	Ti 1.5	V 1.6	Cr 1.6	Mn 1.5	Fe 1.8	Co 1.8	Ni 1.8	Cu 1.9	Zn 1.6	Ga 1.6	Ge 1.8	As 2.0	Se 2.4	Br 2.8
Rb 0.8	Sr 1.0	Y 1.2	Zr 1.4	Nb 1.6	Mo 1.8	Tc 1.9	Ru 2.2	Rh 2.2	Pd 2.2	Ag 1.9	Cd 1.7	In 1.7	Sn 1.8	Sb 1.9	Te 2.1	I 2.5
Cs 0.7	Ba 0.9	La 1.1	Hf 1.3	Ta 1.5	W 1.7	Re 1.9	Os 2.2	Ir 2.2	Pt 2.2	Au 2.4	Hg 1.9	Tl 1.8	Pb 1.8	Bi 1.9	Po 2.0	At 2.2

> **Interactive Example 3-2**   **Electronegativity**
>
> Judging from their relative positions in the Periodic Table, which element in each pair has the larger electronegativity?
>
> (a) Lithium or carbon   (b) Nitrogen or oxygen   (c) Carbon or oxygen
>
> **Strategy**
>
> The elements in each pair are in the second period of the Periodic Table. Within a period, electronegativity increases from left to right across the period.
>
> **Solution**
>
> (a) C > Li   (b) O > N   (c) O > C
>
> **Problem 3-2**
>
> Judging from their relative positions in the Periodic Table, which element in each pair has the larger electronegativity?
>
> (a) Lithium or potassium   (b) Nitrogen or phosphorus
> (c) Carbon or silicon

3-5   What Is an Ionic Bond?

A. Forming Ionic Bonds

According to the Lewis model of bonding, an ionic bond forms by the transfer of one or more valence-shell electrons from an atom of lower electronegativity to the valence shell of an atom of higher electronegativity. The more electronegative atom gains one or more valence electrons and becomes an anion; the less electronegative atom loses one or more valence electrons and becomes a cation. The compound formed by the electrostatic attraction of positive and negative ions is called an **ionic compound**.

As a guideline, we say that this type of electron transfer to form an ionic compound is most likely to occur if the difference in electronegativity between two atoms is approximately 1.9 or greater. A bond is more likely to be covalent if this difference is less than 1.9. You should be aware that the value of 1.9 for the formation of an ionic bond is somewhat arbitrary. Some chemists prefer a slightly larger value, others a slightly smaller value. The essential point is that the value of 1.9 gives us a guidepost against which to decide if a bond is more likely to be ionic or more likely to be covalent. Section 3-7 discusses covalent bonding.

An example of an ionic compound is that formed between the metal sodium (electronegativity 0.9) and the nonmetal chlorine (electronegativity 3.0). The difference in electronegativity between these two elements is 2.1. In forming the ionic compound NaCl, the single $3s$ valence electron of a sodium atom is transferred to the partially filled valence shell of a chlorine atom.

$$\text{Na } (1s^22s^22p^63s^1) + \text{Cl } (1s^22s^22p^63s^23p^5) \longrightarrow \text{Na}^+ (1s^22s^22p^6) + \text{Cl}^- (1s^22s^22p^63s^23p^6)$$

     Sodium atom   Chlorine atom   Sodium ion   Chloride ion

In the following equation, we use a single-headed curved arrow to show the transfer of one electron from sodium to chlorine.

$$\text{Na}\cdot + \cdot\ddot{\underset{\cdot\cdot}{\text{Cl}}}: \longrightarrow \text{Na}^+ \; :\ddot{\underset{\cdot\cdot}{\text{Cl}}}:^-$$

66 Chapter 3 Chemical Bonds

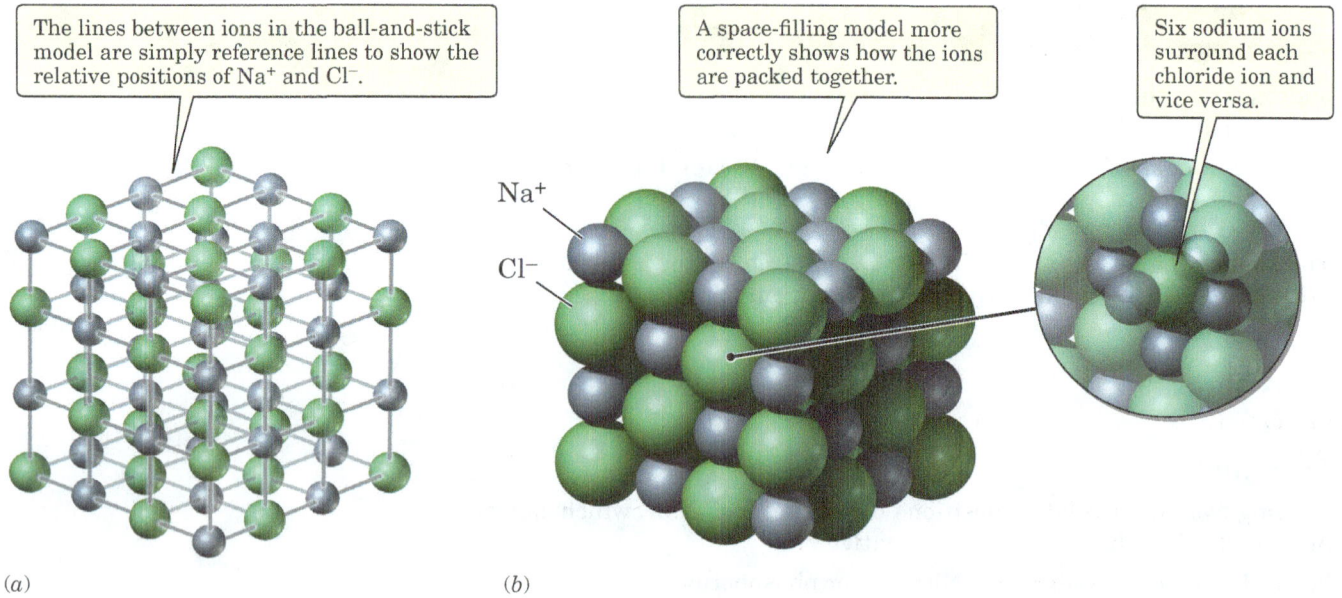

FIGURE 3-1 The structure of a sodium chloride crystal. (a) Ball-and-stick models show the relative positions of the ions. (b) Space-filling models show the relative sizes of the ions.

The ionic bond in solid sodium chloride results from the force of electrostatic attraction between positive sodium ions and negative chloride ions. In its solid (crystalline) form, sodium chloride consists of a three-dimensional array of Na^+ and Cl^- ions arranged as shown in Figure 3-1.

Although ionic compounds do not consist of molecules, they do have a definite ratio of one kind of ion to another; their formulas give this ratio. For example, NaCl represents the simplest ratio of sodium ions to chloride ions—namely, 1:1.

B. Predicting Formulas of Ionic Compounds

Ions are charged particles, but the matter we see all around us and deal with every day is electrically neutral (uncharged). If ions are present in any sample of matter, the total number of positive charges must equal the total number of negative charges. Therefore, we cannot have a sample containing only Na^+ ions. Any sample that contains Na^+ ions must also contain negative ions, such as Cl^-, Br^-, or S^{2-}, and the sum of the positive charges must equal the sum of the negative charges.

EXAMPLE 3-3 Formulas of Ionic Compounds

Write the formulas for the ionic compounds formed from the following ions:
(a) Lithium ion and bromide ion
(b) Barium ion and iodide ion
(c) Aluminum ion and sulfide ion

Strategy

The formula of an ionic compound shows the simplest whole-number ratio between cations and anions. In an ionic compound, the total number of positive charges of the cations and the total number of negative charges of the anions must be equal. Therefore, to predict the formula of an ionic compound, you must know the charges of the ions involved.

Solution

(a) Table 3-1 shows that the charge on a lithium ion is +1, and Table 3-3 shows that the charge on a bromide ion is −1. Therefore, the formula for lithium bromide is LiBr.

(b) The charge on a barium ion is +2 (Table 3-1) and the charge on an iodide ion is −1 (Table 3-3). Two I$^-$ ions are required to balance the charge of one Ba^{2+} ion. Therefore, the formula for barium iodide is BaI$_2$.

(c) The charge on an aluminum ion is +3 (Table 3-1) and the charge on a sulfide ion is −2 (Table 3-3). For the compound to have an overall charge of zero, the ions must combine in the ratio of two aluminum ions to three sulfur ions. The formula of aluminum sulfide is Al$_2$S$_3$.

Problem 3-3

Write the formulas for the ionic compounds formed from the following ions:

(a) Potassium ion and chloride ion   (b) Calcium ion and fluoride ion
(c) Iron(III) ion and oxide ion

Remember that the subscripts in the formulas for ionic compounds represent the ratio of the ions. Thus, a crystal of BaI$_2$ has twice as many iodide ions as barium ions. For ionic compounds, when both charges are 2, as in the compound formed from Ba^{2+} and O^{2-}, we must "reduce to lowest terms." That is, barium oxide is BaO, not Ba$_2$O$_2$. The reason is that we are looking at ratios only, and the ratio of ions in barium oxide is 1:1.

3-6 How Do We Name Ionic Compounds?

To name an ionic compound, we give the name of the cation first, followed by the name of the anion.

A. Binary Ionic Compounds of Metals That Form Only One Positive Ion

A **binary compound** contains only two elements. In a **binary ionic compound**, both of the elements are present as ions. The name of the compound consists of the name of the metal from which the cation (positive ion) was formed, followed by the name of the anion (negative ion). We generally ignore subscripts in naming binary ionic compounds. For example, AlCl$_3$ is named aluminum chloride. We know this compound contains three chloride ions because the positive and negative charges in the compound must be equal—that is, one Al^{3+} ion must combine with three Cl$^-$ ions to balance the charges.

Interactive Example 3-4   Binary Ionic Compounds

Name these binary ionic compounds:

(a) LiBr   (b) Ag$_2$S   (c) NaBr

Strategy

The name of an ionic compound consists of two words: name of the cation followed by the name of the anion.

Solution

(a) Lithium bromide (b) Silver sulfide (c) Sodium bromide

Problem 3-4

Name these binary ionic compounds:

(a) MgO (b) BaI_2 (c) KCl

Interactive Example 3-5 — Binary Ionic Compounds

Write the formulas for these binary ionic compounds:

(a) Barium hydride (b) Sodium fluoride (c) Calcium oxide

Strategy

Write the formula of the positive ion and then the formula of the negative ion. Remember that the number of positive and negative charges must be equal. Show the ratio of each ion in the formula of the compound by subscripts. Where only one of either ion is present, do not show a subscript.

Chemical Compound	Ions Present	Analysis
Barium hydride	Ba^{2+} and H^-	One Ba^{2+} ion must combine with two H^- ions to balance the charges.
Sodium fluoride	Na^+ and F^-	One Na^+ ion must combine with one F^- ion to balance the charges.
Calcium oxide	Ca^{2+} and O^{2-}	One Ca^{2+} ion must combine with one O^{2-} ion to balance the charges.

Solution

(a) BaH_2 (b) NaF (c) CaO

Problem 3-5

Write the formulas for these binary ionic compounds:

(a) Magnesium chloride (b) Aluminum oxide (c) Lithium iodide

B. Binary Ionic Compounds of Metals That Form More Than One Positive Ion

Table 3-2 shows that many transition metals form more than one positive ion. For example, copper forms both Cu^+ and Cu^{2+} ions. For systematic names, we use Roman numerals in the name to show the charge. For common names, we use the *-ous, -ic* system.

EXAMPLE 3-6 — Binary Ionic Compounds

Give each binary ionic compound a systematic name and a common name.

(a) CuO (b) Cu_2O

Strategy

The name of a binary ionic compound consists of two words. First is the name of the cation followed by the name of the anion. Because transition metals typically form more than one cation, the charge on the cation must be indicated by a Roman numeral in parentheses following the name of the transition metal or by using the suffix *-ic* to show the higher of the two possible cation charges or the suffix *-ous* to show the lower of the two possible cation charges.

Chemical Compound	Ions Present	Analysis
CuO	Cu^{2+} and O^{2-}	One Cu^{2+} ion must combine with one O^{2-} ion to balance the charges.
Cu_2O	Cu^+ and O^{2-}	Two Cu^+ ions must combine with one O^{2-} ion to balance the charges.

Solution

(a) Systematic name: copper(II) oxide. Common name: cupric oxide.
(b) Systematic name: copper(I) oxide. Common name: cuprous oxide.

Remember in answering part (b) that we ignore subscripts in naming binary ionic compounds. Therefore, the 2 in Cu_2O is not indicated in the name. You know that two copper(I) ions are present because two positive charges are needed to balance the two negative charges on an O^{2-} ion.

Problem 3-6

Give each binary compound a systematic name and a common name.
(a) FeO   (b) Fe_2O_3

C. Ionic Compounds That Contain Polyatomic Ions

To name ionic compounds containing polyatomic ions, name the positive ion first and then the negative ion, each as a separate word. Remember to refer to the names of the common polyatomic ions found in Table 3-4.

Interactive Example 3-7   Polyatomic Ions

Name these ionic compounds, each of which contains a polyatomic ion:
(a) $NaNO_3$ (b) $CaCO_3$ (c) $(NH_4)_2SO_3$ (d) NaH_2PO_4

Strategy

To name ionic compounds containing polyatomic ions (Table 3-4), name the positive ion first and then the negative ion, each as a separate word.

Solution

(a) Sodium nitrate   (b) Calcium carbonate
(c) Ammonium sulfite   (d) Sodium dihydrogen phosphate

Problem 3-7

Name these ionic compounds, each of which contains a polyatomic ion:
(a) K_2HPO_4 (b) $Al_2(SO_4)_3$ (c) $FeCO_3$

CHEMICAL CONNECTIONS 3B

Ionic Compounds in Medicine

Many ionic compounds have medical uses, some of which are shown in the table.

Formula	Name	Medical Use
$AgNO_3$	Silver nitrate	Antibiotic
$BaSO_4$	Barium sulfate	Radiopaque medium for X-ray work
$CaSO_4$	Calcium sulfate	Plaster of Paris casts
$FeSO_4$	Iron(II) sulfate	Treatment of iron deficiency
$KMnO_4$	Potassium permanganate	Anti-infective (external)
KNO_3	Potassium nitrate (saltpeter)	Diuretic
Li_2CO_3	Lithium carbonate	Treatment of bipolar disorder
$MgSO_4$	Magnesium sulfate (Epsom salts)	Cathartic
$NaHCO_3$	Sodium bicarbonate (baking soda)	Antacid
NaI	Sodium iodide	Iodine for thyroid hormones
NH_4Cl	Ammonium chloride	Acidification of the digestive system
$(NH_4)_2CO_3$	Ammonium carbonate	Expectorant
SnF_2	Tin(II) fluoride	To strengthen teeth (external)
ZnO	Zinc oxide	Astringent (external)

Drinking a "barium cocktail," which contains barium sulfate, makes the intestinal tract visible on an X-ray.

Test your knowledge with Problems 3-93, 3-94, and 3-95.

3-7 What Is a Covalent Bond?

A. Formation of a Covalent Bond

A covalent bond forms when electron pairs are shared between two atoms whose difference in electronegativity is less than 1.9. As we have already mentioned, the most common covalent bonds occur between two nonmetals or between a nonmetal and a metalloid.

According to the Lewis model, a pair of electrons in a covalent bond functions in two ways simultaneously: the two atoms share it, and it fills the valence shell of each atom. The simplest example of a covalent bond is that in a hydrogen molecule, H_2. When two hydrogen atoms bond, the single electrons from each atom combine to form an electron pair. A bond formed by sharing a pair of electrons is called a **single bond** and is represented by a single line between the two atoms. The electron pair shared between the two hydrogen atoms in H_2 completes the valence shell of each hydrogen. Thus, in H_2, each hydrogen has, in effect, two electrons in its valence shell and an electron configuration like that of helium, the noble gas nearest to it in atomic number.

Single bond A bond formed by sharing one pair of electrons and represented by a single line between two atoms

Table 3-6 Classification of Chemical Bonds

Electronegativity Difference Between Bonded Atoms	Type of Bond	Most Likely Formed Between
Less than 0.5	Nonpolar covalent	Two nonmetals or a nonmetal and a metalloid
0.5 to 1.9	Polar covalent	
Greater than 1.9	Ionic	A metal and a nonmetal

B. Nonpolar and Polar Covalent Bonds

Although all covalent bonds involve the sharing of electrons, they differ widely in the degree of sharing. We classify covalent bonds into two categories, **nonpolar covalent** and **polar covalent**, depending on the difference in electronegativity between the bonded atoms. In a nonpolar covalent bond, electrons are shared equally. In a polar covalent bond, they are shared unequally. It is important to realize that no sharp line divides these two categories, nor, for that matter, does a sharp line divide polar covalent bonds and ionic bonds. Nonetheless, the rule-of-thumb guidelines given in Table 3-6 will help you decide whether a given bond is more likely to be nonpolar covalent, polar covalent, or ionic.

An example of a polar covalent bond is that in H—Cl, in which the difference in electronegativity between the bonded atoms is $3.0 - 2.1 = 0.9$. A covalent bond between carbon and hydrogen is classified as nonpolar covalent because the difference in electronegativity between these two atoms is only $2.5 - 2.1 = 0.4$. You should be aware, however, that there is some slight polarity to a C—H bond, but because it is quite small, we arbitrarily say that a C—H bond is nonpolar. Increasing difference in electronegativity are related to increasing bond polarity.

Nonpolar covalent A covalent bond between two atoms whose difference in electronegativity is less than 0.5

Polar covalent A covalent bond between two atoms whose difference in electronegativity is between 0.5 and 1.9

EXAMPLE 3-8 Classification of Chemical Bonds

Classify each bond as nonpolar covalent, polar covalent, or ionic.

(a) O—H (b) N—H (c) Na—F (d) C—Mg (e) C—S

Strategy
Using Table 3-5, determine the difference in electronegativity between bonded atoms. Then, use the values given in Table 3-6 to classify the type of bond formed.

Solution

Bond	Difference in Electronegativity	Type of Bond
(a) O—H	$3.5 - 2.1 = 1.4$	Polar covalent
(b) N—H	$3.0 - 2.1 = 0.9$	Polar covalent
(c) Na—F	$4.0 - 0.9 = 3.1$	Ionic
(d) C—Mg	$2.5 - 1.2 = 1.3$	Polar covalent
(e) C—S	$2.5 - 2.5 = 0.0$	Nonpolar covalent

Problem 3-8
Classify each bond as nonpolar covalent, polar covalent, or ionic.

(a) S—H (b) P—H (c) C—F (d) C—Cl

Dipole A chemical species in which there is a separation of charge; there is a positive pole in one part of the species and a negative pole in another part

FIGURE 3-2 HCl is a polar covalent molecule. In the electron density map of HCl, red indicates a region of high electron density and blue indicates a region of low electron density.

An important consequence of the unequal sharing of electrons in a polar covalent bond is that the more electronegative atom gains a greater fraction of the shared electrons and acquires a partial negative charge, indicated by the symbol δ− (read "delta minus"). The less electronegative atom has a lesser fraction of the shared electrons and acquires a partial positive charge, indicated by the symbol δ+ (read "delta plus"). This separation of charge produces a **dipole** (two poles). We commonly show the presence of a bond dipole by an arrow, with the head of the arrow near the negative end of the dipole and a cross on the tail of the arrow near the positive end (Figure 3-2).

We can also show the polarity of a covalent bond by an electron density map. In this type of molecular model, a blue color shows the presence of a δ+ charge and a red color shows the presence of a δ− charge. Figure 3-2 also shows an electron density map of HCl. The ball-and-stick model in the center of the electron density map shows the orientation of the atomic nuclei in space. The transparent surface surrounding the ball-and-stick model shows the relative sizes of the atoms (equivalent to the size shown by a space-filling model). Colors on the surface show the distribution of electron density. We see by the blue color that hydrogen bears a δ+ charge and by the red color that chlorine bears a δ− charge.

EXAMPLE 3-9 Polarity of A Covalent Bond

Using the symbols δ− and δ+, indicate the polarity in each polar covalent bond.

(a) C—O   (b) N—H   (c) C—Mg

Strategy

The more electronegative atom of a covalent bond bears a partial negative charge, and the less electronegative atom bears a partial positive charge.

Solution

For (a), C and O are both in period 2 of the Periodic Table. Because O is farther to the right than C, it is more electronegative than C. For (c), Mg is a metal located to the far left in the Periodic Table and C is a nonmetal located to the right. All nonmetals, including H, have a greater electronegativity than do the metals in columns 1A and 2A. The electronegativity of each element is given below the symbol of the element.

$$\begin{array}{ccc} \overset{\delta+ \quad \delta-}{(a)\ \text{C—O}} & \overset{\delta- \quad \delta+}{(b)\ \text{N—H}} & \overset{\delta- \quad \delta+}{(c)\ \text{C—Mg}} \\ \quad 2.5\ \ 3.5 & \quad 3.0\ \ 2.1 & \quad 2.5\ \ 1.2 \end{array}$$

Problem 3-9

Using the symbols δ− and δ+, indicate the polarity in each polar covalent bond.

(a) C—N   (b) N—O   (c) C—Cl

Lewis structures Formulas for molecules or ions showing all pairs of bonding electrons as single, double, or triple bonds and all nonbonded electrons as pairs of Lewis dots

C. Drawing Lewis Structures of Covalent Compounds

The ability to draw **Lewis structures** for covalent molecules is a fundamental skill for the study of chemistry. The following How To box will help you with this task.

HOW TO...

Draw Lewis Structures

1. **Determine the number of valence electrons in the molecule.**
   Add up the number of valence electrons contributed by each atom. To determine the number of valence electrons, you only need to know the number of each kind of atom in the molecule. For each unit of negative charge on the ion, add one electron. For each unit of positive charge, subtract one electron.
   Example: The Lewis structure for formaldehyde, CH_2O, must show 12 valence electrons:

   $$4 \text{ (from C)} + 2 \text{ (from the two H)} + 6 \text{ (from O)} = 12$$

2. **Determine the connectivity of the atoms (which atoms are bonded to each other) and connect bonded atoms by single bonds.**
   Determining the connectivity of the atoms is often the most challenging part of drawing a Lewis structure. For some molecules, we ask you to propose connectivity. For most, however, we give you the experimentally determined connectivity and ask you to complete the Lewis structure.
   Example: The atoms in formaldehyde are bonded in the following order. *Note that we do not attempt at this point to show bond angles or the three-dimensional shape of the molecule; we just show what is bonded to what.*

   $$\begin{array}{c} O \\ | \\ H-C-H \end{array}$$

   This partial structure shows six valence electrons in the three single bonds. In it, we have accounted for six of the 12 valence electrons.

3. **Arrange the remaining electrons so that each atom has a complete outer shell.**
   Each hydrogen atom must be surrounded by two electrons. Each carbon, nitrogen, oxygen, and halogen atom must be surrounded by eight valence electrons. The remaining valence electrons may be shared between atoms in bonds or may be unshared pairs on a single atom. A pair of electrons involved in a covalent bond (**bonding electrons**) is shown as a single line; an unshared pair of electrons (**nonbonding electrons**) is shown as a pair of Lewis dots.

Lewis structure

Ball-and-stick model of formaldehyde

Unshared pairs of electrons are not shown in ball-and-stick models

By placing two pairs of bonding electrons between C and O, we give carbon a complete octet. By placing the remaining four electrons on oxygen as two Lewis dot pairs, we give oxygen eight valence electrons and a complete octet (octet rule). Note that we placed the two pairs of bonding electrons between C and O before we assigned the unshared pairs of electrons on the oxygen.

Bonding electrons Valence electrons involved in forming a covalent bond; that is, shared electrons

Nonbonding electrons Valence electrons not involved in forming covalent bonds; that is, unshared electrons

Double bond A bond formed by sharing two pairs of electrons and represented by two lines between the two bonded atoms

Triple bond A bond formed by sharing three pairs of electrons and represented by three lines between the two bonded atoms

As a check on this structure, verify (1) that each atom has a complete valence shell (which each does) and (2) that the Lewis structure has the correct number of valence electrons (12, which it does).

4. In a **double bond**, two atoms share two pairs of electrons; we represent a double bond by two lines between the bonded atoms. Double bonds are most common between atoms of C, N, O, and S. In the organic and biochemistry chapters in particular, we shall see many examples of C=C and C=O double bonds.

5. In a **triple bond**, two atoms share three pairs of electrons; we show a triple bond by three lines between the bonded atoms. Triple bonds are most common between atoms of C and N, for example in —C≡C— and —C≡N: triple bonds.

Table 3-7 gives Lewis structures and names for several small molecules. Notice that each hydrogen is surrounded by two valence electrons and that each carbon, nitrogen, oxygen, and chlorine is surrounded by eight valence electrons. Furthermore, each carbon has four bonds, each nitrogen has three bonds and one unshared pair of electrons, each oxygen has two bonds and two unshared pairs of electrons, and chlorine (as well as the other halogens) has one bond and three unshared pairs of electrons.

Table 3-7 Lewis Structures for Several Small Molecules

(The number of valence electrons in each molecule is given in parentheses after the molecular formula of the compound.)

EXAMPLE 3-10 Lewis Structures of Covalent Compounds

State the number of valence electrons in each molecule and draw a Lewis structure for each:

(a) Hydrogen peroxide, H_2O_2
(b) Methanol, CH_3OH
(c) Acetic acid, CH_3COOH

Strategy

To determine the number of valence electrons in a molecule, add the number of valence electrons contributed by each kind of atom in the molecule. To draw a Lewis structure, determine the connectivity of the atoms and connect bonded atoms by single bonds. Then arrange the remaining valence electrons so that each atom has a complete outer shell.

Solution

(a) A Lewis structure for hydrogen peroxide, H_2O_2, must show the 14 valence electrons—six from each oxygen and the one from each hydrogen, for a total of $12 + 2 = 14$ valence electrons. We know that hydrogen forms only one covalent bond, so the connectivity of atoms must be as follows:

$$H—O—O—H$$

The three single bonds account for six valence electrons. The remaining eight valence electrons must be placed on the oxygen atoms to give each a complete octet:

Ball-and-stick models show only nuclei, covalent bonds, and the shape of the molecule; they do not show unshared pairs of electrons

(b) A Lewis structure for methanol, CH_3OH, must show the four valence electrons from carbon, one from each hydrogen, and the six from oxygen for a total of $4 + 4 + 6 = 14$ valence electrons. The connectivity of atoms in methanol is given on the left, below. The five single bonds in this partial structure account for ten valence electrons. The remaining four valence electrons must be placed on oxygen as two Lewis dot pairs to give it a complete octet.

The order of attachment of atoms   Lewis dot structure

(c) A molecule of acetic acid, CH_3COOH, must contain the four valence electrons from each carbon, the six from each oxygen, and the one from each hydrogen for a total of $8 + 12 + 4 = 24$ valence electrons. The connectivity of atoms, shown on the left below, contains seven single bonds, which account for 14 valence electrons. The remaining ten electrons must be added in such a way that each carbon and oxygen atom has a complete outer shell of eight electrons. This can be done in only one way, which creates a double bond between carbon and one of the oxygens.

The order of attachment of atoms   Lewis dot structure   (Unshared electron pairs not shown)

In this Lewis structure, each carbon has four bonds: one carbon has four single bonds, and the other carbon has two single bonds and one double bond. Each oxygen has two bonds and two unshared pairs of electrons: one oxygen has one double bond and two unshared pairs of electrons, and the other oxygen has two single bonds and two unshared pairs of electrons.

Problem 3-10

Draw a Lewis structure for each molecule. Each has only one possible order of attachment of its atoms, which is left for you to determine.

(a) Ethane, C_2H_6
(b) Chloromethane, CH_3Cl
(c) Hydrogen cyanide, HCN

EXAMPLE 3-11 Covalent Bonding of Carbon

Why does carbon have four bonds and no unshared pairs of electrons in some covalent compounds?

Strategy

In answering this question, you need to consider the electron configuration of carbon, the number of electrons its valence shell can hold, and the orbitals available to it for sharing electrons to form covalent bonds.

Solution

In forming covalent compounds, carbon reacts to obtain a filled valence shell; that is, a complete octet in its valence shell and an electron configuration resembling that of neon, the noble gas nearest it in atomic number.

Carbon is a second-period element and can contain no more than eight electrons in its valence shell; that is, in its one $2s$ and three $2p$ orbitals. When carbon has four bonds, it has a complete valence shell and a complete octet. With eight electrons, its $2s$ and $2p$ orbitals are now completely occupied and can hold no more electrons. Adding an additional pair of electrons would place ten electrons in the valence shell of carbon and violate the octet rule.

Problem 3-11

Draw a Lewis structure of a covalent compound in which carbon has:

(a) Four single bonds
(b) Two single bonds and one double bond
(c) Two double bonds
(d) One single bond and one triple bond

D. Exceptions to the Octet Rule

The Lewis model of covalent bonding focuses on valence electrons and the necessity for each atom other than hydrogen to have a completed valence shell containing eight electrons. Although most molecules formed by main-group elements (Groups 1A–7A) have structures that satisfy the octet rule, some important exceptions exist.

One exception involves molecules that contain an atom with more than eight electrons in its valence shell. Atoms of period 2 elements use one $2s$ and three $2p$ orbitals for bonding. These four orbitals can contain only eight valence electrons—hence the octet rule. Atoms of period 3 elements, however, have one $3s$ orbital, three $3p$ orbitals, and five $3d$ orbitals; they can accommodate more than eight electrons in their valence shells (Section 2-6A). In phosphine, PH_3, phosphorus has eight electrons in its valence shell and obeys the octet rule. The phosphorus atoms in phosphorus pentachloride, PCl_5, and phosphoric acid, H_3PO_4, have ten electrons in their valence shells and, therefore, are exceptions to the octet rule.

Sulfur, another period 3 element, forms compounds in which it has 8, 10, and even 12 electrons in its valence shell. The sulfur atom in H_2S has 8 electrons in its valence shell and obeys the octet rule. The sulfur atoms in SO_2 and H_2SO_4 have 10 and 12 electrons, respectively, in their valence shells and are exceptions to the octet rule.

3-8 How Do We Name Binary Covalent Compounds?

A **binary covalent compound** is a binary (two-element) compound in which all bonds are covalent. In naming a binary covalent compound:

1. First name the less electronegative element (see Table 3-5). Note that the less electronegative element is also generally written first in the formula.
2. Then name the more electronegative element. To name it, add *-ide* to the stem name of the element. Chlorine, for example, becomes chloride and oxygen becomes oxide (Table 3-3).
3. Use the prefixes *di-, tri-, tetra-*, and so on, to show the number of atoms of each element. The prefix *mono-* is omitted when it refers to the first atom named, and it is rarely used with the second atom. An exception to this rule is CO, which is named carbon monoxide.

The name is then written as two words.

Name of the first element in the formula; use prefixes *di-* and so forth if necessary	Name of the second element; use prefixes *mono-* and so forth if necessary

Note that the use of prefixes is only for binary covalent compounds and should not be used to name ionic compounds.

Interactive Example 3-12 — Binary Covalent Compounds

Name these binary covalent compounds:

(a) NO (b) SF_2 (c) N_2O

Strategy

The systematic name of a binary covalent compound consists of two words. The first word gives the name of the element that appears first in the formula. A prefix (*di-*, *tri-*, *tetra-*, and so forth) is used to show the number of atoms of that element in the formula. The second word consists of (1) a suffix designating the number of atoms of the second element, (2) the stem name of the second element, and (3) the suffix *-ide*.

Solution

(a) Nitrogen oxide (more commonly called nitric oxide)
(b) Sulfur difluoride
(c) Dinitrogen oxide (more commonly called nitrous oxide or laughing gas)

Problem 3-12

Name these binary covalent compounds:

(a) NO_2 (b) PBr_3 (c) SCl_2 (d) BF_3

3-9 What Is Resonance?

As chemists developed a deeper understanding of covalent bonding in organic and inorganic compounds, it became obvious that for a great many molecules and ions, no single Lewis structure provides a truly accurate representation. For example, Figure 3-3 shows three Lewis structures for the carbonate ion, CO_3^{2-}. In each structure, carbon is bonded to three oxygen atoms by a combination of one double bond and two single bonds. Each Lewis structure implies that one carbon-oxygen bond is different from the other two. However, this is not the case. It has been determined experimentally that all three carbon-oxygen bonds are identical.

FIGURE 3-3 Three Lewis structures for the carbonate ion.

The problem for chemists, then, is how to describe the structure of molecules and ions for which no single Lewis structure is adequate and yet still retain Lewis structures. As an answer to this problem, Linus Pauling proposed the theory of resonance.

Resonance A theory that many molecules and ions are best described as a hybrid of two or more Lewis contributing structures

Contributing structure Representations of a molecule or ion that differ only in the distribution of valence electrons

A. Theory of Resonance

According to the theory of **resonance**, many molecules and ions are best described by writing two or more Lewis structures and considering the real molecule or ion to be a hybrid of these structures. An individual Lewis structure is called a **contributing structure**. They are also sometimes referred

CHEMICAL CONNECTIONS 3C

Nitric Oxide: Air Pollutant and Biological Messenger

Nitric oxide, NO, is a colorless gas whose importance in the environment has been known for several decades but whose biological importance is only now being fully recognized. This molecule has 11 valence electrons. Because its number of electrons is odd, it is not possible to draw a structure for NO that obeys the octet rule; there must be one unpaired electron, here shown on the less electronegative nitrogen atom.

An unpaired electron → :N=O:
Nitric oxide

The importance of NO in the environment arises from the fact that it forms as a by-product during the combustion of fossil fuels. Under the temperature conditions of internal combustion engines and other combustion sources, nitrogen and oxygen of the air react to form small quantities of NO:

$$N_2 + O_2 \xrightarrow{heat} 2NO$$
Nitric oxide

When inhaled, NO passes from the lungs into the bloodstream. There it interacts with the iron in hemoglobin, decreasing its ability to carry oxygen. What makes nitric oxide so hazardous in the environment is that it reacts almost immediately with oxygen to form NO_2. When dissolved in water, NO_2 reacts with water to form nitric acid and nitrous acid, which are major acidifying components of acid rain.

$$2NO + O_2 \longrightarrow 2NO_2$$
Nitric oxide   Nitrogen dioxide

$$NO_2 + H_2O \longrightarrow HNO_3 + HNO_2$$
Nitrogen dioxide   Nitric acid   Nitrous acid

Imagine the surprise when it was discovered within the last two decades that this highly reactive, seemingly hazardous compound is synthesized in humans and plays a vital role as a signaling molecule in the cardiovascular system (Chemical Connections 24F).

Colorless nitric oxide, NO, coming from the tank, bubbles through the water. When it reaches the air, it is oxidized to brown nitrogen dioxide, NO_2.

Test your knowledge with Problem 3-97.

---

to as **resonance structures** or **resonance contributors**. We show that the real molecule or ion is a **resonance hybrid** of the various contributing structures by interconnecting them with **double-headed arrows**. Do not confuse the double-headed arrow with the double arrow used to show chemical equilibrium. As we explain shortly, resonance structures are not in equilibrium with each other.

Figure 3-4 shows three contributing structures for the carbonate ion. These contributing structures are said to be equivalent. All three have identical patterns of covalent bonding.

The use of the term "resonance" for this theory of covalent bonding appears to suggest that bonds and electron pairs are constantly changing back and forth from one position to another over time. This notion is not

Resonance hybrid A molecule or ion described as a composite or hybrid of a number of contributing structures

Double-headed arrows Symbols used to show that the structures on either side of it are resonance-contributing structures

FIGURE 3-4 The carbonate ion represented as a hybrid of three equivalent contributing structures. Curved arrows (in red) show how electron pairs are redistributed from one contributing structure to the next.

HOW TO...

Draw Curved Arrows and Push Electrons

Notice in Figure 3-4 that the only difference among contributing structures (a), (b), and (c) is the position of the valence electrons. To generate one resonance structure from another, chemists use a **curved arrow**. The arrow indicates where a pair of electrons originates (the tail of the arrow) and where it is repositioned in an alternative contributing structure (the head of the arrow).

A curved arrow is nothing more than a bookkeeping symbol for keeping track of electron pairs or, as some call it, **electron pushing**. Do not be misled by its simplicity. Electron pushing will help you see the relationship among contributing structures.

Following are contributing structures for the nitrite and acetate ions. Curved arrows show how the contributing structures are interconverted. For each ion, the contributing structures are equivalent. They have the same bonding patterns.

Nitrite ion
(equivalent contributing structures)

Acetate ion
(equivalent contributing structures)

A common mistake is to use curved arrows to indicate the movement of atoms or positive charges. This is never correct. Curved arrows are used only to show the repositioning of electron pairs when a new contributing structure is generated.

at all correct. The carbonate ion, for example, has one—and only one—real structure. The problem is ours. How do we represent that one real structure? The resonance method offers a way to represent the real structure while simultaneously retaining Lewis structures with electron-pair bonds and showing all nonbonding pairs of electrons. Thus, although we realize that the carbonate ion is not accurately represented by any one contributing structure shown in Figure 3-4, we continue to represent it by one of them for convenience. We understand, of course, that we are referring to the resonance hybrid.

Resonance, when it exists, is a stabilizing factor—that is, a resonance hybrid is more stable than any one of its hypothetical contributing structures. We will see three particularly striking illustrations of the stability

EXAMPLE 3-13 Resonance

Draw the contributing structure indicated by the curved arrows. Be certain to show all valence electrons and all charges.

(a)

(b)

(c)

Strategy
Curved arrows show the repositioning of a pair of electrons either from a bond to an adjacent atom as in parts (a) and (b) or from an atom to an adjacent bond as in parts (b) and (c).

Solution

(a)   (b)   (c)

Problem 3-13
Draw the contributing structure indicated by the curved arrows. Be certain to show all valence electrons and all charges.

(a)   (b)   (c)

of resonance hybrids when we consider the unusual chemical properties of benzene and aromatic hydrocarbons in Chapter 13, the acidity of carboxylic acids in Chapter 18, and the geometry of the amide bonds in proteins in Chapter 19.

B. Writing Acceptable Contributing Structures

Certain rules must be followed to write acceptable contributing structures:

1. All contributing structures must have the same number of valence electrons.
2. All contributing structures must obey the rules of covalent bonding. In particular, no contributing structure may have more than two electrons in the valence shell of hydrogen or more than eight electrons in the valence shell of a second-period element. Third-period elements, such as phosphorus and sulfur, may have more than eight electrons in their valence shells.
3. The positions of all atomic nuclei must be the same in all resonance structures; that is, contributing structures differ only in the distribution of valence electrons.

EXAMPLE 3-14 Resonance Contributing Structures

Which sets are valid pairs of contributing structures?

$$\underset{(a)}{CH_3-\underset{\underset{\|}{\overset{\overset{:O:}{\|}}{C}}}{}-CH_3 \quad \text{and} \quad CH_3-\underset{+}{\overset{\overset{:\ddot{O}:^-}{|}}{C}}-CH_3} \qquad \underset{(b)}{CH_3-\overset{\overset{:O:}{\|}}{C}-CH_3 \quad \text{and} \quad CH_2=\overset{\overset{:\ddot{O}-H}{|}}{C}-CH_3}$$

Strategy
The guideline tested in this example is that contributing structures involve only the redistribution of valence electrons. The position of all atoms remains the same.

Solution
(a) These are valid contributing structures. They differ only in the distribution (location) of valence electrons.
(b) These are not valid contributing structures. They differ in the arrangement of their atoms.

Problem 3-14
Which sets are valid pairs of contributing structures?

(a) $CH_3-C(\overset{:\ddot{O}:}{}{})(\underset{:\ddot{O}:^-}{})$ and $CH_3-\underset{+}{C}(\overset{:\ddot{O}:^-}{})(\underset{:\ddot{O}:}{})$

(b) $CH_3-C(\overset{:\ddot{O}:}{}{})(\underset{:\ddot{O}:^-}{})$ and $CH_3-C(\overset{:\ddot{O}:^-}{})(\underset{:\ddot{O}:}{})$

A final note: do not confuse resonance contributing structures with equilibration among different species. A molecule described as a resonance hybrid is not equilibrating among the individual electron configurations of the contributing structures. Rather, the molecule has only one structure, which is best described as a hybrid of its various contributing structures. The color wheel provides a good analogy. Purple is not a primary color; the primary colors blue and red are mixed to make purple. You can think of molecules represented by resonance hybrids as being purple. Purple is not sometimes blue and sometimes red: purple is purple. In an analogous way, a molecule described as a resonance hybrid is not sometimes one contributing structure and sometimes another: it is a single structure all the time.

3-10 How Do We Predict Bond Angles in Covalent Molecules?

In Section 3-7, we used a shared pair of electrons as the fundamental unit of covalent bonds and drew Lewis structures for several small molecules containing various combinations of single, double, and triple bonds (see, for example, Table 3-7). We can predict the degree of a **bond angle** in these and other molecules by using the **valence-shell electron-pair repulsion (VSEPR) model**.

According to this model, the valence electrons of an atom may be involved in the formation of single, double, or triple bonds or they may be unshared. Each combination creates a negatively charged region of electron density around a nucleus. Because like charges repel each other, the various regions

Bond angle The angle between two bonded atoms and a central atom

(a) Linear

(b) Trigonal planar

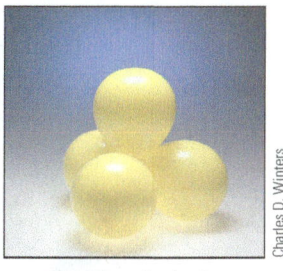
(c) Tetrahedral

FIGURE 3-5 Inflated balloon models to predict bond angles. (a) Two balloons assume a linear shape with a bond angle of 180° about the tie point. (b) Three balloons assume a trigonal planar shape with bond angles of 120° about the tie point. (c) Four balloons assume a tetrahedral shape with bond angles of 109.5° about the tie point.

of electron density around a nucleus spread out so that each is as far away as possible from the others.

You can demonstrate the bond angles predicted by this model in a very simple way. Imagine that an inflated balloon represents a region of electron density. Two inflated balloons tied together by their ends assume the shape shown in Figure 3-5(a). The point where they are tied together represents the atom about which you want to predict a bond angle, and the balloons represent regions of electron density about that atom.

We use the VSEPR model and the balloon model analogy in the following way to predict the shape of a molecule of methane, CH_4. The Lewis structure for CH_4 shows a carbon atom surrounded by four regions of electron density. Each region contains a pair of electrons forming a single covalent bond to a hydrogen atom. According to the VSEPR model, the four regions point away from carbon so that they are as far away from one another as possible. The maximum separation occurs when the angle between any two regions of electron density is 109.5°. Therefore, we predict all H—C—H bond angles to be 109.5° and the shape of the molecule to be **tetrahedral** [Figure 3-5(c) and 3-6]. The H—C—H bond angles in methane have been measured experimentally and found to be 109.5°. Thus, the bond angles and shape of methane predicted by the VSEPR model are identical to those observed experimentally.

We can predict the shape of an ammonia molecule, NH_3, in the same way. The Lewis structure of NH_3 shows nitrogen surrounded by four regions of electron density. Three regions contain single pairs of electrons that form covalent bonds with hydrogen atoms. The fourth region contains an unshared pair of electrons [Figure 3-7(a)]. Using the VSEPR model, we predict that the four regions are arranged in a tetrahedral manner and that the three H—N—H bond angles in this molecule are 109.5°. The observed bond angles are 107.3°. We can explain this small difference between the predicted and the observed angles by proposing that the unshared pair of electrons on nitrogen repels adjacent bonding electron pairs more strongly than the bonding pairs repel one another.

The geometry of an ammonia molecule is described as **pyramidal**; that is, the molecule is shaped like a triangular-based pyramid with the three hydrogens located at the base and the single nitrogen located at the apex.

Figure 3-8 shows a Lewis structure and a ball-and-stick model of a water molecule. In H_2O, oxygen is surrounded by four regions of electron density. Two of these regions contain pairs of electrons used to form single covalent bonds to hydrogens; the remaining two regions contain unshared electron pairs. Using the VSEPR model, we predict that the four regions of electron density around oxygen are arranged in a tetrahedral manner and that the

▶ Ammonia gas is drilled into the soil of a farm field. Most of the ammonia manufactured in the world is used as fertilizer because ammonia supplies the nitrogen needed by green plants.

(a)

(b)

FIGURE 3-6 The shape of a methane molecule, CH_4, is tetrahedral. (a) Lewis structure and (b) ball-and-stick model. The hydrogens occupy the four corners of a regular tetrahedron, and all H—C—H bond angles are 109.5°.

(a)

(b)

FIGURE 3-7 The shape of an ammonia molecule, NH_3, is pyramidal. (a) Lewis structure and (b) ball-and-stick model. The H—N—H bond angles are 107.3°, slightly smaller than the H—C—H bond angles of methane.

(a)

(b)

FIGURE 3-8 The shape of a water molecule, H_2O, is bent. (a) Lewis structure and (b) ball-and-stick model.

H—O—H bond angle is 109.5°. Experimental measurements show that the actual H—O—H bond angle in a water molecule is 104.5°, a value smaller than that predicted. We can explain this difference between the predicted and the observed bond angle by proposing, as we did for NH_3, that unshared pairs of electrons repel adjacent pairs more strongly than bonding pairs do. Note that the distortion from 109.5° is greater in H_2O, which has two unshared pairs of electrons, than it is in NH_3, which has only one unshared pair. Therefore, the actual geometry of a water molecule is described as **bent.**

A general prediction emerges from this discussion. If a Lewis structure shows four regions of electron density around an atom, the VSEPR model predicts a tetrahedral distribution of electron density and bond angles of approximately 109.5°.

In many of the molecules we will encounter, three regions of electron density surround an atom. Figure 3-9 shows Lewis structures and ball-and-stick models for molecules of formaldehyde, CH_2O, and ethylene, C_2H_4.

FIGURE 3-9 The shapes of formaldehyde, CH_2O, and ethylene, C_2H_4, are trigonal planar.

Carbon dioxide

:O=C=O:

Side view   End view

Acetylene

H—C≡C—H

Side view   End view

FIGURE 3-10 The shapes of carbon dioxide, CO_2, and acetylene, C_2H_2, are linear.

In the VSEPR model, we treat a double bond as a single region of electron density. In formaldehyde, three regions of electron density surround carbon. Two regions contain single pairs of electrons, each of which forms a single bond to a hydrogen; the third region contains two pairs of electrons, which form a double bond to oxygen. In ethylene, three regions of electron density also surround each carbon atom; two contain single pairs of electrons, and the third contains two pairs of electrons.

Three regions of electron density about an atom are farthest apart when they lie in a plane and make angles of 120° with one another. Thus, the predicted H—C—H and H—C—O bond angles in formaldehyde and the H—C—H and H—C—C bond angles in ethylene are all 120°. Furthermore, all atoms in each molecule lie in a plane. Thus, both formaldehyde and ethylene are planar molecules. The geometry about an atom surrounded by three regions of electron density, as in formaldehyde and ethylene, is described as **trigonal planar**.

In still other types of molecules, two regions of electron density surround a central atom. Figure 3-10 shows Lewis structures and ball-and-stick models of molecules of carbon dioxide, CO_2, and acetylene, C_2H_2.

In carbon dioxide, two regions of electron density surround carbon; each contains two pairs of electrons and forms a double bond to an oxygen atom. In acetylene, two regions of electron density also surround each carbon; one contains a single pair of electrons and forms a single bond to a hydrogen atom, and the other contains three pairs of electrons and forms a triple bond to a carbon atom. In each case, the two regions of electron density are farthest apart if they form a straight line through the central atom and create an angle of 180°. Both carbon dioxide and acetylene are linear molecules.

Table 3-8 summarizes the predictions of the VSEPR model. In this table, three-dimensional shapes are shown using a solid wedge to represent a bond coming toward you, out of the plane of the paper. A broken wedge represents a bond going away from you, behind the plane of the paper. A solid line represents a bond in the plane of the paper.

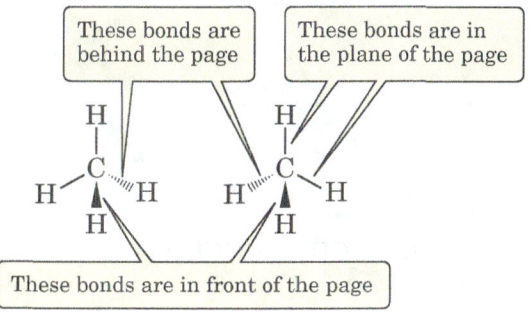

Table 3-8 Predicted Molecular Shapes (VSEPR model)

Regions of Electron Density Around Central Atom	Predicted Distribution of Electron Density	Predicted Bond Angles	Examples (Shape of the Molecule)
4	Tetrahedral	109.5°	Methane (tetrahedral), Ammonia (pyramidal), Water (bent)
3	Trigonal planar	120°	Ethylene (planar), Formaldehyde (planar)
2	Linear	180°	Carbon dioxide (linear), Acetylene (linear)

EXAMPLE 3-15 Predicting Bond Angles In Covalent Compounds

Predict all bond angles and the shape of each molecule:

(a) CH_3Cl (b) $CH_2=CHCl$

Strategy

To predict bond angles, first draw a correct Lewis structure for the compound. Be certain to show all unpaired electrons. Then determine the number of regions of electron density (either 2, 3, or 4) around each atom and use that number to predict bond angles (either 109.5°, 120°, or 180°).

Solution

(a) The Lewis structure for CH_3Cl shows that four regions of electron density surround carbon. Therefore, we predict that the distribution of electron pairs about carbon is tetrahedral, all bond angles are 109.5°, and the shape of CH_3Cl is tetrahedral.

(b) In the Lewis structure for $CH_2 = CHCl$, three regions of electron density surround each carbon. Therefore, we predict that all bond angles are 120° and that the molecule is planar. The bonding about each carbon is trigonal planar.

(Side view)   (Viewed along the C=C bond)

Problem 3-15

Predict all bond angles for these molecules:

(a) CH_3OH (b) CH_2Cl_2 (c) H_2CO_3 (carbonic acid)

3-11 How Do We Determine If a Molecule Is Polar?

In Section 3-7B, we used the terms "polar" and "dipole" to describe a covalent bond in which one atom bears a partial positive charge and the other bears a partial negative charge. We also saw that we can use the difference in electronegativity between bonded atoms to determine the polarity of a covalent bond and the direction of its dipole. We can now combine our understanding of bond polarity and molecular geometry (Section 3-10) to predict the polarity of molecules. To discuss the physical and chemical properties of a molecule, it is essential to have an understanding of polarity. Many chemical reactions, for example, are driven by the interaction of the positive part of one molecule with the negative part of another molecule.

A molecule will be polar if (1) it has polar bonds and (2) its centers of partial positive charge and partial negative charge lie at different places within the molecule. Consider first carbon dioxide, CO_2, a molecule with two polar carbon–oxygen double bonds. The oxygen on the left pulls electrons of the O=C bond toward it, giving it a partial negative charge. Similarly, the oxygen on the right pulls electrons of the C=O bond toward it by the same amount, giving it the same partial negative charge as the oxygen on the left. Carbon bears a partial positive charge. We can show the polarity of these bonds by using the symbols $\delta+$ and $\delta-$. Alternatively, we can show that each carbon–oxygen bond has a dipole by using an arrow, where the head of the arrow points to the negative end of the dipole and the crossed tail is positioned at the positive end of the dipole. Because carbon dioxide is a linear molecule, its centers of negative and positive partial charge coincide. Therefore, CO_2 is a nonpolar molecule; that is, it has no dipole.

$\overset{\delta-}{\ddot{O}}=\overset{\delta+}{C}=\overset{\delta-}{\ddot{O}}$

Carbon dioxide
(a nonpolar molecule)

In a water molecule, each O—H bond is polar. Oxygen, the more electronegative atom, bears a partial negative charge, and each hydrogen bears a partial positive charge. The center of partial positive charge in a water molecule is located halfway between the two hydrogen atoms, and the center of partial negative charge is on the oxygen atom. Thus, a water molecule has polar bonds and because of its geometry is a polar molecule.

Water
(a polar molecule)

The center of partial positive charge is midway between the two hydrogen atoms

Ammonia has three polar N—H bonds. Because of its geometry, the centers of partial positive and partial negative charges are found at different places within the molecule. Thus, ammonia has polar bonds and because of its geometry is a polar molecule.

Ammonia
(a polar molecule)

The center of partial positive charge is midway between the three hydrogen atoms

EXAMPLE 3-16 Polarity of Covalent Molecules

Which of these molecules are polar? Show the direction of the molecular dipole by using an arrow with a crossed tail.

(a) CH_2Cl_2 (b) CH_2O (c) C_2H_2

Strategy

To determine whether a molecule is polar, first determine if it has polar bonds, and if it does, determine if the centers of positive and negative charge lie at the same or different places within the molecule. If they lie at the same place, the molecule is nonpolar; if they lie at different places, the molecule is polar.

Solution

Both dichloromethane, CH_2Cl_2, and formaldehyde, CH_2O, have polar bonds and because of their geometry are polar molecules. Because acetylene, C_2H_2, contains no polar bonds, it is a nonpolar molecule.

Problem 3-16
Which of these molecules are polar? Show the direction of the molecular dipole by using an arrow with a crossed tail.

(a) H_2S (b) HCN   (c) C_2H_6

Summary of Key Questions

Section 3-2 What Is the Octet Rule?
- The **octet rule** states that elements of Groups 1A–7A tend to gain or lose electrons so as to achieve an outer shell containing eight valence electrons and the same electron configuration as that of the noble gas nearest to it in atomic number.
- An atom with almost eight valence electrons tends to gain the needed electrons to have eight electrons in its valence shell—that is, to achieve the same electron configuration as the noble gas nearest to it in atomic number. In gaining electrons, the atom becomes a negatively charged ion called an **anion**.
- An atom with only one or two valence electrons tends to lose the number of electrons required to have eight valence electrons in its next lower shell—that is, to have the same electron configuration as the noble gas nearest to it in atomic number. In losing electrons, the atom becomes a positively charged ion called a **cation**.

Section 3-3 How Do We Name Anions and Cations?
- For metals that form only one type of cation, the name of the cation is the name of the metal followed by the word "ion."
- For metals that form more than one type of cation, we show the charge on the ion by placing a Roman numeral enclosed in parentheses immediately following the name of the metal. Alternatively, for some elements, we use the suffix *-ous* to show the lower positive charge and *-ic* to show the higher positive charge.
- A **monatomic anion** is named by adding *-ide* to the stem part of the name.
- A **polyatomic ion** contains more than one type of atom.

Section 3-4 What Are the Two Major Types of Chemical Bonds?
- The two major types of chemical bonds are ionic bonds and covalent bonds.
- According to the Lewis model of chemical bonding, atoms bond together in such a way that each atom participating in the bond acquires a valence-shell electron configuration matching that of the noble gas nearest to it in atomic number.
- **Electronegativity** is a measure of the force of attraction that an atom exerts on electrons it shares in a chemical bond. It increases from left to right across a row and from bottom to top in a column of the Periodic Table.
- An **ionic bond** forms between two atoms if the difference in electronegativity between them is greater than 1.9.
- A **covalent bond** forms if the difference in electronegativity between the bonded atoms is 1.9 or less.

Section 3-5 What Is an Ionic Bond?
- An **ionic bond** forms by the transfer of valence-shell electrons from an atom of lower electronegativity to the valence shell of an atom of higher electronegativity.
- In an ionic compound, the total number of positive charges must equal the total number of negative charges.

Section 3-6 How Do We Name Ionic Compounds?

- For a **binary ionic compound**, name the cation first, followed by the name of the anion. Where a metal ion may form different cations, use a Roman numeral to show its positive charge. To name an ionic compound that contains one or more polyatomic ions, name the cation first, followed by the name of the anion.

Section 3-7 What Is a Covalent Bond?

- According to the Lewis model, a **covalent bond** forms when pairs of electrons are shared between two atoms whose difference in electronegativity is 1.9 or less.
- A pair of electrons in a covalent bond is shared by two atoms and at the same time fills the valence shell of each atom.
- A **nonpolar covalent bond** is a covalent bond in which the difference in electronegativity between bonded atoms is less than 0.5. A **polar covalent bond** is a covalent bond in which the difference in electronegativity between bonded atoms is between 0.5 and 1.9. In a polar covalent bond, the more electronegative atom bears a partial negative charge ($\delta-$) and the less electronegative atom bears a partial positive charge ($\delta+$). This separation of charge produces a **dipole**.
- A **Lewis structure** for a covalent compound must show (1) the correct arrangement of atoms, (2) the correct number of valence electrons, (3) no more than two electrons in the outer shell of hydrogen, and (4) no more than eight electrons in the outer shell of any second-period element.
- Exceptions to the octet rule include compounds of third-period elements, such as phosphorus and sulfur, which may have as many as 10 and 12 electrons, respectively, in their valence shells.

Section 3-8 How Do We Name Binary Covalent Compounds?

- To name a **binary covalent compound**, name the less electronegative element first, followed by the name of the more electronegative element. The name of the more electronegative element is derived by adding *-ide* to the stem name of the element. Use the prefixes *di-*, *tri-*, *tetra-*, and so on, to show the presence of two or more atoms of the same kind.

Section 3-9 What Is Resonance?

- According to the theory of resonance, a molecule or ion for which no single Lewis structure is adequate is best described by writing two or more **resonance contributing structures** and considering the real molecule or ion to be a hybrid of these contributing structures. To show how pairs of valence electrons are redistributed from one contributing structure to the next, we use curved arrows. A curved arrow extends from where a pair of electrons is initially shown (on an atom or in a covalent bond) to its new location (on an adjacent atom or an adjacent covalent bond).

Section 3-10 How Do We Predict Bond Angles in Covalent Molecules?

- The **valence-shell electron-pair repulsion (VSEPR) model** predicts bond angles of 109.5° about atoms surrounded by four regions of electron density, angles of 120° about atoms surrounded by three regions of electron density, and angles of 180° about atoms surrounded by two regions of electron density.

Section 3-11 How Do We Determine If a Molecule Is Polar?

- A molecule is polar (has a dipole) if it has polar bonds and the centers of its partial positive and partial negative charges do not coincide.
- If a molecule has polar bonds but the centers of its partial positive and negative charges coincide, the molecule is nonpolar (it has no dipole).

Problems

Orange-numbered problems are applied.

Section 3-2 What Is the Octet Rule?

3-17 Answer true or false.
   (a) The octet rule refers to the chemical bonding patterns of the first eight elements of the Periodic Table.
   (b) The octet rule refers to the tendency of certain elements to react in such a way that they achieve an outer shell of eight valence electrons.
   (c) In gaining electrons, an atom becomes a positively charged ion called a cation.
   (d) When an atom forms an ion, only the number of valence electrons changes; the number of protons and neutrons in the nucleus does not change.
   (e) In forming ions, Group 2A elements typically lose two electrons to become cations with a charge of +2.
   (f) In forming an ion, a sodium atom ($1s^2 2s^2 2p^6 3s^1$) completes its valence shell by adding one electron to fill its 3s shell ($1s^2 2s^2 2p^6 3s^2$).
   (g) The elements of Group 6A typically react by accepting two electrons to become anions with a charge of -2.
   (h) With the exception of hydrogen, the octet rule applies to all elements in periods 1, 2, and 3.
   (i) Atoms and the ions derived from them have very similar physical and chemical properties.

3-18 How many electrons must each atom gain or lose to acquire an electron configuration identical to the noble gas nearest to it in atomic number?
(a) Li (b) Cl (c) P (d) Al
(e) Sr (f) S (g) Si (h) O

3-19 Show how each chemical change obeys the octet rule.
(a) Lithium forms Li^+ (b) Oxygen forms O^{2-}

3-20 Show how each chemical change obeys the octet rule.
(a) Hydrogen forms H^- (hydride ion)
(b) Aluminum forms Al^{3+}

3-21 Write the formula for the most stable ion formed by each element.
(a) Mg (b) F (c) Al
(d) S (e) K (f) Br

3-22 Why is Li^- not a stable ion?

3-23 Predict which ions are stable:
(a) I^- (b) Se^{2+} (c) Na^+ (d) S^{2-} (e) Li^{2+} (f) Ba^{3+}

3-24 Predict which ions are stable:
(a) Br^{2-} (b) C^{4-} (c) Ca^+
(d) Ar^+ (e) Na^+ (f) Cs^+

3-25 Why are carbon and silicon reluctant to form ionic bonds?

3-26 Table 3-2 shows the following ions of copper: Cu^+ and Cu^{2+}. Do these violate the octet rule? Explain.

Section 3-3 How Do We Name Anions and Cations?

3-27 Answer true or false.
(a) For Group 1A and Group 2A elements, the name of the ion each forms is simply the name of the element followed by the word ion; for example, Mg^{2+} is named magnesium ion.
(b) H^+ is named hydronium ion, and H^- is named hydride ion.
(c) The nucleus of H^+ consists of one proton and one neutron.
(d) Many transition and inner transition elements form more than one positively charged ion.
(e) In naming metal cations with two different charges, the suffix *-ous* refers to the ion with a charge of +1 and *-ic* refers to the ion with a charge of +2.
(f) Fe^{3+} may be named either iron(III) ion or ferric ion.
(g) The anion derived from a bromine atom is named bromine ion.
(h) The anion derived from an oxygen atom is named oxide ion.
(i) HCO_3^- is named hydrogen carbonate ion.
(j) The prefix *bi-* in the name "bicarbonate" ion indicates that this ion has a charge of −2.
(k) The hydrogen phosphate ion has a charge of +1, and the dihydrogen phosphate ion has a charge of +2.
(l) The phosphate ion is PO_3^{4-}.
(m) The nitrite ion is NO_2^-, and the nitrate ion is NO_3^-.
(n) The carbonate ion is CO_3^{2-}, and the hydrogen carbonate ion is HCO_3^-.

3-28 Name each polyatomic ion.
(a) HCO_3^- (b) NO_2^- (c) SO_4^{2-}
(d) HSO_4^- (e) $H_2PO_4^-$

Section 3-4 What Are the Two Major Types of Chemical Bonds?

3-29 Answer true or false.
(a) According to the Lewis model of bonding, atoms bond together in such a way that each atom participating in the bond acquires an outer-shell electron configuration matching that of the noble gas nearest to it in atomic number.
(b) Atoms that lose electrons to achieve a filled valence shell become cations and form ionic bonds with anions.
(c) Atoms that gain electrons to achieve filled valence shells become anions and form ionic bonds with cations.
(d) Atoms that share electrons to achieve filled valence shells form covalent bonds.
(e) Ionic bonds tend to form between elements on the left side of the Periodic Table, and covalent bonds tend to form between elements on the right side of the Periodic Table.
(f) Ionic bonds tend to form between a metal and a nonmetal.
(g) When two nonmetals combine, the bond between them is usually covalent.
(h) Electronegativity is a measure of an atom's attraction for the electrons it shares in a chemical bond with another atom.
(i) Electronegativity generally increases with atomic number.
(j) Electronegativity generally increases with atomic weight.
(k) Electronegativity is a periodic property.
(l) Fluorine, in the upper-right corner of the Periodic Table, is the most electronegative element; hydrogen, in the upper-left corner, is the least electronegative element.
(m) Electronegativity depends on both the nuclear charge and the distance of the valence electrons from the nucleus.
(n) Electronegativity generally increases from left to right across a period of the Periodic Table.
(o) Electronegativity generally increases from top to bottom in a column of the Periodic Table.

3-30 Why does electronegativity generally increase going up a column (group) of the Periodic Table?

3-31 Why does electronegativity generally increase going from left to right across a row of the Periodic Table?

3-32 Judging from their relative positions in the Periodic Table, which element in each pair has the larger electronegativity?
(a) F or Cl (b) O or S (c) C or N (d) C or F

3-33 Toward which atom are the bonding electrons shifted in a covalent bond between each of the following pairs:
(a) H and Cl (b) N and O
(c) C and O (d) Cl and Br
(e) C and S (f) P and S (g) H and O

3-34 Which of these bonds is the most polar? The least polar?
(a) C—N (b) C—C (c) C—O

3-35 Classify each bond as nonpolar covalent, polar covalent, or ionic.
(a) C—Cl (b) C—Li (c) C—N

3-36 Classify each bond as nonpolar covalent, polar covalent, or ionic.
(a) C—Br (b) S—Cl (c) C—P

Section 3-5 What Is an Ionic Bond?

3-37 Answer true or false.
(a) An ionic bond is formed by the combination of positive and negative ions.
(b) An ionic bond between two atoms forms by the transfer of one or more valence electrons from the atom of higher electronegativity to the atom of lower electronegativity.
(c) As a rough guideline, we say that an ionic bond will form if the difference in electronegativity between two atoms is approximately 1.9 or greater.
(d) In forming NaCl from sodium and chlorine atoms, one electron is transferred from the valence shell of sodium to the valence shell of chlorine.
(e) The formula of sodium sulfide is Na_2S.
(f) The formula of calcium hydroxide is CaOH.
(g) The formula of aluminum sulfide is AlS.
(h) The formula of iron(III) oxide is Fe_3O_2.
(i) Barium ion is Ba^{2+}, and oxide ion is O^{2-}; therefore, the formula of barium oxide is Ba_2O_2.

3-38 Complete the chart by writing formulas for the compounds formed:

	Br^-	MnO_4^-	O^{2-}	NO_3^-	SO_4^{2-}	PO_4^{3-}	OH^-
Li^+							
Ca^{2+}							
Co^{3+}							
K^+							
Cu^{2+}							

3-39 Write a formula for the ionic compound formed from each pair of elements.
(a) Sodium and bromine (b) Sodium and oxygen
(c) Aluminum and chlorine (d) Barium and chlorine
(e) Magnesium and oxygen

3-40 Although not a transition metal, lead can form Pb^{2+} and Pb^{4+} ions. Write the formula for the compound formed between each of these lead ions and the following anions:
(a) Chloride ion (b) Hydroxide ion
(c) Oxide ion

3-41 Describe the structure of sodium chloride in the solid state.

3-42 What is the charge on each ion in these compounds?
(a) CaS (b) MgF_2 (c) Cs_2O
(d) $ScCl_3$ (e) Al_2S_3

3-43 Write the formula for the compound formed from the following pairs of ions:
(a) Iron(III) ion and hydroxide ion
(b) Barium ion and chloride ion
(c) Calcium ion and phosphate ion
(d) Sodium ion and permanganate ion

3-44 Write the formula for the ionic compound formed from the following pairs of ions:
(a) Iron(II) ion and chloride ion
(b) Calcium ion and hydroxide ion
(c) Ammonium ion and phosphate ion
(d) Tin(II) ion and fluoride ion

3-45 Which formulas are not correct? For each that is not correct, write the correct formula.
(a) Ammonium phosphate; $(NH_4)_2PO_4$
(b) Barium carbonate; Ba_2CO_3
(c) Aluminum sulfide; Al_2S_3
(d) Magnesium sulfide; MgS

3-46 Which formulas are not correct? For each that is not correct, write the correct formula.
(a) Calcium oxide; CaO_2
(b) Lithium oxide; LiO
(c) Sodium hydrogen phosphate; $NaHPO_4$
(d) Ammonium nitrate; NH_4NO_3

Section 3-6 How Do We Name Ionic Compounds?

3-47 Answer true or false.
(a) The name of a binary ionic compound consists of the name of the positive ion followed by the name of the negative ion.
(b) In naming binary ionic compounds, it is necessary to state the number of each ion present in the compound.
(c) The formula of aluminum oxide is Al_2O_3.
(d) Both copper(II) oxide and cupric oxide are acceptable names for CuO.
(e) The systematic name for Fe_2O_3 is iron(II) oxide.
(f) The systematic name for $FeCO_3$ is iron carbonate.
(g) The systematic name for NaH_2PO_4 is sodium dihydrogen phosphate.
(h) The systematic name for K_2HPO_4 is dipotassium hydrogen phosphate.

(i) The systematic name for Na₂O is sodium oxide.
(j) The systematic name for PCl₃ is potassium chloride.
(k) The formula of ammonium carbonate is NH₄CO₃.

3-48 Potassium chloride and potassium bicarbonate are used as potassium dietary supplements. Write the formula of each compound.

3-49 Potassium nitrite has been used as a vasodilator and as an antidote for cyanide poisoning. Write the formula of this compound.

3-50 Name the polyatomic ion(s) in each compound.
(a) Na₂SO₃ (b) KNO₃ (c) Cs₂CO₃
(d) NH₄OH (e) K₂HPO₄ (f) Ca(ClO₄)₂

3-51 Write the formulas for the ions present in each compound.
(a) NaBr (b) FeSO₃ (c) Mg₃(PO₄)₂
(d) KH₂PO₄ (e) NaHCO₃ (f) Ba(NO₃)₂

3-52 Name these ionic compounds:
(a) NaF (b) MgS (c) Al₂O₃
(d) BaCl₂ (e) Ca(HSO₃)₂ (f) KI
(g) Sr₃(PO₄)₂ (h) Fe(OH)₂ (i) NaH₂PO₄
(j) Pb(CH₃COO)₂ (k) BaH₂ (l) (NH₄)₂HPO₄

3-53 Write formulas for the following ionic compounds:
(a) Potassium bromide (b) Calcium oxide
(c) Mercury(II) oxide (d) Copper(II) phosphate
(e) Lithium sulfate (f) Iron(III) sulfide

3-54 Write formulas for the following ionic compounds:
(a) Ammonium hydrogen sulfite
(b) Magnesium acetate
(c) Strontium dihydrogen phosphate
(d) Silver carbonate
(e) Strontium chloride
(f) Barium permanganate
(g) Aluminum perchlorate

Section 3-7 What Is a Covalent Bond?

3-55 Answer true or false.
(a) A covalent bond is formed between two atoms whose difference in electronegativity is less than 1.9.
(b) If the difference in electronegativity between two atoms is zero (they have identical electronegativities), then the two atoms will not form a covalent bond.
(c) A covalent bond formed by sharing two electrons is called a double bond.
(d) In the hydrogen molecule (H₂), the shared pair of electrons completes the valence shell of each hydrogen.
(e) In the molecule CH₄, each hydrogen has an electron configuration like that of helium and carbon has an electron configuration like that of neon.
(f) In a polar covalent bond, the more electronegative atom has a partial negative charge (δ−) and the less electronegative atom has a partial positive charge (δ+).
(g) These bonds are arranged in order of *increasing* polarity C—H < N—H < O—H.
(h) These bonds are arranged in order of *increasing* polarity H—F < H—Cl < H—Br.
(i) A polar bond has a dipole with the negative end located at the more electronegative atom.
(j) In a single bond, two atoms share one pair of electrons; in a double bond, they share two pairs of electrons; and in a triple bond, they share three pairs of electrons.
(k) The Lewis structure for ethane, C₂H₆, must show eight valence electrons.
(l) The Lewis structure for formaldehyde, CH₂O, must show 12 valence electrons.
(m) The Lewis structure for the ammonium ion, NH₄⁺, must show nine valence electrons.
(n) Atoms of third-period elements can hold more than eight electrons in their valence shells.

3-56 How many covalent bonds are normally formed by each element?
(a) N (b) F (c) C (d) Br (e) O

3-57 What is:
(a) A single bond? (b) A double bond?
(c) A triple bond?

3-58 In Section 2-3B, we saw that there are seven diatomic elements.
(a) Draw Lewis structures for each of these diatomic elements.
(b) Which diatomic elements are gases at room temperature? Which are liquids? Which are solids?

3-59 Draw a Lewis structure for each covalent compound.
(a) CH₄ (b) C₂H₂ (c) C₂H₄
(d) BF₃ (e) CH₂O (f) C₂Cl₆

3-60 What is the difference between a molecular formula, a structural formula, and a Lewis structure?

3-61 State the total number of valence electrons in each molecule.
(a) NH₃ (b) C₃H₆ (c) C₂H₄O₂ (d) C₂H₆O
(e) CCl₄ (f) HNO₂ (g) CCl₂F₂ (h) O₂

3-62 Draw a Lewis structure for each of the following molecules and ions. In each case, the atoms can be connected in only one way.
(a) Br₂ (b) H₂S (c) N₂H₄ (d) N₂H₂
(e) CN⁻ (f) NH₄⁺ (g) N₂ (h) O₂

3-63 What is the difference between (a) a bromine atom, (b) a bromine molecule, and (c) a bromide ion? Draw the Lewis structure for each.

3-64 Acetylene (C₂H₂), hydrogen cyanide (HCN), and nitrogen (N₂) each contain a triple bond. Draw a Lewis structure for each molecule. Which of these are polar molecules, and which are nonpolar molecules?

3-65 Why can't hydrogen have more than two electrons in its valence shell?

3-66 Why can't second-row elements have more than eight electrons in their valence shells? That is, why does the octet rule work for second-row elements?

3-67 Why does nitrogen have three bonds and one unshared pair of electrons in covalent compounds?

3-68 Draw a Lewis structure of a covalent compound in which nitrogen has:
(a) Three single bonds and one unshared pair of electrons
(b) One single bond, one double bond, and one unshared pair of electrons
(c) One triple bond and one unshared pair of electrons

3-69 Why does oxygen have two bonds and two unshared pairs of electrons in covalent compounds?

3-70 Draw a Lewis structure of a covalent compound in which oxygen has:
(a) Two single bonds and two unshared pairs of electrons
(b) One double bond and two unshared pairs of electrons

3-71 The ion O^{6+} has a complete outer shell. Why is this ion not stable?

3-72 Draw a Lewis structure for a molecule in which a carbon atom is bonded by a double bond to (a) another carbon atom, (b) an oxygen atom, and (c) a nitrogen atom.

3-73 Which of the following molecules have an atom that does not obey the octet rule (not all of these are stable molecules)?
(a) BF_3 (b) CF_2 (c) BeF_2 (d) C_2H_4
(e) CH_3 (f) N_2 (g) NO

Section 3-8 How Do We Name Binary Covalent Compounds?

3-74 Answer true or false.
(a) A binary covalent compound contains two kinds of atoms.
(b) The two types of atoms in a binary covalent compound are named in this order: first the more electronegative element and then the less electronegative element.
(c) The name for SF_2 is sulfur difluoride.
(d) The name for CO_2 is carbon dioxide.
(e) The name for CO is carbon oxide.
(f) The name for HBr is hydrogen bromide.
(g) The name for CCl_4 is carbon tetrachloride.

3-75 Name these binary covalent compounds.
(a) SO_2 (b) SO_3 (c) PCl_3 (d) CS_2

Section 3-9 What Is Resonance?

3-76 Write two acceptable contributing structures for the bicarbonate ion, HCO_3^- and show by the use of curved arrows how the first contributing structure is converted to the second.

3-77 Ozone, O_3, is an unstable blue gas with a characteristic pungent odor. In an ozone molecule, the connectivity of the atoms is O—O—O and both O—O bonds are equivalent.
(a) How many valence electrons must be present in an acceptable Lewis structure for an ozone molecule?

(b) Write two equivalent resonance contributing structures for ozone. Be certain to show any positive or negative charges that may be present in your contributing structures. By equivalent contributing structures, we mean that each has the same pattern of bonding.
(c) Show by the use of curved arrows how the first of your contributing structures may be converted to the second.
(d) Based on your contributing structures, predict the O—O—O bond angle in ozone.
(e) Explain why the following is not an acceptable contributing structure for an ozone molecule:

$$\ddot{\text{O}}=\ddot{\text{O}}=\ddot{\text{O}}$$

3-78 Nitrous oxide, N_2O, laughing gas, is a colorless, nontoxic, tasteless, and odorless gas. It is used as an inhalation anesthetic in dental and other surgeries. Because nitrous oxide is soluble in vegetable oils (fats), it is used commercially as a propellant in whipped toppings.

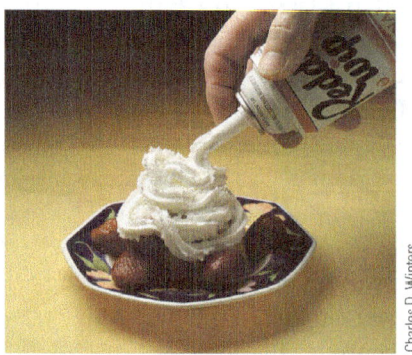

Nitrous oxide dissolves in fats. The gas is added under pressure to cans of whipped topping. When the valve is opened, the gas expands, thus expanding (whipping) the topping and forcing it out of the can.

(a) How many valence electrons are present in a molecule of N_2O?
(b) Write two equivalent contributing structures for this molecule. The connectivity in nitrous oxide is N—N—O.
(c) Explain why the following is not an acceptable contributing structure:

$$:\text{N}≡\text{N}=\ddot{\text{O}}$$

Section 3-10 How Do We Predict Bond Angles in Covalent Molecules?

3-79 Answer true or false.
(a) The letters VSEPR stand for valence-shell electron-pair repulsion.
(b) In predicting bond angles about a central atom in a covalent molecule, the VSEPR model considers only shared electron pairs (electron pairs involved in forming covalent bonds).

(c) The VSEPR model treats the two electron pairs of a double bond as one region of electron density and the three electron pairs of a triple bond as one region of electron density.

(d) In carbon dioxide, O═C═O, carbon is surrounded by four pairs of electrons and the VSEPR model predicts 109.5° for the O—C—O bond angle.

(e) For a central atom surrounded by three regions of electron density, the VSEPR model predicts bond angles of 120°.

(f) The geometry about a carbon atom surrounded by three regions of electron density is described as trigonal planar.

(g) For a central atom surrounded by four regions of electron density, the VSEPR model predicts bond angles of 360°/4 = 90°.

(h) For the ammonia molecule, NH_3, the VSEPR model predicts H—N—H bond angles of 109.5°.

(i) For the ammonium ion, NH_4^+, the VSEPR model predicts H—N—H bond angles of 109.5°.

(j) The VSEPR model applies equally well to covalent compounds of carbon, nitrogen, and oxygen.

(k) In water, H—O—H, the oxygen atom forms covalent bonds to two other atoms, and therefore, the VSEPR model predicts an H—O—H bond angle of 180°.

(l) If you fail to consider unshared pairs of valence electrons when you use the VSEPR model, you will arrive at an incorrect prediction.

(m) Given the assumptions of the VSEPR model, the only bond angles it predicts for compounds of carbon, nitrogen, and oxygen are 109.5°, 120°, and 180°.

3-80 State the shape of a molecule whose central atom is surrounded by:
(a) Two regions of electron density
(b) Three regions of electron density
(c) Four regions of electron density

3-81 Hydrogen and oxygen combine in different ratios to form H_2O (water) and H_2O_2 (hydrogen peroxide).
(a) How many valence electrons are found in H_2O? In H_2O_2?
(b) Draw Lewis structures for each molecule in part (a). Be certain to show all valence electrons.
(c) Using the VSEPR model, predict the bond angles about the oxygen atom in water and about each oxygen atom in hydrogen peroxide.

3-82 Hydrogen and nitrogen combine in different ratios to form three compounds: NH_3 (ammonia), N_2H_4 (hydrazine), and N_2H_2 (diimide).
(a) How many valence electrons must the Lewis structure of each molecule show?
(b) Draw a Lewis structure for each molecule.
(c) Predict the bond angles about the nitrogen atom(s) in each molecule.

3-83 Predict the shape of each molecule.
(a) CH_4 (b) PH_3 (c) CHF_3 (d) SO_2
(e) SO_3 (f) CCl_2F_2 (g) NH_3 (h) PCl_3

3-84 Predict the shape of each ion.
(a) NO_2^- (b) NH_4^+ (c) CO_3^{2-}

Section 3-11 How Do We Determine If a Molecule Is Polar?

3-85 Answer true or false.
(a) To predict whether a covalent molecule is polar or nonpolar, you must know both the polarity of each bond and the geometry (shape) of the molecule.
(b) A molecule may have two or more polar bonds and still be nonpolar.
(c) All molecules with polar bonds are polar.
(d) If water were a linear molecule with an H—O—H bond angle of 180°, water would be a nonpolar molecule.
(e) H_2O and NH_3 are polar molecules, but CH_4 is nonpolar.
(f) In methanol, CH_3OH, the O—H bond is more polar than the C—O bond.
(g) Dichloromethane, CH_2Cl_2, is polar, but tetrachloromethane, CCl_4, is nonpolar.
(h) Ethanol, CH_3CH_2OH, the alcohol of alcoholic beverages, has polar bonds, has a net dipole, and is a polar molecule.

3-86 Both CO_2 and SO_2 have polar bonds. Account for the fact that CO_2 is nonpolar and SO_2 is polar.

3-87 Consider the molecule boron trifluoride, BF_3.
(a) Write a Lewis structure for BF_3.
(b) Predict the F—B—F bond angles using the VSEPR model.
(c) Does BF_3 have polar bonds? Is it a polar molecule?

3-88 Is it possible for a molecule to have polar bonds and yet have no dipole? Explain.

3-89 Is it possible for a molecule to have no polar bonds and yet have a dipole? Explain.

3-90 In each case, tell whether the bond is ionic, polar covalent, or nonpolar covalent.
(a) Br_2 (b) BrCl (c) HCl (d) SrF_2
(e) SiH_4 (f) CO (g) N_2 (h) CsCl

3-91 Account for the fact that chloromethane, CH_3Cl, which has only one polar C—Cl bond, is a polar molecule, but carbon tetrachloride, CCl_4, which has four polar C—Cl bonds, is a nonpolar molecule.

Chemical Connections

3-92 (Chemical Connections 3A) What are the three main inorganic components of one dry mixture currently used to create synthetic bone?

3-93 (Chemical Connections 3B) Why is sodium iodide often present in the table salt we buy at the grocery store?

3-94 (Chemical Connections 3B) What is a medical use of barium sulfate?

3-95 (Chemical Connections 3B) What is a medical use of potassium permanganate?

3-96 (Chemical Connections 3A) What is the most prevalent metal ion in bone and tooth enamel?

3-97 (Chemical Connections 3C) In what way does the gas nitric oxide, NO, contribute to the acidity of acid rain?

Additional Problems

3-98 Explain why argon does not form either (a) ionic bonds or (b) covalent bonds.

3-99 Knowing what you do about covalent bonding in compounds of carbon, nitrogen, and oxygen and given the fact that silicon is just below carbon in the Periodic Table, phosphorus is just below nitrogen, and sulfur is just below oxygen, predict the molecular formula for the compound formed by (a) silicon and chlorine, (b) phosphorus and hydrogen, and (c) sulfur and hydrogen.

3-100 Use the valence-shell electron-pair repulsion model to predict the shape of a molecule in which a central atom is surrounded by five regions of electron density—as, for example, in phosphorus pentafluoride, PF_5. (Hint: Use molecular models or if you do not have a set handy, use marshmallows or gumdrops and toothpicks.)

3-101 Use the valence-shell electron-pair repulsion model to predict the shape of a molecule in which a central atom is surrounded by six regions of electron density, as, for example, in sulfur hexa-fluoride, SF_6.

3-102 Chlorine dioxide, ClO_2, is a yellow to reddish yellow gas at room temperature. This strong oxidizing agent is used for bleaching cellulose, paper pulp, and textiles and for water purification. It was the gas used to kill anthrax spores in the anthrax-contaminated Hart Senate Office Building.
(a) How many valence electrons are present in ClO_2?
(b) Draw a Lewis structure for this molecule. (Hint: The order of attachment of atoms in this molecule is O—Cl—O. Chlorine is a third-period element, and its valence shell may contain more than eight electrons.)

3-103 Using the information in Figure 2-16, estimate the H—O and H—S distances (the atom—atom distances) in H_2O and H_2S, respectively.

3-104 Arrange the single covalent bonds within each set in order of increasing polarity.
(a) C—H, O—H, N—H   (b) C—H, C—Cl, C—I
(c) C—C, C—O, C—N

3-105 Consider the structure of Vitamin E shown below, which is found most abundantly in wheat germ oil, sunflower, and safflower oils:

Vitamin E

(a) Identify the various types of geometries present in each central atom using VSEPR theory.
(b) Determine the various relative bond angles associated with each central atom using VSEPR theory.
(c) Which is the most polar bond in Vitamin E?
(d) Would you predict Vitamin E to be polar or nonpolar?

3-106 Consider the structure of Penicillin G shown below, an antibiotic used to treat bacterial infections caused by gram-positive organisms, derived from Penicillium fungi:

Penicillin G

(a) Identify the various types of geometries present in each central atom using VSEPR theory.
(b) Determine the various relative bond angles associated with each central atom using VSEPR theory.
(c) Which is the most polar bond in Penicillin G?
(d) Would you predict Penicillin G to be polar or nonpolar?

3-107 Ephedrine, a molecule at one time found in the dietary supplement ephedra, has been linked to adverse health reactions, such as heart attacks, strokes, and heart palpitations. The use of ephedra in dietary supplements is now banned by the FDA.

Ephedrine

(a) Which is the most polar bond in ephedra?
(b) Would you predict ephedra to be polar or nonpolar?

3-108 Allene, C_3H_4, has the structural formula $CH_2=C=CH_2$.
(a) Describe the shape of this molecule.
(b) Is allene polar or nonpolar?

3-109 Until several years ago, the two chlorofluorocarbons (CFCs) most widely used as heat transfer media in refrigeration systems were Freon-11 (trichlorofluoromethane, CCl_3F) and Freon-12 (dichlorodifluoromethane, CCl_2F_2). Draw a three-dimensional representation of each molecule and indicate the direction of its polarity.

Reading Labels

3-110 Name and write the formula for the fluorine-containing compound present in fluoridated toothpastes and dental gels.

3-111 If you read the labels of sun-blocking lotions, you will find that a common UV-blocking agent is a compound containing zinc. Name and write the formula of this zinc-containing compound.

3-112 On packaged table salt, it is common to see a label stating that the salt "supplies iodide, a necessary nutrient." Name and write the formula of the iodine-containing nutrient compound found in iodized salt.

3-113 We are constantly warned about the dangers of "lead-based" paints. Name and write the formula for a lead-containing compound found in lead-based paints.

3-114 If you read the labels of several liquid and tablet antacid preparations, you will find that in many of them, the active ingredients are compounds containing hydroxide ions. Name and write formulas for these hydroxide ion–containing compounds.

3-115 Iron forms Fe^{2+} and Fe^{3+} ions. Which ion is found in over-the-counter preparations intended to treat "iron-poor blood"?

3-116 Read the labels of several multivitamin/multimineral formulations. Among their components, you will find a number of so-called trace minerals—minerals required in the diet of a healthy adult in amounts less than 100 mg per day or present in the body in amounts less than 0.01% of total body weight. Following are 18 trace minerals. Name at least one form of each trace mineral present in multivitamin formulations.

(a) Phosphorus (b) Magnesium
(c) Potassium (d) Iron
(e) Calcium (f) Zinc
(g) Manganese (h) Titanium
(i) Silicon (j) Copper
(k) Boron (l) Molybdenum
(m) Chromium (n) Iodine
(o) Selenium (p) Vanadium
(q) Nickel (r) Tin

3-117 Write formulas for these compounds.

(a) Calcium sulfite, which is used in preserving cider and other fruit juices
(b) Calcium hydrogen sulfite, which is used in dilute aqueous solutions for washing casks in brewing to prevent souring and cloudiness of beer and to prevent secondary fermentation
(c) Calcium hydroxide, which is used in mortar, plaster, cement, and other building and paving materials
(d) Calcium hydrogen phosphate, which is used in animal feeds and as a mineral supplement in cereals and other foods

3-118 Many paint pigments contain transition metal compounds. Name the compounds in these pigments using a Roman numeral to show the charge on a transition metal ion.

(a) Yellow, CdS (b) Green, Cr_2O_3
(c) White, TiO_2 (d) Purple, $Mn_3(PO_4)_2$
(e) Blue, Co_2O_3 (f) Ochre, Fe_2O_3

Looking Ahead

3-119 Perchloroethylene, which is a liquid at room temperature, is one of the most widely used solvents for commercial dry cleaning. It is sold for this purpose under several trade names, including Perclene®. Does this molecule have polar bonds? Is it a polar molecule? Does it have a dipole?

$$\underset{\text{Perchloroethylene}}{\overset{Cl}{\underset{Cl}{\diagdown}}C=C\overset{Cl}{\underset{Cl}{\diagup}}}$$

3-120 Vinyl chloride is the starting material for the production of poly(vinyl chloride), abbreviated PVC. Its recycling code is "V". The major use of PVC is for tubing in residential and commercial construction (Section 12-7).

$$\underset{\text{Vinyl chloride}}{\overset{H}{\underset{H}{\diagdown}}C=C\overset{Cl}{\underset{H}{\diagup}}}$$

(a) Complete the Lewis structure for vinyl chloride by showing all unshared pairs of electrons.
(b) Predict the H—C—H, H—C—C, and Cl—C—H bond angles in this molecule.
(c) Does vinyl chloride have polar bonds? Is it a polar molecule? Does it have a dipole?

3-121 Tetrafluoroethylene is the starting material for the production of poly(tetrafluoroethylene), PTFE, a polymer that is widely used for the preparation of nonstick coatings on kitchenware (Section 12-7). The most widely known trade name for this product is Teflon®.

Tetrafluoroethylene

(a) Complete the Lewis structure for tetrafluoroethylene by showing all unshared pairs of electrons.
(b) Predict the F—C—F and F—C—C bond angles in this molecule.
(c) Does tetrafluoroethylene have polar bonds? Is it a polar molecule? Does it have a dipole?

3-122 Some of the following structural formulas are incorrect because they contain one or more atoms that do not have their normal number of covalent bonds. Which structural formulas are incorrect, and which atom or atoms in each have the incorrect number of bonds?

(a)
$$\text{Cl}-\underset{\underset{H}{|}}{\overset{\overset{H}{|}}{C}}=\underset{}{\overset{\overset{H}{|}}{C}}-H$$

(b)
$$H-O-\underset{\underset{H}{|}}{\overset{\overset{H}{|}}{C}}-\underset{\underset{H}{|}}{\overset{\overset{H}{|}}{C}}-N-\underset{\underset{H}{|}}{\overset{\overset{H}{|}}{C}}-H$$

(c)
$$H-\underset{\underset{Br}{|}}{C}=C=\underset{\underset{H}{|}}{C}-O-\underset{\underset{H}{|}}{\overset{\overset{H}{|}}{C}}-H$$

(d)
$$H-C\equiv C-\underset{\underset{H}{|}}{\overset{\overset{H}{|}}{C}}-\underset{\underset{H}{|}}{C}=C-H$$

3-123 Sodium borohydride, $NaBH_4$, has found wide use as a reducing agent in organic chemistry. It is an ionic compound composed of one sodium ion, Na^+, and one borohydride ion, BH_4^-.

(a) How many valence electrons are present in the borohydride ion?
(b) Draw a Lewis structure for the borohydride ion.
(c) Predict the H—B—H bond angles in the borohydride ion.

3-124 Given your answer to Problem 3-123 and knowing that aluminum is immediately below boron in column 3A of the Periodic Table, propose a structure for lithium aluminum hydride, another widely used reducing agent in organic chemistry.

3-125 In Chapter 27, you will learn that adenosine 5'-triphosphate (ATP) serves as a common currency into which the energy gained from food is converted and stored for use during muscle contraction. Consider the structure of ATP (▼ refer to the structure at bottom of page).

▼ **Chemical structure for problem 3-125**

(a) Identify the various types of geometries present in each central atom using VSEPR theory.
(b) Determine the various relative bond angles associated with each central atom using VSEPR theory.
(c) What is the most polar bond in ATP?
(d) Would you predict ATP to be polar or nonpolar?

3-126 Androstenedione, a muscle-building dietary supplement that is allowed in baseball but is banned in professional football, college athletics, and the Olympic sports (see Chemical Connections 21C), has the following formula:

Androstenedione

(a) Identify the various types of geometries present in each central atom using VSEPR theory.
(b) Determine the various relative bond angles associated with each central atom using VSEPR theory.
(c) What is the most polar bond in androstenedione?
(d) Would you predict androstenedione to be polar or nonpolar?

ATP

3-127 Amoxicillin is an antibiotic used to treat bacterial infections caused by susceptible microorganisms. Consider the skeletal structure of amoxicillin (▼ refer to the structure at bottom of page). where all the bonded atoms are shown but double bonds, triple bonds, and/or lone pairs are missing:
(a) Complete the structure of amoxicillin.
(b) Identify the various types of geometries present in each central atom using VSEPR theory.
(c) Determine the various relative bond angles associated with each central atom using VSEPR theory.
(d) What is the most polar bond in Amoxicillin?
(e) Would you predict amoxicillin to be polar or nonpolar?
(f) Is amoxicillin expected to possess resonance? Explain why or why not.

▼ **Chemical structure for problem 3-127**

3-128 Cyclopropane, an anesthetic with extreme reactivity under normal conditions, consists of three carbon atoms linked to each other to form a ring with a formula C_3H_6.
(a) Based on this description, draw a Lewis structure for this molecule.
(b) Identify the geometry present in each central carbon atom using VSEPR theory.
(c) What is the predicted bond angle associated with each central carbon atom using VSEPR theory?
(d) What do you predict is the actual observed C-C-C bond angle, given the shape and size of the ring?
(e) Explain why cyclopropane is considerably less stable than other three-carbon compounds that do not contain a ring.

Amoxicillin skeletal structure

Chemical Reactions

4

Fireworks are spectacular displays of chemical reactions.

KEY QUESTIONS

- 4-1 What Is a Chemical Reaction?
- 4-2 How Do We Balance Chemical Equations?

  How To . . . Balance a Chemical Equation
- 4-3 How Can We Predict If Ions in Aqueous Solution Will React with Each Other?
- 4-4 What Are Oxidation and Reduction?
- 4-5 What Are Formula Weights and Molecular Weights?
- 4-6 What Is a Mole and How Do We Use It to Calculate Mass Relationships?
- 4-7 How Do We Calculate Mass Relationships in Chemical Reactions?
- 4-8 What Is Heat of Reaction?

4-1 What Is a Chemical Reaction?

In Chapter 1, we learned that chemistry is mainly concerned with two things: the structure of matter and the transformations from one form of matter to another. In Chapters 2 and 3, we discussed the first of these topics, and now we are ready to turn our attention to the second. In a chemical change, also called a chemical reaction, one or more reactants (starting materials) are converted into one or more products. Chemical reactions occur all around us. They fuel and keep alive the cells of living tissues; they occur when we light a match, cook a dinner, start a car, listen to a CD player, or watch television. Most of the world's manufacturing processes involve chemical reactions; they include petroleum refining and food processing as well as the manufacture of drugs, plastics, synthetic fibers, fertilizers, explosives, and many other materials.

In this chapter, we discuss four aspects of chemical reactions: (1) how to write and balance chemical equations, (2) types of chemical reactions, (3) mass relationships in chemical reactions, and (4) heat gains and losses.

4-2 How Do We Balance Chemical Equations?

When propane, which is the major component in bottled gas or LPG (liquefied petroleum gas), burns in air, it reacts with the oxygen in the air. These two reactants are converted to the products carbon dioxide and water in a chemical reaction called **combustion**. We can write this chemical reaction in the form of a **chemical equation**, using chemical formulas for the reactants and products and an arrow to indicate the direction in which the reaction proceeds. In addition, it is important to show the state of each reactant and product; that is, whether it is a gas, liquid, or solid. We use the symbol (g) for gas, (ℓ) for liquid, (s) for solid, and (aq) for a substance dissolved in water (aqueous). We place the appropriate symbol immediately following each reactant and product. In our combustion equation, propane, oxygen, and carbon dioxide are gases, and the flame produced when propane burns is hot enough so that the water that forms is a gas (steam). ◀

Combustion Burning in air

Chemical equation A representation using chemical formulas of the process that occurs when reactants are converted to products

$$C_3H_8(g) + O_2(g) \longrightarrow CO_2(g) + H_2O(g)$$
Propane   Oxygen   Carbon dioxide   Water

The equation we have written is incomplete, however. While it tells us the formulas of the starting materials and products (which every chemical equation must do) and the physical state of each reactant and product, it does not give the amounts correctly. It is not balanced, which means that the number of atoms on the left side of the equation is not the same as the number of atoms on the right side. From the law of conservation of mass (Section 2-3A), we know that atoms are neither destroyed nor created in chemical reactions; they merely shift from one substance to another. Thus, all of the atoms present at the start of the reaction (on the left side of the equation) must still be present at the end (on the right side of the equation). In the equation we have just written, three carbon atoms are on the left but only one is on the right.

▶ Propane burning in air

Charles D. Winters

HOW TO . . .

Balance a Chemical Equation

To balance an equation, we place numbers in front of the formulas until the number of each kind of atom in the products is the same as that in the starting materials. These numbers are called coefficients. As an example, let us balance our propane equation:

$$C_3H_8(g) + O_2(g) \longrightarrow CO_2(g) + H_2O(g)$$
Propane   Oxygen   Carbon dioxide   Water

To balance an equation:

1. Begin with atoms that appear in only one compound on the left and only one compound on the right. In the equation for the reaction of propane and oxygen, begin with either carbon or hydrogen.
2. If an atom occurs as a free element—as, for example O_2, in the reaction of propane with oxygen—balance this element last.
3. You can change only coefficients in balancing an equation; you cannot change chemical formulas. For example, if you have H_2O on the left side of an equation but need two oxygens, you can add the coefficient "2" to read $2H_2O$. You cannot, however, get two oxygens by changing

the formula to H_2O_2. Doing so would change the chemical composition of the expected product from water, H_2O, to hydrogen peroxide, H_2O_2.

In the equation for the combustion (burning) of propane with oxygen, we can begin with carbon. Three carbon atoms appear on the left and one on the right. If we put a 3 in front of the CO_2 (indicating that three CO_2 molecules are formed), three carbons will appear on each side and the carbons will be balanced:

Three C on each side

$$C_3H_8(g) + O_2(g) \longrightarrow 3CO_2(g) + H_2O(g)$$

Next, we look at the hydrogens. There are eight on the left and two on the right. If we put a 4 in front of the H_2O, there will be eight hydrogens on each side and the hydrogens will be balanced:

Eight H on each side

$$C_3H_8(g) + O_2(g) \longrightarrow 3CO_2(g) + 4H_2O(g)$$

The only atom still unbalanced is oxygen. Notice that we saved this reactant for last (rule 2). There are two oxygen atoms on the left and ten on the right. If we put a 5 in front of the O_2 on the left, we both balance the oxygen atoms and arrive at the balanced equation:

Ten O on each side

$$C_3H_8(g) + 5O_2(g) \longrightarrow 3CO_2(g) + 4H_2O(g)$$

At this point, the equation ought to be balanced, but we should always check, just to make sure. In a balanced equation, there must be the same number of atoms of each element on both sides. A check of our work shows three C, ten O, and eight H atoms on each side. The equation is indeed balanced.

Interactive Example 4-1   Balancing a Chemical Equation

Balance this equation:

$$Ca(OH)_2(s) + HCl(g) \longrightarrow CaCl_2(s) + H_2O(\ell)$$

Calcium hydroxide   Hydrogen chloride   Calcium chloride

Strategy

To balance an equation, we place numbers in front of the formulas until there are identical numbers of atoms on each side of the equation. Begin with atoms that appear in only one compound on the left and only one compound on the right.

Solution

The calcium is already balanced—there is one Ca on each side. There is one Cl on the left and two on the right. To balance chlorine, we add the coefficient 2 in front of HCl:

$$Ca(OH)_2(s) + 2HCl(g) \longrightarrow CaCl_2(s) + H_2O(\ell)$$

Looking at hydrogens, we see that there are four hydrogens on the left but only two on the right. Placing the coefficient 2 in front of H_2O balances the hydrogens. It also balances the oxygens and completes the balancing of the equation:

$$Ca(OH)_2(s) + 2HCl(g) \longrightarrow CaCl_2(s) + 2H_2O(\ell)$$

Problem 4-1

Following is an unbalanced equation for photosynthesis, the process by which green plants convert carbon dioxide and water to glucose and oxygen. Balance this equation:

$$CO_2(g) + H_2O(\ell) \xrightarrow{\text{Photosynthesis}} \underset{\text{Glucose}}{C_6H_{12}O_6(aq)} + O_2(g)$$

▶ A pocket lighter contains butane in both the liquid and gaseous state.

EXAMPLE 4-2 Balancing a Chemical Equation

Balance this equation for the combustion of butane, the fluid most commonly used in pocket lighters: ◀

$$\underset{\text{Butane}}{C_4H_{10}(g)} + O_2(g) \longrightarrow CO_2(g) + H_2O(g)$$

Strategy

The equation for the combustion of butane is very similar to the one we examined at the beginning of this section for the combustion of propane. To balance an equation, we place numbers in front of the formulas until there are identical numbers of atoms on each side of the equation.

Solution

To balance carbons, put a 4 in front of the CO_2 (because there are four carbons on the left). Then to balance hydrogens, place a 5 in front of the H_2O to give ten hydrogens on each side of the equation.

$$C_4H_{10}(g) + O_2(g) \longrightarrow 4CO_2(g) + 5H_2O(g)$$

When we count the oxygens, we find 2 on the left and 13 on the right. We can balance their numbers by putting 13/2 in front of the O_2.

$$C_4H_{10}(g) + \frac{13}{2}O_2(g) \longrightarrow 4CO_2(g) + 5H_2O(g)$$

Although chemists sometimes have good reason to write equations with fractional coefficients, it is common practice to use only whole-number coefficients. We accomplish that by multiplying everything by 2, which gives the balanced equation:

$$2C_4H_{10}(g) + 13O_2(g) \longrightarrow 8CO_2(g) + 10H_2O(g)$$

Problem 4-2

Balance this equation:

$$C_6H_{14}(g) + O_2(g) \longrightarrow CO_2(g) + H_2O(g)$$

EXAMPLE 4-3 Balancing a Chemical Equation

Balance this equation:

$$Na_2SO_3(aq) + H_3PO_4(aq) \longrightarrow H_2SO_3(aq) + Na_3PO_4(aq)$$

Sodium sulfite · Phosphoric acid · Sulfurous acid · Sodium phosphate

Strategy
The key to balancing equations like this one is to realize that polyatomic ions such as SO_3^{2-} and PO_4^{3-} remain intact on both sides of the equation.

Solution
We can begin by balancing the Na^+ ions. We put a 3 in front of Na_2SO_3 and a 2 in front of Na_3PO_4, giving us six Na^+ ions on each side:

Six Na on each side

$$3Na_2SO_3(aq) + H_3PO_4(aq) \longrightarrow H_2SO_3(aq) + 2Na_3PO_4(aq)$$

There are now three SO_3^{2-} units on the left and only one on the right, so we put a 3 in front of H_2SO_3:

Three SO_3^{2-} units on each side

$$3Na_2SO_3(aq) + H_3PO_4(aq) \longrightarrow 3H_2SO_3(aq) + 2Na_3PO_4(aq)$$

Now let's look at the PO_4^{3-} units. There are two PO_4^{3-} units on the right but only one on the left. To balance them, we put a 2 in front of H_3PO_4. In so doing, we balance not only the PO_4^{3-} units but also the hydrogens and arrive at the balanced equation:

Two PO_4^{3-} units on each side

$$3Na_2SO_3(aq) + 2H_3PO_4(aq) \longrightarrow 3H_2SO_3(aq) + 2Na_3PO_4(aq)$$

Problem 4-3
Balance this equation:

$$K_2C_2O_4(aq) + Ca_3(AsO_4)_2(s) \longrightarrow K_3AsO_4(aq) + CaC_2O_4(s)$$

Potassium oxalate · Calcium arsenate · Potassium arsenate · Calcium oxalate

---

One final point about balancing chemical equations. The following equation for the combustion of propane is correctly balanced.

$$C_3H_8(g) + 5O_2(g) \longrightarrow 3CO_2(g) + 4H_2O(g)$$

Propane

Would it be correct if we doubled all the coefficients?

$$2C_3H_8(g) + 10O_2(g) \longrightarrow 6CO_2(g) + 8H_2O(g)$$

Propane

Yes, this revised equation is mathematically and scientifically correct, but chemists do not normally write equations with coefficients that are all divisible by a common number. A correctly balanced equation is almost always written with the coefficients expressed in the lowest set of whole numbers.

4-3 How Can We Predict If Ions in Aqueous Solution Will React with Each Other?

Many ionic compounds are soluble in water. As we saw in Section 3-5, ionic compounds always consist of both positive and negative ions. When they dissolve in water, the positive and negative ions separate from each other. We call such separation **dissociation**. For example,

$$NaCl(s) \xrightarrow{H_2O} Na^+(aq) + Cl^-(aq)$$

What happens when we mix **aqueous solutions** of two different ionic compounds? Does a reaction take place between the ions? The answer depends on the ions. For example, if any of the negative and positive ions come together to form a water-insoluble compound, then a reaction takes place and a precipitate forms. This is sometimes referred to as a **precipitation reaction**.

Suppose we prepare one solution by dissolving sodium chloride, NaCl, in water and a second solution by dissolving silver nitrate, $AgNO_3$, in water.

$$\text{Solution 1} \quad NaCl(s) \xrightarrow{H_2O} Na^+(aq) + Cl^-(aq)$$
$$\text{Solution 2} \quad AgNO_3(s) \xrightarrow{H_2O} Ag^+(aq) + NO_3^-(aq)$$

If we now mix the two solutions, four ions are present in the solution: Ag^+, Na^+, Cl^-, and NO_3^-. Two of these ions, Ag^+ and Cl^-, react to form the compound AgCl (silver chloride), which is insoluble in water. A reaction therefore takes place, forming a white precipitate of AgCl that slowly sinks to the bottom of the container (Figure 4-1). We write this reaction as follows:

$$Ag^+(aq) + NO_3^-(aq) + Na^+(aq) + Cl^-(aq) \longrightarrow AgCl(s) + Na^+(aq) + NO_3^-(aq)$$
$$\text{Silver ion} \quad \text{Nitrate ion} \quad \text{Sodium ion} \quad \text{Chloride ion} \quad \text{Silver chloride}$$

Notice that the Na^+ and NO_3^- ions do not participate in a reaction, but merely remain dissolved in the water. Ions that do not participate in a reaction are called **spectator ions**, certainly an appropriate name.

We can simplify the equation for the formation of silver chloride by omitting all spectator ions:

$$\text{Net ionic equation:} \quad Ag^+(aq) + Cl^-(aq) \longrightarrow AgCl(s)$$
$$\text{Silver ion} \quad \text{Chloride ion} \quad \text{Silver chloride}$$

This kind of equation that we write for ions in solution is called a **net ionic equation**. Like all other chemical equations, net ionic equations must be balanced. We balance them in the same way we do other equations, except now we must make sure that charges balance as well as atoms.

Net ionic equations show only the ions that react—no spectator ions are shown. For example, consider the net ionic equation for the precipitation of arsenic(III) sulfide from aqueous solution:

$$\text{Net ionic equation:} \quad 2As^{3+}(aq) + 3S^{2-}(aq) \longrightarrow As_2S_3(s)$$

Not only are there two arsenic and three sulfur atoms on each side, but the total charge on the left side is the same as the total charge on the right side; they are both zero.

In general, ions in solution react with each other only when one of these four things can happen:

1. Two ions form a solid that is insoluble in water, known as a precipitation reaction. AgCl is one example, as shown in Figure 4-1.

Aqueous solutions Solutions in which the solvent is water

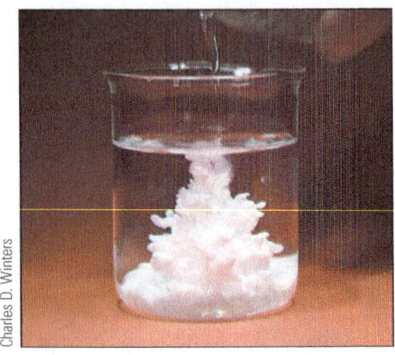

FIGURE 4-1 Adding Cl^- ions to a solution of Ag^+ ions produces a white precipitate of silver chloride, AgCl.

Precipitation reaction When positive and negative ions react to form an insoluble compound

Spectator ions Ions that appear unchanged on both sides of a chemical equation

Net ionic equation A chemical equation that does not contain spectator ions, where both atoms and charges are balanced

2. Two ions form a gas that escapes from the reaction mixture as bubbles. An example is the reaction of sodium bicarbonate, $NaHCO_3$, with HCl to form the gas carbon dioxide, CO_2 (Figure 4-2). The net ionic equation for this reaction, where the expected reactant ion $H^+(aq)$ from HCl is written as a hydrated ion or $H_3O^+(aq)$, is written as:

Net ionic equation: $HCO_3^-(aq) + H_3O^+(aq) \longrightarrow CO_2(g) + 2H_2O(\ell)$
  Bicarbonate ion   Carbon dioxide

3. An acid neutralizes a base to form water. Acid-base reactions are so important that we devote Chapter 8 to them.

4. One of the ions can oxidize another. We discuss this type of reaction in Section 4-4.

In many cases, no reaction takes place when we mix solutions of ionic compounds because none of these situations holds. For example, if we mix solutions of copper(II) nitrate, $Cu(NO_3)_2$, and potassium sulfate, K_2SO_4, we merely have a mixture containing Cu^{2+}, K^+, NO_3^-, and SO_4^{2-} ions dissolved in water. None of these ions react with each other; therefore, we see nothing happening (Figure 4-3).

FIGURE 4-2 When aqueous solutions of $NaHCO_3$ and HCl are mixed, a reaction between HCO_3^- and H_3O^+ ions produces CO_2 gas, which can be seen as bubbles.

(a)

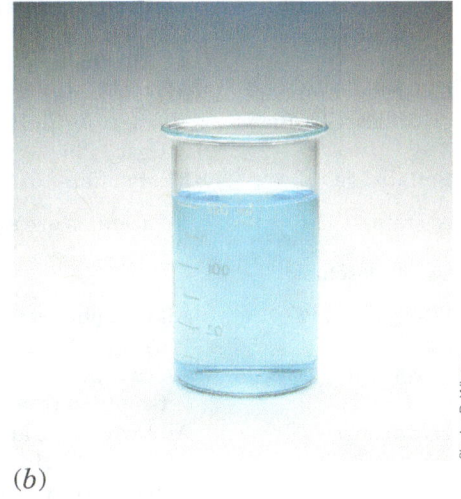
(b)

FIGURE 4-3 (a) The beaker on the left contains a solution of potassium sulfate (colorless), and the beaker on the right contains a solution of copper(II) nitrate (blue). (b) When the two solutions are mixed, the blue color becomes lighter because the copper(II) nitrate is less concentrated, but no other chemical reaction occurs.

EXAMPLE 4-4 Net Ionic Equation

When a solution of barium chloride, $BaCl_2$, is added to a solution of sodium sulfate, Na_2SO_4, a white precipitate of barium sulfate, $BaSO_4$, forms. Write the net ionic equation for this reaction. ◀

Strategy

The net ionic equation shows only those ions that combine to form a precipitate.

Solution

Because both barium chloride and sodium sulfate are ionic compounds, each exists in water as its dissociated ions:

$$Ba^{2+}(aq) + 2Cl^-(aq) + 2Na^+(aq) + SO_4^{2-}(aq)$$

We are told that a precipitate of barium sulfate forms:

$$Ba^{2+}(aq) + 2Cl^-(aq) + 2Na^+(aq) + SO_4^{2-}(aq) \longrightarrow$$
$$BaSO_4(s) + 2Na^+(aq) + 2Cl^-(aq)$$
  Barium sulfate

▶ The mixing of solutions of barium chloride, $BaCl_2$, and sodium sulfate, Na_2SO_4, forms a white precipitate of barium sulfate, $BaSO_4$.

Because Na⁺ and Cl⁻ ions appear on both sides of the equation (they are spectator ions), we cancel them and are left with the following net ionic equation:

Net ionic equation: $Ba^{2+}(aq) + SO_4^{2-}(aq) \longrightarrow BaSO_4(s)$

Problem 4-4

When a solution of copper(II) chloride, $CuCl_2$, is added to a solution of potassium sulfide, K_2S, a black precipitate of copper(II) sulfide, CuS, forms. Write the net ionic equation for the reaction.

Of the four ways for ions to react in water, the formation of an insoluble compound via a precipitation reaction is one of the most common. We can predict when this result will happen if we know the solubilities of the ionic compounds. Some useful guidelines for the solubility of ionic compounds in water are given in Table 4-1.

Table 4-1 Solubility Rules for Ionic Compounds

Usually Soluble	
Li^+, Na^+, K^+, Rb^+, Cs^+, NH_4^+	All Group 1A (alkali metal) and ammonium salts are soluble.
Nitrates, NO_3^-	All nitrates are soluble.
Chlorides, bromides, iodides, Cl^-, Br^-, I^-	All common chlorides, bromides, and iodides are soluble except AgCl, Hg_2Cl_2, $PbCl_2$, AgBr, Hg_2Br_2, $PbBr_2$, AgI, Hg_2I_2, PbI_2
Sulfates, SO_4^{2-}	Most sulfates are soluble except $CaSO_4$, $SrSO_4$, $BaSO_4$, $PbSO_4$
Acetates, CH_3COO^-	All acetates are soluble.
Usually Insoluble	
Phosphates, PO_4^{3-}	All phosphates are insoluble except those of NH_4^+ and Group 1A (the alkali metal) cations.
Carbonates, CO_3^{2-}	All carbonates are insoluble except those of NH_4^+ and Group 1A (the alkali metal) cations. ◀
Hydroxides, OH^-	All hydroxides are insoluble except those of NH_4^+ and Group 1A (the alkali metal) cations. $Sr(OH)_2$, $Ba(OH)_2$, and $Ca(OH)_2$ are only slightly soluble. ◀
Sulfides, S^{2-}	All sulfides are insoluble except those of NH_4^+ and Group 1A (the alkali metal) and Group 2A cations. MgS, CaS, and BaS are only slightly soluble.

▶ Sea animals of the mollusk family often use insoluble calcium carbonate, $CaCO_3$ to construct their shells.

▶ Both $Fe(OH)_3$, iron(III) hydroxide, and $CuCO_3$, copper(II) carbonate, are insoluble in water.

CHEMICAL CONNECTIONS 4A

Solubility and Tooth Decay

The outermost protective layer of a tooth is the enamel, which is composed of approximately 95% hydroxyapatite, $Ca_5(PO_4)_3(OH)$, and 5% collagen (Figure 22-13). Like most other phosphates and hydroxides, hydroxyapatite is insoluble in water. In acidic media, however, it dissolves to a slight extent, yielding Ca^{2+}, PO_4^{3-}, and OH^- ions. This loss of enamel creates pits and cavities in the tooth.

Acidity in the mouth is produced by bacterial fermentation of remnants of food, especially carbohydrates.

Once pits and cavities form in the enamel, bacteria can hide there and cause further damage in the underlying softer material called dentin. The fluoridation of water brings F^- ions to the hydroxyapatite. There, F^- ions take the place of OH^- ions, forming the considerably less acid-soluble fluoroapatite, $Ca_5(PO_4)_3F$. Fluoride-containing toothpastes enhance this exchange process and provide protection against tooth decay.

Test your knowledge with Problems 4-78 and 4-79.

4-4 What Are Oxidation and Reduction?

Oxidation-reduction is one of the most important and common types of chemical reactions. **Oxidation** is the loss of electrons. **Reduction** is the gain of electrons. An **oxidation-reduction reaction** (often called a **redox reaction**) involves the transfer of electrons from one species to another. An example is the oxidation of zinc by copper ions, the net ionic equation for which is:

$$Zn(s) + Cu^{2+}(aq) \longrightarrow Zn^{2+}(aq) + Cu(s)$$

When we put a piece of zinc metal into a beaker containing copper(II) ions in aqueous solution, three things happen (Figure 4-4):

1. Some of the zinc metal dissolves and goes into solution as Zn^{2+}.
2. Copper metal deposits on the surface of the zinc metal.
3. The blue color of the Cu^{2+} ions gradually disappears.

Zinc atoms lose electrons to copper ions and become zinc ions:

$$Zn(s) \longrightarrow Zn^{2+}(aq) + 2e^- \qquad \text{Zn is oxidized}$$

At the same time, Cu^{2+} ions gain electrons from the zinc atoms. The copper ions are reduced:

$$Cu^{2+}(aq) + 2e^- \longrightarrow Cu(s) \qquad Cu^{2+} \text{ is reduced}$$

Oxidation The loss of electrons; the gain of oxygen atoms and/or the loss of hydrogen atoms

Reduction The gain of electrons; the loss of oxygen atoms and/or the gain of hydrogen atoms

Redox reaction An oxidation-reduction reaction

FIGURE 4-4 When a piece of zinc is added to a solution containing Cu^{2+} ions, Zn is oxidized by Cu^{2+} ions and Cu^{2+} ions are reduced by the Zn.

Oxidizing agent An entity that accepts electrons in an oxidation-reduction reaction

Reducing agent An entity that donates electrons in an oxidation-reduction reaction

Oxidation and reduction are not independent reactions. That is, a species cannot gain electrons from nowhere, nor can a species lose electrons to nothing. In other words, no oxidation can occur without an accompanying reduction, and vice versa. In the preceding reaction, Cu^{2+} oxidizes Zn. We call Cu^{2+} an **oxidizing agent**. Similarly, Zn reduces Cu^{2+}, and we call Zn a **reducing agent**.

We summarize these oxidation-reduction relationships for the reaction of zinc metal with Cu^{2+} ion in the following way:

Note here that a curved arrow from Zn(s) to Cu^{2+} shows the transfer of two electrons from zinc to copper ion. Refer to Section 8-6B for additional examples of redox reactions which feature active metals with strong acids.

Although the definitions we have given for oxidation (loss of electrons) and reduction (gain of electrons) are easy to apply in many redox reactions, they are not so easy to apply in other cases. For example, another redox reaction is the combustion (burning) of methane, CH_4, in which CH_4 is oxidized to CO_2 while O_2 is reduced to H_2O.

$$CH_4(g) + 2O_2(g) \longrightarrow CO_2(g) + 2H_2O(g)$$
Methane

It is not easy to see the electron loss and gain in such a reaction, so chemists developed another definition of oxidation and reduction, one that is easier to apply in many cases, especially where organic (carbon-containing) compounds are involved:

Oxidation: The gain of oxygen atoms and/or the loss of hydrogen atoms

Reduction: The loss of oxygen atoms and/or the gain of hydrogen atoms

Applying these alternative definitions to the reaction of methane with oxygen, we find the following:

$$CH_4(g) \quad + \quad 2O_2(g) \longrightarrow CO_2(g) + 2H_2O(g)$$
Gains O and loses H; is oxidized   Gains H; is reduced
Is the reducing agent   Is the oxidizing agent

In fact, this second definition is much older than the one involving electron transfer; it is the definition given by Lavoisier when he first discovered oxidation and reduction more than 200 years ago. Note that we could not apply this definition to our zinc-copper example.

EXAMPLE 4-5 Oxidation-Reduction

In each equation, identify the substance that is oxidized, the substance that is reduced, the oxidizing agent, and the reducing agent.
(a) $Al(s) + Fe^{3+}(aq) \longrightarrow Al^{3+}(aq) + Fe(s)$
(b) $CH_3OH(g) + O_2(g) \longrightarrow HCOOH(g) + H_2O(g)$
   Methanol   Formic acid

Strategy

The substance that is oxidized loses electrons and is a reducing agent. The substance that gains electrons is the oxidizing agent and is reduced. For organic compounds, oxidation involves the gain of oxygen atoms and/or the loss of hydrogen atoms. Reduction involves the gain of hydrogen atoms and/or the loss of oxygen atoms. ◄

Solution

(a) Al(s) loses three electrons and becomes Al^{3+}; therefore, aluminum is oxidized. In the process of being oxidized, Al(s) gives its electrons to Fe^{3+}, and so Al(s) is the reducing agent. Fe^{3+} gains three electrons and becomes Fe(s) and is reduced. In the process of being reduced, Fe^{3+} accepts three electrons from Al(s), and so Fe^{3+} is the oxidizing agent. To summarize:

(b) Because it is not easy to see the loss or gain of electrons in this example, we apply the second set of definitions. In converting CH_3OH to HCOOH, CH_3OH both gains oxygen atoms and loses hydrogen atoms; it is oxidized. In being converted to H_2O, O_2 gains hydrogen atoms; it is reduced. The compound oxidized is the reducing agent; CH_3OH is the reducing agent. The compound reduced is the oxidizing agent; O_2 is the oxidizing agent. To summarize:

$$CH_3OH(g) + O_2(g) \longrightarrow HCOOH(g) + H_2O(g)$$

Is oxidized; methanol is the reducing agent

Is reduced; oxygen is the oxidizing agent

Problem 4-5

In each equation, identify the substance that is oxidized, the substance that is reduced, the oxidizing agent, and the reducing agent:
(a) $Ni^{2+}(aq) + Cr(s) \longrightarrow Ni(s) + Cr^{2+}(aq)$
(b) $CH_2O(g) + H_2(g) \longrightarrow CH_3OH(g)$
   Formaldehyde   Methanol

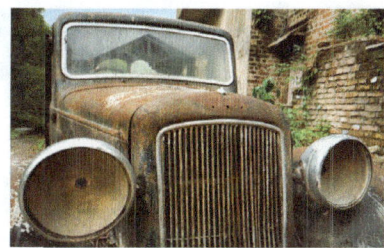

▶ The rusting of iron and steel can be a serious problem in an industrial society. In rusting, iron is oxidized.

We have said that redox reactions are extremely common. Here are some important categories:

1. **Combustion** All combustion (burning) reactions are redox reactions in which the compounds or mixtures that are burned are oxidized by oxygen, O_2. They include the burning of gasoline, diesel oil, fuel oil, natural gas, coal, wood, and paper. All these materials contain carbon, and all except coal also contain hydrogen. If the combustion is complete, carbon is oxidized to CO_2 and oxygen is reduced to H_2O. In an incomplete combustion, these elements are converted to other compounds, many of which cause air pollution. ◄

Unfortunately, much of today's combustion that takes place in gasoline and diesel engines and in furnaces is incomplete and so contributes

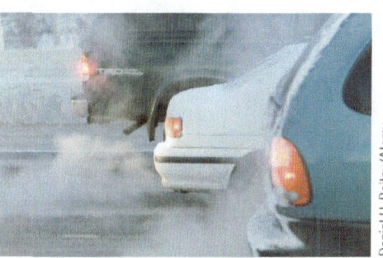

▶ Air pollution is caused by incomplete fuel combustion.

CHEMICAL CONNECTIONS 4B

Voltaic Cells

In Figure 4-4, we see that when a piece of zinc metal is put in a solution containing Cu^{2+} ions, zinc atoms give electrons to Cu^{2+} ions. We can change the experiment by putting the zinc metal in one beaker and the Cu^{2+} ions in another and then connecting the two beakers by a length of wire and a salt bridge (see the accompanying figure). A reaction still takes place; that is, zinc atoms still give electrons to Cu^{2+} ions, but now the electrons must flow through the wire to get from the Zn to the Cu^{2+}. This flow of electrons produces an electric current, and the electrons keep flowing until either the Zn or the Cu^{2+} is used up. In this way, the apparatus generates an electric current by using a redox reaction. We call this device a **voltaic cell** or, more commonly, a battery.

The electrons produced at the zinc end carry negative charges. This end of the battery is a negative electrode (called the **anode**). The electrons released at the anode as zinc is oxidized go through an outside circuit and, in doing so, produce the battery's electric current. At the other end of the battery via the positively charged electrode (called the **cathode**), electrons are consumed as Cu^{2+} ions are reduced to copper metal.

To see why a salt bridge is necessary, we must look at the Cu^{2+} solution. Because we cannot have positive charges in any place without an equivalent number of negative charges, negative ions must be in the beaker as well—perhaps sulfate, nitrate, or some other anion. When electrons come over the wire, the Cu^{2+} is converted to Cu:

$$Cu^{2+}(aq) + 2e^- \longrightarrow Cu(s)$$

This reaction diminishes the number of Cu^{2+} ions, but the number of negative ions remains unchanged. The salt bridge is necessary to carry some of these negative ions to the other beaker, where they are needed to balance the Zn^{2+} ions being produced by the following reaction:

$$Zn(s) \longrightarrow Zn^{2+}(aq) + 2e^-$$

Voltaic cell. The electron flow over the wire from Zn to Cu^{2+} is an electric current that causes the lightbulb to glow.

Test your knowledge with Problem 4-80.

---

to air pollution. In the incomplete combustion of methane, for example, carbon is oxidized to carbon monoxide, CO, because there is not a sufficient supply of oxygen to complete its oxidation to CO_2:

$$2CH_4(g) + 3O_2(g) \longrightarrow 2CO(g) + 4H_2O(g)$$
Methane

CHEMICAL CONNECTIONS 4C

Artificial Pacemakers and Redox

An artificial pacemaker is a small electrical device that uses electrical impulses, delivered by electrodes contacting the heart muscles, to regulate the beating of the heart. The primary purpose of a pacemaker is to maintain an adequate heart rate, either because the heart's native pacemaker does not beat fast enough, or perhaps there is a blockage in the heart's electrical conduction system. When a pacemaker detects that the heart is beating too slowly, it sends an electrical signal to the heart, generated via a redox reaction, so that the heart muscle beats faster. Modern pacemakers are externally programmable and allow a cardiologist to select the optimum pacing modes for individual patients.

Early pacemakers generated an electrical impulse via the following redox reaction:

$$Zn + Hg^{2+} \longrightarrow Zn^{2+} + Hg$$

The zinc atom is oxidized to Zn^{2+}, and Hg^{2+} is reduced to Hg. Many contemporary artificial pacemakers contain a lithium-iodine battery, which has a longer battery life (10 years or more). Consider the unbalanced redox reaction for the lithium-iodine battery:

$$Li + I_2 \longrightarrow LiI$$

The lithium atom is oxidized to Li^+, and the I_2 molecule is reduced to I^-. When the pacemaker fails to sense a heartbeat within a normal beat-to-beat time period, an electrical signal produced from these reactions is initiated, stimulating the ventricle of the heart. This sensing and stimulating activity continues on a beat-by-beat basis. More complex systems include the ability to stimulate both the atrial and ventricular chambers.

A pacemaker is a medical device that uses electrical impulses, delivered by electrodes contacting the heart muscles, to regulate the beating of the heart.

Test your knowledge with Problem 4-81.

2. **Respiration** Humans and animals get their energy by respiration. The oxygen in the air we breathe oxidizes carbon-containing compounds in our cells to produce CO_2 and H_2O. Note that respiration is equivalent to combustion, except that it takes place more slowly and at a much lower temperature. We discuss respiration more fully in Chapter 27. The important product of respiration is not CO_2 (which the body eliminates) or H_2O, but energy.

3. **Rusting** We all know that when iron or steel objects are left out in the open air, they eventually rust (steel is mostly iron but contains certain other elements as well). In rusting, iron is oxidized to a mixture of iron oxides. We can represent the main reaction by the following equation:

$$4Fe(s) + 3O_2(g) \longrightarrow 2Fe_2O_3(s)$$

FIGURE 4-5 (a) Dry cell batteries. (b) A lead storage battery.

4. **Bleaching** Most bleaching involves oxidation, and common bleaches are oxidizing agents. The colored compounds being bleached are usually organic compounds; oxidation converts them to colorless compounds. ◂

5. **Batteries** A voltaic cell (Chemical Connections 4B) is a device in which electricity is generated from a chemical reaction. Such cells are often called batteries (Figure 4-5). We are all familiar with batteries in our cars and in such portable devices as radios, flashlights, cell phones, and computers. In all cases, the reaction that takes place in the battery is a redox reaction.

4-5 What Are Formula Weights and Molecular Weights?

We begin our study of mass relationships with a discussion of formula weight. The **formula weight (FW)** of a compound is the sum of the atomic weights in atomic mass units (amu) of all the atoms in the compound's formula. The term "formula weight" can be used for both ionic and molecular compounds and tells nothing about whether the compound is ionic or molecular.

Another term, **molecular weight (MW)**, is strictly correct only when used for covalent compounds. In this book, we use "formula weight" for both ionic and covalent compounds and "molecular weight" only for covalent compounds.

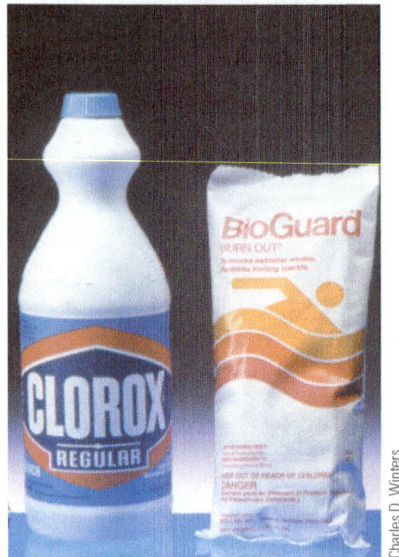

▸ These household bleaches are oxidizing agents.

A table of atomic weights is given on the inside front cover. Atomic weights can also be found in the Periodic Table on the inside front cover.

Molecular weight (MW) The sum of the atomic weights of all atoms in a molecular compound expressed in atomic mass units (amu)

EXAMPLE 4-6 Molecular Weight

What is the molecular weight of (a) glucose, $C_6H_{12}O_6$, and (b) urea, $(NH_2)_2CO$?

Strategy

Molecular weight is the sum of the atomic weights of all atoms in the molecular formula expressed in atomic mass units (amu).

Solution

(a) Glucose, $C_6H_{12}O_6$
    C $6 \times 12.0 = 72.0$
    H $12 \times 1.0 = 12.0$
    O $6 \times 16.0 = 96.0$
    $C_6H_{12}O_6 = 180.0$ amu

(b) Urea, $(NH_2)_2CO$
    N $2 \times 14.0 = 28.0$
    H $4 \times 1.0 = 4.0$
    C $1 \times 12.0 = 12.0$
    O $1 \times 16.0 = 16.0$
    $(NH_2)_2CO = 60.0$ amu

> **Problem 4-6**
> What is (a) the molecular weight of ibuprofen, $C_{13}H_{18}O_2$, and (b) the formula weight of barium phosphate, $Ba_3(PO_4)_2$?

4-6 What Is a Mole and How Do We Use It to Calculate Mass Relationships?

Atoms and molecules are so tiny (Section 2-4F) that chemists are seldom able to deal with them one at a time. When we weigh even a very small quantity of a compound, huge numbers of formula units (perhaps 10^{19}) are present. The formula unit may be atoms, molecules, or ions. To overcome this problem, chemists long ago defined a unit called the **mole (mol)**. A mole is the amount of substance that contains as many atoms, molecules, or ions as there are atoms in exactly 12 g of carbon-12. The important point here is that whether we are dealing with a mole of iron atoms, a mole of methane molecules, or a mole of sodium ions, a mole always contains the same number of formula units. We are accustomed to scale-up factors in situations where there are large numbers of units involved in counting. We count eggs by the dozen and pencils by the gross. Just as the dozen (12 units) is a useful scale-up factor for eggs and the gross (144 units) a useful scale-up factor for pencils, the mole is a useful scale-up factor for atoms and molecules. We are soon going to see that the number of units is much larger for a mole than for a dozen or a gross.

Mole (mol) The formula weight of a substance expressed in grams

The number of formula units in a mole is called **Avogadro's number** after the Italian physicist Amadeo Avogadro (1776–1856), who first proposed the concept of a mole but was not able to experimentally determine the number of units it represented. Note that Avogadro's number is not a defined value, but rather a value that must be determined experimentally. Its value is now known to nine significant figures.

Avogadro's number 6.02×10^{23}, is the number of formula units per mole

$$\text{Avogadro's number} = 6.02214199 \times 10^{23} \text{ formula units per mole}$$

For most calculations in this text, we round this number to three significant figures to 6.02×10^{23} formula units per mole.

A mole of hydrogen atoms is 6.02×10^{23} hydrogen atoms, a mole of sucrose (table sugar) molecules is 6.02×10^{23} sugar molecules, a mole of apples is 6.02×10^{23} apples, and a mole of sodium ions is 6.02×10^{23} sodium ions. Just as we call 12 of anything a dozen, 20 a score, and 144 a gross, we call 6.02×10^{23} of anything a mole.

The **molar mass** of any substance (the mass of one mole of the substance) is the formula weight of the substance expressed in grams per mole. For instance, the formula weight of glucose, $C_6H_{12}O_6$ (Example 4-6), is 180.0 amu; therefore, 180.0 g of glucose is one mole of glucose. Likewise, the formula weight of urea, $(NH_2)_2CO$, is 60.0 amu and, therefore, one mole of urea is 60.0 grams of urea. For atoms, one mole is the atomic weight expressed in grams: 12.0 g of carbon is one mole of carbon atoms; 32.1 g of sulfur is one mole of sulfur atoms; and so on. As you see, the important point here is that to talk about the mass of a mole, we need to know the chemical formula of the substance we are considering. Figure 4-6 shows one-mole quantities of several compounds.

Molar mass The mass of one mole of a substance expressed in grams; the formula weight of a compound expressed in grams

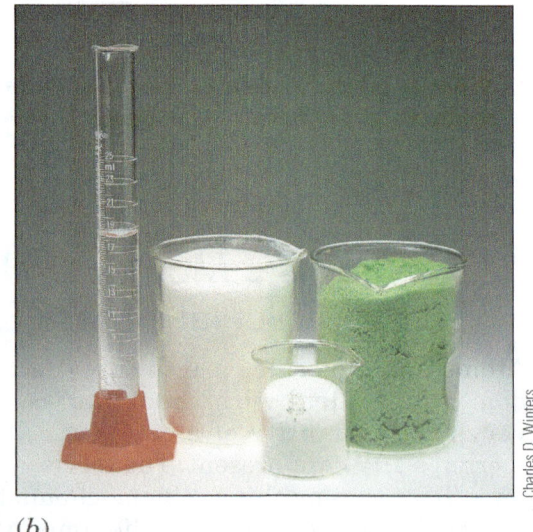

(a) (b)

FIGURE 4-6 One-mole quantities of (a) six metals and (b) four compounds. (a) Top row (left to right): Cu beads (63.5 g), Al foil (27.0 g), and Pb shot (207.2 g). Bottom row (left to right): S powder (32.1 g), Cr chunks (52.0 g), and Mg shavings (24.4 g). (b) H_2O (18.0 g); small beaker, NaCl (58.4 g); large left beaker, aspirin, $C_9H_8O_4$ (180.2 g); and large right beaker, green $NiCl_2 \cdot 6H_2O$ (237.7 g).

Now that we know the relationship between moles and molar mass (g/mol), we can use molar mass as a conversion factor to convert from grams to moles and from moles to grams. For this calculation, we use molar mass as a conversion factor.

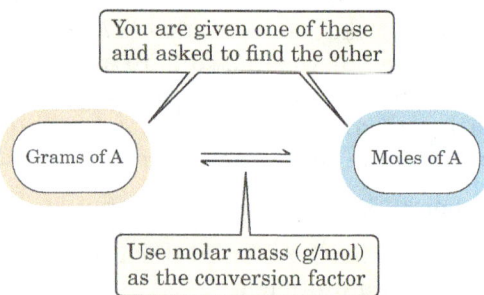

Suppose we want to know the number of moles of water in a graduated cylinder that contains 36.0 g of water. We know that the molar mass of water is 18.0 g/mol. If 18.0 g of water is one mole of water, then 36.0 g must be two moles of water.

$$36.0 \text{ g } H_2O \times \frac{1 \text{ mol } H_2O}{18.0 \text{ g } H_2O} = 2.00 \text{ mol } H_2O$$

Molar mass can also be used to convert from moles to grams. Suppose you have a beaker that contains 0.753 mol of sodium chloride and you want to calculate the number of grams of sodium chloride in the beaker. As a conversion factor, use the fact that the molar mass of NaCl is 58.5 g/mol.

$$0.753 \text{ mol NaCl} \times \frac{58.5 \text{ g NaCl}}{1 \text{ mol NaCl}} = 44.1 \text{ g NaCl}$$

Interactive Example 4-7 — Moles

We have 27.5 g of sodium fluoride, NaF, the form of fluoride ions most commonly used in fluoride toothpastes and dental gels. How many moles of NaF is this?

Strategy

The formula weight of NaF = 23.0 + 19.0 = 42.0 amu. Thus, each mole of NaF has a mass of 42.0 g, allowing us to use the conversion factor 1 mol NaF = 42.0 g NaF.

Solution

$$27.5 \text{ g NaF} \times \frac{1 \text{ mol NaF}}{42.0 \text{ g NaF}} = 0.655 \text{ mol NaF}$$

Problem 4-7

A person drinks 1500. g of water per day. How many moles is this?

Interactive Example 4-8 — Moles

We wish to weigh 3.41 mol of ethanol, C_2H_6O. How many grams is this?

Strategy

The formula weight of C_2H_6O is 2(12.0) + 6(1.0) + 16.0 = 46.0 amu, so the conversion factor is 1 mol C_2H_6O = 46.0 g C_2H_6O.

Solution

$$3.41 \text{ mol } C_2H_6O \times \frac{46.0 \text{ g } C_2H_6O}{1.00 \text{ mol } C_2H_6O} = 157 \text{ g } C_2H_6O$$

Problem 4-8

We wish to weigh 2.84 mol of sodium sulfide, Na_2S. How many grams is this?

Interactive Example 4-9 — Moles

How many moles of nitrogen atoms and oxygen atoms are in 21.4 mol of the explosive trinitrotoluene (TNT), $C_7H_5N_3O_6$?

Strategy

The molecular formula $C_7H_5N_3O_6$ tells us that each molecule of TNT contains three nitrogen atoms and six oxygen atoms. It also tells us that each mole of TNT contains three moles of N atoms and six moles of O atoms. Therefore, we have the following conversion factors: 1 mol TNT = 3 mol N atoms, and 1 mol TNT = 6 mol O atoms.

Solution

The number of moles of N atoms in 21.4 moles of TNT is:

$$21.4 \text{ mol TNT} \times \frac{3 \text{ mol N atoms}}{1 \text{ mol TNT}} = 64.2 \text{ mol N atoms}$$

The number of moles of O atoms in 21.4 moles of TNT is:

$$21.4 \text{ mol TNT} \times \frac{6 \text{ mol O atoms}}{1 \text{ mol TNT}} = 128 \text{ mol O atoms}$$

Note that we give the answer to three significant figures because we were given the number of moles to three significant figures. The ratio of moles of O atoms to moles of TNT is an exact number.

Problem 4-9

How many moles of C atoms, H atoms, and O atoms are in 2.5 mol of glucose, $C_6H_{12}O_6$?

Interactive Example 4-10 — Moles

How many moles of sodium ions, Na^+, are in 5.63 g of sodium sulfate, Na_2SO_4?

Strategy

The formula weight of Na_2SO_4 is $2(23.0) + 32.1 + 4(16.0) = 142.1$ amu. In the conversion of grams Na_2SO_4 to moles, we use the conversion factors 1 mol Na_2SO_4 = 142.1 g Na_2SO_4 and 1 mole Na_2SO_4 = 2 moles Na^+.

Solution

First, we need to find out how many moles of Na_2SO_4 are in the sample.

$$5.63 \text{ g Na}_2\text{SO}_4 \times \frac{1 \text{ mol Na}_2\text{SO}_4}{142.1 \text{ g Na}_2\text{SO}_4} = 0.0396 \text{ mol Na}_2\text{SO}_4$$

The number of moles of Na^+ ions in 0.0396 mol of Na_2SO_4 is:

$$0.0396 \text{ mol Na}_2\text{SO}_4 \times \frac{2 \text{ mol Na}^+}{1 \text{ mol Na}_2\text{SO}_4} = 0.0792 \text{ mol Na}^+$$

Problem 4-10

How many moles of copper(I) ions, Cu^+, are there in 0.062 g of copper(I) nitrate, $CuNO_3$?

EXAMPLE 4-11 — Molecules per Gram

An aspirin tablet, $C_9H_8O_4$, contains 0.360 g of aspirin. (The rest of the tablet is starch or other fillers.) How many molecules of aspirin are present in this tablet?

Strategy

The formula weight of aspirin is $9(12.0) + 8(1.0) + 4(16.0) = 180.0$ amu, which gives us the conversion factor 1 mol aspirin = 180.0 g aspirin. To convert moles of aspirin to molecules of aspirin, we use the conversion factor 1 mole aspirin = 6.02×10^{23} molecules aspirin.

Solution

First, we need to find out how many moles of aspirin are in 0.360 g:

$$0.360 \text{ g aspirin} \times \frac{1 \text{ mol aspirin}}{180.0 \text{ g aspirin}} = 0.00200 \text{ mol aspirin}$$

The number of molecules of aspirin in a tablet is:

$$0.00200 \text{ mol} \times 6.02 \times 10^{23} \frac{\text{molecules}}{\text{mol}} = 1.20 \times 10^{21} \text{ molecules}$$

Problem 4-11

How many molecules of water, H_2O, are in a glass of water (235 g)?

4-7 How Do We Calculate Mass Relationships in Chemical Reactions?

A. Stoichiometry

As we saw in Section 4-2, a balanced chemical equation tells us not only which substances react and which are formed, but also the molar ratios in which they react. For example, using the molar ratios in a balanced chemical equation, we can calculate the mass of starting materials needed to produce a particular mass of a product. The quantitative relationship between the amounts of reactants consumed and products formed in chemical reactions as expressed by a balanced chemical equation is called **stoichiometry**.

In Section 4-2, we saw that the coefficients in an equation represent numbers of molecules. Because moles are proportional to molecules (Section 4-6), the coefficients in an equation also represent numbers of moles. Let us look once again at the balanced equation for the combustion of propane:

$$C_3H_8(g) + 5O_2(g) \longrightarrow 3CO_2(g) + 4H_2O(g)$$
Propane

Stoichiometry The quantitative relationship between reactants and products in a chemical reaction as expressed by a balanced chemical equation

This equation tells us that propane and oxygen are converted to carbon dioxide and water and that 1 mol of propane combines with 5 mol of oxygen to produce 3 mol of carbon dioxide and 4 mol of water; that is, we know the molar ratios involved. The same is true for any other balanced equation. This fact allows us to answer questions such as the following:

1. How many moles of any particular product are formed if we start with a given mass of a starting material?
2. How many grams (or moles) of one starting material are necessary to react completely with a given number of grams (or moles) of another starting material?
3. How many grams (or moles) of starting material are needed if we want to form a certain number of grams (or moles) of a certain product?
4. How many grams (or moles) of a product are obtained when a certain amount of another product is formed?

It might seem as if we have four different types of problems here. In fact, we can solve them all by the simple procedure summarized in the following diagram:

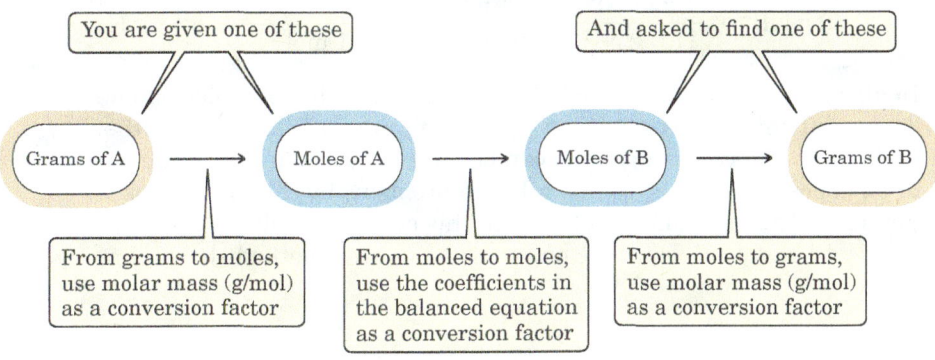

You will always need a conversion factor that relates moles to moles. You will also need the conversion factors for grams to moles and from moles to grams according to the way the problem is asked; you may need one or both in some problems and not in others. It is easy to weigh a given number of grams, but the molar ratio determines the amount of substance involved in a reaction.

Interactive Example 4-12   Stoichiometry

Ammonia is produced on an industrial scale by the reaction of nitrogen gas with hydrogen gas (the Haber process) according to this balanced equation:

$$N_2(g) + 3H_2(g) \longrightarrow 2NH_3(g)$$
$$\text{Ammonia}$$

How many grams of N_2 are necessary to produce 7.50 g of NH_3?

Strategy

The coefficients in an equation refer to the relative numbers of moles, not grams. Therefore, we must first find out how many moles of NH_3 are in 7.50 g of NH_3. To convert grams of NH_3 to moles of NH_3, we use the conversion factor 17.0 g NH_3 = 1 mol NH_3. We see from the balanced chemical equation that 2 mol NH_3 are produced from 1 mol of N_2, which gives us the conversion factor 2 mol NH_3 = 1.0 mol N_2. Finally, we convert moles of N_2 to grams of N_2, using the conversion factor 1 mol N_2 = 28.0 g N_2. Thus, solving this example requires three steps and three conversion factors.

Solution

Step 1: Convert 7.50 grams of NH_3 to moles of NH_3.

$$7.50 \text{ g } NH_3 \times \frac{1 \text{ mol } NH_3}{17.0 \text{ g } NH_3} = 0.441 \text{ mol } NH_3$$

Step 2: Convert moles of NH_3 to moles of N_2.

$$0.441 \text{ mol } NH_3 \times \frac{1 \text{ mol } N_2}{2 \text{ mol } NH_3} = 0.221 \text{ mol } N_2$$

Step 3: Convert moles N_2 to grams of N_2.

$$0.221 \text{ mol } N_2 \times \frac{28.0 \text{ g } N_2}{1 \text{ mol } N_2} = 6.18 \text{ g } N_2$$

Alternatively, one could perform the calculations in one continuous step.

$$7.50 \text{ g } NH_3 \times \frac{1 \text{ mol } NH_3}{17.0 \text{ g } NH_3} \times \frac{1 \text{ mol } N_2}{2 \text{ mol } NH_3} \times \frac{28.0 \text{ g } N_2}{1 \text{ mol } N_2} = 6.18 \text{ g } N_2$$

In all such problems, we are given a mass (or number of moles) of one compound and asked to find the mass (or number of moles) of another compound. The two compounds can be on the same side of the equation or on opposite sides. We can do all such problems by the three steps we just used. The remaining problems in this chapter involving multiple steps will be solved using the one continuous step method shown directly above.

Problem 4-12

Pure aluminum is prepared by the electrolysis of aluminum oxide according to this equation:

$$Al_2O_3(s) \xrightarrow{Electrolysis} Al(s) + O_2(g)$$
Aluminum oxide

(a) Balance this equation.
(b) What mass of aluminum oxide is required to prepare 27 g (1 mol) of aluminum?

Interactive Example 4-13 — Stoichiometry

Silicon to be used in computer chips is manufactured in a process represented by the following reaction:

$$SiCl_4(s) + 2Mg(s) \longrightarrow Si(s) + 2MgCl_2(s)$$
Silicon tetrachloride · · · · · · · · · Magnesium chloride

A sample of 225 g of silicon tetrachloride, $SiCl_4$, is reacted with an excess (more than necessary) of Mg. How many moles of Si are produced?

Strategy

To solve this example, we first convert grams of $SiCl_4$ to moles of $SiCl_4$, then moles of $SiCl_4$ to moles of Si.

Solution

Step 1: First, we convert grams of $SiCl_4$ to moles of $SiCl_4$. For this calculation, we use the conversion factor 1 mol $SiCl_4$ = 170 g $SiCl_4$.

Step 2: To convert moles of $SiCl_4$ to moles of Si, use the conversion factor 1 mol $SiCl_4$ = 1 mol Si, which we obtain from the balanced chemical equation. Now we do the arithmetic and obtain the answer:

$$225 \text{ g SiCl}_4 \times \frac{1 \text{ mol SiCl}_4}{170 \text{ g SiCl}_4} \times \frac{1 \text{ mol Si}}{1 \text{ mol SiCl}_4} = 1.32 \text{ mol Si}$$

▶ As microprocessor chips become increasingly smaller, the purity of the silicon becomes more important, because impurities can prevent the circuit from working properly.

Problem 4-13

In the industrial synthesis of acetic acid, methanol is reacted with carbon monoxide. How many moles of CO are required to produce 16.6 mol of acetic acid?

$$CH_3OH(g) + CO(g) \longrightarrow CH_3COOH(\ell)$$
Methanol · · · Carbon monoxide · · · Acetic acid

EXAMPLE 4-14 — Stoichiometry

When urea, $(NH_2)_2CO$, is acted on by the enzyme urease in the presence of water, ammonia and carbon dioxide are produced. Urease, the catalyst, is placed over the reaction arrow.

$$(NH_2)_2CO(aq) + H_2O(\ell) \xrightarrow{Urease} 2NH_3(aq) + CO_2(g)$$
Urea · Ammonia

If excess water is present (more than necessary for the reaction), how many grams each of CO_2 and NH_3 are produced from 0.83 mol of urea?

Strategy

We are given moles of urea and asked for grams of CO_2. First, we use the conversion factor 1 mol urea = 1 mol CO_2 to find the number of moles of CO_2 that will be produced and then convert moles of CO_2 to grams of CO_2. We use the same strategy to find the number of grams of NH_3 produced.

Solution

For grams of CO_2:

Step 1: We first convert moles of urea to moles of carbon dioxide using the conversion factor derived from the balanced chemical equation, 1 mol urea = 1 mol carbon dioxide.

Step 2: Use the conversion factor 1 mol CO_2 = 44 g CO_2 and then do the math to give the answer:

$$0.83 \text{ mol urea} \times \frac{1 \text{ mol } CO_2}{1 \text{ mol urea}} \times \frac{44 \text{ g } CO_2}{1 \text{ mol } CO_2} = 37 \text{ g } CO_2$$

For grams of NH_3:

Steps 1 and 2 combine into one equation, where we follow the same procedure as for CO_2 but use different conversion factors:

$$0.83 \text{ mol urea} \times \frac{2 \text{ mol } NH_3}{1 \text{ mol urea}} \times \frac{17 \text{ g } NH_3}{1 \text{ mol } NH_3} = 28 \text{ g } NH_3$$

Problem 4-14

Ethanol is produced industrially by the reaction of ethylene with water in the presence of an acid catalyst. How many grams of ethanol are produced from 7.24 mol of ethylene? Assume that excess water is present.

$$\underset{\text{Ethylene}}{C_2H_4(g)} + H_2O(\ell) \longrightarrow \underset{\text{Ethanol}}{C_2H_6O(\ell)}$$

B. Limiting Reagents

Frequently, reactants are mixed in molar proportions that differ from those that appear in a balanced equation. It often happens that one reactant is completely used up but one or more other reactants are not all used up. At times, we deliberately choose to have an excess of one reagent over another. As an example, consider an experiment in which NO is prepared by mixing five moles of N_2 with one mole of O_2. Only one mole of N_2 will react, consuming the one mole of O_2. The oxygen is used up completely, and four moles of nitrogen remain. These molar relationships are summarized under the balanced equation:

	$N_2(g)$	+ $O_2(g)$	$\longrightarrow$ 2NO(g)
Before reaction (moles)	5.0	1.0	0
After reaction (moles)	4.0	0	2.0

Limiting reagent The reactant that is consumed, leaving an excess of another reagent or reagents unreacted

The **limiting reagent** is the reactant that is used up first. In this example, O_2 is the limiting reagent, because it governs how much NO can form. The other reagent, N_2, is in excess.

EXAMPLE 4-15 Limiting Reagent

Suppose 12 g of C is mixed with 64 g of O_2 and the following reaction takes place:

$$C(s) + O_2(g) \longrightarrow CO_2(g)$$

(a) Which reactant is the limiting reagent, and which reactant is in excess?
(b) How many grams of CO_2 will be formed?

Strategy

Determine how many moles of each reactant are present initially. Because C and O_2 react in a 1:1 molar ratio, the reactant present in the smaller molar amount is the limiting reagent and determines how many moles and, therefore, how many grams of CO_2 can be formed.

Solution

(a) We use the molar mass of each reactant to calculate the number of moles of each compound present before reaction.

$$12 \text{ g C} \times \frac{1 \text{ mol C}}{12 \text{ g C}} = 1.0 \text{ mol C}$$

$$64 \text{ g O}_2 \times \frac{1 \text{ mol O}_2}{32 \text{ g O}_2} = 2.0 \text{ mol O}_2$$

According to the balanced equation, reaction of one mole of C requires one mole of O_2. But two moles of O_2 are present at the start of the reaction. Therefore, C is the limiting reagent and O_2 is in excess.

(b) To calculate the number of grams of CO_2 formed, we use the conversion factor 1 mol CO_2 = 44 g CO_2.

$$12 \text{ g C} \times \frac{1 \text{ mol C}}{12 \text{ g C}} \times \frac{1 \text{ mol CO}_2}{1 \text{ mol C}} \times \frac{44 \text{ g CO}_2}{1 \text{ mol CO}_2} = 44 \text{ g CO}_2$$

We can summarize these numbers in the following table. Note that as required by the law of conservation of mass, the sum of the masses of the material present after reaction is the same as the amount present before any reaction took place, namely 76 g of material.

	C	+	O_2	$\longrightarrow$	CO_2
Before reaction	12 g		64 g		0
Before reaction	1.0 mol		2.0 mol		0
After reaction	0		1.0 mol		1.0 mol
After reaction	0		32.0 g		44.0 g

Problem 4-15

Assume that 6.0 g of C and 2.1 g of H_2 are mixed and react to form methane according to the following balanced equation:

$$C(s) + 2H_2(g) \longrightarrow CH_4(g)$$
$$\text{Methane}$$

(a) Which is the limiting reagent, and which reactant is in excess?
(b) How many grams of CH_4 are produced in the reaction?

C. Percent Yield

When carrying out a chemical reaction, we often get less of a product than we might expect from the type of calculation we discussed earlier in this section. For example, suppose we react 32.0 g (1 mol) of CH_3OH with excess CO to form acetic acid:

$$CH_3OH + CO \longrightarrow CH_3COOH$$
Methanol   Carbon monoxide   Acetic acid

If we calculate the expected yield based on the stoichiometry of the balanced equation, we find that we should get 1 mol (60.0 g) of acetic acid. Suppose we get only 57.8 g of acetic acid. Does this result mean that the law of conservation of mass is being violated? No, it does not. We get less than 60.0 g of acetic acid because some of the CH_3OH does not react or because some of it reacts in another way or perhaps because our laboratory technique is not perfect and we lose a little in transferring it from one container to another.

At this point, we need to define three terms, all of which relate to yield of product in a chemical reaction:

Actual yield The mass of product actually formed or isolated in a chemical reaction.

Theoretical yield The mass of product that should form in a chemical reaction according to the stoichiometry of the balanced equation.

Percent yield The actual yield divided by the theoretical yield times 100.

$$\text{Percent yield} = \frac{\text{actual yield}}{\text{theoretical yield}} \times 100\%$$

We summarize the data for the preceding preparation of acetic acid in the following table:

	CH_3OH	+	CO	$\longrightarrow$	CH_3COOH
Before reaction	32.0 g		Excess		0
Before reaction	1.00 mol		Excess		0
Theoretical yield					1.00 mol
Theoretical yield					60.0 g
Actual yield					57.8 g

We calculate the percent yield in this experiment as follows:

$$\text{Percent yield} = \frac{57.8 \text{ acetic acid}}{60.0 \text{ g acetic acid}} \times 100\% = 96.3\%$$

Occasionally, the percent yield is greater than 100%. For example, if a chemist fails to dry a product completely before weighing it, the product weighs more than it should because it also contains water. In such cases, the actual yield may be larger than expected and the percent yield may be greater than 100%.

EXAMPLE 4-16   Percent Yield

In an experiment forming ethanol, the theoretical yield is 50.5 g. The actual yield is 46.8 g. What is the percent yield?

Strategy

Percent yield is the actual yield divided by the theoretical yield times 100.

Solution

$$\% \text{ Yield} = \frac{46.8 \text{ g}}{50.5 \text{ g}} \times 100\% = 92.7\%$$

Problem 4-16

In an experiment to prepare aspirin, the theoretical yield is 153.7 g. If the actual yield is 124.3 g, what is the percent yield?

Why is it important to know the percent yield of a chemical reaction or a series of reactions? The most important reason often relates to cost. If the yield of commercial product is, say, only 10%, the chemists will probably be sent back to the lab to vary experimental conditions in an attempt to improve the yield. As an example, consider a reaction in which starting material A is converted first to compound B, then to compound C, and finally to compound D.

$$A \longrightarrow B \longrightarrow C \longrightarrow D$$

Suppose the yield is 50% at each step. In this case, the yield of compound D is 13% based on the mass of compound A. If, however, the yield at each step is 90%, the yield of compound D increases to 73%; and if the yield at each step is 99%, the yield of compound D is 97%. These numbers are summarized in the following table.

If the Percent Yield per Step Is	The Percent Yield of Compound D Is
50%	0.50 × 0.50 × 0.50 × 100 = 13%
90%	0.90 × 0.90 × 0.90 × 100 = 73%
99%	0.99 × 0.99 × 0.99 × 100 = 97%

4-8 What Is Heat of Reaction?

In almost all chemical reactions, not only are starting materials converted to products, but heat is also either given off or absorbed. For example, when one mole of carbon is oxidized by oxygen to produce one mole of CO_2, 94.0 kilocalories of heat is given off per mole of carbon:

$$C(s) + O_2(g) \longrightarrow CO_2(g) + 94.0 \text{ kcal}$$

The heat given off or gained in a reaction is called the **heat of reaction**. A reaction that gives off heat is **exothermic**; a reaction that absorbs heat is **endothermic**. The amount of heat given off or absorbed is proportional to the amount of material. For example, when 2 mol of carbon is oxidized by oxygen to give carbon dioxide, 2 × 94.0 = 188 kcal of heat is given off.

The energy changes accompanying a chemical reaction are not limited to heat. In some reactions, such as in voltaic cells (Chemical Connections 4B), the energy given off takes the form of electricity. In other reactions, such as photosynthesis (the reaction whereby plants convert water and carbon dioxide to carbohydrates and oxygen), the energy absorbed is in the form of light.

An example of an endothermic reaction is the decomposition of mercury(II) oxide: ◀

$$2HgO(s) + 43.4 \text{ kcal} \longrightarrow 2Hg(\ell) + O_2(g)$$

Mercury(II) oxide
(Mercuric oxide)

Heat of reaction The heat given off or absorbed in a chemical reaction

Exothermic A chemical reaction that gives off heat

Endothermic A chemical reaction that absorbs heat

▶ Mercury (II) oxide, a red compound, decomposes into two elements when heated: mercury (a metal) and oxygen (a nonmetal). Mercury vapor condenses on the cooler upper portion of the test tube.

This equation tells us that if we want to decompose 2 mol of mercury(II) oxide into the elements Hg(ℓ) and O_2(g), we must add 43.4 kcal of energy to HgO. Incidentally, the law of conservation of energy tells us that the reverse reaction, the oxidation of mercury, must give off exactly the same amount of heat:

$$2Hg(\ell) + O_2(g) \longrightarrow 2HgO(s) + 43.4 \text{ kcal}$$

Especially important are the heats of reaction for combustion reactions. As we saw in Section 4-4, combustion reactions are the most important heat-producing reactions, because most of the energy required for modern society to function is derived from them. All combustions are exothermic. The heat given off in a combustion reaction is called the **heat of combustion**.

Summary of Key Questions

Section 4-2 How Do We Balance Chemical Equations?
- A **chemical equation** is an expression showing which reactants are converted to which products. A balanced chemical equation shows how many moles of each starting material are converted to how many moles of each product according to the law of conservation of mass.

Section 4-3 How Can We Predict If Ions in Aqueous Solution Will React with Each Other?
- When ions are mixed in aqueous solution, they react with one another only if (1) a precipitate forms, (2) a gas forms, (3) an acid neutralizes a base, or (4) an oxidation-reduction takes place.
- Ions that do not react are called **spectator ions**.
- A **net ionic equation** shows only those ions that react. In a net ionic equation, both the charges and the number (mass) of atoms must be balanced.

Section 4-4 What Are Oxidation and Reduction?
- **Oxidation** is the loss of electrons; **reduction** is the gain of electrons. These two processes must take place together; you cannot have one without the other. The joint process is often called a redox reaction.
- Oxidation can also be defined as the gain of oxygens and/or the loss of hydrogens; reduction can also be defined as the loss of oxygens and/or the gain of hydrogens.

Section 4-5 What Are Formula Weights and Molecular Weights?
- The **formula weight (FW)** of a compound is the sum of the atomic weights of all atoms in the compound expressed in atomic mass units (amu). Formula weight applies to both ionic and molecular compounds.
- The term **molecular weight**, also expressed in amu, applies to only molecular compounds.

Section 4-6 What Is a Mole and How Do We Use It to Calculate Mass Relationships?
- A **mole (mol)** of any substance is defined as Avogadro's number (6.02×10^{23}) of formula units of the substance.
- The **molar mass** of a substance is its formula weight expressed in grams.

Section 4-7 How Do We Calculate Mass Relationships in Chemical Reactions?
- **Stoichiometry** is the study of the mass relationships in chemical reactions.
- The reagent that is used up first in a reaction is called the **limiting reagent**.
- The **percent yield** for a reaction equals the **actual yield** divided by the **theoretical yield** multiplied by 100.

Section 4-8 What Is Heat of Reaction?
- Almost all chemical reactions are accompanied by either a gain or a loss of heat. This heat is called the heat of reaction.
- Reactions that give off heat are exothermic; those that absorb heat are endothermic.
- The heat given off in a combustion reaction is called the **heat of combustion**.

Problems

Orange-numbered problems are applied.

Section 4-2 How Do We Balance Chemical Equations?

4-17 Balance each equation.
(a) $HI + NaOH \longrightarrow NaI + H_2O$
(b) $Ba(NO_3)_2 + H_2S \longrightarrow BaS + HNO_3$
(c) $CH_4 + O_2 \longrightarrow CO_2 + H_2O$
(d) $C_4H_{10} + O_2 \longrightarrow CO_2 + H_2O$
(e) $Fe + CO_2 \longrightarrow Fe_2O_3 + CO$

4-18 Balance each equation.
(a) $H_2 + I_2 \longrightarrow HI$
(b) $Al + O_2 \longrightarrow Al_2O_3$
(c) $Na + Cl_2 \longrightarrow NaCl$
(d) $Al + HBr \longrightarrow AlBr_3 + H_2$
(e) $P + O_2 \longrightarrow P_2O_5$

4-19 If you blow carbon dioxide gas into a solution of calcium hydroxide, a milky-white precipitate of calcium carbonate forms. Write a balanced equation for the formation of calcium carbonate in this reaction.

4-20 Calcium oxide is prepared by heating limestone (calcium carbonate, $CaCO_3$) to a high temperature, at which point it decomposes to calcium oxide and carbon dioxide. Write a balanced equation for this preparation of calcium oxide.

4-21 The brilliant white light in some firework displays is produced by burning magnesium in air. The magnesium reacts with oxygen in the air to form magnesium oxide. Write a balanced equation for this reaction.

4-22 The rusting of iron is a chemical reaction of iron with oxygen in the air to form iron(III) oxide. Write a balanced equation for this reaction.

4-23 When solid carbon burns in a limited supply of oxygen gas, the gas carbon monoxide, CO, forms. This gas is deadly to humans because it combines with hemoglobin in the blood, making it impossible for the blood to transport oxygen. Write a balanced equation for the formation of carbon monoxide.

4-24 Solid ammonium carbonate, $(NH_4)_2CO_3$, decomposes at room temperature to form gaseous ammonia, carbon dioxide, and water. Because of the ease of decomposition and the penetrating odor of ammonia, ammonium carbonate can be used as smelling salts. Write a balanced equation for this decomposition.

4-25 In the chemical test for arsenic, the gas arsine, AsH_3, is prepared. When arsine is decomposed by heating, arsenic metal deposits as a mirror-like coating on the surface of a glass container and hydrogen gas, H_2, is given off. Write a balanced equation for the decomposition of arsine.

4-26 When a piece of aluminum metal is dropped into hydrochloric acid, HCl, hydrogen is released as a gas and a solution of aluminum chloride forms. Write a balanced equation for the reaction.

4-27 In the industrial chemical preparation of chlorine, Cl_2, electric current is passed through an aqueous solution of sodium chloride to give $Cl_2(g)$ and $H_2(g)$. The other product of this reaction is sodium hydroxide. Write a balanced equation for this reaction.

Section 4-3 How Can We Predict If Ions in Aqueous Solution Will React with Each Other?

4-28 Answer true or false.
(a) A net ionic equation shows only those ions that undergo chemical reaction.
(b) In a net ionic equation, the number of moles of starting material must equal the number of moles of product.
(c) A net ionic equation must be balanced by both mass and charge.
(d) As a generalization, all lithium, sodium, and potassium salts are soluble in water.
(e) As a generalization, all nitrate (NO_3^-) salts are soluble in water.
(f) As a generalization, most carbonate (CO_3^{2-}) salts are insoluble in water.
(g) Sodium carbonate, Na_2CO_3, is insoluble in water.
(h) Ammonium carbonate, $(NH_4)_2CO_3$, is insoluble in water.
(i) Calcium carbonate, $CaCO_3$, is insoluble in water.
(j) Sodium dihydrogen phosphate, NaH_2PO_4, is insoluble in water.
(k) Sodium hydroxide, NaOH, is soluble in water.
(l) Barium hydroxide, $Ba(OH)_2$, is soluble in water.

4-29 Balance these net ionic equations.
(a) $Ag^+(aq) + Br^-(aq) \longrightarrow AgBr(s)$
(b) $Cd^{2+}(aq) + S^{2-}(aq) \longrightarrow CdS(s)$
(c) $Sc^{3+}(aq) + SO_4^{2-}(aq) \longrightarrow Sc_2(SO_4)_3(s)$
(d) $Sn^{2+}(aq) + Fe^{2+}(aq) \longrightarrow Sn(s) + Fe^{3+}(aq)$
(e) $K(s) + H_2O(\ell) \longrightarrow K^+(aq) + OH^-(aq) + H_2(g)$

4-30 In the equation

$$2Na^+(aq) + CO_3^{2-}(aq) + Sr^{2+}(aq) + 2Cl^-(aq) \longrightarrow SrCO_3(s) + 2Na^+(aq) + 2Cl^-(aq)$$

(a) Identify the spectator ions.
(b) Write the balanced net ionic equation.

4-31 Predict whether a precipitate will form when aqueous solutions of the following compounds are mixed. If a precipitate will form, write its formula and write a net ionic equation for its formation. To make your predictions, use the solubility generalizations in Section 4-3.
(a) $CaCl_2(aq) + K_3PO_4(aq) \longrightarrow$
(b) $KCl(aq) + Na_2SO_4(aq) \longrightarrow$
(c) $(NH_4)_2CO_3(aq) + Ba(NO_3)_2(aq) \longrightarrow$
(d) $FeCl_2(aq) + KOH(aq) \longrightarrow$
(e) $Ba(NO_3)_2(aq) + NaOH(aq) \longrightarrow$
(f) $Na_2S(aq) + SbCl_3(aq) \longrightarrow$
(g) $Pb(NO_3)_2(aq) + K_2SO_4(aq) \longrightarrow$

4-32 When a solution of ammonium chloride is added to a solution of lead(II) nitrate, $Pb(NO_3)_2$, a white precipitate, lead(II) chloride, forms. Write a balanced net ionic equation for this reaction. Both ammonium chloride and lead nitrate exist as dissociated ions in aqueous solution.

4-33 When a solution of hydrochloric acid, HCl, is added to a solution of sodium sulfite, Na_2SO_3, sulfur dioxide gas is released from the solution. Write a net ionic equation for this reaction. An aqueous solution of HCl contains H^+ and Cl^- ions, and Na_2SO_3 exists as dissociated ions in aqueous solution.

4-34 When a solution of sodium hydroxide is added to a solution of ammonium carbonate, H_2O is formed and ammonia gas, NH_3, is released when the solution is heated. Write a net ionic equation for this reaction. Both NaOH and $(NH_4)_2CO_3$ exist as dissociated ions in aqueous solution.

4-35 Using the solubility generalizations given in Section 4-3, predict which of these ionic compounds are soluble in water.
(a) KCl (b) NaOH (c) $BaSO_4$
(d) Na_2SO_4 (e) Na_2CO_3 (f) $Fe(OH)_2$

4-36 Using the solubility generalizations given in Section 4-3, predict which of these ionic compounds are soluble in water.
(a) $MgCl_2$ (b) $CaCO_3$ (c) Na_2SO_3
(d) NH_4NO_3 (e) $Pb(OH)_2$

Section 4-4 What Are Oxidation and Reduction?

4-37 Answer true or false.
(a) When a substance is oxidized, it loses electrons.
(b) When a substance gains electrons, it is reduced.
(c) In a redox reaction, the oxidizing agent becomes reduced.
(d) In a redox reaction, the reducing reagent becomes oxidized.
(e) When Zn is converted to Zn^{2+} ion, zinc is oxidized.
(f) Oxidation can also be defined as the loss of oxygen atoms and/or the gain of hydrogen atoms.
(g) Reduction can also be defined as the gain of oxygen atoms and/or the loss of hydrogen atoms.
(h) When oxygen, O_2, is converted to hydrogen peroxide, H_2O_2, we say that O_2 is reduced.
(i) Hydrogen peroxide, H_2O_2, is an oxidizing agent.
(j) All combustion reactions are redox reactions.
(k) The products of complete combustion (oxidation) of hydrocarbon fuels are carbon dioxide, water, and heat.
(l) In the combustion of hydrocarbon fuels, oxygen is the oxidizing agent and the hydrocarbon fuel is the reducing agent.
(m) Incomplete combustion of hydrocarbon fuels can produce significant amounts of carbon monoxide.
(n) Most common bleaches are oxidizing agents.

4-38 In the reaction

$$Pb(s) + 2Ag^+(aq) \longrightarrow Pb^{2+}(aq) + 2Ag(s)$$

(a) Which species is oxidized and which is reduced?
(b) Which species is the oxidizing agent and which is the reducing agent?

4-39 In the reaction

$$C_7H_{12}(\ell) + 10O_2(g) \longrightarrow 7CO_2(g) + 6H_2O(\ell)$$

(a) Which species is oxidized and which is reduced?
(b) Which species is the oxidizing agent and which is the reducing agent?

4-40 When a piece of sodium metal is added to water, hydrogen is evolved as a gas and a solution of sodium hydroxide is formed.
(a) Write a balanced equation for this reaction.
(b) What is oxidized in this reaction? What is reduced?

Section 4-5 What Are Formula Weights and Molecular Weights?

4-41 Answer true or false.
(a) Formula weight is the mass of a compound expressed in grams.
(b) 1 atomic mass unit (amu) is equal to 1 gram (g).
(c) The formula weight of H_2O is 18 amu.
(d) The molecular weight of H_2O is 18 amu.
(e) The molecular weight of a covalent compound is the same as its formula weight.

4-42 Calculate the formula weight of:
(a) KCl (b) Na_3PO_4 (c) $Fe(OH)_2$
(d) $NaAl(SO_3)_2$ (e) $Al_2(SO_4)_3$ (f) $(NH_4)_2CO_3$

4-43 Calculate the molecular weight of:
(a) Sucrose, $C_{12}H_{22}O_{11}$ (b) Glycine, $C_2H_5NO_2$
(c) DDT, $C_{14}H_9Cl_5$

Section 4-6 What Is a Mole and How Do We Use It to Calculate Mass Relationships?

4-44 Answer true or false.
(a) The mole is a counting unit, just as a dozen is a counting unit.
(b) Avogadro's number is the number of formula units in one mole.
(c) Avogadro's number, to three significant figures, is 6.02×10^{23} formula units per mole.
(d) 1 mol of H_2O contains $3 \times 6.02 \times 10^{23}$ formula units.
(e) 1 mol of H_2O has the same number of molecules as 1 mol of H_2O_2.
(f) The molar mass of a compound is its formula weight expressed in amu.
(g) The molar mass of H_2O is 18 g/mol.
(h) 1 mol of H_2O has the same molar mass as 1 mol of H_2O_2.
(i) 1 mol of ibuprofen, $C_{13}H_{18}O_2$, contains 33 mol of atoms.
(j) To convert moles to grams, multiply by Avogadro's number.
(k) To convert grams to moles, divide by molar mass.
(l) 1 mol of H_2O contains 1 mol of hydrogen atoms and one mol of oxygen atoms.

(m) 1 mol of H_2O contains 2 g of hydrogen atoms and 1 g of oxygen atoms.
(n) 1 mole of H_2O contains 18.06×10^{23} atoms.

4-45 Calculate the number of moles in:
(a) 32 g of methane, CH_4
(b) 345.6 g of nitric oxide, NO
(c) 184.4 g of chlorine dioxide, ClO_2
(d) 720. g of glycerin, $C_3H_8O_3$

4-46 Calculate the number of grams in:
(a) 1.77 mol of nitrogen dioxide, NO_2
(b) 0.84 mol of 2-propanol, C_3H_8O (rubbing alcohol)
(c) 3.69 mol of uranium hexafluoride, UF_6
(d) 0.348 mol of galactose, $C_6H_{12}O_6$
(e) 4.9×10^{-2} mol of vitamin C, $C_6H_8O_6$

4-47 Calculate the number of moles of:
(a) O atoms in 18.1 mol of formaldehyde, CH_2O
(b) Br atoms in 0.41 mol of bromoform, $CHBr_3$
(c) O atoms in 3.5×10^3 mol of $Al_2(SO_4)_3$
(d) Hg atoms in 87 g of HgO

4-48 Calculate the number of moles of:
(a) S^{2-} ions in 6.56 mol of Na_2S
(b) Mg^{2+} ions in 8.320 mol of $Mg_3(PO_4)_2$
(c) acetate ions, CH_3COO^-, in 0.43 mol of $Ca(CH_3COO)_2$

4-49 Calculate the number of:
(a) nitrogen atoms in 25.0 g of TNT, $C_7H_5N_3O_6$
(b) carbon atoms in 40.0 g of ethanol, C_2H_6O
(c) oxygen atoms in 500. mg of aspirin, $C_9H_8O_4$
(d) sodium atoms in 2.40 g of sodium dihydrogen phosphate, NaH_2PO_4

4-50 How many molecules are in each of the following?
(a) 2.9 mol of TNT, $C_7H_5N_3O_6$
(b) one drop (0.0500 g) of water
(c) 3.1×10^{-1} g of aspirin, $C_9H_8O_4$

4-51 What is the mass in grams of each number of molecules of formaldehyde, CH_2O?
(a) 100. molecules (b) 3000. molecules
(c) 5.0×10^6 molecules (d) 2.0×10^{24} molecules

4-52 The molecular weight of hemoglobin is about 68,000 amu. What is the mass in grams of a single molecule of hemoglobin?

4-53 A typical deposit of cholesterol, $C_{27}H_{46}O$, in an artery might have a mass of 3.9 mg. How many molecules of cholesterol are in this mass?

Section 4-7 How Do We Calculate Mass Relationships in Chemical Reactions?

4-54 Answer true or false.
(a) Stoichiometry is the study of mass relationships in chemical reactions.
(b) To determine mass relationships in a chemical reaction, you first need to know the balanced chemical equation for the reaction.
(c) To convert from grams to moles and vice versa, use Avogadro's number as a conversion factor.
(d) To convert from grams to moles and vice versa, use molar mass as a conversion factor.
(e) A limiting reagent is the reagent that is used up first.
(f) Suppose a chemical reaction between A and B requires 1 mol of A and 2 mol of B. If 1 mol of each is present, then B is the limiting reagent.
(g) Theoretical yield is the yield of product that should be obtained according to the balanced chemical equation.
(h) Theoretical yield is the yield of product that should be obtained if all limiting reagent is converted to product.
(i) Percent yield is the number of grams of product divided by the number of grams of the limiting reagent times 100.
(j) To calculate percent yield, divide the mass of product formed by the theoretical yield and multiply by 100.

4-55 For the reaction:
$$2N_2(g) + 3O_2(g) \longrightarrow 2N_2O_3(g)$$
(a) How many moles of N_2 are required to react completely with 1 mole of O_2?
(b) How many moles of N_2O_3 are produced from the complete reaction of 1 mole of O_2?
(c) How many moles of O_2 are required to produce 8 moles of N_2O_3?

4-56 Magnesium reacts with sulfuric acid according to the following equation. How many moles of H_2 are produced by the complete reaction of 230. mg of Mg with sulfuric acid?
$$Mg(s) + H_2SO_4(aq) \longrightarrow MgSO_4(aq) + H_2(g)$$

4-57 Chloroform, $CHCl_3$, is prepared industrially by the reaction of methane with chlorine. How many grams of Cl_2 are needed to produce 1.50 moles of chloroform?
$$CH_4(g) + 3Cl_2(g) \longrightarrow CHCl_3(\ell) + 3HCl(g)$$
Methane   Chloroform

4-58 At one time, acetaldehyde was prepared industrially by the reaction of ethylene with air in the presence of a copper catalyst. How many grams of acetaldehyde can be prepared from 81.7 g of ethylene?
$$2C_2H_4(g) + O_2(g) \xrightarrow{\text{Catalyst}} 2C_2H_4O(g)$$
Ethylene   Acetaldehyde

4-59 Chlorine dioxide, ClO_2, is used for bleaching paper. It is also the gas used to kill the anthrax spores that contaminated the Hart Senate Office Building in the fall of 2001. Chlorine dioxide is prepared by treating sodium chlorite with chlorine gas.
$$NaClO_2(aq) + Cl_2(g) \longrightarrow ClO_2(g) + NaCl(aq)$$
Sodium   Chlorine
chlorite   dioxide

(a) Balance the equation for the preparation of chlorine dioxide.

(b) Calculate the weight of chlorine dioxide that can be prepared from 5.50 kg of sodium chlorite.

4-60 Ethanol, C_2H_6O, is added to gasoline to produce "gasohol," a fuel for automobile engines. How many grams of O_2 are required for complete combustion of 421 g of ethanol?

$$C_2H_5OH(\ell) + 3O_2(g) \longrightarrow 2CO_2(g) + 3H_2O$$
Ethanol

4-61 In photosynthesis, green plants convert CO_2 and H_2O to glucose, $C_6H_{12}O_6$. How many grams of CO_2 are required to produce 5.1 g of glucose?

$$6CO_2(g) + 6H_2O(\ell) \xrightarrow{Photosynthesis} C_6H_{12}O_6(aq) + 6O_2(g)$$
Glucose

4-62 Iron ore is converted to iron by heating it with coal (carbon), and oxygen according to the following equation:

$$2Fe_2O_3(s) + 6C(s) + 3O_2(g) \longrightarrow 4Fe(s) + 6CO_2(g)$$

If the process is run until 3940. g of Fe is produced, how many grams of CO_2 will also be produced?

4-63 Given the reaction in Problem 4-62, how many grams of C are necessary to react completely with 0.58 g of Fe_2O_3?

4-64 Aspirin is made by the reaction of salicylic acid with acetic anhydride. How many grams of aspirin are produced if 85.0 g of salicylic acid is treated with excess acetic anhydride?

(C_7H_6O_3) Salicyclic acid (s) + (C_4H_6O_3) Acetic anhydride (ℓ) → (C_9H_8O_4) Aspirin (s) + CH_3COOH (C_2H_4O_2) Acetic acid (ℓ)

4-65 Suppose the preparation of aspirin from salicylic acid and acetic anhydride (Problem 4-64) gives a yield of 75.0% of aspirin. How many grams of salicylic acid must be used to prepare 50.0 g of aspirin?

4-66 Benzene reacts with bromine to produce bromobenzene according to the following equation:

$$C_6H_6(\ell) + Br_2(\ell) \longrightarrow C_6H_5Br(\ell) + HBr(g)$$
Benzene   Bromine   Bromobenzene   Hydrogen bromide

If 60.0 g of benzene is mixed with 135 g of bromine,
(a) Which is the limiting reagent?
(b) How many grams of bromobenzene are formed in the reaction?

4-67 Ethyl chloride is prepared by the reaction of chlorine with ethane according to the following balanced equation.

$$C_2H_6(g) + Cl_2(g) \longrightarrow C_2H_5Cl(\ell) + HCl(g)$$
Ethane   Ethyl chloride

When 5.6 g of ethane is reacted with excess chlorine, 8.2 g of ethyl chloride forms. Calculate the percent yield of ethyl chloride.

4-68 Diethyl ether is made from ethanol according to the following reaction:

$$2C_2H_5OH(\ell) \longrightarrow (C_2H_5)_2O(\ell) + H_2O(\ell)$$
Ethanol   Diethyl ether

In an experiment, 517 g of ethanol gave 391 g of diethyl ether. What was the percent yield in this experiment?

Section 4-8 What Is Heat of Reaction?

4-69 Answer true or false.
(a) Heat of reaction is the heat given off or absorbed by a chemical reaction.
(b) An endothermic reaction is one that gives off heat.
(c) If a chemical reaction is endothermic, the reverse reaction is exothermic.
(d) All combustion reactions are exothermic.
(e) If the reaction of glucose ($C_6H_{12}O_6$) and O_2 in the body to give CO_2 and H_2O is an exothermic reaction, then photosynthesis in green plants (the reaction of CO_2 and H_2O to give glucose and O_2) is an endothermic process.
(f) The energy required to drive photosynthesis comes from the sun in the form of electromagnetic radiation.

4-70 What is the difference between exothermic and endothermic reactions?

4-71 Which of these reactions are exothermic, and which are endothermic?
(a) $2NH_3(g) + 22.0 \text{ kcal} \longrightarrow N_2(g) + 3H_2(g)$
(b) $H_2(g) + F_2(g) \longrightarrow 2HF(g) + 124 \text{ kcal}$
(c) $C(s) + O_2(g) \longrightarrow CO_2(g) + 94.0 \text{ kcal}$
(d) $H_2(g) + CO_2(g) + 9.80 \text{ kcal} \longrightarrow H_2O(g) + CO(g)$
(e) $C_3H_8(g) + 5O_2(g) \longrightarrow 3CO_2(g) + 4H_2O(g) + 531 \text{ kcal}$

4-72 In the following reaction, 9.80 kcal is absorbed per mole of CO_2 undergoing reaction. How much heat is given off if two moles of water are reacted with two moles of carbon monoxide?

$$H_2(g) + CO_2(g) + 9.80 \text{ kcal} \longrightarrow H_2O(g) + CO(g)$$

4-73 Following is the equation for the combustion of acetone:

$$2C_3H_6O(\ell) + 8O_2(g) \longrightarrow 6CO_2(g) +$$
Acetone
$$6H_2O(g) + 853.6 \text{ kcal}$$

How much heat will be given off if 0.37 mol of acetone is burned completely?

4-74 The oxidation of glucose, $C_6H_{12}O_6$, to carbon dioxide and water is exothermic. The heat liberated is the same whether glucose is metabolized in the body or burned in air.

$$C_6H_{12}O_6 + 6O_2 \longrightarrow 6CO_2 + 6H_2O + 670 \text{ kcal/mol}$$
Glucose

Calculate the heat liberated when 15.0 g of glucose is metabolized to carbon dioxide and water in the body.

4-75 The heat of combustion of glucose, $C_6H_{12}O_6$, is 670 kcal/mol. The heat of combustion of ethanol, C_2H_6O, is 327 kcal/mol. The heat liberated by oxidation of each compound is the same whether it is burned in air or metabolized in the body. On a kcal/g basis, metabolism of which compound liberates more heat?

4-76 A plant requires approximately 4178 kcal for the production of 1.00 kg of starch (Chapter 20) from carbon dioxide and water.
(a) Is the production of starch in a plant an exothermic process or an endothermic process?
(b) Calculate the energy in kilocalories required by a plant for the production of 6.32 g of starch.

4-77 To convert 1 mol of iron(III) oxide to its elements requires 196.5 kcal:

$$Fe_2O_3(s) + 196.5 \text{ kcal} \longrightarrow 2Fe(s) + \frac{3}{2}O_2(g)$$

How many grams of iron can be produced if 156.0 kcal of heat is absorbed by a large-enough sample of iron(III) oxide?

Chemical Connections

4-78 (Chemical Connections 4A) How does fluoride ion protect the tooth enamel against decay?

4-79 (Chemical Connections 4A) What ions are present in hydroxyapatite?

4-80 (Chemical Connections 4B) A voltaic cell is represented by the following equation:

$$Fe(s) + Zn^{2+}(aq) \longrightarrow Fe^{2+}(aq) + Zn(s)$$

Which electrode is the anode, and which is the cathode?

4-81 (Chemical Connections 4C) Balance the lithium-iodine battery redox reaction described in this section and identify the oxidizing and reducing agents present.

Additional Problems

4-82 When gaseous dinitrogen pentoxide, N_2O_5, is bubbled into water, nitric acid, HNO_3, forms. Write a balanced equation for this reaction.

4-83 In a certain reaction, Cu^+ is converted to Cu^{2+}. Is Cu^+ ion oxidized or reduced in this reaction? Is Cu^+ ion an oxidizing agent or a reducing agent in this reaction?

4-84 Using the equation:

$$Fe_2O_3(s) + 3CO(g) \longrightarrow 2Fe(s) + 3CO_2(g)$$

(a) Show that this is a redox reaction. Which species is oxidized, and which is reduced?
(b) How many moles of Fe_2O_3 are required to produce 38.4 mol of Fe?
(c) How many grams of CO are required to produce 38.4 mol of Fe?

4-85 Methyl tertiary butyl ether (or MTBE), a chemical compound with molecular formula $C_5H_{12}O$, is an additive used as an oxygenate to raise the octane number of gas, although its use has declined in the last few years in response to environmental and health concerns. Write the balanced molecular equation for the reaction involving the complete burning of liquid MTBE in air.

4-86 When an aqueous solution of Na_3PO_4 is added to an aqueous solution of $Cd(NO_3)_2$, a precipitate forms. Write a net ionic equation for this reaction and identify the spectator ions.

4-87 The active ingredient in an analgesic tablet is 488 mg of aspirin, $C_9H_8O_5$. How many moles of aspirin does the tablet contain?

4-88 Chlorophyll, the compound responsible for the green color of leaves and grasses, contains one atom of magnesium in each molecule. If the percentage by weight of magnesium in chlorophyll is 2.72%, what is the molecular weight of chlorophyll?

4-89 If 7.0 kg of N_2 is added to 11.0 kg of H_2 to form NH_3, which reactant is in excess?

$$N_2(g) + 3H_2(g) \longrightarrow 2NH_3(g)$$

4-90 Lead(II) nitrate and aluminum chloride react according to the following equation:

$$3Pb(NO_3)_2 + 2AlCl_3 \longrightarrow 3PbCl_2 + 2Al(NO_3)_3$$

In an experiment, 8.00 g of lead nitrate reacted with 2.67 g of aluminum chloride to give 5.55 g of lead chloride.
(a) Which reactant was the limiting reagent?
(b) What was the percent yield?

4-91 Assume that the average red blood cell has a mass of 2×10^{-8} g and that 20% of its mass is hemoglobin (a protein whose molar mass is 68,000). How many molecules of hemoglobin are present in one red blood cell?

4-92 Reaction of pentane, C_5H_{12}, with oxygen, O_2, gives carbon dioxide and water.
(a) Write a balanced equation for this reaction.
(b) In this reaction, what is oxidized and what is reduced?
(c) What is the oxidizing agent, and what is the reducing agent?

4-93 Ammonia is prepared industrially by the reaction of nitrogen and hydrogen according to the following equation:

$$N_2(g) + 3H_2(g) \longrightarrow 2NH_3(g)$$
Ammonia

If 29.7 kg of N_2 is added to 3.31 kg of H_2,
(a) Which reactant is the limiting reagent?
(b) How many grams of the other reactant are left over?
(c) How many grams of NH_3 are formed if the reaction goes to completion?

Tying It Together

4-94 2,3,7,8-Tetrachlorodibenzo-*p*-dioxin (TCDD) is a potent poison with the chemical formula $C_{12}H_4Cl_4O_2$. The average lethal dose in humans is approximately 2.9×10^{-2} mg per kg of body weight. How many molecules of TCDD constitute a lethal dose for an 82-kg individual?

4-95 Furan, an organic compound used in the synthesis of nylon and referenced in Section 20-2, has the molecular formula C_4H_4O.
(a) Determine the number of moles of furan in a 441 mg sample.
(b) If the density of furan is known to be 0.936 g/mL, how many carbon atoms are present in 0.060 L of furan?
(c) Calculate the mass in grams of 9.86×10^{25} molecules of furan.

4-96 A sample of gold consisting of 8.68×10^{23} atoms with a density of 19.3 g/mL is hammered into a sheet that covers an area of 1.00×10^2 ft^2. Determine the thickness of the sheet in centimeters.

4-97 Consider the production of $KClO_4(aq)$ via the three balanced sequential reactions below, where the percentage yield of each reaction is written above the reaction arrows:

$$Cl_2(g) + 2KOH(aq) \xrightarrow{92.1\%} KCl(aq) + KClO(aq) + H_2O(\ell)$$

$$3KClO(aq) \xrightarrow{86.7\%} 2KCl(aq) + KClO_3(aq)$$

$$4KClO_3(aq) \xrightarrow{75.3\%} 3KClO_4(aq) + KCl(aq)$$

Determine the mass in grams of $KClO_4(aq)$ produced at the end of the three-step reaction sequence if a student begins with 966 kg of $Cl_2(g)$.

4-98 Elemental chlorine is commonly used to kill microorganisms in drinking water supplies as well as to remove sulfides. For example, noxious-smelling hydrogen sulfide gas is removed from water via the following unbalanced chemical equation:

$$H_2S(aq) + Cl_2(aq) \longrightarrow HCl(aq) + S_8(s)$$

(a) Write a balanced equation for this reaction.
(b) Determine the mass in grams of elemental sulfur, S_8, which is produced when 50.0 L of water containing 1.5×10^{-5} g of H_2S per liter is treated with 1.0 g of Cl_2.
(c) Calculate the percent yield of the reaction if 5.8×10^{-4} g of S_8 is generated.

Looking Ahead

4-99 The two major sources of energy in our diets are fats and carbohydrates. Palmitic acid, one of the major components of both animal fats and vegetable oils, belongs to a group of compounds called fatty acids. The metabolism of fatty acids is responsible for the energy from fats. The major carbohydrates in our diets are sucrose (table sugar; Section 20-4A) and starch (Section 20-5A). Both starch and sucrose are first converted in the body to glucose, and then glucose is metabolized to produce energy. The heat of combustion of palmitic acid is 2385 kcal/mol, and that of glucose is 670. kcal/mol. Below are unbalanced equations for the metabolism of each body fuel:

$$C_{16}H_{32}O_2(aq) + O_2(g) \longrightarrow$$
Palmitic acid
(256 g/mol)

$$CO_2(g) + H_2O(\ell) + 2385 \text{ kcal/mol}$$

$$C_6H_{12}O_6(aq) + O_2(g) \longrightarrow$$
Glucose
(180. g/mol)

$$CO_2(g) + H_2O(\ell) + 670 \text{ kcal/mol}$$

(a) Balance the equation for the metabolism of each fuel.
(b) Calculate the heat of combustion of each in kcal/g.
(c) In terms of kcal/mol, which of the two is the better source of energy for the body?
(d) In terms of kcal/g, which of the two is the better source of energy for the body?

4-100 The heat of combustion of methane, CH_4, the major component of natural gas, is 213 kcal/mol. The heat of combustion of propane, C_3H_8, the major component of LPG, or bottled gas, is 530. kcal/mol.
(a) Write a balanced equation for the complete combustion of each to CO_2 and H_2O.
(b) On a kcal/mol basis, which of these two fuels is the better source of heat energy?
(c) On a kcal/g basis, which of these two fuels is the better source of heat energy?

Challenge Problems

4-101 An automobile with gasoline consisting of octane, $C_8H_{18}(\ell)$, has a density of 0.69 g/mL. If the automobile travels 168 miles with a gas mileage of 21.2 mi/gal, how many kg of CO_2 are produced assuming complete combustion of octane and excess oxygen?

4-102 Aspartame, an artificial sweetener used as a sugar substitute in some foods and beverages, has the molecular formula $C_{14}H_{18}N_2O_5$.

(a) How many mg of aspartame are present in 3.72×10^{26} molecules of aspartame?

(b) Imagine you obtain 25.0 mL of aspartame, which is known to have a density of 1.35 g/mL. How many molecules of aspartame are present in this volume?

(c) How many hydrogen atoms are present in 1.00 mg of aspartame?

(d) Complete the skeletal structure of aspartame, where all the bonded atoms are shown but double bonds, triple bonds, and/or lone pairs are missing.

Aspartame skeletal structure

(e) Identify the various types of geometries present in each central atom of aspartame using VSEPR theory.

(f) Determine the various relative bond angles associated with each central atom of aspartame using VSEPR theory.

(g) What is the most polar bond in aspartame?

(h) Would you predict aspartame to be polar or nonpolar?

(i) Is aspartame expected to possess resonance? Explain why or why not.

(j) Consider the combustion of aspartame, which results in formation of $NO_2(g)$ as well as other expected products. Write a balanced chemical equation for this reaction.

(k) Calculate the weight of $CO_2(g)$ that can be prepared from 1.62 g of aspartame mixed with 2.11 g of oxygen gas.

4-103 Caffeine, a central nervous system stimulant, has the molecular formula $C_8H_{10}N_4O_2$.

(a) How many moles of caffeine are present in 6.19×10^{25} molecules of caffeine?

(b) Imagine you dissolve caffeine in water to a volume of 100.0 mL, which is known to have a density of 1.23 g/mL. How many molecules of caffeine are present in this volume?

(c) How many nitrogen atoms are present in 3.5 mg of caffeine?

(d) Complete the skeletal structure of caffeine, where all the bonded atoms are shown but double bonds, triple bonds, and/or lone pairs are missing.

Caffeine skeletal structure

(e) Identify the various types of geometries present in each central atom of caffeine using VSEPR theory.

(f) Determine the various relative bond angles associated with each central atom of caffeine using VSEPR theory.

(g) What is the most polar bond in caffeine?

(h) Would you predict caffeine to be polar or nonpolar?

(i) Consider the combustion of caffeine, which results in formation of $NO_2(g)$ as well as other expected products. Write a balanced chemical equation for this reaction.

(j) The heat of combustion for caffeine is 2211 kcal/mol. How much heat will be given off if 0.81 g of caffeine is burned completely?

(k) Calculate the weight of $H_2O(g)$ that can be prepared from 8.00 g of caffeine mixed with 20.3 g of oxygen gas.

Gases, Liquids, and Solids

5

Hot-air balloon, Utah

KEY QUESTIONS

- 5-1 What Are the Three States of Matter?
- 5-2 What Is Gas Pressure and How Do We Measure It?
- 5-3 What Are the Laws That Govern the Behavior of Gases?
- 5-4 What Are Avogadro's Law and the Ideal Gas Law?
- 5-5 What Is Dalton's Law of Partial Pressures?
- 5-6 What Is the Kinetic Molecular Theory?
- 5-7 What Types of Intermolecular Attractive Forces Exist Between Molecules?
- 5-8 How Do We Describe the Behavior of Liquids at the Molecular Level?
- 5-9 What Are the Characteristics of the Various Types of Solids?
- 5-10 What Is a Phase Change and What Energies Are Involved?

5-1 What Are the Three States of Matter?

Various forces hold matter together causing it to take different forms. In an atomic nucleus, very strong forces of attraction keep the protons and neutrons together (Chapter 2). In an atom itself, there are attractions between the positive nucleus and the negative electrons that surround it. Within molecules, atoms are attracted to each other by covalent bonds, the arrangement of which causes the molecules to assume a particular shape. Within an ionic crystal, three-dimensional shapes arise because of electrostatic attractions between ions.

In addition to these forces, there are intermolecular attractive forces. These forces, which are the subject of this chapter, are weaker than any of the forces already mentioned; nevertheless, they help determine whether a particular compound is a solid, a liquid, or a gas at any given temperature.

Intermolecular attractive forces help hold matter together; in effect, they counteract another form of energy—kinetic energy—that tends to lead to a number of different ways for molecules to arrange themselves. Intermolecular attractive forces counteract the kinetic energy that molecules possess, which keeps them constantly moving in random, disorganized ways. Kinetic energy increases with increasing temperature. Therefore, the higher the temperature, the greater the tendency of particles to have more possible arrangements. The total energy remains the same, but it is more widely dispersed. This dispersal of energy will have some important consequences, as we will see shortly.

Gas
- Molecules far apart and disordered
- Negligible interactions between molecules

Liquid
- Intermediate situation

Solid
- Molecules close together and ordered
- Strong interactions between molecules

FIGURE 5-1 The three states of matter. A gas has no definite shape, and its volume is the volume of the container. A liquid has a definite volume but no definite shape. A solid has a definite shape and a definite volume.

Pressure The force per unit area exerted against a surface

The physical state of matter thus depends on a balance between the kinetic energy of particles, which tends to keep them apart, and the intermolecular attractive forces between them, which tend to bring them together.

At high temperatures, molecules possess a high kinetic energy and move so fast that the intermolecular attractive forces between them are too weak to hold them together. This situation is called the **gaseous state**. At lower temperatures, molecules move more slowly, to the point where the forces of attraction between them become important. When the temperature is low enough, a gas condenses to form a **liquid state**. Molecules in the liquid state still move past each other, but they travel much more slowly than they do in the gaseous state. When the temperature is even lower, molecules no longer have enough energy to move past each other. In the **solid state**, each molecule has a certain number of nearest neighbors, and these neighbors do not change (Figure 5-1).

The intermolecular attractive forces are the same in all three states. The difference is that in the gaseous state (and to a lesser degree in the liquid state), the kinetic energy of the molecules is great enough to overcome the attractive forces between them.

Most substances can exist in any of the three states. Typically a solid, when heated to a sufficiently high temperature, melts and becomes a liquid. The temperature at which this change takes place is called the melting point. Further heating causes the temperature to rise to the point at which the liquid boils and becomes a gas. This temperature is called the boiling point. Not all substances, however, can exist in all three states. For example, wood and paper cannot be melted. Upon heating, they either decompose or burn (depending on whether air is present), but they do not melt. Another example is sugar, which does not melt when heated but rather forms a dark substance called caramel.

5-2 What Is Gas Pressure and How Do We Measure It?

On the Earth, we live under a blanket of air that presses down on us and on everything else around us. As we know from weather reports, the **pressure** of the atmosphere varies from day to day.

A gas consists of molecules in rapid, random motion. The pressure a gas exerts on a surface, such as the walls of a container, results from the continual bombardment on the walls of the container by the rapidly moving gas molecules. We use an instrument called a **barometer** (Figure 5-2) to measure atmospheric pressure. One type of barometer consists of a long glass tube that is completely filled with mercury and then inverted into a pool of mercury in a dish. Because there is no air at the top of the mercury column inside the tube (there is no way air could get in), no gas pressure is exerted on the mercury column. The entire atmosphere, however, exerts its pressure on the mercury in the open dish. The difference in the heights of the two mercury levels is a measure of the atmospheric pressure.

Pressure is most commonly measured in **millimeters of mercury (mm Hg)**. Pressure is also measured in **torr**, a unit named in honor of the Italian physicist and mathematician Evangelista Torricelli (1608–1647), who invented the barometer. At sea level, the average pressure of the atmosphere is 760 mm Hg. We use this number to define still another unit of pressure, the **atmosphere (atm)**.

There are several other units with which to measure pressure. The SI unit is the pascal, and meteorologists report pressure in inches of mercury and bars. In this book, we use only mm Hg and atm.

A barometer is adequate for measuring the pressure of the atmosphere, but to measure the pressure of a gas in a container, we use a simpler instrument called a **manometer**. One type of manometer consists of a U-shaped tube containing mercury (Figure 5-3). Arm A has been evacuated and sealed and has zero pressure. Arm B is connected to the container in which the gas sample is enclosed. The pressure of the gas depresses the level of the mercury in arm B. The difference between the two mercury levels gives the pressure directly in mm Hg. If more gas is added to the sample container, the mercury level in B will be pushed down and that in A will rise as the pressure in the bulb increases.

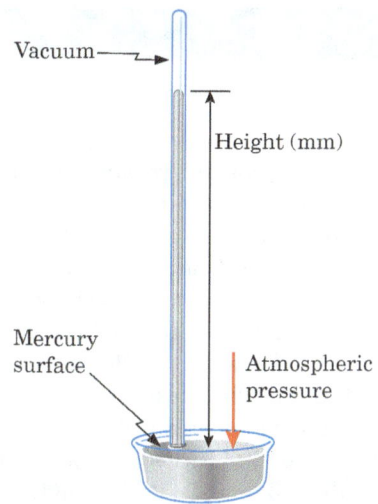

FIGURE 5-2 A mercury barometer.

5-3 What Are the Laws That Govern the Behavior of Gases?

By observing the behavior of gases under different sets of temperatures and pressures, scientists have established a number of relationships. In this section, we study three of the most important of these. The gas laws we describe below hold not only for pure gases but also for mixtures of gases.

$$1 \text{ atm} = 760 \text{ mm Hg}$$
$$= 760 \text{ torr}$$
$$= 101,325 \text{ Pa}$$
$$= 29.92 \text{ in. Hg}$$
$$= 1.01325 \text{ bars}$$

A. Boyle's Law and the Pressure–Volume Relationship

Boyle's law states that for a fixed mass of an ideal gas at a constant temperature, the volume of the gas is inversely proportional to the applied pressure. If the pressure doubles, for example, the volume decreases by one-half. This law can be stated mathematically in the following equation, where P_1 and V_1 are the initial pressure and volume and P_2 and V_2 are the final pressure and volume:

$$PV = \text{constant} \quad \text{or} \quad P_1V_1 = P_2V_2$$

This relationship between pressure and volume is illustrated in Figure 5-4.

FIGURE 5-3 A mercury manometer.

FIGURE 5-4 Boyle's law. Boyle's law experiment showing the compressibility of gases.

CHEMICAL CONNECTIONS 5A

Breathing and Boyle's Law

Under normal resting conditions, we breathe about 12 times per minute, each time inhaling and exhaling about 500. mL of air. When we inhale, we lower the diaphragm or raise the rib cage, either of which increases the volume of the chest cavity. In accord with Boyle's law, as the volume of the chest cavity increases, the pressure within it decreases and becomes lower than the outside pressure. As a result, air flows from the higher-pressure area outside the body into the lungs. While the difference in these two pressures is only about 3 mm Hg, it is enough to cause air to flow into the lungs. In exhaling, we reverse the process: We raise the diaphragm or lower the rib cage. The resulting decrease in volume increases the pressure inside the chest cavity, causing air to flow out of the lungs.

In certain diseases, the chest becomes paralyzed and the affected person cannot move either the diaphragm or the rib cage. In such a case, a respirator is used to help the person breathe. The respirator first pushes down on the chest cavity and forces air out of the lungs. The pressure of the respirator is then lowered below atmospheric pressure, causing the rib cage to expand and draw air into the lungs.

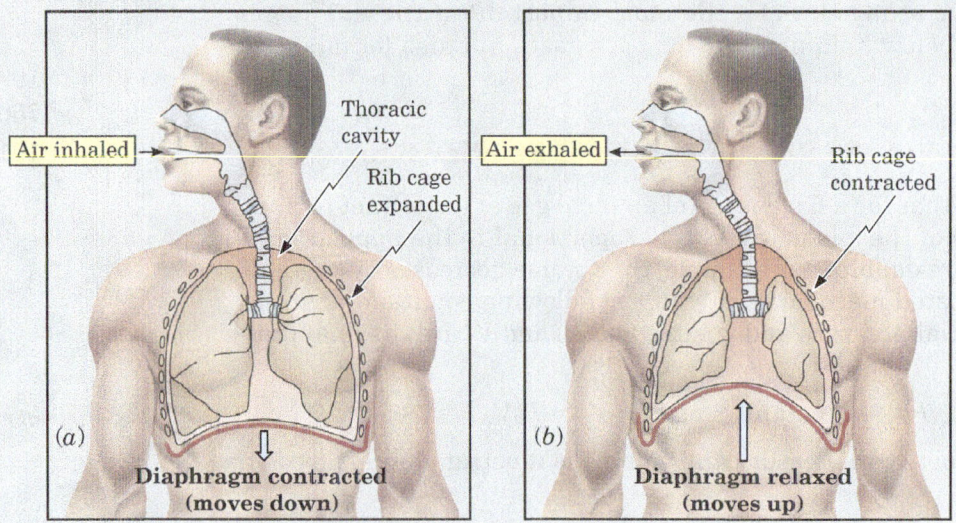

Schematic drawing of the chest cavity. (a) The lungs fill with air. (b) Air empties from the lungs.

Test your knowledge with Problem 5-87.

B. Charles's Law and the Temperature–Volume Relationship

Charles's law states that the volume of a fixed mass of an ideal gas at a constant pressure is directly proportional to the temperature in kelvins (K). In other words, as long as the pressure on a gas remains constant, increasing the temperature of the gas causes an increase in the volume occupied by the gas. Charles's law can be stated mathematically this way:

$$\frac{V}{T} = \text{constant} \quad \text{or} \quad \frac{V_1}{T_1} = \frac{V_2}{T_2}$$

This relationship between volume and temperature is the basis of the hot-air balloon operation (Figure 5-5). When using the gas laws, temperature must be expressed in kelvins (K). The zero in this scale is the lowest possible temperature.

C. Gay-Lussac's Law and the Temperature–Pressure Relationship

Gay-Lussac's law states that, for a fixed mass of a gas at constant volume, the pressure is directly proportional to the temperature in kelvins (K):

$$\frac{P}{T} = \text{constant} \quad \text{or} \quad \frac{P_1}{T_1} = \frac{P_2}{T_2}$$

As the temperature of the gas increases, the pressure increases proportionately. Consider, for example, what happens inside an autoclave. Steam generated inside an autoclave at 1 atm pressure has a temperature of 100°C. As the steam is heated further, the pressure within the autoclave increases. A valve controls the pressure inside the autoclave; if the pressure exceeds the designated maximum, the valve opens, releasing the steam. At maximum pressure, the temperature may reach 120°C to 150°C. All microorganisms in the autoclave are destroyed at such high temperatures. Table 5-1 shows mathematical expressions of these three gas laws.

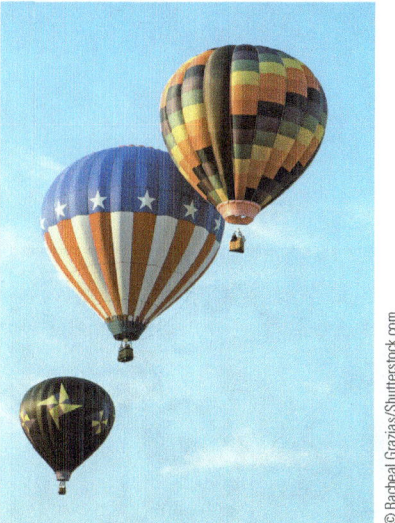

FIGURE 5-5 Charles's law illustrated in a hot-air balloon. Because the balloon can stretch, the pressure inside it remains constant. When the air in the balloon is heated, its volume increases, expanding the balloon. As the air in the balloon expands, it becomes less dense than the surrounding air, providing the lift for the balloon. (Charles was one of the first balloonists.)

Table 5-1 Mathematical Expressions of the Three Gas Laws for a Fixed Mass of Gas

Name	Expression	Constant
Boyle's law	$P_1V_1 = P_2V_2$	T
Charles's law	$\frac{V_1}{T_1} = \frac{V_2}{T_2}$	P
Gay-Lussac's law	$\frac{P_1}{T_1} = \frac{P_2}{T_2}$	V

The three gas laws can be combined and expressed by a mathematical equation called the **combined gas law**:

$$\frac{PV}{T} = \text{constant} \quad \text{or} \quad \frac{P_1V_1}{T_1} = \frac{P_2V_2}{T_2}$$

Combined gas law The pressure, volume, and temperature in kelvins of two samples of the same gas are related by the equation $P_1V_1/T_1 = P_2V_2/T_2$

EXAMPLE 5-1 The Combined Gas Law

A gas occupies 3.00 L at 2.00 atm pressure. Calculate its volume when we increase the pressure to 10.15 atm at the same temperature.

Strategy

First, we identify the known quantities. Because T_1 and T_2 are the same in this example and consequently cancel each other, we don't need to know the temperature. We use the relationship $P_1V_1 = P_2V_2$ and solve the combined gas law for V_2.

Solution

Initial: $P_1 = 2.00$ atm $V_1 = 3.00$ L

Final: $P_2 = 10.15$ atm $V_2 = ?$

$$V_2 = \frac{P_1V_1\cancel{T_2}}{\cancel{T_1}P_2} = \frac{(2.00 \,\cancel{\text{atm}})(3.00 \text{ L})}{10.15 \,\cancel{\text{atm}}} = 0.591 \text{ L}$$

Problem 5-1

A gas occupies 3.8 L at 0.70 atm pressure. If we expand the volume at constant temperature to 6.5 L, what is the final pressure?

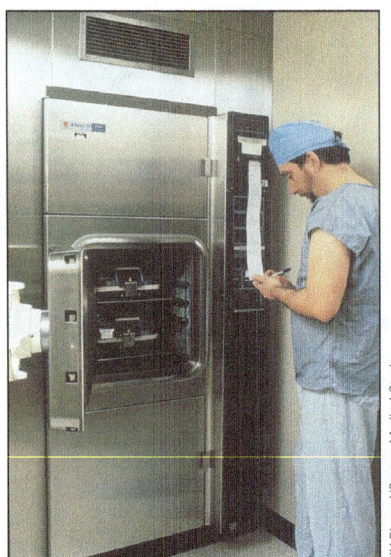

▶ An autoclave used to sterilize hospital equipment.

EXAMPLE 5-2 The Combined Gas Law

In an autoclave, steam at 100°C is generated at 1.00 atm. After the autoclave is closed, the steam is heated at constant volume until the pressure gauge indicates 1.13 atm. What is the final temperature in the autoclave? ◀

Strategy

All temperatures in gas law calculations must be in kelvins; therefore, we must first convert the Celsius temperature to kelvins. Then we identify the known quantities. Because V_1 and V_2 are the same in this example and consequently cancel each other, we don't need to know the volume of the autoclave.

Solution

Step 1: Convert from degrees C to K.

$$100°C = 100 + 273 = 373 \text{ K}$$

Step 2: Identify the known quantities.

Initial: $P_1 = 1.00$ atm $T_1 = 373$ K
Final: $P_2 = 1.13$ atm $T_2 = ?$

Step 3: Solve the combined gas law equation for T_2, the new temperature.

$$T_2 = \frac{P_2 V_2 T_1}{P_1 V_1} = \frac{(1.13 \text{ atm})(373 \text{ K})}{1.00 \text{ atm}} = 421 \text{ K}$$

The final temperature is 421K, or 421 − 273 = 148°C.

Problem 5-2

A constant volume of oxygen gas, O_2, is heated from 120.°C to 212°C. The final pressure is 20.3 atm. What was the initial pressure?

Interactive Example 5-3 The Combined Gas Law

A gas in a flexible container has a volume of 0.50 L and a pressure of 1.0 atm at 393 K. When the gas is heated to 500. K, its volume expands to 3.0 L. What is the new pressure of the gas in the flexible container?

Strategy

We identify the known quantities and then solve the combined gas law for the new pressure.

Solution

Step 1: The known quantities are:

Initial: $P_1 = 1.0$ atm $V_1 = 0.50$ L $T_1 = 393$ K

Final: $P_2 = ?$ $V_2 = 3.0$ L $T_2 = 500.$ K

Step 2: Solving the combined gas law for P_2, we find:

$$P_2 = \frac{P_1 V_1 T_2}{T_1 V_2} = \frac{(1.0 \text{ atm})(0.50 \text{ L})(500.\text{ K})}{(3.0 \text{ L})(393 \text{ K})} = 0.21 \text{ atm}$$

Problem 5-3

A gas is expanded from an initial volume of 20.5 L at 0.92 atm at room temperature (23.0°C) to a final volume of 340.6 L. During the expansion, the gas cools to 12.0°C. What is the new pressure?

5-4 What Are Avogadro's Law and the Ideal Gas Law?

The relationship between the mass of gas present and its volume is described by **Avogadro's law**, which states that equal volumes of gases at the same temperature and pressure contain equal numbers of molecules. Thus, if the temperature, pressure, and volumes of two gases are the same, then the two gases contain the same number of molecules, regardless of their identity (Figure 5-6). Avogadro's law is valid for all gases, no matter what they are.

The actual temperature and pressure at which we compare two or more gases do not matter. It is convenient, however, to select one temperature and one pressure as standard, and chemists have chosen 1 atm as the standard pressure and 0°C (273 K) as the standard temperature. These conditions are called **standard temperature and pressure (STP)**.

All gases at STP or at any other combination of temperature and pressure contain the same number of molecules in any given volume. But how many molecules is that? In Chapter 4, we saw that one mole contains 6.02×10^{23} formula units. What volume of a gas at STP contains one mole of molecules? This quantity has been measured experimentally and found to be 22.4 L. Thus, one mole of any gas at STP occupies a volume of 22.4 L.

Avogadro's law allows us to write a gas law that is valid not only for any pressure, volume, and temperature, but also for any quantity of gas. This law, called the **ideal gas law**, is:

$$PV = nRT$$

where P = pressure of the gas in atmospheres (atm)
          V = volume of the gas in liters (L)
          n = amount of the gas in moles (mol)
          T = temperature of the gas in kelvins (K)
          R = a constant for all gases, called the **ideal gas constant**

We can find the value of R by using the fact that one mole of any gas at STP occupies a volume of 22.4 L:

$$R = \frac{PV}{nT} = \frac{(1.00 \text{ atm})(22.4 \text{ L})}{(1.00 \text{ mol})(273 \text{ K})} = 0.0821 \frac{\text{L} \cdot \text{atm}}{\text{mol} \cdot \text{K}}$$

$T, P,$ and V are equal in both containers

FIGURE 5-6 Avogadro's law. Two tanks of gas of equal volume at the same temperature and pressure contain the same number of molecules.

Avogadro's law Equal volumes of gases at the same temperature and pressure contain the same number of molecules

Standard temperature and pressure (STP) 0°C (273 K) and one atmosphere pressure

Ideal gas law $PV = nRT$

Ideal gas A gas whose physical properties are described accurately by the ideal gas law

Ideal gas constant (R) 0.0821 L · atm · mol^{-1} · K^{-1}

▶ Molar volume. The box has a volume of 22.4 L, which is the volume of one mole of gas at STP (standard temperature and pressure). The basketball is shown for comparison.

The ideal gas law holds for all ideal gases at any temperature, pressure, and volume. But the only gases we have around us in the real world are real gases. Real gases behave most like ideal gases at low pressures (1 atm or less) and high temperatures (300 K or higher). How valid is it to apply the ideal gas law to real gases? The answer is that under most experimental conditions, real gases behave sufficiently like ideal gases that we can use the ideal gas law for them with little trouble. Thus, using $PV = nRT$, we can calculate any one quantity—$P, V, T,$ or n—if we know the other three quantities.

EXAMPLE 5-4   Ideal Gas Law

1.00 mole of CH_4 gas occupies 20.0 L at 1.00 atm pressure. What is the temperature of the gas in kelvins?

Strategy

Solve the ideal gas law for T and plug in the given values:

Solution

$$T = \frac{PV}{nR} = \frac{PV}{n} \times \frac{1}{R} = \frac{(1.00 \text{ atm})(20.0 \text{ L})}{(1.00 \text{ mol})} \times \frac{\text{mol} \cdot \text{K}}{0.0821 \text{ L} \cdot \text{atm}} = 244 \text{ K}$$

Note that we calculated the temperature for 1.00 mol of CH_4 gas under these conditions. The answer would be the same for 1.00 mol of CO_2, N_2, NH_3, or any other gas under these conditions. Note also that we have shown the gas constant separately to make it clear what is happening with the units attached to all quantities. We are going to do this throughout.

Problem 5-4

If 2.00 mol of NO gas occupies 10.0 L at 295 K, what is the pressure of the gas in atmospheres?

EXAMPLE 5-5   Ideal Gas Law

If there is 5.0 g of CO_2 gas in a 10. L cylinder at 25°C, what is the gas pressure within the cylinder?

Strategy

We are given the quantity of CO_2 in grams, but to use the ideal gas law, we must express the quantity in moles. Therefore, we must first convert grams of CO_2 to moles CO_2 and then use this value in the ideal gas law. To convert from grams to moles, we use the conversion factor 1.00 mol CO_2 = 44 g CO_2.

Solution

Step 1: Convert grams of CO_2 to moles of CO_2.

$$5.0 \text{ g } CO_2 \times \frac{1 \text{ mol } CO_2}{44 \text{ g } CO_2} = 0.11 \text{ mol } CO_2$$

Step 2: We now use this value in the ideal gas equation to solve for the pressure of the gas. Note that temperature must be expressed in kelvins.

$$P = \frac{nRT}{V}$$

$$= \frac{nT}{V} \times R = \frac{(0.11 \text{ mol } CO_2)(298 \text{ K})}{10. \text{ L}} \times \frac{0.0821 \text{ L} \cdot \text{atm}}{\text{mol} \cdot \text{K}} = 0.27 \text{ atm}$$

Problem 5-5

A certain quantity of neon gas is under 1.05 atm pressure at 303 K in a 10.0 L vessel. How many moles of neon are present?

EXAMPLE 5-6 Ideal Gas Law

If 3.3 g of a gas at 40°C and 1.15 atm pressure occupies a volume of 1.00 L, what is the mass of one mole of the gas?

Strategy

This problem is more complicated than previous ones. We are given grams of gas and P, T, and V values and asked to calculate the mass of one mole of the gas (g/mol). We can solve this problem in two steps: (1) Use the ideal gas law to calculate the number of moles of gas present in the sample. (2) We are given the mass of gas (3.3 grams) and use the ratio grams/mole to determine the mass of one mole of the gas.

Solution

Step 1: Use the P, V, and T measurements and the ideal gas law to calculate the number of moles of gas present in the sample. To use the ideal gas law, we must first convert 40°C to kelvins: $40 + 273 = 313$ K.

$$n = \frac{PV}{RT} = \frac{PV}{T} \times \frac{1}{R} = \frac{(1.15 \text{ atm})(1.00 \text{ L})}{313 \text{ K}} \times \frac{\text{mol} \cdot \text{K}}{0.0821 \text{ L} \cdot \text{atm}}$$

$$= 0.0448 \text{ mol}$$

Step 2: Calculate the mass of one mole of the gas by dividing grams by moles.

$$\text{Mass of one mole} = \frac{3.3 \text{ g}}{0.0448 \text{ mol}} = 74 \text{ g} \cdot \text{mol}^{-1}$$

Problem 5-6

An unknown amount of He gas occupies 30.5 L at 2.00 atm pressure and 300. K. What is the weight of the gas in the container?

5-5 What Is Dalton's Law of Partial Pressures?

In a mixture of gases, each molecule acts independently of all the others, provided that the gases behave as ideal gases and do not interact with each other in any way. For this reason, the ideal gas law works for mixtures of gases as well as for pure gases. **Dalton's law of partial pressures** states that the total pressure, P_T, of a mixture of gases is the sum of the partial pressures of each individual gas:

$$P_T = P_1 + P_2 + P_3 + \cdots$$

A corollary to Dalton's law is that the **partial pressure** of a gas in a mixture is the pressure that the gas would exert if it were alone in the

Partial pressure The pressure that a gas in a mixture of gases would exert if it were alone in the container

CHEMICAL CONNECTIONS 5B

Hyperbaric Medicine

Ordinary air contains 21% oxygen. Under certain conditions, the cells of tissues can become starved for oxygen (hypoxia), and quick oxygen delivery is needed. Increasing the percentage of oxygen in the air supplied to a patient is one way to remedy this situation, but sometimes even breathing pure (100%) oxygen may not be enough. For example, in carbon monoxide poisoning, hemoglobin, which normally carries most of the O_2 from the lungs to the tissues, binds CO and cannot take up any O_2 in the lungs. Without any help, tissues would soon become starved for oxygen and the patient would die. When oxygen is administered under a pressure of 2 to 3 atm, it dissolves in the plasma to such a degree that the tissues receive enough of it to recover without the help of the poisoned hemoglobin molecules. Other conditions for which hyperbaric medicine is used are treatment of gas gangrene, smoke inhalation, cyanide poisoning, skin grafts, thermal burns, and diabetic lesions.

Breathing pure oxygen for prolonged periods, however, is toxic. For example, if O_2 is administered at 2 atm for

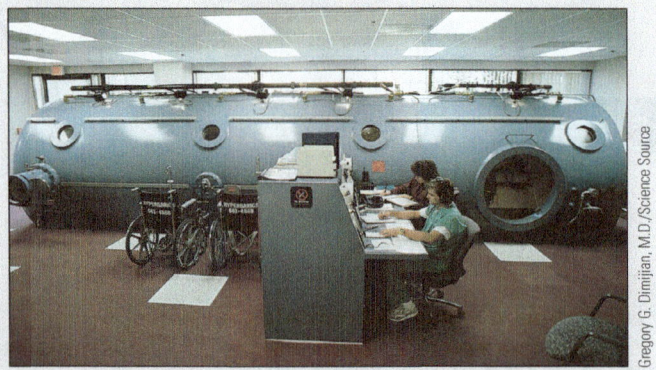

Hyperbaric oxygen chamber at Medical City Dallas Hospital

more than 6 hours, it may damage both lung tissue and the central nervous system. In addition, this treatment may cause nuclear cataract formation, necessitating postrecovery eye surgery. Therefore, recommended exposures to O_2 are 2 hours at 2 atm and 90 minutes at 3 atm. The benefits of hyperbaric medicine must be carefully weighed against these and other contraindications.

Test your knowledge with Problem 5-88.

---

container. The equation holds separately for each gas in the mixture as well as for the mixture as a whole.

Consider a mixture of nitrogen and oxygen illustrated in Figure 5-7. At constant volume and temperature, the total pressure of the mixture is equal to the pressure that the nitrogen alone plus the oxygen alone would exert. The pressure of one gas in a mixture of gases is called the partial pressure of that gas.

FIGURE 5-7 Dalton's law of partial pressures.

> **EXAMPLE 5-7** Dalton's Law of Partial Pressures
>
> To a tank containing N_2 at 2.0 atm and O_2 at 1.0 atm, we add an unknown quantity of CO_2 until the total pressure within the tank is 4.6 atm. What is the partial pressure of the CO_2?
>
> **Strategy**
>
> Dalton's law tells us that the addition of CO_2 does not affect the partial pressures of the N_2 or O_2 already present in the tank. The partial pressures of N_2 and O_2 remain at 2.0 atm and 1.0 atm, respectively, and their sum is 3.0 atm. The final total pressure within the tank, which is 4.6 atm, must be due to the partial pressure of the added CO_2.
>
> **Solution**
>
> If the final pressure is 4.6 atm, the partial pressure of the added CO_2 must be 1.6 atm. Thus, when the final pressure is 4.6 atm, the partial pressures are
>
> $$\underset{\text{Total pressure}}{4.6 \text{ atm}} = \underset{\substack{\text{Partial} \\ \text{pressure} \\ \text{of } N_2}}{2.0 \text{ atm}} + \underset{\substack{\text{Partial} \\ \text{pressure} \\ \text{of } O_2}}{1.0 \text{ atm}} + \underset{\substack{\text{Partial} \\ \text{pressure} \\ \text{of } CO_2}}{1.6 \text{ atm}}$$
>
> ### Problem 5-7
> A vessel under 2.015 atm pressure contains nitrogen, N_2, and water vapor, H_2O. The partial pressure of N_2 is 1.908 atm. What is the partial pressure of the water vapor?

5-6 What Is the Kinetic Molecular Theory?

To this point, we have studied the macroscopic properties of gases—namely, the various laws dealing with the relationships among temperature, pressure, volume, and number of moles of gas in a sample. Now let us examine the behavior of gases at the molecular level and see how we can explain their macroscopic behavior in terms of molecules and the interactions between them.

The relationship between the observed behavior of gases and the behavior of individual gas molecules within the gas can be explained by the **kinetic molecular theory**, which makes the following assumptions about the molecules of a gas:

1. Gases consist of particles, either atoms or molecules, constantly moving through space in straight lines, in random directions, and with various speeds. Because these particles move in random directions, different gases mix readily.

2. The average kinetic energy of gas particles is proportional to the temperature in kelvins. The higher the temperature, the faster they move through space and the greater their kinetic energy.

3. Molecules collide with each other, much as billiard balls do, bouncing off each other and changing directions. Each time they collide, they may exchange kinetic energies (one moves faster than before; the other, slower), but the total kinetic energy of the gas sample remains the same.

4. Gas particles have no volume. Most of the volume taken up by a gas is empty space, which explains why gases can be compressed so easily.

FIGURE 5-8 The kinetic molecular model of a gas. Molecules of nitrogen (blue) and oxygen (red) are in constant motion and collide with each other and with the walls of the container. Collisions of gas molecules with the walls of the container cause gas pressure. In air at STP, 6.02×10^{23} molecules undergo approximately 10 billion collisions per second.

Condensation The change of a substance from the vapor or gaseous state to the liquid state

Solidification The change of a substance from the liquid state to the solid state

5. There are no attractive forces between gas particles. They do not stick together after a collision occurs.
6. Molecules collide with the walls of the container, and these collisions constitute the pressure of the gas (Figure 5-8). The greater the number of collisions per unit time, the greater the pressure. The greater the average kinetic energy of the gas molecules, the greater the pressure.

These six assumptions of the kinetic molecular theory give us an idealized picture of the molecules of a gas and their interactions with one another (Figure 5-8). In real gases, however, forces of attraction between molecules do exist and molecules do occupy some volume. Because of these factors, a gas described by these six assumptions of the kinetic molecular theory is called an **ideal gas**. In reality, there is no ideal gas; all gases are real. At STP, however, most real gases behave in much the same way that an ideal gas would, so we can safely use these assumptions.

5-7 What Types of Intermolecular Attractive Forces Exist Between Molecules?

As noted in Section 5-1, the strength of the intermolecular forces (forces between molecules) in any sample of matter determines whether the sample is a gas, a liquid, or a solid under given conditions of temperature and pressure. In general, the closer the molecules are to each other, the greater the effect of the intermolecular forces. For example, when the temperature of a gas is high (room temperature or higher) and the pressure is low (1 atm or less), molecules of the gas are so far apart that we can effectively ignore attractions between them and treat the gas as ideal. When the temperature decreases, the pressure increases, or both, the distances between molecules decrease so that we can no longer ignore intermolecular forces. In fact, these forces become so important that they cause **condensation** (change from a gas to a liquid) and **solidification** (change from a liquid to a solid). Therefore, before discussing the structures and properties of liquids and solids, we must look at the nature of these intermolecular forces of attraction.

In this section, we discuss three types of intermolecular forces: London dispersion forces, dipole–dipole interactions, and hydrogen bonding. Table 5-2 shows the strengths of these three forces. Also shown for comparison are the strengths of ionic and covalent bonds, both of which are considerably stronger than the other three types of intermolecular forces. Although intermolecular forces are relatively weak compared to the strength of ionic and covalent bonds, it is the intermolecular forces that determine many of the physical properties of molecules, such as melting point, boiling point, and viscosity. As we will see in Chapters 21–31, these forces are also extremely important in influencing the three-dimensional shapes of biomolecules such as proteins and nucleic acids and in affecting how these types of biomolecules recognize and interact with one another.

A. London Dispersion Forces

Intermolecular attractive forces exist between all molecules, whether they are polar or nonpolar. If the temperature falls far enough, even nonpolar molecules such as He, Ne, H_2, and CH_4 can be liquefied. Neon, for example, is a gas at room temperature and atmospheric pressure. It can be liquefied if cooled to $-246°C$. The fact that these and other nonpolar gases can be liquefied means that some sort of interactions must occur between them.

Table 5-2 Forces of Attraction Between Molecules and Ions

	Attractive Force	Example	Typical Energy (k cal/mol)
Intramolecular	Ionic bonds	$Na^+ \cdots Cl^-$, $Mg^{2+} \cdots O^{2-}$	170–970
	Single, double, and triple covalent bonds	C—C C=C C≡C O—H	80–95 175 230 90–120
Intermolecular	Hydrogen bonding	$H-\overset{\delta-}{O}\cdots \overset{\delta+}{H}-O$ (with H's attached)	2–10
	Dipole–dipole interaction	H_3C $\overset{\delta-}{C}=O \cdots \overset{\delta+}{C}=O$ (H_3C groups)	1–6
	London dispersion forces	Ne $\cdots$ Ne	0.01–2.0

These weak intermolecular attractive forces are called **London dispersion forces**, after the American chemist Fritz London (1900–1954), who was the first to explain them.

London dispersion forces have their origin in electrostatic interactions. To visualize the origin of these forces, it is necessary to think in terms of instantaneous distributions of electrons within an atom or a molecule. Consider, for example, a sample of neon atoms, which can be liquefied if cooled to −246°C. Over time, the distribution of electron density in a neon atom is symmetrical, and a neon atom has no permanent dipole; that is, there is no separation of positive and negative charges. However, at any given instant, the electron density in a neon atom may be shifted more toward one part of the atom than another, thus creating a temporary dipole (Figure 5-9). This temporary dipole, which lasts for only tiny fractions of a second, induces temporary dipoles in adjacent neon atoms. These short-lived attractions between the temporary dipoles are the London dispersion forces, inducing a complementary attractive dipole in neighboring neon atoms to form the liquid state.

London dispersion forces exist between all molecules, but they are the only forces of attraction between nonpolar molecules. They range in strength from 0.01 to 2.0 kcal/mol depending on the mass, size, and shape of the interacting molecules. In general, their strength increases as the mass and number of electrons in a molecule increase. Even though London dispersion forces are very weak, they contribute significantly to the attractive forces between large molecules because they act over large surface areas.

B. Dipole–Dipole Interactions

As mentioned in Section 3-7B, many molecules are polar. The attraction between the positive end of one dipole and the negative end of another dipole is called a **dipole–dipole interaction**. These interactions can exist between two identical polar molecules or between two different polar molecules. To see the importance of dipole–dipole interactions, we can look at the differences in boiling points between nonpolar and polar molecules of comparable molecular weight. Butane, C_4H_{10}, with a

London dispersion forces Extremely weak attractive forces between atoms or molecules caused by the electrostatic attraction between temporary induced dipoles

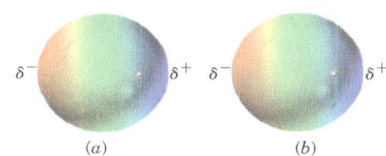

FIGURE 5-9 London dispersion forces. A temporary polarization of electron density in one neon atom creates positive and negative charges, which in turn induce temporary positive and negative charges in an adjacent atom. The intermolecular attractions between the temporary induced positive end of one dipole and the negative end of another temporary dipole are called London dispersion forces.

Dipole–dipole interaction The attraction between the positive end of a dipole of one molecule and the negative end of another dipole in the same or different molecule molecular weight of 58 amu, is a nonpolar molecule with a boiling point of 0.5°C. Acetone, C_3H_6O, with the same molecular weight, has a boiling point of 58°C. Acetone is a polar molecule, and its molecules are held together in the liquid state by dipole–dipole attractions between the negative end of the C=O dipole of one molecule and the positive end of the C=O dipole of another. Because it requires more energy to overcome the dipole–dipole interactions between acetone molecules than it does to overcome the considerably weaker London dispersion forces between butane molecules, acetone has a higher boiling point than butane.

$$CH_3-CH_2-CH_2-CH_3 \qquad CH_3-\overset{\overset{\displaystyle O^{\delta-}}{\|}}{\underset{\delta+}{C}}-CH_3$$

Butane (bp 0.5°C)   Acetone (bp 58°C)

C. Hydrogen Bonding

As we have just seen, the attraction between the positive end of one dipole and the negative end of another results in dipole–dipole attraction. When the positive end of one dipole is a hydrogen atom bonded to an O, N, or F (atoms of high electronegativity; see Table 3-5) and the negative end of the other dipole is an O, N, or F atom, the attractive interaction between dipoles is particularly strong and is given a special name: **hydrogen bonding**.

An example is the hydrogen bonding that occurs between molecules of water in both the liquid and solid states (Figure 5-10).

The strength of hydrogen bonding ranges from 2 to 10 kcal/mol. The strength in liquid water, for example, is approximately 5 kcal/mol. By comparison, the strength of the O—H covalent bond in water is approximately 119 kcal/mol. As can be seen by comparing these numbers, an O—H hydrogen bond is considerably weaker than an O—H covalent bond. Nonetheless, the presence of hydrogen bonds in liquid water has an important effect on the physical properties of water. Because of hydrogen bonding, extra energy is required to separate each water molecule from its neighbors—hence the relatively high boiling point of water. As we will see in later chapters, hydrogen bonds play an especially important role in biological molecules.

Hydrogen bonds are not restricted to water, however. They form between two molecules whenever one molecule has a hydrogen atom covalently bonded to O, N, or F and the other molecule has an O, N, or F atom bearing

Hydrogen bonding An intermolecular force of attraction between the partial positive charge on a hydrogen atom bonded to an atom of high electronegativity, most commonly oxygen or nitrogen, and the partial negative charge on a nearby oxygen or nitrogen

FIGURE 5-10 Two water molecules joined by a hydrogen bond. (a) Structural formulas and (b) ball-and-stick models.

a partial negative charge. We will encounter hydrogen bonding again when we introduce alchohols (Chapter 14), amines (Chapter 16), and carboxylic acids (Chapter 18).

Because oxygen and nitrogen atoms are more commonly encountered in biochemical systems involving hydrogen bonding, we will focus our discussions on these two atoms.

EXAMPLE 5-8 Hydrogen Bonding

Can a hydrogen bond form between:
(a) Two molecules of methanol, CH_3OH?
(b) Two molecules of formaldehyde, CH_2O?
(c) One molecule of methanol, CH_3OH, and one of formaldehyde, CH_2O?

Strategy

Examine the Lewis structure of each molecule and determine if there is a hydrogen atom bonded to either a nitrogen or oxygen atom. That is, determine if there is a polar O—H or N—H bond in one molecule in which hydrogen bears a partial positive charge. In other words, is there a hydrogen bond donor? Then examine the Lewis structure of the other molecule and determine if there is a polar bond in which either oxygen or nitrogen bears a partial negative charge. In other words, is there a potential hydrogen bond acceptor? If both features are present (a hydrogen bond donor and a hydrogen bond acceptor), then hydrogen bonding is possible.

Solution

(a) Yes. Methanol is a polar molecule and has a hydrogen atom covalently bonded to an oxygen atom (a hydrogen bond donor site). The hydrogen bond acceptor site is the oxygen atom of the polar O—H bond.

(b) No. Although formaldehyde is a polar molecule, it does not have a hydrogen covalently bonded to an oxygen or nitrogen atom (it has no hydrogen bond donor site). Its molecules, however, are attracted to each other by dipole–dipole interaction—that is, by the attraction between the negative end of the C=O dipole of one molecule and the positive end of the C=O dipole of another molecule.

$$\begin{array}{c} H \\ \diagdown \\ {}^{\delta+}C={}^{\delta-}O \\ \diagup \\ H \end{array}$$

(c) Yes. Methanol has a hydrogen atom bonded to an oxygen atom (hydrogen bond donor site) and formaldehyde has an oxygen atom bearing a partial negative charge (a hydrogen bond acceptor site).

Problem 5-8

Will the molecules in each set form a hydrogen bond between them?
(a) A molecule of water and a molecule of methanol, CH_3OH
(b) Two molecules of methane, CH_4

5-8 How Do We Describe the Behavior of Liquids at the Molecular Level?

We have seen that we can describe the behavior of gases under most circumstances by the ideal gas law, which assumes that there are no attractive forces between molecules. As pressure increases in a real gas, however, the molecules of the gas become squeezed into a smaller space, with the result that attractions between molecules become increasingly more effective in causing molecules to stick together.

If the distances between molecules decrease so that they touch or almost touch each other, the gas condenses to a liquid. Unlike gases, liquids do not fill all the available space, but they do have a definite volume, irrespective of the container. Because gases have a lot of empty space between molecules, it is easy to compress them into a smaller volume. In contrast, there is very little empty space in liquids; consequently, liquids are difficult to compress. A great increase in pressure is needed to cause even a very small decrease in the volume of a liquid. Thus, liquids, for all practical purposes, are incompressible. In addition, the density of liquids is much greater than that of gases because the same mass occupies a much smaller volume in liquid form than it does in gaseous form.

The brake system in a car is based on hydraulics. The force you exert with the brake pedal is transmitted to the brake via cylinders filled with liquid. This system works very well until an air leak occurs. Once air gets into the brake line, pushing the brake pedal compresses the air and greatly reduces the ability of the fluid to transfer force into pressure.

The positions of molecules in the liquid state are random, and some irregular empty space is available into which molecules can slide. Molecules in the liquid state are, therefore, constantly changing their positions with respect to neighboring molecules. This property causes liquids to be fluid and explains why liquids have a constant volume but not a constant shape.

FIGURE 5-11 Surface tension. Molecules in the interior of a liquid have equal intermolecular attractions in every direction. Molecules at the surface (the liquid–gas interface), however, experience greater attractions toward the interior of the liquid than toward the gaseous state above it. Therefore, molecules on the surface are preferentially pulled toward the center of the liquid. This pull crowds the molecules on the surface, thereby creating a layer, like an elastic skin, that is tough to penetrate.

A. Surface Tension

Unlike gases, liquids have surface properties, one of which is **surface tension** (Figure 5-11). The surface tension of a liquid is directly related to the strength of the intermolecular attraction between its molecules. It is defined as the energy required to increase the surface area of a liquid. Water has a high surface tension because of strong hydrogen bonding among water molecules. As a result, a steel needle can easily be made to float on the surface of water. If, however, the same needle is pushed below the elastic skin into the interior of the liquid, it sinks to the bottom. Similarly, water bugs gliding on the surface of a pond appear to be walking on an elastic skin of water. ◂

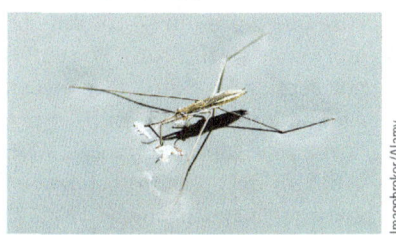

▸ A water-strider standing on water. The surface tension of water supports it.

B. Vapor Pressure

An important property of liquids is their tendency to evaporate. A few hours after a heavy rain, for example, most of the puddles have dried up; the water has evaporated and gone into the air. The same thing occurs if we leave a container of water or any other liquid out in the open. Let us explore how this change occurs.

In any liquid, there is a distribution of velocities among its molecules. Some of the molecules have high kinetic energy and move rapidly. Others have low kinetic energy and move slowly. Whether fast- or slow-moving, molecules in the interior of the liquid cannot go very far before they hit another molecule and have their speed and direction changed by the collision. Molecules at the surface, however, are in a different situation (Figure 5-12). If they are moving slowly (have a low kinetic energy), they cannot escape from the liquid because of the attractions of their neighboring molecules. If they are moving rapidly (have a high kinetic energy) and upward, however, they can escape from the liquid and enter the gaseous space above it.

In an open container, this process continues until all molecules have escaped. If the liquid is in a closed container, as in Figure 5-13, the molecules in the gaseous state cannot diffuse away (as they would do if the container were open). Instead, they remain in the air space above the liquid, where they move rapidly in straight lines until they strike something. Some of these vapor molecules move downward, strike the surface of the liquid, and are recaptured by it.

At this point, we have reached **equilibrium**. As long as the temperature does not change, the number of vapor molecules reentering the liquid

FIGURE 5-12 Evaporation. Some molecules at the surface of the liquid are moving fast enough to escape into the gaseous space.

Vapor A gas

Equilibrium A condition in which two opposing processes occur at an equal rate

CHEMICAL CONNECTIONS 5C

Blood Pressure Measurement

Liquids, like gases, exert a pressure on the walls of their containers. Blood pressure, for example, results from pulsating blood pushing against the walls of the blood vessels. When the heart ventricles contract and push blood out into the arteries, the blood pressure is high (systolic pressure); when the ventricles relax, the blood pressure is lower (diastolic pressure). Blood pressure is usually expressed as a fraction showing systolic over diastolic pressure—for instance, 120/80. The normal range in young adults is 100 to 120 mm Hg systolic and 60 to 80 mm Hg diastolic. In older adults, the corresponding normal ranges are 115 to 135 and 75 to 85 mm Hg, respectively.

A sphygmomanometer—the instrument used to measure blood pressure—consists of a bulb, a cuff, a manometer, and a stethoscope. The cuff is wrapped around the upper arm and inflated by squeezing the bulb (Figure, part a). The inflated cuff exerts a pressure on the arm, which is read on the manometer. When the cuff is sufficiently inflated, its pressure collapses the brachial artery, preventing pulsating blood from flowing to the lower arm (Figure, part b). At this pressure, no sound is heard in the stethoscope because the applied pressure in the cuff is greater than the blood pressure. Next, the cuff is slowly deflated, which decreases the pressure on the arm. The first faint tapping sound is heard when the pressure in the cuff just matches the systolic pressure as the ventricle contracts—that is, when the pressure in the cuff is low enough to allow pulsating blood to begin flowing into the lower arm. As the cuff pressure continues to decrease, the tapping first becomes louder and then begins to fade. At the point when the last faint tapping sound is heard, the cuff pressure matches the diastolic pressure when the ventricle is relaxed, thus allowing continuous blood flow into the lower arm (Figure, part c).

Digital blood pressure monitors are now available for home or office use. In these instruments, the stethoscope and the manometer are combined in a sensory device that records the systolic and diastolic blood pressures together with the pulse rate. The cuff and the inflation bulb are used the same way as in traditional sphygmomanometers.

(a) Measurement of blood pressure (b) Systolic pressure (c) Diastolic pressure

Test your knowledge with Problem 5-89.

Vapor pressure The partial pressure of a gas in equilibrium with its liquid form in a closed container

equals the number escaping from it. At equilibrium, the rate of vaporization equals the rate of liquefaction, and the space above the liquid shown in Figure 5-13 contains both air and vapor molecules, and we can measure the partial pressure of the vapor, called **vapor pressure** of the liquid. Note that we measure the partial pressure of a gas but call it the vapor pressure of the liquid.

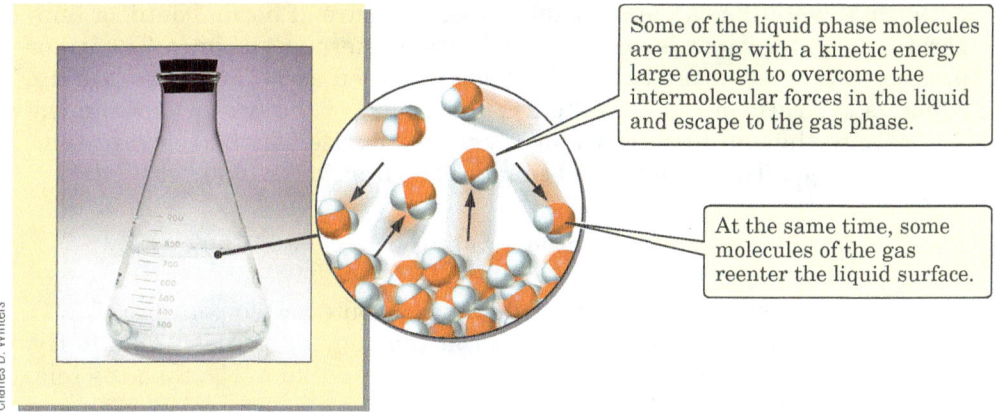

FIGURE 5-13 Evaporation and condensation. In a closed container, molecules of the liquid escape into the vapor phase and the liquid recaptures vapor molecules.

The vapor pressure of a liquid is a physical property of the liquid and a function of temperature (Figure 5-14). As the temperature of a liquid increases, the average kinetic energy of its molecules increases and it becomes easier for molecules to escape from the liquid state to the gaseous state. As the temperature of the liquid increases, its vapor pressure continues to increase until it equals the atmospheric pressure. At this point, bubbles of vapor form under the surface of the liquid and then force their way upward through the surface of the liquid, and the liquid boils.

The molecules that evaporate from a liquid surface are those that have a higher kinetic energy. When they enter the gas phase, the molecules left behind are those with a lower kinetic energy. Because the temperature of a sample is proportional to the average kinetic energy of its molecules, the temperature of the liquid drops as a result of evaporation. This evaporation of a layer of water from your skin produces the cooling effect you feel when you come out of a swimming pool and the water evaporates from your skin.

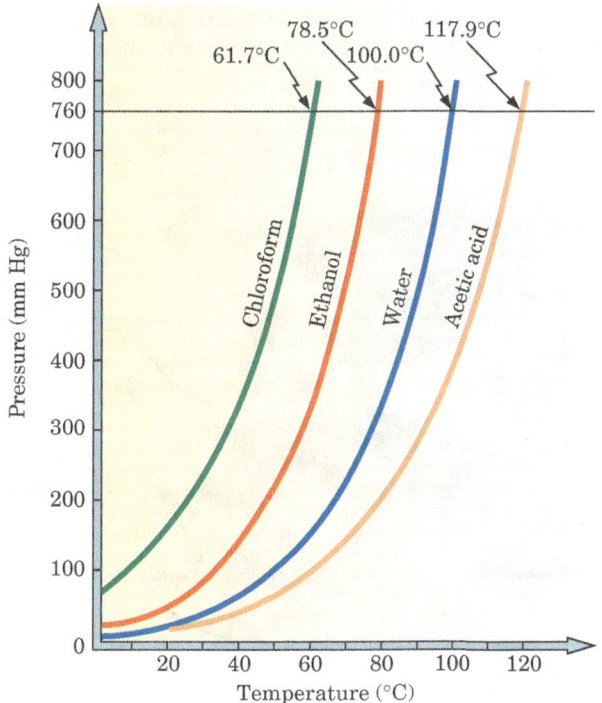

FIGURE 5-14 The change in vapor pressure with temperature for four liquids. The normal boiling point of a liquid is defined as the temperature at which its vapor pressure equals 760 mm Hg.

Because water has an appreciable vapor pressure at normal outdoor temperatures, water vapor is present in the atmosphere at all times. The vapor pressure of water in the atmosphere is expressed as **relative humidity**, which is the ratio of the actual partial pressure of the water vapor in the air, P_{H_2O}, to the equilibrium vapor pressure of water at the relevant temperature, $P°_{H_2O}$. The factor of 100 changes the fraction to a percentage.

$$\text{Relative humidity} = \frac{P_{H_2O}}{P°_{H_2O}} \times 100\%$$

For example, consider a typical warm day with an outdoor temperature of 25°C. The equilibrium vapor pressure of water at this temperature is 23.8 mm Hg. If the actual partial pressure of water vapor were 17.8 mm Hg, then the relative humidity would be 75%.

$$\text{Relative humidity} = \frac{17.8}{23.8} \times 100\% = 75\%$$

C. Boiling Point

Boiling point The temperature at which the vapor pressure of a liquid is equal to the atmospheric pressure

Normal boiling point The temperature at which a liquid boils under a pressure of 1 atm

The **boiling point** of a liquid is the temperature at which its vapor pressure is equal to the pressure of the atmosphere in contact with its surface. The boiling point when the atmospheric pressure is 1 atm is called the **normal boiling point**. For example, 100°C is the normal boiling point of water because that is the temperature at which water boils at 1 atm pressure (Figure 5-15).

The use of a pressure cooker is an example of boiling water at higher temperatures. In this type of pot, food is cooked at, say, 2 atm, at which pressure the boiling point of water is 121°C. Because the food has been raised to a higher temperature, it cooks faster than it would in an open pot, in which boiling water cannot get hotter than 100°C. Conversely, at low pressures, water boils at lower temperatures. For example, in Salt Lake City, Utah, where the average barometric pressure is about 650 mm Hg, the boiling point of water is about 95°C.

D. Factors That Affect Boiling Point

As Figure 5-14 shows, different liquids have different normal boiling points. Table 5-3 gives molecular formulas, molecular weights, and normal boiling points for five liquids.

FIGURE 5-15 Boiling point.

CHEMICAL CONNECTIONS 5D

The Densities of Ice and Water

The hydrogen-bonded superstructure of ice contains empty spaces in the middle of each hexagon because the H_2O molecules in ice are not as closely packed as those in liquid water. For this reason, ice has a lower density (0.917 g/cm^3) than does liquid water (1.00 g/cm^3). As ice melts, some of the hydrogen bonds are broken and the hexagonal superstructure of ice collapses into the more densely packed organization of water. This change explains why ice floats on top of water instead of sinking to the bottom. Such behavior is highly unusual—most substances are denser in the solid state than they are in the liquid state. The lower density of ice keeps fish and microorganisms alive in rivers and lakes that would freeze solid each winter if the ice sank to the bottom. The presence of ice on top insulates the remaining water and keeps it from freezing.

The fact that ice has a lower density than liquid water means that a given mass of ice takes up more space than the same mass of liquid water. This factor explains the damage done to biological tissues by freezing. When parts of the body (usually fingers, toes, nose, and ears) are subjected to extreme cold, they develop a condition called frostbite. Water in cells freezes despite the blood's attempt to keep the temperature at 37°C. As liquid water freezes, it expands and in doing so ruptures the walls of cells containing it, causing damage. In some cases, frostbitten fingers or toes must be amputated.

Cold weather can damage plants and crops in a similar way. Many plants are killed when the air temperature drops below the freezing point of water for several hours. Trees can survive cold winters because they have a low water content inside their trunks and branches. In the state of Florida, occasional harsh winters run the risk of also damaging millions of dollars worth of fruit produce when temperatures reach below 0°C. As such, farmers spray water on their crops in an effort to keep fruits from freezing inside.

Slow freezing is often more damaging to plant and animal tissues than quick freezing. In slow freezing, only a few crystals form, and these can grow to large sizes, rupturing cells. In quick freezing, such as that which can be achieved by cooling in liquid nitrogen (at a temperature of -196°C), many tiny crystals form. Because they do not grow much, tissue damage may be minimal.

(a) In the structure of ice, each water molecule occupies a fixed position in a regular array or lattice. (b) The form of a snowflake reflects the hexagonal arrangement of water molecules within the ice crystal lattice.

Test your knowledge with Problems 5-90 and 5-91.

Table 5-3 Names, Molecular Formulas, Molecular Weights, and Normal Boiling Points for Hexane and the Four Liquids in Figure 5-14

Name	Molecular Formula	Molecular Weight (amu)	Boiling Point (°C)
Chloroform	$CHCl_3$	120	62
Hexane	$CH_3CH_2CH_2CH_2CH_2CH_3$	86	69
Ethanol	CH_3CH_2OH	46	78
Water	H_2O	18	100
Acetic acid	CH_3COOH	60	118

As you study the information in this table, note that chloroform, which has the largest molecular weight of the five compounds, has the lowest boiling point. Water, which has the lowest molecular weight, has the second highest boiling point. From a study of these and other compounds, chemists have determined that the boiling point of covalent compounds depends primarily on three factors:

1. **Intermolecular forces** Water (H_2O, MW 18) and methane (CH_4, MW 16) have about the same molecular weight. The normal boiling point of water is 100°C, while that of methane is −164°C. The difference in boiling points reflects the fact that CH_4 molecules in the liquid state must overcome only the weak London dispersion forces to escape to the vapor state (low boiling point). In contrast, water molecules, being hydrogen-bonded to each other, need more kinetic energy (and a higher boiling temperature) to escape into the vapor phase. Thus the difference in boiling points between these two compounds is due to the greater strength of hydrogen bonding compared with the much weaker London dispersion forces.

2. **Number of sites for intermolecular interaction (surface area)** Consider the boiling points of methane, CH_4, and hexane, C_6H_{14}. Both are nonpolar compounds with no possibility for hydrogen bonding or dipole–dipole interactions between their molecules. The only force of attraction between molecules of either compound is London dispersion forces. The normal boiling point of hexane is 69°C, and that of methane is −164°C. The difference in their boiling points reflects the fact that hexane has more electrons and a larger surface area than methane. Because of its larger surface area, there are more sites for London dispersion forces to arise between hexane molecules than between methane molecules and, therefore, hexane has the higher boiling point.

3. **Molecular shape** When molecules are similar in every way except arrangement of the atoms, the strengths of London dispersion forces determine their relative boiling points. Consider pentane, bp 36.2°C, and 2,2-dimethylpropane, bp 9.5°C (Figure 5-16). Both compounds have the same molecular formula, C_5H_{12}, and the same molecular weight, but the boiling point of pentane is approximately 26° higher than that of 2,2-dimethylpropane. This difference in boiling points is related to the arrangement of the atoms in the following way. The only forces of attraction between these nonpolar molecules are London dispersion forces. Pentane is a roughly straight chain molecule, whereas 2,2-dimethylpropane has a branched arrangement and a smaller surface area than pentane. As surface area decreases, contact between adjacent molecules, the strength of London dispersion forces, and boiling points all decrease. Consequently, London dispersion forces between molecules of 2,2-dimethylpropane are weaker than those between molecules of pentane. Therefore, 2,2-dimethylpropane has a lower boiling point.

Pentane
(bp 36.2°C)

2,2-Dimethylpropane
(bp 9.5°C)

FIGURE 5-16 Pentane and 2,2-dimethylpropane have the same molecular formula, C_5H_{12}, but quite different shapes.

5-9 What Are the Characteristics of the Various Types of Solids?

When liquids are cooled, their molecules come closer together and attractive forces between them become so strong that random motion stops, and a solid forms. Formation of a solid from a liquid is called solidification or, alternatively, **crystallization**. Even in the solid state, molecules and ions do not stop moving completely. They vibrate around fixed points.

All solids have a regular shape that, in many cases, is obvious to the eye (Figure 5-17). This regular shape often reflects the arrangement of the particles within the crystal. In table salt, for example, the Na^+ and Cl^- ions are arranged in a cubic system (Figure 3-1). Metals such as solid gold also consist of particles arranged in a regular crystal lattice, but here the particles are atoms rather than ions. Because the particles in a solid are almost always closer together than they are in the corresponding liquid, solids almost always have a higher density than liquids. A notable exception to this generality is ice (see Chemical Connections 5D).

A fundamental distinction between kinds of solids is that some are crystalline and others are amorphous. **Crystalline solids** are those whose atoms, ions, or molecules have an ordered arrangement extending over a long range. These can be further categorized as ionic, molecular, polymeric, network covalent, or metallic as summarized in Table 5-4. By contrast, **amorphous solids** consist of randomly arranged particles that have no ordered long-range structure.

Ionic solids consist of orderly arrays of ions held together by ionic bonds in a crystal lattice. The strength of an ionic bond depends on the charges of the ions, their relative sizes, and directly impacts the physical properties of the solid. For example, NaCl, in which the ions have charges of 1+ and 1−, has a melting point of 801°C; MgO, in which the charges are 2+ and 2−, melts at 2852°C. Molecular solids consist of atoms or molecules held together by intermolecular forces (London dispersion forces, dipole-dipole interactions, and hydrogen boding). Because molecules are held only by intermolecular forces, which are much weaker than ionic bonds, molecular solids generally have lower melting points than ionic solids.

Other types of solids exist as well. For example, some are extremely large molecules, with each molecule having as many as 10^{23} atoms, all connected by covalent bonds. In such a case, the entire crystal is one big molecule. We call such molecules network covalent solids. A classic example is diamond [Figure 5-18(a)]. When you hold a diamond in your hand, you are holding a gigantic assembly of bonded atoms. Like ionic crystals, network covalent solids have very high melting points—if they can be melted at all. In many cases, they cannot be.

Crystallization The formation of a solid from a liquid

Garnet   Sulfur   Quartz   Pyrite

FIGURE 5-17 Some crystals.

Metallic solids consist of metal atoms surrounded by valence electrons. The bonding in metallic solids is too strong to be due to London dispersion forces, and yet there are not sufficient valence electrons for ordinary covalent bonds to be formed between atoms. Instead, the resulting metallic bond is due to the overlap of valence electrons that are scattered throughout the entire solid. The strength of the bonding increases as the number of electrons available for bonding increases. Table 5-4 further describes the various types of solids.

Table 5-4 Types of Solids

Type	Made Up of	Characteristics	Examples
Ionic	Ions in a crystal lattice	High melting point	NaCl, K_2SO_4
Molecular	Molecules in a crystal lattice	Low melting point	Ice, aspirin
Polymeric	Giant molecules; can be crystalline, semicrystalline, or amorphous	Low melting point or cannot be melted; soft or hard	Rubber, plastics, proteins
Network	A very large number of atoms connected by covalent bonds	Very hard; very high melting point or cannot be melted	Diamond, quartz
Amorphous	Randomly arranged atoms or molecules	Mostly soft, can be made to flow, but no melting point	Soot, tar, glass
Metallic	Metal atoms surrounded by a cloud of electrons	Soft to very hard; low to very high melting point	Au, Fe, Cu

Some elements exist in different forms in the same physical state. These forms have different chemical and physical properties and are known as **allotropes**. The best-known example is the element carbon, which exists in more than 40 known structural forms, five of which are crystalline (Figure 5-18) but most of which are amorphous. Diamond occurs when solidification takes place under very high pressure (thousands of atmospheres). Another form of carbon is the graphite in a pencil. Carbon atoms are packed differently in high-density hard diamonds than they are in low-density soft graphite.

FIGURE 5-18 Solid forms of carbon: (a) diamond, (b) graphite, (c) "buckyball," (d) nanotube, and (e) soot.

In a third form of carbon, each molecule contains 60 carbon atoms arranged in a structure having 12 pentagons and 20 hexagons as faces, resembling a soccer ball [Figure 5-18(c)]. Because the famous architect Buckminster Fuller (1895–1983) invented domes of a similar structure (he called them geodesic domes), the C-60 substance was named buckminsterfullerene, or "buckyball" for short. The discovery of buckyballs has generated a whole new area of carbon chemistry. Similar cage-like structures containing 72, 80, and even larger numbers of carbon have been synthesized. As a group, they are called fullerenes.

New variations on the fullerenes are nanotubes [Figure 5-18(d)]. The *nano-* part of the name comes from the fact that the cross section of each tube is only nanometers in size (1 nm = 10^{-9} m). Nanotubes come in a variety of forms. Single-walled carbon nanotubes can vary in diameter from 1 to 3 nm and are about 20 mm long. These structures have generated great industrial interest because of their optical and electronic properties. They may play a role in the miniaturization of instruments, giving rise to a new generation of nanoscale devices.

Soot is the fifth form of solid carbon. This substance solidifies directly out of carbon vapor and is an amorphous solid [Figure 5-18(e)]. Another example of an amorphous solid is glass. In essence, glass is an immobilized liquid.

5-10 What Is a Phase Change and What Energies Are Involved?

A. The Heating Curve for $H_2O(s)$ to $H_2O(g)$

Imagine the following experiment: We heat a piece of ice that is initially at −20°C. At first, we don't see any difference in its physical state. The temperature of the ice increases, but its appearance does not change. At 0°C, the ice begins to melt and liquid water appears. As we continue heating, more and more of the ice melts, but the temperature stays constant at 0°C until all the ice has melted and only liquid water remains. After all the ice has melted, the temperature of the water again increases as heat is added. At 100°C, the water boils. We continue heating as it continues to evaporate, but the temperature of the remaining liquid water does not change. Only after all the liquid water has changed to gaseous water (steam) does the temperature of the sample rise above 100°C.

These changes in state are called **phase changes**. A **phase** is any part of a system that looks uniform (homogeneous) throughout. Solid water (ice) is one phase, liquid water is another, and gaseous water is still another phase. Table 5-5 summarizes the energies for each step in the conversion of 1.0 g of ice to 1.0 g of steam.

Let us calculate the heat required to raise the temperature of 1.0 g of ice at −20°C to water vapor at 120°C and compare our results with the data in Table 5-5. We begin with ice, whose specific heat is 0.48 cal/g · °C (Table 1-4).

Phase changes Changes from one physical state (gas, liquid, or solid) to another

Table 5-5 Energy Required to Heat 1.0 g of Solid Water at −20°C to 120°C

Physical Change	Energy (cal)	Basis for Calculation of Energy Required
Warming ice from −20°C to 0°C	9.6	Specific heat of ice = 0.48 cal/g · °C
Melting ice; temperature = 0°C	80	Heat of fusion of ice = 80. cal/g
Warming water from 0°C to 100°C	100	Specific heat of liquid water = 1.00 cal/g · °C
Boiling water; temperature = 100°C	540	Heat of vaporization = 540 cal/g
Warming steam from 100°C to 120°C	9.6	Specific heat of steam = 0.48 cal/g · °C

Recall from Section 1-9B that it requires 0.48 × 20 = 9.6 cal to raise the temperature of 1.0 g of ice from −20 to 0°C.

$$0.48 \frac{\text{cal}}{g \cdot °C} \times 1.0\,g \times 20°C = 9.6\text{ cal}$$

After the ice reaches 0°C, additional heat causes a phase change: Solid water melts and becomes liquid water. The heat necessary to melt 1.0 g of any solid is called its **heat of fusion**. The heat of fusion of ice is 80. cal/g. Thus it requires 80 cal to melt 1.0 g of ice—that is, to change 1.0 g of ice at 0°C to liquid water at 0°C.

Only after the ice has completely melted does the temperature of the water rise again. The **specific heat** of liquid water is 1.00 cal/g · °C (Table 1-4). Thus it requires 100 cal to raise the temperature of 1.0 g of liquid water from 0°C to 100°C. Contrast this with the 80 cal required to melt 1.0 g of ice.

When liquid water reaches 100°C, the normal boiling point of water, the temperature of the sample remains constant while another phase change takes place: Liquid water vaporizes to gaseous water. The amount of heat necessary to vaporize 1.0 g of a liquid at its normal boiling point is called its **heat of vaporization**. For water, this value is 540 cal/g. Once all of the liquid water has been vaporized, the temperature again rises as the water vapor (steam) is heated. The specific heat of steam is 0.48 cal/g (Table 1-4). Thus it requires 9.6 cal to heat 1.0 g of steam from 100°C to 120°C. The data for heating 1.0 g of water from −20°C to 120°C can be shown in a graph called a **heating curve** (Figure 5-19). The total energy required for this conversion involves 739 cal of heat. Notice that the heat of vaporization is greater than the heat of fusion because all intermolecular forces must be overcome before vaporization can occur. In particular, while many hydrogen bonds remain in the solid to liquid transition, virtually all hydrogen bonds must be broken in the liquid to gas transition, hence the difference between the heat of fusion (80. cal/g) and heat of vaporization (540 cal/g).

An important aspect of these phase changes is that each one of them is reversible. If we start with liquid water at room temperature and cool it by immersing the container in a dry ice bath (−78°C), the reverse process is observed. The temperature drops until it reaches 0°C, and then ice begins to crystallize. During this phase change, the temperature of the sample stays constant but heat is given off. The amount of heat given off when 1.0 g of liquid water at 0°C freezes is exactly the same as the amount of heat absorbed when the 1.0 g of ice at 0°C melts.

FIGURE 5-19 The heating curve of ice. The graph shows the effect of adding heat to 1.0 g of ice initially at −20°C and raising its temperature to 120°C.

A transition from the solid state directly into the vapor state without going through the liquid state is called **sublimation**. ◀ Solids usually sublime only at reduced pressures (less than 1 atm). At high altitudes, where the atmospheric pressure is low, snow sublimes. Solid CO_2 (dry ice) sublimes at $-78.5°C$ under 1 atm pressure. At 1 atm pressure, CO_2 can exist only as a solid or as a gas, never as a liquid.

EXAMPLE 5-9 Heat of Fusion

The heat of fusion of ice is 80. cal/g. How many calories are required to melt 1.0 mol of ice?

Strategy

We first convert moles of ice to grams of ice using the conversion factor 1 mol of ice = 18 g of ice. We then use the heat of fusion of ice (80. cal/g) to calculate the number of calories required to melt the given quantity of ice.

Solution

1.0 mole of H_2O has a mass of 18 g. We use the factor-label method to calculate the heat required to melt 1.0 mole of ice at 0°C.

$$\frac{80.\,\text{cal}}{\text{g-ice}} \times 18\ \text{g-ice} = 1.4 \times 10^3\ \text{cal}$$

Problem 5-9

What mass of water at 100°C can be vaporized by the addition of 45.0 kcal of heat?

EXAMPLE 5-10 Heat of Fusion and Phase Change

What will be the final temperature if we add 1000. cal of heat to 10.0 g of ice at 0°C?

Strategy

The first thing the added heat does is to melt the ice. So we must first determine if 1000 cal is sufficient to melt the ice completely. If less than 1000. cal is required to melt the ice to liquid water, then the remaining heat will serve to raise the temperature of the liquid water. The specific heat (SH; Section 1-9) of liquid water is 1.00 cal/g · °C (Table 1-4).

Solution

Step 1: This phase change will use 10.0 g × 80. cal/g = 8.0×10^2 cal, which leaves 2.0×10^2 cal to raise the temperature of the liquid water.

Step 2: The temperature of the liquid water is now raised by the remaining heat. The relationship between specific heat, mass, and temperature change is given by the following equation (Section 1-9):

$$\text{Amount of heat} = \text{SH} \times m \times (T_2 - T_1)$$

Solving this equation for $T_2 - T_1$ gives:

$$T_2 - T_1 = \text{amount of heat} \times \frac{1}{\text{SH}} \times \frac{1}{m}$$

$$T_2 - T_1 = 2.0 \times 10^2\ \text{cal} \times \frac{\text{g} \cdot °\text{C}}{1.00\ \text{cal}} \times \frac{1}{10.0\ \text{g}} = 20.°\text{C}$$

▶ These freeze-dried coffee crystals were prepared by subliming water from frozen coffee.

Thus, the temperature of the liquid water will rise by 20°C from 0°C to 20°C.

Problem 5-10

The specific heat of iron is 0.11 cal/g · °C (Table 1-4). The heat of fusion of iron—that is, the heat required to convert iron from a solid to a liquid at its melting point—is 63.7 cal/g. Iron melts at 1530°C. How much heat must be added to 1.0 g of iron at 25°C to completely melt it?

We can show all phase changes for any substance on a **phase diagram**. Figure 5-20 is a phase diagram for water. Temperature is plotted on the *x*-axis and pressure on the *y*-axis. The three areas with different colors are labeled solid, liquid, and vapor. Within these areas water exists either as ice or liquid water or water vapor. The line (A–B) separating the solid phase from the liquid phase contains all the freezing (melting) points of water—for example, 0°C at 1 atm and 0.005°C at 400 mm Hg.

At the melting point, the solid and liquid phases coexist. The line separating the liquid phase from the gas phase (A–C) contains all the boiling points of water—for example, 100°C at 760 mm Hg and 84°C at 400 mm Hg. At the boiling points, the liquid and gas phases coexist.

Finally, the line separating the solid phase from the gas phase (A–D) contains all the sublimation points. At the sublimation points, the solid and gas phases coexist.

At a unique point (A) on the phase diagram, called the **triple point**, all three phases coexist. The triple point for water occurs at 0.01°C and 4.58 mm Hg pressure.

A phase diagram illustrates how one may go from one phase to another. For example, suppose we have water vapor at 95°C and 660 mm Hg (E). We want to condense it to liquid water. We can decrease the temperature to 70°C without changing the pressure (moving horizontally from E to F). Alternatively, we can increase the pressure to 760 mm Hg without changing the temperature (moving vertically from E to G). Or we can change both temperature and pressure (moving from E to H). Any of these processes will condense the water vapor to liquid water, although the resulting liquids will be at different pressures and temperatures. The phase diagram allows us to visualize what will happen to the phase of a substance when we change the experimental conditions from one set of temperature and pressure to another set.

FIGURE 5-20 Phase diagram of water. Temperature and pressure scales are greatly reduced.

EXAMPLE 5-11 Phase Diagram

What will happen to ice at 0°C if the pressure decreases from 1 atm to 0.001 atm?

Strategy
Consult the location on the phase diagram of water (Figure 5-20) and find the point that corresponds to the given temperature and pressure conditions.

Solution
According to Figure 5-20, when the pressure decreases while the temperature remains constant, we move vertically from 1 atm (760 mm Hg) to 0.001 atm (0.76 mm Hg). During this process, we cross the boundary separating the solid phase from the vapor phase. Thus, when the pressure drops to 0.001 atm, ice sublimes and becomes vapor.

Problem 5-11
What will happen to water vapor if it is cooled from 100°C to −30°C while the pressure stays at 1 atm?

▶ In this photo, vapors of gaseous CO_2 are cold enough to cause the moisture in the air above it to condense. The mixture of CO_2 vapors and condensed water vapor is heavier than air and slowly glides along the table or other surface on which the dry ice sits.

CHEMICAL CONNECTIONS 5E

Supercritical Carbon Dioxide

We are conditioned to think that a compound may exist in three phases: solid, liquid, or gas. Under certain pressures and temperatures, however, more phases may exist. A case in point is the plentiful nonpolar substance carbon dioxide. At room temperature and 1 atm pressure, CO_2 is a gas. Even when it is cooled to −78°C, it does not become a liquid but rather goes directly from a gas to a solid, which we call dry ice. ◀ At room temperature, a pressure of 60 atm is necessary to force molecules of CO_2 close enough together that they condense to a liquid.

Much more esoteric is the form of carbon dioxide called supercritical CO_2, which has some of the properties of a gas and some of the properties of a liquid. It has the density of a liquid but maintains its gas-like property of being able to flow with little viscosity or surface tension. What makes supercritical CO_2 particularly useful is that it is an excellent solvent for many organic materials. For example, supercritical CO_2 can extract caffeine from ground coffee beans, and after extraction when the pressure is released, it simply evaporates, leaving no traces behind. Similar processes can be performed with organic solvents, but traces of solvent may be left behind in the decaffeinated coffee and may alter its taste.

To understand the supercritical state, it is necessary to think about the interactions of molecules in the gas and liquid states. In the gaseous state, molecules are far apart, there is very little interaction between them, and most of the volume occupied by the gas is empty space. In the liquid state, molecules are held close together by the attractive forces between their molecules, and there is very little empty space between them. The supercritical state is something between these two states. Molecules are close enough together to give the sample some of the properties of a liquid but at the same time far enough apart to give it some of the properties of a gas.

The critical temperature and pressure for carbon dioxide are 31°C and 73 atm. When supercritical CO_2 is cooled below the critical temperature and/or compressed, there is a phase transition and the gas and liquid coexist. At a critical temperature and pressure, the two phases merge. Above critical conditions, the supercritical fluid exists, which exhibits characteristics that are intermediate between those of a gas and those of a liquid.

The phase diagram for carbon dioxide

Test your knowledge with Problem 5-92.

Summary of Key Questions

Section 5-1 What Are the Three States of Matter?
- Matter can exist in three different states: gas, liquid, and solid.
- Attractive forces between molecules tend to hold matter together, whereas the kinetic energy of the molecules tends to disorganize matter.

Section 5-2 What Is Gas Pressure and How Do We Measure It?
- Gas pressure results from the bombardment of gas particles on the walls of its container.
- The pressure of the atmosphere is measured with a barometer. Three common units of pressure are millimeters of mercury, torr, and atmospheres. 1 mm Hg = 1 torr and 760 mm Hg = 1 atm.

Section 5-3 What Are the Laws That Govern the Behavior of Gases?
- **Boyle's law** states that for a fixed mass of gas at constant temperature, the volume of the gas is inversely proportional to the pressure.
- **Charles's law** states that the volume of a fixed mass of gas at constant pressure is directly proportional to the temperature in kelvins.
- **Gay-Lussac's law** states that for a fixed mass of gas at constant volume, the pressure is directly proportional to the temperature in kelvins.
- These laws are combined and expressed as the **combined gas law**:

$$\frac{P_1 V_1}{T_1} = \frac{P_2 V_2}{T_2}$$

Section 5-4 What Are Avogadro's Law and the Ideal Gas Law?
- **Avogadro's law** states that equal volumes of gases at the same temperature and pressure contain the same number of molecules.
- The ideal gas law, $PV = nRT$, incorporates Avogadro's law into the combined gas law.
- In summary, in problems involving gases, the most important equation is:
  The **ideal gas law,** useful when three of the variables P, V, T, and n are given and you are asked to calculate the fourth variable:

$$PV = nRT$$

Section 5-5 What Is Dalton's Law of Partial Pressures?
- **Dalton's law of partial pressures** states that the total pressure of a mixture of gases is the sum of the partial pressures of each individual gas.

Section 5-6 What Is the Kinetic Molecular Theory?
- The **kinetic molecular theory** explains the behavior of gases. Molecules in the gaseous state move rapidly and randomly, allowing a gas to fill all the available space of its container. Gas molecules have no volume and there are no forces of attraction between them. In their random motion, gas molecules collide with the walls of the container and thereby exert pressure.

Section 5-7 What Types of Intermolecular Attractive Forces Exist Between Molecules?
- **Intermolecular forces of attraction** are responsible for the condensation of gases into the liquid state and for the solidification of liquids to the solid state. In order of increasing strength, the intermolecular forces of attraction are **London dispersion forces, dipole–dipole attractions**, and **hydrogen bonds**.

Section 5-8 How Do We Describe the Behavior of Liquids at the Molecular Level?
- **Surface tension** is the energy required to increase the surface area of a liquid.
- **Vapor pressure** is the pressure of a vapor (gas) above its liquid in a closed container. The vapor pressure of a liquid increases with increasing temperature.
- The **boiling point** of a liquid is the temperature at which its vapor pressure equals the atmospheric pressure. The boiling point of a liquid is determined by (1) the nature and strength of the intermolecular forces between its molecules, (2) the number of sites for intermolecular interaction, and (3) molecular shape.

Section 5-9 What Are the Characteristics of the Various Types of Solids?
- Solids crystallize in well-formed geometrical shapes that include crystalline (ionic, molecular, polymeric, network covalent, or metallic) and amorphous solids.
- The **melting point** is the temperature at which a substance changes from the solid state to the liquid state.
- **Crystallization** is the formation of a solid from a liquid.
- **Allotropes** are elements that exist in different forms while in the same physical state with different chemical and physical properties.

Section 5-10 What Is a Phase Change and What Energies Are Involved?
- A **phase** is any part of a system that looks uniform throughout. A **phase change** involves a change of matter from one physical state to another—that is, from a solid, liquid, or gaseous state to any one of the other two states.
- **Sublimation** is a change from a solid state directly to a gaseous state.
- **Heat of fusion** is the heat necessary to convert 1.0 g of any solid to a liquid.
- **Heat of vaporization** is the heat necessary to convert 1.0 g of any liquid to the gaseous state.
- A **phase diagram** allows the visualization of what happens to the phase of a substance when the temperature or pressure is changed.
- A phase diagram contains all the melting points, boiling points, and sublimation points where two phases coexist.
- A phase diagram also contains a unique triple point where all three phases coexist.

Problems

Orange-numbered problems are applied.

Section 5-2 What Is Gas Pressure and How Do We Measure It?

5-12 A weather report says that the barometric pressure is 29.5 inches of mercury. What is this pressure in atmospheres?

5-13 Use the kinetic molecular theory to explain why, at constant temperature, the pressure of a gas increases as its volume is decreased.

5-14 Use the kinetic molecular theory to explain why the pressure of a gas in a fixed-volume container increases as its temperature is increased.

5-15 Name three ways by which the volume of a gas can be decreased.

Section 5-3 What Are the Laws That Govern the Behavior of Gases?

5-16 Answer true or false.
(a) For a sample of gas at constant temperature, its pressure multiplied by its volume is a constant.
(b) For a sample of gas at constant temperature, increasing the pressure increases the volume.
(c) For a sample of gas at constant temperature, $P_1/V_1 = P_2/V_2$.
(d) As a gas expands at constant temperature, its volume increases.
(e) The volume of a sample of gas at constant pressure is directly proportional to its temperature—the higher its temperature, the greater its volume.
(f) A hot-air balloon rises because hot air is less dense than cooler air.
(g) For a gas sample in a container of fixed volume, an increase in temperature results in an increase in pressure.
(h) For a gas sample in a container of fixed volume, $P \times T$ is a constant.
(i) When steam at 100°C in an autoclave is heated to 120°C, the pressure within the autoclave increases.
(j) When a gas sample in a flexible container at constant pressure at 25°C is heated to 50°C, its volume doubles.
(k) Lowering the diaphragm causes the chest cavity to increase in volume and the pressure of air in the lungs to decrease.
(l) Raising the diaphragm decreases the volume of the chest cavity and forces air out of the lungs.

5-17 A sample of gas has a volume of 6.20 L at 20°C at a pressure of 1.10 atm. What is its volume at the same temperature and at a pressure of 0.925 atm?

5-18 Methane gas is compressed from 20. L to 2.5 L at a constant temperature. The final pressure is 12.2 atm. What was the original pressure?

5-19 A gas syringe at 20°C contains 20.0 mL of CO_2 gas. The pressure of the gas in the syringe is 1.0 atm. What is the pressure in the syringe at 20°C if the plunger is depressed to 10.0 mL?

5-20 Suppose that the pressure in an automobile tire is 2.30 atm at a temperature of 20.0°C. What will the pressure in the tire be if after 10 miles of driving the temperature of the tire increases to 47.0°C?

5-21 A sample of 23.0 L of NH_3 gas at 10.0°C is heated at constant pressure until it fills a volume of 50.0 L. What is the new temperature in °C?

5-22 If a sample of 4.17 L of ethane gas, C_2H_6, at 725°C is cooled to 175°C at constant pressure, what is the new volume?

5-23 A sample of SO_2 gas has a volume of 5.2 L. It is heated at constant pressure from 30. to 90.°C. What is its new volume?

5-24 A sample of B_2H_6 gas in a 35-mL container is at a pressure of 450. mm Hg and a temperature of 625°C. If the gas is allowed to cool at constant volume until the pressure is 375 mm Hg, what is the new temperature in °C?

5-25 A gas in a bulb as in Figure 5-3 registers a pressure of 833 mm Hg in the manometer in which the reference arm of the U-shaped tube (A) is sealed and evacuated. What will the difference in the mercury levels be if the reference arm of the U-shaped tube is open to atmospheric pressure (760 mm Hg)?

5-26 In an autoclave, a constant amount of steam is generated at a constant volume. Under 1.00 atm pressure, the steam temperature is 100.°C. What pressure setting should be used to obtain a 165°C steam temperature for the sterilization of surgical instruments?

5-27 A sample of the inhalation anesthetic gas Halothane, $C_2HBrClF_3$, in a 500-mL cylinder has a pressure of 2.3 atm at 0°C. What will be the pressure of the gas if its temperature is warmed to 37°C (body temperature)?

5-28 Complete this table:

V_1	T_1	P_1	V_2	T_2	P_2
546 L	43°C	6.5 atm	_____	65°C	1.9 atm
43 mL	−56°C	865 torr	_____	43°C	1.5 atm
4.2 L	234 K	0.87 atm	3.2 L	29°C	_____
1.3 L	25°C	740 mm Hg	_____	0°C	1.0 atm

5-29 Complete this table:

V_1	T_1	P_1	V_2	T_2	P_2
6.35 L	10°C	0.75 atm	___	0°C	1.0 atm
75.6 L	0°C	1.0 atm	___	35°C	735 torr
1.06 L	75°C	0.55 atm	3.2 L	0°C	___

5-30 A balloon filled with 1.2 L of helium at 25°C and 0.98 atm pressure is submerged in liquid nitrogen at −196°C. Calculate the final volume of the helium in the balloon.

5-31 A balloon used for atmospheric research has a volume of 1×10^6 L. Assume that the balloon is filled with helium gas at STP and then allowed to ascend to an altitude of 10 km, where the pressure of the atmosphere is 243 mm Hg and the temperature is −33°C. What will the volume of the balloon be under these atmospheric conditions?

5-32 A gas occupies 56.44 L at 2.00 atm and 310. K. If the gas is compressed to 23.52 L and the temperature is lowered to 281 K, what is the new pressure?

5-33 A certain quantity of helium gas is at a temperature of 27°C and a pressure of 1.00 atm. What will the new temperature be if its volume is doubled at the same time that its pressure is decreased to one-half its original value?

5-34 A sample of 30.0 mL of krypton gas, Kr, is at 756 mm Hg and 25.0°C. What is the new volume if the pressure is decreased to 325 mm Hg and the temperature is decreased to −12.5°C?

5-35 A 26.4-mL sample of ethylene gas, C_2H_4, has a pressure of 2.50 atm at 2.5°C. If the volume is increased to 36.2 mL and the temperature is raised to 10°C, what is the new pressure?

Section 5-4 What Are Avogadro's Law and the Ideal Gas Law?

5-36 Answer true or false.
  (a) Avogadro's law states that equal volumes of gas at the same temperature and pressure contain equal numbers of molecules.
  (b) At STP, one mole of uranium hexafluoride (UF_6, MW 352 amu), the gas used in uranium enrichment programs, occupies a volume of 352 L.
  (c) If two gas samples have the same temperature, volume, and pressure, then both contain the same number of molecules.
  (d) The value of Avogadro's number is 6.02×10^{23} g/mol.
  (e) Avogadro's number is valid only for gases at STP.
  (f) The ideal gas law is $PV = nRT$.
  (g) When using the ideal gas law for calculations, temperature must be in degrees Celsius.
  (h) If one mole of ethane (CH_3CH_3) gas occupies 20.0 L at 1.00 atm, the temperature of the gas is 244 K.
  (i) One mole of helium (MW 4.0 amu) gas at STP occupies twice the volume of one mole of hydrogen (MW 2.0 amu).

5-37 A sample of a gas at 77°C and 1.33 atm occupies a volume of 50.3 L.
  (a) How many moles of the gas are present?
  (b) Does your answer depend on knowing what gas it is?

5-38 What is the volume in liters occupied by 1.21 g of Freon-12 gas, CCl_2F_2, at 0.980 atm and 35°C?

5-39 An 8.00-g sample of a gas occupies 22.4 L at 2.00 atm and 273 K. What is the molar mass of the gas?

5-40 What volume is occupied by 5.8 g of propane gas, C_3H_8, at 23°C and 1.15 atm pressure?

5-41 Does the density of a gas increase, decrease, or stay the same as the pressure increases at constant temperature? As the temperature increases at constant pressure?

5-42 What volume in milliliters does 0.275 g of uranium hexafluoride gas, UF_6, occupy at its boiling point of 56°C at 365 torr?

5-43 A hyperbaric chamber has a volume of 200. L.
  (a) How many moles of oxygen are needed to fill the chamber at room temperature (23°C) and 3.00 atm pressure?
  (b) How many grams of oxygen are needed?

5-44 One breath of air has a volume of 2 L at STP. If air contains 20.9% oxygen, how many molecules of oxygen are in one breath?

5-45 An average pair of lungs has a volume of 5.5 L. If the air they contain is 21% oxygen, how many molecules of O_2 do the lungs contain at 1.1 atm and 37°C?

5-46 Calculate the molar mass of a gas if 3.30 g of the gas occupies 660. mL at 735 mm Hg and 27°C.

5-47 The three main components of dry air and the percentage of each are N_2 (78.08%), O_2 (20.95%), and Ar (0.93%).
  (a) Calculate the mass of one mole of air.
  (b) Given the mass of one mole of air, calculate the density of air in g/L at STP.

5-48 The density of Freon-12, CCl_2F_2 at STP is 4.99 g/L, which means that it is approximately four times more dense than air. Show how the kinetic molecular theory of gases accounts for the fact that although Freon-12 is more dense than air, it nevertheless finds its way to the stratosphere, where it is implicated in the destruction of Earth's protective ozone layer.

5-49 Calculate the density in g/L of each of these gases at STP. Which gases are denser than air as calculated in Problem 5-47(b)? Which are less dense than air?
  (a) SO_2 (b) CH_4 (c) H_2
  (d) He   (e) CO_2

5-50 How many molecules of CO are in 100. L of CO at STP?

5-51 The density of liquid octane, C_8H_{18}, is 0.7025 g/mL. If 1.00 mL of liquid octane is vaporized at 100°C and 725 torr, what volume does the vapor occupy?

5-52 The density of acetylene gas, C_2H_2, in a 4-L container at 0°C and 2 atm pressure is 0.02 g/mL. What would be the density of the gas under identical temperature and pressure if the container were partitioned into two 2-L compartments?

5-53 Sodium metal reacts explosively with hydrochloric acid, HCl(aq), as shown in the following chemical equation:

$$2Na(s) + 2HCl(aq) \longrightarrow 2NaCl(aq) + H_2(g)$$

What volume of $H_2(g)$ is produced when 3.50 g of Na(s) is reacted with an excess of hydrochloric acid at a temperature of 18°C and a pressure of 0.995 atm?

5-54 Automobile air bags are inflated by nitrogen gas. When a significant collision occurs, an electronic sensor triggers the decomposition of sodium azide to form nitrogen gas and sodium metal. The nitrogen gas then inflates nylon bags, which protect the driver and front-seat passenger from impact with the dashboard and windshield.

$$2NaN_3(s) \longrightarrow 2Na(s) + 3N_2(g)$$
Sodium azide

What volume of nitrogen gas measured at 1 atm and 27°C is formed by the decomposition of 100. g of sodium azide?

Section 5-5 What Is Dalton's Law of Partial Pressures?

5-55 Answer true or false.
(a) Partial pressure is the pressure that a gas in a container would exert if it were alone in the container.
(b) The units of partial pressure are grams per liter.
(c) Dalton's law of partial pressures states that the total pressure of a mixture of gases is the sum of the partial pressures of each gas.
(d) If 1 mole of CH_4 gas at STP is added to 22.4 L of N_2 at STP, the final pressure in the 22.4 L container will be 1.00 atm.

5-56 The three main components of dry air and the percentage of each are nitrogen (78.08%), oxygen (20.95%), and argon (0.93%).
(a) Calculate the partial pressure of each gas in a sample of dry air at 760 mm Hg.
(b) Calculate the total pressure exerted by these three gases combined.

5-57 Air in the trachea contains oxygen (19.4%), carbon dioxide (0.4%), water vapor (6.2%), and nitrogen (74.0%). If the pressure in the trachea is assumed to be 1.0 atm, what are the partial pressures of these gases in this part of the body?

5-58 The partial pressures of a mixture of gases are as follows: oxygen, 210 mm Hg; nitrogen, 560 mm Hg; and carbon dioxide, 15 mm Hg. The total pressure of the gas mixture was 790 mm Hg. Is there another gas present in the mixture?

Section 5-6 What Is the Kinetic Molecular Theory?

5-59 Answer true or false.
(a) According to the kinetic molecular theory, gas particles have mass but no volume.
(b) According to the kinetic molecular theory, the average kinetic energy of gas particles is proportional to the temperature in degrees Celsius.
(c) According to the kinetic molecular theory, when gas particles collide, they bounce off each other with no change in total kinetic energy.
(d) According to the kinetic molecular theory, there are only weak intramolecular forces of attraction between gas particles.
(e) According to the kinetic molecular theory, the pressure of a gas in a container is the result of collisions of gas particles on the walls of the container.
(f) Warming a gas results in an increase in the average kinetic energy of its particles.
(g) When a gas is compressed, the increase in its pressure is the result of an increase in the number of collisions of its particles on the walls of the container.
(h) The kinetic molecular theory describes the behavior of ideal gases, of which there are only a few.
(i) As the temperature and volume of a gas increase, the behavior of the gas becomes more like the behavior predicted by the ideal gas law.
(j) If the assumptions of the kinetic molecular theory of gases are correct, then there is no combination of temperature and pressure at which a gas would become liquid.

5-60 Compare and contrast Dalton's atomic theory and the kinetic molecular theory.

Section 5-7 What Types of Attractive Forces Exist Between Molecules?

5-61 Answer true or false.
(a) Of the forces of attraction between particles, London dispersion forces are the weakest and covalent bonds are the strongest.
(b) All covalent bonds have approximately the same energy.
(c) London dispersion forces arise because of the attraction of temporary induced dipoles.
(d) In general, London dispersion forces increase as molecular size increases.
(e) London dispersion forces occur only between polar molecules—they do not occur between nonpolar atoms or molecules.
(f) The existence of London dispersion forces accounts for the fact that even small, nonpolar particles such as Ne, He, and H_2 can be liquefied if the temperature is low enough and the pressure is high enough.
(g) For nonpolar gases at STP, the average kinetic energy of its particles is greater than the force of attraction between gas particles.
(h) Dipole–dipole interaction is the attraction between the positive end of one dipole and the negative end of another.
(i) Dipole–dipole interactions exist between CO molecules but not between CO_2 molecules.
(j) If two polar molecules have approximately the same molecular weight, the strength of the dipole–dipole interactions between the molecules of each will be approximately the same.

(k) Hydrogen bonding refers to the single covalent bond between the two hydrogen atoms in H—H.

(l) The strength of hydrogen bonding in liquid water is approximately the same as that of an O—H covalent bond in water.

(m) Hydrogen bonding, dipole–dipole interactions, and London dispersion forces have in common that the forces of attraction between particles are all electrostatic (positive to negative and negative to positive).

(n) Water (H_2O, bp 100°C) has a higher boiling point than hydrogen sulfide (H_2S, bp −61°C) because the hydrogen bonding between H_2O molecules is stronger than that between H_2S molecules.

(o) The hydrogen bonding among molecules containing N—H groups is stronger than that among molecules containing O—H groups.

5-62 Which forces are stronger, intramolecular covalent bonds or intermolecular hydrogen bonds?

5-63 Under which condition does water vapor behave most ideally?
 (a) 0.5 atm, 400 K   (b) 4 atm, 500 K
 (c) 0.01 atm, 500 K

5-64 Can water and dimethyl sulfoxide, $(CH_3)_2S{=}O$, molecules form hydrogen bonds between them?

5-65 What kind of intermolecular interactions take place in (a) liquid CCl_4 and (b) liquid CO? Which will have the highest surface tension?

5-66 Ethanol, C_2H_5OH, and carbon dioxide, CO_2, have approximately the same molecular weight, yet carbon dioxide is a gas at STP and ethanol is a liquid. How do you account for this difference in physical property?

5-67 Can dipole–dipole interactions ever be weaker than London dispersion forces? Explain.

5-68 Which compound has a higher boiling point: butane, C_4H_{10}, or hexane, C_6H_{14}?

Section 5-8 How Do We Describe the Behavior of Liquids at the Molecular Level?

5-69 Answer true or false.
 (a) The ideal gas law assumes that there are no attractive forces between molecules. If this were true, then there would be no liquids.
 (b) Unlike a gas, whose molecules move freely in any direction, molecules in a liquid are locked into fixed positions, giving the liquid a constant shape.
 (c) Surface tension is the force that prevents a liquid from being stretched.
 (d) Surface tension creates an elastic-like layer on the surface of a liquid.
 (e) Water has a high surface tension because H_2O is a small molecule.
 (f) Vapor pressure is proportional to temperature—as the temperature of a liquid sample increases, its vapor pressure also increases.
 (g) When molecules evaporate from a liquid, the temperature of the liquid drops.
 (h) Evaporation is a cooling process because it leaves fewer molecules with high kinetic energy in the liquid state.
 (i) The boiling point of a liquid is the temperature at which its vapor pressure equals the atmospheric pressure.
 (j) As the atmospheric pressure increases, the boiling point of a liquid increases.
 (k) The temperature of boiling water is related to how vigorously it is boiling—the more vigorous the boiling, the higher the temperature of the water.
 (l) The most important factor determining the relative boiling points of liquids is molecular weight—the greater the molecular weight, the higher the boiling point.
 (m) Ethanol (CH_3CH_2OH, bp 78.5°C) has a greater vapor pressure at 25°C than water (H_2O, bp 100°C).
 (n) Hexane ($CH_3CH_2CH_2CH_2CH_2CH_3$, bp 69°C) has a higher boiling point than methane (CH_4, bp −164°C) because hexane has more sites for hydrogen bonding between its molecules than does methane.
 (o) A water molecule can participate in hydrogen bonding through each of its hydrogen atoms and through its oxygen atom.
 (p) For nonpolar molecules of comparable molecular weight, the more compact the shape of the molecule, the higher its boiling point.

5-70 The melting point of chloroethane, CH_3CH_2Cl, is −136°C and its boiling point is 12°C. Is chloroethane a gas, a liquid, or a solid at STP?

Section 5-9 What Are the Characteristics of the Various Types of Solids?

5-71 Answer true or false.
 (a) Formation of a liquid from a solid is called melting; formation of a solid from a liquid is called crystallization.
 (b) Most solids have a higher density than their liquid forms.
 (c) Molecules in a solid are locked into fixed positions.
 (d) Each element has one and only one solid (crystalline) form.
 (e) Diamond and graphite are both crystalline forms of carbon.
 (f) Diamond consists of hexagonal crystals of carbon arranged in a repeating pattern.
 (g) The *nano* in nanotube refers to the structure dimensions, which are in the nanometer (10^{-9} m) range.
 (h) Nanotubes have lengths up to 1 nm.
 (i) A buckyball (C_{60}) has a diameter of 1 nm.
 (j) All solids, if heated to a high enough temperature, can be melted.
 (k) Glass is an amorphous solid.

5-72 Identify the type of crystalline solid (i.e., ionic, molecular, metallic, network covalent, polymeric) formed by each of the following:
(a) glucose, $C_6H_{12}O_6$
(b) silver, Ag
(c) silicon carbide, SiC
(d) bottle containing Arrowhead® brand mountain spring water
(e) potassium iodide, KI
(f) elemental sulfur powder, S_8

Section 5-10 What Is a Phase Change and What Energies Are Involved?

5-73 Answer true or false.
(a) A phase change from solid to liquid is called melting.
(b) A phase change from liquid to gas is called boiling.
(c) If heat is added slowly to a mixture of ice and liquid water, the temperature of the sample gradually increases until all of the ice is melted.
(d) Heat of fusion is the heat required to melt 1 g of a solid.
(e) Heat of vaporization is the heat required to evaporate 1 g of liquid at the normal boiling point of the liquid.
(f) Steam burns are more damaging to the skin than hot-water burns because the specific heat of steam is so much higher than the specific heat of hot water.
(g) The heat of vaporization of water is approximately the same as its heat of fusion.
(h) The specific heat of water is the heat required to raise the temperature of 1 g of water from 0°C to 100°C.
(i) Melting a solid is an exothermic process; crystallization of a liquid is an endothermic process.
(j) Melting a solid is a reversible process; the solid can be converted to a liquid and the liquid back to a solid with no change in composition of the sample.
(k) Sublimation is a phase change from solid directly to gas.

5-74 Calculate the specific heat (Section 1-9) of gaseous Freon-12, CCl_2F_2, if it requires 170. cal to change the temperature of 36.6 g of Freon-12 from 30.°C to 50.°C.

5-75 The heat of vaporization of liquid Freon-12, CCl_2F_2, is 4.71 kcal/mol. Calculate the energy required to vaporize 39.2 g of this compound. The molecular weight of Freon-12 is 120.9 amu.

5-76 The specific heat (Section 1-9) of mercury is 0.0332 cal/g·°C. Calculate the energy necessary to raise the temperature of one mole of liquid mercury by 36°C.

5-77 Using Figure 5-14, estimate the vapor pressure of ethanol at (a) 30°C, (b) 40°C, and (c) 60°C.

5-78 CH_4 and H_2O have about the same molecular weight. Which has the higher vapor pressure at room temperature? Explain.

5-79 The normal boiling point of a substance depends on both the mass of the molecule and the attractive forces between molecules. Arrange the compounds in each set in order of increasing boiling point and explain your answer:
(a) HCl, HBr, HI   (b) O_2, HCl, H_2O_2

5-80 Refer to Figure 5-19. How many calories are required to bring one mole of ice at 0°C to a liquid state at room temperature (23°C)?

5-81 Compare the number of calories absorbed when 100. g of ice at 0°C is changed to liquid water at 37°C with the number of calories absorbed when 100. g of liquid water is warmed from 0°C to 37°C.

5-82 (a) How much energy is released when 10. g of steam at 100°C is condensed and cooled to body temperature (37°C)?
(b) How much energy is released when 100. g of liquid water at 100°C is cooled to body temperature (37°C)?
(c) Why are steam burns more painful than hot-water burns?

5-83 When iodine vapor hits a cold surface, iodine crystals form. Name the phase change that is the reverse of this condensation.

5-84 If a 156-g block of dry ice, CO_2, is sublimed at 25°C and 740 mm Hg, what volume does the gas occupy?

5-85 Trichlorofluoromethane (Freon-11, CCl_3F) as a spray is used to temporarily numb the skin around minor scrapes and bruises. It accomplishes this by reducing the temperature of the treated area, thereby numbing the nerve endings that perceive pain. Calculate the heat in kilocalories that can be removed from the skin by 1.00 mL of Freon-11. The density of Freon-11 is 1.49 g/mL, and its heat of vaporization is 6.42 kcal/mol.

5-86 Using the phase diagram of water (Figure 5-20), describe the process by which you can sublime 1 g of ice at −10°C and at 1 atm pressure to water vapor at the same temperature.

Chemical Connections

5-87 (Chemical Connections 5A) What happens when a person lowers the diaphragm in his or her chest cavity?

5-88 (Chemical Connections 5B) In carbon monoxide poisoning, the hemoglobin is incapable of transporting oxygen to the tissues. How does the oxygen get delivered to the cells when a patient is put into a hyperbaric chamber?

5-89 (Chemical Connections 5C) In a sphygmomanometer one listens to the first tapping sound as the constrictive pressure of the arm cuff is slowly released. What is the significance of this tapping sound?

5-90 (Chemical Connections 5D) Why is the damage by severe frostbite irreversible?

5-91 (Chemical Connections 5D) If you fill a glass bottle with water, cap it, and cool to −10°C, the bottle will crack. Explain.

5-92 (Chemical Connections 5E) In what way does supercritical CO_2 have some of the properties of a gas and some of the properties of a liquid?

Additional Problems

5-93 Why is it difficult to compress a liquid or a solid?

5-94 Explain in terms of the kinetic molecular theory what causes (a) the pressure of a gas and (b) the temperature of a gas.

5-95 The unit of pressure most commonly used for checking the inflation of automobile and bicycle tires is pounds per square inch (lb/in^2), abbreviated psi. The conversion factor between atm and psi is 1.00 atm = 14.7 psi. Suppose an automobile tire is filled to a pressure of 34 psi. What is the pressure in atm in the tire?

5-96 The gas in an aerosol can is at a pressure of 3.0 atm at 23°C. What will the pressure of the gas in the can be if the temperature is raised to 400°C?

5-97 Why do aerosol cans carry the warning "Do not incinerate"?

5-98 Under certain weather conditions (just before rain), the air becomes less dense. How does this change affect the barometric pressure reading?

5-99 An ideal gas occupies 387 mL at 275 mm Hg and 75°C. If the pressure changes to 1.36 atm and the temperature increases to 105°C, what is the new volume?

5-100 Arrange the following solids in order of increasing expected melting points: $CO_2(s)$, $Xe(s)$, $CaO(s)$, $H_2O(s)$, $LiCl(s)$, and $HCl(s)$.

5-101 On the basis of what you have learned about intermolecular forces, predict which liquid has the highest boiling point:
  (a) Pentane, C_5H_{12}
  (b) Chloroform, $CHCl_3$
  (c) Water, H_2O

5-102 A 10-L gas cylinder is filled with N_2 to a pressure of 35 in. Hg. How many moles of N_2 do you have to add to your container to raise the pressure to 60 in. Hg? Assume a constant temperature of 27°C.

5-103 When filled, a typical tank for an outdoor grill contains 20. lb of LP (liquefied petroleum) gas, the major component of which is propane, C_3H_8. For this problem, assume that propane is the only substance present.
  (a) How do you account for the fact that when propane is put under pressure, it can be liquefied?
  (b) How many kilograms of propane does a full tank contain?
  (c) How many moles of propane does a full tank contain?
  (d) If the propane in a full tank was released into a flexible container, what volume would it occupy at STP?

5-104 Explain why many gases are transparent.

5-105 The density of a gas is 0.00300 g/cm^3 at 100.°C and 1.00 atm. What is the mass of one mole of the gas?

5-106 The normal boiling point of hexane, C_6H_{14}, is 69°C, and that of pentane, C_5H_{12}, is 36°C. Predict which of these compounds has a higher vapor pressure at 20°C.

5-107 If 60.0 g of NH_3 occupies 35.1 L under a pressure of 77.2 in. Hg, what is the temperature of the gas, in °C?

5-108 Water is a liquid at STP. Hydrogen sulfide, H_2S, a heavier molecule, is a gas under the same conditions. Explain.

5-109 Why does the temperature of a liquid drop as a result of evaporation?

5-110 What volume of air (21% oxygen) measured at 25°C and 0.975 atm is required to completely oxidize 3.42 g of aluminum to aluminum oxide, Al_2O_3?

Tying It Together

5-111 Diving, particularly SCUBA (Self-Contained Underwater Breathing Apparatus) diving, subjects the body to increased pressure. Each 10. m (approximately 33 ft) of water exerts an additional pressure of 1 atm on the body.
  (a) What is the pressure on the body at a depth of 100. ft?
  (b) The partial pressure of nitrogen gas in air at 1 atm is 593 mm Hg. Assuming a SCUBA diver breathes compressed air, what is the partial pressure of nitrogen entering the lungs from a breathing tank at a depth of 100. ft?
  (c) The partial pressure of oxygen gas in the air at 2 atm is 158 mm Hg. What is the partial pressure of oxygen in the air in the lungs at a depth of 100. ft?
  (d) Why is it absolutely essential to exhale vigorously in a rapid ascent from a depth of 100. ft?

5-112 Consider the mixing of 3.5 L of $CO_2(g)$ and 1.8 L of $H_2O(g)$ at 35°C and 740 mm Hg. Determine the mass of $O_2(g)$ that can be produced from the unbalanced reaction:

$$CO_2(g) + H_2O(g) \longrightarrow C_4H_{10}(\ell) + O_2(g)$$

5-113 Ammonia and gaseous hydrogen chloride react to form ammonium chloride according to the following equation:

$$NH_3(g) + HCl(g) \longrightarrow NH_4Cl(s)$$

If 4.21 L of $NH_3(g)$ at 27°C and 1.02 atm is combined with 5.35 L of $HCl(g)$ at 26°C and 0.998 atm, what mass of $NH_4Cl(s)$ will be generated?

5-114 Carbon dioxide gas, saturated with water vapor, can be produced by the addition of aqueous acid to calcium carbonate based on the following balanced net ionic equation:

$$CaCO_3(s) + 2H^+(aq) \longrightarrow Ca^{2+}(aq) + H_2O(\ell) + CO_2(g)$$

  (a) How many moles of wet $CO_2(g)$, collected at 60.°C and 774 torr total pressure, are produced by the complete reaction of 10.0 g of $CaCO_3$ with excess acid?
  (b) What volume does this wet CO_2 occupy?
  (c) What volume would the CO_2 occupy at 774 torr if a desiccant (a chemical drying agent) were added to remove the water? The vapor pressure of water at 60.°C is 149.4 mm Hg.

5-115 Ammonium nitrite decomposes upon heating to form nitrogen gas and water vapor according to the following unbalanced chemical reaction:

$$NH_4NO_2(s) \longrightarrow N_2(g) + H_2O(g)$$

When a sample is decomposed in a test tube, 511 mL of wet $N_2(g)$ is collected over water at 26°C and 745 torr total pressure. How many grams of dry $NH_4NO_2(s)$ were initially decomposed? The vapor pressure of water at 26°C is 25.2 torr.

5-116 How much total heat in calories is required to raise the temperature of 3.50 g of ice at -10.0°C to water vapor at 115°C? Refer to Table 5-5 for relevant data.

5-117 Determine the total amount of heat lost in calories when 5.75 g of water vapor at 120.°C is cooled to $-20.$°C. Refer to Table 5-5 for relevant data.

Challenge Problems

5-118 Isooctane, which has a chemical formula C_8H_{18}, is the component of gasoline from which the term *octane rating* derives.

(a) Write the balanced chemical equation for the combustion of isooctane.

(b) The density of isooctane is 0.792 g/mL. How many kg of CO_2 are produced each year by the annual U.S. gasoline consumption of 4.6×10^{10} L?

(c) What is the volume in liters of this CO_2 at STP?

(d) The chemical formula for isooctane can be represented by $(CH_3)_3CCH_2CH(CH_3)_2$. Draw a Lewis structure of isooctane.

(e) Another molecule with the same molecular formula is octane, which can be represented by:

When comparing isooctane and octane, one structure is observed to have a boiling point of 99°C, while another is known to have a boiling point of 125°C. Which substance, isooctane or octane, is expected to have the higher boiling point?

(f) Determine whether isooctane or octane is expected to have the greater vapor pressure.

5-119 Consider the decomposition of solid ammonium nitrate to form gaseous dinitrogen oxide and water vapor. A 2.50 g sample of $NH_4NO_3(s)$ is introduced into a 1.75 L flask and heated to 230°C.

(a) Write the balanced chemical equation for this decomposition process.

(b) What is the partial pressure of $N_2O(g)$ and $H_2O(g)$ produced?

(c) Determine the total gas pressure present in the flask at 230°C.

(d) Using VSEPR theory, draw three equivalent resonance structures for $N_2O(g)$.

5-120 A 0.325 g sample of a compound containing carbon and hydrogen only occupies a volume of 193 mL at 749 mm Hg and 26.1°C.

(a) Determine the molecular weight of this compound containing carbon and hydrogen only.

(b) Draw a possible Lewis structure for this compound.

(c) Determine the various relative bond angles associated with each central atom of this compound using VSEPR theory.

(d) Would you predict this compound to be polar or nonpolar?

(e) What types of intermolecular forces are present in a container with this compound?

Solutions and Colloids

Human blood cells in an isotonic solution.

KEY QUESTIONS

- 6-1 What Do We Need to Know as We Begin?
- 6-2 What Are the Most Common Types of Solutions?
- 6-3 What Are the Distinguishing Characteristics of Solutions?
- 6-4 What Factors Affect Solubility?
- 6-5 What Are the Most Common Units for Concentration?
- 6-6 Why Is Water Such a Good Solvent?
- 6-7 What Are Colloids?
- 6-8 What Is a Colligative Property?

6-1 What Do We Need to Know as We Begin?

In Chapter 2, we discussed pure substances—compounds made of two or more elements in a fixed ratio. Such systems are the easiest to study, so it was convenient to begin with them. In our daily lives, however, we more frequently encounter mixtures—systems consisting of more than one component. Air, smoke, seawater, milk, blood, and rocks, for example, are mixtures (Section 2-2C).

If a mixture is uniform throughout at the molecular level, we call it a homogeneous mixture or, more commonly, a solution. Filtered air and seawater, for example, are both solutions. They are clear and transparent. In contrast, in most rocks we can see distinct regions separated from each other by well-defined boundaries. Such rocks are heterogeneous mixtures. Another example is a mixture of sand and sugar. We can easily distinguish between the two components; the mixing does not occur at the molecular level (Figure 2-3). Thus, mixtures are classified on the basis of how they look to the unaided eye.

Some systems, however, fall between homogeneous and heterogeneous mixtures. Cigarette smoke, milk, and blood plasma may look homogeneous, but they do not have the transparency of air or seawater. These mixtures are classified as suspensions. We will deal with such systems in Section 6-7.

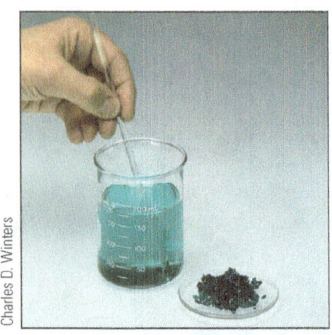

▶ Making a homogeneous solution. A green solid, nickel nitrate, is stirred into water, where it dissolves to form a homogeneous solution.

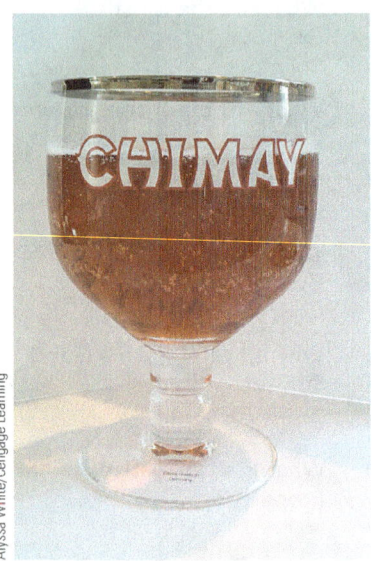

▶ Beer is a solution in which a liquid (alcohol), a solid (malt), and a gas (CO_2) are dissolved in the solvent, water.

▶ Mixtures can be homogeneous, as with brass, which is a solid solution of copper and zinc. Alternatively, they can be heterogeneous, as with granite, which contains discrete regions of different minerals (feldspar, mica, and quartz).

Alloys Homogeneous mixtures of two or more metals

Although mixtures can contain many components, we will generally restrict our discussion to two-component systems, with the understanding that everything we say can be extended to multicomponent systems.

6-2 What Are the Most Common Types of Solutions?

When we think of a solution, we normally think of a liquid. Liquid solutions, such as sugar in water, are the most common kind, but there are also solutions that are gases or solids. In fact, all mixtures of gases are solutions. Because gas molecules are far apart from each other and much empty space separates them, two or more gases can mix with each other in any proportions. Because the mixing takes place at the molecular level, a true solution always forms; that is, there are no heterogeneous mixtures of gases.

With solids, we are at the other extreme. Whenever we mix solids, we almost always get a heterogeneous mixture. Because even microscopic pieces of solid still contain many billions of particles (molecules, ions, or atoms), there is no way to achieve mixing at the molecular level. Homogeneous mixtures of solids (or **alloys**), such as brass, do exist, but we make them by melting the solids, mixing the molten components, and allowing the mixture to solidify.

Table 6-1 lists the five most common types of solutions. Examples of other types (gas in solid, liquid in gas, and so on) are also known but are much less important. ◀

Table 6-1 The Most Common Types of Solutions

Solute		Solvent	Appearance of Solution	Example
Gas	in	Liquid	Liquid	Carbonated water
Liquid	in	Liquid	Liquid	Wine
Solid	in	Liquid	Liquid	Salt water (saline solution)
Gas	in	Gas	Gas	Air
Solid	in	Solid	Solid	14-Carat gold

When a solution consists of a solid or a gas dissolved in a liquid, the liquid is called the **solvent** and the solid or gas is called the **solute**. A solvent may have several solutes dissolved in it, even of different types. A common example is spring water, in which gases (carbon dioxide and oxygen) and solids (salts) are dissolved in the solvent, water.

When one liquid is dissolved in another, a question may arise regarding which is the solvent and which is the solute. The one present in the greater amount is usually called the solvent. We normally do not use the terms "solute" and "solvent" when talking about solutions of gases in gases or solids in solids.

6-3 What Are the Distinguishing Characteristics of Solutions?

The following are some properties of solutions:

1. **The distribution of particles in a solution is uniform.**
   Every part of the solution has exactly the same composition and properties as every other part. That, in fact, is the definition of "homogeneous." As a consequence, we cannot usually tell a solution from a pure solvent

CHEMICAL CONNECTIONS 6A

Acid Rain

The water vapor evaporated by the sun from oceans, lakes, and rivers condenses and forms clouds that eventually fall as rain. The raindrops contain small amounts of CO_2, O_2, and N_2. Table 6A shows that of these gases, CO_2 is the most soluble in water. When CO_2 dissolves in water, it reacts with a water molecule to give carbonic acid, H_2CO_3.

$$CO_2(g) + H_2O(\ell) \longrightarrow \underset{\text{Carbonic acid}}{H_2CO_3(aq)}$$

The acidity caused by the CO_2 is not harmful; however, contaminants that result from industrial pollution may create a serious acid rain problem. Burning coal or oil that contains sulfur generates sulfur dioxide, SO_2, which has a high solubility in water. Sulfur dioxide in the air is oxidized to sulfur trioxide, SO_3. The reaction of sulfur dioxide with water gives sulfurous acid, and the reaction of sulfur trioxide with water gives sulfuric acid.

$$\underset{\text{Sulfur dioxide}}{SO_2} + H_2O \longrightarrow \underset{\text{Sulfurous acid}}{H_2SO_3}$$

$$\underset{\text{Sulfur trioxide}}{SO_3} + H_2O \longrightarrow \underset{\text{Sulfuric acid}}{H_2SO_4}$$

Smelting, which involves melting or fusing an ore as part of the refining (or separation) process, produces other soluble gases as well. In many parts of the world, especially those located downwind from heavily industrialized areas, the result is acid rain that pours down on forests and lakes. It damages vegetation and kills fish. Acid rain has been observed with increasing frequency in the eastern United States, in North Carolina, in the Adirondack Mountains of New York State, and in parts of New England, as well as in eastern Canada.

Trees killed by acid rain at Mt. Mitchell in North Carolina.

Table 6A The Solubility of Some Gases in Water

Gas	Solubility (g/kg H_2O at 20°C and 1 atm)
O_2	0.0434
N_2	0.0190
CO_2	1.688
H_2S	3.846
SO_2	112.80
NO_2	0.0617

Test your knowledge with Problems 6-78 and 6-79.

---

simply by looking at it. A glass of pure water looks the same as a glass of water containing dissolved salt or sugar. In some cases, we can tell by looking—for example, if the solution is colored and we know that the solvent is colorless.

2. **The components of a solution do not separate on standing.**
   A solution of vinegar (acetic acid in water), for example, will never separate.

3. **A solution cannot be separated into its components by filtration.**
   Both the solvent and the solute pass through a filter paper.

4. **For any given solute and solvent, it is possible to make solutions of many different compositions.**
   For example, we can easily make a solution of 1 g of glucose in 100. g of water, or 2 g, or 6 g, or 8.7 g, or any other amount of glucose up to the solubility limit (Section 6-4).

5. **Solutions are almost always transparent.**
   They may be colorless or colored, but we can usually see through them. Solid solutions are exceptions.

6. **Solutions can be separated into pure components.**
   Common separation methods include distillation and chromatography, which we may learn about in the laboratory portion of this course. The separation of a solution into its components is a physical change, not a chemical one.

6-4 What Factors Affect Solubility?

The **solubility** of a solid in a liquid is the maximum amount of the solid that will dissolve in a given amount of a particular solvent at a given temperature. Suppose we wish to make a solution of table salt (NaCl) in water. We take some water, add a few grams of salt, and stir. At first, we see the particles of salt suspended in the water. Soon, however, all the salt dissolves. Now let us add more salt and continue to stir. Again, the salt dissolves. Can we repeat this indefinitely? The answer is no—there is a limit. The solubility of table salt is 36.2 g per 100. g of water at 30°C. If we add more salt than that amount, the excess solid does not dissolve but rather remains suspended as long as we keep stirring; it will sink to the bottom after we stop stirring.

Solubility is a physical constant, like melting point or boiling point. Each solid has a different solubility in every liquid. Some solids have a very low solubility in a particular solvent; we often call these solids *insoluble*. Others have a much higher solubility; we call these *soluble*. However, there is always a solubility limit (see Section 4-3 for some useful solubility generalizations). The same is true for gases dissolved in liquids. Different gases have different solubilities in a solvent (see Chemical Connections 6A). Some liquids are essentially insoluble in other liquids (gasoline in water), whereas others are soluble to a limit. For example, 100. g of water dissolves about 6 g of diethyl ether (another liquid). If we add more ether than that amount, two layers will form (Figure 6-1).

Some liquids, however, are completely soluble in other liquids, no matter how much is present. An example is ethanol, C_2H_6O, and water, which form a solution no matter what quantities of each are mixed. We say that water and ethanol are **miscible** in all proportions.

When a solvent contains all the solute it can hold at a given temperature, we call the solution **saturated.** Any solution containing a lesser amount of solute is **unsaturated.** If we add more solute to a saturated solution at constant temperature, it looks as if none of the additional solid dissolves, because the solution already holds all the solute that it can. Actually, an equilibrium similar to the one discussed in Section 5-8B is at work in this situation. Some particles of the additional solute dissolve, but an equal quantity of dissolved solute comes out of solution. Thus, even though the concentration of dissolved solute does not change, the solute particles themselves are constantly going into and out of solution.

A **supersaturated solution** contains more solute in the solvent than it can normally hold at a given temperature under equilibrium conditions. A supersaturated solution is not stable; when disturbed in any way, such as by stirring or shaking, the excess solute precipitates—thus, the solution returns to equilibrium and becomes merely saturated.

Whether a particular solute dissolves in a particular solvent depends on several factors, as discussed next.

A. Nature of the Solvent and the Solute

The more similar two compounds are, the more likely that one will be soluble in the other. Here the rule is "like dissolves like." This is not an absolute rule, but it does apply in a great many cases.

FIGURE 6-1 Diethyl ether and water form two layers. A separatory funnel permits the bottom layer to be drawn off.

Supersaturated solution A solution that contains more than the equilibrium amount of solute at a given temperature and pressure

When we say "like," we mostly mean similar in terms of polarity. In other words, polar compounds dissolve in polar solvents because the positive end of the dipole of one molecule attracts the negative end of the dipole of the other. Furthermore, nonpolar compounds dissolve in nonpolar solvents. For example, the liquids benzene (C_6H_6) and carbon tetrachloride (CCl_4) are nonpolar compounds. They dissolve in each other, and other nonpolar materials, such as gasoline, dissolve in them. In contrast, ionic compounds such as sodium chloride (NaCl) and polar compounds such as table sugar ($C_{12}H_{22}O_{11}$) are insoluble in these solvents.

The most important polar solvent is water. We have already seen that most ionic compounds are soluble in water, as are small covalent compounds that can form hydrogen bonds with water. It is worth noting that even polar molecules are usually insoluble in water if they are unable either to react with water or to form hydrogen bonds with water molecules. Water as a solvent is discussed in Section 6-6.

B. Temperature

For most solids and liquids that dissolve in liquids, the general rule is that solubility increases with increasing temperature. Sometimes the increase in solubility is great. In other cases it is only moderate. For a few substances, solubility even decreases with increasing temperature (Figure 6-2).

For example, the solubility of glycine, H_2N-CH_2-COOH, a white crystalline solid and a polar building block of proteins, is 52.8 g in 100. g of water at 80°C but only 33.2 g at 30°C. If, for instance, we prepare a saturated solution of glycine in 100. g of water at 80°C, it will hold 52.8 g of glycine. If we then allow the solution to cool to 30°C where the solubility is 33.2 g, we might expect the excess glycine, 19.6 g, to precipitate from solution as crystals. It often does, but on many occasions, it does not. The latter case is an example of a supersaturated solution. Even though the solution contains more glycine than the water can normally hold at 30°C, the excess glycine stays in solution because the molecules need a seed—a surface on which to begin crystallizing. If no such surface is available, no precipitate will form.

Supersaturated solutions are not indefinitely stable, however. If we shake or stir the solution, we may find that the excess solid precipitates at once (Figure 6-3). Another way to crystallize the excess solute is to add a crystal of the solute, a process called **seeding.** The seed crystal provides the surface onto which the solute molecules can converge.

For gases, solubility in liquids almost always decreases with increasing temperature. The effect of temperature on the solubility of gases in water can have important consequences for fish, for example. Oxygen is only slightly soluble in water, and fish need that oxygen to live. When the temperature of a body of water increases, perhaps because of the output from a nuclear power plant, the solubility of oxygen decreases and may become so low that fish die. This situation is called thermal pollution.

C. Pressure

Pressure has little effect on the solubility of liquids or solids. For gases, however, **Henry's law** applies (Figure 6-4): the higher the pressure, the greater the solubility of a gas in a liquid. This concept is the basis of the hyperbaric medicine discussed in Chemical Connections 5B. When the pressure increases, more O_2 dissolves in the blood plasma and reaches tissues at higher-than-normal pressures (2 to 3 atm).

Henry's law also explains why a bottle of beer or any other carbonated beverage foams when it is opened. The bottle is sealed under greater than 1 atm of pressure. When opened at 1 atm, the solubility of CO_2 in the liquid

FIGURE 6-2 The solubilities of some solids in water as a function of temperature. The solubility of glycine increases rapidly, that of NaCl barely increases, and that of Li_2SO_4 decreases with increasing temperature.

FIGURE 6-3 When a supersaturated aqueous solution of sodium acetate ($CH_3COO^-Na^+$) is disturbed, the excess salt crystallizes rapidly.

Henry's law The solubility of a gas in a liquid is directly proportional to the pressure

CHEMICAL CONNECTIONS 6B

The Bends

Deep-sea divers encounter high pressures while under water (see Problem 5-111). For them to breathe properly under such conditions, oxygen must be supplied under pressure. At one time, this goal was achieved with compressed air. As pressure increases, the solubility of gases in the blood increases. This is especially true for nitrogen, which constitutes almost 80% of our air.

When divers come up and the pressure on their bodies decreases, the solubility of nitrogen in their blood decreases as well. As a consequence, the previously dissolved nitrogen in the blood and tissues starts to form small bubbles, especially in the veins. The formation of gas bubbles (called the bends) can hamper blood circulation. If this condition is allowed to develop uncontrolled, a resulting pulmonary embolism can prove fatal.

If the diver's ascent is gradual, regular exhalation and diffusion through the skin remove the dissolved gases. Divers use decompression chambers, where the high pressure is gradually reduced to normal pressure.

If decompression disease develops after a dive, patients are put into a hyperbaric chamber (see Chemical Connections 5B), where they breathe pure oxygen at 2.8 atm pressure. In the standard form of treatment, the pressure is reduced to 1 atm over a period of 6 hours.

Nitrogen also has a narcotic effect on divers when they breathe compressed air at depths greater than 40 m. This effect, called "rapture of the deep," is similar to alcohol-induced intoxication.

Ascending too rapidly will cause dissolved nitrogen bubbles to be released and form bubbles in the blood.

Because of the problem caused by nitrogen, divers' tanks often are charged with a helium–oxygen mixture instead of with air. The solubility of helium in blood is affected less by pressure than is the solubility of nitrogen.

Sudden decompression and ensuing bends are important not only in deep-sea diving but also in high-altitude flight, especially orbital flight.

Test your knowledge with Problems 6-80 and 6-81.

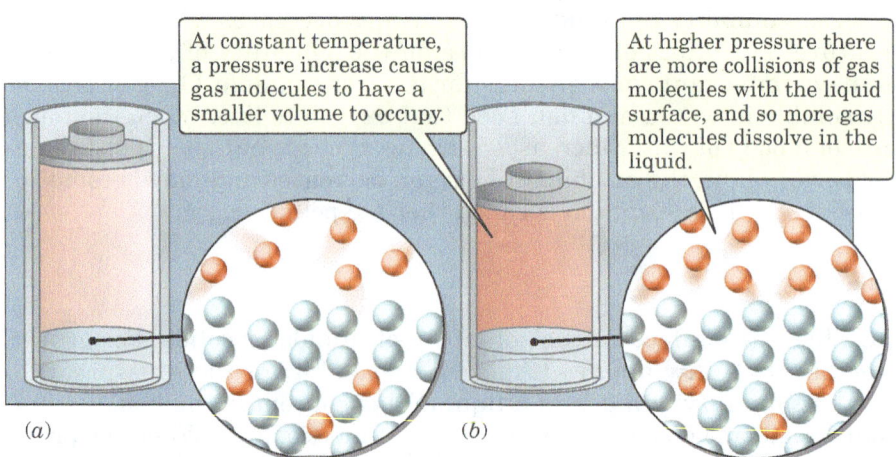

FIGURE 6-4 Henry's law. (a) A gas sample in a liquid under pressure in a closed container. (b) The pressure is increased at constant temperature, causing more gas to dissolve.

decreases. The excess CO_2 is released, forming bubbles, and the gas pushes out some of the liquid. ◄

6-5 What Are the Most Common Units for Concentration?

We can express the amount of a solute dissolved in a given quantity of solvent—that is, the **concentration** of the solution—in a number of ways. Some concentration units are better suited for certain purposes than others are. Sometimes qualitative terms are good enough. For example, we may say that a solution is dilute or concentrated. These terms give us little specific information about the concentration, but we know that a concentrated solution contains more solute than a dilute solution does.

For most purposes, however, we need quantitative concentrations. For example, a nurse must know precisely how much glucose to give to a patient. Many methods of expressing concentration exist, but in this chapter, we deal with just the three most important: percent concentration, molarity, and parts per million (ppm).

► Application of Henry's law. The greater the partial pressure of CO_2 over the soft drink in the bottle, the greater the concentration of dissolved CO_2. When the bottle is opened, the partial pressure of CO_2 drops and CO_2 bubbles out of the solution.

A. Percent Concentration

Chemists represent **percent concentration (% w/v)** in three ways. The most common is mass of solute per volume of solution (w/v):

$$\text{Weight/volume (w/v)\%} = \frac{\text{mass of solute}}{\text{volume of solution}} \times 100$$

Percent concentration (% w/v) The number of grams of solute in 100. mL of solution

If we dissolve 10. g of sucrose (table sugar) in enough water so that the total volume is 100. mL, the concentration is 10.% w/v. Note that here we need to know the total volume of the solution, not the volume of the solvent.

EXAMPLE 6-1 | **Percent Concentration**

The label on a bottle of vinegar says it contains 5.0% w/v acetic acid, CH_3COOH. The bottle contains 240 mL of vinegar. How many grams of acetic acid are in the bottle?

Strategy

We are given the volume of the solution and its weight/volume concentration. To calculate the number of grams of CH_3COOH present in this solution, we use the conversion factor 5.0 g of acetic acid in 100. mL of solution.

Solution

$$240 \text{ mL solution} \times \frac{5.0 \text{ g } CH_3COOH}{100. \text{ mL solution}} = 12 \text{ g } CH_3COOH$$

Problem 6-1

How would we prepare 250 mL of a 4.4% w/v KBr solution in water? Assume that a 250-mL volumetric flask is available.

A second way to represent percent concentration is weight of solute per weight of solution (w/w):

$$\text{Weight/weight (w/w)\%} = \frac{\text{weight solute}}{\text{weight of solution}} \times 100$$

> ### EXAMPLE 6-2 Weight/Volume Percent
>
> If 6.0 g of NaCl is dissolved in enough water to make 300. mL of solution, what is the w/v percent of NaCl?
>
> #### Strategy
>
> To calculate the w/v percent, we divide the weight of the solute by the volume of the solution and multiply by 100:
>
> #### Solution
>
> $$\frac{6.0 \text{ g NaCl}}{300. \text{ mL solution}} \times 100 = 2.0\% \text{ w/v}$$
>
> #### Problem 6-2
>
> If 6.7 g of lithium iodide, LiI, is dissolved in enough water to make 400. mL of solution, what is the w/v percent of LiI?

Calculations of w/w percent are essentially the same as w/v percent calculations, except that we use the weight of the solution instead of its volume. A volumetric flask is not used for these solutions. (Why not?)

Finally, we can represent percent concentration as volume of solute per volume of solution (v/v) percent:

$$\text{Volume/volume (v/v)}\% = \frac{\text{volume solute}}{\text{volume of solution}} \times 100$$

The unit v/v percent is used only for solutions of liquids in liquids—most notably, alcoholic beverages. For example, 40.% v/v ethanol in water means that 40. mL of ethanol has been added to enough water to make 100. mL of solution. This solution might also be called 80 proof, where proof of an alcoholic beverage is twice the v/v percent concentration.

B. Molarity

For many purposes, it is easiest to express concentration by using the weight or volume percentage methods just discussed. When we want to focus on the number of molecules present, however, we need another concentration unit. For example, a 5% solution of glucose in water does not contain the same number of solute molecules as a 5% solution of ethanol in water. That is why chemists often use molarity. **Molarity (M)** is defined as the number of moles of solute dissolved in 1 L of solution. The units of molarity are moles per liter.

$$\text{Molarity } (M) = \frac{\text{moles solute } (n)}{\text{volume of solution (L)}}$$

Thus, in the same volume of solution, a 0.2 M solution of glucose, $C_6H_{12}O_6$, in water contains the same number of molecules of solute as a 0.2 M solution of ethanol, C_2H_6O, in water. In fact, this relationship holds true for equal volumes of any solution, as long as the molarities are the same.

We can prepare a solution of a given molarity in essentially the same way that we prepare a solution of given w/v concentration, except that we use moles instead of grams in our calculations. We can always find out how

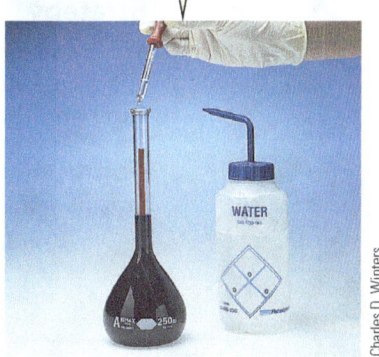

Combine ~240 mL distilled H₂O with 0.395 g (0.00250 mol) KMnO₄ in a 250.0-mL volumetric flask.

Shake the flask to dissolve the KMnO₄.

After the solid dissolves, add sufficient water to fill the flask to the mark etched in the neck, indicating a volume of 250.0 mL, and shake the flask again to thoroughly mix its contents.

FIGURE 6-5 Solution preparation from a solid solute, making 250.0 mL of 0.0100 M aqueous solution of $KMnO_4$.

many moles of solute are in any volume of a solution of known molarity by using the following relationship:

$$\text{Molarity} \times \text{volume in liters} = \text{number of moles}$$

$$\frac{\text{moles}}{\text{liters}} \times \text{liters} = \text{moles}$$

The solution is then prepared as shown in Figure 6-5.

EXAMPLE 6-3 Molarity

How do we prepare 2.0 L of a 0.15 M aqueous solution of sodium hydroxide, NaOH?

Strategy

We are given solid NaOH and want 2.0 L of a 0.15 M solution. First, we find out how many moles of NaOH are present in 2.0 liters of this solution; then we convert this number of moles to grams.

Solution

Step 1: Determine the number of moles of NaOH in 2.0 liters of this solution. For this calculation, we use molarity as a conversion factor.

$$\frac{0.15 \text{ mol NaOH}}{1.0 \text{ L}} \times 2.0 \text{ L} = 0.30 \text{ mol NaOH}$$

Step 2: To convert 0.30 mol of NaOH to grams of NaOH, we use the molar mass of NaOH (40.0 g/mol) as a conversion factor:

$$0.30 \text{ mol NaOH} \times \frac{40.0 \text{ g NaOH}}{1 \text{ mol NaOH}} = 12 \text{ g NaOH}$$

Step 3: To prepare this solution, we place 12 g of NaOH in a 2-L volumetric flask, add some water, swirl until the solid dissolves, and then fill the flask with water to the 2-L mark.

Problem 6-3

How would we prepare 2.0 L of a 1.06 M aqueous solution of KCl?

▶ Blood serum, coagulated blood, and whole blood.

(a) Blood serum

(b) Coagulated blood

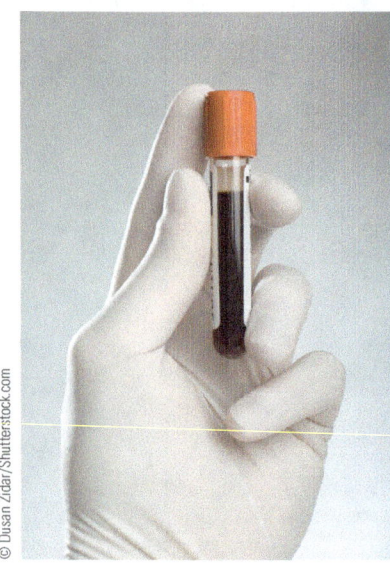

(c) Whole blood

Interactive Example 6-4 — Molarity

If we dissolve 18.0 g of Li_2O (molar mass = 29.9 g/mol) in sufficient water to make 500. mL of solution, what is the molarity of the solution?

Strategy

We are given 18.0 g Li_2O in 500. mL of water and want the molarity of the solution. We first calculate the number of moles of Li_2O in 18.0 g of Li_2O and then convert from moles per 500. mL to moles per liter.

Solution

To calculate the number of moles of Li_2O in a liter of solution, we use two conversion factors: molar mass of Li_2O = 29.9 g and 1000 mL = 1 L.

$$\frac{18.0 \text{ g Li}_2\text{O}}{500. \text{ mL}} \times \frac{1 \text{ mol Li}_2\text{O}}{29.9 \text{ g Li}_2\text{O}} \times \frac{1000 \text{ mL}}{1 \text{ L}} = 1.20\ M$$

Problem 6-4

If we dissolve 0.440 g of KSCN in enough water to make 340. mL of solution, what is the molarity of the resulting solution?

EXAMPLE 6-5 — Molarity

The concentration of sodium chloride in blood serum is approximately 0.14 M. What volume of blood serum contains 2.0 g of NaCl? ◀

Strategy

We are given the concentration in moles per liter and asked to calculate the volume of blood that contains 2.0 g NaCl. To find the volume of blood, we use two conversion factors: the molar mass of NaCl is 58.4 g and the concentration of NaCl in blood is 0.14 M.

Solution

$$2.0 \text{ g NaCl} \times \frac{1 \text{ mol NaCl}}{58.4 \text{ g NaCl}} \times \frac{1 \text{ L}}{0.14 \text{ mol NaCl}} = 0.24 \text{ L} = 2.4 \times 10^2 \text{ mL}$$

Note that the answer in mL must be expressed to no more than two significant figures because the mass of NaCl (2.0 g) is given to only two significant figures. To write the answer as 240. mL would be expressing it to three significant figures. We solve the problem of significant figures by expressing the answer in scientific notation.

Problem 6-5

If a 0.300 M glucose solution is available for intravenous infusion, how many milliliters of this solution are needed to deliver 10.0 g of glucose?

EXAMPLE 6-6 — Molarity

How many grams of HCl are in 225 mL of 6.00 M HCl?

Strategy

We are given 225 mL of 6.00 M HCl and asked to find grams of HCl present. We use two conversion factors—the molar mass of HCl = 36.5 g and 1000 mL = 1 L.

Solution

$$225 \; \cancel{mL} \times \frac{1 \; \cancel{L}}{1000 \; \cancel{mL}} \times \frac{6.00 \; \cancel{\text{mol HCl}}}{1 \; \cancel{L}} \times \frac{36.5 \; \text{g HCl}}{1 \; \cancel{\text{mol HCl}}} = 49.3 \; \text{g HCl}$$

Problem 6-6

A certain wine contains 0.010 M NaHSO$_3$ (sodium bisulfite) as a preservative. How many grams of sodium bisulfite must be added to a 100. gallon barrel of wine to reach this concentration? Assume no change in volume of wine upon addition of the sodium bisulfite.

C. Dilution

We frequently prepare solutions by diluting concentrated solutions rather than by weighing out pure solute (Figure 6-6). Because we add only solvent during dilution, the number of moles of solute remains unchanged. Before dilution, the equation that applies is:

$$M_1 V_1 = \text{moles}$$

After dilution, the volume and molarity have both changed and we have:

$$M_2 V_2 = \text{moles}$$

Because the number of moles of solute is the same both before and after dilution, we can say that:

$$M_1 V_1 = M_2 V_2$$

We can use this handy equation (the units of which are moles = moles) for all dilution problems.

FIGURE 6-6 Solution preparation by dilution. Here 100 mL of 0.100 M potassium dichromate, K$_2$Cr$_2$O$_7$, is diluted to 1.000 L. The result is dilution by a factor of 10.

EXAMPLE 6-7 Dilution

Suppose we have a bottle of concentrated acetic acid (6.0 M). How would we prepare 200. mL of a 3.5 M solution of acetic acid?

Strategy

We are given $M_1 = 6.0\ M$ and asked to calculate V_1. We are also given $M_2 = 3.5\ M$ and $V_2 = 200.$ mL, that is $V_2 = 0.200$ L.

Solution

$$M_1 V_1 = M_2 V_2$$

$$\frac{6.0\ \text{mol}}{1.0\ \text{L}} \times V_1 = \frac{3.5\ \text{mol}}{1.0\ \text{L}} \times 0.200\ \text{L}$$

Solving this equation for V_1 gives:

$$V_1 = \frac{3.5\ \text{mol} \times 0.200\ \text{L}}{6.0\ \text{mol}} = 0.12\ \text{L}$$

To prepare this solution, we place 0.12 L, or 120. mL, of concentrated acetic acid in a 200. mL volumetric flask, add some water and mix, and then fill to the calibration mark with water.

Problem 6-7

We are given a solution of 12.0 M HCl and want to prepare 300. mL of a 0.600 M solution. How would we prepare it?

A similar equation can be used for dilution problems involving percent concentrations:

$$\%_1 V_1 = \%_2 V_2$$

EXAMPLE 6-8 Dilution

Suppose we have a solution of 50.% w/v NaOH on hand. How would we prepare 500. mL of a 0.50% w/v solution of NaOH?

Strategy

We are given 50.% w/v NaOH and asked to prepare 500. mL (V_2) of 0.50% solution V_1. We use the relationship:

$$\%_1 V_1 = \%_2 V_2$$

Solution

$$(50.\%) \times V_1 = (0.50\%) \times 500.\ \text{mL}$$

$$V_1 = \frac{0.50\% \times 500.\ \text{mL}}{50.\%} = 5.0\ \text{mL}$$

To prepare this solution, we add 5.0 mL of the 50.% w/v solution (the concentrated solution) to a 500.-mL volumetric flask, then some water and mix, and then fill to the mark with water. Note that this is a dilution by a factor of 100.

Problem 6-8

A concentrated solution of 15% w/v KOH solution is available. How would we prepare 20.0 mL of a 0.10% w/v KOH solution?

D. Parts per Million

Sometimes we need to deal with very dilute solutions—for example, 0.0001%. In such cases, it is more convenient to use the unit **parts per million (ppm)** to express concentration.

$$\text{ppm} = \frac{\text{g solute}}{\text{g solution}} \times 10^6$$

For example, if drinking water is polluted with lead ions to the extent of 1 ppm, it means that there is 1 mg of lead ions in 1 kg (1 L) of water. When reporting concentration in ppm, the units must be the same for both solute and solution—for example, mg of solute per 10^6 mg of solution, or g of solute per g of solution. Some solutions are so dilute that we use **parts per billion (ppb)** to express their concentrations.

$$\text{ppb} = \frac{\text{g solute}}{\text{g solution}} \times 10^9$$

EXAMPLE 6-9 Parts per Million (ppm)

Verify that 1 mg of lead in 1 kg of drinking water is equivalent to 1 ppm lead.

Strategy

The units we are given are milligrams and kilograms. To report ppm, we must convert them to a common unit, say grams. For this calculation, we use two conversion factors: 1000 mg = 1 g and 1 kg solution = 1000 g solution.

Solution

Step 1: First we find the mass (grams) of lead:

$$1 \text{ mg lead} \times \frac{1 \text{ g lead}}{1000 \text{ mg lead}} = 1 \times 10^{-3} \text{ g lead}$$

Step 2: Next, we find the mass (grams) of the solution:

$$1 \text{ kg solution} \times \frac{1000 \text{ g solution}}{1 \text{ kg solution}} = 1 \times 10^3 \text{ g solution}$$

Step 3: Finally use these values to calculate the concentration of lead in ppm:

$$\text{ppm} = \frac{1 \times 10^{-3} \text{ g lead}}{1 \times 10^3 \text{ g solution}} \times 10^6 = 1 \text{ ppm}$$

Problem 6-9

Sodium hydrogen sulfate, $NaHSO_4$, which dissolves in water to release H^+ ion, is used to adjust the pH of the water in swimming pools. Suppose we add 560. g of $NaHSO_4$ to a swimming pool that contains 4.5×10^5 L of water at 25°C. What is the Na^+ ion concentration in ppm?

Modern methods of analysis allow us to detect such minuscule concentrations. Some substances are harmful even at concentrations measured in ppb. One such substance is dioxin, an impurity in the 2,4,5-T herbicide sprayed by the United States as a defoliant in Vietnam.

6-6 Why Is Water Such a Good Solvent?

Water covers about 75% of the Earth's surface in the form of oceans, ice caps, glaciers, lakes, and rivers. Water vapor is always present in the atmosphere. Life evolved in water and without it life as we know it could not

CHEMICAL CONNECTIONS 6C

Electrolyte Solutions in Body and Intravenous Fluids

Body fluids typically contain a mixture of several electrolytes (see Section 6-6C) such as Na^+, Ca^{2+}, Cl^-, HCO_3^-, and HPO_4^{2-}. The ions present generally originate from more than one source. We measure each individual ion present in terms of an equivalent (Eq), which is the molar amount of an ion equal to one mole of positive or negative electrical charge. For example, 1 mole of Na^+ ion and HCO_3^- ion are each one equivalent because they supply one mole of electrical charge. Ions with a 2+ or 2− charge, such as Ca^{2+} and HPO_4^{2-}, are each two equivalents per one mole of ion.

The concentrations of electrolytes present in body fluids and in intravenous fluids given to a patient are often expressed in milliequivalents per liter (mEq/L) of solution. For example, lactated Ringer's solution is often used for fluid resuscitation after a patient suffers blood loss due to trauma, surgery, or a brain injury. It consists of three cations (130. mEq/L Na^+, 4 mEq/L K^+, and 3 mEq/L Ca^{2+}) and two anions (109 mEq/L Cl^- and 28 mEq/L $C_3H_5O_3^-$, lactate). Notice that the charge balance of the solution is maintained, and the total number of positive charges is equal to the total number of negative charges. The various electrolyte concentrations present

in lactated Ringer's solution are one of many possible intravenous replacement solutions used in a clinical setting. The use of specific intravenous solutions depends on the fluid, electrolytic, and nutritional needs of an individual patient.

Test your knowledge with Problems 6-82 and 6-83.

---

exist. The human body is about 60% water. This water is found both inside the cells of the body (intracellular) and outside the cells (extracellular). Most of the important chemical reactions in living tissue occur in aqueous solution; water serves as a solvent to transport reactants and products from one place in the body to another. Water is also itself a reactant or product in many biochemical reactions. The properties that make water such a good solvent are its polarity and its hydrogen-bonding capacity (Section 5-7C).

A. How Does Water Dissolve Ionic Compounds?

We learned in Section 3-5 that ionic compounds in the solid state are composed of a regular array of ions in a crystal lattice. The crystal is held together by ionic bonds, which are electrostatic attractions between positive and negative ions. Water, of course, is a polar molecule. When a solid ionic compound is added to water, water molecules surround the ions at the surface of the crystal. The negative ions (anions) attract the positive poles of water molecules, and the positive ions (cations) attract the negative poles of water molecules (Figure 6-7). Each ion attracts multiple water molecules. When the combined force of attraction to water molecules is greater than the force of attraction of the ionic bonds that keeps the ions in the crystal, the ions will be completely dislodged. Water molecules now surround the ion removed from the crystal (Figure 6-8). Such ions are said to be **hydrated.** A more general term, covering all solvents, is **solvated.** The solvation layer—that is, the surrounding shell of solvent molecules—acts as a cushion. It prevents a solvated anion from colliding directly with a solvated cation, thereby keeping the solvated ions in solution.

FIGURE 6-7 When water dissolves an ionic compound, water molecules remove anions and cations from the surface of the solid and water molecules surround the ions.

FIGURE 6-8 Anions and cations solvated by water.

Not all ionic solids are soluble in water. Some rules for predicting solubilities were given in Section 4-3.

B. Solid Hydrates

The attraction between ions and water molecules is so strong in some cases that the water molecules are an integral part of the crystal structure of the solids. Water molecules in a crystal are called **water of hydration.** The substances that contain water in their crystals are themselves called **hydrates.** For example, both gypsum and plaster of Paris are hydrates of calcium sulfate: gypsum is calcium sulfate dihydrate, $CaSO_4 \cdot 2H_2O$, and plaster of Paris is calcium sulfate monohydrate, $(CaSO_4)_2 \cdot H_2O$ The dot in the formula $CaSO_4 \cdot 2H_2O$ indicates that H_2O is present in the crystal, but it is not covalently bonded to the Ca^{2+} or SO_4^{2-} ions. Some hydrates hold on to their water molecules tenaciously. To remove them, the crystals must be heated for some time at a high temperature. The crystal without its water of hydration is called **anhydrous.** In many cases, anhydrous crystals attract water so strongly that they absorb from the water vapor in the air. That is, some anhydrous crystals become hydrated upon standing in air. Crystals that do so are called **hygroscopic.**

Hydrated crystals often look different from the anhydrous forms. For example, copper(II) sulfate pentahydrate, $CuSO_4 \cdot 5H_2O$, is blue but the anhydrous form, $CuSO_4$, is white (Figure 6-9).

The difference between hydrated and anhydrous crystals can sometimes have an effect in the body. For example, the compound sodium urate exists in the anhydrous form as spherical crystals, but in the monohydrate form as needle-shaped crystals (Figure 6-10). The deposition of sodium urate monohydrate in the joints (mostly in the big toe) causes gout. If we want a hygroscopic compound to remain anhydrous, we must place it in a sealed container that contains no water vapor.

Hygroscopic A quality of a substance that is able to absorb water vapor from the air

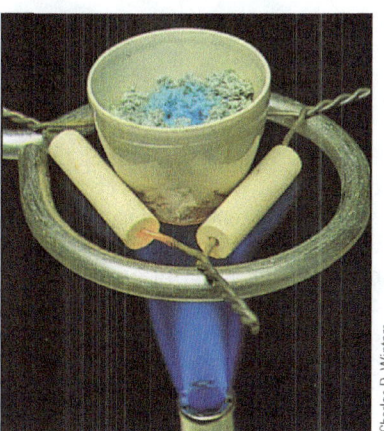

FIGURE 6-9 When blue hydrated copper(II) sulfate, $CuSO_4 \cdot 5H_2O$, is heated and the compound releases its water of hydration, it changes to white anhydrous copper(II) sulfate, $CuSO_4$.

FIGURE 6-10 (a) The needle-shaped sodium urate monohydrate crystals that cause gout. (b) The pain of gout as depicted by an eighteenth-century cartoonist.

C. Electrolytes

Ions in water migrate from one place to another, maintaining their charge in the process. As a consequence, solutions of ions conduct electricity. They can do so because ions in the solution migrate independently of one another. As shown in Figure 6-11, cations migrate to the negative electrode, called the **cathode,** and anions migrate to the positive electrode, called the **anode.** The movement of ions constitutes an electric current. The migration of ions completes the circuit initiated by the battery and can cause an electric bulb to light up (see also Chemical Connections 4B).

A substance, such as potassium chloride, that conducts an electric current when dissolved in water or when in the molten state is called an **electrolyte.** Hydrated K^+ ions carry positive charges, and hydrated Cl^- ions carry negative charges; as a result, the bulb in Figure 6-11 lights brightly if these ions are present. A substance that does not conduct electricity is called a **nonelectrolyte.** Distilled water, for example, is a nonelectrolyte. The light bulb shown in Figure 6-11 does not light up if only distilled water is placed in the beaker. However, with tap water in the beaker, the bulb lights dimly. Tap water contains enough ions to carry electricity, but their concentration is so low that the solution conducts only a small amount of electricity.

As we see, electric conductance depends on the concentration of ions. The higher the ion concentration, the greater the electric conductance of the solution. Nevertheless, differences in electrolytes exist. If we take a 0.1 M aqueous NaCl and compare it with a 0.1 M aqueous acetic acid (CH_3COOH), we find that the NaCl solution lights a bulb brightly, but the acetic acid solution lights it only dimly. We might have expected the two solutions to behave similarly, because each has the same concentration, 0.1 M, and each compound provides two ions, a cation and an anion (Na^+ and Cl^-, H^+ and CH_3COO^-). The reason they behave differently is that, whereas NaCl dissociates completely to two ions (each hydrated and each moving independently), in the case of CH_3COOH, only a few of its molecules dissociate into ions. Most of the acetic acid molecules do not dissociate, and undissociated molecules do not conduct electricity. Compounds that dissociate completely are called **strong electrolytes,** and those that dissociate only partially into ions are called **weak electrolytes.**

Electrolytes are important components of the body because they help to maintain the acid–base balance and the water balance. The most important

FIGURE 6-11 Conductance by an electrolyte. When an electrolyte, such as KCl, is dissolved in water and provides ions that move about, their migration completes an electrical circuit and the lightbulb in the circuit glows. The ions of every KCl unit have dissociated to K^+ and Cl^-. The Cl^- ions move toward the positive electrode and the K^+ ions move toward the negative electrode, thereby transporting electrical charge through the solution.

CHEMICAL CONNECTIONS 6D

Hydrates and Air Pollution: The Decay of Buildings and Monuments

Many buildings and monuments in urban areas throughout the world are decaying, ruined by air pollution. The main culprit in this process is acid rain, an end product of air pollution. The stones most commonly used for buildings and monuments are limestone and marble, both of which are largely calcium carbonate. In the absence of polluted air, these stones can last for thousands of years. Thus, many statues and buildings from ancient times (Babylonian, Egyptian, Greek, and others) survived until recently with little change. Indeed, they remain intact in many rural areas.

In urban areas, however, the air is polluted with SO_2 and SO_3, which come mostly from the combustion of coal and petroleum products containing small amounts of sulfur compounds as impurities (see Chemical Connections 6A). They react with the calcium carbonate at the surface of the stones to form calcium sulfate. When calcium sulfate interacts with rainwater, it forms the dihydrate gypsum.

The problem is that gypsum has a larger volume than the original marble or limestone, and its presence causes the surface of the stone to expand. This activity, in turn, results in flaking. Eventually, statues such as those in the Parthenon (in Athens, Greece) become noseless and later faceless.

$$SO_3(g) + H_2O(g) \longrightarrow H_2SO_4(\ell)$$
Sulfur trioxide   Sulfuric acid

$$CaCO_3(s) + H_2SO_4(\ell) \longrightarrow CaSO_4(s) + H_2O(g) + CO_2(g)$$
Calcium carbonate   Calcium sulfate
(marble, limestone)

$$CaSO_4(s) + 2H_2O(g) \longrightarrow CaSO_4(s) \cdot 2H_2O(s)$$
Calcium sulfate   Calcium sulfate dihydate
                                       (gypsum)

Acid rain damage to stonework on the walls of York Minster, York, England.

Test your knowledge with Problems 6-84 and 6-85.

---

cations in tissues of the human body are Na^+, K^+, Ca^{2+}, and Mg^{2+}. The most important anions in the body are HCO_3^-, Cl^-, HPO_4^{2-}, and $H_2PO_4^-$. ◀

D. How Does Water Dissolve Covalent Compounds?

Water is a good solvent not only for ionic compounds but also for many covalent compounds. In a few cases, the covalent compounds dissolve because they react with water. An example of a covalent compound that dissolves in water is HCl. HCl is a gas (with a penetrating, choking odor) that attacks the mucous membranes of the eyes, nose, and throat. When dissolved in water, HCl molecules react with water to give ions:

$$HCl(g) + H_2O(\ell) \longrightarrow Cl^-(aq) + H_3O^+(aq)$$
Hydrogen   Hydronium ion
chloride

▶ Sports drinks help to maintain the body's electrolyte balance.

Another example is the gas sulfur trioxide, which reacts with water as follows:

$$SO_3(g) + 2H_2O(\ell) \longrightarrow H_3O^+(aq) + HSO_4^-(aq)$$
Sulfur   Hydronium
trioxide   ion

FIGURE 6-12 Solvation of a polar covalent compound by water. The dotted lines represent hydrogen bonds.

Note that H^+ does not exist in aqueous solution; it combines with a water molecule and forms a hydronium ion, H_3O^+. Because HCl and SO_3 are completely converted to ions in dilute aqueous solution, these solutions are ionic solutions and behave just as other electrolytes do (they conduct a current). Nevertheless, HCl and SO_3 are themselves covalent compounds, unlike salts such as NaCl.

Most covalent compounds that dissolve in water do not, in fact, react with water. They dissolve because water molecules surround the entire covalent molecule and solvate it. For example, when methanol, CH_3OH, dissolves in water, the methanol molecules are solvated by the water molecules (Figure 6-12).

There is a simple way to predict which covalent compounds will dissolve in water and which will not. Covalent compounds will dissolve in water if they can form hydrogen bonds with water, provided that the solute molecules are fairly small. Hydrogen bonding is possible between two molecules if one of them contains an O, N, or F atom (a hydrogen bond acceptor) and the other contains an O—H, N—H, or F—H bond (a hydrogen bond donor). Every water molecule contains an O atom and O—H bonds. Therefore, water can form hydrogen bonds with any molecule that also contains an O, N, or F atom or an O—H, N—H, or F—H bond. If these molecules are small enough, they will be soluble in water. How small? In general, they can have no more than three C atoms for each O or N atom.

For example, acetic acid, CH_3COOH, is soluble in water, but benzoic acid, C_6H_5COOH, is not significantly soluble. Similarly, ethanol, C_2H_6O, is soluble in water, but dipropyl ether, $C_6H_{14}O$, is not. Table sugar, $C_{12}H_{22}O_{11}$ (Section 20-4A), is very soluble in water. Although each molecule of sucrose contains a large number (12) of carbon atoms, it has so many oxygen atoms (11) that it forms many hydrogen bonds with water molecules; thus, a sucrose molecule in aqueous solution is very well solvated.

As a generalization, covalent molecules that do not contain O or N atoms are almost always insoluble in water. For example, methanol, CH_3OH, is infinitely soluble in water, but chloromethane, CH_3Cl, is not. The exception to this generalization is the rare case where a covalent compound reacts with water—for instance, HCl.

E. Water in the Body

Water is important in the body not only because it dissolves ionic substances as well as some covalent compounds, but also because it hydrates all polar molecules in the body. In this way, water serves as a vehicle to transport most of the organic compounds, nutrients, and fuels used by the body, as well as waste material. Blood and urine are two examples of aqueous body fluids.

In addition, the hydration of macromolecules such as proteins, nucleic acids, and polysaccharides allows the proper motions within these molecules, which are necessary for such functions as enzyme activity (see Chapter 23).

6-7 What Are Colloids?

Up to now, we have discussed only solutions. The maximum diameter of the solute particles in a true solution is about 1 nm. If the diameter of the solute particles exceeds this size, then we no longer have a true solution—we have a **colloid.** In a colloid (also called a colloidal dispersion or colloidal system), the diameter of the solute particles ranges from about 1 to 1000. nm. The term *colloid* has acquired a new name recently. In Section 5-9, we encountered the term *nanotube*. The "nano" part refers to dimensions in the nanometer range (1 nm = 10^{-9} m), which is the size range of colloids. Thus, when we encounter terms such as "nanoparticle" or "nanoscience," they are equivalent to "colloidal particle" or "colloid science," although the former terms refer mostly to particles with a well-defined geometrical shape (such as tubes), while the latter terms are more general.

Colloidal particles usually have a very large surface area, which accounts for the two basic characteristics of colloidal systems:

1. They scatter light and therefore appear turbid, cloudy, or milky.
2. Although colloidal particles are large, they form stable dispersions—they do not form separate phases that settle out. As with true solutions, colloids can exist in a variety of phases: gas, liquid, or solid (Table 6-2).

All colloids exhibit the following characteristic effect. When we shine light through a colloid and look at the system from a 90° angle, we see the pathway of the light without seeing the colloidal particles themselves (they are too small to see). Rather, we see flashes of the light scattered by the particles in the colloid (Figure 6-13).

The **Tyndall effect** is due to light scattering by colloidal particles. Smoke, serum, and fog, to name a few examples, all exhibit the Tyndall effect. We are all familiar with the sunbeams that can be seen when sunlight passes through dusty air. This, too, is an example of the Tyndall effect. Again, we do not see the particles in dusty air, but only the light scattered by them.

Colloidal systems are stable. Mayonnaise, for example, stays emulsified and does not separate into oil and water. When the size of colloidal particles is larger than about 1000. nm, however, the system is unstable and separates into phases. Such systems are called **suspensions.**

For example, if we take a lump of soil and disperse it in water, we get a muddy suspension. The soil particles are anywhere from 10^3 to 10^9 nm in diameter. The muddy mixture scatters light and, therefore, appears turbid. It is not a stable system, however. If left alone, the soil particles soon settle, with clear water found above the sediment. Therefore, soil in water is a suspension, not a colloidal system.

Table 6-3 summarizes the properties of solutions, colloids, and suspensions.

What makes a colloidal dispersion stable? To answer this question, we must first realize that colloidal particles are in constant motion. Just look at the dust particles dancing in a ray of sunlight that enters a room. Actually, you do not see the dust particles themselves; they are too small. Rather, you see flashes of scattered light. The motion of the dust particles dispersed in air is a random, chaotic motion. This motion of any colloidal particle suspended in a solvent is called **Brownian motion** (Figure 6-14).

The constant buffeting and collisions by solvent molecules cause the colloidal particles to move in random Brownian motion. (In the case of the dust particles, the solvent is air.) This ongoing motion creates favorable conditions for collisions between particles. When such large particles collide, they stick together, combine to give larger particles, and finally settle out of the solution. That is what happens in a suspension.

Table 6-2 Types of Colloidal Systems

Type	Example
Gas in gas	None
Gas in liquid	Whipped cream
Gas in solid	Marshmallows
Liquid in gas	Clouds, fog
Liquid in liquid	Milk, mayonnaise
Liquid in solid	Cheese, butter
Solid in gas	Smoke
Solid in liquid	Jelly
Solid in solid	Dried paint

Tyndall effect Light passing through and scattered by a colloid viewed at a right angle

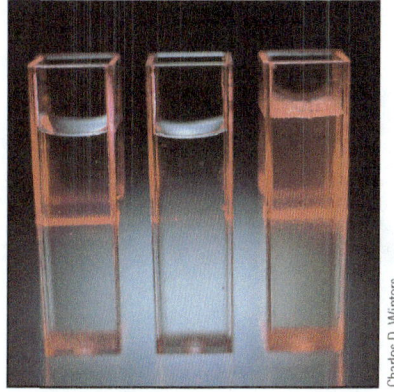

FIGURE 6-13 The Tyndall effect. A narrow beam of light from a laser is passed through a colloidal mixture (left), then a NaCl solution, and finally a colloidal mixture of gelatin and water (right). This illustrates the light-scattering ability of the colloid-sized particles.

FIGURE 6-14 Brownian motion.

Table 6-3 Properties of Three Types of Mixtures

Property	Solutions	Colloids	Suspensions
Particle size (nm)	0.1–1.0	1–1000	>1000
Filterable with ordinary paper	No	No	Yes
Homogeneous	Yes	Borderline	No
Settles on standing	No	No	Yes
Behavior to light	Transparent	Tyndall effect	Translucent or opaque

CHEMICAL CONNECTIONS 6E

Emulsions and Emulsifying Agents

Oil and water do not mix. Even when we stir them vigorously and the oil droplets become dispersed in the water, the two phases separate as soon as we stop stirring. There are, however, a number of stable colloidal systems made of oil and water, known as **emulsions**. For example, the oil droplets in milk are dispersed in an aqueous solution. This is possible because milk contains a protective colloid—the milk protein called casein. Casein molecules surround the oil droplets, and because they are polar and carry a charge, they protect and stabilize the oil droplets. Casein is thus an emulsifying agent.

Another emulsifying agent is egg yolk. This ingredient in mayonnaise coats the oil droplets and prevents them from separating.

Test your knowledge with Problem 6-86.

Emulsions Systems, such as fats in milk, consisting of a liquid with or without an emulsifying agent in an immiscible liquid, usually as droplets of larger than colloidal size

So why do colloidal particles remain in solution despite all the collisions due to their Brownian motion? Two reasons explain this phenomenon:

1. Most colloidal particles carry a large solvation layer. If the solvent is water, as in the case of protein molecules in the blood, the colloidal particles are surrounded by a large number of water molecules, which move together with the colloidal particles and cushion them. When two colloidal particles collide as a result of Brownian motion, they do not actually touch each other; instead, only their solvent layers collide. As a consequence, the particles do not stick together and precipitate. Instead, they stay in solution.

2. The large surface area of colloidal particles acquires charges from the solution. All colloids in a particular solution acquire the same kind of charge—for example, a negative charge. This development leaves a net negative charge in the solvent. When a charged colloidal particle encounters another charged colloidal particle, the two repel each other because of their like charges.

Thus, the combined effects of the solvation layer and the surface charge keep colloidal particles in a stable dispersion. By taking advantage of these effects, chemists can either increase or decrease the stability of a colloidal system. If we want to get rid of a colloidal dispersion, we can remove the solvation layer, the surface charge, or both. For example, proteins in the blood form a colloidal dispersion. If we want to isolate a protein from blood, we may want to precipitate it. We can accomplish this task in two ways: by removing the hydration layer or by removing the surface charges. If we add a solvent such as ethanol or acetone, each of which has great affinity for water, water is removed from the solvation layer of the protein, and when unprotected protein molecules collide, they stick together and form sediment. Similarly, by adding an electrolyte such as NaCl to the solution, we can remove the charges from the surface of the proteins (by a mechanism too complicated to discuss here). Without their protective charges, two protein molecules will no longer repel each other. Instead, when they collide, they stick together and precipitate from the solution. ◂

▶ Freshly made wines are often cloudy because of colloidal particles (left). Removing the particles clarifies the wine (right).

Colligative property A property of a solution that depends only on the number of solute particles and not on the chemical identity of the solute

6-8 What Is a Colligative Property?

A **colligative property** is any property of a solution that depends only on the number of solute particles dissolved in the solvent and not on the nature of the solute particles. Several colligative properties exist, including freezing-point depression, boiling-point elevation, and osmotic pressure. Of these three, osmotic pressure is of paramount importance in biological systems.

A. Freezing-Point Depression

One mole of any particle, whether it is a molecule or ion, dissolved in 1000. g of water lowers the freezing point of the water by 1.86°C. The nature of the solute does not matter, only the number of particles.

$$\Delta T_f = \frac{-1.86°C}{\text{mol}} \times \text{mol of particles}$$

This principle is used in a number of practical ways. In winter, we use salts (sodium chloride and calcium chloride) to melt snow and ice on our streets. The salts dissolve in the melting snow and ice, which lowers the freezing point of the water. ◀ Another application is the use of antifreeze in automobile radiators. Because water expands upon freezing (see Chemical Connections 5D), the ice formed in a car's cooling system when the outside temperature falls below 0°C can crack the engine block. The addition of antifreeze prevents this problem, because it makes the water freeze at a much lower temperature. The most common automotive antifreeze is ethylene glycol, $C_2H_6O_2$.

Freezing-point depression The decrease in the freezing point of a liquid caused by adding a solute

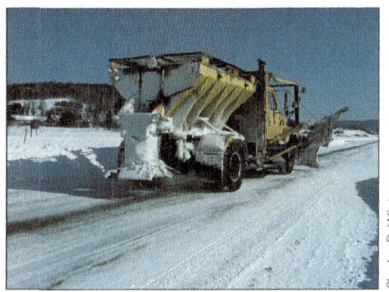

▶ Salting lowers the freezing point of ice.

EXAMPLE 6-10   Freezing-Point Depression

If we add 275 g of ethylene glycol, $C_2H_6O_2$, a nondissociating molecular compound, per 1000. g of water in a car radiator, what will be the freezing point of this solution?

Strategy

We are given 275 g of ethylene glycol (molar mass 62.1 g) per 1000. g water and asked to calculate the freezing point of the solution. We first calculate the moles of ethylene glycol present in the solution and then the freezing-point depression caused by that number of moles.

Solution

$$\Delta T = 275 \text{ g } C_2H_6O_2 \times \frac{1 \text{ mol } C_2H_6O_2}{62.1 \text{ g } C_2H_6O_2} \times \frac{-1.86°C}{1 \text{ mol } C_2H_6O_2} = -8.24°C$$

The freezing point of the water will be lowered from 0°C to −8.24°C, and the radiator will not crack if the outside temperature remains above −8.24°C (17.17°F).

Problem 6-10

If we add 215 g of methanol, CH_3OH, to 1000. g of water, what will be the freezing point of the solution?

If a solute is ionic, then each mole of solute dissociates to more than one mole of particles. For example, if we dissolve one mole (58.5 g) of NaCl in 1000. g of water, the solution contains two moles of solute particles: one mole each of Na^+ and Cl^-. The freezing point of water will be lowered by twice 1.86°C, that is, by 3.72°C per mole of NaCl.

EXAMPLE 6-11   Freezing-Point Depression

What will be the freezing point of the resulting solution if we dissolve one mole of potassium sulfate, K_2SO_4, in 1000. g of water?

Strategy and Solution

One mole of K_2SO_4 dissociates to produce three moles of ions: two moles of K^+ and one mole of SO_4^{2-}. The freezing point will be lowered by $3 \times 1.86°C = 5.58°C$, and the solution will freeze at $-5.58°C$.

Problem 6-11

Which aqueous solution would have the lowest freezing point?
(a) 6.2 M NaCl   (b) 2.1 M Al(NO$_3$)$_3$ (c) 4.3 M K$_2$SO$_3$

B. Boiling-Point Elevation

The boiling point of a substance is the temperature at which the vapor pressure of the substance equals atmospheric pressure. A solution containing a nonvolatile solute has a lower vapor pressure than the pure solvent and must be at a higher temperature before its vapor pressure equals atmospheric pressure and it boils. Thus, the boiling point of a solution containing a nonvolatile solute is higher than that of the pure solvent.

One mole of any molecule or ion dissolved in 1000. g of water raises the boiling point of the water by 0.512°C. The nature of the solute does not matter, only the number of particles.

$$\Delta T_b = \frac{0.512°C}{\text{mol}} \times \text{mol of particles}$$

EXAMPLE 6-12   Boiling-Point Elevation

Calculate the boiling point of a solution prepared by dissolving 275 g of ethylene glycol ($C_2H_6O_2$) in 1000. mL of water.

Strategy

To calculate the boiling point elevation, we must determine the number of moles of ethylene glycol dissolved in 1000. mL of water. We use the conversion factor 1.00 mole of ethylene glycol = 62.1 g of ethylene glycol.

Solution

$$\Delta T = 275 \text{ g } C_2H_6O_2 \times \frac{1 \text{ mol } C_2H_6O_2}{62.1 \text{ g } C_2H_6O_2} \times \frac{0.512°C}{1 \text{ mol } C_2H_6O_2} = 2.27°C$$

The boiling point is raised by 2.27°C. Therefore, the solution boils at 102.3°C.

Problem 6-12

Calculate the boiling point of a solution prepared by dissolving 310. g of ethanol, CH_3CH_2OH, in 1000. mL of water.

C. Osmotic Pressure

To understand osmotic pressure, let us consider the experimental setup shown in Figure 6-15. Suspended in the beaker is a bag containing a 5% solution of sugar in water. The bag is made of a **semipermeable membrane**

FIGURE 6-15 Demonstration of osmotic pressure.

that contains very tiny pores, far too small for us to see but large enough to allow solvent (water) molecules to pass through them but not the larger solvated sugar molecules.

When the bag is submerged in pure water, Figure 6-15(a), water flows into the bag by **osmosis** and raises the liquid level in the tube attached to the bag, Figure 6-15(b). Although sugar molecules are too big to pass through the membrane, water molecules easily move back and forth across it. However, this process cannot continue indefinitely because gravity prevents the difference in levels from becoming too great. Eventually a dynamic equilibrium is achieved. The height of the liquid in the tube remains unchanged and is a measure of osmotic pressure.

The liquid level in the glass tube and the breaker can be made equal again if we apply an external pressure through the glass tube. The amount of external pressure required to equalize the levels is called the **osmotic pressure (Π)**.

Although this discussion assumes that one compartment contains pure solvent and the other a solution, the same principle applies if both compartments contain solutions, as long as their concentrations are different. The solution of higher concentration always has a higher osmotic pressure than the one of lower concentration, which means that the flow of solvent molecules always occurs from the more dilute solution into the more concentrated solution. Of course, the number of particles is the most important consideration. We must remember that in ionic solutions, each mole of solute gives rise to more than one mole of particles. For convenience in calculation, we define a new term, **osmolarity,** which is molarity (M) of the solution multiplied by the number of particles (i) produced by each formula unit of solute.

$$\text{Osmolarity} = M \times i$$

Osmosis The passage of solvent molecules from a less concentrated solution across a semipermeable membrane into a more concentrated solution

Osmotic pressure (Π) The amount of external pressure that must be applied to the more concentrated solution to stop the passage of solvent molecules across a semipermeable membrane

> ### EXAMPLE 6-13 Osmolarity
>
> A 0.89 percent w/v NaCl aqueous solution is referred to as a physiological or isotonic saline solution because it has the same concentration of salts as normal human blood. Although blood contains several salts, saline solution has only NaCl. What is the osmolarity of this solution?
>
> #### Strategy
>
> We are given a 0.89% solution—that is, a solution that contains 0.89 g NaCl per 100. mL of solution. Because osmolarity is based on grams of solute per 1000. grams of solution, we calculate that this solution contains 8.9 g of NaCl per 1000. g of solution. Given this concentration, we can then calculate the molarity of the solution.
>
> #### Solution
>
> $$\frac{0.89 \text{ g NaCl}}{100. \text{ mL}} \times \frac{1000 \text{ mL}}{1 \text{ L}} \times \frac{1 \text{ mol NaCl}}{58.4 \text{ g NaCl}} = \frac{0.15 \text{ mol NaCl}}{1 \text{ L}} = 0.15 \, M$$
>
> Each formula unit of NaCl dissociates into two particles, namely Na^+ and Cl^-; therefore, the osmolarity is two times the molarity.
>
> $$\text{Osmolarity} = 0.15 \times 2 = 0.30 \text{ osmol}$$
>
> #### Problem 6-13
>
> What is the osmolarity of a 3.3% w/v Na_3PO_4 solution?

As noted earlier, osmotic pressure is a colligative property. The osmotic pressure generated by a solution across a semipermeable membrane—the difference between the heights of the two columns in Figure 6-15(b)—depends on the osmolarity of the solution. If the osmolarity increases by a factor of 2, the osmotic pressure will also increase by a factor of 2. Osmotic pressure is very important in biological organisms because cell membranes are semipermeable. For example, red blood cells in the body are suspended in a medium called plasma, which must have the same osmolarity as the red blood cells. Two solutions with the same osmolarity are called **isotonic,** so plasma is said to be isotonic with red blood cells. As a consequence, no osmotic pressure is generated across the cell membrane.

Cell-shriveling by osmosis occurs when vegetables or meats are cured in brine (a concentrated aqueous solution of NaCl). When a fresh cucumber is soaked in brine, water flows from the cucumber cells into the brine, leaving behind a shriveled cucumber, Figure 6-16 (*right*). With the proper spices added to the brine, the cucumber becomes a tasty pickle. A cucumber soaked in pure water is affected very little, as shown in Figure 6-16 (*left*).

What would happen if we suspended red blood cells in distilled water instead of in plasma? Inside the red blood cells, the osmolarity is approximately the same as in a physiological saline solution—0.30 osmol (**an isotonic solution**). Distilled water has zero osmolarity. As a consequence, water flows into the red blood cells. The volume of the cells increases, and they swell, as shown in Figure 6-18(b). The membrane cannot resist the osmotic pressure, and the red blood cells eventually burst, spilling their contents into the water. We call this process **hemolysis.**

Solutions in which the osmolarity (and hence osmotic pressure) is lower than that of suspended cells are called **hypotonic solutions.** Obviously, it

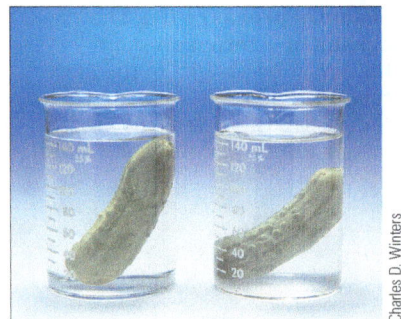

FIGURE 6-16 Osmosis and vegetables.

CHEMICAL CONNECTIONS 6F

Reverse Osmosis and Desalinization

In osmosis, the solvent flows spontaneously from the dilute solution compartment into the concentrated solution compartment. In reverse osmosis, the opposite happens. When we apply pressures greater than the osmotic pressure to the more concentrated solution, solvent flows from it to the more dilute solution by a process we call **reverse osmosis** (Figure 6-17).

Reverse osmosis is used to make drinkable water from seawater or brackish water. In large plants in the Persian Gulf countries, for example, more than 100. atm pressure is applied to seawater containing 35,000. ppm salt. The water that passes through the semipermeable membrane under this pressure contains only 400. ppm salt—well within the limits set by the World Health Organization for drinkable water.

An emergency hand-operated water desalinator that works by reverse osmosis. It can produce 4.5 L of pure water per hour from seawater, which can save someone adrift at sea.

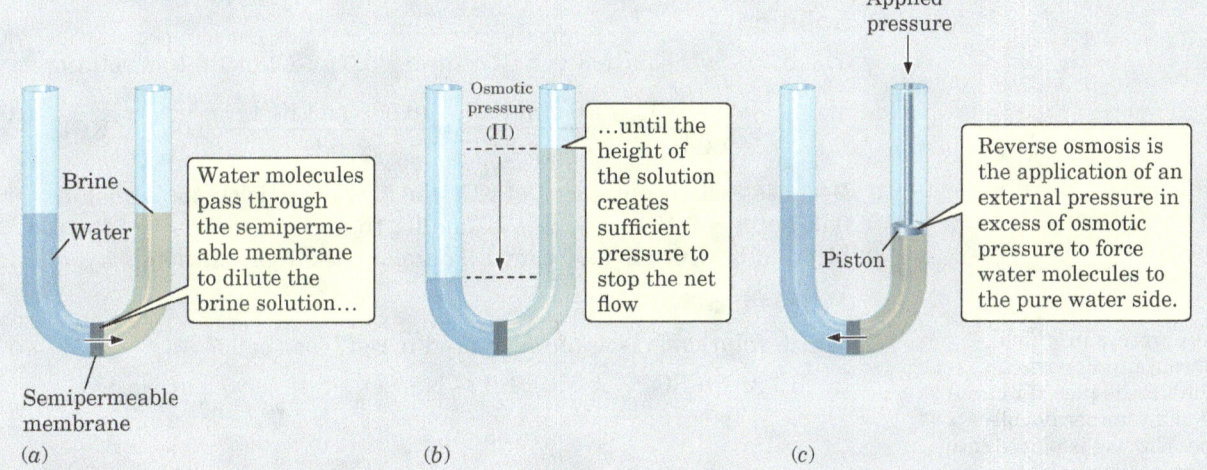

FIGURE 6-17 Normal and reverse osmosis. Normal osmosis is represented in (a) and (b). Reverse osmosis is represented in (c).

Test your knowledge with Problems 6-87 and 6-88.

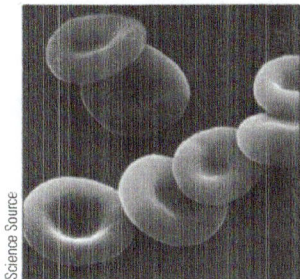
(a) The cells are unaffected.

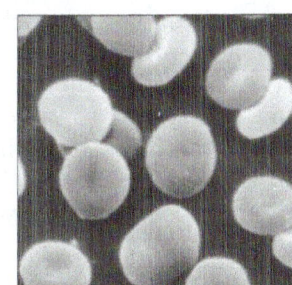
(b) The cells swell by hemolysis.

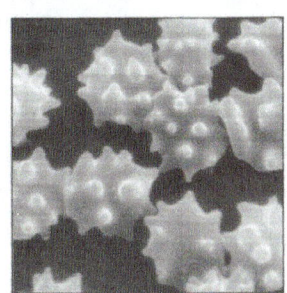
(c) The cells shrink by crenation.

FIGURE 6-18 Red blood cells in solutions of different osmolarity or *tonicity*. Red blood cells in (a) isotonic, (b) hypotonic, and (c) hypertonic solutions.

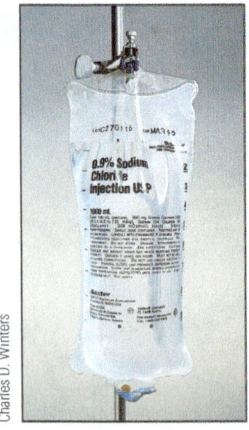

▶ An isotonic saline solution.

is very important that we always use isotonic solutions and never hypotonic solutions in intravenous feeding and blood transfusion. Hypotonic solutions would simply kill the red blood cells by hemolysis. ◀

Equally important, we should not use **hypertonic solutions.** A hypertonic solution has a greater osmolarity (and greater osmotic pressure) than the red blood cells. If red blood cells are placed in a hypertonic solution— for example, 0.5 osmol glucose solution—water flows from the cells into the glucose solution through the semipermeable cell membrane. This process, called **crenation,** shrivels the cells, as shown in Figure 6-18(c).

As already mentioned in Example 6-13, 0.89 w/v% NaCl (physiological saline) is isotonic with red blood cells and is used in intravenous injections.

EXAMPLE 6-14 Toxicity

Is a 0.50% w/v aqueous solution of KCl (a) hypertonic, (b) hypotonic, or (c) isotonic compared to red blood cells?

Strategy

Calculate the osmolarity of the solution, which is its molarity times the number of particles produced by each formula unit of solute.

Solution

The 0.50% w/v solution of KCl contains 5.0 g KCl in 1.0 L of solution:

$$\frac{5.0 \text{ g KCl}}{1.0 \text{ L}} \times \frac{1.0 \text{ mol KCl}}{74.6 \text{ g KCl}} = \frac{0.067 \text{ mol KCl}}{1.0 \text{ L}} = 0.067 \, M \text{ KCl}$$

Because each formula unit of KCl yields two particles, the osmolarity is $0.067 \times 2 = 0.13$ osmol; this is smaller than the osmolarity of the red blood cells, which is 0.30 osmol. Therefore, the KCl solution is hypotonic.

Problem 6-14

Which solution is isotonic compared to red blood cells?
(a) 0.1 M Na$_2$SO$_4$ (b) 1.0 M Na$_2$SO$_4$ (c) 0.2 M Na$_2$SO$_4$

D. Dialysis

Dialysis A process in which a solution containing particles of different sizes is placed in a bag made of a semipermeable membrane. The bag is placed into a solvent or solution containing only small molecules. The solution in the bag reaches equilibrium with the solvent outside, allowing the small molecules to diffuse across the membrane but retaining the large molecules.

An osmotic semipermeable membrane allows only solvent and not solute molecules to pass. If, however, the openings in the membrane are somewhat larger, then small solute molecules can also pass through, but large solute molecules, such as macromolecular and colloidal particles, cannot. This process is called **dialysis.**

For example, ribonucleic acids are important biological molecules that we will study in Chapter 25. When biochemists prepare ribonucleic acid solutions, they must remove small particles, such as NaCl, from the solution to obtain a pure nucleic acid preparation. To do so, they place the nucleic acid solution in a dialysis bag (made of cellophane) of sufficient pore size to allow all the small particles to diffuse and retain only the large nucleic acid molecules. If the dialysis bag is suspended in flowing distilled water, all NaCl and small particles will leave the bag. After a certain amount of time, the bag will contain only the pure nucleic acids dissolved in water. ◀

Our kidneys work in much the same way. The millions of nephrons, or kidney cells, have very large surface areas in which the capillaries of the

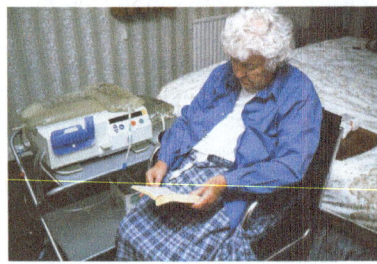

▶ A portable dialysis unit.

CHEMICAL CONNECTIONS 6G

Hemodialysis

The kidneys' main function is to remove toxic waste products from the blood. When the kidneys are not functioning properly, these waste products accumulate and may threaten life. **Hemodialysis** is a process that performs the same filtration function (see the figure).

In hemodialysis, the patient's blood circulates through a long tube of cellophane membrane suspended in an isotonic solution and then returns to the patient's vein. The cellophane membrane retains the large particles (for example, proteins) but allows the small ones, including the toxic wastes, to pass through. In this way, dialysis removes wastes from the blood.

If the cellophane tube were suspended in distilled water, other small molecules, such as glucose, and ions, such as Na^+ and Cl^-, would also be removed from the blood. That is something we don't want to happen. The isotonic solution used in hemodialysis consists of 0.6% NaCl, 0.04% KCl, 0.2% $NaHCO_3$, and 0.72% glucose (all w/v). It ensures that no glucose or Na^+ is lost from the blood.

A patient usually remains on an artificial kidney machine for four to seven hours. During this time, the isotonic bath is changed every two hours. Kidney machines allow people with kidney failure to lead a normal life, although they must take these hemodialysis treatments regularly.

A schematic diagram of the hollow-fiber (or capillary) dialyzer, the most commonly used artificial kidney. During dialysis, blood flows through small tubes constructed of a semipermeable membrane; the tubes themselves are bathed in the dialyzing solution.

Test your knowledge with Problems 6-89 and 6-90.

blood vessels come in contact with the nephrons. The kidneys serve as a gigantic filtering machine. The waste products of the blood dialyse out through semipermeable membranes in the glomeruli and enter collecting tubes that carry the urine to the ureter. The glomeruli of the kidneys are fine capillary blood vessels in which the body's waste products are removed from the blood. Meanwhile, large protein molecules and cells are retained in the blood.

Summary of Key Questions

Section 6-1 What Do We Need to Know as We Begin?
- Systems containing more than one component are **mixtures.**
- **Homogeneous mixtures** are uniform throughout.
- **Heterogeneous mixtures** exhibit well-defined boundaries between phases.

Section 6-2 What Are the Most Common Types of Solutions?
- The most common types of solutions are gas in liquid, liquid in liquid, solid in liquid, gas in gas, and solid in solid.
- When a solution consists of a solid or gas dissolved in a liquid, the liquid acts as the **solvent,** and the solid or

gas is the **solute.** When one liquid is dissolved in another, the liquid present in greater amount is considered to be the solvent.

Section 6-3 What Are the Distinguishing Characteristics of Solutions?

- The distribution of solute particles is uniform throughout.
- The components of a solution do not separate on standing.
- A solution cannot be separated into its components by filtration.
- For any given solute and solvent, it is possible to make solutions of many different compositions.
- Most solutions are transparent.

Section 6-4 What Factors Affect Solubility?

- The **solubility** of a substance is the maximum amount of the substance that dissolves in a given amount of solvent at a given temperature.
- "Like dissolves like" means that polar molecules are soluble in polar solvents and that nonpolar molecules are soluble in nonpolar solvents. The solubility of solids and liquids in liquids usually increases with increasing temperature; the solubility of gases in liquids usually decreases with increasing temperature.

Section 6-5 What Are the Most Common Units for Concentration?

- Percent concentration is given in either weight per unit volume of solution (w/v) or volume per unit volume of solution (v/v).
- Percent weight/volume (w/v%) is the weight of solute per unit volume of solvent multiplied by 100.
- Percent volume/volume (v/v%) is the volume of solute per unit volume of solution multiplied by 100.
- **Molarity (M)** is the number of moles of solute per liter of solution.

Section 6-6 Why Is Water Such a Good Solvent?

- Water is the most important solvent, because it dissolves polar compounds and ions through hydrogen bonding and dipole–dipole interactions. Hydrated ions are surrounded by water molecules (as a solvation layer) that move together with the ion, cushioning it from collisions with other ions. Aqueous solutions of ions and molten salts are **electrolytes** and conduct electricity.

Section 6-7 What Are Colloids?

- Colloids exhibit a chaotic random motion, called **Brownian motion.** Colloids are stable mixtures despite the relatively large size of the colloidal particles (1 to 1000 nm). The stability results from the solvation layer that cushions the colloidal particles from direct collisions and from the electric charge on the surface of colloidal particles.

Section 6-8 What Is a Colligative Property?

- A **colligative property** is a property of a solution that depends only on the number of solute particles present.
- **Freezing-point depression, boiling-point elevation,** and **osmotic pressure** are examples of colligative properties.
- Osmotic pressure operates across an osmotic semipermeable membrane that allows only solvent molecules to pass but screens out all larger particles. In osmotic pressure calculations, concentration is measured in **osmolarity,** which is the molarity of the solution multiplied by the number of particles produced by dissociation of the solute.
- Red blood cells in a **hypotonic solution** swell and burst, a process called **hemolysis.**
- Red blood cells in a **hypertonic solution** shrink, a process called **crenation.** Some semipermeable membranes allow small solute particles to pass through along with solvent molecules.
- In **dialysis,** such membranes are used to separate larger particles from smaller ones.

Problems

Orange-numbered problems are applied.

Section 6-2 What Are the Most Common Types of Solutions?

6-15 Answer true or false.
(a) A solute is the substance dissolved in a solvent to form a solution.
(b) A solvent is the medium in which a solute is dissolved to form a solution.
(c) Some solutions can be separated into their components by filtration.
(d) Acid rain is a solution.

6-16 Answer true or false.
(a) Solubility is a physical property like melting point and boiling point.
(b) All solutions are transparent—that is, you can see through them.
(c) Most solutions can be separated into their components by physical methods such as distillation and chromatography.

6-17 Vinegar is a homogeneous aqueous solution containing 6% acetic acid. Which is the solvent?

6-18 Suppose you prepare a solution by dissolving glucose in water. Which is the solvent, and which is the solute?

6-19 In each of the following, tell whether the solutes and solvents are gases, liquids, or solids.
(a) Bronze (see Chemical Connections 2E)
(b) Cup of coffee
(c) Car exhaust
(d) Champagne

6-20 Give a familiar example of solutions of each of these types:
(a) Liquid in liquid
(b) Solid in liquid
(c) Gas in liquid
(d) Gas in gas

6-21 Are mixtures of gases true solutions or heterogeneous mixtures? Explain.

Section 6-4 What Factors Affect Solubility?

6-22 Answer true or false.
(a) Water is a good solvent for ionic compounds because water is a polar liquid.
(b) Small covalent compounds dissolve in water if they can form hydrogen bonds with water molecules.
(c) The solubility of ionic compounds in water generally increases as temperature increases.
(d) The solubility of gases in liquids generally increases as temperature increases.
(e) Pressure has little effect on the solubility of liquids in liquids.
(f) Pressure has a major effect on the solubility of gases in liquids.
(g) In general, the greater the pressure of a gas over water, the greater the solubility of the gas in water.
(h) Oxygen, O_2, is insoluble in water.

6-23 We dissolved 0.32 g of aspartic acid in 115.0 mL of water and obtained a clear solution. After it stands for two days at room temperature, we notice a white powder at the bottom of the beaker. What may have happened?

6-24 The solubility of a compound is 2.5 g in 100. mL of aqueous solution at 25°C. If we put 1.12 g of the compound in a 50.-mL volumetric flask at 25°C and add sufficient water to fill it to the 50.-mL mark, what kind of solution do we get—saturated or unsaturated? Explain.

6-25 A small amount of solid is added to a separatory funnel containing layers of diethyl ether and water. After shaking the separatory funnel, in which layer will we find each of the following solids?
(a) NaCl   (b) Camphor ($C_{10}H_{16}O$)   (c) KOH

6-26 On the basis of polarity and hydrogen bonding, which solute would be the most soluble in benzene, C_6H_6?
(a) CH_3OH (b) H_2O (c) $CH_3CH_2CH_2CH_3$ (d) H_2SO_4

6-27 Suppose that you discover a stain on an oil painting and want to remove it without damaging the painting. The stain is not water-insoluble. Knowing the polarities of the following solvents, which one would you try first and why?
(a) Benzene, C_6H_6
(b) Isopropyl (rubbing) alcohol, C_3H_7OH
(c) Hexane, C_6H_{14}

6-28 Which pairs of liquids are likely to be miscible?
(a) H_2O and CH_3OH (b) H_2O and C_6H_6
(c) C_6H_{14} and CCl_4 (d) CCl_4 and CH_3OH

6-29 The solubility of aspartic acid in water is 0.500 g in 100. mL at 25°C. If we dissolve 0.251 g of aspartic acid in 50.0 mL of water at 50°C and let the solution cool to 25°C without stirring, shaking, or otherwise disturbing the solution, would the resulting solution be a saturated, unsaturated, or supersaturated solution? Explain.

6-30 Near a power plant, warm water is discharged into a river. Sometimes dead fish are observed in the area. Why do fish die in the warm water?

6-31 If a bottle of beer is allowed to stand for several hours after being opened, it becomes "flat" (it loses CO_2). Explain.

6-32 Would you expect the solubility of ammonia gas in water at 2 atm pressure to be:
(a) greater than,   (b) the same as, or
(c) smaller than at 0.5 atm pressure?

Section 6-5 What Are the Most Common Units for Concentration?

6-33 Verify the following statements.
(a) One part per million corresponds to one minute in two years, or a single penny in $10,000.
(b) One part per billion corresponds to one minute in 2000 years, or a single penny in $10 million.

6-34 Describe how we would make the following solutions:
(a) 500.0 mL of a 5.32% w/w H_2S solution in water
(b) 342.0 mL of a 0.443% w/w benzene solution in toluene
(c) 12.5 mL of a 34.2% w/w dimethyl sulfoxide solution in acetone

6-35 Describe how we would prepare the following solutions:
(a) 280. mL of a 27% v/v solution of ethanol, C_2H_6O, in water
(b) 435 mL of a 1.8% v/v solution of ethyl acetate, $C_4H_8O_2$, in water
(c) 1.65 L of an 8.00% v/v solution of benzene, C_6H_6, in chloroform, $CHCl_3$

6-36 Describe how we would prepare the following solutions:
(a) 250 mL of a 3.6% w/v solution of NaCl in water
(b) 625 mL of a 4.9% w/v solution of glycine, $C_2H_5NO_2$, in water
(c) 43.5 mL of a 13.7% w/v solution of Na_2SO_4 in water
(d) 518 mL of a 2.1% w/v solution of acetone, C_3H_6O, in water

6-37 Calculate the w/v percentage of each of these solutes:
(a) 623 mg of casein in 15.0 mL of milk
(b) 74 mg of vitamin C in 250 mL of orange juice
(c) 3.25 g of sucrose in 186 mL of coffee

6-38 Describe how we would prepare 250 mL of 0.10 M NaOH from solid NaOH and water.

6-39 Assuming that the appropriate volumetric flasks are available, describe how we would make these solutions:
(a) 175 mL of a 1.14 M solution of NH_4Br in water
(b) 1.35 L of a 0.825 M solution of NaI in water

(c) 330 mL of a 0.16 M solution of ethanol, C_2H_6O, in water

6-40 What is the molarity of each solution?
(a) 47 g of KCl dissolved in enough water to give 375 mL of solution
(b) 82.6 g of sucrose, $C_{12}H_{22}O_{11}$, dissolved in enough water to give 725 mL of solution
(c) 9.3 g of ammonium sulfate, $(NH_4)_2SO_4$, dissolved in enough water to give 2.35 L of solution

6-41 A teardrop with a volume of 0.5 mL contains 5.0 mg NaCl. What is the molarity of the NaCl in the teardrop?

6-42 The concentration of stomach acid, HCl, is approximately 0.10 M. What volume of stomach acid contains 0.25 mg of HCl?

6-43 The label on a sparkling cider says it contains 22.0 g glucose ($C_6H_{12}O_6$), 190. mg K^+, and 4.00 mg Na^+ per serving of 240. mL of cider. Calculate the molarities of these ingredients in the sparkling cider.

6-44 If 3.18 g $BaCl_2$ is dissolved in enough solvent to make 500.0 mL of solution, what is the molarity of this solution?

6-45 The label on a jar of jam says it contains 13 g of sucrose, $C_{12}H_{22}O_{11}$ per tablespoon (15 mL). What is the molarity of sucrose in the jam?

6-46 A particular toothpaste contains 0.17 g NaF in 75 mL toothpaste. What are the percent w/v and the molarity of NaF in the toothpaste?

6-47 A student has a bottle labeled 0.750% albumin solution. The bottle contains exactly 5.00 mL. How much water must the student add to make the concentration of albumin become 0.125%?

6-48 How many grams of solute are present in each of the following aqueous solutions?
(a) 575 mL of a 2.00 M solution of nitric acid, HNO_3
(b) 1.65 L of a 0.286 M solution of alanine, $C_3H_7NO_2$
(c) 320 mL of a 0.0081 M solution of calcium sulfate, $CaSO_4$

6-49 A student has a stock solution of 30.0% w/v H_2O_2 (hydrogen peroxide). Describe how the student should prepare 250 mL of a 0.25% w/v H_2O_2 solution.

6-50 To make 5.0 L of a fruit punch that contains 10% v/v ethanol, how much 95% v/v ethanol must be mixed with how much fruit juice?

6-51 A pill weighing 325 mg contains the following. What is the concentration of each in ppm?
(a) 12.5 mg Captopril, a medication for high blood pressure
(b) 22 mg Mg^{2+}
(c) 0.27 mg Ca^{2+}

6-52 One slice of enriched bread weighing 80. g contains 70. μg of folic acid. What is the concentration of folic acid in ppm and ppb?

6-53 Dioxin is considered to be poisonous in concentrations above 2 ppb. If a lake containing 1×10^7 L has been contaminated by 0.1 g of dioxin, did the concentration reach a dangerous level?

6-54 An industrial wastewater contains 3.60 ppb cadmium, Cd^{2+}. How many mg of Cd^{2+} could be recovered from a ton (1016 kg) of this wastewater?

6-55 According to the label on a piece of cheese, one serving of 28 g provides the following daily values: 2% of Fe, 6% of Ca, and 6% of vitamin A. The recommended daily allowance (RDA) of each of these nutrients are as follows: 15 mg Fe, 1200 mg Ca, and 0.800 mg vitamin A. Calculate the concentrations of each of these nutrients in the cheese in ppm.

Section 6-6 Why Is Water Such a Good Solvent?

6-56 Answer true or false.
(a) The properties that make water a good solvent are its polarity and its capacity for hydrogen bonding.
(b) When ionic compounds dissolve in water, their ions become solvated by water molecules.
(c) The term "water of hydration" refers to the number of water molecules that surround an ion in aqueous solution.
(d) The term "anhydrous" means "without water."
(e) An electrolyte is a substance that dissolves in water to give a solution that conducts electricity.
(f) In a solution that conducts electricity, cations migrate toward the cathode and anions migrate toward the anode.
(g) Ions must be present in a solution for the solution to conduct electricity.
(h) Distilled water is a nonelectrolyte.
(i) A strong electrolyte is a substance that dissociates completely into ions in aqueous solution.
(j) All compounds that dissolve in water are electrolytes.

6-57 Considering polarities, electronegativities, and similar concepts learned in Chapter 3, classify each of the following as a strong electrolyte, a weak electrolyte, or a nonelectrolyte.
(a) KCl (b) C_2H_6O (ethanol) (c) NaOH
(d) HCl (e) $C_6H_{12}O_6$ (glucose)

6-58 Which of the following would produce the brightest light in the conductance apparatus shown in Figure 6-11?
(a) 0.1 M KCl (b) 0.1 M $(NH_4)_3PO_4$
(c) 0.5 M sucrose

6-59 Ethanol is very soluble in water. Describe how water dissolves ethanol.

6-60 Predict which of these covalent compounds is soluble in water.
(a) C_2H_6 (b) CH_3OH (c) HF
(d) NH_3 (e) CCl_4

Section 6-7 What Are Colloids?

6-61 Answer true or false.
(a) A colloid is a state of matter intermediate between a solution and a suspension, in which particles are large enough to scatter light but too small to settle out from solution.
(b) Colloidal solutions appear cloudy because the colloidal particles are large enough to scatter visible light.

6-62 A type of car tire is made of synthetic rubber in which carbon black particles of the size of 200–500 nm are randomly dispersed. Because carbon black absorbs light, we do not see any turbidity (that is, a Tyndall effect). Do we consider a tire to be a colloidal system, and if so, what kind? Explain.

6-63 On the basis of Tables 6-1 and 6-2, classify the following systems as homogeneous, heterogeneous, or colloidal mixtures.
(a) Physiological saline solution (b) Orange juice
(c) A cloud (d) Wet sand (e) Soap suds (f) Milk

6-64 Table 6-2 shows no examples of a gas-in-gas colloidal system. Considering the definition of a colloid, explain why.

6-65 A solution of protein is transparent at room temperature. When it is cooled to 10°C, it becomes turbid. What causes this change in appearance?

6-66 What gives nanotubes their unique optical and electrical properties?

Section 6-8 What Is a Colligative Property?

6-67 Calculate the freezing points of solutions made by dissolving 1.00 mole of each of the following ionic solutes in 1000. g of H_2O.
(a) NaCl (b) $MgCl_2$
(c) $(NH_4)_2CO_3$ (d) $Al(HCO_3)_3$

6-68 If we add 175 g of ethylene glycol, $C_2H_6O_2$, per 1000. g of water to a car radiator, what will be the freezing point of the solution?

6-69 Methanol, CH_3OH, is used as an antifreeze. How many grams of methanol would you need per 1000. g of water for an aqueous solution to stay liquid at −20.°C?

6-70 In winter, after a snowstorm, salt (NaCl) is spread to melt the ice on roads. How many grams of salt per 1000. g of ice is needed to make it liquid at −5°C?

6-71 A 4 M acetic acid (CH_3COOH) solution lowers the freezing point by −8°C; a 4 M KF solution yields a −15°C freezing-point depression. What can account for this difference?

Osmosis

6-72 In an apparatus using a semipermeable membrane, a 0.005 M glucose (a small molecule) solution yielded an osmotic pressure of 10 mm Hg. What kind of osmotic pressure change would you expect if instead of a semipermeable membrane you used a dialysis membrane?

6-73 In each case, tell which side (if either) rises and why. The solvent is water.

Osmotic membrane

	A	B
(a)	1% glucose	5% glucose
(b)	0.1 M glucose	0.5 M glucose
(c)	1 M NaCl	1 M glucose
(d)	1 M NaCl	1 M K_2SO_4
(e)	3% NaCl	3% KCl
(f)	1 M NaBr	1 M KCl

6-74 An osmotic semipermeable membrane that allows only water to pass separates two compartments, A and B. Compartment A contains 0.9% NaCl, and compartment B contains 3% glycerol, $C_3H_8O_3$.
(a) In which compartment will the level of solution rise?
(b) Which compartment (if either) has the higher osmotic pressure?

6-75 Calculate the osmolarity of each of the following solutions.
(a) 0.39 M Na_2CO_3 (b) 0.62 M $Al(NO_3)_3$
(c) 4.2 M LiBr (d) 0.009 M K_3PO_4

6-76 Two compartments are separated by a semipermeable osmotic membrane through which only water molecules can pass. Compartment A contains a 0.3 M KCl solution, and compartment B contains a 0.2 M Na_3PO_4 solution. Predict from which compartment the water will flow to the other compartment.

6-77 A 0.9% NaCl solution is isotonic with blood plasma. Which solution would crenate red blood cells?
(a) 0.3% NaCl (b) 0.9 M glucose (MW 180)
(c) 0.9% glucose

Chemical Connections

6-78 (Chemical Connections 6A) Oxides of nitrogen (NO, NO_2, N_2O_3) are also responsible for acid rain. Which acids can be formed from these nitrogen oxides?

6-79 (Chemical Connections 6A) What makes normal rainwater slightly acidic?

6-80 (Chemical Connections 6B) Why do deep-sea divers use a helium–oxygen mixture in their tanks instead of air?

6-81 (Chemical Connections 6B) What is nitrogen narcosis?

6-82 (Chemical Connections 6C) A solution contains 54 mEq/L of Cl^- and 12 mEq/L of HCO_3^-. If Na^+ is the only cation present in the solution, what is the Na^+ concentration in milliequivalents per liter?

6-83 (Chemical Connections 6C) The concentration of Ca^{2+} ion present in a blood sample is found to be 4.6 mEq/L. How many milligrams of Ca^{2+} ion are present in 250.0 mL of the blood?

6-84 (Chemical Connections 6D) What is the chemical formula for the main component of limestone and marble?

6-85 (Chemical Connections 6D) Write balanced equations (two steps) for the conversion of marble to gypsum dihydrate.

6-86 (Chemical Connections 6E) What is the protective colloid in milk?

6-87 (Chemical Connections 6F) What is the minimum pressure on seawater that will force water to flow from the concentrated solution into the dilute solution?

6-88 (Chemical Connections 6F) The osmotic pressure generated across a semipermeable membrane by a solution is directly proportional to its osmolarity. Given the data in Chemical Connections 6F on the purification of seawater, estimate what pressure you would need to apply to purify brackish water containing 5000. ppm salt by reverse osmosis.

6-89 (Chemical Connections 6G) A manufacturing error occurred in the isotonic solution used in hemodialysis. Instead of 0.2% $NaHCO_3$, 0.2% of $KHCO_3$ was added. Did this error change the labeled tonicity of the solution? If so, is the resulting solution hypotonic or hypertonic? Would such an error create an electrolyte imbalance in the patient's blood? Explain.

6-90 (Chemical Connections 6G) The artificial kidney machine uses a solution containing 0.6% w/v NaCl, 0.04% w/v KCl, 0.2% w/v $NaHCO_3$, and 0.72% w/v glucose. Show that this is an isotonic solution.

Additional Problems

6-91 When a cucumber is put into a saline solution to pickle it, the cucumber shrinks; when a prune is put into the same solution, the prune swells. Explain what happens in each case.

6-92 A solution of As_2O_3 has a molarity of 2×10^{-5} M. What is this concentration in ppm? (Assume that the density of the solution is 1.00 g/mL.)

6-93 Two bottles of water are carbonated, with CO_2 gas being added, under 2 atm pressure and then capped. One bottle is stored at room temperature; the other is stored in the refrigerator. When the bottle stored at room temperature is opened, large bubbles escape, along with a third of the water. The bottle stored in the refrigerator is opened without frothing or bubbles escaping. Explain.

6-94 How many grams of ethylene glycol must be added to 1000. g of water to create an automobile radiator coolant mixture that will not freeze at $-15°C$?

6-95 Both methanol, CH_3OH, and ethylene glycol, $C_2H_6O_2$, are used as antifreeze. Which is more efficient—that is, which produces a lower freezing point if equal weights of each are added to the same weight of water?

6-96 We know that a 0.89% saline (NaCl) solution is isotonic with blood. In a real-life emergency, you run out of physiological saline solution and have only KCl as a salt and distilled water. Would it be acceptable to make a 0.89% aqueous KCl solution and use it for intravenous infusion? Explain.

6-97 Carbon dioxide and sulfur dioxide are soluble in water because they react with water. Write possible equations for these reactions.

6-98 A reagent label shows that the reagent contains 0.05 ppm lead as a contaminant. How many grams of lead are present in 5.0 g of the reagent?

6-99 A concentrated nitric acid solution contains 35% HNO_3. How would we prepare 300. mL of 4.5% solution?

6-100 Which will have greater osmotic pressure?
  (a) A 0.9% w/v NaCl solution
  (b) A 25% w/v solution of a nondissociating dextran with a molecular weight of 15,000.

6-101 Government regulations permit a 6 ppb concentration of a certain pollutant. How many grams of pollutant are allowed in 1 ton (1016 kg) of water?

6-102 The average osmolarity of seawater is 1.18 osmol. How much pure water would have to be added to 1.0 mL of seawater for it to achieve the osmolarity of blood (0.30 osmol)?

6-103 A swimming pool containing 20,000. L of water is chlorinated to have a final Cl_2 concentration of 0.00500 M. What is the Cl_2 concentration in ppm? How many kilograms of Cl_2 were added to the swimming pool to reach this concentration?

6-104 The density of a solution that is 20.0% $HClO_4$ is 1.138 g/mL. Calculate the molarity of the solution.

6-105 A 10.0% H_2SO_4 solution has a density of 1.07 g/mL. How many milliliters of solution contain 8.37 g of H_2SO_4?

Looking Ahead

6-106 Synovial fluid that exists in joints is a colloidal solution of hyaluronic acid (Section 20-6A) in water. To isolate hyaluronic acid from synovial fluid, a biochemist adds ethanol, C_2H_6O, to bring the solution to 65% ethanol. The hyaluronic acid precipitates upon standing. What makes the hyaluronic acid solution unstable and causes it to precipitate?

Challenge Problems

6-107 A solution is made by dissolving 25.0 g of magnesium chloride crystals in 1000. g of water.
  (a) What will be the freezing point of the new solution assuming complete dissociation of the $MgCl_2$ salt?
  (b) Determine the boiling point of the new solution assuming complete dissociation of the $MgCl_2$ salt.

6-108 Explain why saltwater fish do not survive when they are suddenly transferred to a freshwater aquarium.

6-109 Consider the reaction of 1.46 g Ca(s) with 115 mL of 0.325 M HBr(aq) according to the following unbalanced chemical equation:

$$Ca(s) + HBr(aq) \longrightarrow CaBr_2(aq) + H_2(g)$$

The hydrogen produced was collected by displacement of water at 22°C with a total pressure of 754 torr.

(a) Which reactant is the limiting reagent? (Chapter 4)

(b) Determine the volume (in L) of hydrogen gas produced if the vapor pressure of water at 22°C is 21 torr. (Chapter 5)

(c) How many grams of the other reactant are left over? (Chapter 4)

6-110 Vitamin B_2, riboflavin, is a nondissociating molecular compound soluble in water. If 370.3 g of riboflavin is dissolved in 1000.0 g of water, the resulting solution has a freezing point of $-1.83°C$.

(a) What is the molar mass of riboflavin? (Chapter 4)

(b) Consider the skeletal structure of riboflavin, where all the bonded atoms are shown but double bonds, triple bonds, and/or lone pairs are missing. Complete the structure as shown below. (Chapter 3)

Riboflavin skeletal structure

6-111 As noted in Section 6-8C, the amount of external pressure that must be applied to a more concentrated solution to stop the passage of solvent molecules across a semipermeable membrane is known as the osmotic pressure (π). The osmotic pressure obeys a law similar in form to the ideal gas law (discussed in Section 5-4), where $PV = nRT$. Substituting π for pressure and solving for osmotic pressures gives the following equation:

$$\pi = \left(\frac{n}{V}\right) RT = MRT,$$ where M is the concentration

or molarity of the solution.

(a) Determine the osmotic pressure at 25°C of a 0.0020 M sucrose ($C_{12}H_{22}O_{11}$) solution.

(b) Seawater contains 3.4 g of salts for every liter of solution. Assuming the solute consists entirely of NaCl (and complete dissociation of the NaCl salt), calculate the osmotic pressure of seawater at 25°C.

(c) The average osmotic pressure of blood is 7.7 atm at 25°C. What concentration of glucose ($C_6H_{12}O_6$) will be isotonic with blood?

(d) Lysozyme is an enzyme that breaks bacterial cell walls. A solution containing 0.150 g of this enzyme in 210. mL of solution has an osmotic pressure of 0.953 torr at 25°C. What is the molar mass of lysozyme?

(e) The osmotic pressure of an aqueous solution of a certain protein was measured in order to determine the protein's molar mass. The solution contained 3.50 mg of protein dissolved in sufficient water to form 5.00 mL of solution. The osmotic pressure of the solution at 25°C was found to be 1.54 torr. Calculate the molar mass of the protein.

6-112 List the following aqueous solutions in order of increasing boiling point: 0.060 M glucose ($C_6H_{12}O_6$), 0.025 M LiBr, and 0.025 M Zn(NO$_3$)$_2$. Assume complete dissociation of any salts.

6-113 List the following aqueous solutions in order of decreasing freezing point: 0.040 M glycerin ($C_3H_8O_3$), 0.025 M NaBr, and 0.015 M Al(NO$_3$)$_3$. Assume complete dissociation of any salts.

Reaction Rates and Chemical Equilibrium

7

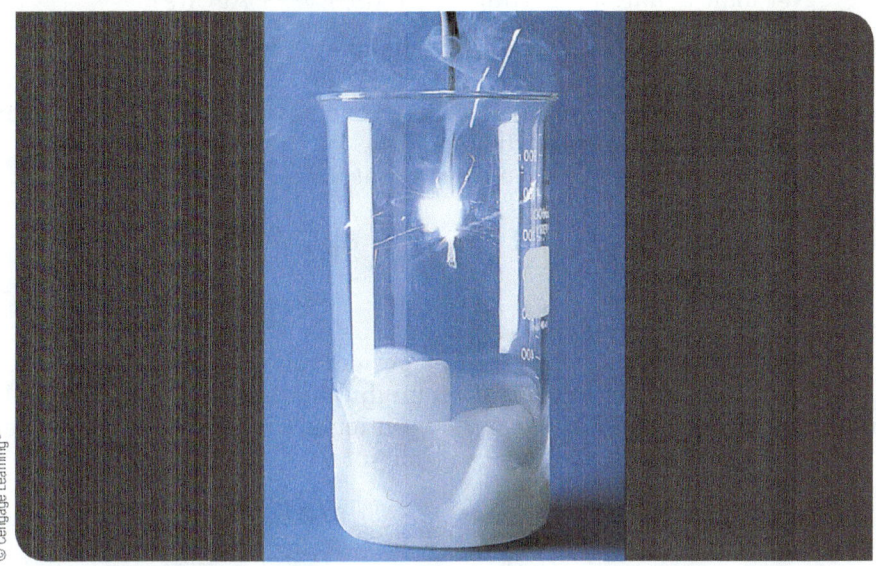

When a glowing ribbon of magnesium is thrust into a beaker of carbon dioxide (from the sublimation of dry ice at the bottom of the beaker), the metal bursts into a brilliant white flame, producing a smoke of magnesium and carbon.

KEY QUESTIONS

- **7-1** How Do We Measure Reaction Rates?
- **7-2** Why Do Some Molecular Collisions Result in Reaction Whereas Others Do Not?
- **7-3** What Is the Relationship Between Activation Energy and Reaction Rate?
- **7-4** How Can We Change the Rate of a Chemical Reaction?
- **7-5** What Does It Mean to Say That a Reaction Has Reached Equilibrium?
- **7-6** What Is an Equilibrium Constant and How Do We Use It?

  How To . . . Interpret the Value of the Equilibrium Constant, K
- **7-7** What Is Le Chatelier's Principle?

7-1 How Do We Measure Reaction Rates?

In this chapter, we are going to look at two closely related topics—reaction rates and chemical equilibrium. Knowing whether a reaction takes place quickly or slowly can give important information about the process in question. If the process has health implications, the information can be especially crucial. Sooner or later, many reactions will appear to stop, but that simply means that two reactions that are the reverse of each other are proceeding at the same rate. When this is the case, the reaction is said to be at equilibrium. The study of chemical equilibrium gives information about how to control reactions, including those that play key roles in life processes. We will address chemical equilibrium later in this chapter.

Some chemical reactions take place rapidly; others are very slow. For example, glucose and oxygen gas react with each other to form water and carbon dioxide:

$$C_6H_{12}O_6(s) + 6O_2(g) \longrightarrow 6CO_2(g) + 6H_2O(\ell)$$
Glucose

This reaction is extremely slow, however. A sample of glucose exposed to O_2 in the air shows no measurable change even after many years.

In contrast, consider what happens when you take one or two aspirin tablets for a slight headache. Very often, the pain disappears in half an hour or so. Thus, the aspirin must have reacted with compounds in the body within that time.

Many reactions occur even faster. For example, if we add a solution of silver nitrate to a solution of sodium chloride (NaCl), a precipitate of silver chloride (AgCl) forms almost instantaneously.

$$\text{Net ionic equation: } Ag^+(aq) + Cl^-(aq) \longrightarrow AgCl(s)$$

The precipitation of AgCl is essentially complete in considerably less than 1 s.

Chemical kinetics The study of the rates of chemical reactions

The study of reaction rates is called **chemical kinetics.** The **rate of a reaction** is the change in concentration of a reactant (or product) per unit time. Every reaction has its own rate, which must be measured in the laboratory. Consider the following reaction carried out in the solvent acetone:

$$CH_3-Cl + I^- \xrightarrow{\text{Acetone}} CH_3-I + Cl^-$$
$$\text{Chloromethane} \qquad\qquad \text{Iodomethane}$$

To determine the reaction rate, we can measure the concentration of the product, iodomethane, in the acetone at periodic time intervals—say, every 10 min. For example, the concentration might increase from 0 to 0.12 mol/L over a period of 30 min. The rate of the reaction is the change in the concentration of iodomethane divided by the time interval:

$$\frac{(0.12 \text{ mol } CH_3I/L) - (0 \text{ mol } CH_3I/L)}{30 \text{ min}} = \frac{0.0040 \text{ mol } CH_3I/L}{\text{min}}$$

This unit is read "0.0040 mole per liter per minute." On average, 0.0040 mol of chloromethane are initially converted to iodomethane for each liter of solution. The rate could also be determined by following the decrease in concentration of CH_3Cl or of I^-, if that is more convenient.

The rate of a reaction is not constant over a long period of time. At the beginning, in most reactions, the change in concentration is directly proportional to time. This period is shown as the linear portion of the graph in Figure 7-1. The rate calculated during this period, called the **initial rate,** is constant during this time interval. Later, as the reactant is used up, the rate of reaction decreases. Figure 7-1 shows a rate determined at a later time as well as the initial rate. The rate determined later is less than the initial rate.

FIGURE 7-1 Changes in the concentration of B in the A → B system with respect to time. The rate (the change in concentration of B per unit time) is largest at the beginning of the reaction and gradually decreases until it reaches zero at the completion of the reaction.

Interactive Example 7-1 — Reaction Rate

Another way to determine the rate of the reaction of chloromethane with iodide ion is to measure the disappearance of I^- from the solution. Suppose that the concentration of I^- was 0.24 mol I^-/L at the start of the reaction. At the end of 20 min, the concentration dropped to 0.16 mol I^-/L. This difference is equal to a change in concentration of 0.08 mol I^-/L. What is the rate of reaction?

Strategy
We use the definition of rate as the change in concentration in a unit of time. We can determine the change in concentration by subtraction. The time interval is given.

Solution
The rate of the reaction is:

$$\frac{(0.16 \text{ mol } I^-/L) - (0.24 \text{ mol } I^-/L)}{20 \text{ min}} = \frac{-0.0040 \text{ mol } I^-/L}{\text{min}}$$

Because the stoichiometry of the components is 1:1 in this reaction, we get the same numerical answer for the rate whether we monitor a reactant or a product. Note, however, that when we measure the concentration of a reactant that disappears with time, the rate of reaction is a negative number.

Problem 7-1
In the reaction

$$2HgO(s) \longrightarrow 2Hg(\ell) + O_2(g)$$

we measure the evolution of oxygen gas to determine the rate of reaction. At the beginning of the reaction (at 0 min), 0.020 L of O_2 is present. After 15 min, the volume of O_2 gas is 0.35 L. What is the rate of reaction?

The rates of chemical reactions—both the ones that we carry out in the laboratory and the ones that take place inside our bodies—are very important. A reaction that goes more slowly than we need may be useless, whereas a reaction that goes too fast may be dangerous. Ideally, we would like to know what causes the enormous variety in reaction rates. In the next three sections, we examine this question.

7-2 Why Do Some Molecular Collisions Result in Reaction Whereas Others Do Not?

For two molecules or ions to react with each other, they must first collide. As we saw in Chapter 5, molecules in gases and liquids are in constant motion and frequently collide with each other. If we want a reaction to take place between two compounds A and B, we allow them to mix if they are gases or dissolve them in a solvent if they are liquids. In either case, the constant motion of the molecules will lead to frequent collisions between molecules of A and B. In fact, we can even calculate how many such collisions will take place in a given period of time. Such calculations indicate that so many collisions occur between A and B molecules that most reactions should be over in considerably less than one second. Because the actual reactions generally proceed much more slowly, we must conclude that most collisions do not result in a reaction. Typically, when a molecule of A collides with a molecule of B, the two simply bounce apart without reacting. Every once in awhile,

Effective collision A collision between two molecules or ions that results in a chemical reaction

though, molecules of A and B collide and react to form a new compound. A collision that results in a reaction between two molecules or ions is called an **effective collision.**

Why are some collisions effective whereas others are not? There are three main reasons:

1. In most cases, for a reaction to take place between A and B, one or more ionic or covalent bonds must be broken in A or B or both, and energy is required for this to happen. The energy comes from the collision between A and B. If the energy of the collision is large enough, bonds will break and a reaction will take place. If the collision energy is too low, the molecules will bounce apart without reacting. The minimum energy necessary for a reaction to occur is called the **activation energy.**

Activation energy The minimum energy necessary to cause a chemical reaction

The energy of any collision depends on the relative speeds (that is, on the relative kinetic energies) of the colliding objects and on their angle of approach. Much greater damage is done in a head-on collision of two cars both going 40 mi/h than in a collision in which a car going 20 mi/h sideswipes one going 10 mi/h. The same consideration applies with molecules, as Figure 7-2 shows.

FIGURE 7-2 The energy of molecular collisions varies. (a) Two fast-moving molecules colliding head-on have a higher collision energy than (b) two slower-moving molecules colliding at an angle.

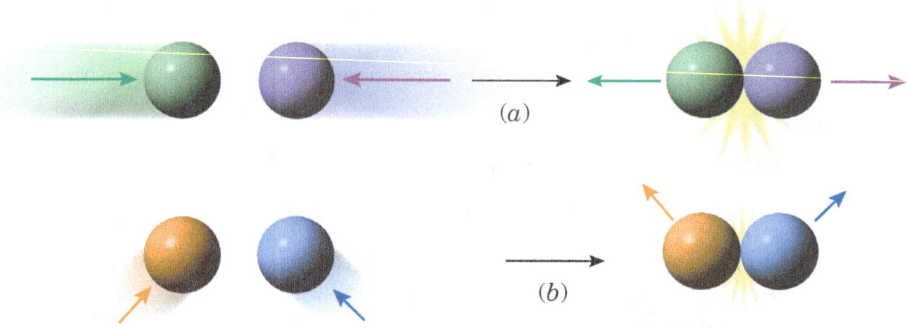

2. Even if two molecules collide with an energy greater than the activation energy, a reaction may not take place if the molecules are not oriented properly when they collide. Consider, for example, the reaction between H_2O and HCl:

$$H_2O(\ell) + HCl(g) \longrightarrow H_3O^+(aq) + Cl^-(aq)$$

For this reaction to take place, the molecules must collide in such a way that the H of the HCl hits the O of the water, as shown in Figure 7-3(a). A collision in which the Cl hits the O, as shown in Figure 7-3(b), cannot lead to a reaction, even if sufficient energy is available.

3. The frequency of collisions is another important factor. If more collisions take place, the chances are that more of them will have sufficient energy and the proper orientation of molecules for a reaction to take place.

FIGURE 7-3 Molecules must be properly oriented for a reaction to take place. (a) HCl and H_2O molecules are oriented so that the H of HCl collides with the O of H_2O, and a reaction takes place. (b) No reaction takes place because Cl, and not H, collides with the O of H_2O. The colored arrows show the path of the molecules.

Returning to the example given at the beginning of this chapter, we can now see why the reaction between glucose and O_2 is so slow. The O_2 molecules are constantly colliding with glucose molecules, but the percentage of effective collisions is extremely tiny at room temperature.

7-3 What Is the Relationship Between Activation Energy and Reaction Rate?

Figure 7-4 shows a typical energy diagram for an exothermic reaction. The products have a lower energy than the reactants; we might, therefore, expect the reaction to take place rapidly. As the curve shows, however, the reactants cannot be converted to products without the necessary activation energy. The activation energy is like a hill. If we are in a mountainous region, we may find that the only way to go from one point to another is to climb over a hill. It is the same in a chemical reaction. Even though the products may have a lower energy than the reactants, the products cannot form unless the reactants "go over the hill" or over a high pass—that is, they must possess the necessary activation energy.

Let us look into this issue more closely. In a typical reaction, existing bonds are broken and new bonds form. For example, when H_2 reacts with N_2 to give NH_3, six covalent bonds (counting a triple bond as three bonds) must break and six new covalent bonds must form.

$$3\text{H—H} + \text{N}\equiv\text{N} \longrightarrow 2\text{H—N}\begin{smallmatrix}\text{H}\\ \text{H}\end{smallmatrix}$$

Ammonia

Breaking a bond requires an input of energy, but a bond forming releases energy. In a "downhill" reaction of the type shown in Figure 7-4, the amount of energy released in creating the new bonds is greater than that required to break the original bonds. In other words, the reaction is exothermic. Yet it may well have a substantial activation energy, or energy barrier, because in most cases, at least one bond must break before any new bonds can form. Thus, energy must be put into the system before we get any back. This is

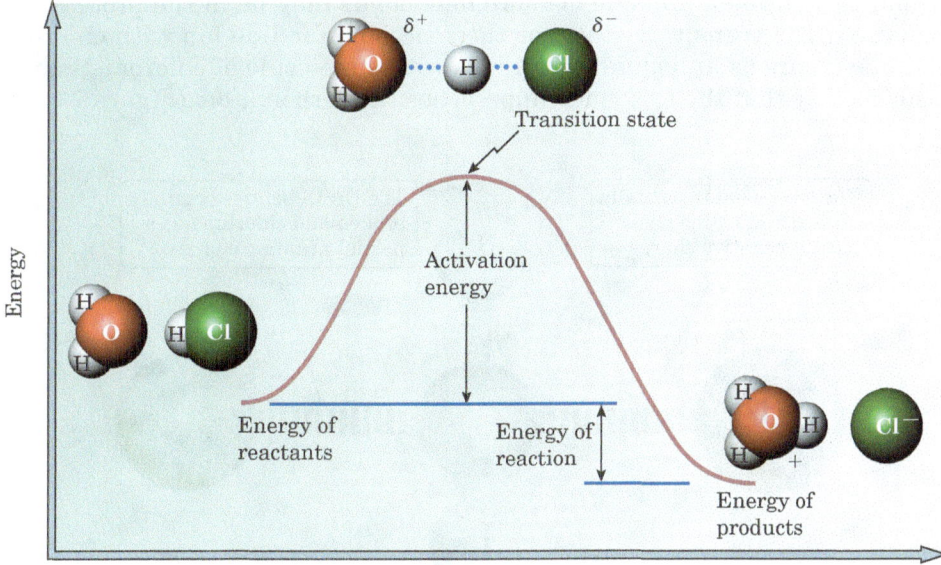

FIGURE 7-4 Energy diagram for the exothermic reaction.

$H_2O(\ell) + HCl(g) \longrightarrow$
$\qquad H_3O^+(aq) + Cl^-(aq)$

The energy of the reactants is greater than the energy of the products. The diagram shows the positions of all atoms before, at, and after the transition state.

FIGURE 7-5 Energy diagram for an endothermic reaction. The energy of the products is greater than that of the reactants.

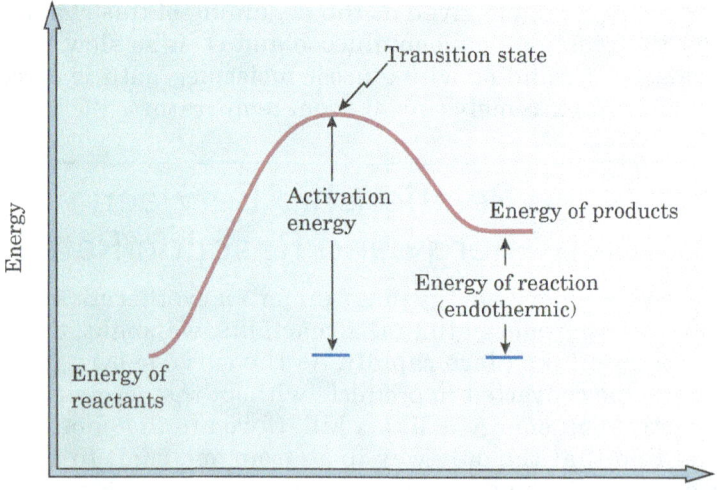

analogous to the following situation: Somebody offers to let you buy into a business from which, for an investment of $10,000, you could get an income of $40,000 per year, beginning in one year. In the long run, you would do very well. First, however, you need to put up the initial $10,000 (the activation energy) to start the business.

Notice that we just used an analogy for discussing energy by comparing energy changes to dollar amounts. Analogies can be useful to a point, but, at times, they are not enough. This point is especially true when we need exact information. Being precise in terminology is highly useful when we talk about energy changes, especially in view of the fact that scientists have developed a number of ways for describing transformations of energy under different conditions.

Every reaction has a different energy diagram. Sometimes, the energy of the products is higher than that of the reactants (Figure 7-5); that is, the reaction is "uphill". For almost all reactions, however, there is an energy "hill"—the activation energy. The activation energy is inversely related to the rate of the reaction. The lower the activation energy, the faster the reaction; the higher the activation energy, the slower the reaction.

The top of the hill on an energy diagram is called the **transition state.** When the reacting molecules reach this point, one or more original bonds are partially broken and one or more new bonds may be in the process of formation. The transition state for the reaction of iodide ion with chloromethane occurs as an iodide ion collides with a molecule of chloromethane in such a way that the iodide ion approaches the carbon atom (Figure 7-6).

FIGURE 7-6 Transition state for the reaction of CH_3Cl with I^-. In the transition state, iodide ion, I^-, attacks the carbon of chloromethane from the side opposite the C—Cl bond. In this transition state, both chlorine and iodine have partial negative charges.

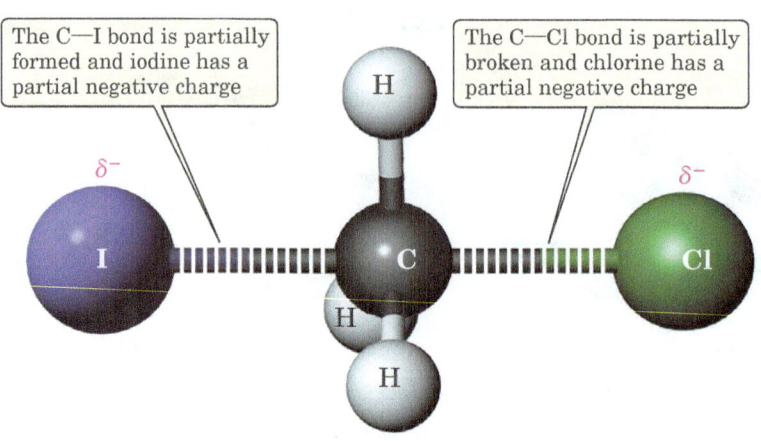

The speed of a reaction is proportional to the probability of effective collisions. In a single-step reaction, the probability that two particles will collide is greater than the probability of a simultaneous collision of five particles. If you consider the net ionic reaction

$$H_2O_2 + 3I^- + 2H^+ \longrightarrow I_3^- + 2H_2O$$

it is highly unlikely that six reactant particles will collide simultaneously; thus, this reaction should be slow. In reality, this reaction is very fast. This fact indicates that the reaction does not occur in one step but rather takes place in multiple steps. In each of those steps, the probability is high for collisions between two particles. Even a simple reaction such as

$$H_2(g) + Br_2(g) \longrightarrow 2HBr(g)$$

occurs in four steps:

Step 1: $Br_2 \xrightarrow{slow} 2Br\cdot$

Step 2: $Br\cdot + H_2 \xrightarrow{fast} HBr + H\cdot$

Step 3: $H\cdot + Br_2 \xrightarrow{fast} HBr + Br\cdot$

Step 4: $Br\cdot + Br\cdot \xrightarrow{fast} Br_2$

The dot ($\cdot$) indicates the single unpaired electron in the atom. The overall rate of the reaction will be controlled by the slowest of the four steps, just as the slowest-moving car controls the flow of traffic on a street. In the preceding reaction, step 1 is the slowest, because it has the highest activation energy.

7-4 How Can We Change the Rate of a Chemical Reaction?

In Section 7-2, we saw that reactions occur as a result of collisions between fast-moving molecules possessing a certain minimum energy (the activation energy). In this section, we examine some of the factors that affect activation energies and reaction rates.

A. Nature of the Reactants

In general, reactions that take place between ions in aqueous solution (Section 4-3) are extremely rapid, occurring almost instantaneously. Activation energies for these reactions are very low because usually no covalent bonds must be broken. As we might expect, reactions between covalent molecules, whether in aqueous solution or not, take place much more slowly. Many of these reactions require 15 min to 24 h or longer for most of the reactants to be converted to the products. Some reactions take a good deal longer, of course, but they are seldom useful.

B. Concentration

Consider the following reaction:

$$A + B \longrightarrow C + D$$

In most cases, the reaction rate increases when we increase the concentration of either or both reactants (Figure 7-7). For many reactions—though

FIGURE 7-7 The reaction of steel wool with oxygen. (*a*) When heated in air, steel wool glows but does not burn rapidly because the concentration of O_2 in the air is only about 20%. (*b*) When the glowing steel wool is put into 100% O_2, it burns vigorously.

(*a*)

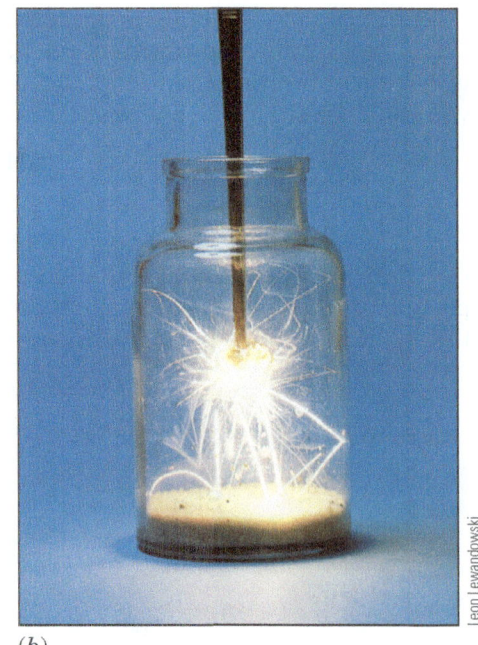
(*b*)

by no means all—a direct relationship exists between concentration and reaction rate; that is, when the concentration of a reactant is doubled, the reaction rate also doubles. This outcome is easily understandable on the basis of collision theory. If we double the concentration of A, there are twice as many molecules of A in the same volume, so the molecules of B in that volume now collide with twice as many A molecules per second than before. Given that the reaction rate depends on the number of effective collisions per second, the rate doubles. In the case where one of the reactants is a solid, the rate is affected by the surface area of the solid. For this reason, a substance in powder form reacts faster than the same substance in the form of large chunks.

We can express the relationship between rate and concentration mathematically. For example, for the reaction

$$2H_2O_2(\ell) \longrightarrow 2H_2O(\ell) + O_2(g)$$

the rate was determined to be -0.01 mol H_2O_2/L/min at a constant temperature when the initial concentration of H_2O_2 was 1 mol/L. In other words, every minute 0.01 mol/L of hydrogen peroxide was used up. Researchers also found that every time the concentration of H_2O_2 was doubled, the rate also doubled. Thus, the rate is directly proportional to the concentration of H_2O_2. We can write this relationship as

$$\text{Rate} = k[H_2O_2]$$

Rate constant A proportionality constant, k, between the molar concentration of reactants and the rate of reaction; rate = k [compound]

where k is a constant, called the **rate constant.** Rate constants are usually calculated from the **initial rates of reaction** and corresponding initial concentrations (Figure 7-1) and are positive values. The brackets [ ] stand for the molar concentration of the chemical species whose formula is between the brackets.

EXAMPLE 7-2 Rate Constants

Calculate the rate constant, k, for the reaction

$$2H_2O_2(\ell) \longrightarrow 2H_2O(\ell) + O_2(g)$$

using the rate and the initial concentration mentioned in the preceding discussion:

$$\frac{-0.01 \text{ mol } H_2O_2}{L \cdot min} \qquad [H_2O_2] = \frac{1 \text{ mol}}{L}$$

Strategy and Solution

We start with the rate equation, solve it for k, and then insert the appropriate experimental values.

$$\text{Rate} = k[H_2O_2]$$

$$k = \frac{\text{Rate}}{[H_2O_2]}$$

$$= \frac{0.01 \text{ mol } H_2O_2}{L \cdot min} \times \frac{L}{1 \text{ mol } H_2O_2}$$

$$= \frac{0.01}{min}$$

Note that all the concentration units cancel and that the rate constant has units that indicate some event in a given time, which makes sense. The answer is also a reasonable number. Also note that the negative sign, used to denote the concentration of the reactant that disappears with time, has been removed, as the rate constant is always positive.

Problem 7-2

Calculate the rate for the reaction in Example 7-2 when the initial concentration of H_2O_2 is 0.36 mol/L.

C. Temperature

In virtually all cases, reaction rates increase with increasing temperature. A rule of thumb for many reactions is that every time the temperature goes up by 10°C, the rate of reaction doubles. This rule is far from exact, but it is not far from the truth in many cases. As you can see, this effect can be quite large. It says, for example, that if we run a reaction at 90°C instead of at room temperature (20°C), the reaction will go about 128 times faster. There are seven 10° increments between 20°C and 90°C, and $2^7 = 128$. Put another way, if it takes 20 h to convert 100 g of reactant A to product C at 20°C, then it would take only 10 min at 90°C. Temperature, therefore, is a powerful tool that lets us increase the rates of reactions that are inconveniently slow. It also lets us decrease the rates of reactions that are inconveniently fast. For example, we might choose to run reactions at low temperatures because explosions might result or the reactions would otherwise be out of control at room temperature.

What causes reaction rates to increase with increasing temperature? Once again, we turn to collision theory. Here temperature has two effects:

1. In Section 5-6, we learned that temperature is related to the average kinetic energy of molecules. When the temperature increases, molecules move more rapidly, which means that they collide more frequently. More frequent collisions mean higher reaction rates. However, this factor is much less important than the second factor.

2. Recall from Section 7-2 that a reaction between two molecules takes place only if an effective collision occurs—a collision with an energy equal to or greater than the activation energy. When the temperature increases, not only is the average speed (kinetic energy) of the molecules greater, but there is also a different distribution of speeds. The number of very fast

CHEMICAL CONNECTIONS 7A

Why High Fever Is Dangerous

Chemical Connections 1B points out that a sustained body temperature of 41.7°C (107°F) is invariably fatal. We can now see why a high fever is dangerous. Normal body temperature is 37°C (98.6°F), and all the many reactions in the body—including respiration, digestion, and the synthesis of various compounds—take place at that temperature. If an increase of 10°C causes the rates of most reactions to approximately double, then an increase of even 1°C makes them go significantly faster than normal.

Fever is a protective mechanism, and a small increase in temperature allows the body to kill germs faster by mobilizing the immune defense mechanism. This increase must be small, however: A rise of 1°C brings the temperature to 38°C (100.4°F); a rise of 3°C brings it to 40°C (104°F). A temperature higher than 104°F increases reaction rates to the danger point.

One can easily detect the increase in reaction rates when a patient has a high fever. The pulse rate increases and breathing becomes faster as the body attempts to supply increased amounts of oxygen for the accelerated reactions. A marathon runner, for example, may become overheated on a hot and humid day. After a time, perspiration can no longer cool his or her body effectively, and the runner may suffer hyperthermia or heat stroke, which, if not treated properly, can cause brain damage.

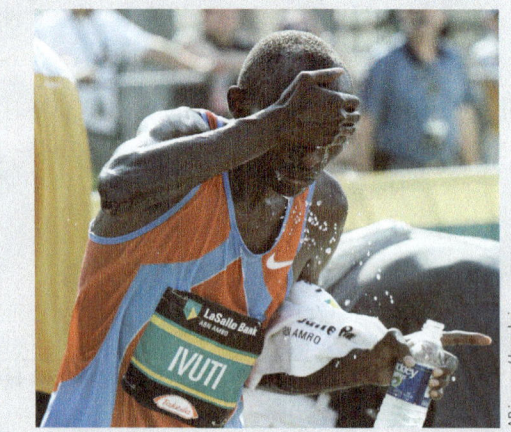

An overheated runner is at risk of serious health problems.

Test your knowledge with Problems 7-42 and 7-43.

---

molecules increases much more than the number with the average speed (Figure 7-8). As a consequence, the number of effective collisions rises even more than the total number of collisions. Not only do more collisions take place, but the percentage of collisions that have an energy greater than the activation energy also rises. This factor is mainly responsible for the sharp increase in reaction rates with increasing temperature.

D. Presence of a Catalyst

Any substance that increases the rate of a reaction without itself being used up is called a **catalyst**. Many catalysts are known—some that increase the rate of only one reaction and others that can affect several reactions. Although we have seen that we can speed up reactions by increasing

Catalyst A substance that increases the rate of a chemical reaction by providing an alternative pathway with a lower activation energy

FIGURE 7-8 Distribution of kinetic energies (molecular velocities) at two temperatures. The kinetic energy on the x-axis designated E_a indicates the energy (molecular velocity) necessary to pass through the activation energy barrier. The shaded areas represent the fraction of molecules that have kinetic energies (molecular velocities) greater than the activation energy.

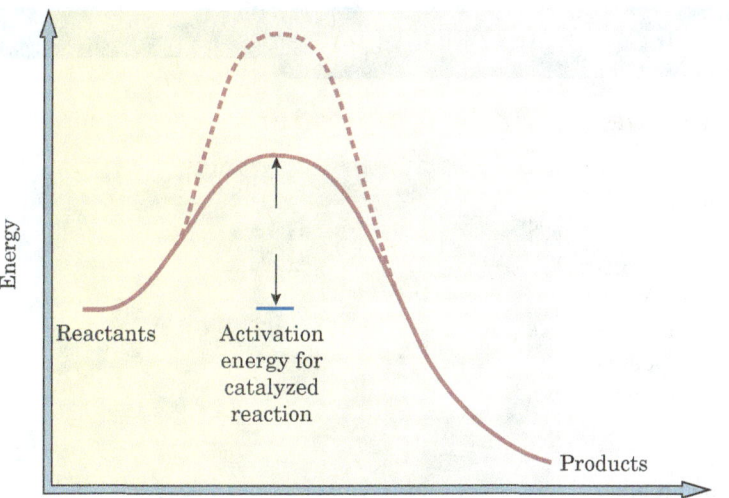

FIGURE 7-9 Energy diagram for a catalyzed reaction. The dashed line shows the energy curve for the uncatalyzed process. The catalyst provides an alternative pathway whose activation energy is lower.

the temperature, in some cases they remain too slow even at the highest temperatures we can conveniently reach. In other cases, it is not feasible to increase the temperature—perhaps because other unwanted reactions would be speeded up, too. In such cases, a catalyst, if we can find the right one for a given reaction, can prove very valuable. Many important industrial processes rely on **heterogeneous catalysts** (see Chemical Connections 7E), and virtually all reactions that take place in living organisms are catalyzed by enzymes (Chapter 22) or **homogeneous catalysts**. ◀

Catalysts work by allowing the reaction to take a different pathway, one with a lower activation energy. Without the catalyst, the reactants would have to get over the higher energy hill shown in Figure 7-9. The catalyst provides a lower hill. As we have seen, a lower activation energy means a faster reaction rate.

Each catalyst has its own way of providing an alternative pathway. Many catalysts provide a surface on which the reactants can meet. For example, the reaction between formaldehyde (HCHO) and hydrogen (H_2) to give methanol (CH_3OH) goes so slowly without a catalyst that it is not practical, even if we increase the temperature to a reasonable level. If the mixture of gases is shaken with finely divided platinum metal, however, the reaction takes place at a convenient rate (Section 12-6D). The formaldehyde and hydrogen molecules meet each other on the surface of the platinum, where the proper bonds can be broken and new bonds form and the reaction can proceed:

Heterogeneous catalysts catalysts in separate phases from the reactants—for example, the solid platinum, Pt(s), in the reaction between $CH_2O(g)$ and $H_2(g)$

Homogeneous catalysts catalysts in the same phase as the reactants—for example, enzymes in body tissues

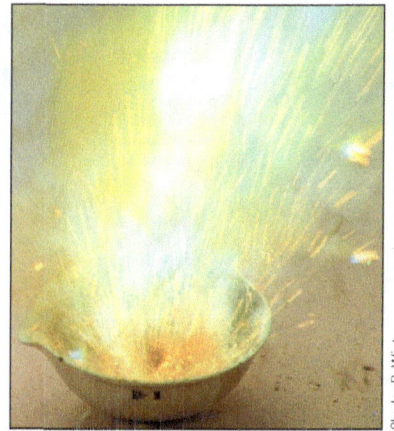

▶ In this dish, chloride ion, Cl^-, acts as a catalyst for the decomposition of NH_4NO_3.

$$\underset{\text{Formaldehyde}}{\overset{H}{\underset{H}{>}}C=O} + H_2 \xrightarrow{Pt} \underset{\text{Methanol}}{H-\overset{H}{\underset{H}{\overset{|}{C}}}-O-H}$$

We often write the catalyst over or under the arrow.

7-5 What Does It Mean to Say That a Reaction Has Reached Equilibrium?

Many reactions are irreversible. When a piece of paper is completely burned, the products are CO_2 and H_2O. Anyone who takes pure CO_2 and H_2O and tries to make them react to give paper and oxygen will not succeed.

CHEMICAL CONNECTIONS 7B

The Effects of Lowering Body Temperature

Like a significant increase in body temperature, a substantial decrease in body temperature below 37°C (98.6°F) can prove harmful because reaction rates are abnormally low. It is sometimes possible to take advantage of this effect. In some heart operations, for example, it is necessary to stop the flow of oxygen to the brain for a considerable time. At 37°C (98.6°F), the brain cannot survive without oxygen for longer than about 5 min without suffering permanent damage. When the patient's body temperature is deliberately lowered to about 28 to 30°C (82.4 to 86°F), however, the oxygen flow can be stopped for a considerable time without causing damage because reaction rates slow down. At 25.6°C (78°F), the body's oxygen consumption is reduced by 50%.

An operating table unit monitors a patient packed in ice.

Test your knowledge with Problems 7-43 and 7-44.

A tree, of course, turns CO_2 and H_2O into wood and oxygen, and we, in sophisticated factories, make paper from the wood. These activities are not the same as directly combining CO_2, H_2O, and energy in a single process to get paper and oxygen, however. Therefore, we can certainly consider the burning of paper to be an irreversible reaction.

Other reactions are reversible. A **reversible reaction** can be made to go in either direction. For example, if we mix carbon monoxide with water in the gas phase at a high temperature, carbon dioxide and hydrogen are produced:

$$CO(g) + H_2O(g) \longrightarrow CO_2(g) + H_2(g)$$

If we desire, we can also make this reaction take place the other way. That is, we can mix carbon dioxide and hydrogen to get carbon monoxide and water vapor:

$$CO_2(g) + H_2(g) \longrightarrow CO(g) + H_2O(g)$$

Let us see what happens when we run a reversible reaction. We will add some carbon monoxide to water vapor in the gas phase. The two compounds begin to react at a certain rate (the forward reaction):

$$CO(g) + H_2O(g) \longrightarrow CO_2(g) + H_2(g)$$

As the reaction proceeds, the concentrations of CO and H_2O gradually decrease because both reactants are being used up. In turn, the rate of the reaction gradually decreases because it depends on the concentrations of the reactants (Section 7-4B).

But what is happening in the other direction? Before we added the carbon monoxide, no carbon dioxide or hydrogen was present. As soon as the forward reaction began, it produced small amounts of these substances, and we now have some CO_2 and H_2. These two compounds will now, of course, begin reacting with each other (the reverse reaction):

$$CO_2(g) + H_2(g) \longrightarrow CO(g) + H_2O(g)$$

At first, the reverse reaction is very slow. As the concentrations of H_2 and CO_2 (produced by the forward reaction) gradually increase, the rate of the reverse reaction also gradually increases.

CHEMICAL CONNECTIONS 7C

Timed-Release Medication

It is often desirable that a particular medicine act slowly and maintain its action evenly in the body for 24 h. We know that a solid in powder form reacts faster than the same weight in pill form because the powder has a greater surface area at which the reaction can take place. To slow the reaction and to have the drug be delivered evenly to the tissues, pharmaceutical companies coat beads of some of their drugs. The coating prevents the drug from reacting for a time. The thicker the coating, the longer it takes the drug to react. A drug with a smaller bead size has more surface area than a drug with a larger bead size; hence, drugs packaged in a smaller bead size will react more rapidly. By combining the proper bead size with the proper amount of coating, the drug can be designed to deliver its effect over a 24-h period. In this way, the patient needs to take only one pill per day.

Coating can also prevent problems related to stomach irritation. For example, aspirin can cause stomach ulceration or bleeding in some people. Enteric (from the Greek *enteron*, which means affecting the intestines) coated aspirin tablets have a polymeric coat that is acid-resistant. Such a drug does not dissolve until it reaches the intestines, where it causes no harm.

A package of timed-release medications.

Test your knowledge with Problem 7-45.

We have a situation, then, in which the rate of the forward reaction gradually decreases, while the rate of the reverse reaction (which began at zero) gradually increases. Eventually the two rates become equal. At this point, the process is in **dynamic equilibrium** (or just **equilibrium**).

$$CO_2(g) + H_2(g) \underset{\text{reverse}}{\overset{\text{forward}}{\rightleftharpoons}} CO(g) + H_2O(g)$$

We use a double arrow to indicate that a reaction is reversible.

What happens in the reaction container once we reach equilibrium? If we measure the concentrations of the substances in the container, we find that no change in concentration takes place after equilibrium is reached (Figure 7-10). Whatever the concentrations of all the substances are at equilibrium, they remain the same forever unless something happens to disturb the equilibrium (as discussed in Section 7-7). This does not mean that all the concentrations must be the same—all of them can, in fact, be different and usually are—but it does mean that, whatever they are, they no longer change once equilibrium has been reached, no matter how long we wait.

Given that the concentrations of all the reactants and products no longer change, can we say that nothing is happening? No, we know that both reactions are occurring; all the molecules are constantly reacting—the CO and H_2O are being changed to CO_2 and H_2, and the CO_2 and H_2 are being changed to CO and H_2O. Because the rates of the forward and reverse reactions are the same, however, none of the concentrations change.

In the example just discussed, we approached equilibrium by adding carbon monoxide to water vapor. Alternatively, we could have added carbon dioxide to hydrogen. In either case, we eventually get an equilibrium mixture containing the same four compounds (Figure 7-11).

Dynamic equilibrium A state in which the rate of the forward reaction equals the rate of the reverse reaction

FIGURE 7-10 Changes in the concentrations of reactants (A and B) and products (C and D) as a system approaches equilibrium. Only A and B are present at the beginning of the reaction.

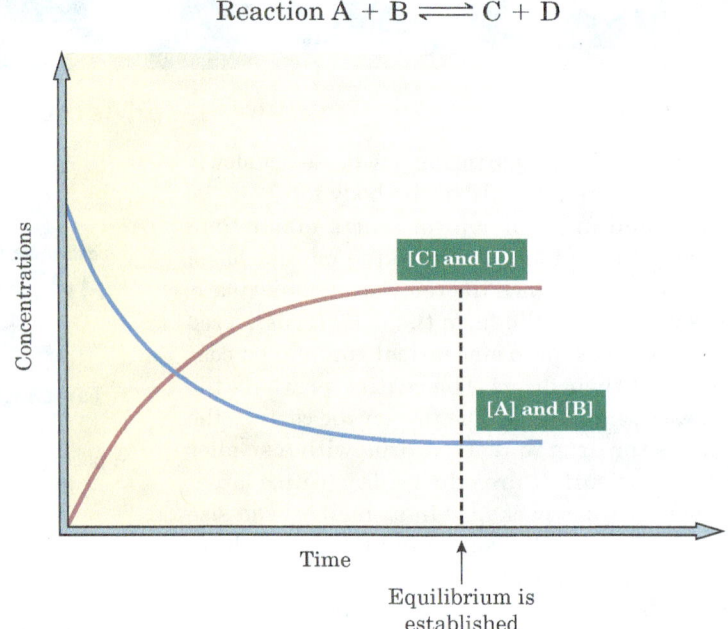

It is not necessary to begin with equal amounts. We could, for example, take 10 moles of carbon monoxide and 0.2 mole of water vapor. We would still arrive at an equilibrium mixture of all four compounds.

7-6 What Is an Equilibrium Constant and How Do We Use It?

Chemical equilibria can be treated by a simple mathematical expression. First, let us write the following reaction as the general equation for all reversible reactions:

$$a\text{A} + b\text{B} \rightleftharpoons c\text{C} + d\text{D}$$

In this equation, the capital letters stand for substances—CO_2, H_2O, CO, and H_2, for instance—and the lowercase letters are the coefficients of the balanced equation. The double arrow shows that the reaction is reversible. In general, any number of substances can be present on either side. Be careful not to use a double-headed single arrow ($\longleftrightarrow$) to denote an equilibrium reaction. This symbol is used to show resonance (Section 3-9).

In the laboratory, we study equilibrium reactions such as the one discussed in the preceding paragraph under carefully controlled conditions. Living things are a far cry from these laboratory conditions. The concept of equilibrium, however, can give useful insight into processes that take place

FIGURE 7-11 An equilibrium can be approached from either direction.

in living organisms, such as humans. The importance of calcium in maintaining bone integrity provides an example.

Bone is primarily calcium phosphate, $Ca_3(PO_4)_2$. This compound is highly insoluble in water, giving bone tissue its stability. Highly insoluble does not mean totally insoluble, or, to put it another way, solubility is not zero. Calcium phosphate solid in water reaches equilibrium with dissolved calcium ions and phosphate ions dissolved in intracellular fluid, which is mostly water.

$$Ca_3(PO_4)_2(s) \rightleftharpoons 3Ca^{2+}(aq) + 2PO_4^{3-}(aq)$$

In bone tissue, calcium phosphate is in contact with dissolved calcium and phosphate ions in intracellular fluid. Dietary calcium increases the concentration of calcium ion in intracellular fluid, favoring the reverse reaction, decreasing solubility, and ultimately increasing bone density.

Once equilibrium is reached, the following equation is valid, where K is a constant called the **equilibrium constant**:

$$K = \frac{[C]^c[D]^d}{[A]^a[B]^b} \quad \text{The equilibrium expression}$$

Equilibrium constant The ratio of product concentrations to reactant concentrations (with exponents that depend on the coefficients of the balanced equation)

Let us examine the equilibrium expression. It is understood that the concentration of a species within brackets is always expressed in moles per liter. The equilibrium expression tells us that when we multiply the equilibrium concentrations of the substances on the right side of the chemical equation and divide this product by the equilibrium concentrations of the substances on the left side (after raising each number to the appropriate power), we get the equilibrium constant, a number that does not change as long as temperature remains constant. In the example involving solid calcium phosphate above, one might expect that the equilibrium constant is:

$$K = \frac{[Ca^{2+}]^3[PO_4^{3-}]^2}{[Ca_3(PO_4)_2]}$$

However, as a general rule, pure solids and pure liquids are not included when writing an equilibrium expression (for reasons that extend beyond the scope of this book). The concentrations of gases and solutes in solution are included because only those concentrations can be varied, and therefore, it is important to state what they are. Therefore, the equilibrium expression actually becomes:

$$K = [Ca^{2+}]^3[PO_4^{3-}]^2$$

Let us look at several examples of how to set up equilibrium expressions.

EXAMPLE 7-3 Equilibrium Expressions

Write the equilibrium expression for the reaction:

$$CO(g) + H_2O(g) \rightleftharpoons CO_2(g) + H_2(g)$$

Strategy and Solution

$$K = \frac{[CO_2][H_2]}{[CO][H_2O]}$$

This expression tells us that at equilibrium, the concentration of carbon dioxide multiplied by the concentration of hydrogen and divided by the concentrations of water and carbon monoxide is a constant, K. Note that no exponent is written in this equation because all of the coefficients of the chemical equation are 1, and by convention, an exponent of 1 is not written.

Problem 7-3
Write the equilibrium expression for the reaction:

$$SO_3(g) + H_2O(\ell) \rightleftharpoons H_2SO_4(aq)$$

This reaction takes place in the atmosphere when water droplets react with the sulfur oxides formed in the combustion of fuels that contain sulfur. The resulting sulfuric acid is a component of acid rain.

EXAMPLE 7-4   Equilibrium Expressions

Write the equilibrium expression for the reaction:

$$C_6H_{12}O_6(s) + 6O_2(g) \rightleftharpoons 6CO_2(g) + 6H_2O(\ell)$$

Strategy and Solution

$$K = \frac{[CO_2]^6}{[O_2]^6}$$

In this case, the chemical equation has coefficients other than unity, so the equilibrium expression contains exponents. Notice that the solid and liquid are excluded from the expression.

Problem 7-4
Write the equilibrium expression for the reaction:

$$2NH_3(g) \rightleftharpoons N_2(g) + 3H_2(g)$$

Now let us see how K is calculated.

EXAMPLE 7-5   Equilibrium Constants

Some H_2 is added to I_2 at 427°C and the following reaction is allowed to come to equilibrium:

$$H_2(g) + I_2(g) \rightleftharpoons 2HI(g)$$

When equilibrium is reached, the concentrations are $[I_2]$ = 0.42 mol/L, $[H_2]$ = 0.025 mol/L, and [HI] = 0.76 mol/L. Calculate K at 427°C.

Strategy
Write the expression for the equilibrium constant, then substitute the values for the concentrations.
The equilibrium expression is:

$$K = \frac{[HI]^2}{[I_2][H_2]}$$

Solution
Substituting the concentrations, we get:

$$K = \frac{[0.76\ M]^2}{[0.42\ M][0.025\ M]} = 55$$

Equilibrium constants are usually written without units. This is current practice among chemists.

> **Problem 7-5**
> What is the equilibrium constant for the following reaction? Equilibrium concentrations are given under the formula of each component.
>
> $$PCl_3 + Cl_2 \rightleftharpoons PCl_5$$
> $1.66\,M 1.66\,M 1.66\,M$

Example 7-5 shows us that the reaction between I_2 and H_2 to give HI has an equilibrium constant of 55. What does this value mean? At constant temperature, equilibrium constants remain the same no matter what concentrations we have. That is, at 427°C, if we begin by adding, say, 5 moles of H_2 to 5 moles of I_2, the forward and then the backward reactions will take place, and equilibrium will eventually be reached. At that point, the value of K will equal 55. If we begin at 427°C with different numbers of moles of H_2 and I_2, perhaps 7 moles of H_2 and 2 moles of I_2, once equilibrium is reached, the value of $[HI]^2/[I_2][H_2]$ will again be 55. It makes no difference what the initial concentrations of the three substances are. At 427°C, as long as all three substances are present and equilibrium has been reached, the concentrations of the three substances will adjust themselves so that the value of the equilibrium constant equals 55.

The equilibrium constant is different for every reaction. Some reactions have a large K; others have a small K. A reaction with a very large K proceeds almost to completion (to the right). For example, K for the following reaction is about 100,000,000, or 10^8 at 25°C:

$$N_2(g) + 3H_2(g) \rightleftharpoons 2NH_3(g)$$

This value of 10^8 for K means that at equilibrium, $[NH_3]$ must be very large and $[N_2]$ and $[H_2]$ must be very small so that $[NH_3]^2/[N_2][H_2]^3 = 10^8$. Thus, if we add N_2 to H_2, we can be certain that when equilibrium is reached, an essentially complete reaction has taken place.

On the other hand, a reaction such as the following, which has a very small K, about 10^{-8} at 25°C, hardly goes forward at all:

$$AgCl(s) \rightleftharpoons Ag^+(aq) + Cl^-(aq)$$

This value of 10^{-8} for K means that at equilibrium, $[Ag^+]$ and $[Cl^-]$ must be very small so that $[Ag^+][Cl^-] = 10^{-8}$.

Equilibrium effects are most obvious in reactions with K values between 10^3 and 10^{-3}. In such cases, the reaction goes part of the way and significant concentrations of all substances are present at equilibrium. An example is the reaction between carbon monoxide and water discussed in Section 7-5, for which K is equal to 10 at 600°C.

HOW TO...

Interpret the Value of the Equilibrium Constant, K

Position of Equilibrium

The first question about the value of an equilibrium constant is whether the number is larger than one or smaller than one. If the number is larger than one, it means the ratio of product concentrations to reactant concentrations favors products. In other words, *the equilibrium lies to the right*. If the number is smaller than one, it means the ratio of product concentrations to reactant concentrations favors reactants. In other words, *the equilibrium lies to the left*.

Numerical Value of K

The next question focuses on the numerical value of the equilibrium constant. As we saw in Section 1-3, we frequently write numbers with exponents, with positive exponents for very large numbers and negative exponents for very small numbers. The sign and the numerical value of the exponent for a given equilibrium constant conveys information about whether the equilibrium lies strongly to the right (reaction goes to completion), strongly to the left (very little product formed), or at some intermediate point with significant amounts of both reactants and products present.

Very Large Values of K (above 10^3)

The conversion of NO gas to NO_2 in the presence of atmospheric oxygen is a reaction of environmental importance. Both these gases are pollutants and play a large role in the formation of smog and of acid rain.

$$2NO(g) + O_2(g) \rightleftharpoons 2NO_2(g)$$

The equilibrium constant for this reaction is 4.2×10^{12} at room temperature. If we start with 10.0 M NO, we find that only $2.2 \times 10^{-6} M$ is found at equilibrium and that the concentration of NO_2 is 10.0 M to within experimental error. Only a negligible amount of NO remains, and we say that the reaction has gone to completion.

Intermediate Values of K (less than 10^3 but more than 10^{-3})

Great care is taken in transporting and handling chlorine gas, especially with regard to fire prevention. Chlorine can react with carbon monoxide (also produced in fires) to produce phosgene ($COCl_2$), one of the poison gases used in World War I.

$$CO(g) + Cl_2(g) \rightleftharpoons COCl_2(g)$$
$$0.50\ M \quad 1.10\ M \quad\quad 0.10\ M$$

The equilibrium constant for this reaction is 0.18 (1.8×10^{-1}) at 600°C. The equilibrium concentrations are given below the formula of each component. They are similar in terms of order of magnitude, but the lower concentration for phosgene is consistent with the equilibrium constant being less than one.

Very Small Values of K (less than 10^{-3})

Barium sulfate is a compound of low solubility widely used to coat the gastrointestinal tract in preparation for X-rays. The solid is in equilibrium with dissolved barium and sulfate ions.

$$BaSO_4(s) \rightleftharpoons Ba^{2+}(aq) + SO_4^{2-}(aq)$$

The equilibrium constant for this reaction is 1.10×10^{-10} at room temperature. The concentrations of barium ion and sulfate ion are each $1.05 \times 10^{-5} M$. This low number implies that very little of the solid has dissolved.

The equilibrium constant for a given reaction remains the same no matter what happens to the concentrations, but the same is not true for changes in temperature.

As pointed out earlier in this section, the equilibrium expression is valid only after equilibrium has been reached. Before that point, there is no equilibrium, and the equilibrium expression is not valid. But how long does it take for a reaction to reach equilibrium? There is no easy answer to this question. Some reactions, if the reactants are well mixed, reach equilibrium in less than one second; others will not get there even after millions of years.

There is no relationship between the rate of a reaction (how long it takes to reach equilibrium) and the value of K. It is possible to have a large K and a slow rate, as in the reaction between glucose and O_2 to give CO_2 and H_2O, which does not reach equilibrium for many years (Section 7-1), or a small K and a fast rate. In other reactions, the rate and K are both large or both small.

7-7 What Is Le Chatelier's Principle?

When a reaction reaches equilibrium, the forward and reverse reactions take place at the same rate, and the equilibrium concentration of the reaction mixture does not change as long as we don't do anything to the system. But what happens if we do? In 1888, Henri Le Chatelier (1850–1936) put forth the statement known as **Le Chatelier's principle:** If an external stress is applied to a system in equilibrium, the system reacts in such a way as to partially relieve that stress. Let us look at five types of stress that can be put on chemical equilibria: adding a reactant or product, removing a reactant or product, and changing the temperature.

Le Chatelier's principle
A principle stating that when a stress is applied to a system in chemical equilibrium, the position of the equilibrium shifts in the direction that will relieve the applied stress

A. Addition of a Reaction Component

Suppose that the reaction between acetic acid and ethanol has reached equilibrium:

$$\underset{\text{Acetic acid}}{CH_3COOH(\ell)} + \underset{\text{Ethanol}}{C_2H_5OH(\ell)} \overset{HCl}{\rightleftharpoons} \underset{\text{Ethyl acetate}}{CH_3COOC_2H_5(\ell)} + H_2O(\ell)$$

This means that the reaction flask contains all four substances (plus the catalyst) and that their concentrations no longer change.

We now disturb the system by adding some acetic acid.

Adding CH_3COOH $\quad\underset{\text{Acetic acid}}{CH_3\overset{O}{\overset{\|}{C}}OH} + \underset{\text{Ethanol}}{HOCH_2CH_3} \overset{HCl}{\rightleftharpoons} \underset{\text{Ethyl acetate}}{CH_3\overset{O}{\overset{\|}{C}}OCH_2CH_3} + H_2O$

Equilibrium shifts to formation of more products →

The result is that the concentration of acetic acid suddenly increases, which increases the rate of the forward reaction. As a consequence, the concentrations of the products (ethyl acetate and water) begin to increase. At the same time, the concentrations of reactants decrease. Now, an increase in the concentrations of the products causes the rate of the reverse reaction to increase, but the rate of the forward reaction is decreasing, so eventually the two rates will be equal again and a new equilibrium will be established. ◄

When that happens, the concentrations are once again constant, but they are not the same as they were before the addition of the acetic acid. The concentrations of ethyl acetate and water are higher now, and the

▶ The tube on the left contains a saturated solution of silver acetate (Ag^+ ions and CH_3COO^- ions) in equilibrium with solid silver acetate. When more silver ions are added in the form of silver nitrate solution, the equilibrium shifts to the right, producing more silver acetate, as can be seen in the tube on the right.
$Ag^-(aq) + CH_3COO^-(aq) \rightleftharpoons$
$\qquad CH_3COOAg(s)$

concentration of ethanol is lower. The concentration of acetic acid is higher because we added some, but it is less than it was immediately after we made the addition.

When we add more of any component to a system in equilibrium, that addition constitutes a stress. The system relieves this stress by increasing the concentrations of the components on the other side of the equilibrium equation. We say that the equilibrium shifts in the opposite direction. The addition of acetic acid, on the left side of the equation, causes the rate of the forward reaction to increase and the reaction to move toward the right: more ethyl acetate and water form, and some of the acetic acid and ethanol are used up. The same thing happens if we add ethanol.

On the other hand, if we add water or ethyl acetate, the rate of the reverse reaction increases and the reaction shifts to the left:

$$CH_3COOH + C_2H_5OH \xrightleftharpoons[]{HCl} CH_3COOC_2H_5 + H_2O \quad \text{Adding ethyl acetate}$$

Acetic acid   Ethanol   Ethyl acetate

←——————— Equilibrium shifts toward formation of reactants

We can summarize by saying that the addition of any component causes the equilibrium to shift to the opposite side.

EXAMPLE 7-6   Le Chatelier's Principle–Effect of Concentration

When dinitrogen tetroxide, a colorless gas, is enclosed in a vessel, a color indicating the formation of brown nitrogen dioxide soon appears (see Figure 7-12 later in this chapter). The intensity of the brown color indicates the amount of nitrogen dioxide formed. The equilibrium reaction is:

$$N_2O_4(g) \rightleftharpoons 2NO_2(g)$$

Dinitrogen tetroxide (colorless)   Nitrogen dioxide (brown)

When more N_2O_4 is added to the equilibrium mixture, the brownish color becomes darker. Explain what happened.

Strategy and Solution

The darker color indicates that more nitrogen dioxide is formed. This happens because the addition of the reactant shifts the equilibrium to the right, forming more product.

Problem 7-6

What happens to the following equilibrium reaction when Br_2 gas is added to the equilibrium mixture?

$$2NOBr(g) \rightleftharpoons 2NO(g) + Br_2(g)$$

B. Removal of a Reaction Component

It is not always as easy to remove a component from a reaction mixture as it is to add one, but there are often ways to do it. The removal of a component, or even a decrease in its concentration, lowers the corresponding reaction rate and changes the position of the equilibrium. If we remove a reactant, the reaction shifts to the left, toward the side from which the reactant was removed. If we remove a product, the reaction shifts to the right, toward the side from which the product was removed.

In the case of the acetic acid–ethanol equilibrium, ethyl acetate has the lowest boiling point of the four components and can be removed by distillation. The equilibrium then shifts to that side so that more ethyl acetate is produced to compensate for the removal. The concentrations of acetic acid and ethanol decrease, and the concentration of water increases. The effect of removing a component is thus the opposite of adding one. The removal of a component causes the equilibrium to shift to the side from which the component was removed.

$$CH_3COOH + C_2H_5OH \xrightleftharpoons{HCl} H_2O + CH_3COOC_2H_5 \quad \text{Removing ethyl acetate}$$

Acetic acid   Ethanol   Ethyl acetate

Equilibrium shifts toward formation of more products →

No matter what happens to the individual concentrations, the value of the equilibrium constant remains unchanged.

EXAMPLE 7-7   Le Chatelier's Principle–Removal of a Reaction Component

The beautiful stone we know as marble is mostly calcium carbonate. When acid rain containing sulfuric acid attacks marble, the following equilibrium reaction can be written: ◀

$$CaCO_3(s) + H_2SO_4(aq) \rightleftharpoons CaSO_4(s) + CO_2(g) + H_2O(\ell)$$

Calcium carbonate   Sulfuric acid   Calcium sulfate   Carbon dioxide

How does the fact that carbon dioxide is a gas influence the equilibrium?

Strategy and Solution

The gaseous CO_2 diffuses away from the reaction site, meaning that this product is removed from the equilibrium mixture. The equilibrium shifts to the right so that the statue continues eroding.

Problem 7-7

Consider the following equilibrium reaction for the decomposition of an aqueous solution of hydrogen peroxide:

$$2H_2O_2(aq) \rightleftharpoons 2H_2O(\ell) + O_2(g)$$

Hydrogen peroxide

Oxygen has limited solubility in water (see the table in Chemical Connections 6A). What happens to the equilibrium after the solution becomes saturated with oxygen?

C. Change in Temperature

The effect of a change in temperature on a reaction that has reached equilibrium depends on whether the reaction is exothermic (gives off heat) or endothermic (requires heat). Let us look first at an exothermic reaction:

$$2H_2(g) + O_2(g) \rightleftharpoons 2H_2O(\ell) + 137{,}000 \text{ cal per mol } H_2O$$

▶ Effects of acid rain on marble. These photographs of marble statues of George Washington located in NYC were taken 60 years apart. Acid rain is dissolving the marble—marble is composed of calcium carbonate, which dissolves in water of low pH.

CHEMICAL CONNECTIONS 7D

Sunglasses and Le Chatelier's Principle

Heat is not the only form of energy that affects equilibria. The statements made in the text regarding endothermic and exothermic reactions can be generalized to reactions involving other forms of energy. A practical illustration of this generalization is the use of sunglasses with adjustable shading. The compound silver chloride, AgCl, is incorporated in the glasses. This compound, upon exposure to sunlight, produces metallic silver, Ag, and chlorine, Cl_2:

$$\text{Light} + 2Ag^+ + 2Cl^- \rightleftharpoons 2Ag(s) + Cl_2$$

The more silver metal produced, the darker the glasses. At night or when the wearer goes indoors, the reaction is reversed according to Le Chatelier's principle. In this case, the addition of energy in the form of sunlight drives the equilibrium to the right; its removal drives the equilibrium to the left.

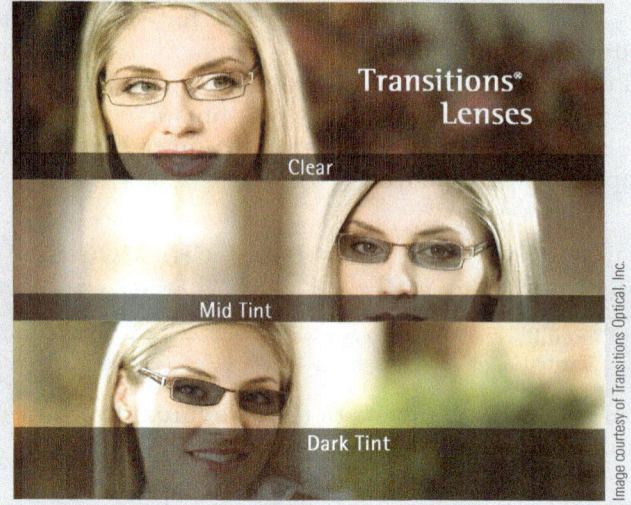

Test your knowledge with Problems 7-46 and 7-47.

If we consider heat to be a product of this reaction, then we can use Le Chatelier's principle and the same type of reasoning as we did before. An increase in temperature means that we are adding heat. Because heat is a product, its addition pushes the equilibrium to the opposite side. We can therefore say that if this exothermic reaction is at equilibrium and we increase the temperature, the reaction goes to the left—the concentrations of H_2 and O_2 increase and that of H_2O decreases. This is true of all exothermic reactions.

- An increase in temperature drives an exothermic reaction toward the reactants (to the left).
- A decrease in temperature drives an exothermic reaction toward the products (to the right).

For an endothermic reaction, of course, the opposite is true.

- An increase in temperature drives an endothermic reaction toward the products (to the right).
- A decrease in temperature drives an endothermic reaction toward the reactants (to the left).

Recall from Section 7-4 that a change in temperature changes not only the position of equilibrium but also the value of K, the equilibrium constant.

EXAMPLE 7-8 Le Chatelier's Principle–Effect of Temperature

The conversion of nitrogen dioxide to dinitrogen tetroxide is an exothermic reaction:

$$2NO_2(g) \rightleftharpoons N_2O_4(g) + 13{,}700 \text{ cal}$$

Nitrogen dioxide (brown) Dinitrogen tetroxide (colorless)

In Figure 7-12 we see that the brown color is darker at 50°C than it is at 0°C. Explain.

Strategy and Solution

To go from 0°C to 50°C, heat must be added. But heat is a product of this equilibrium reaction, as it is written in the question. The addition of heat, therefore, shifts the equilibrium to the left. This shift produces more $NO_2(g)$, leading to the darker brown color.

Problem 7-8

Consider the following equilibrium reaction:

$$A \rightleftharpoons B$$

Increasing the temperature results in an increase in the equilibrium concentration of B. Is the conversion of A to B an exothermic reaction or an endothermic reaction? Explain.

D. Change in Pressure

A change in pressure influences the equilibrium only if one or more components of the reaction mixture are gases. Consider the following equilibrium reaction:

$$N_2O_4(g) \rightleftharpoons 2NO_2(g)$$

Dinitrogen tetroxide (colorless)   Nitrogen dioxide (brown)

In this equilibrium, we have one mole of gas as a reactant and two moles of gas as products. According to Le Chatelier's principle, an increase in pressure shifts the equilibrium in the direction that will decrease the moles in the gas phase and thus decrease the internal pressure. In the preceding reaction, the equilibrium will shift to the left.

- An increase in pressure shifts the reaction toward the side with fewer moles of gas.
- A decrease in pressure shifts the reaction toward the side with more moles of gas.
- When the moles of gas are the same in a balanced reaction, an increase or decrease in pressure results in no shift of the reaction.

Le Châtelier's principle does not apply to pressure increases caused by the addition of a nonreactive (inert) gas to the reaction mixture. This addition has no effect on the equilibrium position.

EXAMPLE 7-9   Le Chatelier's Principle–Effect of Gas Pressure

In the production of ammonia, both reactants and products are gases:

$$N_2(g) + 3H_2(g) \rightleftharpoons 2NH_3(g)$$

What kind of pressure change would increase the yield of ammonia?

Strategy and Solution

There are four moles of gases on the left side and two moles on the right side. To increase the yield of ammonia, we must shift the equilibrium to

FIGURE 7-12 Effect of temperature on the N_2O_4–NO_2 system at equilibrium. (*top*) At 50°C, the deep brown color indicates the predominance of NO_2. (*bottom*) At 0°C, N_2O_4, which is colorless, predominates.

the right. An increase in pressure shifts the equilibrium toward the side with fewer moles—that is, to the right. Thus, an increase in pressure will increase the yield of ammonia.

Problem 7-9

What happens to the following equilibrium reaction when the pressure is increased?

$$O_2(g) + 4ClO_2(g) \rightleftharpoons 2Cl_2O_5(g)$$

E. The Effects of a Catalyst

As we saw in Section 7-4D, a catalyst increases the rate of a reaction without itself being changed. For a reversible reaction, catalysts always increase the rates of both the forward and reverse reactions to the same extent. Therefore, the addition of a catalyst has no effect on the position of equilibrium. However, adding a catalyst to a system not yet at equilibrium causes it to reach equilibrium faster than it would without the catalyst.

CHEMICAL CONNECTIONS 7E

The Haber Process

Both humans and other animals need proteins and other nitrogen-containing compounds to live. Ultimately, the nitrogen in these compounds comes from the plants that we eat. Although the atmosphere contains plenty of N_2, nature converts it to compounds usable by biological organisms in only one way: Certain bacteria have the ability to "fix" atmospheric nitrogen—that is, convert it to ammonia. Most of these bacteria live in the roots of certain plants such as clover, alfalfa, peas, and beans. However, the amount of nitrogen fixed by such bacteria each year is far less than the amount necessary to feed all the humans and animals in the world.

The world today can support its population only by using fertilizers made by artificial fixing, primarily the **Haber process,** which converts N_2 to NH_3.

$$N_2(g) + 3H_2(g) \rightleftharpoons 2NH_3(g) + 22 \text{ kcal}$$

Early workers who focused on the problem of fixing nitrogen were troubled by a conflict between equilibrium and rate. Because the synthesis of ammonia is an exothermic reaction, an increase in temperature drives the equilibrium to the left; so the best results (largest possible yield) should be obtained at low temperatures. At low temperatures, however, the rate is too slow to produce any meaningful amounts of NH_3. In 1908, Fritz Haber (1868–1934) solved this problem when he discovered a catalyst that permits the reaction to take place at a convenient rate at 500°C.

Ammonia is produced by bacteria in these root nodules.

The NH_3 produced by the Haber process is converted to fertilizers, which are used all over the world. Without these fertilizers, food production would diminish so much that widespread starvation would result.

Test your knowledge with Problem 7-48.

Summary of Key Questions

Section 7-1 How Do We Measure Reaction Rates?
- The **rate of a reaction** is the change in concentration of a reactant or a product per unit time. Some reactions are fast; others are slow.

Section 7-2 Why Do Some Molecular Collisions Result in Reaction Whereas Others Do Not?
- The rate of a reaction depends on the number of **effective collisions**—that is, collisions that lead to a reaction.
- The energy necessary for a reaction to take place is the **activation energy.** Effective collisions have (1) more than the activation energy required for the reaction to proceed forward and (2) the proper orientation in space of the colliding particles.

Section 7-3 What Is the Relationship Between Activation Energy and Reaction Rate?
- The lower the activation energy, the faster the reaction.
- An energy diagram shows the progress of a reaction.
- The position at the top of the curve in an energy diagram is called the **transition state.**

Section 7-4 How Can We Change the Rate of a Chemical Reaction?
- Reaction rates generally increase with increasing concentration and temperature; they also depend on the nature of the reactants.
- The rates of some reactions can be increased by adding a **catalyst,** a substance that provides an alternative pathway with a lower activation energy.
- A rate constant gives the relationship between the rate of the reaction and the concentrations of the reactants at a constant temperature.

Section 7-5 What Does It Mean to Say That a Reaction Has Reached Equilibrium?
- Many reactions are reversible and eventually reach equilibrium.
- At **equilibrium,** the forward and reverse reactions take place at equal rates and concentrations do not change.

Section 7-6 What Is an Equilibrium Constant and How Do We Use It?
- Every equilibrium has an **equilibrium expression** and an **equilibrium constant,** K, which does not change when concentrations change but does change when temperature changes.
- There is no relationship between the value of the equilibrium constant, K, and the rate at which equilibrium is reached.

Section 7-7 What Is Le Chatelier's Principle?
- **Le Chatelier's principle** tells us what happens when we put stress on a system in equilibrium.
- The addition of a component causes the equilibrium to shift to the opposite side.
- The removal of a component causes the equilibrium to shift to the side from which the component is removed.
- Increasing the temperature drives an exothermic equilibrium to the side of the reactants; increasing the temperature drives an endothermic equilibrium to the side of the products.
- Increasing the pressure of a mixture shifts the equilibrium in the direction that decreases the moles in the gas phase; decreasing the pressure of a mixture shifts the equilibrium in the direction that increases the moles in the gas phase.
- Addition of a catalyst has no effect on the position of equilibrium.

Problems

Orange-numbered problems are applied.

Section 7-1 How Do We Measure Reaction Rates?

7-10 The rate of disappearance of HCl was measured for the following reaction:

$$CH_3OH + HCl \longrightarrow CH_3Cl + H_2O$$

The initial concentration of HCl is 1.85 M. Its concentration decreases to 1.58 M in 54.0 min. What is the rate of reaction?

7-11 Consider the following reaction:

$$\underset{\text{Chloromethane}}{CH_3-Cl} + I^- \longrightarrow \underset{\text{Iodomethane}}{CH_3-I} + Cl^-$$

Suppose we start the reaction with an initial iodomethane concentration of 0.260 M. This concentration increases to 0.840 M over a period of 1 h 20 min. What is the rate of reaction?

Section 7-2 Why Do Some Molecular Collisions Result in Reaction Whereas Others Do Not?

7-12 Two kinds of gas molecules are reacted at a set temperature. The gases are blown into the reaction vessel from two tubes. In setup A, the two tubes are aligned parallel to each other; in setup B, they are 90° to each other; and in setup C, they are aligned directly opposite each other. Which setup would yield the most effective collisions?

7-13 Why are reactions between ions in aqueous solution generally much faster than reactions between covalent molecules?

Section 7-3 What Is the Relationship Between Activation Energy and Reaction Rate?

7-14 What is the likelihood that the following reaction occurs in a single step? Explain.

$$O_2(g) + 4ClO_2(g) \rightleftharpoons 2Cl_2O_5(g)$$

7-15 A certain reaction is exothermic by 9 kcal/mol and has an activation energy of 14 kcal/mol. Draw an energy diagram for this reaction and label the transition state.

Section 7-4 How Can We Change the Rate of a Chemical Reaction?

7-16 A quart of milk quickly spoils if left at room temperature but keeps for several days in a refrigerator. Explain.

7-17 If a certain reaction takes 16 h to go to completion at 10°C, what temperature should we run it if we want it to go to completion in 1 h?

7-18 In most cases, when we run a reaction by mixing a fixed quantity of substance A with a fixed quantity of substance B, the rate of the reaction begins at a maximum and then decreases as time goes by. Explain.

7-19 If you were running a reaction and wanted it to go faster, what three things might you try to accomplish this goal?

7-20 What factors determine whether a reaction run at a given temperature will be fast or slow?

7-21 Explain how a catalyst increases the rate of a reaction.

7-22 If you add a piece of marble, $CaCO_3$, to a 6 M HCl solution at room temperature, you will see some bubbles form around the marble as gas slowly rises. If you crush another piece of marble and add it to the same solution at the same temperature, you will see vigorous gas formation, so much so that the solution appears to be boiling. Explain.

Section 7-5 What Does It Mean to Say That a Reaction Has Reached Equilibrium?

7-23 Burning a piece of paper is an irreversible reaction. Give some other examples of irreversible reactions.

7-24 Suppose the following reaction is at equilibrium:

$$PCl_3 + Cl_2 \rightleftharpoons PCl_5$$

(a) Are the equilibrium concentrations of PCl_3, Cl_2, and PCl_5 necessarily equal? Explain.

(b) Is the equilibrium concentration of PCl_3 necessarily equal to that of Cl_2? Explain.

Section 7-6 What Is an Equilibrium Constant and How Do We Use It?

7-25 Write equilibrium expressions for these reactions.
(a) $2H_2O_2(g) \rightleftharpoons 2H_2O(g) + O_2(g)$
(b) $2N_2O_5(g) \rightleftharpoons 2N_2O_4(g) + O_2(g)$
(c) $6H_2O(g) + 6CO_2(g) \rightleftharpoons C_6H_{12}O_6(s) + 6O_2(g)$

7-26 Write the chemical equations corresponding to the following equilibrium expressions.

(a) $K = \dfrac{[H_2CO_3]}{[CO_2][H_2O]}$

(b) $K = \dfrac{[P_4][O_2]^5}{[P_4O_{10}]}$

(c) $K = \dfrac{[F_2]^3[PH_3]}{[HF]^3[PF_3]}$

7-27 Consider the following equilibrium reaction. Under each species is its equilibrium concentration. Calculate the equilibrium constant for the reaction.

$$CO(g) + H_2O(g) \rightleftharpoons CO_2(g) + H_2(g)$$
$$0.933\ M \quad 0.720\ M \quad 0.133\ M \quad 3.37\ M$$

7-28 When the following reaction reached equilibrium at 325 K, the equilibrium constant was found to be 172. When a sample was taken of the equilibrium mixture, it was found to contain 0.0714 M NO_2. What was the equilibrium concentration of N_2O_4?

$$2NO_2(g) \rightleftharpoons N_2O_4(g)$$

7-29 The following reaction was allowed to reach equilibrium at 25°C. Under each component is its equilibrium concentration. Calculate the equilibrium constant, K, for this reaction.

$$2NOCl(g) \rightleftharpoons 2NO(g) + Cl_2(g)$$
$$2.6\ M \qquad 1.4\ M \quad 0.34\ M$$

7-30 Write the equilibrium expression for this reaction:

$$HNO_3(aq) + H_2O(\ell) \rightleftharpoons H_3O^+(aq) + NO_3^-(aq)$$

7-31 Here are equilibrium constants for several reactions. Which of them favor the formation of products, and which favor the formation of reactants?
(a) 4.5×10^{-8}
(b) 32
(c) 4.5
(d) 3.0×10^{-7}
(e) 0.0032

7-32 A particular reaction has an equilibrium constant of 1.13 under one set of conditions and an equilibrium constant of 1.72 under a different set of conditions. Which conditions would be more advantageous in an industrial process that sought to obtain the maximum amount of products? Explain.

7-33 If a reaction is very exothermic—that is, if the products have a much lower energy than the reactants—can we be reasonably certain that it will take place rapidly?

7-34 If a reaction is very endothermic—that is, if the products have a much higher energy than the reactants—can we be reasonably certain that it will take place extremely slowly or not at all?

7-35 A reaction has a high rate constant but a small equilibrium constant. What does this mean in terms of producing an industrial product?

Section 7-7 What Is Le Chatelier's Principle?

7-36 Complete the following table showing the effects of changing reaction conditions on the equilibrium and value of the equilibrium constant, K.

Change in Condition	How the Reacting System Changes to Achieve a New Equilibrium	Does the Value of K Increase or Decrease?
Addition of a reactant	Shift to product formation	Neither
Removal of a reactant		
Addition of a product		
Removal of a product		
Increasing pressure		

7-37 Assume that the following exothermic reaction is at equilibrium:

$$H_2(g) + I_2(g) \rightleftharpoons 2HI(g)$$

Tell whether the position of equilibrium will shift to the right or the left if we:
(a) Remove some HI
(b) Add some I_2
(c) Remove some I_2
(d) Increase the temperature
(e) Add a catalyst

7-38 The following reaction is endothermic:

$$3O_2(g) \rightleftharpoons 2O_3(g)$$

If the reaction is at equilibrium, tell whether the equilibrium will shift to the right or the left if we:
(a) Remove some O_3
(b) Remove some O_2
(c) Add some O_3
(d) Decrease the temperature
(e) Add a catalyst
(f) Increase the pressure

7-39 The following reaction is exothermic: After it reaches equilibrium, we add a few drops of Br_2.

$$2NO(g) + Br_2(g) \rightleftharpoons 2NOBr(g)$$

(a) What will happen to the equilibrium?
(b) What will happen to the equilibrium constant?

7-40 Is there any change in conditions that change the equilibrium constant, K, of a given reaction?

7-41 The equilibrium constant at 1127°C for the following endothermic reaction is 571:

$$2H_2S(g) \rightleftharpoons 2H_2(g) + S_2(g)$$

If the mixture is at equilibrium, what happens to K if we:
(a) Add some H_2S?
(b) Add some H_2?
(c) Lower the temperature to 1000°C?

Chemical Connections

7-42 (Chemical Connections 7A) In a bacterial infection, body temperature may rise to 101°F. Does this body defense kill the bacteria directly by heat or by another mechanism? If so, by which mechanism?

7-43 (Chemical Connections 7A and 7B) Why is a high fever dangerous? Why is a low body temperature dangerous?

7-44 (Chemical Connections 7B) Why do surgeons sometimes lower body temperatures during heart operations?

7-45 (Chemical Connections 7C) A painkiller—for example, Tylenol—can be purchased in two forms, each containing the same amount of drug. One form is a solid coated pill, and the other is a capsule that contains tiny beads and has the same coat. Which medication will act faster? Explain.

7-46 (Chemical Connections 7D) What reaction takes place when sunlight hits the compound silver chloride?

7-47 (Chemical Connections 7D) You have a recipe to manufacture sunglasses: 3.5 g AgCl/kg glass. A new order comes in to manufacture sunglasses to be used in deserts like the Sahara. How would you change the recipe?

7-48 (Chemical Connections 7E) If the equilibrium for the Haber process is unfavorable at high temperatures, why do factories nevertheless use high temperatures?

Additional Problems

7-49 In the reaction between H_2 and Cl_2 to give HCl, a 10°C increase in temperature doubles the rate of reaction. If the rate of reaction at 15°C is 2.8 moles of HCl per liter per second, what are the rates at −5°C and at 45°C?

7-50 Draw an energy diagram for an exothermic reaction that yields 75 kcal/mol. The activation energy is 30 kcal/mol.

7-51 Draw a diagram similar to Figure 7-4. Draw a second line of the energy profile starting and ending at the same level as the first but having a smaller peak than the first line. Label them 1 and 2. What may have occurred to change the energy profile of a reaction from 1 to 2?

7-52 For the reaction

$$2NOBr(g) \rightleftharpoons 2NO(g) + Br_2(g)$$

the rate of the reaction was −2.3 mol NOBr/L/h when the initial NOBr concentration was 6.2 mol NOBr/L. What is the rate constant of the reaction?

7-53 The equilibrium constant for the following reaction is 25:

$$2NOBr(g) \rightleftharpoons 2NO(g) + Br_2(g)$$

A measurement made on the equilibrium mixture found that the concentrations of NO and Br_2 were each 0.80 M. What is the concentration of NOBr at equilibrium?

7-54 In the following reaction, the concentration of N_2O_4 in mol/L was measured at the end of the times shown. What is the initial rate of the reaction?

$$N_2O_4(g) \rightleftharpoons 2NO_2(g)$$

Time (s)	[N_2O_4]
0	0.200
10	0.180
20	0.162
30	0.146

7-55 How could you increase the rate of a gaseous reaction without adding more reactants or a catalyst and without changing the temperature?

7-56 In an endothermic reaction, the activation energy is 10.0 kcal/mol. Is the activation energy of the reverse reaction also 10.0 kcal/mol, or would it be more or less? Explain with the aid of a diagram.

7-57 Write the reaction to which the following equilibrium expression applies:

$$K = \frac{[NO_2]^4[H_2O]^6}{[NH_3]^4[O_2]^7}$$

7-58 The rate for the following reaction at 300 K was found to be 0.22 M NO_2/min. What would be the approximate rate at 320 K?

$$N_2O_4(g) \rightleftharpoons 2NO_2(g)$$

7-59 Assume that two different reactions are taking place at the same temperature. In reaction A, two different spherical molecules collide to yield a product. In reaction B, the shape of the colliding molecules is rodlike. Each reaction has the same number of collisions per second and the same activation energy. Which reaction goes faster?

7-60 Is it possible for an endothermic reaction to have zero activation energy?

7-61 In the following reaction, the rate of appearance of I_2 is measured at the times shown. What is the initial rate of the reaction?

$$2HI(g) \rightleftharpoons H_2(g) + I_2(g)$$

Time (s)	[I_2]
0	0
10	0.30
20	0.57
30	0.81

7-62 A reaction occurs in three steps with the following rate constants:

$$A \xrightarrow[\text{Step 1}]{k_1 = 0.3\ M} B \xrightarrow[\text{Step 2}]{k_2 = 0.05\ M} C \xrightarrow[\text{Step 3}]{k_3 = 4.5\ M} D$$

(a) Which step is the slow step?
(b) Which step has the lowest activation energy?

Looking Ahead

7-63 As we shall see in Chapter 8, weak acids such as acetic acid only partially dissociate in solution, as shown in the following simplified net ionic equilibrium reaction.

$$CH_3COOH \rightleftharpoons H^+ + CH_3COO^-$$
Acetic acid

(a) Suppose that initially, only 0.10 M acetic acid is present. Analysis of the equilibrium mixture shows that the concentration of acetic acid reduces to 0.098 M. Determine the equilibrium concentrations of H^+ and CH_3COO^-.

(b) What is the expected equilibrium constant, K, for this reaction?

7-64 As we shall see in Chapter 20, there are two forms of glucose, designated alpha (α) and beta (β), which are in equilibrium in aqueous solution. The equilibrium constant for the reaction is 1.5 at 30°C.

$$\alpha\text{-D-glucose}(aq) \rightleftharpoons \beta\text{-D-glucose}(aq) \quad K = 1.5$$

(a) If you begin with a fresh 1.0 M solution of α-D-glucose in water, what will be its concentration when equilibrium is reached?

(b) Calculate the percentage of α-glucose and of β-glucose present at equilibrium in aqueous solution at 30°C.

7-65 Consider the reaction A $\longrightarrow$ B, for which you wish to determine the rate. You do not have any convenient method for determining the amount of B formed. You do, however, have a method for determining the amount of A left as the reaction proceeds. Does it make any difference if you determine the rate in terms of disappearance of A rather than appearance of B? Why or why not?

7-66 You have a choice of two methods for determining the rate of a reaction. In the first method, you have to extract part of the reaction mixture to test for the amount of product formed. In the second method, you can do continuous monitoring of the amount of product formed. Which method is preferable and why?

7-67 You want to measure reaction rates for some very fast reactions. What sort of technical difficulties do you expect to arise?

7-68 You make five measurements of the rate of a reaction, and for each measurement, you determine the rate constant. The values of four of the rate constants are close to each other (within experimental error). The other is quite different. Is your result likely to represent a different rate or an error in calculation? Why?

Tying It Together

7-69 Pure carbon exists is several forms, two of which are diamond and graphite. The conversion of the diamond form to the graphite form is exothermic to a very slight extent. How is it that jewelers can advertise "Diamonds are forever"?

7-70 You have decided to change the temperature at which you run a certain reaction in hopes of obtaining more product more quickly. You find that you actually get less of the desired product, although you get to the equilibrium state more quickly. What happened?

Challenge Problems

7-71 You have a beaker that contains solid silver chloride (AgCl) and a saturated solution of Ag^+ and Cl^- ions in equilibrium with the solid.

$$AgCl(s) \rightleftharpoons Ag^+(aq) + Cl^-(aq)$$

You add several drops of a sodium chloride solution. What happens to the concentration of silver ions?

7-72 What would happen to the reaction that produces ammonia if water is present in the reaction mixture?

$$N_2(g) + 3H_2(g) \rightleftharpoons 2NH_3(g)$$

Hint: Ammonia is very soluble in water.

7-73 The equilibrium constant, K, is 2.4×10^{-3} for the following reaction at a certain temperature.

$$H_2(g) + F_2(g) \rightleftharpoons 2HF(g)$$

If the concentrations of both $H_2(g)$ and $F_2(g)$ are found to be 0.0021 M at equilibrium, what is the concentration of HF(g) under these conditions?

7-74 It can be shown that a mathematical relationship exists between rate constants and equilibrium constants. For example, consider the following generic reaction, where the rate constant k refers to the rate of the forward reaction and the rate constant k' refers to the rate of the reverse reaction.

$$A + B \underset{k'}{\overset{k}{\rightleftharpoons}} C + D$$

Verify that the equilibrium constant for a reaction is equal to the ratio of the rate constants for the forward and reverse reactions.

$$K = \frac{k}{k'}$$

7-75 The following exothermic reaction is at equilibrium.

$$Zn(s) + 4H^+(aq) + 2NO_3^-(aq) \rightleftharpoons Zn^{2+}(aq) + 2NO_2(g) + 2H_2O(\ell)$$

Consider each of the following changes separately and state the effect (increase, decrease, or no change) that the change from the first column has on the equilibrium value of the quantity listed in the second column.

Change	Quantity	Effect
Increase the pressure	Concentration of NO_3^-	
Add some Zn	Concentration of NO_2	
Decrease the H^+	Concentration of Zn^{2+}	
Add Pt catalyst	Equilibrium constant, K	
Add some Ar gas	Concentration of H^+	
Decrease the Zn^{2+}	Equilibrium constant, K	
Increase the temperature	Concentration of Zn	

7-76 When a 0.10 M solution of glucose-1-phosphate is incubated with a catalytic amount of phosphoglucomutase, the glucose-1-phosphate is transformed into glucose-6-phosphate until equilibrium is established at 25°C.

glucose-1-phosphate $\rightleftharpoons$ glucose-6-phosphate

If the equilibrium concentration of glucose-6-phosphate is found to be 9.6×10^{-2} M, determine the equilibrium constant, K, for this reaction at 25°C.

7-77 A certain endothermic reaction (see Figure 7-5) at equilibrium has an activation energy of 40. kJ. If the energy of reaction is found to be 30. kJ, what is the activation energy for the reverse reaction?

7-78 Consider the equilibrium of phosphorus pentachloride, PCl_5, with its decomposition products, where $K = 6.3 \times 10^{-4}$ at a certain temperature.

$$PCl_5(g) \rightleftharpoons PCl_3(g) + Cl_2(g)$$

At equilibrium, it is found that the concentration of PCl_5 is three times the concentration of PCl_3. Determine the concentration of Cl_2 under these conditions.

7-79 Consider the reaction shown below at a certain temperature.

$$2H_2O(g) \rightleftharpoons 2H_2(g) + O_2(g)$$

The equilibrium constant, K, is equal to 8.7×10^3 at a certain temperature. At equilibrium, it is found that $[H_2] = 1.9 \times 10^{-2}$ M and $[O_2] = 8.0 \times 10^{-2}$ M. What is the concentration of H_2O at equilibrium?

7-80 Consider the equilibrium reaction shown in Problem 7-72. Suppose an equilibrium mixture contains 0.036 M N_2 and 0.15 M H_2. The equilibrium constant, K, is equal to 0.29 at a certain temperature. What is the concentration of NH_3?

7-81 The first step in the industrial synthesis of hydrogen is the reaction of steam and methane to give carbon monoxide and hydrogen at 1400 K.

$$H_2O(g) + CH_4 \rightleftharpoons CO(g) + 3H_2(g)$$

The equilibrium constant, K, is equal to 4.7 at 1400 K. A mixture of reactants and products at 1400 K contains 0.035 M H_2O, 0.050 M CH_4. 0.15 M CO, and 0.20 M H_2.

(a) Is this reaction at equilibrium? Hint: how does your calculated ratio of products to reactants compare to the equilibrium constant noted above.

(b) In which direction does the reaction proceed to reach equilibrium?

7-82 An equilibrium mixture of O_2, SO_2, and SO_3 contains equal concentrations of SO_2 and SO_3 at a certain temperature.

$$2SO_2(g) + O_2(g) \rightleftharpoons 2SO_3(g)$$

The equilibrium constant, K, is equal to 2.7×10^2. Calculate the equilibrium concentration of O_2.

7-83 Consider the following reaction, where the rate constant k refers to the rate of the forward reaction and the rate constant k' refers to the rate of the reverse reaction as shown in Problem 7-74.

$$(CF_3)_2CO(g) + H_2O(g) \underset{k'}{\overset{k}{\rightleftharpoons}} (CF_3)_2C(OH)_2(g)$$

At 76°C, the forward reaction rate constant, k, is 0.13 and the reverse rate constant, k' is 6.2×10^{-4}. What is the value of the equilibrium constant, K?

7-84 Consider the reaction of chloromethane with OH^- in aqueous solution.

$$CH_3Cl(g) + OH^-(aq) \underset{k'}{\overset{k}{\rightleftharpoons}} CH_3OH(aq) + Cl^-(aq)$$

At room temperature, the rate constant for the forward reaction, k, is 6×10^{-6} and the equilibrium constant, K, is equal to 1×10^{16}. Calculate the rate constant for the reverse reaction, k' at room temperature.

8 Acids and Bases

KEY QUESTIONS

- 8-1 What Are Acids and Bases?
- 8-2 How Do We Define the Strength of Acids and Bases?
- 8-3 What Are Conjugate Acid–Base Pairs?
  - *How To* . . . Name Common Acids
- 8-4 How Can We Tell the Position of Equilibrium in an Acid–Base Reaction?
- 8-5 How Do We Use Acid Ionization Constants?
  - *How To* . . . Use Logs and Antilogs
- 8-6 What Are the Properties of Acids and Bases?
- 8-7 What Are the Acidic and Basic Properties of Pure Water?
- 8-8 What Are pH and pOH?
- 8-9 How Do We Use Titrations to Calculate Concentration?
- 8-10 What Are Buffers?
- 8-11 How Do We Calculate the pH of a Buffer?
- 8-12 What Are TRIS, HEPES, and These Buffers with the Strange Names?

Some foods and household products are very acidic, while others are basic. From your prior experiences, can you tell which ones belong to which category?

8-1 What Are Acids and Bases?

We frequently encounter acids and bases in our daily lives. Oranges, lemons, and vinegar are examples of acidic foods, and sulfuric acid is in our automobile batteries. As for bases, we take antacid tablets for heartburn and use household ammonia as a cleaning agent. What do these substances have in common? Why are acids and bases usually discussed together?

In 1884, a young Swedish chemist named Svante Arrhenius (1859–1927) answered the first question by proposing what was then a new definition of acids and bases. According to the Arrhenius definition, an **acid** is a substance that produces H^+ ions in aqueous solution, and a **base** is a substance that produces OH^- ions in aqueous solution.

This definition of acid is a slight modification of the original Arrhenius definition, which stated that an acid produces hydrogen ions, H^+. Today we know that H^+ ions cannot exist in water. An H^+ ion is a bare proton, and a charge of +1 is too concentrated to exist on such a tiny particle

(Section 3-2). Therefore, an H$^+$ ion in water immediately combines with an H$_2$O molecule to give a **hydronium ion**, H$_3$O$^+$.

Hydronium ion The H$_3$O$^+$ ion

$$H^+(aq) + H_2O(\ell) \longrightarrow H_3O^+(aq)$$
Hydronium ion

Apart from this modification, the Arrhenius definitions of acid and base are still valid and useful today, as long as we are talking about aqueous solutions. Although we know that acidic aqueous solutions do not contain H$^+$ ions, we frequently use the terms "H$^+$" and "proton" when we really mean "H$_3$O$^+$". The three terms are generally used interchangeably.

When an acid dissolves in water, it reacts with the water to produce H$_3$O$^+$. For example, hydrogen chloride, HCl, in its pure state is a poisonous gas. When HCl dissolves in water, it reacts with a water molecule to give hydronium ion and chloride ion:

$$H_2O(\ell) + HCl(aq) \longrightarrow H_3O^+(aq) + Cl^-(aq)$$

Thus, a bottle labeled aqueous "HCl" is actually not HCl at all, but rather an aqueous solution of H$_3$O$^+$ and Cl$^-$ ions.

We can show the transfer of a proton from an acid to a base by using a curved arrow. First, we write the Lewis structure (Section 2-6F) of each reactant and product. Then we use curved arrows to show the change in position of electron pairs during the reaction. The tail of the curved arrow is located at the electron pair. The head of the curved arrow shows the new position of the electron pair.

$$H-\overset{..}{\underset{|}{\overset{|}{O}}}: + H-\overset{..}{\underset{..}{Cl}}: \longrightarrow H-\overset{..}{\underset{|}{\overset{+}{O}}}-H + :\overset{..}{\underset{..}{Cl}}:^-$$
$$\phantom{H-\overset{..}{\underset{|}{\overset{|}{O}}}:}\,HH$$

In this equation, the curved arrow on the left shows that an unshared pair of electrons on oxygen forms a new covalent bond with hydrogen. The curved arrow on the right shows that the pair of electrons of the H—Cl bond is given entirely to chlorine to form a chloride ion. Thus, in the reaction of HCl with H$_2$O, a proton is transferred from HCl to H$_2$O, and in the process, an O—H bond forms and an H—Cl bond is broken.

With bases, the situation is slightly different. Many bases are metal hydroxides, such as KOH, NaOH, Mg(OH)$_2$, and Ca(OH)$_2$. When these ionic solids dissolve in water, their ions merely separate, and each ion is solvated by water molecules (Section 6-6A). For example,

$$NaOH(s) \xrightarrow{H_2O} Na^+(aq) + OH^-(aq)$$

Other bases are not hydroxides. Instead, they produce OH$^-$ ions in water by reacting with water molecules. The most important example of this kind of base is ammonia, NH$_3$, a poisonous gas. When ammonia dissolves in water, it reacts with water to produce ammonium ions and hydroxide ions.

$$NH_3(aq) + H_2O(\ell) \rightleftharpoons NH_4^+(aq) + OH^-(aq)$$

As we will see in Section 8-2, ammonia is a weak base, and the position of the equilibrium for its reaction with water lies considerably toward the left. In a 1.0 M solution of NH$_3$ in water, for example, only about 4 molecules of NH$_3$ out of every 1000 react with water to form NH$_4^+$ and OH$^-$. Thus, when ammonia is dissolved in water, it exists primarily as hydrated NH$_3$ molecules. Nevertheless, some OH$^-$ ions are produced, and therefore, NH$_3$ is a base.

Bottles of NH_3 in water are sometimes labeled "ammonium hydroxide" or "NH_4OH," but this gives a false impression of what is really in the bottle. Most of the NH_3 molecules have not reacted with the water, so the bottle contains mostly NH_3 and H_2O and only a little NH_4^+ and OH^-.

We indicate how the reaction of ammonia with water takes place by using curved arrows to show the transfer of a proton from a water molecule to an ammonia molecule. Here, the curved arrow on the left shows that the unshared pair of electrons on nitrogen forms a new covalent bond with a hydrogen of a water molecule. At the same time as the new N—H bond forms, an O—H bond of a water molecule breaks and the pair of electrons forming the H—O bond moves entirely to oxygen, forming OH^-.

$$H-\underset{\underset{H}{|}}{\overset{\overset{H}{|}}{N}}: + H-\overset{..}{\underset{..}{O}}-H \longrightarrow H-\underset{\underset{H}{|}}{\overset{\overset{H}{|}}{N}}\!\!\!{}^+\!\!-H + :\overset{..}{\underset{..}{O}}-H$$

Thus, ammonia produces an OH^- ion by taking H^+ from a water molecule and leaving OH^- behind.

8-2 How Do We Define the Strength of Acids and Bases?

All acids are not equally strong. According to the Arrhenius definition, a **strong acid** is one that reacts completely or almost completely with water to form H_3O^+ ions. Table 8-1 gives the names and molecular formulas for six of the most common strong acids. They are strong acids because when they dissolve in water, they dissociate completely to give H_3O^+ ions.

Strong acid An acid that ionizes completely in aqueous solution

Table 8-1 Strong Acids and Bases

Acid Formula	Name	Base Formula	Name
HCl	Hydrochloric acid	LiOH	Lithium hydroxide
HBr	Hydrobromic acid	NaOH	Sodium hydroxide
HI	Hydroiodic acid	KOH	Potassium hydroxide
HNO_3	Nitric acid	$Ba(OH)_2$	Barium hydroxide
H_2SO_4	Sulfuric acid	$Ca(OH)_2$	Calcium hydroxide
$HClO_4$	Perchloric acid	$Sr(OH)_2$	Strontium hydroxide

Weak acids Acids that are only partially ionized in aqueous solution

Weak acids produce a much smaller concentration of H_3O^+ ions. Acetic acid, for example, is a weak acid. In water it exists primarily as acetic acid molecules; only a few acetic acid molecules (4 out of every 1000) are converted to acetate ions.

$$\underset{\text{Acetic acid}}{CH_3COOH(aq)} + H_2O(\ell) \rightleftharpoons \underset{\text{Acetate ion}}{CH_3COO^-(aq)} + H_3O^+(aq)$$

Strong bases Bases that ionize completely in aqueous solution

Weak base A base that is only partially ionized in aqueous solution

There are six common **strong bases** (Table 8-1), all of which are metal hydroxides. They are strong bases because, when they dissolve in water, they ionize completely to give OH^- ions. Another base, $Mg(OH)_2$, dissociates almost completely once dissolved, but it is very insoluble in water to begin with. We classify it as a **weak base.** As we saw in Section 8-1, ammonia is a weak base because the equilibrium for its reaction with water lies far to the left.

CHEMICAL CONNECTIONS 8A

Some Important Acids and Bases

STRONG ACIDS Sulfuric acid, H_2SO_4, is used in many industrial processes, such as manufacturing fertilizer, dyes and pigments, and rayon. In fact, sulfuric acid is one of the most widely produced single chemicals in the United States.

Hydrochloric acid, HCl, is an important acid in chemistry laboratories. Pure HCl is a gas, and the HCl in laboratories is an aqueous solution. HCl is the acid in the gastric fluid in your stomach, where it is secreted at a strength of about 5% w/v.

Nitric acid, HNO_3, is a strong oxidizing agent. A drop of it causes the skin to turn yellow because the acid reacts with skin proteins. A yellow color upon contact with nitric acid has long been a test for proteins.

WEAK ACIDS Acetic acid, CH_3COOH, is present in vinegar (about 5%). Pure acetic acid is called glacial acetic acid because of its melting point of 17°C, which means that it freezes on a moderately cold day.

Boric acid, H_3BO_3, is a solid. Solutions of boric acid in water were once used as antiseptics, especially for eyes. Boric acid is toxic when swallowed.

Phosphoric acid, H_3PO_4, is one of the strongest of the weak acids. The ions produced from it—$H_2PO_4^-$, HPO_4^{2-}, and PO_4^{3-}—are important in biochemistry (see Section 27-3).

STRONG BASES Sodium hydroxide, NaOH, also called lye, is the most important of the strong bases. It is a solid whose aqueous solutions are used in many industrial processes, including the manufacture of glass and soap.

Potassium hydroxide, KOH, also a solid, is used for many of the same purposes as NaOH.

WEAK BASES Ammonia, NH_3, the most important weak base, is a gas with many industrial uses. One of its chief uses is for fertilizers. A 5% solution is sold in supermarkets as a cleaning agent, and weaker solutions are used as "spirits of ammonia" to revive people who have fainted.

Magnesium hydroxide, $Mg(OH)_2$, is a solid that is insoluble in water. A suspension of about 8% $Mg(OH)_2$ in water is called milk of magnesia and is used as a laxative. $Mg(OH)_2$ is also used to treat wastewater in metal-processing plants and as a flame retardant in plastics.

Test your knowledge with Problem 8-75.

It is important to understand that the strength of an acid or a base is not related to its concentration. HCl is a strong acid, whether it is concentrated or dilute, because it dissociates completely in water to chloride ions and hydronium ions. Acetic acid is a weak acid, whether it is concentrated or dilute, because the equilibrium for its reaction with water lies far to the left. When acetic acid dissolves in water, most of it is present as undissociated CH_3COOH molecules.

$$HCl(aq) + H_2O(\ell) \longrightarrow Cl^-(aq) + H_3O^+(aq)$$

$$\underset{\text{Acetic acid}}{CH_3COOH(aq)} + H_2O(\ell) \rightleftharpoons \underset{\text{Acetate ion}}{CH_3COO^-(aq)} + H_3O^+(aq)$$

In Section 6-6C, we saw that electrolytes (substances that produce ions in aqueous solution) can be strong or weak. The strong acids and bases in Table 8-1 are strong electrolytes. Almost all other acids and bases are weak electrolytes.

8-3 What Are Conjugate Acid–Base Pairs?

The Arrhenius definitions of acid and base are very useful in aqueous solutions. But what if water is not involved? In 1923, the Danish chemist Johannes Brønsted and the English chemist Thomas Lowry independently proposed the following definitions: An **acid** is a proton donor, a **base** is a proton acceptor, and an **acid–base reaction** is a proton-transfer reaction. Furthermore, according to the Brønsted–Lowry definitions, any pair of molecules or ions that can be interconverted by transfer of a proton is called a **conjugate acid–base pair**. When an acid transfers a proton to a base, the acid is converted to its **conjugate base**. When a base accepts a proton, it is converted to its **conjugate acid**.

Conjugate acid–base pair A pair of molecules or ions that are related to one another by the gain or loss of a proton

Conjugate base In the Brønsted-Lowry theory, a substance formed when an acid donates a proton to another molecule or ion

Conjugate acid In the Brønsted–Lowry theory, a substance formed when a base accepts a proton

We can illustrate these relationships by examining the reaction between acetic acid and ammonia:

$$\underset{\text{Acetic acid}\\(\text{Acid})}{CH_3COOH} + \underset{\text{Ammonia}\\(\text{Base})}{NH_3} \rightleftharpoons \underset{\substack{\text{Acetate}\\\text{ion}\\(\text{Conjugate}\\\text{base of}\\\text{acetic acid})}}{CH_3COO^-} + \underset{\substack{\text{Ammonium}\\\text{ion}\\(\text{Conjugate}\\\text{acid of}\\\text{ammonia})}}{NH_4^+}$$

We can use curved arrows to show how this reaction takes place. The curved arrow on the right shows that the unshared pair of electrons on nitrogen becomes shared to form a new H—N bond. At the same time that the H—N bond forms, the O—H bond breaks and the electron pair of the O—H bond moves entirely to oxygen to form —O⁻ of the acetate ion. The result of these two electron-pair shifts is the transfer of a proton from an acetic acid molecule to an ammonia molecule:

$$\underset{\substack{\text{Acetic acid}\\(\text{Proton donor})}}{CH_3-\overset{\overset{\displaystyle :O:}{\|}}{C}-\overset{..}{\underset{..}{O}}-H} + \underset{\substack{\text{Ammonia}\\(\text{Proton acceptor})}}{:\!\!\underset{\underset{H}{|}}{\overset{\overset{H}{|}}{N}}\!\!-H} \rightleftharpoons \underset{\substack{\text{Acetate}\\\text{ion}}}{CH_3-\overset{\overset{\displaystyle :O:}{\|}}{C}-\overset{..}{\underset{..}{O}}\!:^{-}} + \underset{\substack{\text{Ammonium}\\\text{ion}}}{H-\overset{\overset{H}{|}}{\underset{\underset{H}{|}}{N}}\!\!^{+}\!-H}$$

Table 8-2 gives examples of common acids and their conjugate bases. As you study the examples of conjugate acid–base pairs in Table 8-2, note the following points:

1. An acid can be positively charged, neutral, or negatively charged. Examples of these charge types are H_3O^+, H_2CO_3, and $H_2PO_4^-$, respectively.

2. A base can be negatively charged or neutral. Examples of these charge types are PO_4^{3-} and NH_3, respectively.

3. Acids are classified as monoprotic, diprotic, or triprotic depending on the number of protons each may give up. Examples of **monoprotic acids** include HCl, HNO_3, and CH_3COOH. Examples of **diprotic acids**

Monoprotic acids Acids that can give up only one proton

Diprotic acids Acids that can give up two protons

Table 8-2 Some Acids and Their Conjugate Bases

	Acid	Name	Conjugate Base	Name	
Strong Acids ↑	HI	Hydroiodic acid	I^-	Iodide ion	Weak Bases
	HCl	Hydrochloric acid	Cl^-	Chloride ion	
	H_2SO_4	Sulfuric acid	HSO_4^-	Hydrogen sulfate ion	
	HNO_3	Nitric acid	NO_3^-	Nitrate ion	
	H_3O^+	Hydronium ion	H_2O	Water	
	HSO_4^-	Hydrogen sulfate ion	SO_4^{2-}	Sulfate ion	
	H_3PO_4	Phosphoric acid	$H_2PO_4^-$	Dihydrogen phosphate ion	
	CH_3COOH	Acetic acid	CH_3COO^-	Acetate ion	
	H_2CO_3	Carbonic acid	HCO_3^-	Bicarbonate ion	
	H_2S	Hydrogen sulfide	HS^-	Hydrogen sulfide ion	
	$H_2PO_4^-$	Dihydrogen phosphate ion	HPO_4^{2-}	Hydrogen phosphate ion	
	NH_4^+	Ammonium ion	NH_3	Ammonia	
	HCN	Hydrocyanic acid	CN^-	Cyanide ion	
	C_6H_5OH	Phenol	$C_6H_5O^-$	Phenoxide ion	
	HCO_3^-	Bicarbonate ion	CO_3^{2-}	Carbonate ion	
	HPO_4^{2-}	Hydrogen phosphate ion	PO_4^{3-}	Phosphate ion	↓
Weak Acids	H_2O	Water	OH^-	Hydroxide ion	Strong Bases
	C_2H_5OH	Ethanol	$C_2H_5O^-$	Ethoxide ion	

include H_2SO_4 and H_2CO_3. An example of a **triprotic acid** is H_3PO_4. Carbonic acid, for example, loses one proton to become bicarbonate ion and then a second proton to become carbonate ion.

$$H_2CO_3 + H_2O \rightleftharpoons HCO_3^- + H_3O^+$$
Carbonic acid — Bicarbonate ion

$$HCO_3^- + H_2O \rightleftharpoons CO_3^{2-} + H_3O^+$$
Bicarbonate ion — Carbonate ion

4. Several molecules and ions appear in both the acid and conjugate base columns; that is, each can function as either an acid or a base. The bicarbonate ion, HCO_3^-, ◄ for example, can give up a proton to become CO_3^{2-} (in which case it is an acid) or it can accept a proton to become H_2CO_3 (in which case it is a base). A substance that can act as either an acid or a base is called **amphiprotic**. The most important amphiprotic substance in Table 8-2 is water, which can accept a proton to become H_3O^+ or lose a proton to become OH^-.

5. A substance cannot be a Brønsted–Lowry acid unless it contains a hydrogen atom, but not all hydrogen atoms can be given up. For example, acetic acid, CH_3COOH, has four hydrogens but is monoprotic—it gives up only one of them. Similarly, phenol, C_6H_5OH, gives up only one of its six hydrogens:

$$C_6H_5OH + H_2O \rightleftharpoons C_6H_5O^- + H_3O^+$$
Phenol — Phenoxide ion

This is because a hydrogen must be bonded to a strongly electronegative atom, such as oxygen or a halogen, to be acidic.

6. There is an inverse relationship between the strength of an acid and the strength of its conjugate base: The stronger the acid, the weaker its conjugate base. HI, for example, is the strongest acid listed in Table 8-2 and I^-,

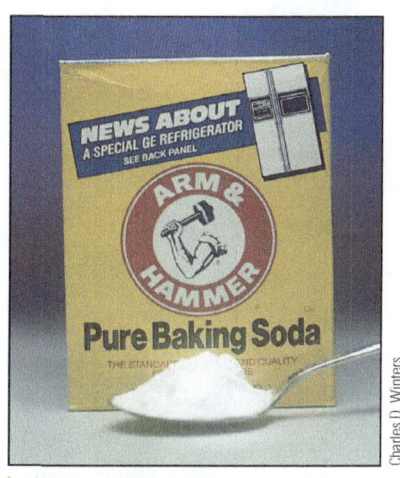

▶ A box of Arm & Hammer baking soda (sodium bicarbonate). Sodium bicarbonate is composed of Na^+ and HCO_3^-, the amphiprotic bicarbonate ion.

Triprotic acid An acid that can give up three protons

Amphiprotic A substance that can act as either an acid or a base its conjugate base, is the weakest base. As another example, CH_3COOH (acetic acid) is a stronger acid than H_2CO_3 (carbonic acid); conversely, CH_3COO^- (acetate ion) is a weaker base than HCO_3^- (bicarbonate ion).

EXAMPLE 8-1   Diprotic Acids

Show how the amphiprotic ion hydrogen sulfate, HSO_4^-, can react as both an acid and a base.

Strategy

For a molecule to act as both an acid and a base, it must be able to both give up a hydrogen ion and accept a hydrogen ion. Therefore, we write two equations, one donating a hydrogen ion and the other accepting one.

Solution

Hydrogen sulfate reacts as an acid in the equation shown below:

$$HSO_4^- + H_2O \rightleftharpoons H_3O^+ + SO_4^{2-}$$

It can react as a base in the equation shown below:

$$HSO_4^- + H_3O^+ \rightleftharpoons H_2O + H_2SO_4$$

Problem 8-1

Draw the acid and base reactions for the amphiprotic ion HPO_4^{2-}.

HOW TO...

Name Common Acids

The names of common acids are derived from the name of the anion that they produce when they dissociate. There are three common endings for these ions: *–ide, –ate,* and *–ite.*

Acids that dissociate into ions with the suffix *–ide* are named
hydro _____ *ic acid*

Cl^-	Chlor*ide* ion	HCl	*hydro*chlor*ic acid*
F^-	Fluor*ide* ion	HF	*hydro*fluor*ic acid*
CN^-	Cyan*ide* ion	HCN	*hydro*cyan*ic acid*

Acids that dissociate into ions with the suffix *–ate* are named
_____ *ic acid*

SO_4^{2-}	Sulf*ate* ion	H_2SO_4	Sulfur*ic acid*
PO_4^{3-}	Phosph*ate* ion	H_3PO_4	Phosphor*ic acid*
NO_3^-	Nitr*ate* ion	HNO_3	Nitr*ic acid*

Acids that dissociate into ions with the suffix *–ite* are named
_____ *ous acid*

SO_3^{2-}	Sulf*ite* ion	H_2SO_3	Sulfur*ous acid*
NO_2^-	Nitr*ite* ion	HNO_2	Nitr*ous acid*

8-4 How Can We Tell the Position of Equilibrium in an Acid–Base Reaction?

We know that HCl reacts with H_2O according to the following equilibrium:

$$HCl + H_2O \rightleftharpoons Cl^- + H_3O^+$$

We also know that HCl is a strong acid, which means the position of this equilibrium lies very far to the right. In fact, this equilibrium lies so far to the right that out of every 10,000 HCl molecules dissolved in water, all but one react with water molecules to give Cl^- and H_3O^+.

For this reason, we usually write the acid reaction of HCl with a unidirectional arrow, as follows:

$$HCl + H_2O \longrightarrow Cl^- + H_3O^+$$

As we have also seen, acetic acid reacts with H_2O according to the following equilibrium:

$$\underset{\text{Acetic acid}}{CH_3COOH} + H_2O \rightleftharpoons \underset{\text{Acetate ion}}{CH_3COO^-} + H_3O^+$$

Acetic acid is a weak acid. Only a few acetic acid molecules react with water to give acetate ions and hydronium ions, and the major species present in equilibrium in aqueous solution are CH_3COOH and H_2O. The position of this equilibrium, therefore, lies very far to the left.

In these two acid–base reactions, water is the base. But what if we have a base other than water as the proton acceptor? How can we determine which are the major species present at equilibrium? That is, how can we determine if the position of equilibrium lies toward the left or toward the right?

As an example, let us examine the acid–base reaction between acetic acid and ammonia to form acetate ion and ammonium ion. As indicated by the question mark over the equilibrium arrow, we want to determine whether the position of this equilibrium lies toward the left or toward the right.

$$\underset{\substack{\text{Acetic acid}\\\text{(Acid)}}}{CH_3COOH} + \underset{\substack{\text{Ammonia}\\\text{(Base)}}}{NH_3} \overset{?}{\rightleftharpoons} \underset{\substack{\text{Acetate ion}\\\text{(Conjugate base}\\\text{of } CH_3COOH)}}{CH_3COO^-} + \underset{\substack{\text{Ammonium ion}\\\text{(Conjugate acid}\\\text{of } NH_3)}}{NH_4^+}$$

In this equilibrium, there are two acids present: acetic acid and ammonium ion. There are also two bases present: ammonia and acetate ion. One way to analyze this equilibrium is to view it as a competition of the two bases, ammonia and acetate ion, for a proton. Which is the stronger base? The information we need to answer this question is found in Table 8-2. We first determine which conjugate acid is the stronger acid and then use this information along with the fact that the stronger the acid, the weaker its conjugate base. From Table 8-2, we see that CH_3COOH is the stronger acid, which means that CH_3COO^- is the weaker base. Conversely, NH_4^+ is the weaker acid, which means that NH_3 is the stronger base. We can now label the relative strengths of each acid and base in this equilibrium:

$$\underset{\substack{\text{Acetic acid}\\\text{(Stronger acid)}}}{CH_3COOH} + \underset{\substack{\text{Ammonia}\\\text{(Stronger base)}}}{NH_3} \overset{?}{\rightleftharpoons} \underset{\substack{\text{Acetate ion}\\\text{(Weaker base)}}}{CH_3COO^-} + \underset{\substack{\text{Ammonium ion}\\\text{(Weaker acid)}}}{NH_4^+}$$

In an acid–base reaction, the equilibrium position always favors reaction of the stronger acid and stronger base to form the weaker acid and weaker base. Thus, at equilibrium, the major species present are the weaker acid and the weaker base. In the reaction between acetic acid and ammonia, therefore, the equilibrium lies to the right and the major species present are acetate ion and ammonium ion:

$$\underset{\substack{\text{Acetic acid}\\\text{(Stronger acid)}}}{CH_3COOH} + \underset{\substack{\text{Ammonia}\\\text{(Stronger base)}}}{NH_3} \rightleftharpoons \underset{\substack{\text{Acetate ion}\\\text{(Weaker base)}}}{CH_3COO^-} + \underset{\substack{\text{Ammonium ion}\\\text{(Weaker acid)}}}{NH_4^+}$$

To summarize, we use the following four steps to determine the position of an acid–base equilibrium:

1. Identify the two acids in the equilibrium; one is on the left side of the equilibrium, and the other is on the right side.
2. Using the information in Table 8-2, determine which acid is the stronger acid and which acid is the weaker acid.
3. Identify the stronger base and the weaker base. Remember that the stronger acid gives the weaker conjugate base and the weaker acid gives the stronger conjugate base.
4. The stronger acid and stronger base react to give the weaker acid and weaker base. The position of equilibrium, therefore, lies on the side of the weaker acid and weaker base.

EXAMPLE 8-2 Acid–Base Pairs

For each acid–base equilibrium, label the stronger acid, the stronger base, the weaker acid, and the weaker base. Then predict whether the position of equilibrium lies toward the right or toward the left.

(a) $H_2CO_3 + OH^- \rightleftharpoons HCO_3^- + H_2O$
(b) $HPO_4^{2-} + NH_3 \rightleftharpoons PO_4^{3-} + NH_4^+$

Strategy

Use Table 8-2 to identify the stronger acid from the weaker acid and the stronger base from the weaker base. Once you have done that, determine in which direction the equilibrium lies. It always lies in the direction of the stronger components moving towards the weaker components.

Solution

Arrows connect the conjugate acid–base pairs, with the red arrows showing the stronger acid. The position of equilibrium in (a) lies toward the right. In (b) it lies toward the left.

(a)
$$\underset{\substack{\text{Stronger}\\\text{acid}}}{H_2CO_3} + \underset{\substack{\text{Stronger}\\\text{base}}}{OH^-} \rightleftharpoons \underset{\substack{\text{Weaker}\\\text{base}}}{HCO_3^-} + \underset{\substack{\text{Weaker}\\\text{acid}}}{H_2O}$$

(b)
$$\underset{\substack{\text{Weaker}\\\text{acid}}}{HPO_4^{2-}} + \underset{\substack{\text{Weaker}\\\text{base}}}{NH_3} \rightleftharpoons \underset{\substack{\text{Stronger}\\\text{base}}}{PO_4^{3-}} + \underset{\substack{\text{Stronger}\\\text{acid}}}{NH_4^+}$$

Problem 8-2

For each acid–base equilibrium, label the stronger acid, the stronger base, the weaker acid, and the weaker base. Then predict whether the position of equilibrium lies toward the right or the left.

(a) $H_3O^+ + I^- \rightleftharpoons H_2O + HI$
(b) $CH_3COO^- + H_2S \rightleftharpoons CH_3COOH + HS^-$

8-5 How Do We Use Acid Ionization Constants?

In Section 8-2, we learned that acids vary in the extent to which they produce H_3O^+ when added to water. Because the ionizations of weak acids in water are all equilibria, we can use equilibrium constants (Section 7-6) to tell us quantitatively just how strong any weak acid is. The reaction that takes place when a weak acid, HA, is added to water is:

$$HA + H_2O \rightleftharpoons A^- + H_3O^+$$

The equilibrium constant expression for this ionization is:

$$K_a = \frac{[A^-][H_3O^+]}{[HA]}$$

The subscript a on K_a shows that it is an equilibrium constant for the ionization of an acid, so K_a is called the **acid ionization constant (K_a)**.

The value of the acid ionization constant for acetic acid, for example, is 1.8×10^{-5}. Because acid ionization constants for weak acids are numbers with negative exponents, we often use an algebraic trick to turn them into numbers that are easier to use. To do so, we take the negative logarithm of the number. Acid strengths are therefore expressed as $-\log K_a$, which we call the pK_a.

$$pK_a = -\log K_a$$

The "p" of anything is just the negative logarithm of that given item. The pK_a of acetic acid is $-\log (1.8 \times 10^{-5})$ which is equal to 4.75. Table 8-3 gives names, molecular formulas, and values of K_a and pK_a for some weak acids. As you study the entries in this table, note the inverse relationship between the values of K_a and pK_a. The weaker the acid, the smaller its K_a, but the larger its pK_a. Also note that for the common logarithm of each measured K_a value, the number of digits after the decimal point equals the number of significant figures in the original number. For example, the K_a of phenol is 1.3×10^{-10}. Because there are two significant figures (Section 1-3) in this

Acid ionization constant (K_a) An equilibrium constant for the ionization of an acid in aqueous solution to H_3O^+ and its conjugate base. K_a is also called an acid dissociation constant

Table 8-3 K_a and pK_a Values for Some Weak Acids

Formula	Name	K_a	pK_a
H_3PO_4	Phosphoric acid	7.5×10^{-3}	2.12
HCOOH	Formic acid	1.8×10^{-4}	3.75
$CH_3CH(OH)COOH$	Lactic acid	1.4×10^{-4}	3.86
CH_3COOH	Acetic acid	1.8×10^{-5}	4.75
H_2CO_3	Carbonic acid	4.3×10^{-7}	6.37
$H_2PO_4^-$	Dihydrogen phosphate ion	6.2×10^{-8}	7.21
H_3BO_3	Boric acid	7.3×10^{-10}	9.14
NH_4^+	Ammonium ion	5.6×10^{-10}	9.25
HCN	Hydrocyanic acid	4.9×10^{-10}	9.31
C_6H_5OH	Phenol	1.3×10^{-10}	9.89
HCO_3^-	Bicarbonate ion	5.6×10^{-11}	10.25
HPO_4^{2-}	Hydrogen phosphate ion	2.2×10^{-13}	12.66

Increasing acid strength ↑

number, the $pK_a = -\log(1.3 \times 10^{-10}) = 9.89$ (two significant figures after the decimal point). We will conform to this rule of significant figures when performing calculations involving logarithms.

One reason for the importance of K_a is that it immediately tells us how strong an acid is. For example, Table 8-3 shows us that although acetic acid, formic acid, and phenol are all weak acids, their strengths as acids are not the same. Formic acid, with a K_a of 1.8×10^{-4}, is stronger than acetic acid, whereas phenol, with a K_a of 1.3×10^{-10}, is much weaker than acetic acid. Phosphoric acid is the strongest of the weak acids. We can tell that an acid is classified as a weak acid by the fact that we list a pK_a for it, and the pK_a is a positive number. If we tried to take the negative logarithm of the K_a for a strong acid, we would get a negative number.

HOW TO...

Use Logs and Antilogs

When dealing with acids, bases, and buffers, we often have to use common or base 10 logarithms (logs). To most people, a logarithm is just a button they push on a calculator. Here we describe briefly how to handle logs and antilogs.

1. What is a logarithm and how is it calculated?
A common logarithm is the power to which you raise 10 to get another number. For example, the log of 100 is 2, because you must raise 10 to the second power to get 100.

$$\log 100 = 2 \text{ because } 10^2 = 100$$

Other examples are:

$$\log 1000 = 3 \text{ because } 10^3 = 1000$$
$$\log 10 = 1 \text{ because } 10^1 = 10$$
$$\log 1 = 0 \text{ because } 10^0 = 1$$
$$\log 0.1 = -1 \text{ because } 10^{-1} = 0.1$$

The common logarithm of a number other than a simple power is usually obtained from a calculator by entering the number and then pressing log. For example,

$$\log 52 = 1.72$$
$$\log 4.5 = 0.65$$
$$\log 0.25 = -0.60$$

Try it now. Enter 100 and then press log. Did you get 2? If so, you did it right. Try again with 52. Enter 52 and press log. Did you get 1.72 (rounded to two decimal places)? Some calculators may have you press log first and then the number. Try it both ways to make sure you know how your calculator works.

2. What are antilogarithms (antilogs)?
An antilog is the reverse of a log. It is also called the inverse log. If you take 10 and raise it to a power, you are taking an antilog. For example,

$$\text{antilog } 5 = 100,000$$

because taking the antilog of 5 means raising 10 to the power of 5 or

$$10^5 = 100,000$$

Try it now on your calculator. What is the antilog of 3? Enter 3 on your calculator. Press INV (inverse) or 2nd (second function), and then press log. The answer should be 1000. Your calculator may be different, but the INV or 2nd function keys are the most common.

3. What is the difference between antilog and −log?
There is a huge and very important difference. Antilog 3 means that we take 10 and raise it to the power of 3, so we get 1000. In contrast, −log 3 means that we take the log of 3, which equals 0.477, and take the negative of it. Thus, −log 3 equals −0.5 For example,

$$\text{antilog } 2 = 100$$
$$-\log 2 = -0.3$$

In the problem below, we will use negative logs to calculate pK_a.

EXAMPLE 8-3 pK_a

K_a for benzoic acid is 6.5×10^{-5}. What is the pK_a of this acid?

Strategy
The pK_a is $-\log K_a$. Thus, use your calculator to find the log of the K_a and then take the negative of it.

Solution
Take the logarithm of 6.5×10^{-5} on your scientific calculator. The answer is -4.19. Because pK_a is equal to $-\log K_a$, you must multiply this value by -1 to get pK_a. The pK_a of benzoic acid is 4.19.

Problem 8-3
K_a for hydrocyanic acid, HCN, is 4.9×10^{-10}. What is its pK_a?

EXAMPLE 8-4   Acid Strength

Which is the stronger acid?
(a) Benzoic acid with a K_a of 6.5×10^{-5} or hydrocyanic acid with a K_a of 4.9×10^{-10}?
(b) Boric acid with a pK_a of 9.14 or carbonic acid with a pK_a of 6.37?

Strategy
Relative acid strength is determined by comparing the K_a values or the pK_a values. If using K_a values, the stronger acid has the larger K_a. If using pK_a values, the stronger acid has the smaller pK_a.

Solution
(a) Benzoic acid is the stronger acid; it has the larger K_a value.
(b) Carbonic acid is the stronger acid; it has the smaller pK_a.

Problem 8-4
Which is the stronger acid?
(a) Carbonic acid, $pK_a = 6.37$, or ascorbic acid (vitamin C), $pK_a = 4.10$?
(b) Aspirin, $pK_a = 3.49$, or acetic acid, $pK_a = 4.75$?

▶ All of these fruits and fruit drinks contain organic acids.

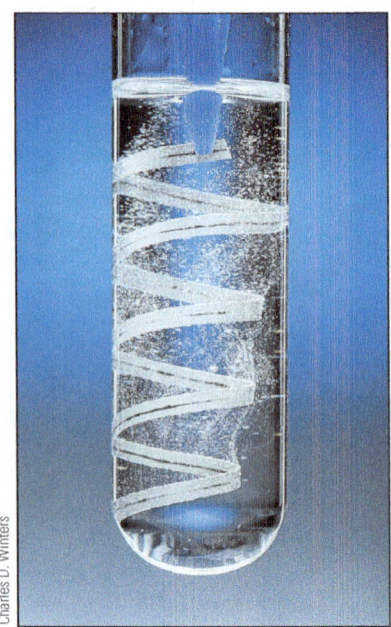

FIGURE 8-1 Acids react with metals. A ribbon of magnesium metal reacts with aqueous HCl to give H_2 gas and aqueous $MgCl_2$.

8-6 What Are the Properties of Acids and Bases?

Today's chemists do not taste the substances they work with, but 200 years ago they routinely did so. That is how we know that acids taste sour and bases taste bitter. The sour taste of lemons, vinegar, and many other foods, for example, is due to the acids they contain.

A. Neutralization

The most important reaction of acids and bases is that they react with each other in a process called neutralization. This name is appropriate because, when a strong corrosive acid such as hydrochloric acid reacts with a strong caustic base such as sodium hydroxide, the product (a solution of ordinary table salt in water) has neither acidic nor basic properties. We call such a solution neutral. Section 8-9 discusses neutralization reactions in detail.

B. Reaction with Metals

Strong acids react with certain metals (called active metals) to produce hydrogen gas, H_2, and a salt. Hydrochloric acid, for example, reacts with magnesium metal to give the salt magnesium chloride and hydrogen gas (Figure 8-1).

$$Mg(s) + 2HCl(aq) \longrightarrow MgCl_2(aq) + H_2(g)$$
Magnesium   Hydrochloric   Magnesium   Hydrogen
             acid   chloride

The reaction of an acid with an active metal to give a salt and hydrogen gas is a redox reaction (Section 4-4). Both the acid and the salt formed are ionized in aqueous solution.

$$Mg(s) + 2H_3O^+(aq) + 2Cl^-(aq) \longrightarrow Mg^{2+}(aq) + 2Cl^-(aq) + H_2(g) + 2H_2O(\ell)$$

The metal is oxidized to a metal ion, H^+ is reduced to H_2, and the spectator ions (Section 4-3) are eliminated as shown in the following net ionic equation:

$$Mg(s) + 2H_3O^+(aq) \longrightarrow Mg^{2+}(aq) + H_2(g) + 2H_2O(\ell)$$

Recall in Section 8-1, we learned that H_3O^+ is commonly written as H^+, although we know that acidic aqueous solutions do not contain H^+ ions. Therefore, the reaction can also be written as:

$$Mg(s) + 2H^+(aq) \longrightarrow Mg^{2+}(aq) + H_2(g)$$

Whether or not a reaction occurs between a metal and an acid depends on how easily each substance is reduced or oxidized. By noting the experimental results obtained from multiple reactions, we construct an **activity series**, which ranks the elements in order of their reducing abilities in aqueous solution. As noted in Table 8-4, the metals located above H_2 give up electrons and are stronger reducing agents, resulting in a reaction between a given metal and an acid. In the preceding example, a ribbon of magnesium metal reacts with aqueous hydrochloric acid because Mg is ranked higher than H_2 on the activity series.

In contrast, metals located below H_2 do not give up electrons as readily and are weaker reducing agents, resulting in no reaction between a given metal and an acid. For example, silver metal will not react with aqueous nitric acid because H_2 is ranked higher than Ag on the activity series.

Table 8-4 Activity Series of Certain Elements

EXAMPLE 8-5   Activity Series

Write the balanced net ionic equation for the reaction of solid chromium with a solution of hydrobromic acid.

Strategy

Whether or not a reaction occurs between a metal and an acid depends on how easily each substance is reduced or oxidized as shown via the activity series (Table 8-4), which ranks the elements in order of their reducing abilities in aqueous solution. Because Cr is ranked higher than H_2 on the activity series, a reaction will result between the metal and the acid.

Solution

As noted in Table 8-4, Cr gives up three electrons to form Cr^{3+}.

$$2Cr(s) + 6HBr(aq) \longrightarrow 2CrBr_3(aq) + 3H_2(g)$$

The acid and the salt formed are ionized in aqueous solution.

$$2Cr(s) + 6H_3O^+(aq) + 6Br^-(aq) \longrightarrow 2Cr^{3+}(aq) + 6Br^-(aq) + 3H_2(g) + 6H_2O(\ell)$$

Spectator ions are eliminated and H_3O^+ can be simplified to H^+, resulting in the following balanced net ionic equation.

$$2Cr(s) + 6H^+(aq) \longrightarrow 2Cr^{3+}(aq) + 3H_2(g)$$

Problem 8-5

Write the balanced net ionic equation for the reaction of lead pellets with a solution of hydroiodic acid.

C. Reaction with Metal Hydroxides

Acids react with metal hydroxides to give a salt and water.

$$\underset{\text{Hydrochloric acid}}{HCl(aq)} + \underset{\text{Potassium hydroxide}}{KOH(aq)} \longrightarrow \underset{\text{Water}}{H_2O(\ell)} + \underset{\text{Potassium chloride}}{KCl(aq)}$$

Both the acid and the metal hydroxide are ionized in aqueous solution. Furthermore, the salt formed is an ionic compound that is present in aqueous solution as anions and cations. Therefore, the actual equation for the reaction of HCl and KOH could be written showing all of the ions present (Section 4-3):

$$H_3O^+ + Cl^- + K^+ + OH^- \longrightarrow 2H_2O + Cl^- + K^+$$

We usually simplify this equation by omitting the spectator ions (Section 4-3), which gives the following equation for the net ionic reaction of any strong acid and strong base to give a soluble salt and water:

$$H_3O^+ + OH^- \longrightarrow 2H_2O$$

D. Reaction with Metal Oxides

Strong acids react with metal oxides to give water and a soluble salt, as shown in the following net ionic equation:

$$2H_3O^+(aq) + \underset{\text{Calcium oxide}}{CaO(s)} \longrightarrow 3H_2O(\ell) + Ca^{2+}(aq)$$

E. Reaction with Carbonates and Bicarbonates

When a strong acid is added to a carbonate such as sodium carbonate, bubbles of carbon dioxide gas are rapidly given off. The overall reaction is a summation of two reactions. In the first reaction, carbonate ion reacts with H_3O^+ to give carbonic acid. Almost immediately, in the second reaction, carbonic acid decomposes to carbon dioxide and water. The following equations show the individual reactions and then the overall reaction:

$$2H_3O^+(aq) + CO_3^{2-}(aq) \longrightarrow H_2CO_3(aq) + 2H_2O(\ell)$$
$$H_2CO_3(aq) \longrightarrow CO_2(g) + H_2O(\ell)$$
$$\overline{2H_3O^+(aq) + CO_3^{2-}(aq) \longrightarrow CO_2(g) + 3H_2O(\ell)}$$

Strong acids also react with bicarbonates such as potassium bicarbonate to give carbon dioxide and water:

$$H_3O^+(aq) + HCO_3^-(aq) \longrightarrow H_2CO_3(aq) + H_2O(\ell)$$
$$H_2CO_3(aq) \longrightarrow CO_2(g) + H_2O(\ell)$$
$$\overline{H_3O^+(aq) + HCO_3^-(aq) \longrightarrow CO_2(g) + 2H_2O(\ell)}$$

To generalize, any acid stronger than carbonic acid will react with carbonate or bicarbonate ion to give CO_2 gas.

The production of CO_2 is what makes bread doughs and cake batters rise. The earliest method used to generate CO_2 for this purpose involved the addition of yeast, which catalyzes the fermentation of carbohydrates to produce carbon dioxide and ethanol (Chapter 28):

$$\underset{\text{Glucose}}{C_6H_{12}O_6} \xrightarrow{\text{Yeast}} 2CO_2 + \underset{\text{Ethanol}}{2C_2H_5OH}$$

The production of CO_2 by fermentation, however, is slow. Sometimes it is desirable to have its production take place more rapidly, in which case bakers use the reaction of $NaHCO_3$ (sodium bicarbonate, also called **baking soda**) and a weak acid. But which weak acid? Vinegar (a 5% solution of acetic acid in water) would work, but it has a potential disadvantage—it imparts a particular flavor to foods. For a weak acid that imparts little or no flavor, bakers use either sodium dihydrogen phosphate, NaH_2PO_4, or potassium dihydrogen phosphate, KH_2PO_4. The two salts do not react when they are dry, but when mixed with water in a dough or batter, they react quite rapidly to produce CO_2. The production of CO_2 is even more rapid in an oven! ◂

$$H_2PO_4^-(aq) + H_2O(\ell) \rightleftharpoons HPO_4^{2-}(aq) + H_3O^+(aq)$$
$$HCO_3^-(aq) + H_3O^+(aq) \longrightarrow CO_2(g) + 2H_2O(\ell)$$
$$\overline{H_2PO_4^-(aq) + HCO_3^-(aq) \longrightarrow HPO_4^{2-}(aq) + CO_2(g) + H_2O(\ell)}$$

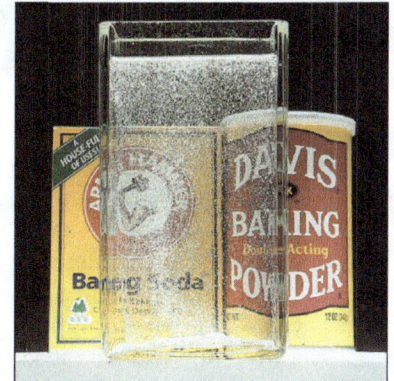

▸ Baking powder contains a weak acid, either sodium or potassium dihydrogen phosphate, and a weak base, sodium or potassium bicarbonate. When they are mixed with water, they react to produce the bubbles of CO_2 seen in this picture.

F. Reaction with Ammonia and Amines

Any acid stronger than NH_4^+ (Table 8-2) is strong enough to react with NH_3 to form a salt. In the following reaction, the salt formed is ammonium chloride, NH_4Cl, which is shown as it would be ionized in aqueous solution:

$$HCl(aq) + NH_3(aq) \longrightarrow NH_4^+(aq) + Cl^-(aq)$$

In Chapter 16, we will meet a family of compounds called amines, which are similar to ammonia except that one or more of the three hydrogen atoms of ammonia are replaced by carbon groups. A typical amine is methylamine, CH_3NH_2. The base strength of most amines is similar to that of NH_3, which means that amines also react with acids to form salts. The salt formed in the reaction of methylamine with HCl is methylammonium chloride, shown here as it would be ionized in aqueous solution:

$$HCl(aq) + \underset{\text{Methylamine}}{CH_3NH_2(aq)} \longrightarrow \underset{\text{Methylammonium ion}}{CH_3NH_3^+(aq)} + Cl^-(aq)$$

The reaction of ammonia and amines with acids to form salts is very important in the chemistry of the body, as we will see in later chapters.

8-7 What Are the Acidic and Basic Properties of Pure Water?

We have seen that an acid produces H_3O^+ ions in water and that a base produces OH^- ions. Suppose that we have absolutely pure water, with no added acid or base. Surprisingly enough, even pure water contains a very small number of H_3O^+ and OH^- ions. They are formed by the transfer of a proton from one molecule of water (the proton donor) to another (the proton acceptor).

$$\underset{\text{Acid}}{H_2O} + \underset{\text{Base}}{H_2O} \rightleftharpoons \underset{\substack{\text{Conjugate} \\ \text{base of } H_2O}}{OH^-} + \underset{\substack{\text{Conjugate} \\ \text{acid of } H_2O}}{H_3O^+}$$

What is the extent of this reaction? We know from the information in Table 8-2 that in this equilibrium, H_3O^+ is the stronger acid and OH^- is the stronger base. Therefore, as shown by the arrows, the equilibrium for this reaction lies far to the left.

CHEMICAL CONNECTIONS 8B

Drugstore Antacids

Stomach fluid is normally quite acidic because of its HCl content. At some time, you probably have gotten "heartburn" caused by excess stomach acidity. To relieve your discomfort, you may have taken an antacid, which, as the name implies, is a substance that neutralizes acids—in other words, a base.

The word "antacid" is a medical term, not one used by chemists. It is, however, found on the labels of many medications available in drugstores and supermarkets. Almost all of them use bases such as $CaCO_3$, $Mg(OH)_2$, $Al(OH)_3$, and $NaHCO_3$ to decrease the acidity of the stomach.

Also in drugstores and supermarkets are nonprescription drugs labeled "acid reducers." Among these brands are Zantac, Tagamet, Pepcid, and Axid. Instead of neutralizing acidity, these compounds reduce the secretion of acid into the stomach. In larger doses (sold only with a prescription), some of these drugs are used in the treatment of stomach ulcers.

Commercial remedies for excess stomach acid.

Test your knowledge with Problem 8-76.

The equilibrium constant for the ionization of water, K_w, is called the **ion product of water**. In pure water at room temperature, K_w has a value of 1.0×10^{-14}.

$$K_w = [H_3O^+][OH^-]$$

$$K_w = 1.0 \times 10^{-14}$$

In pure water, H_3O^+ and OH^- form in equal amounts, so their concentrations must be equal. That is, in pure water:

$[H_3O^+] = 1.0 \times 10^{-7}$ mol/L
$[OH^-] = 1.0 \times 10^{-7}$ mol/L $\Big\}$ In pure water

These are very small concentrations, not enough to make pure water a conductor of electricity. Pure water is not an electrolyte (Section 6-6C).

The equation for the ionization of water is important because it applies not only to pure water but also to any water solution. The product of $[H_3O^+]$ and $[OH^-]$ in any aqueous solution is equal to 1.0×10^{-14}. If, for example, we add 0.010 mol of HCl to 1 L of pure water, it reacts completely to give H_3O^+ ions and Cl^- ions. The concentration of H_3O^+ will be 0.010 M, or $1.0 \times 10^{-2}\,M$. This means that $[OH^-]$ must be $1.0 \times 10^{-14}/1.0 \times 10^{-2} = 1.0 \times 10^{-12}\,M$.

Interactive Example 8-6   Water Equation

The $[OH^-]$ of an aqueous solution is $1.0 \times 10^{-4}\,M$. What is its $[H_3O^+]$?

Strategy

To determine the hydrogen ion concentration when you know the hydroxide ion concentration, you simply divide the $[OH^-]$ into 10^{-14}.

Solution

We substitute into the equation:

$$[H_3O^+][OH^-] = 1.0 \times 10^{-14}$$

$$[H_3O^+] = \frac{1.0 \times 10^{-14}}{1.0 \times 10^{-4}} = 1.0 \times 10^{-10}\ M$$

Problem 8-6

The $[OH^-]$ of an aqueous solution is $1.0 \times 10^{-12}\ M$. What is its $[H_3O^+]$?

Aqueous solutions can have a very high $[H_3O^+]$, but the $[OH^-]$ must then be very low, and vice versa. Any solution with a $[H_3O^+]$ greater than $1.0 \times 10^{-7}\ M$ is acidic. In such solutions, of necessity $[OH^-]$ must be less than $1.0 \times 10^{-7}\ M$. The higher the $[H_3O^+]$, the more acidic the solution. Similarly, any solution with an $[OH^-]$ greater than $1.0 \times 10^{-7}\ M$ is basic. Pure water, in which $[H_3O^+]$ and $[OH^-]$ are equal (they are both $1.0 \times 10^{-7}\ M$), is neutral—that is, neither acidic nor basic.

8-8 What Are pH and pOH?

Because hydronium ion concentrations for most solutions are numbers with negative exponents, these concentrations are more conveniently expressed as pH, where

$$pH = -\log [H_3O^+]$$

similarly to how we expressed pK_a values in Section 8-5.

In Section 8-7, we saw that a solution is acidic if its $[H_3O^+]$ is greater than $1.0 \times 10^{-7}\ M$ and that it is basic if its $[H_3O^+]$ is less than $1.0 \times 10^{-7}\ M$. We can now state the definitions of acidic and basic solutions in terms of pH. ◄

▶ The pH of this soft drink is 3.12. Soft drinks are often quite acidic.

A solution is acidic if its pH is less than 7.00

A solution is basic if its pH is greater than 7.00

A solution is neutral if its pH is equal to 7.00

EXAMPLE 8-7 Calculating pH

(a) The $[H_3O^+]$ of a certain liquid detergent is $1.4 \times 10^{-9}\ M$. What is its pH? Is this solution acidic, basic, or neutral?
(b) The pH of black coffee is 5.3. What is its $[H_3O^+]$? Is it acidic, basic, or neutral?

Strategy

To determine the pH when given the concentration of a hydrogen ion, just take the negative of the log. If it is less than 7, the solution is acidic. If it is greater than 7, it is basic.

If given the pH, you can immediately determine if it is acidic, basic, or neutral according to how the number relates to 7. To convert the pH to the $[H_3O^+]$, take the inverse log of $-pH$.

Solution

(a) On your calculator, take the log of 1.4×10^{-9}. The answer is -8.85. Multiply this value by -1 to give the pH of 8.85. This solution is basic.

(b) Enter 5.3 into your calculator and then press the $+/-$ key to change the sign to minus and give -5.3. Then take the antilog of this number. The $[H_3O^+]$ of black coffee is 5×10^{-6}. This solution is acidic.

Problem 8-7

(a) The $[H_3O^+]$ of an acidic solution is 3.5×10^{-3} M. What is its pH?

(b) The pH of tomato juice is 4.1. What is its $[H_3O^+]$? Is this solution acidic, basic, or neutral?

Just as pH is a convenient way to designate the concentration of H_3O^+, pOH is a convenient way to designate the concentration of OH^-.

$$pOH = -\log[OH^-]$$

As we saw in the previous section, in aqueous solutions, the ion product of water, K_w, is 1.0×10^{-14}, which is equal to the product of the concentration of H^+ and OH^-:

$$K_w = 1.0 \times 10^{-14} = [H^+][OH^-]$$

By taking the logarithm of both sides, and the fact that $-\log(1.0 \times 10^{-14}) = 14.00$ we can rewrite this equation as shown below:

$$14.00 = pH + pOH$$

Thus, once we know the pH of a solution, we can easily calculate the pOH.

EXAMPLE 8-8 Calculating pOH

The $[OH^-]$ of a strongly basic solution is 1.0×10^{-2}. What are the pOH and pH of this solution?

Strategy

When given the $[OH^-]$, determine the pOH by taking the negative logarithm. To calculate the pH, subtract the pOH from 14.

Solution

The pOH is $-\log 1.0 \times 10^{-2}$ or 2.00, and the pH is $14.00 - 2.00 = 12.00$.

Problem 8-8

The $[OH^-]$ of a solution is 1.0×10^{-4} M. What are the pOH and pH of this solution?

All fluids in the human body are aqueous; that is, the only solvent present is water. Consequently, all body fluids have a pH value. Some of them have a narrow pH range; others have a wide pH range. The pH of blood, for example, must be between 7.35 and 7.45 (slightly basic). If it goes outside these limits, illness and even death may result (Chemical Connections 8C). In contrast, the pH of urine can vary from 5.5 to 7.5. Table 8-5 gives pH values for some common materials.

Table 8-5 pH Values of Some Common Materials

Material	pH	Material	pH
Battery acid	0.5	Saliva	6.5–7.5
Gastric juice	1.0–3.0	Pure water	7.0
Lemon juice	2.2–2.4	Blood	7.35–7.45
Vinegar	2.4–3.4	Bile	6.8–7.0
Tomato juice	4.0–4.4	Pancreatic fluid	7.8–8.0
Carbonated beverages	4.0–5.0	Seawater	8.0–9.0
Black coffee	5.0–5.1	Soap	8.0–10.0
Urine	5.5–7.5	Milk of magnesia	10.5
Rain (unpolluted)	6.2	Household ammonia	11.7
Milk	6.3–6.6	Lye (1.0 M NaOH)	14.0

▶ The pH of three household substances. The colors of the acid–base indicators in the flasks show that vinegar is more acidic than club soda and the cleaner is basic.

One thing you must remember when you see a pH value is that because pH is a logarithmic scale, an increase (or decrease) of one pH unit means a tenfold decrease (or increase) in the $[H_3O^+]$. For example, a pH of 3 does not sound very different from a pH of 4. The first, however, means a $[H_3O^+]$ of 10^{-3} M, whereas the second means a $[H_3O^+]$ of 10^{-4} M. The $[H_3O^+]$ of the pH 3 solution is ten times the $[H_3O^+]$ of the pH 4 solution.

There are two ways to measure the pH of an aqueous solution. One way is to use pH paper, which is made by soaking plain paper with a mixture of pH indicators. A pH **indicator** is a substance that changes color at a certain pH. When we place a drop of solution on this paper, the paper turns a certain color. To determine the pH, we compare the color of the paper with the colors on a chart supplied with the paper. ◀

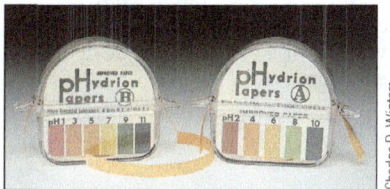

▶ Strips of paper impregnated with indicator are used to find an approximate pH.

One example of an acid–base indicator is the compound methyl orange. When a drop of methyl orange is added to an aqueous solution with a pH of 3.2 or lower, this indicator turns red and the entire solution becomes red. When added to an aqueous solution with a pH of 4.4 or higher, this indicator turns yellow. These particular limits and colors apply only to methyl orange. Other indicators have other limits and colors (Figure 8-2). With pH indicators, the chemical form of the indicator determines its color. The lower pH color is due to the acid form of the indicator, while the higher pH color is associated with the conjugate base form of the indicator.

The second way of determining pH is more accurate and more precise. In this method, we use a pH meter (Figure 8-3). We dip the electrode of the

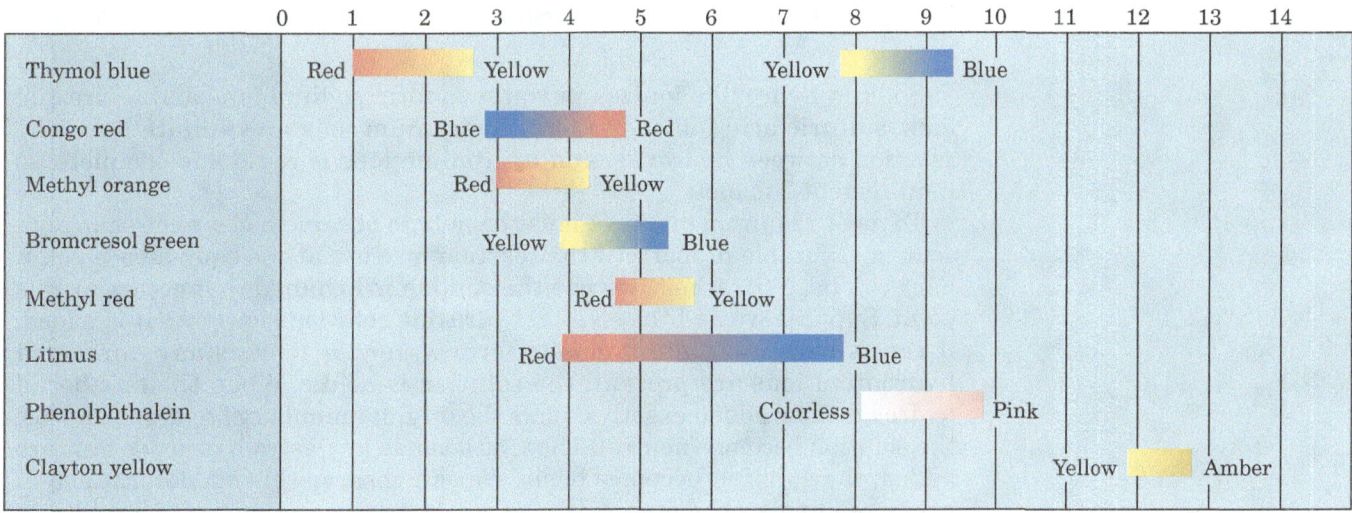

FIGURE 8-2 Some acid–base indicators. Note that some indicators have two color changes.

FIGURE 8-3 A pH meter can rapidly and accurately measure the pH of an aqueous solution.

Titration An analytical procedure whereby we react a known volume of a solution of known concentration with a known volume of a solution of unknown concentration

Equivalence point The point in a titration at which there is a stoichiometrically equal number of moles of each reactant present

pH meter into the solution whose pH is to be measured and then read the pH on a display. The most commonly used pH meters read pH to the nearest hundredth of a unit. It should be mentioned that the accuracy of a pH meter, like that of any instrument, depends on correct calibration.

8-9 How Do We Use Titrations to Calculate Concentration?

Laboratories, whether medical, academic, or industrial, are frequently asked to accurately and precisely determine the concentration of a particular substance in solution, such as the concentration of acetic acid in a given sample of vinegar, or the concentrations of iron, calcium, and magnesium ions in a sample of "hard" water. Determinations of solution concentrations can be made using an analytical technique called a **titration**.

In a titration, we react a known volume of a solution of known concentration with a known volume of a solution of unknown concentration. The solution of unknown concentration may contain an acid (such as stomach acid), a base (such as ammonia), an ion (such as Fe^{2+} ion), or any other substance whose concentration we are asked to determine. If we know the titration volumes and the mole ratio in which the solutes react, we can then calculate the concentration of the second solution.

Titrations must meet several requirements:

1. We must know the equation for the reaction so that we can determine the stoichiometric ratio of reactants to use in our calculations.
2. The reaction must be rapid and complete.
3. When the reactants have combined exactly, there must be a clear-cut change in some measurable property of the reaction mixture. We call the point at which the stoichiometrically correct number of moles of reactants combine exactly the **equivalence point** of the titration.
4. We must have accurate measurements of the amount of each reactant.

Let us apply these requirements to the titration of a solution of sulfuric acid of known concentration with a solution of sodium hydroxide of unknown concentration. We know the balanced equation for this acid–base reaction, so requirement 1 is met.

$$2NaOH(aq) + H_2SO_4(aq) \longrightarrow Na_2SO_4(aq) + 2H_2O(\ell)$$
(Concentration not known)   (Concentration known)

Sodium hydroxide ionizes in water to form sodium ions and hydroxide ions; sulfuric acid ionizes to form hydronium ions and sulfate ions. The reaction between hydroxide and hydronium ions is rapid and complete, so requirement 2 is met.

To meet requirement 3, we must be able to observe a clear-cut change in some measurable property of the reaction mixture at the equivalence point. For acid–base titrations, we use the sudden pH change that occurs at this point. Suppose we add the sodium hydroxide solution slowly. As it is added, it reacts with hydronium ions to form water. As long as any unreacted hydronium ions are present, the solution is acidic. When the number of hydroxide ions added exactly equals the original number of hydronium ions, the solution becomes neutral. Then, as soon as any extra hydroxide ions are added, the solution becomes basic. We can observe this sudden change in pH by reading a pH meter.

Another way to observe the change in pH at the equivalence point is to use an acid–base indicator (Section 8-8). Such an indicator changes color

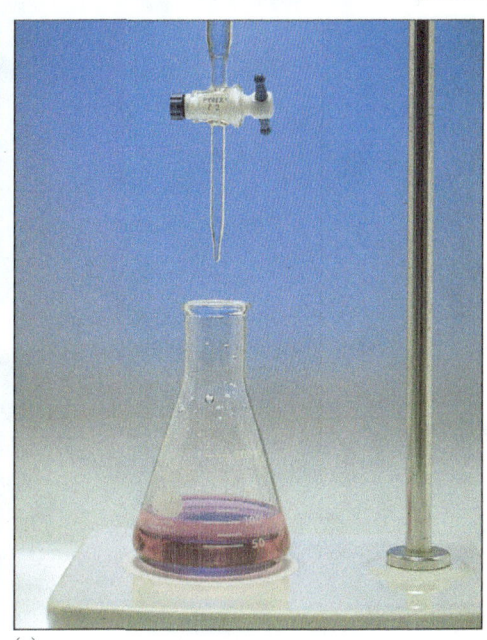

(a) (b) (c)

FIGURE 8-4 An acid–base titration. (a) An acid of known concentration is in the Erlenmeyer flask. (b) When a base is added from the buret, the acid is neutralized. (c) The end point is reached when the color of the indicator changes from colorless to pink.

when the solution changes pH. Phenolphthalein, for example, is colorless in acid solution and pink in basic solution. If this indicator is added to the original sulfuric acid solution, the solution remains colorless as long as excess hydronium ions are present. After enough sodium hydroxide solution has been added to react with all of the hydronium ions, the next drop of base provides excess hydroxide ions and the solution turns pink (Figure 8-4). Thus, we have a clear-cut indication of the equivalence point. The point at which an indicator changes color is called the **end point** of the titration. It is convenient if the end point and the equivalence point are the same, but there are many pH indicators whose end points are not at pH 7. An indicator is chosen that will change as close to the equivalent point as possible.

To meet requirement 4, which is that the volume of each solution used must be known, we use volumetric glassware such as volumetric flasks, burets, and pipets.

Data for a typical acid–base titration are given in Example 8-9. Note that the experiment is run in triplicate, a standard procedure for checking the precision of a titration.

EXAMPLE 8-9 Titrations

Following are data for the titration of 0.108 M H_2SO_4 with a solution of NaOH of unknown concentration. What is the concentration of the NaOH solution?

	Volume of 0.108 M H_2SO_4	Volume of NaOH
Trial I	25.0 mL	33.48 mL
Trial II	25.0 mL	33.46 mL
Trial III	25.0 mL	33.50 mL

$$2\text{NaOH}(aq) + \text{H}_2\text{SO}_4(aq) \longrightarrow \text{Na}_2\text{SO}_4(aq) + 2\text{H}_2\text{O}(\ell)$$

(Concentration not known) (Concentration known)

Strategy

Use the volume of the acid and its concentration to calculate how many moles of hydrogen ions are available to be titrated. At the equivalence point, the moles of base used will equal the moles of H^+ available. Divide the moles of base by the volume of base used in liters to calculate the concentration of the base.

Solution

From the balanced equation for this acid–base reaction, we know the stoichiometry: Two moles of NaOH react with one mole of H_2SO_4. From the three trials, we calculate that the average volume of the NaOH required for complete reaction is 33.48 mL. Because the units of molarity are moles/liter, we must convert volumes of reactants from milliliters to liters. We can then use the factor-label method (Section 1-5) to calculate the molarity of the NaOH solution. What we wish to calculate is the number of moles of NaOH per liter of NaOH.

$$\frac{\text{mol NaOH}}{\text{L NaOH}} = \frac{0.108 \text{ mol } \text{H}_2\text{SO}_4}{1 \text{ L } \text{H}_2\text{SO}_4} \times \frac{0.0250 \text{ L } \text{H}_2\text{SO}_4}{0.03348 \text{ L NaOH}} \times \frac{2 \text{ mol NaOH}}{1 \text{ mol } \text{H}_2\text{SO}_4}$$

$$= \frac{0.161 \text{ mol NaOH}}{\text{L NaOH}} = 0.161\, M$$

Problem 8-9

Calculate the concentration of an acetic acid solution using the following data. Three 25.0-mL samples of acetic acid were titrated to a phenolphthalein end point with 0.121 M NaOH. The volumes of NaOH were 19.96 mL, 19.73 mL, and 19.79 mL.

It is important to understand that a titration is not a method for determining the acidity (or basicity) of a solution. If we want to do that, we must measure the sample's pH, which is the only measurement of solution acidity or basicity. Rather, titration is a method for determining the total acid or base concentration of a solution, which is not the same as the acidity. For example, a 0.1 M solution of HCl in water has a pH of 1.0, but a 0.1 M solution of acetic acid has a pH of 2.9. These two solutions have the same concentration of acid and each neutralizes the same volume of NaOH solution, but they have very different acidities.

8-10 What Are Buffers?

As noted earlier, the body must keep the pH of blood between 7.35 and 7.45. Yet we frequently eat acidic foods such as oranges, lemons, sauerkraut, and tomatoes, and doing so eventually adds considerable quantities of H_3O^+ to the blood. Despite these additions of acidic or basic substances, the body manages to keep the pH of blood remarkably constant. The body manages this feat by using buffers. A **buffer** is a solution whose pH changes very little when small amounts of H_3O^+ or OH^- ions are added to it. In a sense, a pH buffer is an acid or base "shock absorber."

Buffer A solution that resists change in pH when limited amounts of an acid or a base are added to it; the most common example is an aqueous solution containing a weak acid and its conjugate base

The most common buffers consist of approximately equal molar amounts of a weak acid and a salt of the weak acid (or alternatively, a weak base and a salt of the weak base, which we will not consider here). For example, if we dissolve 1.0 mol of acetic acid (a weak acid) and 1.0 mol of its conjugate base (in the form of CH_3COONa, sodium acetate) in 1.0 L of water, we have a good buffer solution. The equilibrium present in this buffer solution is:

$$\underset{\text{Acetic acid}\atop\text{(A weak acid)}}{CH_3COOH} + H_2O \rightleftharpoons \underset{\text{Acetate ion}\atop\text{(Conjugate base}\atop\text{of a weak acid)}}{CH_3COO^-} + H_3O^+$$

(Added as CH_3COOH) (Added as $CH_3COO^-Na^+$)

A. How Do Buffers Work?

A buffer resists a drastic change in pH upon the addition of small quantities of acid or base. To see how, we will use an acetic acid–sodium acetate buffer as an example. If a strong acid such as HCl is added to this buffer solution, the added H_3O^+ ions react with CH_3COO^- ions and are removed from solution.

$$\underset{\text{Acetate ion}\atop\text{(Conjugate base}\atop\text{of a weak acid)}}{CH_3COO^-} + H_3O^+ \longrightarrow \underset{\text{Acetic acid}\atop\text{(A weak acid)}}{CH_3COOH} + H_2O$$

There is a slight increase in the concentration of CH_3COOH as well as a slight decrease in the concentration of CH_3COO^-, but there is no appreciable change in pH. We say that this solution is buffered because it resists a change in pH upon the addition of small quantities of a strong acid.

If NaOH or another strong base is added to the buffer solution, the added OH^- ions react with CH_3COOH molecules and are removed from solution:

$$\underset{\text{Acetic acid}\atop\text{(A weak acid)}}{CH_3COOH} + OH^- \longrightarrow \underset{\text{Acetate ion}\atop\text{(Conjugate base}\atop\text{of a weak acid)}}{CH_3COO^-} + H_2O$$

Here there is a slight decrease in the concentration of CH_3COOH as well as a slight increase in the concentration of CH_3COO^-, but, again, there is no appreciable change in pH.

The important point about this or any other buffer solution is that when the conjugate base of the weak acid removes H_3O^+, it is converted to the undissociated weak acid. Because a substantial amount of weak acid is already present, there is no appreciable change in its concentration, and because H_3O^+ ions are removed from solution, there is no appreciable change in pH. By the same token, when the weak acid removes OH^- ions from solution, it is converted to its conjugate base. Because OH^- ions are removed from solution, there is no appreciable change in pH.

The effect of a buffer can be quite powerful. Addition of either dilute HCl or NaOH to pure water, for example, causes a dramatic change in pH (Figure 8-5).

When HCl or NaOH is added to a phosphate buffer, the results are quite different. Suppose we have a phosphate buffer solution of pH 7.21 prepared by dissolving 0.10 mol NaH_2PO_4 (a weak acid) and 0.10 mol Na_2HPO_4

(a) pH 7.0

(b) pH 2.0

(c) pH 12.0

FIGURE 8-5 The addition of HCl and NaOH to pure water. (a) The pH of pure water is 7.0. (b) The addition of 0.01 mol of HCl to 1 L of pure water causes the pH to decrease to 2.0. (c) The addition of 0.01 mol of NaOH to 1 L of pure water causes the pH to increase to 12.0.

(its conjugate base) in enough water to make 1.00 L of solution. If we add 0.010 mol of HCl to 1.0 L of this solution, the pH decreases to only 7.12. If we add 0.01 mol of NaOH, the pH increases to only 7.30.

Phosphate buffer (pH 7.21) + 0.010 mol HCl   pH 7.21 $\longrightarrow$ 7.12

Phosphate buffer (pH 7.21) + 0.010 mol NaOH   pH 7.21 $\longrightarrow$ 7.30

Had the same amount of acid or base been added to 1 liter of pure water, the resulting pH values would have been 2 and 12, respectively.

Figure 8-6 shows the effect of adding acid to a buffer solution.

(a)

(b)

FIGURE 8-6 Buffer solutions. The solution in the Erlenmeyer flask on the right in both (a) and (b) is a buffer of pH 7.40, the same pH as human blood. The buffer solution also contains bromcresol green, an acid–base indicator that is blue at pH 7.40 (see Figure 8-2). (a) The beaker contains some of the pH 7.40 buffer and the bromcresol green indicator to which has been added 5.0 mL of 0.10 M HCl. After the addition of the HCl, the pH of the buffer solution drops only 0.65 unit to 6.75. (b) The beaker contains pure water and bromcresol green indicator to which has been added 5.0 mL of 0.10 M HCl. After the addition of the HCl, the pH of the unbuffered solution drops to 3.02.

B. Buffer pH

In the previous example, the pH of the buffer containing equal molar amounts of $H_2PO_4^-$ and HPO_4^{2-} is 7.21. From Table 8-3, we see that 7.21 is the pK_a of the acid $H_2PO_4^-$. This is not a coincidence. If we make a buffer solution by mixing equimolar concentrations of any weak acid and its conjugate base, the pH of the solution will equal the pK_a of the weak acid.

This fact allows us to prepare buffer solutions to maintain almost any pH. For example, if we want to maintain a pH of 9.14, we could make a buffer solution from boric acid, H_3BO_3, and sodium dihydrogen borate, NaH_2BO_3, the sodium salt of its conjugate base (see Table 8-3).

EXAMPLE 8-10 Buffers

What is the pH of a buffer solution containing equimolar quantities of:

(a) H_3PO_4 and NaH_2PO_4?   (b) H_2CO_3 and $NaHCO_3$?

Strategy

When there are equimolar quantities of a weak acid and its conjugate base in a buffer solution, the pH is always the same as the pK_a of the weak acid. Look up the pK_a of the weak acid in Table 8-3.

Solution

Because we are adding equimolar quantities of a weak acid and its conjugate base, the pH is equal to the pK_a of the weak acid, which we find in Table 8-3:

(a) pH = 2.12   (b) pH = 6.37

Problem 8-10

What is the pH of a buffer solution containing equimolar quantities of:

(a) NH_4Cl and NH_3?   (b) CH_3COOH and CH_3COONa?

C. Buffer Capacity

Buffer capacity is the amount of hydronium or hydroxide ions that a buffer can absorb without a significant change in its pH. We have already mentioned that a pH buffer is an acid–base "shock absorber." We now ask what makes one solution a better acid–base shock absorber than another solution. The nature of the buffer capacity of a pH buffer depends on both its pH relative to its pK_a and its concentration.

> **pH:** The closer the pH of the buffer is to the pK_a of the weak acid, the more symmetric the buffer capacity, meaning the buffer can resist a pH change with added acid or added base.
>
> **Concentration:** The greater the concentration of the weak acid and its conjugate base, the greater the buffer capacity.

Buffer capacity The extent to which a buffer solution can prevent a significant change in pH of a solution upon addition of a strong acid or a strong base

An effective buffer has a pH equal to the pK_a of the weak acid ±1. For acetic acid, for example, the pK_a is 4.75. Therefore, a solution of acetic acid and sodium acetate functions as an effective buffer within the pH range of approximately 3.75–5.75. When the pH of the buffer solution is equal to the pK_a of the conjugate acid, the solution will have equal capacity with respect to additions of either acid or base. If the pH of the buffer is below the pK_a, the capacity will favor the addition of base. When the pH is above the pK_a, the acid buffer capacity will be greater than the base buffer capacity.

Buffer capacity also depends on concentration. The greater the concentration of the weak acid and its conjugate base, the greater the buffer capacity. We could make a buffer solution by dissolving 1.0 mol each of CH_3COONa and CH_3COOH in 1 L of H_2O, or we could use only 0.10 mol of each. Both solutions have the same pH of 4.75. However, the former has a buffer capacity ten times that of the latter. If we add 0.2 mol of HCl to the former solution, it performs the way we expect—the pH drops to 4.57. If we add 0.2 mol of HCl to the latter solution, however, the pH drops to 1.0 because the buffer has been used up. That is, the amount of H_3O^+ added has exceeded the buffer capacity. The first 0.10 mol of HCl completely neutralizes essentially all the CH_3COO^- present. After that, the solution contains only CH_3COOH and is no longer a buffer, so the second 0.10 mol of HCl decreases the pH to 1.0.

D. Blood Buffers

The average pH of human blood is 7.4. Any change larger than 0.10 pH unit in either direction may cause illness. If the pH goes below 6.8 or above 7.8, death may result. To hold the pH of the blood close to 7.4, the body uses three buffer systems: carbonate, phosphate, and proteins (proteins are discussed in Chapter 22).

The most important of these systems is the carbonate buffer. The weak acid of this buffer is carbonic acid, H_2CO_3; the conjugate base is the bicarbonate ion, HCO_3^-. The pK_a of H_2CO_3 is 6.37 (from Table 8-3). Because the pH of an equal mixture of a weak acid and its salt is equal to the pK_a of the weak acid, a buffer with equal concentrations of H_2CO_3 and HCO_3^- has a pH of 6.37.

Blood, however, has a pH of 7.4. The carbonate buffer can maintain this pH only if $[H_2CO_3]$ and $[HCO_3^-]$ are not equal. In fact, the necessary $[HCO_3^-]/[H_2CO_3]$ ratio is about 10:1. The normal concentrations of these species in blood are about 0.025 M HCO_3^- and 0.0025 M H_2CO_3. This buffer works because any added H_3O^+ is neutralized by the HCO_3^- and any added OH^- is neutralized by the H_2CO_3.

The fact that the $[HCO_3^-]/[H_2CO_3]$ ratio is 10:1 means that this system is a better buffer for acids, which lower the ratio and thus improve buffer effectiveness, than for bases, which raise the ratio and decrease buffer capacity. This is in harmony with the actual functioning of the body because under normal conditions, larger amounts of acidic than basic substances enter the blood. The 10:1 ratio is easily maintained under normal conditions, because the body can very quickly increase or decrease the amount of CO_2 entering the blood.

The second most important buffering system of the blood is a phosphate buffer made up of hydrogen phosphate ion, HPO_4^{2-}, and dihydrogen phosphate ion, $H_2PO_4^-$. In this case, a 1.6:1 $[HPO_4^{2-}]/[H_2PO_4^-]$ ratio is necessary to maintain a pH of 7.4. This ratio is well within the limits of good buffering action.

8-11 How Do We Calculate the pH of a Buffer?

Suppose we want to make a phosphate buffer solution of pH 7.00. The weak acid with a pK_a closest to this desired pH is $H_2PO_4^-$, with a pK_a of 7.21. If we use equal concentrations of NaH_2PO_4 and Na_2HPO_4, however, we will have a buffer of pH 7.21. We want a phosphate buffer that is slightly more acidic than 7.21, so it would seem reasonable to use more of the weak acid,

$H_2PO_4^-$, and less of its conjugate base, HPO_4^{2-}. But what proportions of these two salts do we use? Fortunately, we can calculate these proportions using the **Henderson–Hasselbalch equation**.

The Henderson–Hasselbalch equation is a mathematical relationship between pH, the pK_a of a weak acid, and the concentrations of the weak acid and its conjugate base. The equation is derived in the following way. Assume that we are dealing with a weak acid, HA, and its conjugate base, A^-.

$$HA + H_2O \rightleftharpoons A^- + H_3O^+$$

$$K_a = \frac{[A^-][H_3O^+]}{[HA]}$$

Taking the logarithm of this equation gives:

$$\log K_a = \log\left[\left([H_3O^+]\frac{[A^-]}{[HA]}\right)\right] = \log[H_3O^+] + \log\frac{[A^-]}{[HA]}$$

Rearranging terms gives us a new expression, in which $-\log K_a$ is, by definition, pK_a and $-\log[H_3O^+]$ is, by definition, pH. Making these substitutions gives the Henderson-Hasselbalch equation.

$$-\log[H_3O^+] = -\log K_a + \log\frac{[A^-]}{[HA]}$$

Henderson–Hasselbalch equation $\quad pH = pK_a + \log\frac{[A^-]}{[HA]}$

Because $\frac{[A^-]}{[HA]}$ is a ratio, it doesn't matter if units are given in terms of concentration, moles, or volumes when using the Henderson–Hasselbalch equation, as long as consistent units are used when calculating the ratio.

The Henderson–Hasselbalch equation gives us a convenient way to calculate the pH of a buffer when the concentrations of the weak acid and its conjugate base are not equal.

EXAMPLE 8-11 Buffer pH Calculation

What is the pH of a phosphate buffer solution containing 1.0 mol/L of sodium dihydrogen phosphate, NaH_2PO_4, and 0.50 mol/L of sodium hydrogen phosphate, Na_2HPO_4?

Strategy

Use the Henderson–Hasselbalch equation to determine the pH. You must know either the number of moles of both the conjugate acid and base or the concentrations of the conjugate acid or base. Divide the conjugate base by the conjugate acid, take the log of that ratio, and add it to the pK_a of the conjugate acid.

Solution

The weak acid in this problem is $H_2PO_4^-$; its ionization produces HPO_4^{2-}. The pK_a of this acid is 7.21 (from Table 8-3). Under the weak acid and its conjugate base are shown their concentrations.

$$H_2PO_4^- + H_2O \rightleftharpoons HPO_4^{2-} + H_3O^+ \quad pK_a = 7.21$$
$\quad$ 1.0 mol/L $\qquad\qquad$ 0.50 mol/L

Substituting these values in the Henderson–Hasselbalch equation gives a pH of 6.91.

$$pH = 7.21 + \log \frac{0.50}{1.0}$$

$$= 7.21 - 0.30 = 6.91$$

Problem 8-11

What is the pH of a boric acid buffer solution containing 0.25 mol/L of boric acid, H_3BO_3, and 0.50 mol/L of its conjugate base? See Table 8-3 for the pK_a of boric acid.

Returning to the problem posed at the beginning of this section, how do we calculate the proportions of NaH_2PO_4 and Na_2HPO_4 needed to make up a phosphate buffer of pH 7.00? We know that the pK_a of $H_2PO_4^-$ is 7.21 and that the buffer we wish to prepare has a pH of 7.00. We can substitute these two values in the Henderson–Hasselbalch equation as follows:

$$7.00 = 7.21 + \log \frac{[HPO_4^{2-}]}{[H_2PO_4^-]}$$

Rearranging and solving gives:

$$\log \frac{[HPO_4^{2-}]}{[H_2PO_4^-]} = 7.00 - 7.21 = -0.21$$

$$\frac{[HPO_4^{2-}]}{[H_2PO_4^-]} = 10^{-0.21} = \frac{0.62}{1.0}$$

Thus, to prepare a phosphate buffer of pH 7.00, we can use 0.62 mol of Na_2HPO_4 and 1.0 mol of NaH_2PO_4. Alternatively, we can use any other amounts of these two salts as long as their mole ratio is 0.62 : 1.0.

8-12 What Are TRIS, HEPES, and These Buffers with the Strange Names?

The original buffers used in the lab were made from simple weak acids and bases, such as acetic acid, phosphoric acid, and citric acid. It was eventually discovered that many of these buffers had limitations. For example, they often changed their pH too much if the solution was diluted or if the temperature changed. They often permeated cells in solution, thereby changing the chemistry of the interior of the cell. To overcome these shortcomings, a scientist named N. E. Good developed a series of buffers that consist of zwitterions, molecules with both positive and negative charges. Zwitterions do not readily permeate cell membranes. Zwitterionic buffers are also more resistant to concentration and temperature changes.

Most of the common synthetic buffers used today have complicated formulas, such as 3-[N-morpholino]propanesulfonic acid, which we abbreviate as MOPS. Table 8-5 gives a few examples.

The important thing to remember is that you don't really need to know the structure of these odd-sounding buffers to use them correctly. The important considerations are the pK_a of the buffer and the concentration you want to have. The Henderson–Hasselbalch equation works just fine whether or not you know the structure of the compound in question.

CHEMICAL CONNECTIONS 8C

Respiratory and Metabolic Acidosis

The pH of blood is normally between 7.35 and 7.45. If the pH goes lower than that level, the condition is called **acidosis.** Acidosis leads to depression of the nervous system. Mild acidosis can result in dizziness, disorientation, or fainting; a more severe case can cause coma. If the acidosis persists for a sufficient period of time or if the pH gets too far away from 7.35 to 7.45, death may result.

Acidosis has several causes. One type, called **respiratory acidosis,** results from difficulty in breathing (hypoventilation). An obstruction in the windpipe or diseases such as pneumonia, emphysema, asthma, or congestive heart failure may diminish the amount of oxygen that reaches the tissues and the amount of CO_2 that leaves the body through the lungs. You can even produce mild acidosis by holding your breath. If you ever tried to see how long you could swim underwater in a pool without surfacing, you will have noticed a deep burning sensation in all your muscles when you finally came up for air. The pH of the blood decreases because the CO_2, unable to escape fast enough, remains in the blood, where it lowers the $[HCO_3^-]/[H_2CO_3]$ ratio. Rapid breathing as a result of physical exertion is more about getting rid of CO_2 than it is about breathing in O_2.

Acidosis caused by other factors is called **metabolic acidosis.** Two causes of this condition are starvation (or fasting) and heavy exercise. When the body doesn't get enough food, it burns its own fat, and the products of this reaction are acidic compounds that enter the blood. This problem sometimes happens to people on fad diets. Heavy exercise causes the muscles to produce excessive amounts of lactic acid, which makes muscles feel tired and sore. The lowering of the blood pH due to lactic acid is also what leads to the rapid breathing, dizziness, and nausea that athletes feel at the end of a sprint. In addition, metabolic acidosis is caused by a number of metabolic irregularities. For example, the disease diabetes

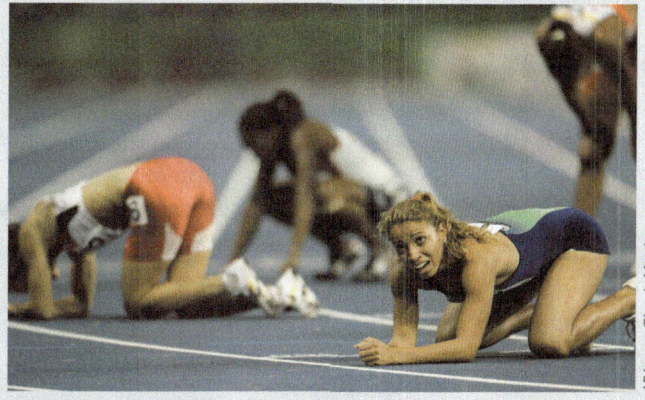

These runners just competed for the gold medal in the 4 × 400 m relay race at the 1996 Olympic Games. The buildup of lactic acid and lowered blood pH caused severe muscle pain and breathlessness.

mellitus produces acidic compounds called ketone bodies (Section 28-6).

Both types of acidosis can be related. When cells are deprived of oxygen, respiratory acidosis results. These cells are unable to produce the energy they need through aerobic (*oxygen-requiring*) pathways that we will learn about in Chapters 27 and 28. To survive, the cells must use the anaerobic (*without oxygen*) pathway called glycolysis. This pathway has lactic acid as an end product, leading to metabolic acidosis. The lactic acid is the body's way of buying time and keeping the cells alive and functioning a little longer. Eventually the lack of oxygen, called an oxygen debt, must be repaid, and the lactic acid must be cleared out. In extreme cases, the oxygen debt is too great, and the individual can die. This was the case of a famous cyclist, Tom Simpson, who died on the slopes of Mont Ventoux during the 1967 Tour de France. Under the influence of amphetamines, he rode so hard that he built up a fatal oxygen debt.

Test your knowledge with Problem 8-77.

EXAMPLE 8-12 Buffer pH Calculation

What is the pH of a solution if you mix 100. mL of 0.20 *M* HEPES in the acid form with 200. mL of 0.20 *M* HEPES in the basic form?

Strategy

To use the Henderson–Hasselbalch equation, you need the ratio of the conjugate base to weak acid forms of the buffer. Because the HEPES solutions have equal concentrations, the ratio of the volumes will give you the ratio of the moles used. Divide the volume of the conjugate base form by the volume of the weak acid form. Take the log of the ratio and add it to the pK_a for HEPES.

Solution

First, we must find the pK_a, which we see from Table 8-6 is 7.55. Then we must calculate the ratio of the conjugate base to the acid. The formula calls for the concentration, but in this situation, the ratio of the concentrations will be the same as the ratio of the moles, which will be the same as the ratio of the volumes, because both solutions had the same starting concentration of 0.20 M. Thus, we can see that the ratio of base to acid is 2:1 because we added twice the volume of base.

$$pH = pK_a + \log([A^-]/[HA]) = 7.55 + \log(2) = 7.85$$

Notice that we did not have to know anything about the structure of HEPES to work out this example.

Problem 8-12

What is the pH of a solution made by mixing 0.2 mol of TRIS acid and 0.05 mol of TRIS base in 500 mL of water?

Table 8-6 Acid and Base Forms of Some Useful Biochemical Buffers

Acid Form (HA)		Base Form (A⁻)	pKa
TRIS—H⁺ (protonated form) $(HOCH_2)_3CNH_3^+$	N—*tris*[hydroxymethyl]aminomethane (TRIS) ⇌	TRIS* (free amine) $(HOCH_2)_3CNH_2$	8.3
⁻TES—H⁺ (zwitterionic form) $(HOCH_2)_3C\overset{+}{N}H_2CH_2CH_2SO_3^-$	N—*tris*[hydroxymethyl]methyl-2-aminoethane sulfonate (TES) ⇌	⁻TES (anionic form) $(HOCH_2)_3CNHCH_2CH_2SO_3^-$	7.55
⁻HEPES—H⁺ (zwitterionic form) HOCH₂CH₂N⁺(H)(CH₂—CH₂)₂NCH₂CH₂SO₃⁻	N—2—hydroxyethylpiperazine-N'-2-ethane sulfonate (HEPES) ⇌	⁻HEPES (anionic form) HOCH₂CH₂N(CH₂—CH₂)₂NCH₂CH₂SO₃⁻	7.55
⁻MOPS—H⁺ (zwitterionic form) O(CH₂—CH₂)₂⁺N(H)CH₂CH₂CH₂SO₃⁻	3—[N—morpholino]propanesulfonic acid (MOPS) ⇌	⁻MOPS (anionic form) O(CH₂—CH₂)₂NCH₂CH₂CH₂SO₃⁻	7.2
²⁻PIPES—H⁺ (protonated dianion) ⁻O₃SCH₂CH₂N(CH₂—CH₂)₂⁺N(H)CH₂CH₂SO₃⁻	Piperazine—N,N'-*bis*[2-ethanesulfonic acid] (PIPES) ⇌	²⁻PIPES (dianion) ⁻O₃SCH₂CH₂N(CH₂—CH₂)₂NCH₂CH₂SO₃⁻	6.8

*Note that TRIS is not a zwitterion.

CHEMICAL CONNECTIONS 8D

Alkalosis and the Sprinter's Trick

Reduced pH is not the only irregularity that can occur in the blood. The pH may also be elevated, a condition called **alkalosis** (blood pH higher than 7.45). It leads to overstimulation of the nervous system, muscle cramps, dizziness, and convulsions. It arises from rapid or heavy breathing, called hyperventilation, which may be caused by fever, infection, the action of certain drugs, or even hysteria. In this case, the excessive loss of CO_2 raises both the ratio of $[HCO_3^-]/[H_2CO_3]$ and the pH.

Athletes who compete in short-distance races that take about a minute to finish have learned how to use hyperventilation to their advantage. By hyperventilating right before the start, they force extra CO_2 out of their lungs. This causes more H_2CO_3 to dissociate into CO_2 and H_2O to replace the lost CO_2. In turn, the loss of the HA form of the bicarbonate blood buffer raises the pH of the blood. When an athlete starts an event with a slightly higher blood pH, he or she can absorb more lactic acid before the blood pH drops to the point where performance is impaired. Of course, the timing of this hyperventilation must be perfect. If the athlete artificially raises blood pH and then the race does not start quickly, the side effect of dizziness will occur.

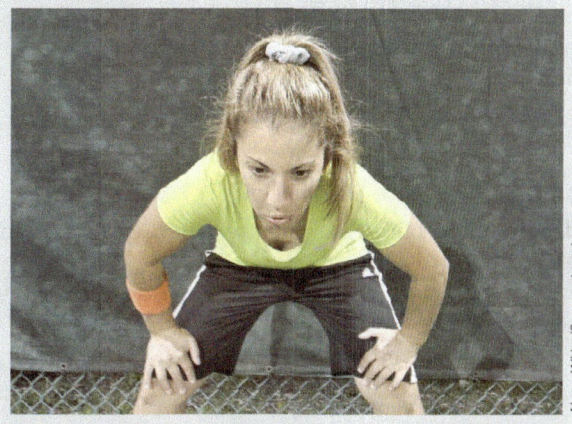

Athletes often hyperventilate before the start of a short distance event. This raises the pH of the blood, allowing it to absorb more H^+ before their performance declines.

Test your knowledge with Problems 8-78 and 8-79.

Summary of Key Questions

Section 8-1 What Are Acids and Bases?

- By the **Arrhenius definitions**, acids are substances that produce H_3O^+ ions in aqueous solution.
- Bases are substances that produce OH^- ions in aqueous solution.

Section 8-2 How Do We Define the Strength of Acids and Bases?

- A strong acid reacts completely or almost completely with water to form H_3O^+ ions.
- A strong base reacts completely or almost completely with water to form OH^- ions.

Section 8-3 What Are Conjugate Acid–Base Pairs?

- The **Brønsted–Lowry definitions** expand the definitions of acid and base beyond water.
- An acid is a proton donor; a base is a proton acceptor.
- Every acid has a **conjugate base**, and every base has a **conjugate acid**. The stronger the acid, the weaker its conjugate base. Conversely, the stronger the base, the weaker its conjugate acid.
- An **amphiprotic substance**, such as water, can act as either an acid or a base.

Section 8-4 How Can We Tell the Position of Equilibrium in an Acid–Base Reaction?

- In an acid–base reaction, the position of equilibrium favors the reaction of the stronger acid and the stronger base to form the weaker acid and the weaker base.

Section 8-5 How Do We Use Acid Ionization Constants?

- The strength of a weak acid is expressed by its **ionization constant**, K_a.
- The larger the value of K_a, the stronger the acid.
- $pK_a = -\log [K_a]$.

Section 8-6 What Are the Properties of Acids and Bases?

- Acids react with metals, metal hydroxides, and metal oxides to give **salts**, which are ionic compounds made up of cations from the base and anions from the acid.
- Acids also react with carbonates, bicarbonates, ammonia, and amines to give salts.

Section 8-7 What Are the Acidic and Basic Properties of Pure Water?

- In pure water, a small percentage of molecules undergo ionization:

$$H_2O + H_2O \rightleftharpoons H_3O^+ + OH^-$$

- As a result, pure water has a concentration of 10^{-7} M for H_3O^+ and 10^{-7} M for OH^-.
- The **ion product of water**, K_w, is equal to 1.0×10^{-14}. $pK_w = 14.00$.

Section 8-8 What Are pH and pOH?

- Hydronium ion concentrations are generally expressed in **pH** units, with $pH = -\log[H_3O^+]$.
- $pOH = -\log[OH^-]$.
- Solutions with pH less than 7 are acidic; those with pH greater than 7 are basic. A **neutral solution** has a pH of 7.
- The pH of an aqueous solution is measured with an acid–base indicator or with a pH meter.

Section 8-9 How Do We Use Titrations to Calculate Concentration?

- We can measure the concentration of aqueous solutions of acids and bases using titration. In an acid–base titration, a base of known concentration is added to an acid of unknown concentration (or vice versa) until an equivalence point is reached, at which point the acid or base being titrated is completely neutralized.

Section 8-10 What Are Buffers?

- A **buffer** does not significantly change its pH when small amounts of either hydronium ions or hydroxide ions are added to it.

- Buffer solutions consist of approximately equal concentrations of a weak acid and its conjugate base.
- The **buffer capacity** depends on both its pH relative to its pK_a and its concentration. The most effective buffer solutions have a pH equal to the pK_a of the weak acid. The greater the concentration of the weak acid and its conjugate base, the greater the buffer capacity.
- The most important buffers for blood are bicarbonate and phosphate.

Section 8-11 How Do We Calculate the pH of a Buffer?

- The **Henderson–Hasselbalch equation** is a mathematical relationship between pH, the pK_a of a weak acid, and the concentrations of the weak acid and its conjugate base:

$$pH = pK_a + \log\frac{[A^-]}{[HA]}$$

Section 8-12 What Are TRIS, HEPES, and These Buffers with the Strange Names?

- Many modern buffers have been designed, and their names are often abbreviated.
- These buffers have qualities useful to scientists, such as not crossing membranes and resisting pH change with dilution or temperature change.
- You do not have to understand the structure of these buffers to use them. The important things to know are the molar mass and the pK_a of the weak acid form of the buffer.

Problems

Orange-numbered problems are applied.

Section 8-1 What Are Acids and Bases?

8-13 Define (a) an Arrhenius acid and (b) an Arrhenius base.

8-14 Write an equation for the reaction that takes place when each acid is added to water.
  (a) HNO_3 (b) HBr
  (c) HCO_3^- (d) NH_4^+

8-15 Write an equation for the reaction that takes place when each base is added to water.
  (a) $LiOH$ (b) $(CH_3)_2NH$
  (c) $Sr(OH)_2$ (d) $CH_3CH_2NH_2$

Section 8-2 How Do We Define the Strength of Acids and Bases?

8-16 For each of the following, tell whether the acid is strong or weak.
  (a) Acetic acid   (b) HCl
  (c) H_3PO_4 (d) H_2SO_4
  (e) HCN   (f) H_2CO_3

8-17 For each of the following, tell whether the base is strong or weak.
  (a) NaOH   (b) Sodium acetate
  (c) KOH   (d) Ammonia
  (e) Water

Section 8-3 What Are Conjugate Acid–Base Pairs?

8-18 Which of these acids are monoprotic, which are diprotic, and which are triprotic? Which are amphiprotic?
  (a) $H_2PO_4^-$ (b) HBO_3^{2-} (c) $HClO_4$ (d) C_2H_5OH
  (e) HSO_3^- (f) HS^- (g) H_2CO_3

8-19 Define (a) a Brønsted–Lowry acid and (b) a Brønsted–Lowry base.

8-20 Write the formula for the conjugate base of each acid.
  (a) H_2SO_4 (b) H_3BO_3 (c) HI
  (d) H_3O^+ (e) NH_4^+ (f) HPO_4^{2-}

8-21 Write the formula for the conjugate base of each acid.
(a) $H_2PO_4^-$ (b) H_2S
(c) HCO_3^- (d) CH_3CH_2OH
(e) H_2O

8-22 Write the formula for the conjugate acid of each base.
(a) OH^- (b) HS^- (c) NH_3
(d) $C_6H_5O^-$ (e) CO_3^{2-} (f) HCO_3^-

8-23 Write the formula for the conjugate acid of each base.
(a) H_2O (b) HPO_4^{2-}
(c) CH_3NH_2 (d) PO_4^{3-}

8-24 Show how the amphiprotic ion hydrogen carbonate, HCO_3^-, can react as both an acid and a base.

8-25 Draw the acid and base reactions for the amphiprotic ion HPO_3^{2-}.

Section 8-4 How Can We Tell the Position of Equilibrium in an Acid–Base Reaction?

8-26 For each equilibrium, label the stronger acid, stronger base, weaker acid, and weaker base. For which reaction(s) does the position of equilibrium lie toward the right? For which does it lie toward the left?
(a) $H_3PO_4 + OH^- \rightleftharpoons H_2PO_4^- + H_2O$
(b) $H_2O + Cl^- \rightleftharpoons HCl + OH^-$
(c) $HCO_3^- + OH^- \rightleftharpoons CO_3^{2-} + H_2O$

8-27 For each equilibrium, label the stronger acid, stronger base, weaker acid, and weaker base. For which reaction(s) does the position of equilibrium lie toward the right? For which does it lie toward the left?
(a) $C_6H_5OH + C_2H_5O^- \rightleftharpoons C_6H_5O^- + C_2H_5OH$
(b) $HCO_3^- + H_2O \rightleftharpoons H_2CO_3 + OH^-$
(c) $CH_3COOH + H_2PO_4^- \rightleftharpoons CH_3COO^- + H_3PO_4$

8-28 Will carbon dioxide be evolved as a gas when sodium bicarbonate is added to an aqueous solution of each compound? Explain.
(a) Sulfuric acid
(b) Ethanol, C_2H_5OH
(c) Ammonium chloride, NH_4Cl

Section 8-5 How Do We Use Acid Ionization Constants?

8-29 Which has the larger numerical value?
(a) The pK_a of a strong acid or the pK_a of a weak acid
(b) The K_a of a strong acid or the K_a of a weak acid

8-30 In each pair, select the stronger acid.
(a) Pyruvic acid ($pK_a = 2.49$) or lactic acid ($pK_a = 3.08$)
(b) Citric acid ($pK_a = 3.08$) or phosphoric acid ($pK_a = 2.10$)
(c) Benzoic acid ($K_a = 6.5 \times 10^{-5}$) or lactic acid ($K_a = 8.4 \times 10^{-4}$)
(d) Carbonic acid ($K_a = 4.3 \times 10^{-7}$) or boric acid ($K_a = 7.3 \times 10^{-10}$)

8-31 Which solution will be more acidic; that is, which will have a lower pH?
(a) $0.10\ M\ CH_3COOH$ or $0.10\ M\ HCl$
(b) $0.10\ M\ CH_3COOH$ or $0.10\ M\ H_3PO_4$
(c) $0.010\ M\ H_2CO_3$ or $0.010\ M\ NaHCO_3$
(d) $0.10\ M\ NaH_2PO_4$ or $0.10\ M\ Na_2HPO_4$
(e) $0.10\ M$ aspirin ($pK_a = 3.47$) or $0.10\ M$ acetic acid

8-32 Which solution will be more acidic; that is, which will have a lower pH?
(a) $0.10\ M\ C_6H_5OH$ (phenol) or $0.10\ M\ C_2H_5OH$ (ethanol)
(b) $0.10\ M\ NH_3$ or $0.10\ M\ NH_4Cl$
(c) $0.10\ M\ NaCl$ or $0.10\ M\ NH_4Cl$
(d) $0.10\ M\ CH_3CH(OH)COOH$ (lactic acid) or $0.10\ M\ CH_3COOH$
(e) $0.10\ M$ ascorbic acid (vitamin C, $pK_a = 4.1$) or $0.10\ M$ acetic acid

Section 8-6 What Are the Properties of Acids and Bases?

8-33 Write an equation for the reaction of HCl with each compound. Which are acid–base reactions? Which are redox reactions?
(a) Na_2CO_3 (b) Mg (c) NaOH (d) Fe_2O_3
(e) NH_3 (f) CH_3NH_2 (g) $NaHCO_3$ (h) Al

8-34 When a solution of sodium hydroxide is added to a solution of ammonium carbonate and then heated, ammonia gas, NH_3, is released. Write a net ionic equation for this reaction. Both NaOH and $(NH_4)_2CO_3$ exist as dissociated ions in aqueous solution.

Section 8-7 What Are the Acidic and Basic Properties of Pure Water?

8-35 Given the following values of $[H_3O^+]$, calculate the corresponding value of $[OH^-]$ for each solution.
(a) $10^{-11}\ M$ (b) $10^{-4}\ M$ (c) $10^{-7}\ M$ (d) $10\ M$

8-36 Given the following values of $[OH^-]$, calculate the corresponding value of $[H_3O^+]$ for each solution.
(a) $10^{-10}\ M$ (b) $10^{-2}\ M$ (c) $10^{-7}\ M$ (d) $10\ M$

Section 8-8 What Are pH and pOH?

8-37 What is the pH of each solution given the following values of $[H_3O^+]$? Which solutions are acidic, which are basic, and which are neutral?
(a) $10^{-8}\ M$ (b) $10^{-10}\ M$ (c) $10^{-2}\ M$
(d) $10^0\ M$ (e) $10^{-7}\ M$

8-38 What is the pH and pOH of each solution given the following values of $[OH^-]$? Which solutions are acidic, which are basic, and which are neutral?
(a) $10^{-3}\ M$ (b) $10^{-1}\ M$ (c) $10^{-5}\ M$ (d) $10^{-7}\ M$

8-39 What is the pH of each solution given the following values of $[H_3O^+]$? Which solutions are acidic, which are basic, and which are neutral?
(a) $3.0 \times 10^{-9}\ M$ (b) $6.0 \times 10^{-2}\ M$
(c) $8.0 \times 10^{-12}\ M$ (d) $5.0 \times 10^{-7}\ M$

Chapter 8 Acids and Bases

8-40 Which is more acidic, a beer with $[H_3O^+] = 3.16 \times 10^{-5}$ or a wine with $[H_3O^+] = 5.01 \times 10^{-4}$?

8-41 What is the $[OH^-]$ and pOH of each solution?
(a) 0.10 M KOH, pH = 13.0
(b) 0.10 M Na_2CO_3, pH = 11.6
(c) 0.10 M Na_3PO_4, pH = 12.0
(d) 0.10 M $NaHCO_3$, pH = 8.4

Section 8-9 How Do We Use Titrations to Calculate Concentration?

8-42 What is the purpose of an acid–base titration?

8-43 What is the molarity of a solution made by dissolving 12.7 g of HCl in enough water to make 1.00 L of solution?

8-44 What is the molarity of a solution made by dissolving 3.4 g of $Ba(OH)_2$ in enough water to make 450 mL of solution? Assume that $Ba(OH)_2$ ionizes completely in water to Ba^{2+} and OH^- ions. What is the pH of the solution?

8-45 Describe how you would prepare each of the following solutions (in each case, assume that the base is a solid).
(a) 400.0 mL of 0.75 M NaOH
(b) 1.0 L of 0.071 M $Ba(OH)_2$
(c) 500.0 mL of 0.1 M KOH
(d) 2.0 L of 0.3 M sodium acetate

8-46 If 25.0 mL of an aqueous solution of H_2SO_4 requires 19.7 mL of 0.72 M NaOH to reach the end point, what is the molarity of the H_2SO_4 solution?

8-47 A sample of 27.0 mL of 0.310 M NaOH is titrated with 0.740 M H_2SO_4. How many milliliters of the H_2SO_4 solution are required to reach the end point?

8-48 A 0.300 M solution of H_2SO_4 was used to titrate 10.00 mL of NaOH; 15.00 mL of acid was required to neutralize the basic solution. What was the molarity of the base?

8-49 A solution of NaOH base was titrated with 0.150 M HCl, and 22.0 mL of acid was needed to reach the end point of the titration. How many moles of the unknown base were in the solution?

8-50 The usual concentration of HCO_3^- ions in blood plasma is approximately 24 millimoles per liter (mmol/L). How would you make up 1.00 L of a solution containing this concentration of HCO_3^- ions?

8-51 What is the end point of a titration?

8-52 Why does a titration not tell us the acidity or basicity of a solution?

Section 8-10 What Are Buffers?

8-53 Write equations to show what happens when, to a buffer solution containing equimolar amounts of CH_3COOH and CH_3COO^-, we add:
(a) H_3O^+ (b) OH^-

8-54 Write equations to show what happens when, to a buffer solution containing equimolar amounts of HPO_4^{2-} and $H_2PO_4^-$, we add
(a) H_3O^+ (b) OH^-

8-55 We commonly refer to a buffer as consisting of approximately equal molar amounts of a weak acid and its conjugate base—for example, CH_3COOH and CH_3COO^-. Is it also possible to have a buffer consisting of approximately equal molar amounts of a weak base and its conjugate acid? Explain.

8-56 What is meant by buffer capacity?

8-57 How can you change the pH of a buffer? How can you change the capacity of a buffer?

8-58 What is the connection between buffer action and Le Chatelier's principle?

8-59 Give two examples of a situation where you would want a buffer to have unequal amounts of the conjugate acid and the conjugate base.

8-60 How is the buffer capacity affected by the ratio of the conjugate base to the conjugate acid?

8-61 Can 100 mL of 0.1 M phosphate buffer at pH 7.2 act as an effective buffer against 20 mL of 1 M NaOH?

Section 8-11 How Do We Calculate the pH of a Buffer?

8-62 What is the pH of a buffer solution made by dissolving 0.10 mol of formic acid, HCOOH, and 0.10 mol of sodium formate, HCOONa, in 1 L of water?

8-63 The pH of a solution made by dissolving 1.0 mol of propanoic acid and 1.0 mol of sodium propanoate in 1.0 L of water is 4.85.
(a) What would the pH be if we used 0.10 mol of each (in 1 L of water) instead of 1.0 mol?
(b) With respect to buffer capacity, how would the two solutions differ?

8-64 Show that when the concentration of the weak acid, [HA], in an acid–base buffer equals that of the conjugate base of the weak acid, $[A^-]$, the pH of the buffer solution is equal to the pK_a of the weak acid.

8-65 Show that the pH of a buffer is 1 unit higher than its pK_a when the ratio of A^- to HA is 10 to 1.

8-66 Calculate the pH of an aqueous solution containing the following:
(a) 0.80 M lactic acid and 0.40 M lactate ion
(b) 0.30 M NH_3 and 1.50 M NH_4^+

8-67 The pH of 0.10 M HCl is 1.0. When 0.10 mol of sodium acetate, CH_3COONa, is added to this solution, its pH changes to 2.9. Explain why the pH changes and why it changes to this particular value.

8-68 If you have 100 mL of a 0.1 M buffer made of NaH_2PO_4 and Na_2HPO_4 that is at pH 6.8 and you add 10 mL of 1 M HCl, will you still have a usable buffer? Why or why not?

Section 8-12 What Are TRIS, HEPES, and These Buffers with the Strange Names?

8-69 Write an equation showing the reaction of TRIS in the acid form with sodium hydroxide (do not write out the chemical formula for TRIS).

8-70 What is the pH of a solution that is 0.1 M in TRIS in the acid form and 0.05 M in TRIS in the basic form?

8-71 Explain why you do not need to know the chemical formula of a buffer compound to use it.

8-72 If you have a HEPES buffer at pH 4.75, will it be a usable buffer? Why or why not?

8-73 Which of the compounds listed in Table 8-6 would be the most effective for making a buffer at pH 8.15? Why?

8-74 Which of the compounds listed in Table 8-6 would be the most effective for making a buffer at pH 7.0?

Chemical Connections

8-75 (Chemical Connections 8A) Which weak base is used as a flame retardant in plastics?

8-76 (Chemical Connections 8B) Name the most common bases used in over-the-counter antacids.

8-77 (Chemical Connections 8C) What causes (a) respiratory acidosis and (b) metabolic acidosis?

8-78 (Chemical Connections 8D) Explain how the sprinter's trick works. Why would an athlete want to raise the pH of his or her blood?

8-79 (Chemical Connections 8D) Another form of the sprinter's trick is to drink a sodium bicarbonate shake before the event. What would be the purpose of doing so? Give the relevant equations.

Additional Problems

8-80 4-Methylphenol, $CH_3C_6H_4OH$ (pK_a = 10.26), is only slightly soluble in water, but its sodium salt, $CH_3C_6H_4O^-Na^+$, is quite soluble in water. In which of the following solutions will 4-methylphenol dissolve more readily than in pure water?
(a) Aqueous NaOH (b) Aqueous $NaHCO_3$
(c) Aqueous NH_3

8-81 Benzoic acid, C_6H_5COOH (pK_a = 4.19), is only slightly soluble in water, but its sodium salt, $C_6H_5COO^-Na^+$, is quite soluble in water. In which of the following solutions will benzoic acid dissolve more readily than in pure water?
(a) Aqueous NaOH (b) Aqueous $NaHCO_3$
(c) Aqueous Na_2CO_3

8-82 Assume that you have a dilute solution of HCl (0.10 M) and a concentrated solution of acetic acid (5.0 M). Which solution is more acidic? Explain.

8-83 Which of the two solutions from Problem 8-82 would take a greater amount of NaOH to hit a phenolphthalein end point assuming you had equal volumes of the two? Explain.

8-84 What is the pH of a solution if you mix 300. mL of 0.30 M TRIS in the base form with 250. mL of 0.15 M TRIS in the acidic form?

8-85 What is the molarity of a solution made by dissolving 0.583 g of the diprotic acid oxalic acid, $H_2C_2O_4$, in enough water to make 1.75 L of solution?

8-86 Following are three organic acids and the pK_a of each: butanoic acid, 4.82; barbituric acid, 5.00; and lactic acid, 3.85.
(a) What is the K_a of each acid?
(b) Which of the three is the strongest acid, and which is the weakest?
(c) What information do you need to predict which of the three acids would require the most NaOH to reach a phenolphthalein end point?

8-87 The pK_a value of barbituric acid is 5.0. If the H_3O^+ and barbiturate ion concentrations are each 0.0030 M, what is the concentration of the undissociated barbituric acid?

8-88 If pure water self-ionizes to give H_3O^+ and OH^- ions, why doesn't pure water conduct an electric current?

8-89 Can an aqueous solution have a pH of zero? Explain your answer using aqueous HCl as your example.

8-90 If an acid, HA, dissolves in water such that the K_a is 1000, what is the pK_a of that acid? Is this scenario possible?

8-91 A scale of K_b values for bases could be set up in a manner similar to that for the K_a scale for acids. However, this setup is generally considered unnecessary. Explain.

8-92 Do a 1.0 M CH_3COOH solution and a 1.0 M HCl solution have the same pH? Explain.

8-93 Do a 1.0 M CH_3COOH solution and a 1.0 M HCl solution require the same amount of 1.0 M NaOH to hit a titration end point? Explain.

8-94 Suppose you wish to make a buffer whose pH is 8.21. You have available 1 L of 0.100 M NaH_2PO_4 and solid Na_2HPO_4. How many grams of the solid Na_2HPO_4 must be added to the stock solution to accomplish this task? (Assume that the volume remains 1 L.)

8-95 In the past, boric acid was used to rinse an inflamed eye. What is the $H_3BO_3/H_2BO_3^-$ ratio in a borate buffer solution that has a pH of 8.40?

8-96 Suppose you want to make a CH_3COOH/CH_3COO^- buffer solution with a pH of 5.60. The acetic acid concentration is to be 0.10 M. What should the acetate ion concentration be?

8-97 For an acid–base reaction, one way to determine the position of equilibrium is to say that the larger of the equilibrium arrow pair points to the acid with the higher value of pK_a. For example,

$$CH_3COOH + HCO_3^- \rightleftharpoons CH_3COO^- + H_2CO_3$$
pK_a = 4.75 $\qquad\qquad\qquad$ pK_a = 6.37

Explain why this rule works.

8-98 When a solution prepared by dissolving 4.00 g of an unknown monoprotic acid in 1.00 L of water is titrated with 0.600 M NaOH, 38.7 mL of the NaOH solution is needed to neutralize the acid. Determine the molarity of the acid solution. What is the molar mass of the unknown acid?

8-99 Write equations to show what happens when, to a buffer solution containing equal amounts of HCOOH and $HCOO^-$, we add:
(a) H_3O^+ (b) OH^-

8-100 If we add 0.10 mol of NH_3 to 0.50 mol of HCl dissolved in enough water to make 1.0 L of solution, what happens to the NH_3? Will any NH_3 remain? Explain.

8-101 Suppose you have an aqueous solution prepared by dissolving 0.050 mol of NaH_2PO_4 in 1 L of water. This solution is not a buffer, but suppose you want to make it into one. How many moles of solid Na_2HPO_4 must you add to this aqueous solution to make it into:

(a) A buffer of pH 7.21
(b) A buffer of pH 6.21
(c) A buffer of pH 8.21

8-102 The pH of a 0.10 M solution of acetic acid is 2.93. When 0.10 mol of sodium acetate, CH_3COONa, is added to this solution, its pH changes to 4.74. Explain why the pH changes and why it changes to this particular value.

8-103 Suppose you have a phosphate buffer ($H_2PO_4^-/HPO_4^{2-}$) of pH 7.21. If you add more solid NaH_2PO_4 to this buffer, would you expect the pH of the buffer to increase, decrease, or remain unchanged? Explain.

8-104 Suppose you have a bicarbonate buffer containing carbonic acid, H_2CO_3, and sodium bicarbonate, $NaHCO_3$, and the pH of the buffer is 6.37. If you add more solid $NaHCO_3$ to this buffer solution, would you expect its pH to increase, decrease, or remain unchanged? Explain.

8-105 A student pulls a bottle of TRIS off a shelf and notes that the bottle says, "TRIS (basic form), $pK_a = 8.3$." The student tells you that if you add 0.1 mol of this compound to 100 mL of water, the pH will be 8.3. Is the student correct? Explain.

Looking Ahead

8-106 Unless under pressure, carbonic acid in aqueous solution breaks down into carbon dioxide and water, and carbon dioxide is evolved as bubbles of gas. Write an equation for the conversion of carbonic acid to carbon dioxide and water.

8-107 Following are pH ranges for several human biological materials. From the pH at the midpoint of each range, calculate the corresponding $[H_3O^+]$. Which materials are acidic, which are basic, and which are neutral?

(a) Milk, pH 6.6–7.6
(b) Gastric contents, pH 1.0–3.0
(c) Spinal fluid, pH 7.3–7.5
(d) Saliva, pH 6.5–7.5
(e) Urine, pH 4.8–8.4
(f) Blood plasma, pH 7.35–7.45
(g) Feces, pH 4.6–8.4
(h) Bile, pH 6.8–7.0

8-108 What is the ratio of $HPO_4^{2-}/H_2PO_4^-$ in a phosphate buffer of pH 7.40 (the average pH of human blood plasma)?

8-109 What is the ratio of $HPO_4^{2-}/H_2PO_4^-$ in a phosphate buffer of pH 7.9 (the pH of human pancreatic fluid)?

Challenge Problems

8-110 A concentrated hydrochloric acid solution contains 36.0% HCl (density = 1.18 g/mL). How many liters are required to produce 10.0 L of a solution that has a pH of 2.05?

8-111 The volume of an adult's stomach ranges from 50 mL when empty to 1 L when full. On a certain day, its volume is 600. mL and its contents have a pH of 2.00.

(a) Determine the number of moles of H^+ present. (Chapter 4)

(b) Assuming that all the H^+ is due to HCl(aq), how many grams of sodium hydrogen carbonate, $NaHCO_3$, will completely neutralize the stomach acid? (Chapter 4)

8-112 Consider an initial 0.040 M hypobromous acid (HOBr) solution at a certain temperature.

$$HOBr(aq) \rightleftharpoons H^+(aq) + OBr^-(aq)$$

At equilibrium after partial dissociation, its pH is found to be 5.05. What is the acid ionization constant, K_a, for hypobromous acid at this temperature?

8-113 A 1.00 L sample of HF gas at 20.0°C and 0.601 atm was dissolved in enough water to make 50.0 mL of hydrofluoric acid solution, HF(aq).

(a) What is the molarity of this solution?
(b) The solution above is allowed to come to equilibrium, and its pH is found to be 1.88. Calculate the acid ionization constant, K_a, for hydrofluoric acid.

8-114 A laboratory student is given an alloy or solid mixture that contains Ag and Pb. The student is directed to separate the two components from one another and decides to treat the mixture with excess concentrated hydrochloric acid. Explain whether this separation will be successful and write any relevant balanced net ionic equations.

8-115 When a solution prepared by dissolving 0.125 g of an unknown diprotic acid in 25.0 mL of water is titrated with 0.200 M NaOH, 30.0 mL of the NaOH solution is needed to neutralize the acid. Determine the molarity of the acid solution. What is the molar mass of the unknown diprotic acid?

8-116 A railroad tank car derails and spills 26 tons of concentrated sulfuric acid (1 ton = 907.185 kg). The acid is 98.0% H_2SO_4 with a density of 1.836 g/mL.

(a) What is the molarity of the acid?
(b) Sodium carbonate, Na_2CO_3, is used to neutralize the acid spill. Determine the kilograms of sodium carbonate required to completely neutralize the acid. (Chapter 4)
(c) How many liters of carbon dioxide at 18°C and 745 mm Hg are produced by this reaction? (Chapter 5)

8-117 Over the past 250 years, the average upper-ocean pH near the Pacific Northwest has decreased by about 0.1 units, from about 8.2 to 8.1. This drop in pH corresponds to an increase in acidity of about 30%. When CO_2 levels in seawater rise, the availability of carbonate ion, CO_3^{2-}, decreases. This makes it more difficult for marine organisms to build and maintain shells and other body parts from calcium carbonate.

(a) Calculate H_3O^+ and OH^- concentrations at pH levels of 8.2 and 8.1.
(b) Demonstrate by calculations that this decrease in pH corresponds to an increase in acidity of about 30%.
(c) Explain the relationship between the pH of seawater and the availability of carbonate ion. Does the change in pH from 8.2 to 8.1 result in an increase or decrease in the availability of carbonate ion?

Nuclear Chemistry

9

The sun's energy is the result of nuclear fusion.

KEY QUESTIONS

- 9-1 How Was Radioactivity Discovered?
- 9-2 What Is Radioactivity?
- 9-3 What Happens When a Nucleus Emits Radioactivity?
  - *How To . . .* Balance a Nuclear Equation
- 9-4 What Is Nuclear Half-Life?
- 9-5 How Do We Detect and Measure Nuclear Radiation?
- 9-6 How Is Radiation Dosimetry Related to Human Health?
- 9-7 What Is Nuclear Medicine?
- 9-8 What Is Nuclear Fusion?
- 9-9 What Is Nuclear Fission and How Is It Related to Atomic Energy?

9-1 How Was Radioactivity Discovered?

Every so often, a scientist makes the kind of discovery that changes the future of the world in some significant way. In 1896, a French physicist, Henri Becquerel (1852–1908), made one of these discoveries. At the time, Becquerel was engaged in a study of phosphorescent materials. In his experiments, he exposed certain salts, among them uranium salts, to sunlight for several hours, whereupon they phosphoresced. He then placed the glowing salts on a photographic plate that had been wrapped in opaque paper. Becquerel observed that by placing a coin or a metal cutout between the phosphorescing salts and the covered plate, he could create photographic images of the coin or metal cutout. He concluded that besides emitting visible light, the phosphorescent materials must have been emitting something akin to X-rays, which William Röntgen had discovered just the previous year. What was even more surprising to Becquerel was that his uranium salts continued to emit this same type of penetrating radiation long after their phosphorescence had ceased. What he had discovered was a type of radiation that Marie Curie was to call radioactivity. For this discovery, Becquerel shared the 1903 Nobel Prize for physics with Pierre and Marie Curie.

In this chapter, we will study the major types of radioactivity, their origin in the nucleus, the uses of radioactivity in the health and biological sciences, and its use as a source of power and energy.

9-2 What Is Radioactivity?

Early experiments identified three kinds of radiation, which were named alpha (α), beta (β), and gamma (γ) rays after the first three letters of the Greek alphabet. Each type of radiation behaves differently when passed between electrically charged plates. When a radioactive material is placed in a lead container that has a small opening, the emitted radiation passes through the opening and then between charged plates (Figure 9-1). One ray (β) is deflected toward the positive plate, indicating that it consists of negatively charged particles. A second ray (α) is deflected toward the negative plate, indicating that it consists of positively charged particles, and a third ray (γ) passes between the charged places without deflection, indicating that it has no charge.

Alpha particles are helium nuclei. Each contains two protons and two neutrons; each has an atomic number of 2 and a charge of +2.

Beta particles are electrons. Each has a charge of −1.

Gamma rays are high-energy electromagnetic radiation. They have no mass or charge.

Gamma rays are only one form of electromagnetic radiation. There are many others, including visible light, radio waves, and cosmic rays. All consist of waves (Figure 9-2).

The only difference between one form of electromagnetic radiation and another is the **wavelength** (λ, Greek letter lambda), which is the distance from one wave crest to the next. The **frequency** (ν, Greek letter nu) of a radiation is the number of crests that pass a given point in one second. Mathematically, wavelength and frequency are related by the following equation, where c is the speed of light (3.0×10^8 m/s):

$$\lambda = \frac{c}{\nu}$$

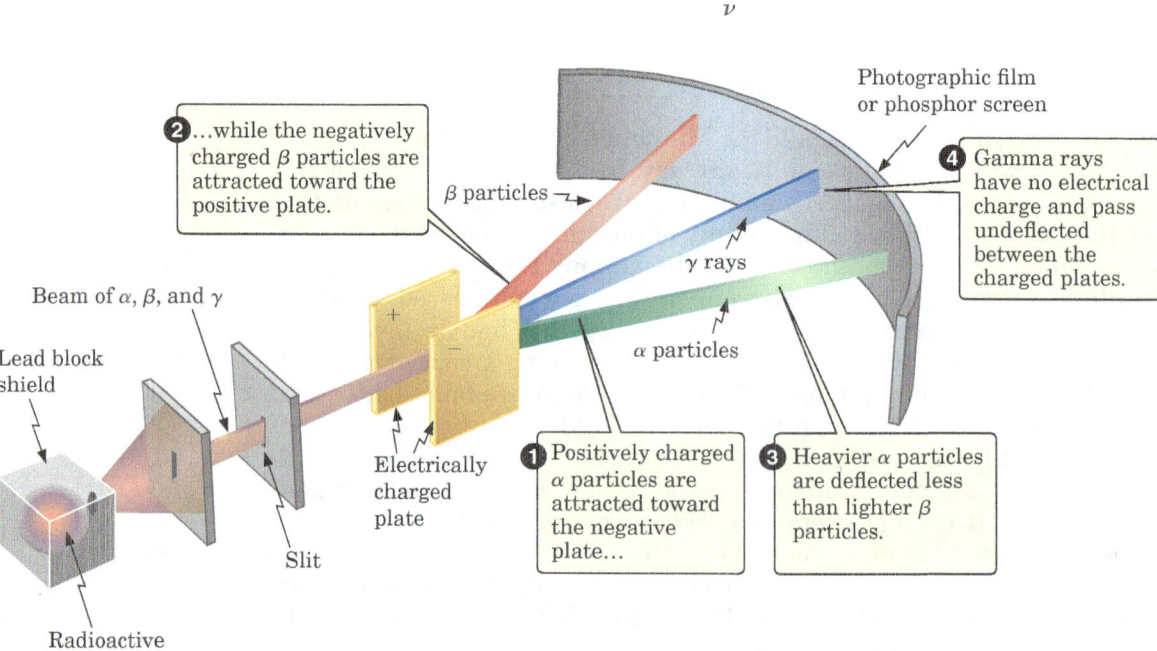

FIGURE 9-1 Electricity and radioactivity. Positively charged alpha (α) particles are attracted to the negative plate and negatively charged beta (β) particles are attracted to the positive plate. Gamma (γ) rays have no charge and are not deflected as they pass between the charged plates. Note that beta particles are deflected more than alpha particles.

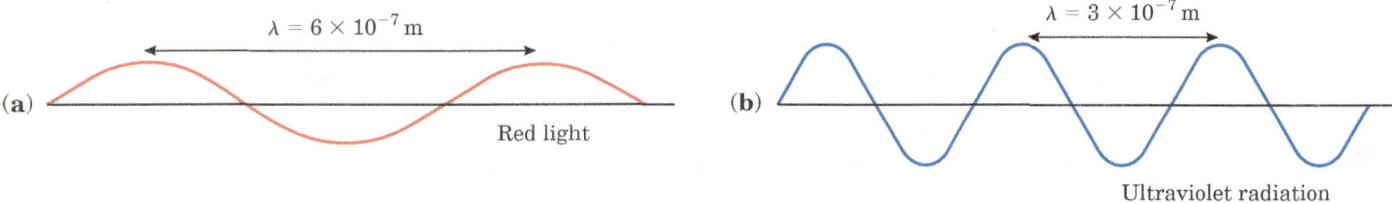

FIGURE 9-2 Two electromagnetic waves with different wavelengths.

As you can see from this relationship, the lower the frequency (ν), the longer the wavelength (λ); or conversely, the higher the frequency, the shorter the wavelength.

A relationship also exists between the frequency (ν) of electromagnetic radiation and its energy; the higher the frequency, the higher its energy. The electron volt (eV) is a non-SI energy unit used frequently in nuclear chemistry. 1 eV = 1.602×10^{-19} J = 3.829×10^{-14} cal. Electromagnetic radiation comes in packets; the smallest units are called **photons**.

Figure 9-3 shows the wavelengths of various types of radiation of the electromagnetic spectrum. Gamma rays are electromagnetic radiation of very high frequency (and high energy). Humans cannot see them because our eyes are not sensitive to waves of this frequency, but instruments (Section 9-5) can detect them. Another kind of radiation, called X-rays, can have higher energies than visible light but less than that of some gamma rays.

Materials that emit radiation (alpha, beta, or gamma) are called **radioactive**. Radioactivity comes from the atomic nucleus and not from the electron cloud that surrounds the nucleus. Table 9-1 summarizes the properties of the particles and rays that come out of radioactive nuclei, along with the properties of some other particles and rays. Note that X-rays are not considered to be a form of radioactivity, because they do not come out of the nucleus but are generated in other ways.

We have said that humans cannot see gamma rays. We cannot see alpha or beta particles either. Likewise, we cannot hear them, smell them, or feel them. They are undetectable by our senses. We can detect radioactivity only by instruments, as discussed in Section 9-5.

9-3 What Happens When a Nucleus Emits Radioactivity?

As mentioned in Section 2-4D, different nuclei consist of different numbers of protons and neutrons. It is customary to indicate these numbers with subscripts and superscripts placed to the left of the atomic symbol. The atomic

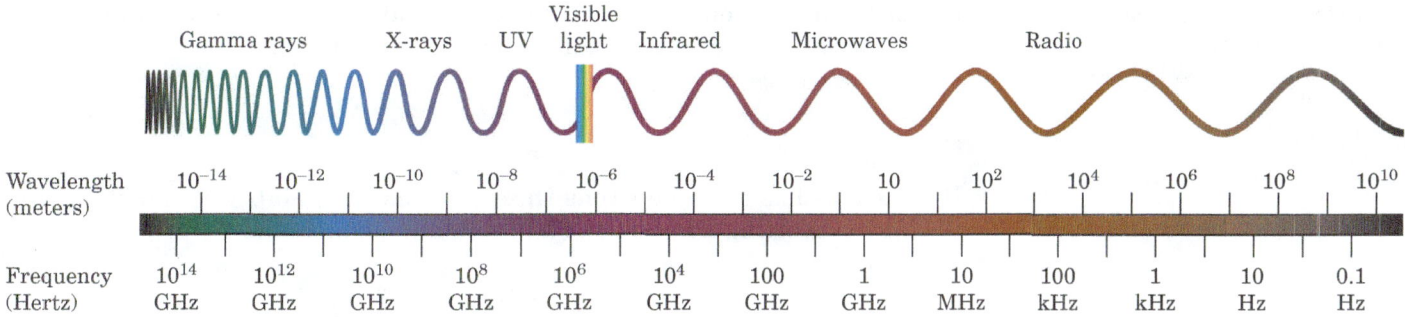

FIGURE 9-3 The electromagnetic spectrum.

Table 9-1 Particles and Rays Frequently Encountered in Radiation

Particle or Ray	Common Name of Radiation	Symbol	Charge	Atomic Mass Units	Penetrating Power[a]	Energy Range[b]
Proton	Proton beam	^1_1H	+1	1	1–3 cm	60 MeV
Electron	Beta particle	$^{\,0}_{-1}e$ or β^-	−1	$0.00055 \left(\frac{1}{1835}\right)$	0–4 mm	1–3 MeV
Neutron	Neutron beam	^1_0n	0	1	—	—
Positron	—	$^{\,0}_{+1}e$ or β^+	+1	0.000555	—	—
Helium nucleus	Alpha particle	^4_2He or α	+2	4	0.02–0.04 mm	3–9 MeV
Energetic radiation	Gamma ray	γ	0	0	1–20 cm	0.1–10 MeV
	X-ray		0	0	0.01–1 cm	0.1–10 MeV

[a]Distance at which half of the radiation has been stopped.
[b]MeV = 1.602×10^{-13} J = 3.829×10^{-14} cal.

number (the number of protons in the nucleus) of an element is shown as a subscript and the mass number (the number of protons and neutrons in the nucleus) as a superscript. Following, for example, are symbols and names for the three known isotopes of hydrogen.

^1_1H hydrogen-1 hydrogen   (not radioactive)
^2_1H hydrogen-2 deuterium (not radioactive)
^3_1H hydrogen-3 tritium   (radioactive)

A. Radioactive and Stable Nuclei

Some isotopes are radioactive, whereas others are stable. Scientists have identified more than 300 naturally occurring isotopes. Of these, 264 are stable, meaning that the nuclei of these isotopes never give off any radioactivity. As far as we can tell, they will last forever. The remainder are **radioactive isotopes (radioisotopes)**—they do give off radioactivity. Furthermore, scientists have made more than 1000 artificial isotopes in laboratories. All artificial isotopes are radioactive.

Radioactive isotopes (radioisotopes) Radiation-emitting isotopes of an element

Isotopes in which the number of protons and neutrons are balanced are stable. In the lighter elements, this balance occurs when the numbers of protons and neutrons are approximately equal. For example, $^{12}_{6}C$ is a stable nucleus (6 protons and 6 neutrons) as are $^{16}_{8}O$ (8 protons and 8 neutrons), $^{20}_{10}Ne$ (10 protons and 10 neutrons), and $^{32}_{16}S$ (16 protons and 16 neutrons). Among the heavier elements, stability requires more neutrons than protons. Lead-206, one of the most stable isotopes of lead, contains 82 protons and 124 neutrons.

If there is a serious imbalance in the proton-to-neutron ratio, either too few or too many neutrons, a nucleus will undergo a **nuclear reaction** to make the ratio more favorable and the nucleus more stable.

Nuclear reaction A reaction that changes the nucleus of an element (usually to the nucleus of another element)

B. Beta Emission

If a nucleus has more neutrons than it needs for stability, it can stabilize itself by converting a neutron to a proton and an electron.

$$^1_0n \longrightarrow {}^1_1H + {}^{\,0}_{-1}e$$

Neutron   Proton   Electron

The proton remains in the nucleus and the electron is emitted from it. The emitted electron is called a beta particle, and the process is called **beta emission**. Phosphorus-32, for example, is a beta emitter:

$$^{32}_{15}P \longrightarrow {}^{32}_{16}S + {}^{0}_{-1}e$$

A phosphorus-32 nucleus has 15 protons and 17 neutrons. The nucleus remaining after an electron has been emitted now has 16 protons and 16 neutrons; its atomic number is increased by 1, but its mass number is unchanged. The new nucleus is, therefore, sulfur-32. Thus, when the unstable phosphorus-32 (15 protons and 17 neutrons) is converted to sulfur-32 (16 protons and 16 neutrons), nuclear stability is achieved.

The changing of one element into another is called **transmutation**. It happens naturally every time an element gives off a beta particle. Every time a nucleus emits a beta particle, it is transformed into another nucleus with the same mass number but an atomic number one unit greater.

HOW TO...

Balance a Nuclear Equation

In writing nuclear equations, we consider only the nucleus and disregard the surrounding electrons. There are two simple rules for balancing nuclear equations:

1. The sum of the mass numbers (superscripts) on both sides of the equation must be equal.
2. The sum of the atomic numbers (subscripts) on both sides of the equation must be equal. For the purposes of determining atomic numbers in a nuclear equation, an electron emitted from the nucleus has an atomic number of −1.

To see how to apply these rules, let us look at the decay of phosphorus-32, a beta emitter.

$$^{32}_{15}P \longrightarrow {}^{32}_{16}S + {}^{0}_{-1}e$$

1. Mass number balance: the total mass number on each side of the equation is 32.
2. Atomic number balance: the atomic number on the left is 15. The sum of the atomic numbers on the right is 16 − 1 = 15.

Thus, we see that in the phosphorus-32 decay equation, mass numbers are balanced (32 and 32) and atomic numbers are balanced (15 and 15); therefore, the nuclear equation is balanced.

EXAMPLE 9-1   Beta Emission

Carbon-14, $^{14}_{6}C$, is a beta emitter. Write an equation for this nuclear reaction and identify the product formed.

$$^{14}_{6}C \longrightarrow ? + {}^{0}_{-1}e$$

Strategy

In beta decay, a neutron is converted to a proton and an electron. The proton remains in the nucleus, and the electron is emitted as a beta particle.

Solution

The $^{14}_{6}C$ nucleus has six protons and eight neutrons. After beta decay, the nucleus has seven protons and seven neutrons:

$$^{14}_{6}C \longrightarrow {^{14}_{7}?} + {^{0}_{-1}e}$$

The sum of the mass numbers on each side of the equation is 14, and the sum of the atomic numbers on each side is 6. We now look in the Periodic Table to determine what element has atomic number 7 and see that it is nitrogen. The product of this nuclear reaction is therefore nitrogen-14, and we can now write a complete equation.

$$^{14}_{6}C \longrightarrow {^{14}_{7}N} + {^{0}_{-1}e}$$

Problem 9-1

Iodine-139 is a beta emitter. Write an equation for this nuclear reaction and identify the product formed.

C. Alpha Emission

For heavy elements, the loss of alpha (α) particles is an especially important stabilization process. For example:

$$^{238}_{92}U \longrightarrow {^{234}_{90}Th} + {^{4}_{2}He}$$

$$^{210}_{84}Po \longrightarrow {^{206}_{82}Pb} + {^{4}_{2}He} + \gamma$$

Note that the radioactive decay of polonium-210 emits both α particles and gamma rays.

A general rule for alpha emission is this: The new nucleus always has a mass number four units lower and an atomic number two units lower than the original.

EXAMPLE 9-2 Alpha Emission

Polonium-218 is an alpha emitter. Write an equation for this nuclear reaction and identify the product formed.

Strategy

An alpha particle has a mass of 4 amu and a charge of +2, so that after alpha emission, the remaining nucleus has an atomic mass that is four units lower and an atomic number that is two units lower.

Solution

The atomic number of polonium is 84, so the partial equation is:

$$^{218}_{84}Po \longrightarrow ? + {^{4}_{2}He}$$

The mass number of the new isotope is 218 − 4 = 214. The atomic number of the new isotope is 84 − 2 = 82. We can now write:

$$^{218}_{84}Po \longrightarrow {^{214}_{82}?} + {^{4}_{2}He}$$

In the Periodic Table, we find that the element with an atomic number of 82 is lead, Pb. Therefore, the product is $^{214}_{82}Pb$, and we can now write the complete equation:

$$^{218}_{84}Po \longrightarrow {^{214}_{82}Pb} + {^{4}_{2}He}$$

> **Problem 9-2**
>
> Thorium-223 is an alpha emitter. Write an equation for this nuclear reaction and identify the product formed.

D. Positron Emission

A positron is a particle that has the same mass as an electron but a charge of $+1$ rather than -1. Its symbol is β^+ or $^{0}_{+1}e$. Positron emission is much rarer than alpha or beta emission. Because a positron has no appreciable mass, the nucleus is transmuted into another nucleus with the same mass number but an atomic number that is one unit less. Carbon-11, for example, is a positron emitter:

$$^{11}_{6}C \longrightarrow \,^{11}_{5}B + \,^{0}_{+1}e$$

In this balanced nuclear equation, the mass numbers on the left and right are 11. The atomic number on the left is 6; on the right the sum of atomic numbers is also 6 ($5 + 1 = 6$).

> **EXAMPLE 9-3 Positron Emission**
>
> Nitrogen-13 is a positron emitter. Write an equation for this nuclear reaction and identify the product.
>
> **Strategy**
>
> A positron has a mass of 0 amu and a charge of $+1$.
>
> **Solution**
>
> We begin by writing the following partial equation:
>
> $$^{13}_{7}N \longrightarrow \,? + \,^{0}_{+1}e$$
>
> Because a positron has no appreciable mass, the mass number of the new isotope is still 13. The sum of the atomic numbers on each side must be 7, which means that the atomic number of the new isotope must be 6. We find in the Periodic Table that the element with atomic number 6 is carbon. Therefore, the new isotope formed in this nuclear reaction is carbon-13 and the balanced nuclear equation is:
>
> $$^{13}_{7}N \longrightarrow \,^{13}_{6}C + \,^{0}_{+1}e$$
>
> **Problem 9-3**
>
> Arsenic-74 is a positron emitter used in locating brain tumors. Write an equation for this nuclear reaction and identify the product.

E. Gamma Emission

Although rare, some nuclei are pure gamma emitters:

$$^{11}_{5}B^* \longrightarrow \,^{11}_{5}B + \gamma$$

Gamma emission often accompanies α and β emissions.

In this equation, $^{11}_{5}B^*$ symbolizes a boron nucleus in a high-energy (excited) state that undergoes gamma emission. In this case, no transmutation takes place. The element is still boron, but its nucleus is in a lower-energy (more stable) state after the emission of excess energy in the form of gamma rays. When all excess energy has been emitted, the nucleus returns to its most stable, lowest-energy state.

F. Electron Capture (E.C.)

In electron capture, an extranuclear electron is absorbed by the nucleus and there reacts with a proton to form a neutron. Thus, electron capture reduces the atomic number of the element, but the mass number is unchanged. Beryllium-7, for example, decays by electron capture to give lithium-7.

$$^{7}_{4}Be + ^{0}_{-1}e \longrightarrow ^{7}_{3}Li$$

> **EXAMPLE 9-4 Electron Capture**
>
> Chromium-51, which is used to image the size and shape of the spleen, decays by electron capture and gamma emission. Write an equation for this nuclear decay and identify the product.
>
> $$^{51}_{24}Cr + ^{0}_{-1}e \longrightarrow ? + \gamma$$
>
> **Strategy and Solution**
>
> Because electron capture results in the conversion of one proton to a neutron and because there is no change in mass number upon gamma emission, the new nucleus has a mass number of 51. The new nucleus, however, has only 23 protons, one less than chromium-51. We find from the Periodic Table that the element with atomic number 23 is vanadium; therefore, the new element formed is vanadium-51. We can now write the complete equation for this nuclear decay.
>
> $$^{51}_{24}Cr + ^{0}_{-1}e \longrightarrow ^{51}_{23}V + \gamma$$
>
> **Problem 9-4**
>
> Thallium-201, a radioisotope used to evaluate heart function in exercise stress tests, decays by electron capture and gamma emission. Write an equation for this nuclear decay and identify the product.

9-4 What Is Nuclear Half-Life?

Suppose we have 40 g of a radioactive isotope—say $^{90}_{38}Sr$. Strontium-90 nuclei are unstable and decay by beta emission to yttrium-90:

$$^{90}_{38}Sr \longrightarrow ^{90}_{39}Y + ^{0}_{-1}\beta$$

Our 40-gram sample of strontium-90 contains about 2.7×10^{23} atoms. We know that these nuclei decay, but at what rate do they decay? Do all of the nuclei decay at once, or do they decay over time? The answer is that they decay over time at a fixed rate. For strontium-90, the decay rate is such that one-half of our original sample (about 1.35×10^{23} atoms) will have decayed in 28.1 years. The time it takes for one-half of any sample of radioactive material to decay is called the **half-life**, $t_{1/2}$.

It does not matter how big or small a sample is. For example, in the case of our 40 g of strontium-90, 20 g will be left at the end of 28.1 years (the rest has been converted to yttrium-90). It will then take another 28.1 years for half of the remainder to decay, so that after 56.2 years, we will have 10 g of strontium-90. If we wait for a third span of 28.1 years, then 5 g will be left. If we had begun with 100 g, then 50 g would be left after the first 28.1-year period.

Figure 9-4 shows the radioactive decay curve of iodine-131. Inspection of this graph shows that at the end of 8 days, half of the original has disappeared. Thus, the half-life of iodine-131 is 8 days. It would take a total of 16 days, or two half-lives, for three-fourths of the original amount of iodine-131 to decay.

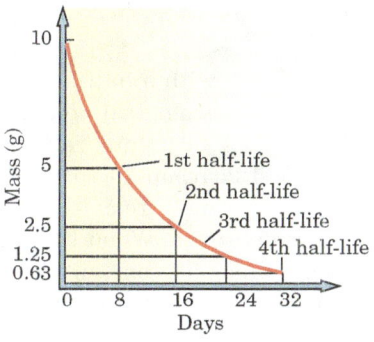

FIGURE 9-4 The decay curve of iodine-131.

EXAMPLE 9-5 Nuclear Half-Life

If 10.0 mg of $^{131}_{53}$I is administered to a patient, how much is left in the body after 32 days?

Strategy and Solution
We know from Figure 9-4 that $t_{1/2}$ of iodine-131 is eight days. The time span of 32 days corresponds to four half-lives. If we start with 10.0 mg, 5.00 mg remains after one half-life, 2.50 mg after two half-lives, 1.25 mg after three half-lives and 0.625 mg after four half-lives.

$$10.0 \text{ mg} \times \overbrace{\frac{1}{2} \times \frac{1}{2} \times \frac{1}{2} \times \frac{1}{2}}^{32 \text{ days (4 half-lives)}} = 0.625 \text{ mg}$$

Problem 9-5
Barium-122 has a half-life of 2 minutes. Suppose you obtained a sample weighing 10.0 g and it takes 10 minutes to set up an experiment in which the barium-122 is to be used. How many grams of barium-122 will remain at the point when you begin the experiment?

It must be noted that in theory, it would take an infinite time period for all of a radioactive sample to decay. In reality, most of the radioactivity decays after ten half-lives, by which time, only 0.098% of the original radioisotope remains.

$$\overbrace{\frac{1}{2} \times \frac{1}{2} \times \frac{1}{2} \times \frac{1}{2} \times \frac{1}{2} \times \frac{1}{2} \times \frac{1}{2} \times \frac{1}{2} \times \frac{1}{2} \times \frac{1}{2}}^{10 \text{ half-lives}} \times 100\% = 0.098\%$$

The half-life of an isotope is independent of temperature and pressure—and, indeed, of all other physical and chemical conditions—and is a property of the particular isotope only. It does not depend on what other kind of atoms surround the particular nucleus (that is, what kind of molecule the nucleus is part of). We do not know any way to speed up radioactive decay or to slow it down.

Table 9-2 gives some half-lives. Even this brief sampling indicates that there are tremendous differences among half-lives. Some isotopes, such as technetium-99m, decay and disappear in a day; others, such as uranium-238, remain radioactive for billions of years. Very short-lived isotopes, especially the artificial heavy elements (Section 9-9) with atomic numbers greater than 100, have half-lives of the order of seconds.

CHEMICAL CONNECTIONS 9A

Radioactive Dating

Carbon-14, with a half-life of 5730 years, can be used to date archeological objects as old as 60,000 years. This dating technique relies on the principle that the carbon-12/carbon-14 ratio of an organism—whether plant or animal—remains constant during the lifetime of the organism. When the organism dies, the carbon-12 level remains constant (carbon-12 is not radioactive), but any carbon-14 present decays by beta emission to nitrogen-14.

$$^{14}_{6}C \longrightarrow \, ^{14}_{7}N + \, ^{0}_{-1}e$$

Using this fact, a scientist can calculate the changed carbon-12/carbon-14 ratio to determine the date of an artifact.

For example, in charcoal made from a tree that has recently died, the carbon-14 gives a radioactive count of 13.70 disintegrations/min per gram of carbon. In a piece of charcoal found in a cave in France near some ancient Cro-Magnon cave paintings, the carbon-14 count was 1.71 disintegrations/min for each gram of carbon. From this information, the cave paintings can be dated. After one half-life, the number of disintegrations/minute per gram is 6.85; after two half-lives, it is 3.42; and after three half-lives, it is 1.71. Therefore, three half-lives have passed since the paintings were created. Given that carbon-14 has a half-life of 5730 years, the paintings are approximately $3 \times 5730 = 17{,}190$ years old.

The famous Shroud of Turin, a piece of linen cloth with the image of a man's head on it, was believed by many to be the original cloth that was wrapped around the body of Jesus Christ after his death. However, radioactive dating showed with 95% certainty that the plants from which the linen was obtained were alive sometime between AD 1260 and 1380, proving that the cloth could not have been the shroud of Christ. Note that it was not necessary to destroy the shroud to perform the tests. In fact, scientists in different laboratories used only a few square centimeters of cloth from its edge.

Rock samples can be dated on the basis of their lead-206 and uranium-238 content. The underlying assumption is that lead-206 comes from the decay of uranium-238, which has a half-life of 4.5 billion years. One of the oldest rocks found on Earth is a granite outcrop in Greenland, dated at 3.7×10^9 years old. On the basis of dating of meteorites, the estimated age of the solar system is 4.6×10^9 years.

Cro-Magnon cave painting.

The Ice Man. This human mummy was found in 1991 in glacial ice high in the Alps. Carbon-14 dating determined that he lived about 5300 years ago. The mummy is exhibited at the South Tyrol Archeological Museum in Bolzano, Italy.

Test your knowledge with Problems 9-61, 9-62, 9-63, and 9-64.

Table 9-2 Half-Lives of Some Radioactive Nuclei

Name	Symbol	Half-Life	Radiation
Hydrogen-3 (tritium)	$^{3}_{1}H$	12.26 years	Beta
Carbon-14	$^{14}_{6}C$	5730 years	Beta
Phosphorus-28	$^{28}_{15}P$	0.28 second	Positron
Phosphorus-32	$^{32}_{15}P$	14.3 days	Beta
Potassium-40	$^{40}_{19}K$	1.28×10^{9} years	Beta+gamma
Scandium-42	$^{42}_{21}Sc$	0.68 second	Positron
Cobalt-60	$^{60}_{27}Co$	5.2 years	Gamma
Strontium-90	$^{90}_{38}Sr$	28.1 years	Beta
Technetium-99m	$^{99m}_{43}Tc$	6.0 hours	Gamma
Indium-116	$^{116}_{49}In$	14 seconds	Beta
Iodine-131	$^{131}_{53}I$	8 days	Beta+gamma
Mercury-197	$^{197}_{80}Hg$	65 hours	Gamma
Polonium-210	$^{210}_{84}Po$	138 days	Alpha
Radon-205	$^{205}_{86}Rn$	2.8 minutes	Alpha
Radon-222	$^{222}_{86}Rn$	3.8 days	Alpha
Uranium-238	$^{238}_{92}U$	4×10^{9} years	Alpha

The usefulness or inherent danger in radioactive isotopes is related to their half-lives. In assessing the long-range health effects of atomic-bomb damage or of nuclear power plant accidents like those at Three Mile Island, Pennsylvania in 1979, Chernobyl (in the former Soviet Union) in 1986 (Chemical Connections 9D), and Fukushima Daiichi (in Japan) in 2011, we can see that radioactive isotopes with long half-lives, such as $^{85}_{36}Kr$ ($t_{1/2}$ = 10 years) or $^{60}_{27}Co$ ($t_{1/2}$ = 5.2 years), are more important than short-lived ones. On the other hand, when a radioactive isotope is used in medical imaging or therapy, short-lived isotopes are more useful because they disappear faster from the body—for example, $^{99m}_{43}Tc$, $^{32}_{15}P$, $^{131}_{53}I$, and $^{197}_{80}Hg$.

9-5 How Do We Detect and Measure Nuclear Radiation?

As already noted, radioactivity is not detectable by our senses. We cannot see it, hear it, feel it, or smell it. How, then, do we know it is there? Alpha, beta, gamma, positron, and X-rays all have a property we can use to detect them. When these types of radiation interact with matter, they knock electrons out of the electron cloud surrounding an atomic nucleus, thereby creating positively charged ions from neutral atoms. For this reason, we call all of these rays **ionizing radiation**.

Ionizing radiation is characterized by two physical measurements: (1) its **intensity** (energy flux), which is the number of particles or photons emerging per unit time, and (2) the **energy** of each particle or photon emitted.

A. Intensity

To measure intensity, we take advantage of the ionizing property of radiation. Instruments such as the **Geiger-Müller counter** (Figure 9-5) and the **proportional counter** contain a gas such as helium or argon. When a radioactive nucleus emits alpha or beta particles or gamma rays, these radiations ionize the gas, and the instrument registers this fact by indicating that an electric current has passed between two electrodes. In this way, the instrument counts radiation particle after particle. ◀

▶ A Geiger-Müller Counter.

FIGURE 9-5 A schematic drawing of a Geiger-Müller counter.

Other measuring devices, such as **scintillation counters**, have a material called a phosphor that emits a unit of light for each alpha or beta particle or gamma ray that strikes it. Once again, the particles are counted one by one. The quantitative measure of radiation intensity can be reported in counts/minute or counts/second.

A common unit of radiation intensity is the **curie** (Ci), named in honor of Marie Curie, whose lifelong work with radioactive materials greatly helped our understanding of nuclear phenomena. One curie is defined as 3.7×10^{10} disintegrations per second (dps). This is radiation of very high intensity, the amount a person would get from exposure to 1.0 g of pure $^{286}_{88}Ra$. This intensity is too high for regular medical use, and the most common units used in the health sciences are small fractions of it. Another, albeit much smaller, unit of radiation activity (intensity) is the **becquerel** (Bq), which is the SI unit. One becquerel is one disintegration per second (dps).

$$1 \text{ becquerel (Bq)} = 1.0 \text{ dps}$$
$$1 \text{ curie (Ci)} = 3.7 \times 10^{10} \text{ dps}$$
$$1 \text{ millicurie (mCi)} = 3.7 \times 10^{7} \text{ dps}$$
$$1 \text{ microcurie } (\mu\text{Ci}) = 3.7 \times 10^{4} \text{ dps}$$

Interactive Example 9-6 — Intensity of Nuclear Radiation

A radioactive isotope with an intensity (activity) of 100. mCi per vial is delivered to a hospital. The vial contains 10. mL of liquid. The instruction is to administer 2.5 mCi intravenously. How many mL of the liquid should be administered?

Strategy and Solution

The intensity (activity) of a sample is directly proportional to the amount present, so:

$$2.5 \text{ mCi} \times \frac{10. \text{ mL}}{100. \text{ mCi}} = 0.25 \text{ mL}$$

9-5 How Do We Detect and Measure Nuclear Radiation?

Problem 9-6

A radioactive isotope in a 9.0-mL vial has an intensity of 300. mCi. A patient is required to take 50. mCi intravenously. How much liquid should be used for the injection?

The intensity of any radiation decreases with the square of the distance. If, for example, the distance (d) from a radiation source doubles, then the intensity (I) of the received radiation decreases by a factor of four.

$$\frac{I_1}{I_2} = \frac{d_2{}^2}{d_1{}^2}$$

EXAMPLE 9-7 Intensity of Nuclear Radiation

If the intensity of radiation is 28 mCi at a distance of 1.0 m, what is the intensity at a distance of 2.0 m?

Strategy

As already noted, the intensity of any radiation decreases with the square of the distance.

Solution

From the preceding equation, we have:

$$\frac{28 \text{ mCi}}{I_2} = \frac{2.0^2}{1.0^2}$$

$$I_2 = \frac{28 \text{ mCi}}{4.0} = 7.0 \text{ mCi}$$

Thus, if the distance from a radioactive source increases by a factor of two, the intensity of the radiation at that distance is decreased by a factor of four.

Problem 9-7

If the intensity of radiation 1.0 cm from a source is 300. mCi, what is the intensity at 3.0 m?

B. Energy

The energies of different particles or photons vary. As shown in Table 9-1, each particle has a certain range of energy. For example, beta particles have an energy range of 1 to 3 MeV (megaelectron volts). This range may overlap with the energy range of some other type of radiation—for example, gamma rays. The penetrating power of a radiation depends on its energy as well on the mass of its particles. Alpha particles are the most massive and the most highly charged and, therefore, the least penetrating; they are stopped by several sheets of ordinary paper, by ordinary clothing, and by the skin. Beta particles have less mass and lower charge than alpha particles and, consequently, have greater penetrating power. They can penetrate several millimeters of bone or tissue. Gamma radiation, which has neither mass nor charge, is the most penetrating of the three types of radiation. Gamma rays can pass completely through the body. Several centimeters of lead or one meter of concrete is required to stop gamma rays (Figure 9-6).

FIGURE 9-6 Penetration of radioactive emissions. Alpha particles, with a charge of +2 and a mass of 4 amu, interact strongly with matter but penetrate the least. They are stopped by several sheets of paper. Beta particles, with less mass and a lower charge than alpha particles, interact less strongly with matter. They easily penetrate paper but are stopped by a 0.5-cm sheet of lead. Gamma rays, with neither mass nor charge, have the greatest penetrating power. It takes 10 cm of lead to stop them.

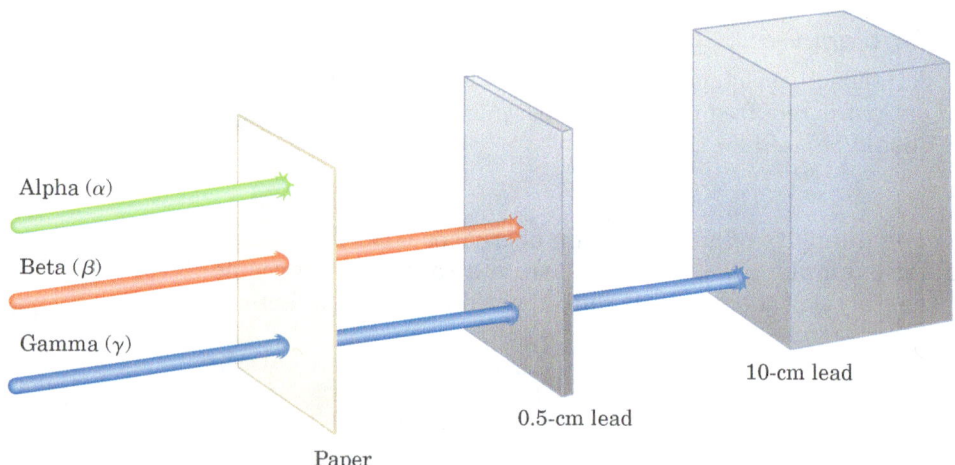

One easy way to protect against ionizing radiation is to wear lead aprons, covering sensitive organs. This practice is followed routinely when diagnostic X-rays are taken. Another way to lessen the damage from ionizing radiation is to move farther away from the source.

9-6 How Is Radiation Dosimetry Related to Human Health?

In studying the effect of radiation on the body, neither the energy of the radiation (in kcal/mol) nor its intensity (in Ci) alone or in combination is of particular importance. Rather, the critical question is what kind of effects such radiation produces in the body. Three different units are used to describe the effects of radiation on the body: roentgens, rads, and rems.

Roentgens (R) Roentgens measure the energy delivered by a radiation source and are, therefore, a measure of exposure to a particular form of radiation. One roentgen is the amount of radiation that produces ions having 2.58×10^{-4} coulomb per kilogram (a coulomb is a unit of electrical charge).

Rads The rad, which stands for *r*adiation *a*bsorbed *d*ose, is a measure of the radiation absorbed from a radiation source. The SI unit is the gray (Gy), where 1 Gy = 100 rad. Roentgens (delivered energy) do not take into account the effect of radiation on tissue and the fact that different tissues absorb different amounts of delivered radiation. Radiation damages body tissue by causing ionization, and for ionization to occur, the tissue must absorb the delivered energy. The relationship between the delivered dose in roentgens and the absorbed dose in rads can be illustrated as follows: Exposure to 1 roentgen yields 0.97 rad of absorbed radiation in water, 0.96 rad in muscle, and 0.93 rad in bone. This relationship holds for high-energy photons. For lower-energy photons, such as "soft" X-rays, each roentgen yields 3 rads of absorbed dose in bone. This principle underlies diagnostic X-rays, wherein soft tissue lets the radiation through to strike a photographic plate but bone absorbs the radiation and casts a shadow on the plate.

Rems The rem, which stands for *r*oentgen *e*quivalent for *m*an, is a measure of the effect of the radiation when a person absorbs 1 roentgen. Other units are the **millirem** (mrem; 1 mrem = 1×10^{-3} rem) and the **sievert** (Sv; 1 Sv = 100 rem). The sievert is the SI unit. The reason for the rem is that tissue damage from 1 rad of absorbed energy depends on the type of radiation. One rad from alpha rays, for example, causes ten times more damage than 1 rad from X-rays or gamma rays. Table 9-3 summarizes the various radiation units and what each measures.

Table 9-3 Radiation Dosimetry

Unit	What the Unit Measures	The SI Unit	Conversion
Roentgen	The amount of radiation delivered from a radiation source	Roentgen (R)	
Rad	The ratio between radiation absorbed by a tissue and that delivered to the tissue	Gray (Gy)	1 rad = 0.01 Gy
Rem	The ratio between the tissue damage caused by a rad of radiation and the type of radiation	Sievert (Sv)	1 rem = 0.01 Sv

Although alpha particles cause more damage than X-rays or gamma rays, they have a very low penetrating power (Table 9-1) and cannot pass through the skin. Consequently, they are not harmful to humans or animals as long as they do not enter the body. If they do get in, however, they can prove quite harmful. They can get inside, for example, if a person swallows or inhales a small particle of a substance that emits alpha particles. Beta particles are less damaging to tissue than alpha particles but penetrate farther and so are generally more harmful. Gamma rays, which can completely penetrate the skin, are by far the most dangerous and harmful form of radiation. Remember, of course, that once alpha particles such as those from radon-222 get in the body, they are very damaging. Therefore, for comparative purposes and for determining exposure from all kinds of sources, the equivalent dose is an important measure. If an organ receives radiation from different sources, the total effect can be summed up in rems (or mrem or Sv). For example, 10 mrem of alpha particles and 15 mrem of gamma radiation give a total of 25 mrem absorbed equivalent dose. Table 9-4 shows the amount of radiation exposure that an average person obtains yearly from both natural and artificial sources.

Table 9-4 Average Exposure to Radiation from Common Sources

Source	Dose (mrem/year)
Naturally Occurring Radiation	
Cosmic rays	27
Terrestrial radiation (rocks, buildings)	28
Inside the human body (K-40 and Ra-226 in the bones)	39
Radon in the air	200
Total	294
Artificial Radiation	
Medical X-rays[a]	39
Nuclear medicine	14
Consumer products	10
Nuclear power plants	0.5
All others	1.5
Total	65
Grand total	359[b]

[a]Individual medical procedures may expose certain parts of the body to much higher levels. For instance, one chest X-ray gives 27 mrem and a diagnostic GI series gives 1970 mrem.

[b]The federal safety standard for allowable occupational exposure is about 5000 mrem/year. It has been suggested that this level be lowered to 4000 mrem/year or even lower to reduce the risk of cancer stemming from low levels of radiation.

SOURCE: National Council on Radiation Protection and Measurements, NCRP Report No. 93 (1993).

Cosmic rays High-energy particles, mainly protons, from outer space bombarding the Earth

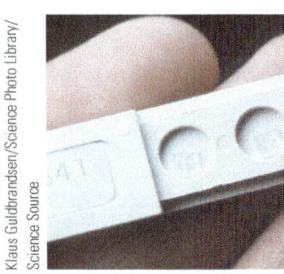

▶ A radiation badge.

The naturally occurring background radiation varies with the geological location. For example, a level that is tenfold higher than the average radiation has been detected in some phosphate mines. People who work in nuclear medicine are, of course, exposed to greater amounts. To ensure that exposures do not get too high, they wear radiation badges. ◀ A single whole-body irradiation of 25 rem causes a noticeable reduction of white blood cells, and 100 rem causes the typical symptoms of radiation sickness, which include nausea, vomiting, a decrease in the white blood cell count, and loss of hair. A dose of 400 rem causes death within one month in 50% of exposed persons, and 600 rem is almost invariably lethal within a short time. It should be noted that as much as 50,000 rem is needed to kill bacteria and as much as 10^6 rem to inactivate viruses.

Fortunately, most of us never get a single dose of more than a few rem and so never suffer from any form of radiation sickness. This does not mean, however, that small doses are totally harmless. The harm may arise in two ways:

1. Small doses of radioactivity over a period of years can cause cancer, especially blood cancers such as leukemia. Repeated exposure to sunlight also carries a risk of tissue damage. Most of the high-energy UV radiation of the sun is absorbed by the Earth's protective ozone layer in the stratosphere. In tanning, however, the repeated overexposure to UV radiation can cause skin cancer (see Chemical Connections 18D).

CHEMICAL CONNECTIONS 9B

The Indoor Radon Problem

Most of our exposure to ionizing radiation comes from natural sources (Table 9-4), with radon gas being the main cause. Radon has more than 20 isotopes, all of which are radioactive. The most important is radon-222, an alpha emitter. Radon-222 is a natural decay product of uranium-238, which is widely distributed in the Earth's crust.

Radon poses a particular health hazard among radioactive elements because it is a gas at normal temperatures and pressures. As a consequence, it can enter our lungs with the air we breathe and become trapped in the mucous lining of the lungs. Radon-222 has a half-life of 3.8 days. It decays naturally and produces, among other isotopes, two harmful alpha emitters: polonium-218 and polonium-214. These polonium isotopes are solids and do not leave the lungs with exhalation. In the long run, they can cause lung cancer.

The U.S. Environmental Protection Agency has set a standard of 4 pCi/L (one picocurie, pCi, is 10^{-12} Ci) as a safe exposure level. A survey of single-family homes in the United States showed that 7% exceeded this level. Most radon seeps into dwellings through cracks in cement foundations and around pipes, then accumulates in basements. The remedy is to ventilate both basements and houses enough to reduce the radiation levels. In a notorious case, a group of houses in Grand Junction, Colorado, was built from bricks made from uranium

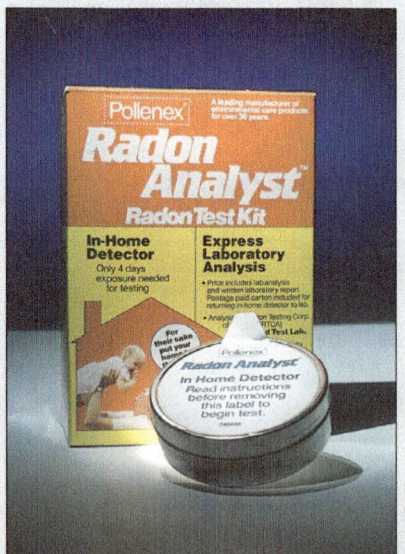

Testing devices are available to determine whether radon is building up in a home.

tailings. Obviously, the radiation levels in these buildings were unacceptably high. Because they could not be controlled, the buildings had to be destroyed. In our modern radiation-conscious age, more and more homebuyers choose to request a certification of radon levels before buying a house.

Test your knowledge with Problem 9-65.

No one knows how many cancers have resulted from this practice, because the doses are so small and continue for so many years that they cannot be measured accurately. Also, because so many other causes of cancer exist, it is difficult or impossible to decide if any particular case is caused by radiation.

2. If any form of radiation strikes an egg or sperm cell, it can cause a change in the genes (see Chemical Connections 25E). Such changes are called mutations. If an affected egg or sperm cell mates, grows, and becomes a new individual, that individual may have mutated characteristics, which are usually harmful and frequently lethal.

Because radiation carries so much potential for harm, it would be nice if we could totally escape it. But can we? Table 9-4 shows that this is impossible. Naturally occurring radiation, called **background radiation**, is present everywhere on Earth. As Table 9-4 shows, this background radiation vastly outstrips the average radiation level from artificial sources (mostly diagnostic X-rays). If we eliminated all forms of artificial radiation, including medical uses, we would still be exposed to the background radiation.

9-7 What Is Nuclear Medicine?

When we think of nuclear chemistry, our first thoughts may well be of nuclear power, atomic bombs, and weapons of mass destruction. True as this may be, it is also true that nuclear chemistry and the use of radioactive elements have become invaluable tools in all areas of science. Nowhere is this more important than in nuclear medicine; that is, in the use of radioactive isotopes as tools for both the diagnosis and treatment of diseases. To describe the full range of medical uses of nuclear chemistry would take far more space than we have in this text. What we have done, instead, is to choose several examples of each use to illustrate the range of applications of nuclear chemistry to the health sciences.

A. Medical Imaging

Medical imaging is the most widely used aspect of nuclear medicine. The goal of medical imaging is to create a picture of a target tissue. To create a useful image requires three things:

- A radioactive element administered in pure form or in a compound that becomes concentrated in the tissue to be imaged.
- A method of detecting radiation from the radioactive source and recording its intensity and location.
- A computer to process the intensity–location data and transform it into a useful image.

Chemically and metabolically, a radioactive isotope in the body behaves in exactly the same way as nonradioactive isotopes of the same element. In the simplest form of imaging, a radioactive isotope is injected intravenously and a technician uses a detector to monitor how the radiation is distributed in the body of the patient. Table 9-5 lists some of the most important radioisotopes used in imaging and diagnosis.

The use of iodine-131, a beta and gamma emitter ($t_{1/2}$ = 8.04 days), to image and diagnose a malfunctioning thyroid gland is a good example. The thyroid gland in the neck produces a hormone, thyroxine, which controls the overall rate of metabolism (use of food) in the body. One molecule of thyroxine contains four iodine atoms. When radioactive iodine-131 is injected into the bloodstream, the thyroid gland takes it up and incorporates it into thyroxine (Chemical Connections 13C). A normally functioning thyroid absorbs about 12% of the administered iodine within a few hours. An overactive

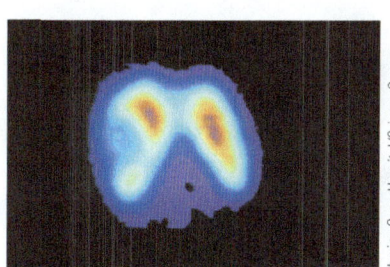

▶ A scan of radiation released by radioactive iodine concentrated in thyroid tissue gives an image of the thyroid gland.

CHEMICAL CONNECTIONS 9C

How Radiation Damages Tissues: Free Radicals

As mentioned earlier, high-energy radiation damages tissue by causing ionization. That is, the radiation knocks electrons out of the molecules that make up the tissue (generally one electron per molecule), thereby forming unstable ions. For example, the interaction of high-energy radiation with water forms H_2O^+, an unstable cation. The positive charge on this cation means that one of the electrons normally present in the water molecule, either one from a covalent bond or from an unshared pair, is missing in this cation; it has been knocked out.

The unpaired electron is on oxygen

$$[H-\ddot{\text{O}}-H]^+$$

Once formed, the H_2O^+ cation is unstable and decomposes to H^+ and a hydroxyl radical:

$$\text{Energy} + H_2O \longrightarrow H_2O^+ + e^-$$

$$H_2O^+ \longrightarrow H^+ + \cdot OH$$

Hydroxyl radical

Though the oxygen atom in the hydroxide ion, OH^-, has a complete octet—it is surrounded by three unshared pairs of electrons and one shared pair—the oxygen atom in the hydroxyl radical is surrounded by only seven valence electrons—two unshared pairs, one shared pair, and one unpaired electron. Compounds that have unpaired electrons are called **free radicals**, or more simply, **radicals**.

Hydroxide ion   Hydroxyl radical

The fact that the oxygen atom of the ·OH radical has an incomplete octet makes this radical extremely reactive. It rapidly interacts with other molecules, causing chemical reactions that damage tissues. These reactions have especially serious consequences if they occur inside cell nuclei and damage genetic material. In addition, they affect rapidly dividing cells more than they do stationary cells. Thus, the damage is greater to embryonic cells, cells of the bone marrow, the intestines, and cells in the lymph. Symptoms of radiation sickness include nausea, vomiting, a decrease in white blood cell count, and loss of hair.

Test your knowledge with Problem 9-66.

Table 9-5 Some Radioactive Isotopes Useful in Medical Imaging

	Isotope	Mode of Decay	Half-Life	Use in Medical Imaging
$^{11}_{6}C$	Carbon-11	β^+, γ	20.3 m	Brain scan to trace glucose metabolism
$^{18}_{9}F$	Fluorine-18	β^+, γ	109 m	Brain scan to trace glucose metabolism
$^{32}_{15}P$	Phosphorus-32	β	14.3 d	Detect eye tumors
$^{51}_{24}Cr$	Chromium-51	E.C., γ	27.7 d	Diagnose albinism; image the spleen and gastrointestinal tract
$^{59}_{26}Fe$	Iron-59	β, γ	44.5 d	Bone marrow function; diagnose anemias
$^{67}_{31}Ga$	Gallium-67	E.C., γ	78.3 h	Whole-body scan for tumors
$^{75}_{34}Se$	Selenium-75	E.C., γ	118 d	Pancreas scan
$^{81m}_{36}Kr$	Krypton-81m	γ	13.3 s	Lung ventilation scan
$^{81}_{38}Sr$	Strontium-81	β	22.2 m	Scan for bone diseases, including cancer
$^{99m}_{43}Tc$	Technetium-99m	γ	6.01 h	Brain, liver, kidney, bone scans; diagnosis of damaged heart muscle
$^{131}_{53}I$	Iodine-131	β, γ	8.04 d	Diagnosis of thyroid malfunction
$^{197}_{80}Hg$	Mercury-197	E.C., γ	64.1 h	Kidney scan
$^{201}_{81}Tl$	Thallium-201	E.C., γ	3.05 d	Heart scan and exercise stress test

thyroid (hyperthyroidism) absorbs and localizes iodine-131 in the gland faster, and an underactive thyroid (hypothyroidism) does so much more slowly. By counting the gamma radiation emitted from the neck, one can determine the rate of uptake of iodine-131 into the thyroid gland and diagnose hyperthyroidism or hypothyroidism.

Most organ scans are similarly based on the preferential uptake of some radioactive isotopes by a particular organ (Figure 9-7).

Another important type of medical imaging is called positron emission tomography (PET). This method is based on the property that certain isotopes (such as carbon-11 and fluorine-18) emit positrons (Section 9-3D). Fluorine-18 decays by positron emission to oxygen-18:

$$^{18}_{9}F \longrightarrow {}^{18}_{8}O + {}^{0}_{+1}e$$

Positrons have very short lives. When a positron and an electron collide, they annihilate each other, resulting in the emission of two gamma rays.

$$\underset{\text{Positron}}{^{0}_{+1}e} + \underset{\text{Electron}}{^{0}_{-1}e} \longrightarrow 2\gamma$$

Because electrons are present in every atom, there are always lots of them around, so positrons generated in the body do not last very long.

A favorite tagged molecule for following the uptake and metabolism of glucose, $C_6H_{12}O_6$, is 18-fluorodeoxyglucose (FDG), a molecule of glucose in which one of glucose's six oxygen atoms is replaced by fluorine-18. When FDG is administered intravenously, the tagged glucose soon enters the blood and from there moves to the brain. Gamma-ray detectors can pick up the signals that come from the areas where the tagged glucose accumulates. In this way, one can see which areas of the brain are involved when we process, for example, visual information (Figure 9-8). Whole body PET scans can be used to diagnose lung, colorectal, head and neck, and esophageal

Normal

Meningioma (brain tumor)

"Brain death"

Scalp tumor

FIGURE 9-7 A comparison of dynamic scan patterns for normal and pathological brains. The studies were performed by injecting technetium-99m into blood vessels.

FIGURE 9-8 Positron emission tomography (PET) brain scans. The upper scans show that 18-fluorodeoxyglucose can cross the blood–brain barrier. The lower scans show that visual stimulation increases blood flow and glucose concentration in certain areas of the brain. These areas are shown in red.

cancers as well as early stages of epilepsy and other diseases that involve abnormal glucose metabolism, such as schizophrenia.

Because tumors have high metabolic rates, PET scans using FDG have become the diagnostic choice for their detection and localization. FDG/PET has been used in the diagnosis of malignant melanoma and malignant lymphoma among other conditions.

Another important use of radioactive isotopes is to learn what happens to an ingested material. The foods and drugs swallowed or otherwise taken in by the body are transformed, decomposed, and excreted. To understand the pharmacology of a drug, it is important to know how and in what part of the body these processes occur. For example, a certain drug may be effective in treating certain bacterial infections. Before beginning a clinical trial of the drug, its manufacturer must prove that the drug is not harmful to humans. In a typical case, the drug is first tested in animal studies. It is synthesized, and some radioactive isotope, such as hydrogen-3, carbon-14, or phosphorus-32, is incorporated into its structure. The drug is administered to the test animals, and after a certain period, the animals are sacrificed. The fate of the drug is then determined by isolating from the body any radioactive compounds formed.

One typical pharmacological experiment studied the effects of tetracycline. This powerful antibiotic tends to accumulate in bones and is not given to pregnant women because it is transferred to the bones of the fetus. A particular tetracycline was tagged with the radioisotope tritium (hydrogen-3), and its uptake in rat bones was monitored in the presence and absence of a sulfa drug. With the aid of a scintillation counter, researchers measured the radiation intensity of maternal and fetal bones. They found that the sulfa drug helped to minimize the accumulation of tetracycline in the fetal bones.

The metabolic fate of essential chemicals in the body can also be followed with radioactive tracers. The use of radioactive isotopes has illuminated a number of normal and pathological body functions as well.

CHEMICAL CONNECTIONS 9D

Magnetic Resonance Imaging

X-rays are commonly used by the medical community to image human bones, muscles, and organs. However, there are several drawbacks to using X-rays for medical imaging. This technique causes adverse biological effects on cells with which they come in contact due to their high-energy radiation; as such, exposure of both the patient and an X-ray technician must be limited. Moreover, because diseased or damaged tissue often yields the same image as healthy tissue, X-rays frequently fail to detect illness or injuries. In the 1980s, a new technique called magnetic resonance imaging (MRI) evolved and is now in widespread use. MRI has none of the disadvantages of X-rays. Diseased tissue appears noticeably different from healthy tissue, and the use of radiofrequency radiation in MRI is not harmful to humans in the doses used.

Certain atomic nuclei have a property known as spin, analogous to the spin associated with electrons as discussed in Section 2-6C. The spin of electrons is responsible for the magnetic properties of atoms. Spinning nuclei behave as tiny magnets, producing magnetic fields. MRI is based on the absorption of energy when certain nuclei are excited by a strong magnetic field. Because hydrogen is a major constituent of aqueous body fluids and fatty tissue, the hydrogen nucleus is the most convenient one for study by MRI. Recall in Section 2-6 we learned that the nucleus of ^1H (a proton) has two possible spin states, either positive or negative.

In the presence of an external magnetic field, a spinning nucleus can align itself either in a lower (parallel) energy state or a higher (antiparallel) energy state. If the spinning nucleus is irradiated with

CHEMICAL CONNECTIONS 9D

Magnetic Resonance Imaging (continued)

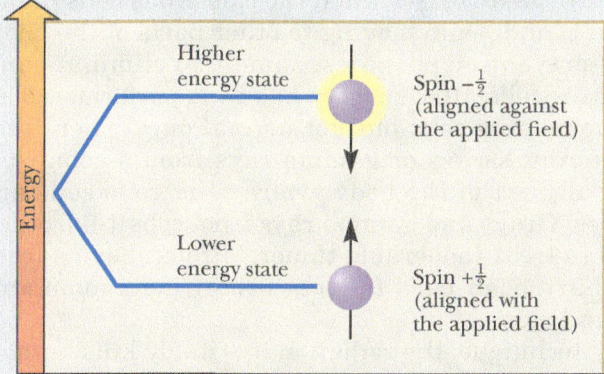

electromagnetic radiation of the proper energy, the nuclear spin can be flipped from the lower energy state to the higher energy state, and this flipping generates a signal that can be detected by sophisticated electronic equipment. This difference in energy corresponds to the radiofrequency portion of the electromagnetic spectrum. The stimulated nuclei give off a signal that can be measured, interpreted, and correlated with their environment in the body. Because hydrogen atoms in the body are in different chemical environments, frequencies of different energies are absorbed. By irradiating the body with pulses of radiofrequency radiation and using sophisticated detection techniques, tissue can be imaged at specific depths within the body, providing pictures with extraordinary detail. The energies absorbed are calculated and converted to three-dimensional color images of the body.

MRI is noninvasive to the body and is quick, safe, and painless. A person is placed in a cavity surrounded by a magnetic field, and an image is generated based on the extent of radiofrequency energy absorption. Differences between normal and malignant tissue, as well as other problems, may be seen. For example, the new generation of MRI instruments is so sensitive that it is able to detect a chemical change in the brain resulting from an external stimulus. A response to a question or the observation of a flash of light produces a measurable signal that can be detected using MRI.

Test your knowledge with Problem 9-67

B. Radiation Therapy

The main use of radioactive isotopes in therapy is the selective destruction of pathological cells and tissues. Recall that radiation, whether from gamma rays, X-rays, or other sources, is detrimental to cells. Ionizing radiation damages cells, especially those that divide rapidly. This damage may be great enough to destroy diseased cells or to sufficiently alter the genes in them so that multiplication of the cells slows down.

In therapy applications, cancerous cells are the main targets for ionizing radiation. Radiation is typically used when a cancer is well localized; it may also be employed when the cancerous cells spread and are in a metastatic state. A metastatic state exists when the cancerous cells break off from their primary site(s) and begin moving to other parts of the body. In addition, it is used for preventive purposes, namely to eliminate any possible remaining cancerous cells after surgery has been performed. The idea, of course, is to kill cancerous cells but not normal ones. Therefore, radiation such as high-energy X-rays or gamma rays from a cobalt-60 source is focused on a small part of the body where cancerous cells are suspected to reside. Besides X-rays and gamma rays from cobalt-60, other ionizing radiation is used to treat inoperable tumors. Proton beams from cyclotrons, for instance, have been used to treat ocular melanoma and tumors of the skull base and spine.

Despite this pinpointing technique, the radiation inevitably kills normal cells along with the cancerous cells. Because the radiation is most effective against rapidly dividing cancer cells rather than normal cells and because the radiation is aimed at a specific location, the damage to healthy tissues is minimized.

Another way to localize radiation damage in therapy is to use specific radioactive isotopes. In the case of thyroid cancer, large doses of iodine-131 are administered, which are taken up by the thyroid gland. The isotope, which has high radioactivity, kills all the types of cells of the gland (cancerous as well as normal ones), but does not appreciably damage other organs.

Another radioisotope, iodine-125, is used in the treatment of prostate cancer. Seeds of iodine-125, a gamma emitter, are implanted in the cancerous area of the prostate gland while being imaged with ultrasound. The seeds deliver 160 Gy (16,000 rad) over their lifetime.

A newer form of prostate cancer treatment with great potential relies on actinium-225, an alpha emitter. As discussed in Section 9-6, alpha particles cause more damage to the tissues than any other form of radiation, but they have low penetrating power. Researchers have developed a very clever way to deliver actinium-225 to the cancerous region of the prostate gland without damaging healthy tissues. The prostate tumor has a high concentration of prostate-specific antigen (PSA) on its surface. A monoclonal antibody (Section 30-4) homes in on the PSA and interacts with it. A single actinium-225 atom attached to such a monoclonal antibody can deliver the desired radiation, thereby destroying the cancer. Actinium-225 is especially effective because it has a half-life of ten days and it decays to three nuclides, themselves alpha emitters. In clinical trials, a single injection of antibody with an intensity in the kBq range (nanocuries) provided tumor regression without toxicity.

9-8 What Is Nuclear Fusion?

An estimated 98% of all matter in the universe is made up of hydrogen and helium. The "big bang" theory of the formation of the universe postulates that our universe started with an explosion (big bang) in which matter was formed out of energy and that at the beginning, only the lightest element, hydrogen, was in existence. Later, as the universe expanded, stars were born when hydrogen clouds collapsed under gravitational forces. In the cores of these stars, hydrogen nuclei fused to form helium.

The fusion of two hydrogen nuclei into a helium nucleus liberates a very large amount of energy in the form of photons, largely by the following reaction:

$$^{2}_{1}H + ^{3}_{1}H \longrightarrow ^{4}_{2}He + ^{1}_{0}n + 5.3 \times 10^{8} \text{ kcal/mol He}$$

Hydrogen-2 (Deuterium)   Hydrogen-3 (Tritium)

This process, called **nuclear fusion**, is how the sun makes its energy. Uncontrolled fusion is employed in the "hydrogen bomb." If we can ever achieve a controlled version of this fusion reaction (which is unlikely to happen in the near term), we should be able to solve our energy problems.

As we have just seen, the fusion of deuterium and tritium nuclei to a helium nucleus gives off a very large amount of energy. What is the source of this energy? When we compare the mass of the reactants and products, we see that there is a loss of $5.0302 - 5.0113 = 0.0189$ g for each mole of helium formed:

$$^{2}_{1}H + ^{3}_{1}H \longrightarrow ^{4}_{2}He + ^{1}_{0}n$$

2.01410 g   3.0161 g   4.0026 g   1.0087 g

5.0302 g   5.0113 g

Nuclear fusion The joining together of atomic nuclei to form a new nucleus heavier than either starting nuclei

When the deuterium and tritium nuclei are converted to helium and a neutron, the extra mass has to go somewhere. Where does it go? The answer is that the missing mass is converted to energy. We even know, from the equation developed by Albert Einstein (1879–1955), how much energy we can get from the conversion of any amount of mass:

$$E = mc^{2}$$

This equation says that the mass (m), in kilograms, that is lost multiplied by the square of the velocity of light (c^{2}, where $c = 3.0 \times 10^{8}$ m/s), in meters squared per second squared (m^{2}/s^{2}), is equal to the amount of energy created (E), in joules. For example, 1 g of matter completely converted to energy would produce 8.8×10^{13} J, which is enough energy to boil 34,000,000 L of water initially at 20°C. This is equivalent to the amount of water in an Olympic-size swimming pool. As you can see, we get a tremendous amount of energy from a little bit of mass.

All of the **transuranium elements** (elements with atomic numbers greater than 92) are artificial and have been prepared by a fusion process in which heavy nuclei are bombarded with light ones. Many, as their names indicate, were first prepared at the Lawrence Laboratory of the University of California, Berkeley, by Glenn Seaborg (1912–1999; Nobel laureate in chemistry, 1951) and his colleagues:

$$^{244}_{96}Cm + ^{4}_{2}He \longrightarrow ^{245}_{97}Bk + ^{1}_{1}H + 2\,^{1}_{0}n$$

$$^{238}_{92}U + ^{12}_{6}C \longrightarrow ^{246}_{98}Cf + 4\,^{1}_{0}n$$

$$^{252}_{98}Cf + ^{10}_{5}B \longrightarrow ^{257}_{103}Lr + 5\,^{1}_{0}n$$

These transuranium elements are unstable, and most have very short half-lives. For example, the half-life of Lawrencium-257 is 0.65 second. Many of the new superheavy elements have been obtained by bombarding lead isotopes with calcium-48 or nickel-64. So far, the creation of elements 110, 111, and 112–116 has been reported, even though their detection was based on the observation of the decay of a single atom.

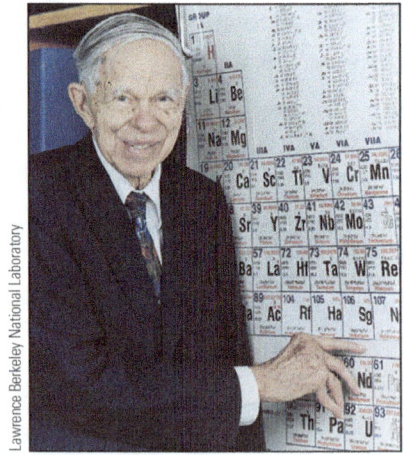

▶ Glenn Seaborg (1912–1999).

A pioneer in developing radioisotopes for medical use, Glenn Seaborg, was the first to produce iodine-131, used subsequently to treat his mother's abnormal thyroid condition. As a result of Seaborg's further research, it became possible to predict accurately the properties of many of the as-yet-undiscovered transuranium elements. In a remarkable 21-year span (1940–1961), Seaborg and his colleagues synthesized ten new transuranium elements (plutonium to lawrencium). He received the Nobel Prize in 1951 for his creation of new elements. In the 1990s, Seaborg was honored by having element 106 named for him. ◀

9-9 What Is Nuclear Fission and How Is It Related to Atomic Energy?

In the 1930s, Enrico Fermi (1901–1954) and his colleagues in Rome and Otto Hahn (1879–1968), Lise Meitner (1878–1968), and Fritz Strassman (1902–1980) in Germany tried to produce new transuranium elements by bombarding uranium-235 with neutrons. To their surprise, they found that, rather than fusion, they obtained **nuclear fission** (fragmentation of large nuclei into smaller pieces):

$$^{235}_{92}U + ^{1}_{0}n \longrightarrow ^{141}_{56}Ba + ^{92}_{36}Kr + 3\,^{1}_{0}n + \gamma + \text{energy}$$

In this reaction, a uranium-235 nucleus first absorbs a neutron to become uranium-236 and then breaks into two smaller nuclei. The most important product of this nuclear decay is energy, which is produced because the products have less mass than the starting materials. This form of energy, called **atomic energy**, has been used for both war (with the atomic bomb) and peace.

With uranium-235, each fission produces three neutrons, which in turn can generate more fissions by colliding with other uranium-235 nuclei. If even one of these neutrons produces a new fission, the process becomes a self-propagating **chain reaction** (Figure 9-9) that continues at a constant

FIGURE 9-9 A chain reaction begins when a neutron collides with a nucleus of uranium-235.

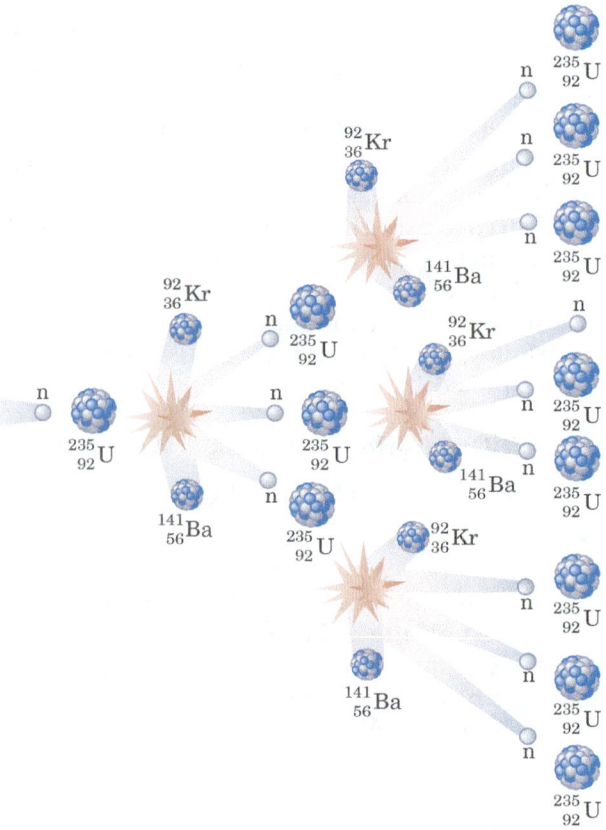

rate. If all three neutrons are allowed to produce new fission, the rate of the reaction increases constantly and eventually culminates in a nuclear explosion. In nuclear power plants, the rate of reaction is controlled by inserting boron control rods into the reactor to absorb neutrons and thereby dampen the rate of fission.

In nuclear power plants, the energy produced by fission is sent to heat exchangers and used to generate steam, which drives a turbine to produce electricity (Figure 9-10). Today, such plants supply more than 15% of the electrical energy in the United States. The opposition to nuclear plants is based on safety considerations and on the unsolved problems of waste disposal. Although nuclear plants in general have good safety records, accidents such as those at Fukushima Daiichi, Chernobyl (Chemical Connections 9D), and Three Mile Island have caused concern. ◀

▶ Nuclear power plant in Salem, New Jersey.

FIGURE 9-10 Schematic diagram of a nuclear power plant.

Nuclear waste disposal is a long-range problem. The fission products of nuclear reactors are highly radioactive themselves, with long half-lives. Spent fuel contains these high-level fission products as nuclear wastes, together with uranium and plutonium that can be recovered and reused as mixed oxide (MOX) fuel. Reprocessing is costly: although done routinely in Europe and Russia, it is not practiced by nuclear plants in the United States for economic reasons. However, this situation may change because cleaner extraction processes have been developed that use supercritical carbon dioxide (Chemical Connections 5E), thereby eliminating the need to dispose of solvent.

The United States has about 50,000 metric tons of spent fuel, stored under water and in dry casks at power plants. The Department of Energy stores the additional nuclear wastes from the nuclear weapons program, research reactors, and other sources in three major sites. After 40 years, the level of radioactivity that the wastes had immediately after their removal from the reactor is reduced a thousandfold. Such nuclear waste is a good candidate for underground burial. For example, the U.S. federal government gave its approval to store nuclear waste at Yucca Mountain, Nevada. However, in light of Japan's recent Fukushima Daiichi nuclear power plant disaster, opponents to the ongoing Yucca Mountain waste disposal are determined to shut the project down, resulting in modern legal and political disputes. ◀

Environmental concerns persist, however. The site cannot be guaranteed to stay dry for centuries. Moisture may corrode the steel cylinders and even

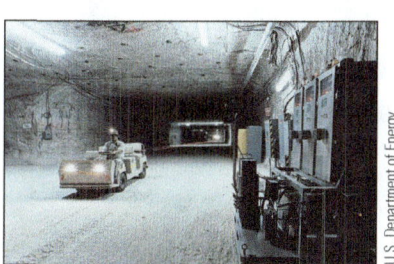

▶ Storage of nuclear wastes in a storage room carved out of an underground salt mine.

CHEMICAL CONNECTIONS 9E

Radioactive Fallout from Nuclear Accidents

On April 26, 1986, an accident occurred at the nuclear reactor in the town of Chernobyl in the former Soviet Union. It was a clear reminder of the dangers involved in this industry and of the far-reaching contamination that such accidents can produce. In Sweden, more than 500 miles away from the accident, the radioactive cloud increased the background radiation from 4 to 15 times the normal level. The radioactive cloud reached England, about 1300 miles away, one week later. There, it increased the natural background radiation by 15%. The radioactivity from iodine-131 was measured at 400 Bq/L in milk and 200 Bq/kg in leafy vegetables. Even some 4000 miles away in Spokane, Washington, an iodine-131 activity of 242 Bq/L was found in rainwater; smaller activities—1.03 Bq/L of ruthenium-103 and 0.66 Bq/L of cesium-137—were recorded as well. These levels are not harmful.

Closer to the source of the nuclear accident, in neighboring Poland, potassium iodide pills were given to children. This step was taken to prevent radioactive iodine-131 (which might come from contaminated food) from concentrating in their thyroid glands, which could lead to cancer. In the wake of the September 11, 2001 terrorist attacks, Massachusetts became the first state to authorize the storage of KI pills in case of nuclear-related terrorist activity.

Map showing those areas most affected by the Chernobyl accident.

Test your knowledge with Problem 9-68.

the inner glass/ceramic cylinders surrounding the nuclear waste. Some fear that leaked materials from such storage tanks may escape if carbon-14 is oxidized to radioactive carbon dioxide or, less likely, that other radioactive nuclides may contaminate the groundwater, which lies far below the desert rock of Yucca Mountain.

To keep these problems in perspective, one must remember that most other ways of generating large amounts of electrical power have their own environmental problems. For example, burning coal or oil contributes to the accumulation of CO_2 in the atmosphere and to acid rain (see Chemical Connections 6A).

Summary of Key Questions

Section 9-1 How Was Radioactivity Discovered?
- Henri Becquerel discovered radioactivity in 1896.

Section 9-2 What Is Radioactivity?
- The four major types of radioactivity are **alpha particles** (helium nuclei), **beta particles** (electrons), **gamma rays** (high-energy photons), and **positrons** (positively charged electrons).

Section 9-3 What Happens When a Nucleus Emits Radioactivity?
- When a nucleus emits a **beta particle**, the new element has the same mass number but an atomic number one unit greater.
- When a nucleus emits an **alpha particle**, the new element has an atomic number two units lower and a mass number four units smaller.

- When a nucleus emits a **positron** (positive electron), the new element has the same mass number but an atomic number one unit smaller.
- In **gamma emission**, no transmutation takes place; only the energy of the nucleus is lowered.
- In **electron capture**, the new element has the same mass number but an atomic number one unit smaller.

Section 9-4 What Is Nuclear Half-Life?

- Each radioactive isotope decays at a fixed rate described by its **half-life**, which is the time required for half of the sample to decay.

Section 9-5 How Do We Detect and Measure Nuclear Radiation?

- Radiation is detected and counted by devices such as **Geiger-Müller counters**.
- The main unit of intensity of radiation is the **curie (Ci)**, which is equal to 3.7×10^{10} disintegrations per second. Other common units are the millicurie (mCi), the microcurie (μCi), and the becquerel (Bq).

Section 9-6 How Is Radiation Dosimetry Related to Human Health?

- For medical purposes and to measure potential radiation damage, the absorbed dose is measured in **rads**. Different particles damage body tissues differently; the **rem** is a measure of relative damage caused by the type of radiation.

Section 9-7 What Is Nuclear Medicine?

- Nuclear medicine is the use of radionuclei for diagnostic imaging and therapy.

Section 9-8 What Is Nuclear Fusion?

- **Nuclear fusion** is the combining (fusing) of two lighter nuclei to form a heavier nucleus. Helium is synthesized in the interiors of stars by the fusion of hydrogen nuclei. The energy released in this process is the energy of our sun.

Section 9-9 What Is Nuclear Fission and How Is It Related to Atomic Energy?

- **Nuclear fission** is the splitting of a heavier nucleus into two or more smaller nuclei. Nuclear fission releases large amounts of energy, which can be either controlled (nuclear reactors) or uncontrolled (nuclear weapons).

Summary of Key Reactions

1. **Beta (β) emission (Section 9-3B)** When a nucleus decays by beta emission, the new element has the same mass number but an atomic number one unit greater.

$$^{32}_{15}P \longrightarrow {}^{32}_{16}S + {}^{0}_{-1}e$$

2. **Alpha (α) emission (Section 9-3C)** When a nucleus decays by alpha emission, the new nucleus has a mass number four units smaller and an atomic number two units smaller.

$$^{238}_{92}U \longrightarrow {}^{234}_{90}Th + {}^{4}_{2}He$$

3. **Positron ($\beta+$) emission (Section 9-3D)** When a nucleus decays by positron emission, the new element has the same mass number but an atomic number one unit smaller.

$$^{11}_{6}C \longrightarrow {}^{11}_{5}B + {}^{0}_{+1}e$$

4. **Gamma (γ) emission (Section 9-3E)** When a nucleus emits gamma radiation, there is no change in either mass number or atomic number of the nucleus.

$$^{11}_{5}B^{*} \longrightarrow {}^{11}_{5}B + \gamma$$

5. **Electron capture (Section 9-3F)** When a nucleus decays by electron capture, the product nucleus has the same mass number but an atomic number one unit smaller.

$$^{7}_{4}Be + {}^{0}_{-1}e \longrightarrow {}^{7}_{3}Li$$

6. **Nuclear Fusion (Section 9-8)** In nuclear fusion, two or more nuclei react to form a larger nucleus. In the process, there is a slight decrease in mass; the sum of the masses of the fusion products is less than the sum of the masses of the starting nuclei. The lost mass appears as energy.

$$^{2}_{1}H + {}^{3}_{1}H \longrightarrow {}^{4}_{2}He + {}^{1}_{0}n + 5.3 \times 10^{8} \text{ kcal/mol He}$$

7. **Nuclear Fission (Section 9-9)** In nuclear fission, a nucleus captures a neutron to form a nucleus with a mass number increased by one unit. The new nucleus then splits into two smaller nuclei.

$$^{235}_{92}U + {}^{1}_{0}n \longrightarrow {}^{141}_{56}Ba + {}^{92}_{36}Kr + 3\,{}^{1}_{0}n + \gamma + \text{energy}$$

Problems

Orange-numbered problems are applied.

Section 9-2 What Is Radioactivity?

9-8 What is the relationship between frequency and wavelength? frequency and energy? wavelength and energy?

9-9 What is the difference between an alpha particle and a proton?

9-10 Microwaves are a form of electromagnetic radiation that is used for the rapid heating of foods. What is the frequency of a microwave with a wavelength of 5.8 cm?

9-11 In each case, given the frequency, give the wavelength in centimeters or nanometers and tell what kind of radiation it is.
(a) 7.5×10^{14}/s
(b) 1.0×10^{10}/s
(c) 1.1×10^{15}/s
(d) 1.5×10^{18}/s

9-12 Red light has a wavelength of 650 nm. What is its frequency?

9-13 Which has the longest wavelength: (a) infrared, (b) ultraviolet, or (c) X-rays? Which has the highest energy?

9-14 Write the symbol for a nucleus with the following components:
(a) 9 protons and 10 neutrons
(b) 15 protons and 17 neutrons
(c) 37 protons and 50 neutrons

9-15 In each pair, tell which isotope is more likely to be radioactive:
(a) Nitrogen-14 and nitrogen-13
(b) Phosphorus-31 and phosphorus-33
(c) Lithium-7 and lithium-9
(d) Calcium-39 and calcium-40

9-16 Which isotope of boron is the most stable: boron-8, boron-10, or boron-12?

9-17 Which isotope of oxygen is the most stable: oxygen-14, oxygen-16, or oxygen-18?

Section 9-3 What Happens When a Nucleus Emits Radioactivity?

9-18 Answer true or false.
(a) The majority (greater than 50%) of the more than 300 naturally occurring isotopes are stable.
(b) More artificial isotopes have been created in the laboratory than there are naturally occurring stable isotopes.
(c) All artificial isotopes created in the laboratory are radioactive.
(d) The terms "beta particle," "beta emission," and "beta ray" all refer to the same type of radiation.
(e) When balancing a nuclear equation, the sum of the mass numbers and the sum of the atomic numbers on each side of the equation must be the same.
(f) The symbol of a beta particle is $_{-1}^{0}\beta$.
(g) When a nucleus emits a beta particle, the new nucleus has the same mass number but an atomic number one unit higher.
(h) When iron-59 ($_{26}^{59}$Fe) emits a beta particle, it is converted to cobalt-59 ($_{27}^{59}$Co).
(i) When a nucleus emits a beta particle, it first captures an electron from outside the nucleus and then emits it.
(j) For the purposes of determining atomic numbers in a nuclear equation, an electron is assumed to have a mass number of zero and an atomic number of -1.
(k) The symbol for an alpha particle is $_{2}^{4}$He.
(l) When a nucleus emits an alpha particle, the new nucleus has an atomic number two units higher and a mass number four units higher.
(m) When uranium-238 ($_{92}^{238}$U) undergoes alpha emission, the new nucleus is thorium-234 ($_{90}^{234}$Th).
(n) The symbol of a positron is $_{+1}^{0}\beta$.
(o) A positron is sometimes referred to as a positive electron.
(p) When a nucleus emits a positron, the new nucleus has the same mass number but an atomic number one unit lower.
(q) When carbon-11 ($_{6}^{11}$C) emits a positron, the new nucleus formed is boron-11 ($_{5}^{11}$B).
(r) Alpha emission and positron emission both result in the formation of a new nucleus with a lower atomic number.
(s) The symbol for gamma radiation is γ.
(t) When a nucleus emits gamma radiation, the new nucleus formed has the same mass number and the same atomic number.
(u) When a nucleus captures an extranuclear electron, the new nucleus formed has the same atomic number but a mass number one unit lower.
(v) When gallium-67 ($_{31}^{67}$Ga) undergoes electron capture, the new nucleus formed is germanium-67 ($_{32}^{67}$Ge).

9-19 Samarium-151 is a beta emitter. Write an equation for this nuclear reaction and identify the product nucleus.

9-20 The following nuclei turn into new nuclei by emitting beta particles. Write an equation for each nuclear reaction and identify the product nucleus.
(a) $^{159}_{63}\text{Eu}$ (b) $^{141}_{56}\text{Ba}$ (c) $^{242}_{95}\text{Am}$

9-21 Chromium-51 is used in diagnosing the pathology of the spleen. The nucleus of this isotope captures an electron according to the following equation. What is the transmutation product?

$$^{51}_{24}\text{Cr} + ^{0}_{-1}\text{e} \longrightarrow ?$$

9-22 The following nuclei decay by emitting alpha particles. Write an equation for each nuclear reaction and identify the product nucleus.
(a) $^{210}_{83}\text{Bi}$ (b) $^{238}_{94}\text{Pu}$ (c) $^{174}_{72}\text{Hf}$

9-23 Curium-248 was bombarded, yielding antimony-116 and cesium-160. What was the bombarding nucleus?

9-24 Phosphorus-29 is a positron emitter. Write an equation for this nuclear reaction and identify the product nucleus.

9-25 For each of the following, write a balanced nuclear equation and identify the radiation emitted.
(a) Beryllium-10 changes to boron-10
(b) Europium-151^m changes to europium-151
(c) Thallium-195 changes to mercury-195
(d) Plutonium-239 changes to uranium-235

9-26 In the first three steps in the decay of uranium-238, the following isotopic species appear: uranium-238 decays to thorium, which then decays to protactinium-234, which then decays to uranium-234. What kind of emission occurs in each step?

9-27 What kind of emission does *not* result in transmutation?

9-28 Complete the following nuclear reactions.
(a) $^{16}_{8}\text{O} + ^{16}_{8}\text{O} \longrightarrow ? + ^{4}_{2}\text{He}$
(b) $^{235}_{92}\text{U} + ^{1}_{0}\text{n} \longrightarrow ^{90}_{38}\text{Sr} + ? + 3\,^{1}_{0}\text{n}$
(c) $^{13}_{6}\text{C} + ^{4}_{2}\text{He} \longrightarrow ^{16}_{8}\text{O} + ?$
(d) $^{210}_{83}\text{Bi} \longrightarrow ? + ^{0}_{-1}\text{e}$
(e) $^{12}_{6}\text{C} + ^{1}_{1}\text{H} \longrightarrow ? + \gamma$

9-29 Americium-240 is made by bombarding plutonium-239 with α particles. In addition to americium-240, a proton and two neutrons are also formed. Write a balanced equation for this nuclear reaction.

Section 9-4 What Is Nuclear Half-Life?

9-30 Answer true or false.
(a) Half-life is the time it takes one-half of a radioactive sample to decay.
(b) The concept of half-life refers to nuclei undergoing alpha, beta, and positron emission; it does not apply to nuclei undergoing gamma emission.
(c) At the end of two half-lives, one-half of the original radioactive sample remains; at the end of three half-lives, one-third of the original sample remains.
(d) If the half-life of a particular radioactive sample is 12 minutes, a time of 36 minutes represents three half-lives.
(e) At the end of three half-lives, only 12.5% of an original radioactive sample remains.

9-31 Iodine-125 emits gamma rays and has a half-life of 60 days. If a 20-mg pellet of iodine-125 is implanted into a prostate gland, how much iodine-125 remains there after one year?

9-32 Polonium-218, a decay product of radon-222 (see Chemical Connections 9B), has a half-life of 3 min. What percentage of the polonium-218 formed will remain in the lung 9 min after inhalation?

9-33 A rock containing 1 mg of plutonium-239 per kg of rock is found in a glacier. The half-life of plutonium-239 is 25,000 years. If this rock was deposited 100,000 years ago during an ice age, how much plutonium-239 per kilogram of rock was in the rock at that time?

9-34 The element radium is extremely radioactive. If you converted a piece of radium metal to radium chloride (with the weight of the radium remaining the same), would it become less radioactive?

9-35 In what ways can we increase the rate of radioactive decay? Decrease it?

9-36 Suppose 50.0 mg of potassium-45, a beta emitter, was isolated in pure form. After one hour, only 3.1 mg of the radioactive material was left. What is the half-life of potassium-45?

9-37 A patient receives 200 mCi of iodine-131, which has a half-life of eight days.
(a) If 12% of this amount is taken up by the thyroid gland after two hours, what will be the activity of the thyroid after two hours, in millicuries and in counts per minute?
(b) After 24 days, how much activity will remain in the thyroid gland?

Section 9-5 How Do We Detect and Measure Nuclear Radiation?

9-38 Answer true or false.
(a) Ionizing radiation refers to any radiation that interacts with neutral atoms or molecules to create positive ions.
(b) Ionizing radiation creates positive ions by striking a nucleus and knocking one or more electrons from the nucleus.
(c) Ionizing radiation creates positive ions by knocking one or more extranuclear electrons from a neutral atom or molecule.
(d) The curie (Ci) and becquerel (Bq) are both units by which we report radiation intensity.
(e) The units of a curie (Ci) are disintegrations per second (dps).
(f) A microcurie (μCi) is a smaller unit than a curie (Ci).

(g) The intensity of radiation is inversely related to the square of the distance from the radiation source; for example, the intensity at three meters from the source is 1/9 of what it is at the source.

(h) Alpha particles are the most massive and highly charged type of nuclear radiation and, therefore, are the most penetrating type of nuclear radiation.

(i) Beta particles have both a smaller mass and a smaller charge than alpha particles and, therefore, are more penetrating than alpha particles.

(j) Gamma rays, with neither mass nor charge, are the least penetrating type of nuclear radiation.

(k) After one half-life, the mass of a radioactive sample remaining is approximately 50% of the original mass.

9-39 If you work in a lab containing radioisotopes emitting all kinds of radiation, from which emission should you seek the most protection?

9-40 What do Geiger-Müller counters measure: (a) the intensity or (b) the energy of radiation?

9-41 It is known that radioactivity is being emitted with an intensity of 175 mCi at a distance of 1.0 m from the source. How far in meters from the source should you stand if you wish to be subjected to no more than 0.20 mCi?

Section 9-6 How Is Radiation Dosimetry Related to Human Health?

9-42 Briefly contrast the three different units used to describe the effects of radiation on the body.

9-43 Does a curie (Ci) measure radiation intensity or energy?

9-44 What property is measured with each of the following terms?
(a) Rad   (b) Rem   (c) Roentgen
(d) Curie (e) Gray (f) Becquerel
(g) Sievert

9-45 A radioactive isotope with an activity (intensity) of 80.0 mCi per vial is delivered to a hospital. The vial contains 7.0 cc of liquid. The instruction is to administer 7.2 mCi intravenously. How many cubic centimeters of liquid should be used for one injection?

9-46 Why does exposure of a hand to alpha rays not cause serious damage to the person, whereas entry of an alpha emitter into the lung as an aerosol produces very serious damage to the person's health?

9-47 A certain radioisotope has an intensity of 10^6 Bq at 1-cm distance from the source. What would be the intensity at 20 cm? Give your answer in both Bq and μCi units.

9-48 Assuming the same amount of absorbed radiation, in rads from three sources, which would be the most damaging to the tissues: alpha particles, beta particles, or gamma rays?

9-49 In an accident involving radioactive exposure, person A received 3.0 Sv while person B received 0.50 mrem exposure. Who was hurt more seriously?

Section 9-7 What Is Nuclear Medicine?

9-50 Answer true or false.

(a) Of the radioisotopes listed in Table 9-5, the majority decay by beta emission.

(b) Isotopes that decay by alpha emission are rarely if ever used in nuclear imaging because alpha emitters are rare.

(c) Gamma emitters are so widely used in medical imaging because gamma radiation is penetrating and, therefore, can easily be measured by radiation detectors outside the body.

(d) When selenium-75 ($^{75}_{34}$Se) decays by electron capture and gamma emission, the new nucleus formed is arsenic-75 ($^{75}_{34}$As).

(e) When iodine-131 ($^{131}_{53}$I) decays by beta and gamma emission, the new nucleus formed is xenon-131 ($^{131}_{54}$Xe).

(f) In positron emission tomography (a PET scan), the detector counts the number of positrons emitted by a tagged material and the location within the body where the tagged material accumulates.

(g) The use of 18-fluorodeoxyglucose (FDG) in PET scans of the brain depends on the fact that FDG behaves in the body as does glucose.

(h) A goal of radiation therapy is to destroy pathological cells and tissues without at the same time damaging normal cells and tissues.

(i) In external beam radiation, radiation from an external source is directed at a tissue either on the surface of the body or within the body.

(j) In internal beam radiation, a radioactive material is implanted in a target tissue to destroy cells in the target tissue without doing appreciable damage to surrounding normal tissues.

9-51 In 1986, the nuclear reactor in Chernobyl had an accident and spewed radioactive nuclei that were carried by the winds for hundreds of miles. Today, among the child survivors of the event, the most common damage is thyroid cancer. What radioactive nucleus do you expect to be responsible for these cancers?

9-52 Cobalt-60, with a half-life of 5.26 years, is used in cancer therapy. The energy of the radiation from cobalt-62 is even higher (half-life = 14 minutes). Why isn't cobalt-62 also used for cancer therapy?

9-53 Match the radioactive isotope with its proper use:
_____ (a) Cobalt-60   1. Heart scan during exercise
_____ (b) Thallium-201 2. Measure water content of body
_____ (c) Tritium   3. Kidney scan
_____ (d) Mercury-197   4. Cancer therapy

Section 9-8 What Is Nuclear Fusion?

9-54 Answer true or false.
(a) In nuclear fusion, two nuclei combine to form a new nucleus.
(b) The energy of the sun is derived from the fusion of two hydrogen-1 (^1_1H) nuclei to form a helium-4 (^4_2He) nucleus.
(c) The energy of the sun occurs because once two hydrogen nuclei fuse, the two positive charges no longer repel each other.
(d) Fusion of hydrogen nuclei in the sun results in a small decrease in mass, which appears as an equivalent amount of energy.
(e) Einstein's famous $E = mc^2$ equation refers to the energy released when two particles of the same mass collide with the speed of light.
(f) Nuclear fusion occurs only in the sun.
(g) Nuclear fusion can be carried out and controlled in the laboratory.

9-55 What are the products of the fusion of hydrogen-2 and hydrogen-3 nuclei?

9-56 Assuming that one proton and two neutrons will be produced in an alpha-bombardment fusion reaction, what target nucleus would you use to obtain berkelium-249?

9-57 Element 109 was first prepared in 1982. A single atom of this element ($^{266}_{109}Mt$), with a mass number of 266, was made by bombarding a bismuth-209 nucleus with an iron-58 nucleus. What other products, if any, must have been formed besides $^{266}_{109}Mt$?

9-58 A new element was formed when lead-208 was bombarded by krypton-86. One could detect four neutrons as the product of the fusion. Identify the new element.

9-59 Boron-10 is used in control rods for nuclear reactors. This nucleus absorbs a neutron and then emits an alpha particle. Write an equation for each nuclear reaction and identify each product nucleus.

9-60 The most abundant isotope of uranium, ^{238}U, does not undergo fission. Instead, it captures a neutron and emits two β particles to make a fissionable isotope of plutonium, which can then be used as fuel in a nuclear reactor. Write an equation for the nuclear reaction and identify the product nucleus.

Chemical Connections

9-61 (Chemical Connections 9A) Why is it accurate to assume that the carbon-14 to carbon-12 ratio in a living plant is constant over the lifetime of the plant?

9-62 (Chemical Connections 9A) In a recent archeological dig in the Amazon region of Brazil, charcoal paintings were found in a cave. The carbon-14 content of the charcoal was one-fourth of what is found in charcoal prepared from that year's tree harvest. How long ago was the cave settled?

9-63 (Chemical Connections 9A) Carbon-14 dating of the Shroud of Turin indicated that the plant from which the shroud was made was alive around AD 1350. To how many half-lives does this time period correspond?

9-64 (Chemical Connections 9A) The half-life of carbon-14 is 5730 years. The wrapping of an Egyptian mummy gave off 7.5 counts per minute per gram of carbon. A piece of linen purchased today would give an activity of 15 counts per minute per gram of carbon. How old is the mummy?

9-65 (Chemical Connections 9B) How does radon-222 produce polonium-218?

9-66 (Chemical Connections 9C) Why is high-energy radiation exposure of water in the body dangerous to rapidly dividing cells?

9-67 (Chemical Connections 9D) How is the presence of the hydrogen atom in the body used in MRI?

9-68 (Chemical Connections 9E) In a nuclear accident, one of the radioactive nuclei that concerns people is iodine-131. Iodine is easily vaporized and can be carried by the winds to locations that are hundreds—even thousands—of miles away. Why is iodine-131 especially harmful?

Additional Problems

9-69 Phosphorus-32 ($t_{1/2}$ = 14.3 h) is used in the medical imaging and diagnosis of eye tumors. Suppose a patient is given 0.010 mg of this isotope. Prepare a graph showing the mass in milligrams remaining in the patient's body after one week. (Assume that none is excreted from the body.)

9-70 During the bombardment of argon-40 with protons, one neutron is emitted for each proton absorbed. What new element is formed?

9-71 Neon-19 and sodium-20 are positron emitters. What products result in each case?

9-72 The half-life of nitrogen-16 is 7 seconds. How long does it take for 100 mg of nitrogen-16 to be reduced to 6.25 mg?

9-73 Do the curie and the becquerel measure the same or different properties of radiation?

9-74 Selenium-75 has a half-life of 120.4 days, so it would take 602 days (five half-lives) to diminish to 3% of the original quantity. Yet this isotope is used for pancreatic scans without any fear that the radioactivity will cause undue harm to the patient. Suggest a possible explanation.

9-75 Use Table 9-4 to determine the percentage of annual radiation we receive from the following sources:
  (a) Naturally occurring sources
  (b) Diagnostic medical sources
  (c) Nuclear power plants

9-76 $^{225}_{89}$Ac is an alpha emitter. In its decay process, it produces three more alpha emitters in succession. Identify each of the decay products.

9-77 Which radiation will cause more ionization, X-rays or radar?

9-78 You have an old wristwatch that still has radium paint on its dial. Measurement of the radioactivity of the watch shows a beta-ray count of 0.50 count/s. If 1.0 microcurie of this sort of radiation produces 1000 mrem/year, how much radiation in mrem do you expect from the wristwatch if you wear it for one year?

9-79 Americium-241, which is used in some smoke detectors, has a half-life of 432 years and is an alpha emitter. What is the decay product of americium-241, and approximately what percentage of the original americium-241 will be still around after 1000 years?

9-80 On rare occasions, a nucleus will capture a beta particle instead of emitting one. Berkelium-246 is such a nucleus. What is the product of this nuclear transmutation?

9-81 A patient is reported to have been irradiated by a dose of 1 sievert in a nuclear accident. Is he in mortal danger?

9-82 What is the ground state of a nucleus?

9-83 Explain the following:
  (a) It is impossible to have a completely pure sample of any radioactive isotope.
  (b) Beta emission of a radioactive isotope creates a new isotope with an atomic number one unit higher than that of the radioactive isotope.

9-84 Yttrium-90, which emits beta particles, is used in radiotherapy. What is the decay product of yttrium-90?

9-85 The half-lives of some oxygen isotopes are as follows:

Oxygen-14 = 71 s   Oxygen-15 = 124 s

Oxygen-19 = 29 s   Oxygen-20 = 14 s

Oxygen-16 is the stable, nonradioactive isotope. Do the half-lives indicate anything about the stability of the other oxygen isotopes?

9-86 $^{225}_{89}$Ac is effective in prostate cancer therapy when administered at kBq levels. If an antibody tagged with $^{225}_{89}$Ac has an intensity of 2 million Bq/mg and if a solution contains 5 mg/L tagged antibody, how many milliliters of the solution should you use for an injection to administer 1 kBq intensity?

9-87 When $^{208}_{82}$Pb is bombarded with $^{64}_{28}$Ni, a new element and six neutrons are produced. Identify the new element.

9-88 Americium-241, the isotope used in smoke detectors, has a half-life of 432 years, which is sufficiently long to allow for handling it in large quantities. This isotope is prepared in the laboratory by bombarding plutonium-239 with α particles. In this reaction, plutonium-239 absorbs two neutrons and then decays by emission of a β particle. Write an equation for this nuclear reaction and identify the isotope formed as an intermediate between plutonium-239 and americium-241.

9-89 Boron-10, an effective absorber of neutrons, is used in control rods of uranium-235 fission reactors (see Figure 9-10) to absorb neutrons and thereby control the rate of reaction. Boron-10 absorbs a neutron and then emits an α particle. Write a balanced equation for this nuclear reaction and identify the nucleus formed as an intermediate between boron-10 and the final nuclear product.

9-90 Tritium, $^{3}_{1}$H, is a beta emitter widely used as a radioactive tracer in chemical and biochemical research. Tritium is prepared by the bombardment of lithium-6 with neutrons. Complete the following nuclear equation:

$$^{6}_{3}\text{Li} + ^{1}_{0}\text{n} \longrightarrow ^{3}_{1}\text{H} + ?$$

9-91 Radon-222 decays to a stable nucleus by a series of three alpha and two beta decays. Determine the stable nucleus that is formed.

9-92 Neptunium-237 decays by a series of steps to bismuth-209. How many alpha and beta particles are produced by this overall decay process?

9-93 Thorium-232 decays by a 10-step process, ultimately yielding lead-208. How many alpha particles and how many beta particles are emitted?

10 Organic Chemistry

KEY QUESTIONS

- 10-1 What Is Organic Chemistry?
- 10-2 Where Do We Obtain Organic Compounds?
- 10-3 How Do We Write Structural Formulas of Organic Compounds?
- 10-4 What Is a Functional Group?

The bark of the Pacific yew contains paclitaxel, a substance that has proven effective in treating certain types of ovarian and breast cancer (see Chemical Connections 10A).

10-1 What Is Organic Chemistry?

Organic chemistry is the chemistry of the compounds of carbon. As you study Chapters 10–19 (organic chemistry) and 20–31 (biochemistry), you will see that organic compounds are everywhere around us. They are in our foods, flavors, and fragrances; in our medicines, toiletries, and cosmetics; in our plastics, films, fibers, and resins; in our paints, varnishes, and glues; and, of course, in our bodies and the bodies of all other living organisms.

Perhaps the most remarkable feature of organic compounds is that they involve the chemistry of carbon and only a few other elements—chiefly, hydrogen, oxygen, and nitrogen. While the majority of organic compounds contain carbon and just these three elements, many also contain sulfur, a halogen (fluorine, chlorine, bromine, or iodine), and phosphorus.

As of the writing of this text, there are 118 known elements. Organic chemistry concentrates on carbon, just one of the 118. The chemistry of the other 117 elements comes under the field of inorganic chemistry. As we see in Figure 10-1, carbon is far from being among the most abundant elements in the Earth's crust. In terms of elemental abundance, approximately 75% of the Earth's crust is composed of just two elements: oxygen and silicon. These two elements are the components of silicate minerals, clays, and sand. In fact, carbon is not even among the ten most abundant elements. Instead, it is merely one of the elements making up the remaining 0.9% of the Earth's crust. Why, then, do we pay such special attention to just one element from among 117?

The first reason is largely historical. In the early days of chemistry, scientists thought organic compounds were those produced by living organisms and that inorganic compounds were those found in rocks and other nonliving matter. At that time, they believed that a "vital force," possessed only by living organisms, was necessary to produce organic compounds. In other words, chemists believed that they could not synthesize any organic compound by starting with only inorganic compounds. This theory was very easy to disprove if, indeed, it was wrong. It required only one experiment in which an organic compound was made from inorganic compounds. In 1828, Friedrich Wöhler (1800–1882) carried out just such an experiment. He heated an aqueous solution of ammonium chloride and silver cyanate, both inorganic compounds and—to his surprise—obtained urea, an "organic" compound found in urine.

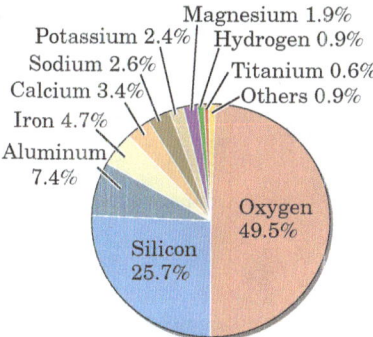

FIGURE 10-1 Abundance of the elements in the Earth's crust.

Although this single experiment of Wöhler's was sufficient to disprove the "doctrine of vital force," it took several years and a number of additional experiments for the entire scientific community to accept the fact that organic compounds could be synthesized in the laboratory. This discovery meant that the terms "organic" and "inorganic" no longer had their original meanings because, as Wöhler demonstrated, organic compounds could be obtained from inorganic materials. A few years later, August Kekulé (1829–1896) put forth a new definition—organic compounds are those containing carbon—and his definition has been accepted ever since.

A second reason for the study of carbon compounds as a separate discipline is the sheer number of organic compounds. Chemists have discovered or synthesized more than 10 million of them, and an estimated 10,000 new ones are reported each year. By comparison, chemists have discovered or synthesized an estimated 1.7 million inorganic compounds. Thus, approximately 85% of all known compounds are organic compounds.

A third reason—and one particularly important for those of you going on to study biochemistry—is that biochemicals of interest to us, including carbohydrates, lipids, proteins, enzymes, nucleic acids (DNA and RNA), hormones, vitamins, and almost all other important chemicals in living systems are organic compounds. Furthermore, their reactions are often strikingly similar to those occurring in test tubes. For this reason, knowledge of organic chemistry is essential for an understanding of biochemistry.

One final point about organic compounds. They generally differ from inorganic compounds in many of their properties, some of which are shown in Table 10-1. Most of these differences stem from the fact that the bonding in organic compounds is almost entirely covalent, while most inorganic compounds have ionic bonds. Thus these two types of compounds differ in their properties because they differ in their structure and composition—not

Table 10-1 A Comparison of Properties of Organic and Inorganic Compounds

Organic Compounds	Inorganic Compounds
Bonding is almost entirely covalent.	Most have ionic bonds.
Many are gases, liquids, or solids with low melting points (less than 360°C).	Most are solids with high melting points.
Most are insoluble in water.	Many are soluble in water.
Most are soluble in organic solvents such as diethyl ether, toluene, and dichloromethane.	Almost all are insoluble in organic solvents.
Aqueous solutions do not conduct electricity.	Aqueous solutions form ions that conduct electricity.
Almost all burn and decompose.	Very few burn.
Reactions are usually slow.	Reactions are often very fast.

because they obey different natural laws. One set of natural laws applies to all compounds. Of course, the entries in Table 10-1 are generalizations, but they are largely true for the vast majority of compounds of both types.

10-2 Where Do We Obtain Organic Compounds?

Chemists obtain organic compounds in two principal ways: isolation from nature and synthesis in the laboratory.

A. Isolation from Nature

Living organisms are "chemical factories." Each terrestrial, marine, and freshwater plant (flora) and animal (fauna)—even microorganisms such as bacteria—makes thousands of organic compounds by a process called biosynthesis. One way, then, to get organic compounds is to extract, isolate, and purify them from biological sources. In this book, we will encounter many compounds that are or have been isolated in this way. Some important examples include vitamin E, the penicillins, table sugar, insulin, quinine, and the anticancer drug paclitaxel (Taxol, see Chemical Connections 10A). Nature also supplies us with three other important sources of organic compounds: natural gas, petroleum, and coal. We will discuss them in Section 11-4.

B. Synthesis in the Laboratory

Ever since Wöhler synthesized urea, organic chemists have sought to develop more ways to synthesize the same compounds or design derivatives of those found in nature. In recent years, the methods for doing so have become so sophisticated that there are few natural organic compounds, no matter how complicated, that chemists cannot synthesize in the laboratory.

Compounds made in the laboratory are identical in both chemical and physical properties to those found in nature—assuming, of course, that each is 100% pure. There is no way that anyone can tell whether a sample of any particular compound was made by chemists or obtained from nature. As a consequence, pure ethanol made by chemists has exactly the same physical and chemical properties as pure ethanol prepared by distilling wine. The same is true for ascorbic acid (vitamin C). There is no advantage, therefore, in paying more money for vitamin C obtained from a natural source than for synthetic vitamin C, because the two are identical in every way. ◀

▶ The vitamin C in an orange is identical to its synthetic tablet form.

CHEMICAL CONNECTIONS 10A

Taxol: A Story of Search and Discovery

In the early 1960s, the National Cancer Institute undertook a program to analyze samples of native plant materials in the hope of discovering substances that would prove effective in the fight against cancer. Among the materials tested was an extract of the bark of the Pacific yew, *Taxus brevifolia*, a slow-growing tree found in the old-growth forests of the Pacific Northwest. This biologically active extract proved to be remarkably effective in treating certain types of ovarian and breast cancer, even in cases where other forms of chemotherapy failed. The structure of the cancer-fighting component of yew bark was determined in 1962, and the compound was named paclitaxel (Taxol).

Paclitaxel (Taxol)

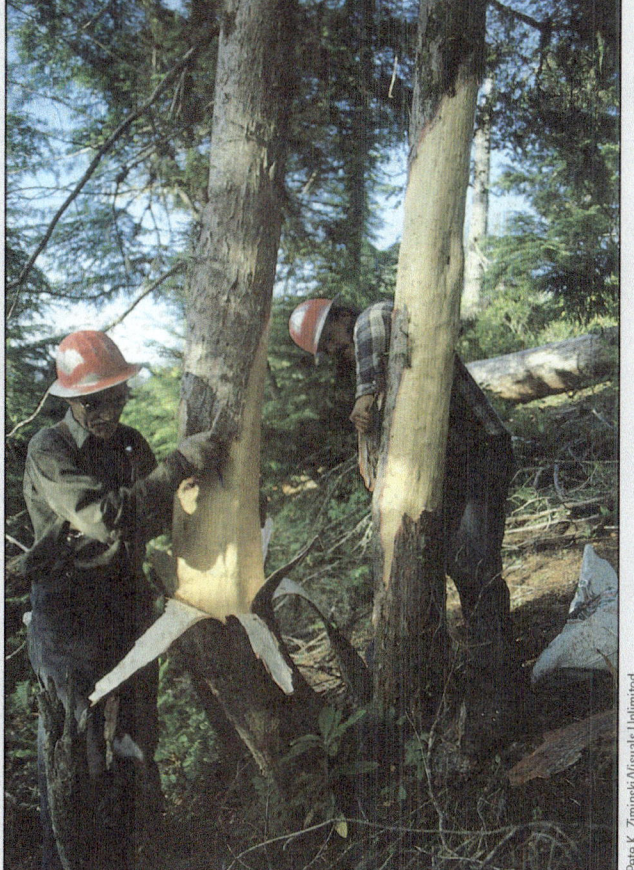

Pacific yew bark being stripped for Taxol extraction

Unfortunately, the bark of a single 100-year-old yew tree yields only about 1 g of Taxol, not enough for effective treatment of even one cancer patient. Furthermore, isolating Taxol means stripping the bark from trees, which kills them. In 1994, chemists succeeded in synthesizing Taxol in the laboratory, but the cost of the synthetic drug was far too high to be economical. Fortunately, an alternative natural source of the drug was found. Researchers in France discovered that the needles of a related plant, *Taxus baccata*, contain a compound that can be converted in the laboratory to Taxol. Because the needles can be gathered without harming the plant, it is not necessary to kill trees to obtain the drug.

Taxol inhibits cell division by acting on microtubules, a key component of the scaffolding of cells. Before cell division can take place, the cell must disassemble these microtubule units, and Taxol prevents this disassembly. Because cancer cells divide faster than normal cells, the drug effectively controls their spread.

The remarkable success of Taxol in the treatment of breast and ovarian cancer has stimulated research efforts to isolate and/or synthesize other substances that will act upon the human body in the same way and that may be even more effective anticancer agents than Taxol.

Test your knowledge with Problems 10-39 and 10-40.

Organic chemists, however, have not been satisfied with merely duplicating nature's compounds. They have also designed and synthesized compounds not found in nature. In fact, the majority of the more than 10 million known organic compounds are purely synthetic and do not exist in living organisms. For example, many modern drugs—Valium, albuterol, Prozac, Zantac, Zoloft, Lasix, Viagra, and Enovid—are synthetic organic compounds not found in nature. Even the over-the-counter drugs aspirin and ibuprofen are synthetic organic compounds not found in nature.

10-3 How Do We Write Structural Formulas of Organic Compounds?

A structural formula shows all the atoms present in a molecule as well as the bonds that connect the atoms to each other. The structural formula for ethanol, whose molecular formula is C_2H_6O, for example, shows all nine atoms and the eight bonds that connect them:

$$\begin{array}{c} H \quad H \\ | \quad | \\ H-C-C-O-H \\ | \quad | \\ H \quad H \end{array}$$
Ethanol

The Lewis model of bonding (Section 3-7C) enables us to see how carbon forms four covalent bonds that may be various combinations of single, double, and triple bonds. Furthermore, the valence-shell electron-pair repulsion (VSEPR) model (Section 3-10) tells us that the most common bond angles about carbon atoms in covalent compounds are approximately 109.5°, 120°, and 180°, for tetrahedral, trigonal planar, and linear geometries, respectively.

Table 10-2 shows several covalent compounds containing carbon bonded to hydrogen, oxygen, nitrogen, and chlorine. From these examples, we see the following:

- Carbon normally forms four covalent bonds and has no unshared pairs of electrons.
- Nitrogen normally forms three covalent bonds and has one unshared pair of electrons.
- Oxygen normally forms two covalent bonds and has two unshared pairs of electrons.
- Hydrogen forms one covalent bond and has no unshared pairs of electrons.
- A halogen (fluorine, chlorine, bromine, and iodine) normally forms one covalent bond and has three unshared pairs of electrons.

Table 10-2 Single, Double, and Triple Bonds in Compounds of Carbon. Bond angles and geometries for carbon are predicted using the VSEPR model.

EXAMPLE 10-1 Writing Structural Formulas

Following are structural formulas for acetic acid, CH_3COOH, and ethylamine, $CH_3CH_2NH_2$.

$$
\begin{array}{cc}
\text{H} \quad \text{O} & \text{H} \quad \text{H} \\
| \quad \;\; \| & | \quad\;\; | \\
\text{H}-\text{C}-\text{C}-\text{O}-\text{H} & \text{H}-\text{C}-\text{C}-\text{N}-\text{H} \\
| & | \quad\;\; | \quad\;\; | \\
\text{H} & \text{H} \quad \text{H} \quad \text{H} \\
\text{Acetic acid} & \text{Ethylamine}
\end{array}
$$

(a) Complete the Lewis structure for each molecule by adding unshared pairs of electrons so that each atom of carbon, oxygen, and nitrogen has a complete octet.

(b) Using the VSEPR model (Section 3-10), predict all bond angles in each molecule.

Strategy and Solution

(a) Each carbon atom must be surrounded by eight valence electrons to have a complete octet. Each oxygen must have two bonds and two unshared pairs of electrons to have a complete octet. Each nitrogen must have three bonds and one unshared pair of electrons to have a complete octet.

(b) To predict bond angles about a carbon, nitrogen, or oxygen atom, count the number of regions of electron density (lone pairs and bonding pairs of electrons about it). If four regions of electron density surround the atom, the predicted bond angles are 109.5°. If three regions surround it, the predicted bond angles are 120°. If two regions surround it, the predicted bond angle is 180°.

Acetic acid

Ethylamine

Problem 10-1

The structural formulas for ethanol, CH_3CH_2OH, and propene, $CH_3CH=CH_2$, are:

$$
\begin{array}{cc}
\text{H} \quad \text{H} & \text{H} \\
| \quad\;\; | & | \\
\text{H}-\text{C}-\text{C}-\text{O}-\text{H} & \text{H}-\text{C}-\text{C}=\text{C}-\text{H} \\
| \quad\;\; | & | \quad\;\; | \quad\;\; | \\
\text{H} \quad \text{H} & \text{H} \quad \text{H} \quad \text{H} \\
\text{Ethanol} & \text{Propene}
\end{array}
$$

(a) Complete the Lewis structure for each molecule showing all valence electrons.

(b) Using the VSEPR model, predict all bond angles in each molecule.

Functional group An atom or group of atoms within a molecule that shows a characteristic set of predictable physical and chemical behaviors

10-4 What Is a Functional Group?

As noted earlier in this chapter, more than 10 million organic compounds have been discovered and synthesized by organic chemists. It might seem an almost impossible task to learn the physical and chemical properties of so many compounds. Fortunately, the study of organic compounds is not as formidable a task as you might think. While organic compounds can undergo a wide variety of chemical reactions, only certain portions of their structures undergo chemical transformations. We call the atoms or groups of atoms of an organic molecule that undergo predictable chemical reactions a **functional group**. As we will see, the same functional group, in whatever organic molecule it occurs, undergoes the same types of chemical reactions. Therefore, we do not have to study the chemical reactions of even a fraction of the 10 million known organic compounds. Instead, we need to identify only a few characteristic functional groups and then study the chemical reactions that each undergoes.

Functional groups are also important because they are the units by which we divide organic compounds into families of compounds. For example, we group those compounds that contain an —OH (hydroxyl) group bonded to a tetrahedral carbon into a family called alcohols; compounds containing a —COOH (carboxyl group) belong to a family called carboxylic acids. Table 10-3 introduces seven of the most common functional groups. A complete list of all functional groups that we will study appears on the inside back cover of the text.

Table 10-3 Seven Common Functional Groups

Family	Functional Group	Example	Name
Alcohol	—OH	CH_3CH_2OH	Ethanol
Amine	—NH_2	$CH_3CH_2NH_2$	Ethylamine
Aldehyde	$-\overset{\overset{O}{\|\|}}{C}-H$	$CH_3\overset{\overset{O}{\|\|}}{C}H$	Acetaldehyde
Ketone	$-\overset{\overset{O}{\|\|}}{C}-$	$CH_3\overset{\overset{O}{\|\|}}{C}CH_3$	Acetone
Carboxylic acid	$-\overset{\overset{O}{\|\|}}{C}-OH$	$CH_3\overset{\overset{O}{\|\|}}{C}OH$	Acetic acid
Ester	$-\overset{\overset{O}{\|\|}}{C}-OR$	$CH_3\overset{\overset{O}{\|\|}}{C}OCH_2CH_3$	Ethyl acetate
Amide	$-\overset{\overset{O}{\|\|}}{C}-NH_2$	$CH_3-\overset{\overset{O}{\|\|}}{C}-NH_2$	Acetamide

At this point, our concern is simply pattern recognition—that is, the ability to recognize and identify one of these seven common functional groups when you see it and how to draw structural formulas of molecules containing it. We will have more to say about the physical and chemical properties of these and several other functional groups in Chapters 11–19.

Functional groups also serve as the basis for naming organic compounds. Ideally, each of the 10 million or more organic compounds must have a unique name different from the name of every other organic compound. We will show how these names are derived in Chapters 11–19 as we study individual functional groups in detail.

To summarize, functional groups:

- Are sites of predictable chemical reactions—a particular functional group, in whatever compound it is found, undergoes the same types of chemical reactions.
- Determine in large measure the physical properties of a compound.
- Serve as the units by which we classify organic compounds into families.
- Serve as a basis for naming organic compounds.

A. Alcohols

As previously mentioned, the functional group of an **alcohol** is an **—OH (hydroxyl) group** bonded to a tetrahedral carbon atom (a carbon having bonds to four atoms). In the general formula of an alcohol (shown below on the left), the symbol R— indicates a carbon group. The important point of the general structure is the —OH group bonded to a tetrahedral carbon atom.

Alcohol A compound containing an —OH (hydroxyl) group bonded to a tetrahedral carbon atom

—OH (hydroxyl) group An —OH group bonded to a tetrahedral carbon atom

Functional group

R = H or carbon group

Structural formula

An alcohol (Ethanol)

Condensed structural formula

Here we represent the alcohol as a **condensed structural formula**, CH_3CH_2OH. In a condensed structural formula, CH_3 indicates a carbon bonded to three hydrogens, CH_2 indicates a carbon bonded to two hydrogens, and CH indicates a carbon bonded to one hydrogen. Unshared pairs of electrons are generally not shown in condensed structural formulas.

Alcohols are classified as **primary (1°), secondary (2°), or tertiary (3°)**, depending on the number of carbon atoms bonded to the carbon bearing the —OH group.

$$CH_3-\underset{\underset{H}{|}}{\overset{\overset{H}{|}}{C}}-OH \qquad CH_3-\underset{\underset{CH_3}{|}}{\overset{\overset{H}{|}}{C}}-OH \qquad CH_3-\underset{\underset{CH_3}{|}}{\overset{\overset{CH_3}{|}}{C}}-OH$$

A 1° alcohol   A 2° alcohol   A 3° alcohol

> **EXAMPLE 10-2** Drawing Structural Formulas of Alcohols
>
> Draw Lewis structures and condensed structural formulas for the two alcohols with the molecular formula C_3H_8O. Classify each as primary, secondary, or tertiary.
>
> ### Strategy and Solution
>
> Begin by drawing the three carbon atoms in a chain. The oxygen atom of the hydroxyl group may be bonded to the carbon chain at two different positions on the chain: either to an end carbon or to the middle carbon.
>
> $$C-C-C \qquad C-C-C-OH \qquad C-\underset{\underset{}{|}}{\overset{\overset{OH}{|}}{C}}-C$$
>
> Carbon chain   The two locations for the —OH group
>
> Finally, add seven more hydrogens, giving a total of eight as shown in the molecular formula. Show unshared electron pairs on the Lewis structures but not on the condensed structural formulas.

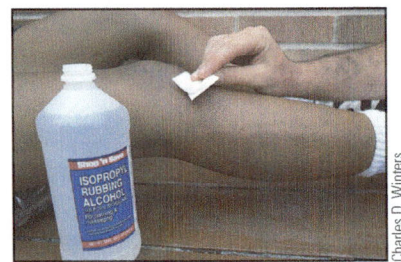

▶ 2-Propanol (isopropyl alcohol) is used to disinfect cuts and scrapes.

The secondary alcohol 2-propanol, whose common name is isopropyl alcohol, is the cooling, soothing component in rubbing alcohol. ◀

Problem 10-2

Draw Lewis structures and condensed structural formulas for the four alcohols with the molecular formula $C_4H_{10}O$. Classify each alcohol as primary, secondary, or tertiary. (*Hint*: First consider the connectivity of the four carbon atoms; they can be bonded either four in a chain or three in a chain with the fourth carbon as a branch on the middle carbon. Then consider the points at which the —OH group can be bonded to each carbon chain.)

B. Amines

Amine An organic compound in which one, two, or three hydrogens of ammonia are replaced by carbon groups; RNH_2, R_2NH, or R_3N.

Amino group A nitrogen atom bonded to one, two, or three R groups, RNH_2, R_2NH, or R_3N, where the symbol R represents a carbon group.

The functional group of an **amine** is an **amino group**—a nitrogen atom bonded to one, two, or three carbon atoms. In a **primary (1°) amine**, nitrogen is bonded to two hydrogens and one carbon group. In a **secondary (2°) amine**, it is bonded to one hydrogen and two carbon groups. In a **tertiary (3°) amine**, it is bonded to three carbon groups. The second and third structural formulas can be written in a more abbreviated form by collecting the CH_3 groups and writing them as $(CH_3)_2NH$ and $(CH_3)_3N$, respectively. The latter are known as condensed structural formulas.

EXAMPLE 10-3 Drawing Structural Formulas of Amines

Draw condensed structural formulas for the two primary amines with the molecular formula C_3H_9N.

Strategy and Solution

For a primary amine, draw a nitrogen atom bonded to two hydrogens and one carbon.

C—C—C—NH₂   C—C—C(NH₂)—C   CH₃CH₂CH₂NH₂   CH₃CH(NH₂)CH₃

The three carbons may be bonded to nitrogen in two ways

Add seven hydrogens to give each carbon four bonds and give the correct molecular formula

Problem 10-3
Draw structural formulas for the three secondary amines with the molecular formula $C_4H_{11}N$.

C. Aldehydes and Ketones

Both aldehydes and ketones contain a C=O (**carbonyl group**). In formaldehyde, the simplest **aldehyde**, the carbonyl group is bonded to two hydrogens. In all other aldehydes the carbonyl group is bonded to one hydrogen and one carbon group. In a condensed structural formula, the aldehyde group may be written showing the carbon-oxygen double bond as CH=O or, alternatively, it may be written —CHO. The functional group of a **ketone** is a carbonyl group bonded to two carbon atoms. In the general structural formula of each functional group, we use the symbol R to represent other groups bonded to carbon to complete the tetravalence of carbon.

Carbonyl group A C=O group

Aldehyde A compound containing a carbonyl group bonded to a hydrogen; a —CHO group

Ketone A compound containing a carbonyl group bonded to two carbon groups

R—C(R)(R)—C(=O)—H   Functional group

CH₃—C(=O)—H   Acetaldehyde (an aldehyde)

R—C(R)(R)—C(=O)—C(R)—   Functional group

CH₃—C(=O)—CH₃   Acetone (a ketone)

EXAMPLE 10-4 Drawing Structural Formulas of Aldehydes

Draw condensed structural formulas for the two aldehydes with the molecular formula C_4H_8O.

Strategy and Solution
First draw the functional group of an aldehyde and then add the remaining carbons. These may be bonded in two ways. Then add seven hydrogens to complete the tetravalence of each carbon.

CH₃CH(CH₃)CHO   or   CH₃CH(CH₃)CHO

$$\underset{\text{O}}{\text{CH}_3\text{CH}_2\text{CH}_2\overset{\|}{\text{C}}\text{H}} \quad \text{or} \quad \text{CH}_3\text{CH}_2\text{CH}_2\text{CHO}$$

Problem 10-4
Draw condensed structural formulas for the three ketones with the molecular formula $C_5H_{10}O$.

D. Carboxylic Acids

Carboxylic acid A compound containing a —COOH group

—COOH (carboxyl: carbonyl + hydroxyl) group A —COOH group

The functional group of a **carboxylic acid** is a —**COOH** (**carb**onyl + hydr**oxyl**) group. In a condensed structural formula, a carboxyl group may also be written —CO_2H.

$$\underset{\substack{\text{Functional} \\ \text{group}}}{\text{R}\overset{:\text{O}:}{\overset{\|}{\text{C}}}\ddot{\text{O}}\text{H}} \qquad \underset{\substack{\text{Acetic acid} \\ \text{(a carboxylic acid)}}}{\text{CH}_3\overset{\text{O}}{\overset{\|}{\text{C}}}\text{OH}}$$

EXAMPLE 10-5 Drawing Structural Formulas of Carboxylic Acids

Draw a condensed structural formula for the single carboxylic acid with the molecular formula $C_3H_6O_2$.

Strategy and Solution

The only way the carbon atoms can be written is three in a chain, and the —COOH group must be on an end carbon of the chain.

$$\text{CH}_3\text{CH}_2\overset{\text{O}}{\overset{\|}{\text{C}}}\text{OH} \quad \text{or} \quad \text{CH}_3\text{CH}_2\text{CO}_2\text{H}$$

Problem 10-5
Draw condensed structural formulas for the two carboxylic acids with the molecular formula $C_4H_8O_2$.

E. Carboxylic Esters

Carboxylic ester A derivative of a carboxylic acid in which a carbon group replaces the H of the carboxyl group

A **carboxylic ester**, commonly referred to as simply an **ester**, is a derivative of a carboxylic acid in which a carbon group replaces the hydrogen of the carboxyl group. The ester group is written —COOR or —CO_2R in this text.

$$\underset{\text{Functional group}}{-\overset{:\text{O}:}{\underset{||}{\text{C}}}-\ddot{\text{O}}-\overset{|}{\underset{|}{\text{C}}}-} \qquad \underset{\substack{\text{Methyl acetate}\\\text{(an ester)}}}{\text{CH}_3-\overset{\text{O}}{\underset{||}{\text{C}}}-\text{O}-\text{CH}_3 \quad \text{or} \quad \text{CH}_3\text{COOCH}_3}$$

EXAMPLE 10-6   Drawing Structural Formulas of Esters

The molecular formula of methyl acetate is $C_3H_6O_2$. Draw the structural formula of another ester with the same molecular formula.

Strategy and Solution

There is only one other ester with this molecular formula. Its structural formula is:

$$\underset{\text{Ethyl formate}}{\text{H}-\overset{\text{O}}{\underset{||}{\text{C}}}-\text{O}-\text{CH}_2-\text{CH}_3}$$

Problem 10-6

Draw structural formulas for the four esters with the molecular formula $C_4H_8O_2$.

F. Amides

An amide is a functional derivative of a carboxylic acid in which an amino group replaces the —OH of the carboxyl group. Alternatively, amide functional groups contain a carbonyl group bonded to a nitrogen atom.

$$\underset{\substack{\text{Acetamide}\\\text{(a 1° amide)}}}{\text{CH}_3\overset{\text{O}}{\underset{||}{\text{C}}}\text{NH}_2} \qquad \underset{\substack{N\text{-Methylacetamide}\\\text{(a 2° amide)}}}{\text{CH}_3\overset{\text{O}}{\underset{||}{\text{C}}}\text{NHCH}_3} \qquad \underset{\substack{N,N\text{-Dimethylformamide}\\\text{(a 3° amide)}}}{\text{H}\overset{\text{O}}{\underset{||}{\text{C}}}\text{N}(\text{CH}_3)_2}$$

The amide group is critically important in nature because it is the link by which amino acids are joined together to form proteins, which are polymers containing many amino acids joined by amide bonds. We will have much more to say about amino acids and proteins in Chapter 22.

Amino acid 1   Amino acid 2   Two amino acids joined by an amide bond

Summary of Key Questions

Section 10-1 What Is Organic Chemistry?
- Organic chemistry is the study of compounds containing carbon.

Section 10-2 Where Do We Obtain Organic Compounds?
- Chemists obtain organic compounds either by isolation from plant and animal sources or by synthesis in the laboratory.

Section 10-3 How Do We Write Structural Formulas of Organic Compounds?
- Carbon normally forms four bonds and has no unshared pairs of electrons. Its four bonds may be four single bonds, two single bonds and one double bond, or one single bond and one triple bond.
- Nitrogen normally forms three bonds and has one unshared pair of electrons. Its bonds may be three single bonds, one single bond and one double bond, or one triple bond.
- Oxygen normally forms two bonds and has two unshared pairs of electrons. Its bonds may be two single bonds or one double bond.

Section 10-4 What Is a Functional Group?
- A **functional group** is a site of chemical reactivity; a particular functional group, in whatever compound it is found, always undergoes the same types of chemical reactions.
- In addition, functional groups are the characteristic structural units by which we both classify and name organic compounds. Important functional groups include the **hydroxyl group** of 1°, 2°, and 3° alcohols; the **amino group** of 1°, 2°, and 3° amines; the **carbonyl group** of aldehydes and ketones; the **carboxyl group** of carboxylic acids; and **ester** and **amide groups.**

Problems

Orange-numbered problems are applied.

Section 10-1 What Is Organic Chemistry?

10-7 Answer true or false.
(a) All organic compounds contain one or more atoms of carbon.
(b) The majority of organic compounds are built from carbon, hydrogen, oxygen, and nitrogen.
(c) By number of atoms, carbon is the most abundant element in the Earth's crust.
(d) Most organic compounds are soluble in water.

Section 10-2 Where Do We Obtain Organic Compounds?

10-8 Answer true or false.
(a) Organic compounds can only be synthesized in living organisms.
(b) Organic compounds synthesized in the laboratory have the same chemical and physical properties as those synthesized in living organisms.
(c) Chemists have synthesized many organic compounds that are not found in nature.

10-9 Is there any difference between vanillin made synthetically and vanillin extracted from vanilla beans?

10-10 Suppose that you are told that organic substances are produced only by living organisms. How would you rebut this assertion?

10-11 What important experiment did Wöhler carry out in 1828?

Section 10-3 How Do We Write Structural Formulas of Organic Compounds?

10-12 Answer true or false.
(a) In organic compounds, carbon normally has four bonds and no unshared pairs of electrons.
(b) When found in organic compounds, nitrogen normally has three bonds and one unshared pair of electrons.
(c) The most common bond angles about carbon in organic compounds are approximately 109.5° and 180°.

10-13 List the four principal elements that make up organic compounds and give the number of bonds each typically forms.

10-14 Think about the types of substances in your immediate environment and make a list of those that are organic—for example, textile fibers. We will ask you to return to this list later in the course and to refine, correct, and possibly expand it.

10-15 How many electrons are in the valence shell of each of the following atoms? Write a Lewis structure for an atom of each element. (*Hint*: Use the Periodic Table.)
(a) Carbon (b) Oxygen
(c) Nitrogen (d) Fluorine

10-16 What is the relationship between the number of electrons in the valence shell of each of the following atoms and the number of covalent bonds each forms?
(a) Carbon (b) Oxygen
(c) Nitrogen (d) Hydrogen

10-17 Write Lewis structures for these compounds. Show all valence electrons. None of them contains a ring of atoms. (*Hint*: Remember that carbon has four bonds, nitrogen has three bonds and one unshared pair of electrons, oxygen has two bonds and two unshared pairs of electrons, and each halogen has one bond and three unshared pairs of electrons.)
(a) H_2O_2 (b) N_2H_4
   Hydrogen peroxide   Hydrazine
(c) CH_3OH (d) CH_3SH
   Methanol   Methanethiol
(e) CH_3NH_2 (f) CH_3Cl
   Methylamine   Chloromethane

10-18 Write Lewis structures for these compounds. Show all valence electrons. None of them contains a ring of atoms.
(a) CH_3OCH_3 (b) C_2H_6
   Dimethyl ether   Ethane
(c) C_2H_4 (d) C_2H_2
   Ethylene   Acetylene
(e) CO_2 (f) CH_2O
   Carbon dioxide   Formaldehyde
(g) H_2CO_3 (h) CH_3COOH
   Carbonic acid   Acetic acid

10-19 Write Lewis structures for these ions.
(a) HCO_3^- (b) CO_3^{2-}
   Bicarbonate ion   Carbonate ion
(c) CH_3COO^- (d) Cl^-
   Acetate ion   Chloride ion

10-20 Why are the following molecular formulas impossible?
(a) CH_5 (b) C_2H_7

Section 10-3 Review of the VSEPR Model

10-21 Explain how to use the valence-shell electron-pair repulsion (VSEPR) model to predict bond angles and geometry about atoms of carbon, oxygen, and nitrogen.

10-22 Suppose you forget to take into account the presence of the unshared pair of electrons on nitrogen in the molecule NH_3. What would you then predict for the H—N—H bond angles and the geometry (bond angles and shape) of ammonia?

10-23 Suppose you forget to take into account the presence of the two unshared pairs of electrons on the oxygen atom of ethanol, CH_3CH_2OH. What would you then predict for the C—O—H bond angle and the geometry of ethanol?

10-24 Use the VSEPR model to predict the bond angles and geometry about each highlighted atom. (*Hint*: Remember to take into account the presence of unshared pairs of electrons.)

(a)
```
    H H
    | |
H—C—C—O—H
    | |
    H H
```

(b)
```
H—C=C—Cl
    | |
    H H
```

(c)
```
    H
    |
H—C—C≡C—H
    |
    H
```

10-25 Use the VSEPR model to predict the bond angles about each highlighted atom.

(a)

(b)
```
    H
    |
H—C—N—H
    | |
    H H
```

(c)

Section 10-4 What Is a Functional Group?

10-26 Answer true or false.
(a) A functional group is a group of atoms in an organic molecule that undergoes a predictable set of chemical reactions.
(b) The functional group of an alcohol, an aldehyde, and a ketone have in common the fact that each contains a single oxygen atom.
(c) A primary alcohol has one —OH group, a secondary alcohol has two —OH groups, and a tertiary alcohol has three —OH groups.
(d) There are two alcohols with the molecular formula C_3H_8O.
(e) There are three amines with the molecular formula C_3H_9N.
(f) Aldehydes, ketones, carboxylic acids, and esters all contain a carbonyl group.
(g) A compound with the molecular formula of C_3H_6O may be either an aldehyde, a ketone, or a carboxylic acid.
(h) Bond angles about the carbonyl carbon of an aldehyde, a ketone, a carboxylic acid, and an ester are all approximately 109.5°.
(i) The molecular formula of the smallest aldehyde is C_3H_6O, and that of the smallest ketone is also C_3H_6O.
(j) The molecular formula of the smallest carboxylic acid is $C_2H_4O_2$.

10-27 What is meant by the term *functional group*?

10-28 List three reasons why functional groups are important in organic chemistry.

10-29 Draw Lewis structures for each of the following functional groups. Show all valence electrons in each functional group.
(a) A carbonyl group
(b) A carboxyl group
(c) A hydroxyl group
(d) A primary amino group
(e) An ester group

10-30 Complete the following structural formulas by adding enough hydrogens to complete the tetravalence of each carbon. Then write the molecular formula of each compound.

(a) C—C=C—C(—C)—C

(b) C—C—C—C(=O)—OH

(c) C—C—C(=O)—C

(d) C—C(—C)—C(=O)—H

(e) C—C(—C)(—C)—C—C—NH$_2$

(f) C—C(—NH$_2$)—C(=O)—OH

(g) C—C(—OH)—C—C—C

(h) C—C(—OH)—C—C(=O)—OH

(i) C=C—C—OH

10-31 What is the meaning of the term *tertiary* (3°) when it is used to classify alcohols?

10-32 Draw a structural formula for the one tertiary (3°) alcohol with the molecular formula $C_4H_{10}O$.

10-33 What is the meaning of the term *tertiary* (3°) when it is used to classify amines?

10-34 Draw condensed structural formulas for all compounds with the molecular formula C_4H_8O that contain a carbonyl group (there are two aldehydes and one ketone).

10-35 Draw structural formulas for each of the following:
(a) The four primary (1°) alcohols with the molecular formula $C_5H_{12}O$.
(b) The three secondary (2°) alcohols with the molecular formula $C_5H_{12}O$.
(c) The one tertiary (3°) alcohol with the molecular formula $C_5H_{12}O$.

10-36 Draw structural formulas for the six ketones with the molecular formula $C_6H_{12}O$.

10-37 Draw structural formulas for the eight carboxylic acids with the molecular formula $C_6H_{12}O_2$.

10-38 Draw structural formulas for each of the following:
(a) The four primary (1°) amines with the molecular formula $C_4H_{11}N$.
(b) The three secondary (2°) amines with the molecular formula $C_4H_{11}N$.
(c) The one tertiary (3°) amine with the molecular formula $C_4H_{11}N$.

Chemical Connections

10-39 (Chemical Connections 10A) How was Taxol discovered?

10-40 (Chemical Connections 10A) In what way does Taxol interfere with cell division?

Additional Problems

10-41 Use the VSEPR model to predict the bond angles about each atom of carbon, nitrogen, and oxygen in these molecules. (*Hint*: First, add unshared pairs of electrons as necessary to complete the valence shell of each atom and then predict the bond angles.)

(a) $CH_3CH_2CH_2OH$
(b)
(c) $CH_3CH=CH_2$
(d) $CH_3C\equiv CCH_3$
(e) CH_3COCH_3 (with C=O)
(f) CH_3NCH_3 (with CH$_3$ branch)

10-42 Silicon is immediately below carbon in Group 4A of the Periodic Table. Predict the C—Si—C bond angles in tetramethylsilane, $(CH_3)_4Si$.

10-43 Phosphorus is immediately below nitrogen in Group 5A of the Periodic Table. Predict the C—P—C bond angles in trimethylphosphine, $(CH_3)_3P$.

10-44 Draw the structure for a compound with the molecular formula:
(a) C_2H_6O that is an alcohol
(b) C_3H_6O that is an aldehyde
(c) C_3H_6O that is a ketone
(d) $C_3H_6O_2$ that is a carboxylic acid

10-45 Draw structural formulas for the eight aldehydes with the molecular formula $C_6H_{12}O$.

10-46 Draw structural formulas for the three tertiary (3°) amines with the molecular formula $C_5H_{13}N$.

10-47 Which of these covalent bonds are polar, and which are nonpolar? (*Hint*: Review Section 3-7B.)
(a) C—C (b) C=C (c) C—H (d) C—O
(e) O—H (f) C—N (g) N—H (h) N—O

10-48 Of the bonds in Problem 10-47, which is the most polar? Which is the least polar?

10-49 Using the symbol $\delta+$ to indicate a partial positive charge and $\delta-$ to indicate a partial negative charge, indicate the polarity of the most polar bond (or bonds if two or more have the same polarity) in each of the following molecules.
(a) CH_3OH
(b) CH_3NH_2
(c) $HSCH_2CH_2NH_2$

(d) CH_3CCH_3
(e)
HCH
(f) CH_3COH

Looking Ahead

10-50 Following is a structural formula and a ball-and-stick model of benzene, C_6H_6.

(a) Predict each H—C—C and C—C—C bond angle in benzene.
(b) Predict the shape of a benzene molecule.

10-51 Following is a structural formula for naphthalene. It was first obtained by heating coal to a high temperature in the absence of air (oxygen). At one time, it was used in "mothballs."

Naphthalene

(a) Predict the shape of naphthalene.
(b) Is a naphthalene molecule polar or nonpolar?

10-52 Identify the functional group(s) in each compound.

(a)
2-Butanone
(a solvent for paints and lacquers)

(b)
Hexanedioic acid
(the second component of nylon-66)

(c)
Lysine
(one of the 20 amino acid building blocks of proteins)

(d)
Dihydroxyacetone
(a component of several artificial tanning lotions)

10-53 Consider molecules with the molecular formula $C_4H_8O_2$. Write the structural formula for a molecule with this molecular formula that contains:
(a) A carboxyl group
(b) An ester group
(c) A ketone group and a 2° alcohol
(d) An aldehyde and a 3° alcohol

10-54 Urea, $(NH_2)_2CO$, is used in plastics and in fertilizers. It is also the primary nitrogen-containing substance excreted by humans.

Urea

(a) Complete the Lewis structure of urea, showing all valence electrons.
(b) Predict the bond angle about each C and N.
(c) Which is the most polar bond in the molecule?
(d) Is urea polar or nonpolar?

10-55 The compound drawn here is lactic acid, a natural compound found in sour milk.

$$\begin{array}{c} H\ H\ O \\ |\ |\ \| \\ H-C-C-C-O-H \\ |\ | \\ H\ O \\ \ \ \ \ \backslash H \end{array}$$

Lactic acid

(a) What is the molecular formula of lactic acid?
(b) Name the two functional groups in lactic acid.
(c) Predict the bond angles about each carbon atom.

(d) Which bonds are polar, and which are nonpolar?
(e) Would you predict that lactic acid is polar or nonpolar?

Tying It Together

10-56 Following is the structural formula of acetylsalicylic acid, better known by its common name aspirin.

Acetylsalicylic acid
(Aspirin)

(a) Name the two oxygen-containing functional groups in aspirin.
(b) What is the molecular formula of aspirin?

10-57 Aspirin is prepared by the reaction of salicylic acid with acetic anhydride as shown in the following equation. The stoichiometry of the reaction is given in the equation. Acetic acid is a by-product of the reaction and must be separated and removed so that aspirin can then be sold as a pure product. How many grams of aspirin can be prepared from 120 grams of salicylic acid? Assume that there is an excess of acetic anhydride. (Chapter 4)

Salicylic acid + Acetic anhydride

↓

Acetylsalicylic acid (Aspirin) + Acetic acid

10-58 Following is the structural formula of acetamide.

Acetamide
(an amide)

(a) Complete the Lewis structure for acetamide, showing all valence electrons.
(b) Use the valence-shell electron-pair repulsion (VSEPR) model (Section 3-10) to predict all bond angles in acetamide.
(c) Which is the most polar bond in acetamide?

10-59 The amide group, in acetamide as well as in all other amides, is best represented as a resonance hybrid (Section 3-9). Following are two contributing structures for the hybrid.

(a)   (b)

(a) Show by the use of curved arrows how contributing structure (a) is converted into contributing structure (b).
(b) Notice that structure (b) contains an oxygen atom with one bond and three unshared pairs of electrons and that this oxygen bears a negative charge. Compare the Lewis structure of this oxygen with the oxygen atom in the hydroxide ion.
(c) Notice that the nitrogen atom of structure (b) has four bonds and bears a positive charge. Compare the Lewis structure and bonding of this nitrogen with the nitrogen in the ammonium ion, NH_4^+.
(d) If the acetamide hybrid is best represented by contributing structure (a), predict the H—N—H bond angle.
(e) If, on the other hand, the acetamide hybrid is best represented by contributing structure (b), predict the H—N—H bond angle.
(f) Proteins are molecules that can be described as polyamides (Chapter 22). Linus Pauling, in his pioneering studies on the structure of proteins, discovered that the actual H—N—H bond angle in each amide bond of a protein is 120°. What does this fact tell you about the relative importance of contributing structures (a) and (b) in the resonance hybrid?

Alkanes

Bunsen burners burn natural gas, which is primarily methane, with small amounts of ethane, propane, and butane.

KEY QUESTIONS

- 11-1 What Are Alkanes?
- 11-2 How Do We Write Structural Formulas of Alkanes?
- 11-3 What Are Constitutional Isomers?
- 11-4 How Do We Name Alkanes?
- 11-5 Where Do We Obtain Alkanes?
- 11-6 What Are Cycloalkanes?
- 11-7 What Are the Shapes of Alkanes and Cycloalkanes?
- *How To . . .* Draw Alternative Chair Conformations of Cyclohexane
- 11-8 What Is *Cis-Trans* Isomerism in Cycloalkanes?
- 11-9 What Are the Physical Properties of Alkanes and Cycloalkanes?
- 11-10 What Are the Characteristic Reactions of Alkanes?
- 11-11 What Are Some Important Haloalkanes?

11-1 What Are Alkanes?

In this chapter, we examine the physical and chemical properties of **alkanes**, the simplest type of organic compounds. Actually, alkanes are members of a larger class of organic compounds called hydrocarbons. A **hydrocarbon** is a compound composed of only carbon and hydrogen. Figure 11-1 shows the four classes of hydrocarbons, along with the characteristic type of bonding between carbon atoms in each class. Alkanes are **saturated hydrocarbons**; that is, they contain only carbon–carbon single bonds. Saturated in this context means that each carbon in the hydrocarbon has the maximum number of hydrogens bonded to it. A hydrocarbon that contains one or more carbon–carbon double bonds, triple bonds, or benzene rings is classified as an **unsaturated hydrocarbon**. We study alkanes (saturated hydrocarbons) in this chapter, and alkenes, alkynes, and arenes (unsaturated hydrocarbons) in Chapters 12 and 13.

We often refer to alkanes as **aliphatic hydrocarbons** because the physical properties of the higher members of this class resemble those of the long carbon-chain molecules we find in animal fats and plant oils (Greek: *aleiphar*, fat or oil).

Alkanes Saturated hydrocarbons whose carbon atoms are arranged in a chain

Hydrocarbon A compound that contains only carbon and hydrogen atoms

Saturated hydrocarbons Hydrocarbons that contain only carbon–carbon single bonds

Aliphatic hydrocarbons Alkanes

FIGURE 11-1 The four classes of hydrocarbons.

11-2 How Do We Write Structural Formulas of Alkanes?

Methane, CH_4, and ethane, C_2H_6, are the first two members of the alkane family. Lewis structures and ball-and-stick models for these molecules are shown below. The shape of methane is tetrahedral, and all H—C—H bond angles are 109.5°. Each carbon atom in ethane is also tetrahedral, and the bond angles in it are all approximately 109.5° as well. Although the three-dimensional shapes of larger alkanes are more complex than those of methane and ethane, the four bonds about each carbon atom are still arranged in a tetrahedral manner, and all bond angles are still approximately 109.5°.

Line-angle formula An abbreviated way to draw structural formulas in which each vertex and line terminus represents a carbon atom and each line represents a bond

The next members of the alkane family are propane, butane, and pentane. In the following representations, these hydrocarbons are drawn as condensed structural formulas, which show all carbons and hydrogens. They can also be drawn in a more abbreviated form called a **line-angle formula**. In this type of representation, a line represents a carbon–carbon bond and a vertex represents a carbon atom. A line ending in space represents a —CH_3 group. To count hydrogens from a line-angle formula, you simply add enough hydrogens in your mind to give each carbon its required four bonds. Chemists use line-angle formulas because they are easier and faster to draw than condensed structural formulas.

Structural formulas for alkanes can be written in yet another condensed form. For example, the structural formula of pentane contains three CH_2 (**methylene**) groups in the middle of the chain. We can group them together and write the structural formula $CH_3(CH_2)_3CH_3$. Table 11-1 gives names and molecular formulas for the first ten alkanes with unbranched chains. Note that the names of all these alkanes end in -*ane*. We will have more to say about naming alkanes in Section 11-4.

	Ball-and-stick model		
Line-angle formula			
Condensed structural formula	$CH_3CH_2CH_3$ Propane	$CH_3CH_2CH_2CH_3$ Butane	$CH_3CH_2CH_2CH_2CH_3$ Pentane

Table 11-1 The First Ten Alkanes with Unbranched Chains

Name	Molecular Formula	Condensed Structural Formula	Name	Molecular Formula	Condensed Structural Formula
methane	CH_4	CH_4	hexane	C_6H_{14}	$CH_3(CH_2)_4CH_3$
ethane	C_2H_6	CH_3CH_3	heptane	C_7H_{16}	$CH_3(CH_2)_5CH_3$
propane	C_3H_8	$CH_3CH_2CH_3$	octane	C_8H_{18}	$CH_3(CH_2)_6CH_3$
butane ◄	C_4H_{10}	$CH_3(CH_2)_2CH_3$	nonane	C_9H_{20}	$CH_3(CH_2)_7CH_3$
pentane	C_5H_{12}	$CH_3(CH_2)_3CH_3$	decane	$C_{10}H_{22}$	$CH_3(CH_2)_8CH_3$

EXAMPLE 11-1 Drawing Line-Angle Formulas

Table 11-1 gives the condensed structural formula for hexane. Draw a line-angle formula for this alkane and number the carbons on the chain beginning at one end and proceeding to the other end.

Strategy and Solution

Hexane contains six carbons in a chain. Its line-angle formula is:

Problem 11-1

Following is a line-angle formula for an alkane. What is the name and the molecular formula of this alkane?

▶ Butane, $CH_3CH_2CH_2CH_3$, is the fuel in this lighter. Butane molecules are present in both the liquid and gaseous states in the lighter.

11-3 What Are Constitutional Isomers?

Constitutional isomers are compounds that have the same molecular formula but different structural formulas. By "different structural formulas," we mean that they differ in the kinds of bonds (single, double, or triple) and/or in the connectivity of their atoms. For the molecular formulas CH_4, C_2H_6, and C_3H_8, only one connectivity of their atoms is possible, so there are no constitutional isomers for these molecular formulas. For the molecular formula C_4H_{10}, two structural formulas are possible: in butane, the four carbons are bonded in a chain; in 2-methylpropane, three carbons are bonded in a chain and the fourth carbon is a branch on the chain. The

Constitutional isomers Compounds with the same molecular formula but a different order of attachment (connectivity) of their atoms. Constitutional isomers have also been called structural isomers, an older term that is still in use.

two constitutional isomers with the molecular formula C_4H_{10} are drawn here both as condensed structural formulas and as line-angle formulas. Also shown are ball-and-stick models of each.

CH₃CH₂CH₂CH₃
Butane
(bp −0.5°C)

$$\begin{array}{c} CH_3 \\ | \\ CH_3CHCH_3 \end{array}$$

2-Methylpropane
(bp −11.6°C)

Butane and 2-methylpropane are different compounds and have different physical and chemical properties. Their boiling points, for example, differ by approximately 11°C.

In Section 10-4, we encountered several examples of constitutional isomers, although we did not call them that at the time. We saw, for example, that there are two alcohols with the molecular formula C_3H_8O, two primary amines with the molecular formula C_3H_9N, two aldehydes with the molecular formula C_4H_8O, and two carboxylic acids with the molecular formula $C_4H_8O_2$.

To determine whether two or more structural formulas represent constitutional isomers, write the molecular formula of each and then compare them. All compounds that have the same molecular formula but different structural formulas are constitutional isomers.

EXAMPLE 11-2 Constitutional Isomerism

Do the structural formulas in each of the following sets represent the same compound or constitutional isomers? (*Hint*: You will find it helpful to redraw each molecule as a line-angle formula, which will make it easier for you to see similarities and differences in molecular structure.)

(a) CH₃CH₂CH₂CH₂CH₂CH₃   and   CH₃CH₂CH₂
                                           |
                                           CH₂CH₂CH₃   (Each is C_6H_{14})

(b)
$$\begin{array}{c} CH_3 \\ | \\ CH_3CHCH_2CH \\ | \quad\quad | \\ CH_3 \quad CH_3 \end{array}$$
and
$$\begin{array}{c} CH_3 \\ | \\ CH_3CH_2CHCHCH_3 \\ | \\ CH_3 \end{array}$$
(Each is C_7H_{16})

Strategy

First, find the longest chain of carbon atoms. It makes no difference whether the chain is drawn as straight or bent; as structural formulas are drawn in this problem, there is no attempt to show three-dimensional shapes. Second, number the longest chain from the end nearest the first branch. Third, compare the lengths of the two chains and the sizes and locations of any branches. Structural formulas that have the same molecular formula and the same connectivity of their atoms represent the same compound; those that have the same molecular formula but different connectivities of their atoms represent constitutional isomers.

Solution

(a) Each structural formula has an unbranched chain of six carbons; they are identical and represent the same compound.

$$CH_3CH_2CH_2CH_2CH_2CH_3 \quad \text{and} \quad CH_3CH_2CH_2CH_2CH_2CH_3$$

(b) Each structural formula has the same molecular formula, C_7H_{16}. In addition, each has a chain of five carbons with two CH_3 branches. Although the branches are identical, they are at different locations on the chains. Therefore, these structural formulas represent constitutional isomers.

Problem 11-2

Do the line-angle formulas in each of the following sets represent the same compound or constitutional isomers?

(a) [line-angle structures] and [line-angle structures]

(b) [line-angle structures] and [line-angle structures]

EXAMPLE 11-3 Constitutional Isomerism

Draw line-angle formulas for the five constitutional isomers with the molecular formula C_6H_{14}.

Strategy

In solving problems of this type, you should devise a strategy and then follow it. Here is one possible strategy. First, draw a line-angle formula for the constitutional isomer with all six carbons in an unbranched chain. Then, draw line-angle formulas for all constitutional isomers with five carbons in a chain and one carbon as a branch on the chain. Finally, draw line-angle formulas for all constitutional isomers with four carbons in a chain and two carbons as branches.

Solution

Here are line-angle formulas for all constitutional isomers with six, five, and four carbons in the longest chain. No constitutional isomers for C_6H_{14} having only three carbons in the longest chain are possible.

Six carbons in an unbranched chain

Five carbons in a chain; one carbon as a branch

Four carbons in a chain; two carbons as branches

Problem 11-3

Draw structural formulas for the three constitutional isomers with the molecular formula C_5H_{12}.

Molecular Formula	Number of Constitutional Isomers
CH_4	1
C_5H_{12}	3
$C_{10}H_{22}$	75
$C_{15}H_{32}$	4347
$C_{25}H_{52}$	36,797,588
$C_{30}H_{62}$	4,111,846,763

The ability of carbon atoms to form strong, stable bonds with other carbon atoms results in a staggering number of constitutional isomers, as the table shows.

Thus, for even a small number of carbon and hydrogen atoms, a very large number of constitutional isomers is possible. In fact, the potential for structural and functional group individuality among organic molecules made from just the basic building blocks of carbon, hydrogen, nitrogen, and oxygen is practically limitless.

11-4 How Do We Name Alkanes?

A. The IUPAC System

Ideally, every organic compound should have a name from which its structural formula can be drawn. For this purpose, chemists have adopted a set of rules established by the **International Union of Pure and Applied Chemistry (IUPAC)**.

The IUPAC name for an alkane with an unbranched chain of carbon atoms consists of two parts: (1) a prefix that shows the number of carbon atoms in the chain and (2) the suffix **-ane**, which shows that the compound is a saturated hydrocarbon. Table 11-2 gives the prefixes used to show the presence of 1 to 20 carbon atoms.

The IUPAC chose the first four prefixes listed in Table 11-2 because they were well established long before the nomenclature was systematized. For example, the prefix *but-* appears in the name butyric acid, a compound of four carbon atoms formed by the air oxidation of butterfat (Latin: *butyrum*,

Table 11-2 Prefixes Used in the IUPAC System to Show the Presence of 1 to 20 Carbons in an Unbranched Chain

Prefix	Number of Carbon Atoms	Prefix	Number of Carbon Atoms	Prefix	Number of Carbon Atoms	Prefix	Number of Carbon Atoms
meth-	1	hex-	6	undec-	11	hexadec-	16
eth-	2	hept-	7	dodec-	12	heptadec-	17
prop-	3	oct-	8	tridec-	13	octadec-	18
but-	4	non-	9	tetradec-	14	nonadec-	19
pent-	5	dec-	10	pentadec-	15	eicos-	20

butter). The prefixes that denote five or more carbons are derived from Latin numbers. Table 11-1 gives the names, molecular formulas, and condensed structural formulas for the first ten alkanes with unbranched chains.

IUPAC names of alkanes with branched chains consist of a parent name that shows the longest chain of carbon atoms and substituent names that indicate the groups bonded to the parent chain. For example:

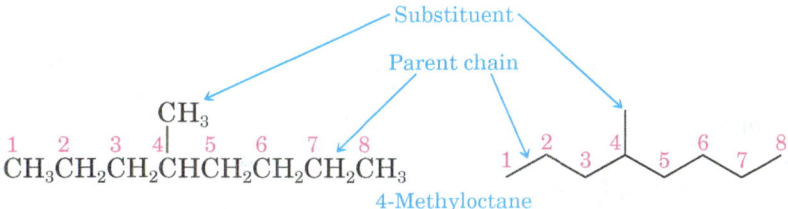

4-Methyloctane

A substituent group derived from an alkane by removal of a hydrogen atom is called an **alkyl group** and is commonly represented by the symbol **R—**. Alkyl groups are named by dropping the *-ane* from the name of the parent alkane and adding the suffix *-yl*. Table 11-3 gives the names and condensed structural formulas for eight of the most common alkyl groups. The prefix *sec-* is an abbreviation for "secondary," meaning a carbon bonded to two other carbons. The prefix *tert-* is an abbreviation for "tertiary," meaning a carbon bonded to three other carbons.

Alkyl group A group derived by removing a hydrogen from an alkane; given the symbol R—

R— A symbol used to represent an alkyl group

Table 11-3 Names of the Eight Most Common Alkyl Groups

Name	Condensed Structural Formula	Name	Condensed Structural Formula
methyl	—CH₃	butyl	—CH₂CH₂CH₂CH₃
ethyl	—CH₂CH₃	isobutyl	—CH₂CHCH₃ | CH₃
propyl	—CH₂CH₂CH₃	sec-butyl	—CHCH₂CH₃ | CH₃
isopropyl	—CHCH₃ | CH₃	tert-butyl	CH₃ | —CCH₃ | CH₃

The rules of the IUPAC system for naming alkanes are as follows:

1. The name for an alkane with an unbranched chain of carbon atoms consists of a prefix showing the number of carbon atoms in the parent chain and the suffix *-ane*.
2. For branched-chain alkanes, take the longest chain of carbon atoms as the parent chain and its name becomes the root name.
3. Give each substituent on the parent chain a name and a number. The number shows the carbon atom of the parent chain to which the substituent is bonded. Use a hyphen to connect the number to the name.

CH₃
|
CH₃CHCH₃
2-Methylpropane

4. If there is one substituent, number the parent chain from the end that gives the substituent the lower number.

$$\underset{\substack{\text{2-Methylpentane}\\ \text{(not 4-methylpentane)}}}{CH_3CH_2CH_2\overset{\overset{\displaystyle CH_3}{|}}{C}HCH_3}$$

5. If the same substituent occurs more than once, number the parent chain from the end that gives the lower number to the substituent encountered first. Indicate the number of times the substituent occurs by a prefix *di-, tri-, tetra-, penta-, hexa-*, and so on. Use a comma to separate position numbers.

$$\underset{\substack{\text{2,4-Dimethylhexane}\\ \text{(not 3,5-dimethylhexane)}}}{CH_3CH_2\overset{\overset{\displaystyle CH_3}{|}}{C}HCH_2\overset{\overset{\displaystyle CH_3}{|}}{C}HCH_3}$$

6. If there are two or more different substituents, list them in alphabetical order and number the chain from the end that gives the lower number to the substituent encountered first. If there are different substituents in equivalent positions on opposite ends of the parent chain, give the substituent first in alphabetical order the lower number.

$$\underset{\substack{\text{3-Ethyl-5-methylheptane}\\ \text{(not 3-methyl-5-ethylheptane)}}}{CH_3CH_2\overset{\overset{\displaystyle CH_3}{|}}{C}HCH_2\underset{\underset{\displaystyle CH_2CH_3}{|}}{C}HCH_2CH_3}$$

7. Do not include the prefixes *di-, tri-, tetra-*, and so on, or the hyphenated prefixes *sec-* and *tert-* in alphabetizing. Alphabetize the names of substituents first and then insert these prefixes. In the following example, the alphabetizing parts are ethyl and methyl, not *ethyl* and *dimethyl*.

$$\underset{\substack{\text{4-Ethyl-2,2-dimethylhexane}\\ \text{(not 2,2-dimethyl-4-ethylhexane)}}}{\underset{\underset{\displaystyle CH_3}{|}}{CH_3\overset{\overset{\displaystyle CH_3}{|}}{C}CH_2\overset{\overset{\displaystyle CH_2CH_3}{|}}{C}HCH_2CH_3}}$$

Interactive Example 11-4   IUPAC Names of Alkanes

Write the molecular formula and IUPAC name for each alkane.

(a)   (b)

Strategy

If there is only one substituent on the parent chain, as in (a), number the parent chain from the end that gives the substituent the lowest possible number. If there are two or more substituents on the parent chain, as in (b), number the parent chain from the end that gives the substituent first in alphabetical order the lowest possible number.

Solution

The molecular formula of (a) is C_5H_{12}, and that of (b) is $C_{11}H_{24}$. In (a), number the longest chain from the end that gives the methyl substituent the lower number (rule 4). In (b), list isopropyl and methyl substituents in alphabetical order (rule 6).

(a) 2-Methylbutane

(b) 4-Isopropyl-2-methylheptane

Problem 11-4

Write the molecular formula and IUPAC name for each alkane.

B. Common Names

In the older system of **common nomenclature**, the total number of carbon atoms in an alkane, regardless of their arrangement, determines the name.

The first three alkanes are methane, ethane, and propane. All alkanes with the molecular formula C_4H_{10} are called butanes, all those with the molecular formula C_5H_{12} are called pentanes, and all those with the molecular formula C_6H_{14} are called hexanes. For alkanes beyond propane, **iso** indicates that one end of an otherwise unbranched chain terminates in a $(CH_3)_2CH-$group. Following are examples of common names:

$CH_3CH_2CH_2CH_3$ Butane

$CH_3\overset{\overset{\displaystyle CH_3}{|}}{C}HCH_3$ Isobutane

$CH_3CH_2CH_2CH_2CH_3$ Pentane

$CH_3CH_2\overset{\overset{\displaystyle CH_3}{|}}{C}HCH_3$ Isopentane

This system of common names has no way of handling other branching patterns; therefore, for more complex alkanes, we must use the more flexible IUPAC system.

In this book, we concentrate on IUPAC names. From time to time, however, we also use common names, especially when chemists and biochemists use them almost exclusively in everyday discussions. When the text gives both IUPAC and common names for a compound, we will always give the IUPAC name first, followed by the common name in parentheses. In this way, you should have no doubt about which name is which.

▶ A petroleum fractional distillation tower.

Cycloalkane A saturated hydrocarbon that contains carbon atoms bonded to form a ring

11-5 Where Do We Obtain Alkanes?

The two major sources of alkanes are natural gas and petroleum. **Natural gas** consists of approximately 90 to 95% methane, 5 to 10% ethane, and a mixture of other relatively low-boiling alkanes—chiefly propane, butane, and 2-methylpropane.

Petroleum is a thick, viscous liquid mixture of thousands of compounds, most of them hydrocarbons, formed from the decomposition of marine plants and animals. Petroleum and petroleum-derived products fuel automobiles, aircraft, and trains. They provide most of the greases and lubricants required for the machinery utilized by our highly industrialized society. Furthermore, petroleum, along with natural gas, provides nearly 90% of the organic raw materials for the synthesis and manufacture of synthetic fibers, plastics, detergents, drugs, dyes, and a multitude of other products. ◀

The fundamental separation process in refining petroleum is fractional distillation (Figure 11-2). Practically all crude petroleum that enters a refinery goes to distillation units, where it is heated to temperatures as high as 370 to 425°C and separated into fractions. Each fraction contains a mixture of hydrocarbons that boils within a particular range.

11-6 What Are Cycloalkanes?

A hydrocarbon that contains carbon atoms joined to form a ring is called a **cyclic hydrocarbon**. When all carbons of the ring are saturated (only carbon–carbon single bonds are present), the hydrocarbon is called a **cycloalkane**. Cycloalkanes with ring sizes ranging from 3 to more than 30 carbon atoms are found in nature, and in principle there is no limit to ring size. Five-membered (cyclopentane) and six-membered (cyclohexane) rings are especially abundant in nature; for this reason, we concentrate on them in this book.

FIGURE 11-2 Fractional distillation of petroleum. The lighter, more volatile fractions are removed from higher up the column; the heavier, less volatile fractions are removed from lower down.

FIGURE 11-3 Examples of cycloalkanes.

H₂C—CH₂
| | or ☐
H₂C—CH₂

Cyclobutane

H₂C⟨CH₂⟩CH₂ or ⬠
H₂C—CH₂

Cyclopentane

H₂C⟨CH₂⟩CH₂ or ⬡
H₂C⟨CH₂⟩CH₂

Cyclohexane

Organic chemists rarely show all carbons and hydrogens when writing structural formulas for cycloalkanes. Rather, we use line-angle formulas to represent cycloalkane rings and represent each ring by a regular polygon having the same number of sides as there are carbon atoms in the ring. For example, we represent cyclobutane by a square, cyclopentane by a pentagon, and cyclohexane by a hexagon (Figure 11-3).

To name a cycloalkane, prefix the name of the corresponding open-chain alkane with *cyclo-* and name each substituent on the ring. If there is only one substituent on the ring, there is no need to give it a location number. If there are two or more substituents, number the ring beginning with the substituent first in alphabetical order.

EXAMPLE 11-5   IUPAC Names of Cycloalkanes

Write the molecular formula and IUPAC name for each cycloalkane.

(a)   (b)

Strategy

For cycloalkanes, the parent name of the ring is the prefix *cyclo-*, plus the name of the alkane with the same number of carbon atoms as are in the ring. If there is only one substituent on the ring, as in (a), there is no need to give it a number. If there are two or more substituents on the ring, as in (b), number the carbon atoms of the ring beginning at the carbon with the substituent first in alphabetical order. If there are three or more substituents, number the atoms of the ring so as to give the substituents the smallest set of numbers and then list them in alphabetical order.

Solution

(a) The molecular formula of this compound is C_8H_{16}. Because there is only one substituent, there is no need to number the atoms of the ring. The IUPAC name of this cycloalkane is isopropylcyclopentane.

(b) The molecular formula is $C_{11}H_{22}$. To name this compound, first number the atoms of the cyclohexane ring beginning with *tert*-butyl, the substituent first in alphabetical order (remember, alphabetical order here is determined by the *b* of butyl and not by the *t* of *tert*-). The name of this cycloalkane is 1-*tert*-butyl-4-methylcyclohexane.

Problem 11-5

Write the molecular formula and IUPAC name for each cycloalkane.

(a) [cyclopentane with isobutyl group]   (b) [cycloheptane with sec-butyl group]   (c) [cyclopropane with tert-butyl group]

11-7 What Are the Shapes of Alkanes and Cycloalkanes?

In this section, we concentrate on ways to visualize molecules as three-dimensional objects and to visualize bond angles and relative distances between various atoms and functional groups within a molecule. At this point, you should review Section 3-10 and use of the valence-shell electron-pair repulsion (VSEPR) model to predict bond angles and shapes of molecules. We urge you to build molecular models of these compounds and to study and manipulate those models. Organic molecules are three-dimensional objects, and it is essential that you become comfortable in dealing with them as such.

A. Alkanes

Although the VSEPR model gives us a way to predict the geometry about each carbon atom in an alkane, it provides us with no information about the three-dimensional shape of an entire molecule. There is, in fact, free rotation about each carbon–carbon bond in an alkane. As a result, even a molecule as simple as ethane has an infinite number of possible three-dimensional shapes, or **conformations**.

Figure 11-4 shows three conformations for a butane molecule. Conformation (a) is the most stable because the methyl groups at the ends of the four-carbon chain are farthest apart. Conformation (b) is formed by a rotation of 120° about the single bond joining carbons 2 and 3. In this conformation, some crowding of groups occurs because the two methyl groups are closer together than they are in conformation (a). Rotation about the C_2—C_3 single bond by another 60° gives conformation (c), which is the most crowded because the two methyl groups face each other.

Figure 11-4 shows only three of the possible conformations for a butane molecule. In fact, there are an infinite number of possible conformations that differ only in the angles of rotation about the various C—C bonds within the molecule. In an actual sample of butane, the conformation of each molecule constantly changes as a result of the molecule's collisions with other butane molecules and with the walls of the container. Even so, at any given time, a majority of butane molecules are in the most stable, fully extended conformation. There are the fewest butane molecules in the most crowded conformation.

To summarize, for any alkane (except, of course, for methane), there are an infinite number of conformations. The majority of molecules in any sample will be in the least crowded conformation; the fewest will be in the most crowded conformation.

Conformations Any three-dimensional arrangements of atoms in a molecule that result from rotation about a single bond

(a) Least crowded conformation; methyl groups are farthest apart.

(b) Intermediate crowding; methyl groups are closer to each other.

(c) Most crowded conformation; methyl groups are closest to each other.

FIGURE 11-4 Three conformations of a butane molecule.

B. Cycloalkanes

We limit our discussion to the conformations of cyclopentanes and cyclohexanes because they are the carbon rings most commonly found in nature. Nonplanar or "puckered" conformations are favored in all cycloalkanes larger than cyclopropane.

Cyclopentane

The most stable conformation of cyclopentane is the **envelope conformation** shown in Figure 11-5. In it, four carbon atoms are in a plane and the fifth carbon atom is bent out of the plane, like an envelope with its flap bent upward. All bond angles in cyclopentane are approximately 109.5°.

FIGURE 11-5 The most stable conformation of cyclopentane.

Cyclohexane

The most stable conformation of cyclohexane is the **chair conformation**, in which all bond angles are approximately 109.5°.

In a chair conformation, the 12 C—H bonds are arranged in two different orientations. Six of them are **axial** bonds, and the other six are **equatorial** orientations. One way to visualize the difference between these two types of bonds is to imagine an axis running through the center of the chair (Figure 11-6). Axial bonds are oriented parallel to this axis. Three of the axial bonds point up; the other three point down. Notice also that axial bonds alternate, first up and then down, as you move from one carbon to the next.

Equatorial bonds are oriented approximately perpendicular to the imaginary axis of the ring and also alternate first slightly up and then slightly down as you move from one carbon to the next. Notice also that if the axial bond on a carbon points upward, the equatorial bond on that carbon points slightly downward. Conversely, if the axial bond on a particular carbon points downward, the equatorial bond on that carbon points slightly upward.

Finally, notice that each equatorial bond is oriented parallel to two ring bonds on opposite sides of the ring. The three pairs of parallel C—H bonds are shown in the following structural formulas, along with the two ring bonds to which each C—H pair is parallel.

Axial A position on a chair conformation of a cyclohexane ring that extends from the ring parallel to the imaginary axis of the ring

Equatorial A position on a chair conformation of a cyclohexane ring that extends from the ring roughly perpendicular to the imaginary axis of the ring

(a) Ball-and-stick model showing all 12 hydrogens

(b) The six equatorial C—H bonds shown in red

(c) The six axial C—H bonds shown in blue

FIGURE 11-6 Chair conformation of cyclohexane showing equatorial and axial C—H bonds.

HOW TO...

Draw Alternative Chair Conformations of Cyclohexane

You will be asked frequently to draw three-dimensional representations of chair conformations of cyclohexane and to show spatial relationships among atoms and groups of atoms bonded to the ring. Here are four steps that will help you to draw them. With a little practice, you will find it easy to draw them.

Step 1: Draw two sets of parallel lines, one line in each set offset from the other in the set as shown.

Step 2: Complete the chair by drawing the head and foot pieces, one up and the other down.

Step 3: Draw the equatorial bonds using ring bonds as a guide. Remember that each equatorial bond is parallel to two ring bonds and that equatorial bonds on opposite carbons of the ring are parallel to one another. Sets of parallel equatorial bonds are shown here in color.

Equatorial bonds

Step 4: Draw the six axial bonds as vertical lines. Remember that all axial bonds are parallel to each other. Sets of parallel axial bonds are shown in color.

Axial bonds

EXAMPLE 11-6 Chair Conformations of Cyclohexanes

Following is a chair conformation of methylcyclohexane showing a methyl group and one hydrogen. Indicate by a label whether each is equatorial or axial.

Strategy

Equatorial bonds are approximately perpendicular to the imaginary axis of the ring and form an equator about the ring. Axial bonds are parallel to the imaginary axis of the ring.

Solution

The methyl group is axial, and the hydrogen is equatorial.

Problem 11-6

Following is a chair conformation of cyclohexane with carbon atoms numbered 1 through 6. Draw methyl groups that are equatorial on carbons 1, 2, and 4.

Suppose that —CH_3 or another group on a cyclohexane ring can occupy either an equatorial or an axial position. Chemists have discovered that a six-membered ring is more stable when the maximum number of substituent groups are equatorial. Perhaps the simplest way to confirm this

---

CHEMICAL CONNECTIONS 11A

The Poisonous Puffer Fish

Nature is by no means limited to carbon in six-membered rings. Tetrodotoxin, one of the most potent toxins known, is composed of a set of interconnected six-membered rings, each in a chair conformation. All but one of these rings contains atoms other than carbon.

Tetrodotoxin is produced in the liver and ovaries of many species of *Tetraodontidae*, one of which is the puffer fish, so called because it inflates itself to an almost spherical spiny ball when alarmed. It is evidently a species highly preoccupied with defense, but the Japanese are not put off by its prickly appearance. They regard the puffer, called *fugu* in Japanese, as a delicacy. To serve it in a public restaurant, a chef must be registered as sufficiently skilled in removing the toxic organs so as to make the flesh safe to eat.

Tetrodotoxin blocks the sodium ion channels, which are essential for neurotransmission (Section 24-3). This

A puffer fish with its body inflated.

blockage prevents communication between neurons and muscle cells and results in weakness, paralysis, and eventual death.

Tetrodotoxin

Test your knowledge with Problem 11-58.

(a) Equatorial methylcyclohexane (b) Axial methylcyclohexane

FIGURE 11-7 Methylcyclohexane. The three hydrogens of the methyl group are shown in green to make them stand out more clearly.

relationship is to examine molecular models. Figure 11-7(a) shows a space-filling model of methylcyclohexane with the methyl group in an equatorial position. In this position, the methyl group is as far away as possible from other atoms of the ring. When methyl is axial [Figure 11-7(b)], it quite literally bangs into two axial hydrogen atoms on the same side of the ring. Thus, the more stable conformation of a substituted cyclohexane ring has a maximum number of its substituents groups in equatorial positions.

11-8 What Is *Cis-Trans* Isomerism in Cycloalkanes?

Cis-trans isomers Isomers that have the same connectivity of their atoms but a different arrangement of their atoms in space due to the presence of either a ring or a carbon–carbon double bond (Chapter 12)

Cycloalkanes with substituents on two or more carbons of the ring show a type of isomerism called **cis-trans isomerism**. Cycloalkane **cis-trans isomers** have (1) the same molecular formula and (2) the same connectivity of their atoms, but (3) a different arrangement of their atoms in space because of restricted rotation around the carbon–carbon single bonds of the ring. We study *cis-trans* isomerism in cycloalkanes in this chapter and that of alkenes in Chapter 12.

We can illustrate *cis-trans* isomerism in cycloalkanes by using 1,2-dimethylcyclopentane as an example. Planar representations of five- and six-membered rings are not spatially accurate because these rings normally exist as envelope and chair conformations. Planar representations are, however, adequate for showing *cis-trans* isomerism. In the following structural formulas, the cyclopentane ring is drawn as a planar pentagon viewed through the plane of the ring. Carbon–carbon bonds of the ring projecting toward you are shown as heavy lines. When viewed from this perspective, substituents bonded to the cyclopentane ring project above and below the plane of the ring. In one isomer of 1,2-dimethylcyclopentane, the methyl groups are on the same side of the ring (either both above or both below the plane of the ring); in the other isomer, they are on opposite sides of the ring (one above and one below the plane of the ring).

The prefix *cis* (Latin: on the same side) indicates that the substituents are on the same side of the ring; the prefix *trans* (Latin: across) indicates that they are on opposite sides of the ring.

cis-1,2-Dimethyl-cyclopentane

trans-1,2-Dimethyl-cyclopentane

Alternatively, we can view the cyclopentane ring as a regular pentagon seen from above, with the ring in the plane of the page. Substituents on the ring then either project toward you (that is, they project up above the page) and are shown by solid wedges or project away from you (project down below the page) and are shown by broken wedges. In the following structural formulas, we show only the two methyl groups; we do not show hydrogen atoms of the ring.

cis-1,2-Dimethyl-cyclopentane

trans-1,2-Dimethyl-cyclopentane

We say that 1,2-dimethylcyclopentane has two stereocenters. A **stereocenter** is a tetrahedral atom, most commonly carbon, at which exchange of two groups produces a stereoisomer. Both carbons 1 and 2 of 1,2-dimethylcyclopentane, for example, are stereocenters; in this molecule, exchange of H and CH₃ groups at either stereocenter converts a *trans* isomer to a *cis* isomer, or vice versa.

Alternatively, we refer to the stereoisomers of 1,2-dimethylcyclopentane as having either a *cis* or a *trans* configuration. **Configuration** refers to the arrangement of atoms about a stereocenter. We say, for example, that exchange of groups at either stereocenter in the *cis* configuration gives the isomer with the *trans* configuration.

Cis and *trans* isomers are also possible for 1,4-dimethylcyclohexane. We can draw a cyclohexane ring as a planar hexagon and view it through the plane of the ring. Alternatively, we can view it as a regular hexagon viewed from above with substituent groups pointing toward us, shown by solid wedges, or pointing away from us, shown by broken wedges.

trans-1,4-Dimethylcyclohexane

cis-1,4-Dimethylcyclohexane

Because *cis-trans* isomers differ in the orientation of their atoms in space, they are **stereoisomers**. *Cis-trans* isomerism is one type of stereoisomerism. We will study another type of stereoisomerism, called enantiomerism, in Chapter 15.

Stereocenter A tetrahedral atom, most commonly carbon, at which exchange of two groups produces a stereoisomer

Configuration Refers to the arrangement of atoms about a stereocenter—that is, to the relative arrangement of parts of a molecule in space

Stereoisomers Isomers that have the same connectivity of their atoms but a different orientation of their atoms in space

EXAMPLE 11-7 *Cis-Trans* Isomerism in Cycloalkanes

Which of the following cycloalkanes show *cis-trans* isomerism? For each that does, draw both isomers.

(a) Methylcyclopentane (b) 1,1-Dimethylcyclobutane
(c) 1,3-Dimethylcyclobutane

Strategy

For a cycloalkane to show *cis-trans* isomerism, it must have at least two substituents, each on a different carbon of the ring.

Solution

(a) Methylcyclopentane does not show *cis-trans* isomerism; it has only one substituent on the ring.
(b) 1,1-Dimethylcyclobutane does not show *cis-trans* isomerism because only one arrangement is possible for the two methyl groups. Because both methyl groups are bonded to the same carbon, they must be *trans* to each other—one above the ring, the other below it.
(c) 1,3-Dimethylcyclobutane shows *cis-trans* isomerism. The two methyl groups may be *cis*, or they may be *trans*.

cis-1,3-Dimethylcyclobutane   *trans*-1,3-Dimethylcyclobutane

Problem 11-7

Which of the following cycloalkanes show *cis-trans* isomerism? For each that does, draw both isomers.

(a) 1,3-Dimethylcyclopentane
(b) Ethylcyclopentane
(c) 1,3-Dimethylcyclohexane

11-9 What Are the Physical Properties of Alkanes and Cycloalkanes?

The most important physical property of alkanes and cycloalkanes is their almost complete lack of polarity. We saw in Section 3-4B that the electronegativity difference between carbon and hydrogen is 2.5−2.1 = 0.4 on the Pauling scale. Given this small difference, we classify a C—H bond as nonpolar covalent. Therefore, alkanes are nonpolar compounds and the only interactions between their molecules are the very weak London dispersion forces (Section 5-7A).

A. Melting and Boiling Points

The boiling points of alkanes are lower than those of almost any other type of compound with the same molecular weight. In general, both boiling and melting points of alkanes increase with increasing molecular weight (Table 11-4).

Alkanes containing 1 to 4 carbons are gases at room temperature. Alkanes containing 5 to 17 carbons are colorless liquids. High-molecular-weight alkanes (those containing 18 or more carbons) are white, waxy solids. Several plant waxes are high-molecular-weight alkanes. The wax found in apple skins, for example, is an unbranched alkane with the molecular formula $C_{27}H_{56}$. ◄ Paraffin wax, a mixture of high-molecular-weight alkanes, is used for wax candles, in lubricants, and to seal home-canned jams, jellies, and other preserves. Petrolatum, so named because it is derived from petroleum refining, is a liquid mixture of high-molecular-weight alkanes. It is sold as mineral oil and Vaseline and is used as an ointment base in pharmaceuticals and cosmetics and as a lubricant and rust preventive.

► Paraffin wax and mineral oil are mixtures of alkanes.

Table 11-4 Physical Properties of Some Unbranched Alkanes

Name	Condensed Structural Formula	Molecular Weight (amu)	Melting Point (°C)	Boiling Point (°C)	Density of Liquid (g/mL at 0°C)*
methane	CH_4	16.0	−182	−164	(a gas)
ethane	CH_3CH_3	30.1	−183	−88	(a gas)
propane	$CH_3CH_2CH_3$	44.1	−190	−42	(a gas)
butane	$CH_3(CH_2)_2CH_3$	58.1	−138	0	(a gas)
pentane	$CH_3(CH_2)_3CH_3$	72.2	−130	36	0.626
hexane	$CH_3(CH_2)_4CH_3$	86.2	−95	69	0.659
heptane	$CH_3(CH_2)_5CH_3$	100.2	−90	98	0.684
octane	$CH_3(CH_2)_6CH_3$	114.2	−57	126	0.703
nonane	$CH_3(CH_2)_7CH_3$	128.3	−51	151	0.718
decane	$CH_3(CH_2)_8CH_3$	142.3	−30	174	0.730

*For comparison, the density of H_2O is 1.000 g/mL at 4°C.

Alkanes that are constitutional isomers are different compounds and have different physical and chemical properties. Table 11-5 lists the boiling points of the five constitutional isomers with the molecular formula of C_6H_{14}. The boiling point of each branched-chain isomer is lower than that of hexane itself; the more branching, the lower the boiling point. These differences in boiling points are related to molecular shape in the following way. As branching increases, the alkane molecule becomes more compact and its surface area decreases. As we learned in Section 5-7A, as surface area decreases, London dispersion forces act over a smaller surface area. Hence, the attraction between molecules decreases and boiling point decreases. Thus, for any group of alkane constitutional isomers, the least-branched isomer generally has the highest boiling point and the most-branched isomer generally has the lowest boiling point.

Table 11-5 Boiling Points of the Five Isomeric Alkanes with the Molecular Formula C_6H_{14}

Name	bp (°C)
hexane	68.7
3-methylpentane	63.3
2-methylpentane	60.3
2,3-dimethylbutane	58.0
2,2-dimethylbutane	49.7

Hexane (bp 68.7°)

Larger surface area, an increase in London dispersion forces, and a higher boiling point

2,2-Dimethylbutane (bp 49.7°)

Smaller surface area, a decrease in London dispersion forces, and a lower boiling point

B. Solubility: A Case of "Like Dissolves Like"

Because alkanes are nonpolar compounds, they are not soluble in water, which dissolves only ionic and polar compounds. Recall that water is a polar substance and that its molecules associate with one another through hydrogen bonding (Section 6-6D). Alkanes do not dissolve in water because they cannot form hydrogen bonds with water. Alkanes, however, are soluble in each other, an example of "like dissolves like" (Section 6-4A). Alkanes are also soluble in other nonpolar organic compounds, such as toluene and diethyl ether.

C. Density

The average density of the liquid alkanes listed in Table 11-4 is about 0.7 g/mL—that of higher-molecular-weight alkanes is about 0.8 g/mL. All liquid and solid alkanes are less dense than water (1.000 g/mL), and because they are insoluble in water, they float on water.

> **EXAMPLE 11-8** **Physical Properties of Alkanes**
>
> Arrange the alkanes in each set in order of increasing boiling point.
> (a) Butane, decane, and hexane
> (b) 2-Methylheptane, octane, and 2,2,4-trimethylpentane
>
> **Strategy**
>
> The compounds in each set are alkanes, and the only forces of attraction between alkane molecules are very weak London dispersion forces. As the number of carbons in a hydrocarbon chain increases, London dispersion forces between chains increase; therefore, the boiling point also increases (Section 5-7A). For alkanes that are constitutional isomers, the strength of London dispersion forces between molecules depends on shape. The more compact the shape, the weaker the intermolecular forces of attraction and the lower the boiling point.
>
> **Solution**
>
> (a) All three compounds are unbranched alkanes. Decane has the longest carbon chain, the strongest London forces between its molecules, and the highest boiling point. Butane has the shortest carbon chain and the lowest boiling point.
>
> Butane Hexane Decane
> bp −0.5°C bp 69°C bp 174°C
>
> (b) These three alkanes are constitutional isomers with the molecular formula C_8H_{18}. 2,2,4-Trimethylpentane is the most highly branched isomer and therefore has the smallest surface area and the lowest boiling point. Octane, the unbranched isomer, has the largest surface area and the highest boiling point.
>
>
>
> 2,2,4-Trimethylpentane 2-Methylheptane Octane
> (bp 99°C) (bp 118°C) (bp 126°C)
>
> **Problem 11-8**
>
> Arrange the alkanes in each set in order of increasing boiling point.
> (a) 2-Methylbutane, pentane, and 2,2-dimethylpropane
> (b) 3,3-Dimethylheptane, nonane, and 2,2,4-trimethylhexane

11-10 What Are the Characteristic Reactions of Alkanes?

The most important chemical property of alkanes and cycloalkanes is their inertness. They are quite unreactive under the normal ionic reaction conditions we studied in Chapters 5 and 8. Under certain conditions, however, alkanes react with oxygen, O_2. By far, their most important reaction with oxygen is oxidation (combustion) to form carbon dioxide and water. They also react with fluorine, bromine, and chlorine to form halogenated hydrocarbons.

A. Reaction with Oxygen: Combustion

Oxidation of hydrocarbons, including alkanes and cycloalkanes, is the basis for their use as energy sources for heat [natural gas, liquefied petroleum gas (LPG), and fuel oil] and power (gasoline, diesel, and aviation fuel). Following are balanced equations for the complete combustion of methane, the major component of natural gas, and propane, the major component of LPG or bottled gas. The heat liberated when an alkane is oxidized to carbon dioxide and water is called its heat of combustion (Section 4-8).

$$CH_4 + 2O_2 \longrightarrow CO_2 + 2H_2O + 212 \text{ kcal/mol}$$
Methane

$$CH_3CH_2CH_3 + 5O_2 \longrightarrow 3CO_2 + 4H_2O + 530 \text{ kcal/mol}$$
Propane

CHEMICAL CONNECTIONS 11B

Octane Rating: What Those Numbers at the Pump Mean

Gasoline is a complex mixture of C_6 to C_{12} hydrocarbons. The quality of gasoline as a fuel for internal combustion engines is expressed in terms of an octane rating. Engine knocking occurs when a portion of the air–fuel mixture explodes prior to the piston reaching the top of its stroke (usually as a result of heat developed during the compression) and independent of ignition by the spark plug. The resulting shockwave of the piston against the cylinder wall reverberates, creating a characteristic metallic "pinging" sound.

Two compounds were selected as reference fuels for rating gasoline quality. One of these, 2,2,4-trimethylpentane (isooctane) has very good antiknock properties and was assigned an octane rating of 100. Heptane, the other reference compound, has poor antiknock properties and was assigned an octane rating of 0.

The **octane rating** of a particular gasoline is the percent of 2,2,4-trimethylpentane in a mixture of 2,2,4-trimethylpentane and heptane that has antiknock properties equivalent to that of the test gasoline. For example, the antiknock properties of 2-methylhexane are equivalent to those of a mixture of 42% 2,2,4-trimethylpentane and 58% heptane; therefore, the octane rating of 2-methylhexane is 42. Ethanol, which is added to gasoline to produce gasohol, has an octane rating of 105. Octane itself has an octane rating of −20.

2,2,4-Trimethylpentane (octane rating 100)

Heptane (octane rating 0)

Typical octane ratings of commonly available gasolines.

Test your knowledge with Problems 11-59 through 11-61.

B. Reaction with Halogens: Halogenation

If we mix methane with chlorine or bromine in the dark at room temperature, nothing happens. If, however, we heat the mixture to 100°C or higher or expose it to light, a reaction begins at once. The products of the reaction between methane and chlorine are chloromethane and hydrogen chloride. What occurs is a substitution reaction—in this case, the substitution of chlorine for hydrogen in methane.

$$CH_4 + Cl_2 \xrightarrow{\text{heat or light}} CH_3Cl + HCl$$
$$\text{Methane} \qquad\qquad\qquad \text{Chloromethane}$$
$$\text{(Methyl chloride)}$$

If chloromethane is allowed to react with more chlorine, further chlorination produces a mixture of dichloromethane, trichloromethane, and tetrachloromethane.

$$CH_3Cl + Cl_2 \xrightarrow{\text{heat}} CH_2Cl_2 + HCl$$
$$\text{Chloromethane} \qquad\qquad \text{Dichloromethane}$$
$$\text{(Methyl chloride)} \qquad\qquad \text{(Methylene chloride)}$$

$$CH_2Cl_2 \xrightarrow[\text{heat}]{Cl_2} CHCl_3 \xrightarrow[\text{heat}]{Cl_2} CCl_4$$
$$\text{Dichloromethane} \quad \text{Trichloromethane} \quad \text{Tetrachloromethane}$$
$$\text{(Methylene chloride)} \quad \text{(Chloroform)} \quad \text{(Carbon tetrachloride)}$$

In the last equation, the reagent Cl_2 is placed over the reaction arrow and the equivalent amount of HCl formed is not shown. Placing reagents over reaction arrows and omitting by-products is commonly done to save space.

We derive IUPAC names of haloalkanes by naming the halogen atom as a substituent (*fluoro-, chloro-, bromo-,* and *iodo-*) and alphabetizing it along with other substituents. Common names consist of the common name of the alkyl group followed by the name of the halogen (chloride, bromide, and so forth) as a separate word. Dichloromethane (methylene chloride) is a widely used solvent for organic compounds.

EXAMPLE 11-9 Halogenation of Alkanes

Write a balanced equation for the reaction of ethane with chlorine to form chloroethane, C_2H_5Cl.

Strategy
The reaction of ethane with chlorine results in the substitution of one of the hydrogen atoms of ethane with a chlorine atom.

Solution

$$CH_3CH_3 + Cl_2 \xrightarrow{\text{heat or light}} CH_3CH_2Cl + HCl$$
$$\text{Ethane} \qquad\qquad\qquad \text{Chloroethane}$$
$$\text{(Ethyl chloride)}$$

Problem 11-9
Reaction of propane with chlorine gives two products, each with the molecular formula C_3H_7Cl. Draw structural formulas for these two compounds and give each an IUPAC name and a common name.

11-11 What Are Some Important Haloalkanes?

One of the major uses of haloalkanes is as intermediates in the synthesis of other organic compounds. Just as we can replace a hydrogen atom of an alkane, we can, in turn, replace the halogen atom with a number of other functional groups. In this way, we can construct more complex molecules. In contrast, alkanes that contain several halogens are often quite unreactive, a fact that has proved especially useful in the design of several classes of consumer products.

A. Chlorofluorocarbons

Of all the haloalkanes, the **chlorofluorocarbons (CFCs)** manufactured under the trade name Freons are the most widely known. CFCs are nontoxic, nonflammable, odorless, and noncorrosive. Originally, they seemed to be ideal replacements for hazardous compounds such as ammonia and sulfur dioxide formerly used as heat-transfer agents in refrigeration systems. Among the CFCs most widely used for this purpose were trichlorofluoromethane (CCl_3F, Freon-11) and dichlorodifluoromethane (CCl_2F_2, Freon-12). CFCs also found wide use as industrial cleaning solvents to prepare surfaces for coatings, to remove cutting oils and waxes from millings, and to remove protective coatings. In addition, they were employed as propellants for aerosol sprays.

B. Solvents

Several low-molecular-weight haloalkanes are excellent solvents in which to carry out organic reactions and to use as cleaners and degreasers. Carbon tetrachloride ("carbon tet") was the first of these compounds to find wide

---

CHEMICAL CONNECTIONS 11C

The Environmental Impact of Freons

Concern about the environmental impact of CFCs arose in the 1970s, when researchers found that more than 4.5×10^5 kg/yr of these compounds were being emitted into the atmosphere. In 1974, Sherwood Rowland and Mario Molina, both of the United States, announced their theory, which has since been amply confirmed, that these compounds destroy the stratospheric ozone layer. When released into the air, CFCs escape to the lower atmosphere. Because of their inertness, however, they do not decompose there. Slowly they find their way to the stratosphere, where they absorb ultraviolet radiation from the Sun and then decompose. As they do so, they set up chemical reactions that lead to the destruction of the stratospheric ozone layer, which shields the Earth against short-wavelength ultraviolet radiation from the Sun. An increase in short-wavelength ultraviolet radiation reaching the Earth is believed to promote the destruction of certain crops and agricultural species and to increase the incidence of skin cancer in light-skinned individuals.

This concern prompted two conventions, one in Vienna in 1985 and one in Montreal in 1987, held by the United Nations Environmental Program. The 1987 meeting produced the Montreal Protocol, which set limits on the production and use of ozone-depleting CFCs and urged the complete phase-out of their production by 1996. This phase-out resulted in enormous costs for manufacturers and is not yet complete in developing countries.

Rowland, Molina, and Paul Crutzen (a Dutch chemist at the Max Planck Institute for Chemistry in Germany) were awarded the 1995 Nobel Prize for chemistry. As noted in the award citation by the Royal Swedish Academy of Sciences, "By explaining the chemical mechanisms that affect the thickness of the ozone layer, these three researchers have contributed to our salvation from a global environmental problem that could have catastrophic consequences."

CHEMICAL CONNECTIONS 11C

The Environmental Impact of Freons (continued)

The chemical industry has responded to this crisis by developing replacement refrigerants that have much lower ozone-depleting potential. The most prominent of these replacements are the hydrofluorocarbons (HFCs) and hydrochlorofluorocarbons (HCFCs).

```
    F  F              H  Cl
    |  |              |  |
F—C—C—H          H—C—C—F
    |  |              |  |
    F  H              H  Cl
  HFC-134a          HCFC-141b
```

These compounds are much more chemically reactive in the atmosphere than the Freons and are destroyed before they reach the stratosphere. However, they are not compatible in the air conditioners of cars manufactured before 1995.

Using Data from infrared spectroscopy scientists can calculate the global warming potential (GWP), Table 11-6, for scores of halocarbons and related compounds, some of which are given in the table. Among them HFO-1234yf has the lowest lifetime in the atmosphere as well as the lowest GWP and thus is an environmentally friendly alternative to the Freons. For these reasons HFO-1234yf has become the refrigerant gas of choice for home and commercial cooling systems. Six of these new halocarbons along with Freon-12 are listed in the table. As shown in Table 11-6, HFO-1234yf has the lowest atmospheric lifetime and the lowest global warming potential.

Table 11-6 Global Warming Potential for some refrigerant gases.

Compound	Formula	Atmospheric lifetime years	Global Warming Potential
CFC-11	CCl_3F	45	4,600
CFC-12	CCl_2F_2	100	10,200
CFC-113	CCl_2FCClF_2	85	5820
CFC-22	$CHClF_2$	11.9	1760
HFC-135a	CH_2FCF_3	13.4	1300
HFO-1234yf	$CF_3CF=CH_2$	0.03	<1

Values of GWP are relative to CO_2, which has a value of 1

Test your knowledge with Problems 11-62 through 11-64.

application, but its use for this purpose has since been discontinued because it is now known that carbon tet is both toxic and a carcinogen. Today, the most widely used haloalkane solvent is dichloromethane, CH_2Cl_2.

Summary of Key Questions

Section 11-1 What Are Alkanes?
- **Alkanes** are saturated hydrocarbons whose carbon atoms are arranged in an open chain.
- A **hydrocarbon** is a compound composed of only carbon and hydrogen.
- A **saturated hydrocarbon** contains only carbon–carbon single bonds.
- An **unsaturated hydrocarbon** is a hydrocarbon that contains one or more carbon–carbon double or triple bonds or benzene rings.

Section 11-2 How Do We Write Structural Formulas of Alkanes?
- In a **line-angle formula**, a line represents a carbon–carbon bond and a vertex represents a carbon atom. To count hydrogens from a line-angle formula, add enough hydrogens to give each carbon its required four bonds.

Section 11-3 What Are Constitutional Isomers?
- **Constitutional isomers** have the same molecular formula but a different connectivity of their atoms.

Section 11-4 How Do We Name Alkanes?

- Alkanes are named according to a set of rules developed by the **International Union of Pure and Applied Chemistry (IUPAC)**.
- The IUPAC name of an alkane consists of two parts: a prefix that tells the number of carbon atoms in the parent chain and the ending **-ane**. Substituents derived from alkanes by removal of a hydrogen atom are called **alkyl groups** and are denoted by the symbol **R—**.

Section 11-5 Where Do We Obtain Alkanes?

- **Natural gas** consists of 90 to 95% methane with lesser amounts of ethane and other lower-molecular-weight hydrocarbons.
- **Petroleum** is a liquid mixture of thousands of different hydrocarbons.

Section 11-6 What Are Cycloalkanes?

- A **cycloalkane** is an alkane that contains carbon atoms bonded to form a ring.
- To name a cycloalkane, prefix the name of the open-chain alkane with cyclo-.

Section 11-7 What Are the Shapes of Alkanes and Cycloalkanes?

- A **conformation** is any three-dimensional arrangement of the atoms of a molecule that results from rotation about a single bond.
- The lowest-energy conformation of cyclopentane is an **envelope conformation**.
- The lowest-energy conformation of cyclohexane is a **chair conformation**. In a chair conformation, six C—H bonds are **axial** and six C—H bonds are **equatorial**. A substituent on a six-membered ring is more stable when it is equatorial than when it is axial.

Section 11-8 What Is Cis-Trans Isomerism in Cycloalkanes?

- **Configuration** refers to the arrangement of atoms about a stereocenter.
- **Cis-trans** isomers of cycloalkanes have (1) the same molecular formula and (2) the same connectivity of their atoms, but (3) a different orientation of their atoms in space because of the restricted rotation around the C—C bonds of the ring.
- For *cis-trans* isomers of cycloalkanes, *cis* means that substituents are on the same side of the ring; *trans* means that they are on opposite sides of the ring.

Section 11-9 What Are the Physical Properties of Alkanes and Cycloalkanes?

- Alkanes are nonpolar compounds, and the only forces of attraction between their molecules are London dispersion forces.
- At room temperature, low-molecular-weight alkanes are gases, higher-molecular-weight alkanes are liquids, and very-high-molecular-weight alkanes are waxy solids.
- For any group of alkane constitutional isomers, the least-branched isomer generally has the highest boiling point and the most-branched isomer generally has the lowest boiling point.
- Alkanes are insoluble in water but soluble in each other and in other nonpolar organic solvents such as toluene. All liquid and solid alkanes are less dense than water.

Section 11-10 What Are the Characteristic Reactions of Alkanes?

- The most important chemical property of alkanes and cycloalkanes is their inertness. They are quite unreactive under normal ionic reaction conditions.
- By far, the most important reaction of alkanes and cycloalkanes is their reaction with oxygen (**combustion**) in which they are converted to carbon dioxide and water.
- Combustion of alkanes and cycloalkanes is the basis for their use as energy sources and power.
- Alkanes and cycloalkanes also react with chlorine, Cl_2, and bromine, Br_2, by **substitution**, in which an atom of chlorine or bromine is substituted for a hydrogen of the alkane or cycloalkane. The products of this type of substitution reaction are haloalkanes and halocycloalkanes.

Section 11-11 What Are Some Important Haloalkanes?

- Haloalkanes are often used as intermediates in the synthesis of other organic compounds. Just as we can replace a hydrogen atom with a halogen atom, we can in turn replace the halogen atom with other functional groups.
- Low-molecular-weight haloalkanes such as dichloromethane, CH_2Cl_2, are excellent solvents in which to carry out organic reactions and to use as degreasers and solvents for cleaning.

Summary of Key Reactions

1. Oxidation of Alkanes (Section 11-10A) Oxidation of alkanes to carbon dioxide and water, an exothermic reaction, is the basis for our use of them as sources of heat and power.

$$CH_3CH_2CH_3 + 5O_2 \longrightarrow 3CO_2 + 4H_2O + 530 \text{ kcal/mol}$$
Propane

2. Halogenation of Alkanes (Section 11-10B) Reaction of an alkane with chlorine or bromine results in the substitution of a halogen atom for a hydrogen.

$$CH_3CH_3 + Cl_2 \xrightarrow{\text{heat or light}} CH_3CH_2Cl + HCl$$
Ethane   Chloroethane
           (Ethyl chloride)

Problems

Orange-numbered problems are applied.

Section 11-2 How Do We Write Structural Formulas of Alkanes?

11-10 Answer true or false.
(a) A hydrocarbon is composed only of the elements carbon and hydrogen.
(b) Alkanes are saturated hydrocarbons.
(c) The general formula of an alkane is C_nH_{2n+2}, where n is the number of carbons in the alkane.
(d) Alkenes and alkynes are unsaturated hydrocarbons.

11-11 Define:
(a) Hydrocarbon
(b) Alkane
(c) Saturated hydrocarbon

11-12 Why is it not accurate to describe an unbranched alkane as a "straight-chain" hydrocarbon?

11-13 What is meant by the term *line-angle formula* as applied to alkanes and cycloalkanes?

11-14 For each condensed structural formula, write a line-angle formula.

(a) $\text{CH}_3\text{CH}_2\overset{\overset{\displaystyle\text{CH}_2\text{CH}_3}{|}}{\text{CH}}\text{CHCH}_2\overset{\overset{\displaystyle\text{CH}_3}{|}}{\text{CH}}\text{CH}_3$
$\quad\quad\quad\quad\quad\quad\underset{\underset{\displaystyle\text{CH}(\text{CH}_3)_2}{|}}{\,}$

(b) $\text{CH}_3\overset{\overset{\displaystyle\text{CH}_3}{|}}{\underset{\underset{\displaystyle\text{CH}_3}{|}}{\text{C}}}\text{CH}_3$

(c) $(\text{CH}_3)_2\text{CHCH}(\text{CH}_3)_2$

(d) $\text{CH}_3\text{CH}_2\overset{\overset{\displaystyle\text{CH}_2\text{CH}_3}{|}}{\underset{\underset{\displaystyle\text{CH}_2\text{CH}_3}{|}}{\text{C}}}\text{CH}_2\text{CH}_3$

(e) $(\text{CH}_3)_3\text{CH}$

(f) $\text{CH}_3(\text{CH}_2)_3\text{CH}(\text{CH}_3)_2$

11-15 Write the molecular formula for each alkane.

(a) [line-angle structure]
(b) [line-angle structure]
(c) [line-angle structure]

Section 11-3 What Are Constitutional Isomers?

11-16 Answer true or false.
(a) Constitutional isomers have the same molecular formulas and the same connectivity of their atoms.
(b) There are two constitutional isomers with the molecular formula C_3H_8.
(c) There are four constitutional isomers with the molecular formula C_4H_{10}.
(d) There are five constitutional isomers with the molecular formula C_5H_{12}.

11-17 Which statements are true about constitutional isomers?
(a) They have the same molecular formula.
(b) They have the same molecular weight.
(c) They have the same connectivity of their atoms.
(d) They have the same physical properties.

11-18 Each member of the following set of compounds is an alcohol; that is, each contains an —OH (hydroxyl group; see Section 10-4A). Which structural formulas represent the same compound, and which represent constitutional isomers?

11-19 Each member of the following set of compounds is an amine; that is, each contains a nitrogen atom bonded to one, two, or three carbon groups (Section 10-4B). Which structural formulas represent the same compound, and which represent constitutional isomers?

11-20 Each member of the following set of compounds is either an aldehyde or a ketone (Section 10-4C). Which structural formulas represent the same compound, and which represent constitutional isomers?

11-21 In the six following sets, which pairs of structural formulas represent constitutional isomers?

11-22 Draw line-angle formulas for the nine constitutional isomers with the molecular formula C_7H_{16}.

Section 11-4 How Do We Name Alkanes?

11-23 Answer true or false.
  (a) The parent name of an alkane is the name of the longest chain of carbon atoms in the alkane.
  (b) Propyl and isopropyl groups are constitutional isomers.
  (c) There are four alkyl groups with the molecular formula C_4H_9.

11-24 Name these alkyl groups:

  (a) CH_3CH_2-
  (b) $CH_3\overset{\overset{\displaystyle CH_3}{|}}{CH}-$
  (c) $CH_3\overset{\overset{\displaystyle CH_3}{|}}{CH}CH_2-$
  (d) $CH_3\overset{\overset{\displaystyle CH_3}{|}}{\underset{\underset{\displaystyle CH_3}{|}}{C}}-$

11-25 Among the ingredients listed in one commercial foam shaving gel are isobutane and isopentane.
  (a) Write the IUPAC name of each hydrocarbon.
  (b) Why are these two hydrocarbons added to the formulation of the shaving gel?

Section 11-5 Where Do We Obtain Alkanes?

11-26 Answer true or false.
(a) The two major sources of alkanes the world over are petroleum and natural gas.
(b) The octane number of a particular gasoline is the number of grams of octane per liter of the fuel.
(c) Octane and 2,2,4-trimethylpentane are constitutional isomers and have the same octane number.

Section 11-6 What Are Cycloalkanes?

11-27 Answer true or false.
(a) Cycloalkanes are saturated hydrocarbons.
(b) Hexane and cyclohexane are constitutional isomers.
(c) The parent name of a cycloalkane is the name of the unbranched alkane with the same number of carbon atoms as are in the cycloalkane ring.

11-28 Write the IUPAC names for these alkanes and cycloalkanes.

(a) $CH_3CHCH_2CH_2CH_3$
    $\quad\;\;|$
    $\quad CH_3$

(b) $CH_3CHCH_2CH_2CHCH_3$
    $\quad\;\;|\quad\quad\quad\;\;|$
    $\quad CH_3\quad\quad CH_3$

(c) $CH_3(CH_2)_4CHCH_2CH_3$
    $\quad\quad\quad\quad\quad|$
    $\quad\quad\quad\quad CH_2CH_3$

(d)

(e)

(f)

11-29 Write line-angle formulas for these alkanes and cycloalkanes.
(a) 2,2,4-Trimethylhexane
(b) 2,2-Dimethylpropane
(c) 3-Ethyl-2,4,5-trimethyloctane
(d) 5-Butyl-2,2-dimethylnonane
(e) 4-Isopropyloctane
(f) 3,3-Dimethylpentane
(g) *trans*-1,3-Dimethylcyclopentane
(h) *cis*-1,2-Diethylcyclobutane

Section 11-7 What Are the Shapes of Alkanes and Cycloalkanes?

11-30 Answer true or false.
(a) Conformations have the same molecular formula and the same connectivity but differ in the three-dimensional arrangement of their atoms in space.
(b) In all conformations of ethane, propane, butane, and higher alkanes, all C—C—C and H—C—H bond angles are approximately 109.5°.
(c) In a cyclohexane ring, if an axial bond is above the plane of the ring on a particular carbon, axial bonds on the two adjacent carbons are below the plane of the ring.
(d) In a cyclohexane ring, if an equatorial bond is above the plane of the ring on a particular carbon, equatorial bonds on the two adjacent carbons are below the plane of the ring.
(e) The more stable chair conformation of a cyclohexane ring has more substituent groups in equatorial positions.

11-31 The condensed structural formula of butane is $CH_3CH_2CH_2CH_3$. Explain why this formula does not show the geometry of the real molecule.

11-32 Calculate the actual C—C—C bond angles in planar (a) cyclopropane and (b) cyclopentane and compare them with optimal bond angles.

Section 11-8 What Is Cis-Trans Isomerism in Cycloalkanes?

11-33 Answer true or false.
(a) *Cis*- and *trans*-cycloalkanes have the same molecular formula but a different connectivity of their atoms.
(b) A *cis* isomer of a cycloalkane can be converted to its *trans* isomer by rotation about an appropriate carbon–carbon single bond.
(c) A *cis* isomer of a cycloalkane can be converted to its *trans* isomer by exchange of two groups at a stereocenter in the *cis*-cycloalkane.
(d) Configuration refers to the arrangement in space of the atoms or groups of atoms at a stereocenter.
(e) *Cis*-1,4-dimethylcyclohexane and *trans*-1,4-dimethylcyclohexane are classified as conformations.

11-34 What structural feature of cycloalkanes makes *cis-trans* isomerism in them possible?

11-35 Is *cis-trans* isomerism possible in alkanes?

11-36 Name and draw structural formulas for the *cis* and *trans* isomers of 1,2-dimethylcyclopropane.

11-37 Name and draw structural formulas for the six cycloalkanes with the molecular formula C_5H_{10}. Include *cis-trans* isomers as well as constitutional isomers.

11-38 Why is equatorial methylcyclohexane more stable than axial methylcyclohexane?

11-39 Following is a structural formula and a ball-and-stick model of cholestanol, a close relative of cholesterol.

Cholestanol
$C_{27}H_{48}O$

(a) Describe the conformation of each six-membered ring and the one five-membered ring.
(b) Is the —OH group on ring **A** in an axial or equatorial position?
(c) Is the —CH$_3$ group between rings **A** and **B** in an axial or an equatorial position?

11-40 Consider a cyclohexane ring substituted with one methyl group and one hydroxyl group. Draw the structural formula for a compound with this composition that:
(a) Does not show cis/trans isomerism.
(b) Does show cis/trans isomerism.

Section 11-9 What Are the Physical Properties of Alkanes and Cycloalkanes?

11-41 Answer true or false.
(a) Boiling points among alkanes with unbranched chains increase as the number of carbons in the chain increases.
(b) Alkanes that are liquid at room temperature are more dense than water.
(c) *Cis* and *trans* isomers have the same molecular formula, the same connectivity, and the same physical properties.
(d) Among alkane constitutional isomers, the least branched isomer generally has the lowest boiling point.
(e) Alkanes and cycloalkanes are insoluble in water.
(f) Liquid alkanes are soluble in each other.

11-42 In Problem 11-22, you drew structural formulas for the nine constitutional isomers with molecular formula C_7H_{16}. Predict which isomer has the lowest boiling point and which has the highest boiling point.

11-43 Which unbranched alkane (Table 11-4) has about the same boiling point as water? Calculate the molecular weight of this alkane and compare it with the molecular weight of water.

11-44 What generalizations can you make about the densities of alkanes relative to the density of water?

11-45 What generalization can you make about the solubility of alkanes in water?

11-46 Suppose that you have samples of hexane and octane. Could you tell the difference by looking at them? What color would each be? How could you tell which is which?

11-47 As you can see from Table 11-4, each CH$_2$ group added to the carbon chain of an alkane increases its boiling point. This increase is greater going from CH_4 to C_2H_6 and from C_2H_6 to C_3H_8 than it is going from C_8H_{18} to C_9H_{20} or from C_9H_{20} to $C_{10}H_{22}$. What do you think is the reason for this difference?

11-48 How are the boiling points of hydrocarbons during petroleum refining related to their molecular weight?

Section 11-10 What Are the Characteristic Reactions of Alkanes?

11-49 Answer true or false.
(a) Combustion of alkanes is an endothermic reaction.
(b) The products of complete combustion of an alkane are carbon dioxide and water.
(c) Halogenation of an alkane converts it to a haloalkane.

11-50 Write balanced equations for the combustion of each of the following hydrocarbons. Assume that each is converted completely to carbon dioxide and water.
(a) Hexane
(b) Cyclohexane
(c) 2-Methylpentane

11-51 The heat of combustion of methane, a component of natural gas, is 212 kcal/mol. That of propane, a component of LP gas, is 530 kcal/mol. On a gram-for-gram basis, which hydrocarbon is the better source of heat energy?

11-52 Draw structural formulas for these haloalkanes.
(a) Bromomethane
(b) Chlorocyclohexane
(c) 1,2-Dibromoethane
(d) 2-Chloro-2-methylpropane
(e) Dichlorodifluoromethane (Freon-12)

11-53 The reaction of chlorine with pentane gives a mixture of three chloroalkanes, each with the molecular formula $C_5H_{11}Cl$. Write a line-angle formula and the IUPAC name for each chloroalkane.

11-54 Complete and balance the equation for the complete combustion of each hydrocarbon.
(a) Hexane (b) Cyclohexane
(c) 2-Methylpentane

11-55 Name and draw structural formulas for all possible monochlorination products that might be formed in each reaction.

(a) [cyclohexane] + Cl$_2$ $\xrightarrow{\text{heat}}$

(b) [pentane] + Cl$_2$ $\xrightarrow{\text{heat}}$

(c) [2,3-dimethylbutane] + Cl$_2$ $\xrightarrow{\text{heat}}$

11-56 There are three constitutional isomers with the molecular formula C$_5$H$_{12}$. When heated with chlorine at 300°C, isomer A gives a mixture of four monochlorination products. Under the same experimental conditions, isomer B gives a mixture of three monochlorination products and isomer C gives only one monochlorination product. From this information, deduce the structural formulas of isomers A, B, and C.

Section 11-11 What Are Some Important Haloalkanes?

11-57 Answer true or false.
(a) Freons are members of a class of organic compounds called chlorofluorocarbons (CFCs).
(b) An advantage of Freons as heat-transfer agents in refrigeration systems, propellants in aerosol sprays, and solvents for industrial cleaning is that they are nontoxic, nonflammable, odorless, and noncorrosive.
(c) Freons in the stratosphere interact with ultraviolet radiation and thereby set up chemical reactions that lead to the destruction of the stratospheric ozone layer.
(d) Alternative names for the important laboratory and industrial solvent CH$_2$Cl$_2$ are dichloromethane, methylene chloride, and chloroform.

Chemical Connections

11-58 (Chemical Connections 11A) How many rings in tetrodotoxin contain only carbon atoms? How many contain nitrogen atoms? How many contain two oxygen atoms?

11-59 (Chemical Connections 11B) What is an "octane rating"? What two reference hydrocarbons are used for setting the scale of octane ratings?

11-60 (Chemical Connections 11B) Octane has an octane rating of −20. Will it produce more or less engine knocking than heptane does?

11-61 (Chemical Connections 11B) Ethanol is added to gasoline to produce E-15 and E-85. It promotes more complete combustion of the gasoline and is an octane booster. Compare the heats of combustion of 2,2,4-trimethylpentane (1304 kcal/mol) and ethanol (327 kcal/mol). Which has the higher heat of combustion in kcal/mol? In kcal/g?

11-62 (Chemical Connections 11C) What are Freons? Why were they considered ideal compounds to use as heat-transfer agents in refrigeration systems? Give structural formulas of two Freons used for this purpose.

11-63 (Chemical Connections 11C) In what way do Freons negatively affect the environment?

11-64 (Chemical Connections 11C) What are HFCs and HCFCs? How does their use in refrigeration systems prevent the environmental problems associated with the use of Freons?

Additional Problems

11-65 Tell whether the compounds in each set are constitutional isomers.

(a) CH$_3$CH$_2$OH and CH$_3$OCH$_3$

(b) CH$_3$$\overset{\overset{\text{O}}{\|}}{\text{C}}CH_3$ and CH$_3$CH$_2$$\overset{\overset{\text{O}}{\|}}{\text{C}}$H

(c) CH$_3$$\overset{\overset{\text{O}}{\|}}{\text{C}}OCH_3$ and CH$_3$CH$_2$$\overset{\overset{\text{O}}{\|}}{\text{C}}$OH

(d) CH$_3$$\overset{\overset{\text{OH}}{|}}{\text{CH}}CH_2CH_3$ and CH$_3$$\overset{\overset{\text{O}}{\|}}{\text{C}}CH_2CH_3$

(e) [cyclopentane] and CH$_3$CH$_2$CH$_2$CH$_2$CH$_2$CH$_3$

(f) [cyclopentane] and CH$_2$=CHCH$_2$CH$_2$CH$_3$

11-66 Explain why each of the following is an incorrect IUPAC name. Write the correct IUPAC name for the compound.
(a) 1,3-Dimethylbutane
(b) 4-Methylpentane
(c) 2,2-Diethylbutane
(d) 2-Ethyl-3-methylpentane
(e) 2-Propylpentane
(f) 2,2-Diethylheptane
(g) 2,2-Dimethylcyclopropane
(h) 1-Ethyl-5-methylcyclohexane

11-67 Which of the following compounds can exist as *cis-trans* isomers? For each that can, draw both isomers using solid and dashed wedges to show the orientation in space of the —OH and —CH₃ groups.

(a) [cyclohexane with OH and CH₃]
(b) [cyclohexane with OH and CH₃]
(c) [cyclohexane with HO and CH₃]

11-68 Tetradecane, $C_{14}H_{30}$, is an unbranched alkane with a melting point of 5.9°C and a boiling point of 254°C. Is tetradecane a solid, liquid, or gas at room temperature?

11-69 Dodecane, $C_{12}H_{26}$, is an unbranched alkane. Predict the following:
(a) Will it dissolve in water?
(b) Will it dissolve in hexane?
(c) Will it burn when ignited?
(d) Is it a liquid, solid, or gas at room temperature and atmospheric pressure?
(e) Is it more or less dense than water?

Looking Ahead

11-70 Following is a structural formula for 2-isopropyl-5-methylcyclohexanol:

[structure: 2-Isopropyl-5-methylcyclohexanol]

Using a planar hexagon representation for the cyclohexane ring, draw a structural formula for the *cis-trans* isomer with isopropyl *trans* to —OH and methyl *cis* to —OH. If you answered this part correctly, you have drawn the isomer found in nature and given the name menthol.

11-71 On the left is a representation of the glucose molecule. Convert this representation to the alternative representations using the rings on the right. (We discuss the structure and chemistry of glucose in Chapter 20.)

[glucose structures — Planar hexagon representation; Chair conformation]

11-72 On the left is a representation for 2-deoxy-D-ribose. This molecule is the "D" of DNA. Convert this representation to the alternative representation using the ring on the right. (We discuss the structure and chemistry of this compound in more detail in Chapter 20.)

[2-Deoxy-D-ribose structures]

11-73 As stated in Section 11-9, the wax found in apple skins is an unbranched alkane with the molecular formula $C_{27}H_{56}$. Explain how the presence of this alkane in apple skins prevents the loss of moisture from within the apple.

11-74 Following are structural formulas for dimethyl ether and ethanol. These compounds are constitutional isomers with the molecular formula C_2H_6O.

CH_3OCH_3 CH_3CH_2OH
Dimethyl ether   Ethanol

One of these compounds has a boiling point of 78°C; the other has a boiling point of −24°C.

(a) Which compound has which boiling point? Explain your reasoning. (Chapter 5)
(b) One compound is soluble in water in all proportions, that is, it is infinitely soluble. The other has a water solubility of only 7.8 g/100 mL. Which compound has which water solubility? How do you account for this difference in water solubility? We discuss the physical properties of alcohols and ethers in detail in Chapter 14.

12 Alkenes and Alkynes

KEY QUESTIONS

- 12-1 What Are Alkenes and Alkynes?
- 12-2 What Are the Structures of Alkenes and Alkynes?
- 12-3 How Do We Name Alkenes and Alkynes?
- 12-4 What Are the Physical Properties of Alkenes and Alkynes?
- 12-5 What Are the Characteristic Reactions of Alkenes?
- 12-6 What Are the Important Polymerization Reactions of Ethylene and Substituted Ethylenes?

Alkenes Unsaturated hydrocarbons that contain a carbon–carbon double bond

Alkynes Unsaturated hydrocarbons that contains a carbon–carbon triple bond

Carotene is a naturally occurring polyene in carrots and tomatoes (Problems 12-70 and 12-71).

12-1 What Are Alkenes and Alkynes?

In this chapter, we begin our study of unsaturated hydrocarbons. Recall from Section 11-1 that unsaturated compounds contain one or more carbon–carbon double bonds, triple bonds, or benzene-like rings. In this chapter, we study **alkenes** and **alkynes**. **Alkenes** are unsaturated hydrocarbons that contain one or more carbon–carbon double bonds. The simplest alkene is ethylene.

Ethylene (an alkene)   Acetylene (an alkyne)

Alkynes are unsaturated hydrocarbons that contain one or more carbon–carbon triple bonds. The simplest alkyne is acetylene. Because alkynes are not widespread in nature and have little importance in biochemistry, we will not study their chemistry in depth.

Compounds containing carbon–carbon double bonds are especially widespread in nature. Furthermore, several low-molecular-weight

> ### CHEMICAL CONNECTIONS 12A
>
> ### Ethylene: A Plant Growth Regulator
>
> As we noted below, ethylene occurs only in trace amounts in nature. True enough, but scientists have discovered that this small molecule is a natural ripening agent for fruits. Thanks to this knowledge, fruit growers can now pick fruit while it is still green and less susceptible to bruising. Then, when the growers are ready to pack the fruit for shipment, they can treat it with ethylene gas. Alternatively, the fruit can be treated with ethephon (Ethrel), which slowly releases ethylene and initiates fruit ripening.
>
> $$Cl-CH_2-CH_2-\overset{\overset{O}{\|}}{\underset{\underset{OH}{|}}{P}}-OH$$
>
> Ethephon
>
> The next time you see ripe bananas in the market, you might wonder when they were picked and whether their ripening was artificially induced.
>
> Text your knowledge with Problem 12-61.

alkenes, including ethylene and propene, have enormous commercial importance in our modern industrialized society. The organic chemical industry worldwide produces more pounds of ethylene than any other organic chemical. Annual production in the United States alone exceeds 55 billion pounds.

What is unusual about ethylene is that it occurs only in trace amounts in nature. The enormous amounts of it required to meet the needs of the chemical industry are derived the world over by thermal cracking of hydrocarbons. In the United States and other areas of the world with vast reserves of natural gas, the major process for the production of ethylene is thermal cracking of the small quantities of ethane extracted from natural gas. In **thermal cracking**, a saturated hydrocarbon is converted to an unsaturated hydrocarbon plus H_2. Ethane is thermally cracked by heating it in a furnace to 800–900°C for a fraction of a second.

$$CH_3CH_3 \xrightarrow[\text{(thermal cracking)}]{800-900°C} CH_2{=}CH_2 + H_2$$
$$\text{Ethane} \qquad\qquad\qquad \text{Ethylene}$$

Europe, Japan, and other parts of the world with limited supplies of natural gas depend almost entirely on the thermal cracking of petroleum for their ethylene.

From the perspective of the chemical industry, the single most important reaction of ethylene and other low-molecular-weight alkenes is polymerization, which we discuss in Section 12-6. The crucial point to recognize is that ethylene and all of the commercial and industrial products synthesized from it are derived from either natural gas or petroleum—both nonrenewable natural resources!

12-2 What Are the Structures of Alkenes and Alkynes?

A. Alkenes

Using the VSEPR model (Section 3-10), we predict bond angles of 120° about each carbon in a double bond. The observed H—C—C bond angle in ethylene, for example, is 121.7°, close to the predicted value. In other alkenes, deviations from the predicted angle of 120° may be somewhat larger because of interactions between alkyl groups bonded to the doubly bonded carbons. The C—C—C bond angle in propene, for example, is 124.7°.

CHEMICAL CONNECTIONS 12B

Cis Double Bonds in Unsaturated Fatty Acids

Fatty acids consist of a carboxylic acid group on one end and a fatty chain on the other. The most common of these have between 12 and 20 carbons in the unbranched hydrocarbon chain. They are the major building blocks of all fats and oils, including animal fats and vegetable oils (Section 21-3). Those that have no carbon–carbon double bond in their hydrocarbon chain are called saturated fatty acids. Those with a single carbon–carbon double bond are called monounsaturated fatty acids, and those with two or more carbon–carbon double bonds in their hydrocarbon chain are called polyunsaturated fatty acids. Following are structural formulas for three C_{18} fatty acids. Notice that all carbon–carbon double bonds in oleic and linolenic acids (both unsaturated fatty acids) have the *cis* configuration. In addition, note that the melting points decrease as the number of double bonds increases. We will describe the physical state of fatty acids in more detail in Chapter 18.

Stearic acid
mp 70°C
(a C_{18} saturated fatty acid)

Oleic acid
mp 13°C
(a C_{18} monounsaturated fatty acid)

Linolenic acid
mp −17°C
(a C_{18} polyunsaturated fatty acid)

Test your knowledge with Problem 12-62.

Ethylene   Propene

If we look at a molecular model of ethylene, we see that the two carbons of the double bond and the four hydrogens bonded to them all lie in the same plane—that is, ethylene is a flat or planar molecule. Furthermore, chemists have discovered that under normal conditions, no rotation is possible about the carbon–carbon double bond of ethylene or, for that matter, of any other alkene. Whereas free rotation occurs about each carbon–carbon single bond in an alkane (Section 11-7A), rotation about the carbon–carbon double bond in an alkene does not normally take place because the bond is so rigid. For an important exception to this generalization about carbon–carbon double bonds, see Chemical Connections 12D on *cis-trans* isomerism in vision.

B. Cis-Trans Stereoisomerism in Alkenes

Cis-trans isomers Isomers that have the same connectivity of their atoms but a different arrangement of their atoms in space. Specifically, *cis* and *trans* stereoisomers result from the presence of either a ring or a carbon–carbon double bond.

Because of the restricted rotation about a carbon–carbon double bond, an alkene in which each carbon of the double bond has two different groups bonded to it shows *cis-trans* isomerism (a type of stereoisomerism). For example, 2-butene has two **cis-trans isomers**. In *cis*-2-butene, the two methyl groups are located on the same side of the double bond and the two hydrogens are on the other side. In *trans*-2-butene, the two methyl groups are located on opposite sides of the double bond. *Cis*-2-butene and *trans*-2-butene are different compounds and have different physical and chemical properties.

cis-2-Butene
mp –139°C, bp 4°C

trans-2-Butene
mp –106°C, bp 1°C

12-3 How Do We Name Alkenes and Alkynes?

Alkenes and alkynes are named using the IUPAC system of nomenclature. As we will see, some are still referred to by their common names.

A. IUPAC Names

The key to the IUPAC system of naming alkenes is the ending **-ene**. Just as the ending *-ane* tells us that a hydrocarbon chain contains only carbon–carbon single bonds, the ending *-ene* tells us that it contains a carbon–carbon double bond. To name an alkene:

1. Find the longest carbon chain that includes the double bond. Indicate the length of the parent chain by using a prefix that tells the number of carbon atoms in it (see Table 11-2) and the suffix *-ene* to show that it is an alkene.
2. Number the chain from the end that gives the lower set of numbers to the carbon atoms of the double bond. Designate the position of the double bond by the number of its first carbon.
3. Branched alkenes are named in a manner similar to alkanes; substituent groups are located and named.

CH₃CH₂CH₂CH₂CH=CH₂
1-Hexene

CH₃CH₂CHCH₂CH=CH₂
          |
         CH₃
4-Methyl-1-hexene

CH₃CH₂CHC=CH₂
       |
      CH₂CH₃
with CH₂CH₃
2,3-Diethyl-1-pentene

Note that although 2,3-diethyl-1-pentene has a six-carbon chain, the longest chain that contains the double bond has only five carbons. The parent alkene is, therefore, a pentene rather than a hexene, and the molecule is named as a disubstituted 1-pentene.

The key to the IUPAC name of an alkyne is the ending **-yne**, which shows the presence of a carbon–carbon triple bond. Thus, HC≡CH is ethyne (or acetylene) and CH₃C≡CH is propyne. In higher alkynes, number the longest carbon chain that contains the triple bond from the end that gives the lower set of numbers to the triply bonded carbons. Indicate the location of the triple bond by the number of its first carbon atom.

$$\underset{\underset{CH_3}{|}}{\overset{4\ \ 3\ \ 2\ \ 1}{CH_3CHC\equiv CH}} \qquad \qquad \underset{\underset{CH_3}{|}}{\overset{1\ \ 2\ \ 3\ \ \ \ 4\ 5\ \ \ 6|\ 7}{CH_3CH_2C\equiv CCH_2\overset{\overset{CH_3}{|}}{C}CH_3}}$$

3-Methyl-1-butyne   6,6-Dimethyl-3-heptyne

EXAMPLE 12-1 IUPAC Names of Alkenes and Alkynes

Write the IUPAC name of each unsaturated hydrocarbon.

(a) $CH_2=CH(CH_2)_5CH_3$ (b) $\underset{H_3C}{\overset{H_3C}{\diagdown}}C=C\underset{H}{\overset{CH_3}{\diagup}}$ (c) $CH_3(CH_2)_2C\equiv CCH_3$

Strategy
Step 1: Locate the parent chain—the longest chain of carbon atoms that contains the carbon–carbon double or triple bond.
Step 2: Number the parent chain from the direction that gives the carbons of the double or triple bond the lower set of numbers. Show the presence of the multiple bond by the suffix *-ene* (for a double bond) or *-yne* (for a triple bond). Indicate the location of the multiple bond by the number of its first carbon atom.
Step 3: Name and locate all substituents on the parent chain. List them in alphabetical order.

Solution
(a) The parent chain contains eight carbons; thus, the parent alkene is octene. To show the presence of the carbon–carbon double bond, use the suffix *-ene*. Number the chain beginning with the first carbon of the double bond. This alkene is 1-octene.
(b) Because there are four carbon atoms in the chain containing the carbon–carbon double bond, the parent alkene is butene. The double bond is between carbons 2 and 3 of the chain, and there is a methyl group on carbon 2. This alkene is 2-methyl-2-butene.
(c) There are six carbons in the parent chain, with the triple bond between carbons 2 and 3. This alkyne is 2-hexyne.

Problem 12-1
Write the IUPAC name of each unsaturated hydrocarbon.

B. Common Names

Despite the precision and universal acceptance of IUPAC nomenclature, some alkenes and alkynes—particularly those of low molecular weight—are known almost exclusively by their common names. Three examples follow:

	$CH_2=CH_2$	$CH_3CH=CH_2$	$\underset{\underset{}{}}{\overset{\overset{CH_3}{|}}{CH_3C=CH_2}}$
IUPAC name:	Ethene	Propene	2-Methylpropene
Common name:	Ethylene	Propylene	Isobutylene

We derive common names for alkynes by prefixing the names of the substituents on the carbon–carbon triple bond to the name *acetylene*:

	HC≡CH	CH₃C≡CH	CH₃C≡CCH₃
IUPAC name:	Ethyne	Propyne	2-Butyne
Common name:	Acetylene	Methylacetylene	Dimethylacetylene

C. *Cis* and *Trans* Configurations of Alkenes

The orientation of the carbon atoms of the parent chain determines whether an alkene is *cis* or *trans*. If the carbons of the parent chain are on the same side of the double bond, the alkene is *cis*; if they are on opposite sides, it is a *trans* alkene. In the first example below, they are on opposite sides and the compound is a *trans* alkene. In the second example, they are on the same side and the compound is a *cis* alkene.

trans-3-Hexene

cis-3,4-Dimethyl-2-pentene

EXAMPLE 12-2 Naming Alkene *Cis* and *Trans* Isomers

Name each alkene and specify its configuration by indicating *cis* or *trans* where appropriate.

(a) CH₃CH₂CH₂   H
         \\   /
          C=C
         /   \\
        H   CH₂CH₃

(b) CH₃CH₂CH₂   CH₂CH₃
         \\   /
          C=C
         /   \\
        CH₃   H

Strategy

For alkenes that show *cis-trans* isomerism, use the designator *cis* to show that the carbon atoms of the parent chain are on the same side of the double bond and *trans* to show that they are on opposite sides of the double bond.

Solution

(a) The chain contains seven carbon atoms and is numbered from the right to give the lower number to the first carbon of the double bond. The carbon atoms of the parent chain are on opposite sides of the double bond. This alkene is *trans*-3-heptene.

(b) The longest chain contains seven carbon atoms and is numbered from the right so that the first carbon of the double bond is carbon 3 of the chain. The carbon atoms of the parent chain are on the same side of the double bond. This alkene is *cis*-4-methyl-3-heptene.

Problem 12-2

Name each alkene and specify its configuration.

(a) (b)

D. Cycloalkenes

In naming **cycloalkenes**, number the carbon atoms of the ring double bond 1 and 2 in the direction that gives the substituent encountered first the lower number. It is not necessary to use a location number for the carbons of the double bond, because according to the IUPAC system of nomenclature, the carbon atoms of the double bond of a cycloalkene will always be carbons 1 and 2. Number substituents and list them in alphabetical order.

3-Methylcyclopentene
(not 5-methylcyclopentene)

4-Ethyl-1-methylcyclohexene
(not 5-ethyl-2-methylcyclohexene)

Interactive Example 12-3 — Naming Cycloalkenes

Write the IUPAC name for each cycloalkene.

(a)   (b)   (c)

Strategy

In naming cycloalkenes, the carbon atoms of the double bond are always numbered 1 and 2 in the direction that gives the substituent encountered first the lowest possible number. If there are multiple substituents, list them in alphabetical order.

Solution

(a) 3,3-Dimethylcyclohexene
(b) 1,2-Dimethylcyclopentene
(c) 4-Isopropyl-1-methylcyclohexene

Problem 12-3

Write the IUPAC name for each cycloalkene.

(a)   (b)   (c)

CHEMICAL CONNECTIONS 12C

The Case of the Iowa and New York Strains of the European Corn Borer

Although humans communicate largely by sight and sound, the vast majority of other species in the animal world communicate by chemical signals. Often, communication within a species is specific for one of two or more stereoisomers. For example, a member of a given species might respond to the *cis* isomer of a chemical but not to the *trans* isomer. Alternatively, it might respond to a precise ratio of *cis* and *trans* isomers but not to other ratios of these same stereoisomers.

Several groups of scientists have studied the components of the sex **pheromones** of both the Iowa and New York strains of the European corn borer. Females of these closely related species secrete the sex attractant 11-tetradecenyl acetate. Males of the Iowa strain show their maximum response to a mixture containing 96% of the *cis* isomer and 4% of the *trans* isomer. When the pure *cis* isomer is used alone, males are only weakly attracted. Males of the New York strain show an entirely different response pattern: They respond maximally to a mixture containing 3% of the *cis* isomer and 97% of the *trans* isomer.

trans-11-Tetradecenyl acetate

cis-11-Tetradecenyl acetate

Evidence suggests that an optimal response to a narrow range of stereoisomers, as we see here, is widespread in nature and that most insects maintain species isolation for mating and reproduction by the stereochemical mixtures of their pheromones.

Test your knowledge with Problems 12-63 through 12-65.

E. Dienes, Trienes, and Polyenes

We name alkenes that contain more than one double bond as alkadienes, alkatrienes, and so on. We often refer to those that contain several double bonds more generally as polyenes (Greek: *poly*, many). Following are three dienes:

$CH_2=CHCH_2CH=CH_2$
1,4-Pentadiene

$CH_2=CCH=CH_2$ with CH_3 substituent
2-Methyl-1,3-butadiene
(Isoprene)

1,3-Cyclopentadiene

We saw earlier that for an alkene with one carbon–carbon double bond that can show *cis-trans* isomerism, two stereoisomers are possible. For an alkene with n carbon–carbon double bonds, each of which can show *cis-trans* isomerism, 2^n stereoisomers are possible.

EXAMPLE 12-4 | *Cis-Trans* Isomerism

How many stereoisomers are possible for 2,4-heptadiene?

$$CH_3-CH=CH-CH=CH-CH_2-CH_3$$
2,4-Heptadiene

Strategy

To show *cis-trans* isomerism, each carbon of the double bond must have two different groups bonded to it.

Solution

This molecule has two carbon–carbon double bonds, each of which shows *cis-trans* isomerism. As shown in the following table, $2^2 = 4$ stereoisomers are possible. Line-angle formulas for two of these dienes are drawn here.

	Double-Bond	
	C_2–C_3	C_4–C_5
(1)	trans	trans
(2)	trans	cis
(3)	cis	trans
(4)	cis	cis

trans,trans-2,4-Heptadiene   *trans,cis*-2,4-Heptadiene

Problem 12-4

Draw structural formulas for the other two stereoisomers of 2,4-heptadiene.

EXAMPLE 12-5 Drawing Alkene *Cis-Trans* Isomers

Draw all stereoisomers that are possible for the following unsaturated alcohol.

$$CH_3C=CHCH_2CH_2C=CHCH_2OH$$
(with CH_3 groups on each of the double-bond carbons)

Strategy

To show *cis-trans* isomerism, each carbon of the double bond must have two different groups bonded to it. If a molecule has n double bonds about which *cis-trans* isomerism is possible, then 2^n isomers are possible, where n is the number of double bonds that show *cis-trans* isomerism.

Solution

Cis-trans isomerism is possible only about the double bond between carbons 2 and 3 of the chain. It is not possible for the double bond between carbons 6 and 7 because carbon 7 has two identical groups on it (review Section 12-2B). Thus, $2^1 = 2$ stereoisomers (one *cis-trans* pair) are possible. The *trans* isomer of this alcohol, named geraniol, is a major component of the oils of rose, citronella, and lemon grass. ◀

The *trans* isomer   The *cis* isomer

▶ Lemon grass.

Problem 12-5

How many stereoisomers are possible for the following unsaturated alcohol?

$$CH_3C=CHCH_2CH_2C=CHCH_2CH_2C=CHCH_2OH$$
(with CH_3 groups on each of the three double-bond carbons)

An example of a biologically important polyunsaturated alcohol for which a number of *cis-trans* stereoisomers are possible is vitamin A. Each of the four carbon–carbon double bonds in the chain of carbon atoms bonded to the substituted cyclohexene ring has the potential for *cis-trans* isomerism. There are, therefore, $2^4 = 16$ stereoisomers possible for this structural formula. Vitamin A, the stereoisomer shown here, is the all-*trans* isomer.

Vitamin A (retinol)

CHEMICAL CONNECTIONS 12D

Cis-Trans Isomerism in Vision

The retina, the light-detecting layer in the back of our eyes, contains reddish compounds called visual pigments. Their name, rhodopsin, is derived from the Greek word meaning "rose-colored." Each rhodopsin molecule is a combination of one molecule of a protein called opsin and one molecule of 11-*cis*-retinal, a derivative of vitamin A in which the CH_2OH group of carbon 15 is converted to an aldehyde group, $-CH=O$.

When rhodopsin absorbs light energy, the less stable 11-*cis* double bond is converted to the more stable 11-*trans* double bond. This isomerization changes the shape of the rhodopsin molecule, which in turn causes the neurons of the optic nerve to fire and produce a visual image.

The retinas of vertebrates contain two kinds of rhodopsin-containing cells: rods and cones. Cones function in bright light and are used for color vision; they are concentrated in the central portion of the retina, called the macula, and are responsible for the greatest visual acuity. The remaining area of the retina consists mostly of rods, which are used for peripheral and night vision. 11-*cis*-retinal is present in both cones and rods. Rods have one kind of opsin, whereas cones have three kinds: one for blue, one for green, and one for red color vision.

Vitamin A (retinol) → (enzyme-catalyzed oxidation) → Vitamin A aldehyde (11-*trans*-retinal)

11-*cis*-Retinal + H_2N—opsin, $-H_2O$ → Rhodopsin (visual purple)

enzyme-catalyzed isomerization of the 11-*trans* double bond to 11-*cis*

11-*trans*-Retinal ← H_2O, opsin removed ←

1. light strikes rhodopsin
2. the 11-*cis* double bond is isomerized to 11-*trans*
3. a nerve impulse is sent via the optic nerve to the visual cortex

Test your knowledge with Problems 12-66 and 12-67.

12-4 What Are the Physical Properties of Alkenes and Alkynes?

Alkenes and alkynes are nonpolar compounds, and the only attractive forces between their molecules are very weak London dispersion forces (Section 5-7A). Their physical properties, therefore, are similar to those of alkanes with the same carbon skeletons. Alkenes and alkynes that are liquid at room temperature have densities less than 1.0 g/mL (they float on water). They are insoluble in water but are soluble in one another and in other nonpolar organic liquids.

12-5 What Are the Characteristic Reactions of Alkenes?

The most characteristic reaction of alkenes is an addition to their carbon–carbon double bond: the double bond is broken, and in its place, single bonds form between the carbons and two new atoms or groups of atoms. Table 12-1 shows several examples of alkene addition reactions along with the descriptive name(s) associated with each reaction.

At this point, you might ask why a carbon–carbon double bond is a site of chemical reactivity, whereas carbon–carbon single bonds are quite unreactive under most experimental conditions. One way to answer this question is to focus on the changes in bonding that occur as a result of an alkene addition reaction. Consider, for example, the addition of hydrogen (H_2) to ethylene. As a result of this addition, one double bond and one single bond (the H—H bond of H_2) are replaced by three single bonds, giving a net conversion of one bond of the double bond to two single bonds.

$$CH_2=CH_2 + H-H \longrightarrow H-CH_2-CH_2-H + \text{heat}$$

one double bond and one single bond   —are replaced by→   three single bonds

Table 12-1 Characteristic Addition Reactions of Alkenes

Reaction	Descriptive Name(s)
$\text{C=C} + HCl \longrightarrow -\overset{H}{\underset{|}{C}}-\overset{Cl}{\underset{|}{C}}-$	hydrochlorination
$\text{C=C} + H_2O \longrightarrow -\overset{H}{\underset{|}{C}}-\overset{OH}{\underset{|}{C}}-$	hydration
$\text{C=C} + Br_2 \longrightarrow -\overset{Br}{\underset{|}{C}}-\overset{Br}{\underset{|}{C}}-$	bromination
$\text{C=C} + H_2 \longrightarrow -\overset{H}{\underset{|}{C}}-\overset{H}{\underset{|}{C}}-$	hydrogenation (reduction)

This and almost all other addition reactions of alkenes are exothermic, which means that the products are more stable (have lower energy) than the reactants. Just because an alkene addition reaction is exothermic, however, doesn't mean that it occurs rapidly. The rate of a chemical reaction depends on its activation energy, not on how exothermic or endothermic it is (Section 7-3). In fact, the addition of H_2 to an alkene is immeasurably slow at room temperature but, as we will see soon, proceeds quite rapidly in the presence of a suitable transition metal catalyst.

A. Addition of Hydrogen Halides (Hydrohalogenation)

Haloalkanes (alkyl halides) are formed when a hydrogen halide (HCl, HBr, and HI) is added to an alkene. Addition of HCl to ethylene, for example, gives chloroethane (ethyl chloride):

$$CH_2=CH_2 + HCl \longrightarrow \underset{\underset{H}{|}}{CH_2}-\underset{\underset{Cl}{|}}{CH_2}$$

Ethylene → Chloroethane (Ethyl chloride)

Addition of HCl to propene gives 2-chloropropane (isopropyl chloride); hydrogen adds to carbon 1 of propene and chlorine adds to carbon 2. If the orientation of addition were reversed, 1-chloropropane (propyl chloride) would form. The observed result is that almost no 1-chloropropane forms. Because 2-chloropropane is the observed product, we say that addition of HCl to propene is **regioselective**.

$$\underset{\text{Propene}}{CH_3\overset{2}{C}H=\overset{1}{C}H_2} + HCl \longrightarrow \underset{\text{2-Chloropropane}}{CH_3\underset{\underset{Cl}{|}}{CH}-\underset{\underset{H}{|}}{CH_2}} \quad \underset{\substack{\text{1-Chloropropane} \\ \text{(not formed)}}}{\cancel{CH_3\underset{\underset{H}{|}}{CH}-\underset{\underset{Cl}{|}}{CH_2}}}$$

Regioselective A reaction in which one direction of bond forming or bond breaking occurs in preference to all other directions

This regioselectivity was noted by Vladimir Markovnikov (1838–1904), who made the following generalization, known as **Markovnikov's rule**: in the addition of HX (where X = halogen) to an alkene, hydrogen adds to the doubly bonded carbon that already has the greater number of hydrogens bonded to it; halogen adds to the other carbon. Markovnikov's rule is often paraphrased as "the rich get richer."

Markovnikov's rule In the addition of HX or H_2O to an alkene, hydrogen adds to the carbon of the double bond having the greater number of hydrogens

EXAMPLE 12-6 Addition of HX to an Alkene

Draw a structural formula for the product of each alkene addition reaction.

(a) $CH_3\underset{\underset{CH_3}{|}}{C}=CH_2 + HI \longrightarrow$

(b) [cyclopentene with CH_3 substituent] + HCl ⟶

Strategy

Use Markovnikov's rule to predict the structural formula for the product of each reaction. The H from the HI and HCl will add to the carbon of the double bond that already has the greater number of H atoms bonded to it.

Solution

(a) Markovnikov's rule predicts that the hydrogen of HI adds to carbon 1 and iodine adds to carbon 2 to give 2-iodo-2-methylpropane.

(b) H adds to carbon 2 of the ring and Cl adds to carbon 1 to give 1-chloro-1-methylcyclopentane.

(a) 2-Iodo-2-methylpropane: CH₃C(CH₃)(I)CH₃ — structure shown with CH₃ on top, I on bottom, of central C bonded to two CH₃ groups.

(b) 1-Chloro-1-methylcyclopentane: cyclopentane ring with Cl and CH₃ on carbon 1 (carbon 2 labeled).

Problem 12-6

Draw a structural formula for the product of each alkene addition reaction.

(a) $CH_3CH=CH_2 + HBr \longrightarrow$

(b) cyclohexylidene $=CH_2 + HBr \longrightarrow$

Markovnikov's rule tells us what happens when we add HCl, HBr, or HI to a carbon–carbon double bond. We know that in the addition of HCl or another halogen acid, one bond of the double bond and the H—Cl bond are broken and that new C—H and C—Cl bonds form. But chemists also want to know how this conversion happens. Are the C=C and H—X bonds broken and both new covalent bonds formed all at the same time? Or does this reaction take place in a series of steps? If the latter, what are these steps and in what order do they take place?

Before we tackle the issue of the steps by which alkene addition reactions take place, let us take a moment for an overview of the most common reaction steps we will encounter and see how these steps build on the chemistry already presented.

Pattern 1: Add a proton. In Section 8-1, we learned that "an acid is a proton donor, a base is a proton acceptor, and an acid–base reaction is a proton-transfer reaction." We also saw that we can use curved arrows to show how a proton-transfer reaction takes place as, for example, in the acid–base reaction between acetic acid and ammonia to form acetate ion and ammonium ion.

$$CH_3-\overset{\overset{\displaystyle :O:}{\|}}{C}-\ddot{O}-H \; + \; :NH_3 \; \rightleftharpoons \; CH_3-\overset{\overset{\displaystyle :O:}{\|}}{C}-\ddot{O}:^- \; + \; H-\overset{+}{N}H_3$$

Acetic acid (proton donor)   Ammonia (proton acceptor)   Acetate ion   Ammonium ion

Pattern 2: Take a proton away. If we run the above reaction in reverse, then it corresponds to taking a proton away from the ammonium ion and transferring it to the acetate ion. We can also use curved arrows to show the flow of electron pairs in this type of reaction.

Pattern 3: Reaction of an electrophile and a nucleophile to form a new covalent bond. Another characteristic pattern is the reaction between an **electrophile** (an electron-poor species that can accept a pair of electrons to form a new covalent bond) and a **nucleophile** (an electron-rich species that can donate a pair of electrons to form a new covalent bond). An example of this type of reaction is that between a carbocation and a halide ion.

The driving force behind this reaction is the strong electrostatic attraction between the positive and negative charges of the reacting species and

Electrophile An electron-poor species that can accept a pair of electrons to form a new covalent bond.

Nucleophile An electron-rich species that can donate a pair of electrons to form a new covalent bond.

the energy released when the covalent bond forms. The following equation shows the flow of electron pairs in this type of reaction.

$$\overset{+}{CH_2CH}-CH_2-CH_3 \ + \ :\!\ddot{\underset{..}{Cl}}\!:^- \ \longrightarrow \ CH_3\overset{\overset{\displaystyle :\!\ddot{\underset{..}{Cl}}\!:}{|}}{CH}-CH_2-CH_3$$

A carbocation (an electrophile) Chloride ion (a nucleophile)

Pattern 4: Reaction of a proton donor with a carbon–carbon double bond to form a new covalent bond. The double bond (in this case a nucleophile), provides the pair of electrons that forms a new covalent bond. This pattern is typical in all alkene reactions in which the reaction is catalyzed by an acid. As you study it, note that in this step the carbon–carbon double bond serves as the nucleophile that provides the electron pair to form the new covalent bond. Remember that in a carbon–carbon double bond, two pairs of electrons are shared between the two carbons. An acid–base reaction in which a double bond provides the pair of electrons for the hydrogen transfer creates a carbocation. And remember that, as shown in Section 8-1, a proton, H^+, does not exist as such in aqueous solution. Instead, it immediately combines with a water molecule to form the hydronium ion, H_3O^+. The following reaction shows the flow of electrons in this type of reaction.

$$CH_3-CH=CH-CH_3 \ + \ H-\overset{+}{\underset{\underset{H}{|}}{\ddot{O}}}-H \ \longrightarrow \ \overset{+}{CH_2CH}-CH_2-CH_3 \ + \ H-\ddot{\underset{..}{O}}-H$$

An alkene (proton acceptor)   Hydronium ion (proton donor)   A carbocation (an electrophile)

While the above equation is the most accurate way to show the proton transfer in an aqueous solution, we will simplify the equation to show just the proton and the formation of the new covalent bond.

$$CH_3-CH=CH-CH_3 \ + \ H^+ \ \longrightarrow \ \overset{+}{CH_2CH}-CH_2-CH_3$$

An alkene (proton acceptor)   A proton   A carbocation (an electrophile)

When we come to analyze reactions taking place in biological systems, we will find that the reaction mediums are rich mixtures of proton donors and proton acceptors. Many of these proton donors and acceptors are present on the enzymes that catalyze biochemical reactions. In fact, some biochemists even talk about wraparound enzymes, which have within their three-dimensional structures both proton acceptors and proton donors. It is the presence of these proton donors and acceptors that gives enzymes their remarkable catalytic ability.

Chemists account for the addition of HX to an alkene by defining a two-step **reaction mechanism**, which we illustrate for the reaction of 2-butene with hydrogen chloride to give 2-chlorobutane. Step 1 is the addition of H^+ to 2-butene. To show this addition, we use a **curved arrow** that shows the repositioning of an electron pair from its origin (the tail of the arrow) to its new location (the head of the arrow). Recall that we used curved arrows in Section 8-1 to show bond breaking and bond formation in proton-transfer reactions. We now use curved arrows in the same way to show bond breaking and bond formation in a reaction mechanism.

Step 1 results in the formation of an organic cation. One carbon atom in this cation has only six electrons in its valence shell, so it carries a charge of +1. A species containing a positively charged carbon atom is called a **carbocation** (*carb*on + *cation*). Carbocations are classified as primary (1°),

Reaction mechanism A step-by-step description of how a chemical reaction occurs

Carbocation A species containing a carbon atom with only three bonds to it and bearing a positive charge secondary (2°), or tertiary (3°), depending on the number of carbon groups bonded to the carbon bearing the positive charge.

Mechanism: Addition of HCl to 2-Butene

Step 1: Add a proton. Reaction of the carbon–carbon double bond of the alkene with H^+ forms a 2° carbocation intermediate. In forming this intermediate, one bond of the double bond breaks and its pair of electrons is used to form a new covalent bond with H^+. One carbon of the double bond is then left with only six electrons in its valence shell and therefore has a positive charge.

$$CH_3CH=CHCH_3 + H^+ \longrightarrow CH_3\overset{+}{C}H-\underset{\underset{H}{|}}{C}HCH_3$$

A 2° carbocation intermediate

Step 2: Reaction of an electrophile and a nucleophile to form a new covalent bond. Reaction of the 2° carbocation intermediate with a chloride ion completes the valence shell of carbon and gives 2-chlorobutane.

$$:\ddot{\underset{..}{Cl}}:^- + CH_3\overset{+}{C}HCH_2CH_3 \longrightarrow CH_3\underset{\underset{:\ddot{Cl}:}{|}}{C}HCH_2CH_3$$

Chloride ion   A 2° carbocation intermediate   2-Chlorobutane (sec-Butyl chloride)

EXAMPLE 12-7   Mechanism of Addition of HX to an Alkene

Propose a two-step mechanism for the addition of HI to methylenecyclohexane to give 1-iodo-1-methylcyclohexane.

Methylenecyclohexane   1-Iodo-1-methylcyclohexane

Strategy

The mechanism for the addition of HI to an alkene is similar to the two-step mechanism proposed for the addition of HCl to 2-butene.

Solution

Step 1: Add a proton. In this step, the carbon–carbon double bond is a nucleophile and H^+ is the electrophile. Reaction of H^+ with the carbon–carbon double bond forms a new C—H bond with the carbon bearing the greater number of hydrogens and gives a 3° carbocation intermediate.

A 3° carbocation intermediate

Step 2: Reaction of an electrophile and a nucleophile to form a new covalent bond. Reaction of the 3° carbocation intermediate with iodide ion completes the valence shell of carbon and gives the product.

Problem 12-7

Propose a two-step mechanism for the addition of HBr to 1-methylcyclohexene to give 1-bromo-1-methylcyclohexane.

B. Addition of Water: Acid-Catalyzed Hydration

In the presence of an acid catalyst, most commonly concentrated sulfuric acid, water adds to the carbon–carbon double bond of an alkene to give an alcohol. Addition of water is called **hydration**. In the case of simple alkenes, hydration follows Markovnikov's rule: H of H_2O adds to the carbon of the double bond with the greater number of hydrogens, and OH of H_2O adds to the carbon with the smaller number of hydrogens. Most industrial ethanol is made by the acid-catalyzed hydration of ethylene.

Hydration Addition of water

$$CH_2{=}CH_2 + H_2O \xrightarrow{H_2SO_4} CH_2{-}CH_2$$
$$\phantom{CH_2{=}CH_2 + H_2O \xrightarrow{H_2SO_4} }\;\;\;\;|\;\;\;\;|$$
$$\phantom{CH_2{=}CH_2 + H_2O \xrightarrow{H_2SO_4} CH_2}H\;\;OH$$

Ethylene → Ethanol

$$CH_3CH{=}CH_2 + H_2O \xrightarrow{H_2SO_4} CH_3CH{-}CH_2$$

Propene → 2-Propanol (OH on CH, H on CH_2)

$$CH_3\underset{\underset{CH_3}{|}}{C}{=}CH_2 + H_2O \xrightarrow{H_2SO_4} CH_3\underset{\underset{HO\;\;H}{|\;\;\;|}}{\overset{\overset{CH_3}{|}}{C}}{-}CH_2$$

2-Methylpropene → 2-Methyl-2-propanol

EXAMPLE 12-8 Acid-Catalyzed Hydration of an Alkene

Draw a structural formula for the alcohol formed by the acid-catalyzed hydration of 1-methylcyclohexene.

Strategy

Markovnikov's rule predicts that H adds to the double-bonded carbon with the greater number of hydrogens.

Solution

H adds to carbon 2 of the cyclohexene ring, and OH then adds to carbon 1.

1-Methylcyclohexene + H_2O $\xrightarrow{H_2SO_4}$ 1-Methylcyclohexanol

Problem 12-8

Draw a structural formula for the alcohol formed by acid-catalyzed hydration of each alkene:
(a) 2-Methyl-2-butene (b) 2-Methyl-1-butene

The mechanism for the acid-catalyzed hydration of an alkene is similar to what we proposed for the addition of HCl, HBr, and HI to an alkene and

314 Chapter 12 Alkenes and Alkynes

is illustrated by the hydration of propene. This mechanism is consistent with the fact that acid is a catalyst. One H^+ is consumed in Step 1, but another is generated in Step 3.

Mechanism: Acid-Catalyzed Hydration of Propene

Step 1: Add a proton. Addition of H^+ to the carbon of the double bond with the greater number of hydrogens gives a 2° carbocation intermediate.

$$CH_3CH=CH_2 + H^+ \longrightarrow CH_3\overset{+}{C}HCH_2H$$

A 2° carbocation intermediate

Step 2: Reaction of an electrophile and a nucleophile to form a new covalent bond. The carbocation intermediate completes its valence shell by forming a new covalent bond with an unshared pair of electrons of the oxygen atom of H_2O to give an **oxonium ion**.

Oxonium ion An ion in which oxygen is bonded to three other atoms and bears a positive charge

$$CH_3\overset{+}{C}HCH_3 + :\overset{..}{O}-H \longrightarrow CH_3CHCH_3 \overset{H\ \ \ \ H}{\underset{\overset{|}{\overset{+}{O}}}{}}$$
$$\phantom{CH_3\overset{+}{C}HCH_3 + :\overset{..}{O}-H \longrightarrow}\ \ \ \ \ \ \text{An oxonium ion}$$

Step 3: Take a proton away. Loss of H^+ from the oxonium ion gives the alcohol and generates a new H^+ catalyst.

$$CH_3CHCH_3 \overset{H\ \ \ \ H}{\underset{\overset{|}{\overset{+}{O}}}{}} \longrightarrow CH_3CHCH_3 \overset{:\overset{..}{O}H}{\underset{|}{}} + H^+$$

EXAMPLE 12-9 | Acid-Catalyzed Hydration of an Alkene

Propose a three-step reaction mechanism for the acid-catalyzed hydration of methylenecyclohexane to give 1-methylcyclohexanol.

Strategy

The reaction mechanism for the acid-catalyzed hydration of methylenecyclohexane is similar to the three-step mechanism proposed for the acid-catalyzed hydration of propene.

Solution

Step 1: Add a proton. Reaction of the carbon–carbon double bond with H^+ gives a 3° carbocation intermediate.

⟨⟩=CH_2 + H^+ ⟶ ⟨⟩⁺—CH_3

Methylenecyclohexane   A 3° carbocation intermediate

Step 2: Reaction of the carbocation intermediate with water completes the valence shell of carbon and gives an oxonium ion.

⟨⟩⁺—CH_3 + :Ö—H ⟶ ⟨⟩—Ö⁺(H)—CH_3
               |
               H

An oxonium ion

Step 3: Take a proton away. Loss of H⁺ from the oxonium ion completes the reaction and generates a new H⁺ catalyst.

Problem 12-9
Propose a three-step reaction mechanism for the acid-catalyzed hydration of 1-methylcyclohexene to give 1-methylcyclohexanol.

C. Addition of Bromine and Chlorine (Halogenation)

Chlorine, Cl_2, and bromine, Br_2, react with alkenes at room temperature by addition of halogen atoms to the carbon atoms of the double bond. This reaction is generally carried out by mixing the pure reagents together or by mixing them in an inert solvent such as dichloromethane, CH_2Cl_2.

$$CH_3CH=CHCH_3 + Br_2 \xrightarrow{CH_2Cl_2} CH_3CH(Br)-CH(Br)CH_3$$
2-Butene → 2,3-Dibromobutane

Cyclohexene + Br_2 → 1,2-Dibromocyclohexane

Addition of bromine is a useful qualitative test for the presence of an alkene. If we dissolve bromine in carbon tetrachloride, the solution is red. In contrast, alkenes and dibromoalkanes are colorless. If we mix a few drops of the red bromine solution with an unknown sample suspected of being an alkene, disappearance of the red color as bromine adds to the double bond tells us that an alkene is, indeed, present.

EXAMPLE 12-10 Addition of Halogens to an Alkene

Complete these reactions.

(a) cyclopentene + Br_2 $\xrightarrow{CH_2Cl_2}$

(b) 1-methylcyclohexene + Cl_2 $\xrightarrow{CH_2Cl_2}$

Strategy

When Br_2 or Cl_2 are added to a cycloalkene, one halogen adds to each carbon of the double bond.

Solution

(a) cyclopentene + Br_2 $\xrightarrow{CH_2Cl_2}$ 1,2-dibromocyclopentane

(b) 1-methylcyclohexene + Cl_2 $\xrightarrow{CH_2Cl_2}$ 1,2-dichloro-1-methylcyclohexane

> **Problem 12-10**
>
> Complete these reactions.
>
> (a) $\text{CH}_3\underset{\underset{\text{CH}_3}{|}}{\overset{\overset{\text{CH}_3}{|}}{\text{C}}}\text{CH}=\text{CH}_2 + \text{Br}_2 \xrightarrow{\text{CH}_2\text{Cl}_2}$
>
> (b) cyclohexane=CH$_2$ + Cl$_2$ $\xrightarrow{\text{CH}_2\text{Cl}_2}$

D. Addition of Hydrogen: Reduction (Hydrogenation)

Virtually all alkenes react quantitatively with molecular hydrogen, H_2, in the presence of a transition metal catalyst to give alkanes. Commonly used transition metal catalysts include platinum, palladium, ruthenium, and nickel. Because the conversion of an alkene to an alkane involves reduction by hydrogen in the presence of a catalyst, the process is called **catalytic reduction** or, alternatively, **catalytic hydrogenation**. In Section 21-3, we will see how catalytic hydrogenation is used to solidify liquid vegetable oils so that margarines and semisolid cooking fats are formed.

$$\underset{trans\text{-2-Butene}}{\overset{\text{H}_3\text{C}}{\underset{\text{H}}{>}}\text{C}=\text{C}\overset{\text{H}}{\underset{\text{CH}_3}{<}}} + \text{H}_2 \xrightarrow[25°\text{C, 3 atm}]{\text{Pd}} \underset{\text{Butane}}{\text{CH}_3\text{CH}_2\text{CH}_2\text{CH}_3} \xleftarrow[25°\text{C, 3 atm}]{\text{Pd}} \text{H}_2 + \underset{cis\text{-2-Butene}}{\overset{\text{H}}{\underset{\text{CH}_3}{>}}\text{C}=\text{C}\overset{\text{H}}{\underset{\text{CH}_3}{<}}}$$

$$\underset{\text{Cyclohexene}}{\bigcirc} + \text{H}_2 \xrightarrow[25°\text{C, 3 atm}]{\text{Pd}} \underset{\text{Cyclohexane}}{\bigcirc}$$

The metal catalyst is used in the form of a finely powdered solid. The reaction is carried out by dissolving the alkene in ethanol or another non-reacting organic solvent, adding the solid catalyst, and exposing the mixture to hydrogen gas at pressures ranging from 1 to 150 atm.

Mechanism: Catalytic Reduction

The transition metal catalysts used in catalytic hydrogenation are able to adsorb large quantities of hydrogen onto their surfaces, probably by forming metal–hydrogen bonds. Similarly, alkenes are adsorbed on metal surfaces with the formation of carbon–metal bonds. It is noted here that carbon–metal bonds are formed, but this is not really addressed in Figure 12-1. Addition of hydrogen atoms to an alkene occurs in two steps (Figure 12-1).

FIGURE 12-1 The addition of hydrogen to an alkene involving a transition metal catalyst. (a) Hydrogen and the alkene are adsorbed on the metal surface, and (b) one hydrogen atom is transferred to the alkene, forming one new C—H bond. The other carbon remains adsorbed on the metal surface. (c) A second C—H bond is formed, and the alkene is released.

12-6 What Are the Important Polymerization Reactions of Ethylene and Substituted Ethylenes?

A. Structure of Polyethylenes

From the perspective of the chemical industry, the single most important reaction of alkenes is the formation of chain-growth **polymers** (Greek: *poly*, many, and *meros*, part). In the presence of certain compounds called initiators, many alkenes form polymers made by the stepwise addition of **monomers** (Greek: *mono*, one, and *meros*, part) to a growing polymer chain, as illustrated by the formation of polyethylene from ethylene. In alkene polymers of industrial and commercial importance, n is a large number, typically several thousand.

$$n\,CH_2{=}CH_2 \xrightarrow[\text{(polymerization)}]{\text{initiator}} {-}(CH_2CH_2)_n{-}$$

Ethylene → Polyethylene

Polymers From the Greek *poly*, many, and *meros*, part; any long-chain molecules synthesized by bonding together many single parts called monomers

Monomers From the Greek *mono*, single, and *meros*, part; the simplest nonredundant units from which a polymer is synthesized

To show the structure of a polymer, we place parentheses around the repeating monomer unit. The structure of an entire polymer chain can be reproduced by repeating this enclosed structure in both directions. A subscript n is placed outside the parentheses to indicate that this unit is repeated n times, as illustrated for the conversion of propylene to polypropylene.

Propene → Part of an extended polymer chain of polypropylene / The repeating unit

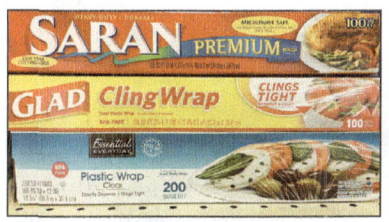

(a)

The most common method of naming a polymer is to attach the prefix **poly-** to the name of the monomer from which the polymer is synthesized—for example, polyethylene and polystyrene. When the name of the monomer consists of two words (for example, the monomer vinyl chloride), its name is enclosed in parentheses. ◀

Vinyl chloride → Poly(vinyl chloride) (PVC)

(b)

▶ Some items made from chain-growth polymers. (*a*) Saran Wrap, a copolymer of vinyl chloride and 1,1-dichloroethylene. (*b*) Items made from polystyrene.

Table 12-2 lists several important polymers derived from ethylene and substituted ethylene, along with their common names and most important uses.

B. Low-Density Polyethylene (LDPE)

The first commercial process for the polymerization of ethylene used **peroxide** initiators at 500°C and 1000 atm and yielded a tough, transparent polymer known as **low-density polyethylene (LDPE)**. At the molecular

Peroxide Any compound that contains an —O—O— bond as, for example, hydrogen peroxide, H—O—O—H

CHEMICAL CONNECTIONS 12E

Recycling Plastics

Plastics are polymers that can be molded when hot and that retain their shape when cooled. Because they are durable and lightweight, plastics are probably the most versatile synthetic materials in existence. In fact, the current production of plastics in the United States exceeds the U.S. production of steel. Plastics have come under criticism, however, for their role in the solid waste crisis. They account for approximately 21% of the volume and 8% of the weight of solid wastes, with most plastic waste consisting of disposable packaging and wrapping.

Six types of plastics are commonly used for packaging applications. In 1988, manufacturers adopted recycling code letters developed by the Society of the Plastics Industry as a means of identifying them.

Currently, only poly(ethylene terephthalate) (PET) and high-density polyethylene (HDPE) are recycled in large quantities. In fact, bottles made of these plastics account for more than 99% of the plastics recycled in the United States.

The synthesis and structure of PET, a polyester, is described in Section 19-6B.

The process for recycling most plastics is simple, with separation of the plastic from other contaminants being the most labor-intensive step. For example, PET soft drink bottles usually have a paper label and adhesive that must be removed before the PET can be reused. Recycling begins with hand or machine sorting, after which the bottles are chopped into small chips. Any ferrous metals are removed by magnets. Any nonferrous metal contaminants are removed by electric eddy currents that cause them to jump like fleas into a bin as they move down a conveyor belt during the sorting process. An air cyclone then removes paper and other lightweight materials. After any remaining labels and adhesives are eliminated with a detergent wash, the PET chips are dried. The PET produced by this method is 99.9% free of contaminants and sells for about half the price of the virgin material. The biggest market for recycled PET in 2005 was fibers. The carpet maker Mohawk Industries, for example, starts with about 250 million lb of recycled PET bottles per year and ends up with 80 to 100 million sq yards of carpet. The largest domestic use of recycled HDPE resins in 2005 was bottles.

Code	Polymer	Common Uses
1 PET	poly(ethylene terephthalate)	soft drink bottles, household chemical bottles, films, textile fibers
2 HDPE	high-density polyethylene	milk and water jugs, grocery bags, squeezable bottles
3 V	poly(vinyl chloride), PVC	shampoo bottles, pipes, shower curtains, vinyl siding, wire insulation, floor tiles
4 LDPE	low-density polyethylene	shrink wrap, trash and grocery bags, sandwich bags, squeeze bottles
5 PP	polypropylene	plastic lids, clothing fibers, bottle caps, toys, diaper linings
6 PS	polystyrene	Styrofoam cups, egg cartons, disposable utensils, packaging materials, appliances
7	all other plastics	various

These students are wearing jackets made from recycled PET soda bottles.

Test your knowledge with Problems 12-68 and 12-69.

Table 12-2 Polymers Derived from Ethylene and Substituted Ethylenes, Along with their Common Names and Most Important Uses.

Monomer Formula	Common Name	Polymer Name(s) and Common Uses
$CH_2{=}CH_2$	ethylene	polyethylene, Polythene; break-resistant containers and packaging materials
$CH_2{=}CHCH_3$	propylene	polypropylene, Herculon; textile and carpet fibers
$CH_2{=}CHCl$	vinyl chloride	poly(vinyl chloride), PVC; construction tubing
$CH_2{=}CCl_2$	1,1-dichloroethylene	poly(1,1-dichloroethylene); Saran Wrap is a copolymer with vinyl chloride
$CH_2{=}CHCN$	acrylonitrile	polyacrylonitrile, Orlon; acrylics and acrylates
$CF_2{=}CF_2$	tetrafluoroethylene	polytetrafluoroethylene, PTFE; Teflon, nonstick coatings
$CH_2{=}CHC_6H_5$	styrene	polystyrene, Styrofoam; insulating materials
$CH_2{=}CHCOOCH_2CH_3$	ethyl acrylate	poly(ethyl acrylate), latex paint
$CH_2{=}C(CH_3)COOCH_3$	methyl methacrylate	poly(methyl methacrylate), Lucite; Plexiglas; glass substitutes

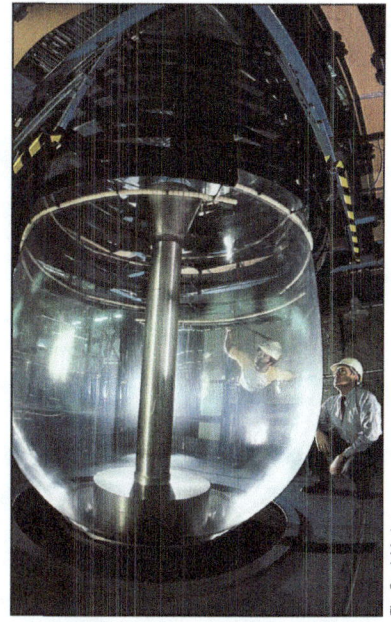

▶ Polyethylene films are produced by extruding molten plastic through a ring-like gap and inflating the film into a balloon.

level, LDPE chains are highly branched, with the result that they do not pack well together and the London dispersion forces (Section 5-7A) between them are weak. LDPE softens and melts at about 115°C, which means that it cannot be used in products that will be exposed to boiling water.

Today, approximately 65% of all LDPE is used for the manufacture of films by the blow-molding technique illustrated in Figure 12-2. LDPE film is inexpensive, which makes it ideal for packaging such consumer items as baked goods and vegetables and for the manufacture of trash bags. ◀

C. High-Density Polyethylene (HDPE)

In the 1950s, Karl Ziegler of Germany and Giulio Natta of Italy developed an alternative method for the polymerization of alkenes, which does not rely on peroxide initiators. Polyethylene from Ziegler-Natta systems, termed **high-density polyethylene (HDPE)**, has little chain branching. Consequently, its chains pack together more closely than those of LDPE, with the result that the London dispersion forces between chains of HDPE are stronger than those in LDPE.

Linear polyethylene (high density)

HDPE has a higher melting point than LDPE and is three to ten times stronger.

FIGURE 12-2 Fabrication of LDPE film. A tube of melted LDPE along with a jet of compressed air is forced through an opening and blown into a gigantic, thin-walled bubble. The film is then cooled and taken up onto a roller. This double-walled film can be slit down the side to give LDPE film or sealed at points along its length to make LDPE bags.

Approximately 45% of all HDPE products are made by the blow-molding process shown in Figure 12-3. HDPE is used for consumer items such as milk and water jugs, grocery bags, and squeezable bottles.

FIGURE 12-3 Blow molding an HDPE container. (a) A short length of HDPE tubing is placed in an open die and the die is closed, sealing the bottom of the tube. (b) Compressed air is forced into the hot polyethylene/die assembly, and the tubing is literally blown up to take the shape of the mold. (c) After the assembly cools, the die is opened, and there is the container!

Summary of Key Questions

Section 12-1 What Are Alkenes and Alkynes?

- An **alkene** is an unsaturated hydrocarbon that contains a carbon–carbon double bond.
- An **alkyne** is an unsaturated hydrocarbon that contains a carbon–carbon triple bond.

Section 12-2 What Are the Structures of Alkenes and Alkynes?

- The structural feature that makes *cis-trans* **stereoisomerism** possible in alkenes is restricted rotation about the two carbons of the double bond.
- The *cis* or *trans* configuration of an alkene is determined by the orientation of the atoms of the parent chain about the double bond.
- If atoms of the parent chain are located on the same side of the double bond, the configuration of the alkene is *cis*; if they are located on opposite sides, the configuration is *trans*.

Section 12-3 How Do We Name Alkenes and Alkynes?

- In IUPAC names, the presence of a carbon–carbon double bond is indicated by a prefix showing the number of carbons in the parent chain and the ending **-ene.** Substituents are numbered and named in alphabetical order.
- The presence of a carbon–carbon triple bond is indicated by a prefix that shows the number of carbons in the parent chain and the ending **-yne.**
- The carbon atoms of the double bond of a cycloalkene are numbered 1 and 2 in the direction that gives the smaller number to the first substituent.
- Compounds containing two double bonds are called **dienes**, those with three double bonds are called **trienes**, and those containing four or more double bonds are called **polyenes**.

Section 12-4 What Are the Physical Properties of Alkenes and Alkynes?

- Because alkenes and alkynes are nonpolar compounds and the only interactions between their molecules are London dispersion forces, their physical properties are similar to those of alkanes with similar carbon skeletons.

Section 12-5 What Are the Characteristic Reactions of Alkenes?

- A characteristic reaction of alkenes is addition to the double bond.
- In addition, the double bond breaks and bonds to two new atoms or groups of atoms form in its place.
- A **reaction mechanism** is a step-by-step description of how a chemical reaction occurs, including the role of the catalyst (if one is present).
- Among the most common steps in a reaction mechanism are:
  - Add a proton.
  - Reaction of an electrophile and a nucleophile to form a new covalent bond.
  - Take a proton away.
- A **carbocation** contains a carbon with only six electrons in its valence shell and bears a positive charge.

Section 12-6 What Are the Important Polymerization Reactions of Ethylene and Substituted Ethylenes?

- **Polymerization** is the process of bonding together many small **monomers** into large, high-molecular-weight **polymers.**

Summary of Key Reactions

1. **Addition of HX (Hydrohalogenation) (Section 12-6A)** Addition of HX to the carbon–carbon double bond of an alkene follows Markovnikov's rule. The reaction occurs in two steps and involves formation of a carbocation intermediate.

2. **Acid-Catalyzed Hydration (Section 12-6B)** Addition of H_2O to the carbon–carbon double bond of an alkene follows Markovnikov's rule. Reaction occurs in three steps and involves formation of carbocation and oxonium ion intermediates.

$$CH_3C=CH_2 + H_2O \xrightarrow{H_2SO_4} CH_3\underset{\underset{OH}{|}}{\overset{\overset{CH_3}{|}}{C}}CH_3$$

3. **Addition of Bromine and Chlorine (Halogenation) (Section 12-6C)** Addition to a cycloalkene gives a 1,2-dihalocycloalkane.

Cyclohexene → 1,2-Dibromocyclohexane

4. **Reduction: Formation of Alkanes (Hydrogenation) (Section 12-6D)** Catalytic reduction involves addition of hydrogen to form two new C—H bonds.

5. **Polymerization of Ethylene and Substituted Ethylenes (Section 12-7A)** In polymerization of alkenes, monomer units bond together without the loss of any atoms.

$$nCH_2=CH_2 \xrightarrow{\text{initiator}} -(CH_2CH_2)_n-$$

Problems

Orange-numbered problems are applied.

Section 12-1 What Are Alkenes and Alkynes?

12-11 Answer true or false.
(a) There are two classes of unsaturated hydrocarbons—alkenes and alkynes.
(b) The bulk of the ethylene used by the chemical industry worldwide is obtained from renewable resources.
(c) Ethylene and acetylene are constitutional isomers.
(d) Cyclohexane and 1-hexene are constitutional isomers.

Section 12-2 What Are the Structures of Alkenes and Alkynes?

12-12 Answer true or false.
(a) Both ethylene and acetylene are planar molecules.
(b) An alkene in which each carbon of the double bond has two different groups bonded to it will show *cis-trans* isomerism.
(c) *Cis-trans* isomers have the same molecular formula but a different connectivity of their atoms.
(d) *Cis*-2-butene and *trans*-2-butene can be interconverted by rotation about the carbon–carbon double bond.
(e) *Cis-trans* isomerism is possible only among appropriately substituted alkenes.
(f) Both 2-hexene and 3-hexene can exist as pairs of *cis-trans* isomers.
(g) Cyclohexene can exist as a pair of *cis-trans* isomers.
(h) 1-Chloropropene can exist as a pair of *cis-trans* isomers.

12-13 What is the difference in structure between a saturated hydrocarbon and an unsaturated hydrocarbon?

12-14 There are three compounds with the molecular formula $C_2H_2Br_2$. Two of these are polar and have a dipole, and one has no dipole (it is not polar). Draw structural formulas for these three compounds and explain why two are polar and the third is not.

12-15 Name and draw structural formulas for all alkenes with the molecular formula C_5H_{10}. As you draw these alkenes, remember that *cis* and *trans* isomers are different compounds and must be counted separately.

12-16 Name and draw structural formulas for alkenes with the molecular formula C₆H₁₂ that have the following carbon skeletons (remember *cis* and *trans* isomers).

(a)
```
        C
        |
   C—C—C—C—C
```

(b)
```
     C   C
     |   |
   C—C—C—C
```

(c)
```
        C
        |
   C—C—C—C
        |
        C
```

12-17 Draw a structural formula for at least one bromoalkene with the molecular formula C₅H₉Br that (a) shows *cis/trans* isomerism and (b) does not show *cis/trans* isomerism.

12-18 Each carbon atom in ethane and in ethylene is surrounded by eight valence electrons and has four bonds to it. Explain how the VSEPR model (Section 3-10) predicts a bond angle of 109.5° about each carbon in ethane but an angle of 120° about each carbon in ethylene.

12-19 Predict all bond angles about each highlighted carbon atom.

(a) [cyclopentene structure] (b) [cyclohexene—CH₂OH structure]

(c) HC≡C—CH=CH₂   (d) [isoprene structure]

12-20 Predict all bond angles about each highlighted carbon atom.

(f) 1,3-Butadiene has two carbon–carbon double bonds and $2^2 = 4$ stereoisomers are possible for it.

12-22 Draw a structural formula for each compound.
(a) *trans*-2-Methyl-3-hexene
(b) 2-Methyl-3-hexyne
(c) 2-Methyl-1-butene
(d) 3-Ethyl-3-methyl-1-pentyne
(e) 2,3-Dimethyl-2-pentene

12-23 Draw a structural formula for each compound.
(a) 3-Chloropropene
(b) 3-Methylcyclohexene
(c) 1,2-Dimethylcyclohexene
(d) *trans*-3,4-Dimethyl-3-heptene
(e) Cyclopropene
(f) 3-Hexyne

12-24 Write the IUPAC name for each unsaturated hydrocarbon.
(a) $CH_2=CH(CH_2)_4CH_3$

(b)

(c) [cyclohexene with CH₃ groups]

(d) $(CH_3)_2CHCH=C(CH_3)_2$
(e) $CH_3(CH_2)_5C\equiv CH$
(f) $CH_3CH_2C\equiv CC(CH_3)_3$

(g) [cyclodecyne structure]

12-25 Write the IUPAC name for each unsaturated hydrocarbon.

(a) [branched alkene]   (b) [methylcyclopentene]

(c) $CH_3CH_2\overset{\overset{CH_2}{\|}}{C}CH_3$

(d)
$$CH_3CH_2CH_2\!\!\diagdown\!\!C=CH_2 \atop CH_3CH_2CH_2\!\!\diagup$$

Section 12-3 How Do We Name Alkenes and Alkynes?

12-21 Answer true or false.
(a) The IUPAC name of an alkene is derived from the name of the longest carbon chain that contains the carbon–carbon double bond.
(b) The IUPAC name of CH₃CH=CHCH₃ is 1,2-dimethylethene.
(c) 2-Methyl-2-butene shows *cis-trans* isomerism.
(d) 1,2-dimethylcyclohexene shows *cis-trans* isomerism.
(e) The IUPAC name of CH₂=CHCH=CHCH₃ is 1,3-pentadiene.

12-26 Explain why each name is incorrect and then write a correct name. Strategy: First draw the structural formula suggested by the incorrect name. Then identify the error, for example, failure to locate the longest chain that contains the alkene functional group etc., and then write the correct name.
(a) 1-Methylpropene   (b) 3-Pentene
(c) 2-Methylcyclohexene   (d) 3,3-Dimethylpentene
(e) 4-Hexyne   (f) 2-Isopropyl-2-butene

12-27 Explain why each name is incorrect and then write a correct name.
(a) 2-Ethyl-1-propene
(b) 5-Isopropylcyclohexene
(c) 4-Methyl-4-hexene
(d) 2-sec-Butyl-1-butene
(e) 6,6-Dimethylcyclohexene
(f) 2-Ethyl-2-hexene

12-28 What structural feature in alkenes makes *cis-trans* isomerism in them possible? What structural feature in cycloalkanes makes *cis-trans* isomerism in them possible? What do these two structural features have in common?

12-29 Which of these alkenes show *cis-trans* isomerism? For each that does, draw structural formulas for both isomers.
(a) 1-Hexene (b) 2-Hexene
(c) 3-Hexene (d) 2-Methyl-2-hexene
(e) 3-Methyl-2-hexene (f) 2,3-Dimethyl-2-hexene

12-30 Which of these alkenes shows *cis-trans* isomerism? For each that does, draw structural formulas for both isomers.
(a) 1-Pentene
(b) 2-Pentene
(c) 3-Ethyl-2-pentene
(d) 2,3-Dimethyl-2-pentene
(e) 2-Methyl-2-pentene
(f) 2,4-Dimethyl-2-pentene

12-31 Cyclodecene exists as both *cis* and *trans* isomers. Draw line-angle formulas for each isomer, showing the configuration of the double bond in each.

12-32 Arachidonic acid is a naturally occurring C_{20} polyunsaturated fatty acid. Draw a line-angle formula for arachidonic acid showing the *cis* configuration about each double bond.

$$CH_3(CH_2)_4(CH=CHCH_2)_4CH_2CH_2COOH$$
Arachidonic acid

12-33 Following is the structural formula of a naturally occurring unsaturated fatty acid.

$$CH_3(CH_2)_7CH=CH(CH_2)_7COOH$$

The *cis* stereoisomer is named oleic acid, and the *trans* isomer is named elaidic acid. Draw a line-angle formula of each acid, showing clearly the configuration of the carbon–carbon double bond in each.

12-34 If you examine the structural formulas for the following cycloalkenes, you will see that the configuration of the double bond is *cis* in each.

Cyclopentene Cyclohexene Cycloheptene

All attempts to synthesize these cycloalkenes in which the double bond has a *trans* configuration have failed. Apparently, it is impossible to have a *trans* configuration in these cycloalkenes. Offer an explanation for why this is so.

12-35 For each molecule that shows *cis-trans* isomerism, draw the *cis* isomer.

12-36 Name and draw structural formulas for all compounds with the molecular formula C_5H_{10} that are:
(a) Alkenes that do not show *cis-trans* isomerism
(b) Alkenes that show *cis-trans* isomerism
(c) Cycloalkanes that do not show *cis-trans* isomerism
(d) Cycloalkanes that show *cis-trans* isomerism

12-37 β-Ocimene, a triene found in the fragrance of cotton blossoms and several essential oils, has the IUPAC name *cis*-3,7-dimethyl-1,3,6-octatriene. (*Cis* refers to the configuration of the double bond between carbons 3 and 4, the only double bond in this molecule about which *cis-trans* isomerism is possible.) Draw a structural formula for β-ocimene.

Section 12-4 What Are the Physical Properties of Alkenes and Alkynes?

12-38 Answer true or false.
(a) Alkenes and alkynes are nonpolar molecules.
(b) The physical properties of alkenes are similar to those of alkanes of the same carbon skeletons.
(c) Alkenes that are liquid at room temperature are insoluble in water and when added to water, will float on water.

Section 12-5 What Are the Characteristic Reactions of Alkenes?

12-39 Answer true or false.
(a) Complete combustion of an alkene gives carbon dioxide and water.
(b) Addition reactions of alkenes involve breaking one of the bonds of the carbon–carbon double bond and formation of two new single bonds in its place.
(c) Markovnikov's rule refers to the regioselectivity of addition reactions of carbon–carbon double bonds.
(d) According to Markovnikov's rule, in the addition of HCl, HBr, or HI to an alkene, hydrogen adds to the carbon of the double bond that already has the greater number of hydrogen atoms bonded to it and the halogen adds to the carbon that has the lesser number of hydrogens bonded to it.
(e) A carbocation is a carbon atom with four bonds that bears a positive charge.
(f) The carbocation derived from ethylene is $CH_3CH_2^+$.
(g) The reaction mechanism for the addition of a halogen acid (HX) to an alkene is divided into two steps, (1) formation of a carbocation and

(2) reaction of the carbocation with halide ion, which complete the reaction.

(h) Acid-catalyzed addition of H_2O to an alkene is called *hydration*.

(i) If a compound fails to react with Br_2, it is unlikely that the compound contains a carbon–carbon double bond.

(j) Addition of H_2 to a double bond is a reduction reaction.

(k) Catalytic reduction of cyclohexene gives hexane.

(l) According to the mechanism presented in the text for acid-catalyzed hydration of an alkene, the H and —OH groups added to the carbon–carbon double bond both arise from the same molecule of H_2O.

(m) The conversion of ethylene, $CH_2{=}CH_2$, to ethanol, CH_3CH_2OH, is an oxidation reaction.

(n) Acid-catalyzed hydration of 1-butene gives 1-butanol. Acid-catalyzed hydration of 2-butene gives 2-butanol.

12-40 Define *alkene addition reaction*. Write an equation for an addition reaction of propene.

12-41 What reagent and/or catalysts are necessary to bring about each conversion?

(a) $CH_3CH{=}CHCH_3 \longrightarrow CH_3CH_2CHBrCH_3$

(b) $CH_3C(CH_3){=}CH_2 \longrightarrow CH_3C(CH_3)(OH)CH_3$

(c) cyclopentene $\longrightarrow$ iodocyclopentane

(d) $CH_3C(CH_3){=}CH_2 \longrightarrow CH_3C(CH_3)(Br)CH_2Br$

12-42 Complete these equations.

(a) (cyclopentyl)$-CH_2CH_3 + HCl \longrightarrow$

(b) (cyclopentyl)$-CH_2CH_3 + H_2O \xrightarrow{H_2SO_4}$

(c) $CH_3(CH_2)_5CH{=}CH_2 + HI \longrightarrow$

(d) (cyclohexyl)$-C(CH_3){=}CH_2 + HCl \longrightarrow$

(e) $CH_3CH{=}CHCH_2CH_3 + H_2O \xrightarrow{H_2SO_4}$

(f) $CH_2{=}CHCH_2CH_2CH_3 + H_2O \xrightarrow{H_2SO_4}$

12-43 Draw structural formulas for all possible carbocations formed by the reaction of each alkene with HCl. Label each carbocation as primary, secondary, or tertiary.

(a) $CH_3CH_2C(CH_3){=}CHCH_3$

(b) $CH_3CH_2CH{=}CHCH_3$

(c) 1-methylcyclopentene

(d) methylenecyclohexane

12-44 Draw a structural formula for the product formed by treatment of 2-methyl-2-pentene with each reagent.

(a) HCl   (b) H_2O in the presence of H_2SO_4

12-45 Draw a structural formula for the product of each reaction.

(a) 1-Methylcyclohexene + Br_2

(b) 1,2-Dimethylcyclopentene + Cl_2

12-46 Draw a structural formula for an alkene with the indicated molecular formula that gives the compound shown as the major product. Note that more than one alkene may give the same compound as the major product.

(a) $C_5H_{10} + H_2O \xrightarrow{H_2SO_4} CH_3C(CH_3)(OH)CH_2CH_3$

(b) $C_5H_{10} + Br_2 \longrightarrow CH_3CH(CH_3)CHBrCH_2Br$

(c) $C_7H_{12} + HCl \longrightarrow$ 1-chloro-1-methylcyclohexane

12-47 Draw a structural formula for an alkene with the molecular formula C_5H_{10} that reacts with Br_2 to give each product.

(a) $CH_3C(CH_3)(Br)CHBrCH_3$ (b) $CH_2BrC(CH_3)(Br)CH_2CH_3$

(c) $CH_2BrCHBrCH_2CH_2CH_3$

12-48 Draw a structural formula for an alkene with the molecular formula C_5H_{10} that reacts with HCl to give the indicated chloroalkane as the major product. More than one alkene may give the same compound as the major product.

(a) CH$_3$C(CH$_3$)(Cl)CH$_2$CH$_3$ (b) CH$_3$CH(CH$_3$)CH(Cl)CH$_3$

(c) CH$_3$CH(Cl)CH$_2$CH$_2$CH$_3$

12-49 With the notable exception of ethanol, the acid-catalyzed hydration of alkenes cannot be used to prepare primary alcohols. It can only be used to prepare secondary and tertiary alcohols from alkenes in good yield. Explain why this is so and illustrate your reasoning with specific examples.

12-50 Draw the structural formula of an alkene that undergoes acid-catalyzed hydration to give the indicated alcohol as the major product. More than one alkene may give each alcohol as the major product.
(a) 3-Hexanol
(b) 1-Methylcyclobutanol
(c) 2-Methyl-2-butanol
(d) 2-Propanol

12-51 Acid-catalyzed hydration of 2-pentene gives a mixture of two alcohols, each with the molecular formula C$_5$H$_{12}$O. Draw the structural formula for both alcohols. Similar treatment of 3-hexene gives only one alcohol with the molecular formula C$_6$H$_{14}$O. Draw the structural formula for this alcohol.

12-52 Terpin, C$_{10}$H$_{20}$O$_2$, is prepared commercially by the acid-catalyzed hydration of limonene.

Limonene

(a) Propose a structural formula for terpin.
(b) How many *cis-trans* isomers are possible for the structural formula you propose?
(c) Terpin hydrate, the isomer of terpin in which the methyl and isopropyl groups are *trans* to each other, is used as an expectorant in cough medicines. Draw a structural formula for terpin hydrate showing the *trans* orientation of these groups.

12-53 Following is the structural formula of hexabromocyclododecane. This compound was at one time used as a flame retardant in polystyrene foam insulation in buildings, but has now been banned because it had been shown to be toxic to aquatic organisms, and can disrupt thyroid hormone levels in laboratory test animals. In addition, it persists in the environment.

Show how this flame retardant could be synthesized from a cycloalkatriene and write the structural formula and IUPAC name of this starting material.

Hexabromocyclododecane

12-54 Propose an explanation for the following experimental observations:
1. Acid-catalyzed hydration of 1-hexene gives a single alcohol in high yield.
2. Acid-catalyzed hydration of *cis*- or *trans*-2-hexene gives a mixture of two alcohols in approximately equal amounts.
3. Acid-catalyzed hydration of *cis*- or *trans*-3-hexene gives a single alcohol in high yield.

12-55 There are nine alkenes with the molecular formula C$_6$H$_{12}$. Five of these were given in Problem 12-54. Draw structural formulas for the other four, and predict the product(s) of the acid-catalyzed hydration of each.

12-56 Draw the product formed by treatment of each alkene with H$_2$/Ni.

12-57 Hydrocarbon A, C$_5$H$_8$, reacts with 2 moles of Br$_2$ to give 1,2,3,4-tetrabromo-2-methylbutane. What is the structure of hydrocarbon A?

12-58 Show how to convert ethylene to these compounds.
(a) Ethane   (b) Ethanol
(c) Bromoethane   (d) 1,2-Dibromoethane
(e) Chloroethane

12-59 Show how to convert 1-butene to these compounds.
(a) Butane   (b) 2-Butanol
(c) 2-Bromobutane   (d) 1,2-Dibromobutane

Section 12-6 What Are the Important Polymerization Reactions of Ethylene and Substituted Ethylenes?

12-60 Answer true or false.
(a) Ethylene contains one carbon–carbon double bond, and polyethylene contains many carbon–carbon double bonds.
(b) All C—C—C bond angles in both LDPE and HDPE are approximately 120°.

Chapter 12 Alkenes and Alkynes

(c) Low-density polyethylene (LDPE) is a highly branched polymer.

(d) High-density polyethylene (HDPE) consists of carbon chains with little branching.

(e) The density of polyethylene polymers is directly related to the degree of chain branching; the greater the branching, the lower the density of the polymer.

(f) PS and PVC are currently recycled.

Chemical Connections

12-61 (Chemical Connections 12A) What is one function of ethylene as a plant growth regulator?

12-62 (Chemical Connections 12B) Linolenic acid is a C_{18} polyunsaturated fatty acid. The configuration of all carbon–carbon double bonds in linolenic acid is *cis*. How many *cis/trans* isomers are possible for this fatty acid?

12-63 (Chemical Connections 12C) What is the meaning of the term *pheromone*?

12-64 (Chemical Connections 12C) What is the molecular formula of 11-tetradecenyl acetate? What is its molecular weight?

12-65 (Chemical Connections 12C) Assume that 1×10^{-12} g of 11-tetradecenyl acetate is secreted by a single corn borer. How many molecules is this?

12-66 (Chemical Connections 12D) What different functions are performed by the rods and cones in the eye?

12-67 (Chemical Connections 12D) In which isomer of retinal is the end-to-end distance longer, the all-*trans* isomer or the 11-*cis* isomer?

12-68 (Chemical Connections 12E) What types of consumer products are made of high-density polyethylene? What types of products are made of low-density polyethylene? One type of polyethylene is currently recyclable, and the other is not. Which is which?

12-69 (Chemical Connections 12E) In recycling codes, what do these abbreviations stand for?
(a) V   (b) PP   (c) PS

Additional Problems

12-70 Write line-angle formulas for all compounds with the molecular formula C_4H_8. Which are sets of constitutional isomers? Which are sets of *cis-trans* isomers?

12-71 Refer to the structure of lycopene at the bottom of the page ▼. Lycopene, $C_{40}H_{56}$, is a deep-red compound that is partially responsible for the red color of ripe fruits, especially tomatoes. Approximately 20 mg of lycopene can be isolated from 1 kg of fresh ripe tomatoes. How many of the carbon–carbon double bonds in lycopene have the possibility for *cis-trans* isomerism? Lycopene is the all-*trans* isomer.

12-72 As you might suspect, β-carotene, $C_{40}H_{56}$, a precursor to vitamin A, was first isolated from carrots. Dilute solutions of β-carotene are yellow—hence its use as a food coloring. In plants, this compound is almost always present in combination with chlorophyll to assist in the harvesting of the energy of sunlight. As tree leaves die in the fall, the green of their chlorophyll molecules is replaced by the yellows and reds of carotene and carotene-related molecules (▼ see β-carotene skeleton at bottom of page).

Compare the carbon skeletons of β-carotene and lycopene. What are the similarities? What are the differences?

12-73 Draw the structural formula for a cycloalkene with the molecular formula C_6H_{10} that reacts with Cl_2 to give each compound.

(a) 1,2-dichlorocyclohexane
(b) 1,2-dichloro-1-methylcyclopentane
(c) 1-chloro-2-chloro-1-methylcyclopentane (with H₃C and Cl on same carbon)
(d) 1-chloro-1-(chloromethyl)cyclopentane

▼ Chemical structures for problems 12-71 and 12-72

Lycopene

β-Carotene

12-74 Propose a structural formula for the product(s) when each of the following alkenes is treated with H_2O/H_2SO_4. Why are two products formed in part (b), but only one in parts (a) and (c)?
  (a) 1-Hexene gives one alcohol with a molecular formula of $C_6H_{14}O$.
  (b) 2-Hexene gives two alcohols, each with a molecular formula of $C_6H_{14}O$.
  (c) 3-Hexene gives one alcohol with a molecular formula of $C_6H_{14}O$.

12-75 *cis*-3-Hexene and *trans*-3-hexene are different compounds and have different physical and chemical properties. Yet, when treated with H_2O/H_2SO_4, each gives the same alcohol. What is this alcohol, and how do you account for the fact that each alkene gives the same one?

12-76 Draw the structural formula of an alkene that undergoes acid-catalyzed hydration to give each of the following alcohols as the major product. More than one alkene may give each compound as the major product.

(a) cyclohexanol
(b) 1-methylcyclohexan-1-ol
(c) 1,2-dimethylcyclopentan-1-ol
(d) 2-methylbutan-2-ol

12-77 Show how to convert cyclopentene into these compounds.
  (a) 1,2-Dibromocyclopentane
  (b) Cyclopentanol
  (c) Iodocyclopentane
  (d) Cyclopentane

Looking Ahead

12-78 In Chapter 21 on the biochemistry of lipids, we will study the three long-chain unsaturated carboxylic acids shown below. Each has 18 carbons and is a component of animal fats, vegetable oils, and biological membranes. Because they consist of a carboxylic acid group on one end and a fatty hydrocarbon chain on the other, they are called fatty acids. How many *cis/trans* isomers are possible for each?

Oleic acid $\quad CH_3(CH_2)_7CH=CH(CH_2)_7COOH$
Linoleic acid $\quad CH_3(CH_2)_4(CH=CHCH_2)_2(CH_2)_6COOH$
Linolenic acid $\quad CH_3CH_2(CH=CHCH_2)_3(CH_2)_6COOH$

12-79 The fatty acids in Problem 12-78 occur in animal fats, vegetable oils, and biological membranes almost exclusively as the all-*cis* isomers. Draw line-angle formulas for each fatty acid showing the *cis* configuration about each carbon–carbon double bond.

12-80 In omega-3 fatty acids, the last carbon of the last double bond of the hydrocarbon chain ends three carbons from the methyl terminal end of the chain. The last carbon of the chain is called the omega carbon, hence, the designation omega-3. Eicosapentaenoic acid is a common omega-3 fatty acid found in cold water fatty fish and health food supplements.

$$CH_3(CH_2CH=CH)_5CH_2CH_2CH_2CO_2H$$

Eicosapentaenoic acid
a C_{20} polyunsaturated fatty acid

  (a) How many *cis-trans* isomers are possible for this fatty acid?
  (b) Draw a line-angle formula for eicosapentaenoic acid, showing the *cis* configuration of all carbon–carbon double bonds in the hydrocarbon chain.

12-81 The following series of three reactions occurs in the metabolic pathway known as the β-oxidation of fatty acids (Section 28-5). Fatty acids are metabolized by this pathway to produce energy. In the following structural formulas, the symbol CoA stands for coenzyme A, an organic molecule involved as a cofactor in many biological reactions. Note that coenzyme A is bound to the fatty acid chain as a thioester, that is, an ester in which a sulfur atom replaces an oxygen atom. Name the type of reaction that occurs in each step and suggest a reason why the sequence is called β-oxidation.

$$-CH_2CH_2CH_2CH_2-\overset{O}{\underset{\|}{C}}-S-CoA \xrightarrow{(1)} -CH_2CH_2CH=CH-\overset{O}{\underset{\|}{C}}-S-CoA \xrightarrow{(2)}$$

A fatty acid thioester

$$-CH_2CH_2\overset{OH}{\underset{|}{C}}HCH_2-\overset{O}{\underset{\|}{C}}-S-CoA \xrightarrow{(3)} -CH_2CH_2\overset{O}{\underset{\|}{C}}CH_2-\overset{O}{\underset{\|}{C}}-S-CoA$$

13 Benzene and Its Derivatives

KEY QUESTIONS

13-1 What Is the Structure of Benzene?

13-2 How Do We Name Aromatic Compounds?

13-3 What Are the Characteristic Reactions of Benzene and Its Derivatives?

13-4 What Are Phenols?

Peppers of the Capsicum family (see Chemical Connections 13F).

Aromatic compounds Benzene or one of its derivatives

Arene A compound containing one or more benzene-like rings

Aryl group A group derived from an arene by removal of a H atom from an arene and given the symbol Ar—

Ar— The symbol used for an aryl group

13-1 What Is the Structure of Benzene?

So far we have described three classes of hydrocarbons—alkanes, alkenes, and alkynes—all called aliphatic hydrocarbons. More than 150 years ago, organic chemists realized that yet another class of hydrocarbons exists, one whose properties are quite different from those of aliphatic hydrocarbons. Because some of these new hydrocarbons have pleasant odors, they were called **aromatic compounds**. Today we know that not all aromatic compounds share this characteristic. Some do have pleasant odors, but some have no odor at all, and others have downright unpleasant odors. A more appropriate definition of an aromatic compound is any compound that has one or more benzene-like rings.

Benzene is an important compound in both the chemical industry and the laboratory, but it must be handled carefully. Not only is it poisonous if ingested in liquid form, but its vapor is also toxic and can be absorbed either by breathing or through the skin. Long-term inhalation can cause liver damage and cancer.

We use the term **arene** to describe aromatic hydrocarbons. Just as a group derived by removal of an H from an alkane is called an alkyl group and given the symbol R—, a group derived by removal of an H from an arene is called an **aryl group** and given the symbol **Ar—**.

Benzene, the simplest aromatic hydrocarbon, was discovered in 1825 by Michael Faraday (1791–1867). The structure of benzene presented an immediate problem to chemists of the day. Benzene has the molecular formula C_6H_6, and a compound with so few hydrogens for its six carbons (compare hexane, C_6H_{14}, and cyclohexane C_6H_{12}), chemists argued, should be unsaturated. But benzene does not behave like an alkene (the only class of unsaturated hydrocarbons known at that time). Though 1-hexene, for example, reacts instantly with Br_2 (Section 12-6C), benzene does not react at all with this reagent. Nor does benzene react with HBr, H_2O/H_2SO_4, or H_2/Pd—all reagents that normally add to carbon–carbon double bonds.

A. Kekulé's Structure of Benzene

The first structure for benzene was proposed by Friedrich August Kekulé in 1872 and consisted of a six-membered ring with alternating single and double bonds, with one hydrogen bonded to each carbon.

A Kekulé structure showing all atoms

A Kekulé structure as a line-angle drawing

Although Kekulé's proposal was consistent with many of the chemical properties of benzene, it was contested for years. The major objection was its failure to account for the unusual chemical behavior of benzene. If benzene contains three double bonds, Kekulé's critics asked, why doesn't it undergo reactions typical of alkenes?

The structure of benzene posed a major problem for chemists because of the limitations of structural theory in 1825 when benzene was discovered. What was clear to chemists was that benzene contained six carbons and six hydrogens but the question was how were they arranged. Following are four structures proposed as alternatives to the Kekulé structure. Notice that these four contain various combinations of rings and carbon–carbon single, double, and triple bonds, and all, including the Kekulé structure, maintain the tetravalence of carbon.

$$H_3C-C\equiv C-C\equiv C-CH_3$$

I

2,4-Hexandiyne

II
"Dewar" structure

III
Prismane

IV

B. Resonance Structure of Benzene

The concept of resonance, developed by Linus Pauling in the 1930s, provided the first adequate description of the structure of benzene. According to the theory of resonance, certain molecules and ions are best described

Resonance hybrid A molecule best described as a hybrid of two or more Lewis contributing structures

by writing two or more Lewis structures and considering the real molecule or ion to be a **resonance hybrid** of these structures. Each individual Lewis structure is called a **contributing structure**. To show that the real molecule is a resonance hybrid of the two Lewis contributing structures, we position a double-headed arrow between them. The two contributing structures for benzene are often called Kekulé structures.

Alternative Lewis contributing structures for benzene

A note about resonance hybrids. Do not confuse resonance contributing structures with equilibration among different chemical species. A molecule described as a resonance hybrid does not equilibrate among the electron configurations of the various contributing structures. Rather, the molecule has only one structure, which is best described as a hybrid of its various contributing structures. A molecule described as a resonance hybrid is not sometimes one contributing structure and sometimes another; it is a single structure all the time.

The resonance hybrid has some of the characteristics of each Lewis contributing structure. For example, the carbon–carbon bonds are neither single nor double but rather something intermediate between the two extremes. It has been determined experimentally that the length of the carbon–carbon bond in benzene is not as long as a carbon–carbon single bond nor as short as a carbon–carbon double bond, but rather is midway between the two. The closed loop of six electrons (two from the second bond of each double bond) characteristic of a benzene ring is sometimes called an **aromatic sextet**.

Wherever we find resonance, we find stability. The real structure is generally more stable than any of its hypothetical Lewis contributing structures. The benzene ring is greatly stabilized by resonance, which explains why it does not undergo the addition reactions typical of alkenes.

13-2 How Do We Name Aromatic Compounds?

A. One Substituent

Monosubstituted alkylbenzenes are named as derivatives of benzene—for example, ethylbenzene. The IUPAC system retains certain common names for several of the simpler monosubstituted alkylbenzenes, including toluene and styrene.

Ethylbenzene   Toluene   Styrene

The IUPAC system also retains common names for the following compounds:

| Phenol | Anisole | Aniline | Benzaldehyde | Benzoic acid |

The substituent group derived by loss of an H from benzene is called a **phenyl group**, C₆H₅—, the common symbol for which is **Ph—**. In molecules containing other functional groups, phenyl groups are often named as substituents.

Phenyl group C₆H₅— the aryl group derived by removing a hydrogen atom from benzene. The name is derived from *phene*, an earlier name for benzene.

Phenyl group (C₆H₅—; Ph—)   1-Phenylcyclohexene   4-Phenyl-1-butene

B. Two Substituents

When two substituents occur on a benzene ring, three isomers are possible. We locate the substituents either by numbering the atoms of the ring or by using the locators *ortho (o), meta (m),* and *para (p)*. The numbers 1,2- are equivalent to *ortho* (Greek: straight); 1,3- to *meta* (Greek: after); and 1,4- to *para* (Greek: beyond).

1,2- or *ortho*-   1,3- or *meta*   1,4- or *para*

When one of the two substituents on the ring imparts a special name to the compound (for example, —CH₃, —OH, —NH₂, or —COOH), we name the compound as a derivative of that parent molecule and assume that the substituent occupies ring position number 1. The IUPAC system retains the common name xylene for the three isomeric dimethylbenzenes. One example is *p*-xylene, a starting material for the synthesis of poly(ethylene terephthalate). Consumer products derived from this polymer include Dacron polyester fibers and Mylar films (Section 19-6B). Where neither substituent imparts a special name, we locate the two substituents and list them in alphabetical order before the ending "benzene." The carbon of the benzene ring with the substituent first in alphabetical order is numbered C—1.

4-Bromobenzoic acid (*p*-Bromobenzoic acid)   3-Chloroaniline (*m*-Chloroaniline)   1,3-Dimethylbenzene (*m*-Xylene)   1-Chloro-4-ethylbenzene (*p*-Chloroethylbenzene)

C. Three or More Substituents

When three or more substituents are present on a benzene ring, specify their locations by numbers. If one of the substituents imparts a special name, then name the molecule as a derivative of that parent molecule. If none of the substituents imparts a special name, then locate the substituents, number them to give the smallest set of numbers, and list them in alphabetical order before the ending "benzene." In the following examples, the first compound is a derivative of toluene and the second is a derivative of phenol. Because no substituent in the third compound imparts a special name, list its three substituents in alphabetical order followed by the word "benzene."

4-Chloro-2-nitrotoluene   2,4,6-Tribromophenol   2-Bromo-1-ethyl-4-nitrobenzene

Interactive Example 13-1 — Naming Aromatic Compounds

Write names for these compounds.

(a) (b) (c)

Strategy
First check to see if one of the substituents on the benzene ring imparts a special name. If one of them does, then name the compound as a derivative of that parent molecule.

Solution
(a) The parent is toluene, and the compound is 3-iodotoluene or *m*-iodotoluene.
(b) The parent is benzoic acid, and the compound is 3,5-dibromobenzoic acid.
(c) The parent is aniline, and the compound is 4-chloroaniline or *p*-chloroaniline.

Problem 13-1
Write names for these compounds.

(a) (b) (c)

CHEMICAL CONNECTIONS 13A

DDT: A Boon and a Curse

Probably the best-known insecticide worldwide is *dichlorodiphenyltrichloroethane* (not an IUPAC name), commonly abbreviated DDT.

Cl—⟨C₆H₄⟩—CH(CCl₃)—⟨C₆H₄⟩—Cl

Dichlorodiphenyltrichloroethane (DDT)

This compound was first prepared in 1874, but it was not until the late 1930s that its potential as an insecticide was recognized. First used for this purpose in 1939, it proved extremely effective in ridding large areas of the world of the insect hosts that transmit malaria and typhus. In addition, it was so effective in killing crop-destroying insect pests that crop yields in many areas of the world increased dramatically.

Widespread use of DDT, however, has proven to be a double-edged sword. Despite DDT's well-known benefits, it has an enormous disadvantage. Because it resists biodegradation, it remains in the soil for years—and this persistence in the environment creates the problem. Scientists estimate that the tissues of adult humans contain, on average, five to ten parts per million of DDT.

Rachel Carson dramatically portrayed the dangers associated with the persistence of DDT in the environment

DDT was sprayed on crops in the United States until its use was banned in the 1970s.

in her 1962 book, *Silent Spring*, which documented serious declines in the populations of eagles and other raptors as well as many other kinds of birds and songbirds. Scientists discovered soon thereafter that DDT inhibits the mechanism by which these birds incorporate calcium into their eggshells. As a result, the shells become so thin and weak that they break during incubation, killing the embryo inside.

Because of these problems, almost all nations have now banned the use of DDT for agricultural purposes. It is still used, however, in some areas to control the population of disease-spreading insects.

Test your knowledge with Problems 13-34 through 13-38.

D. Polynuclear Aromatic Hydrocarbons

Polynuclear aromatic hydrocarbons (PAHs) contain two or more benzene rings, with each pair of rings sharing two adjacent carbon atoms. Naphthalene, anthracene, and phenanthrene, the most common PAHs, and substances derived from them are found in coal tar and high-boiling petroleum residues.

Polynuclear aromatic hydrocarbons (PAHs) Hydrocarbons containing two or more benzene rings, each of which share two carbon atoms with another benzene ring

Naphthalene   Anthracene   Phenanthrene

At one time, naphthalene was used in mothballs and as an insecticide to preserve woolens and furs, but its use decreased after the introduction of chlorinated hydrocarbons such as *p*-dichlorobenzene.

CHEMICAL CONNECTIONS 13B

Carcinogenic Polynuclear Aromatic Hydrocarbons and Smoking

A **carcinogen** is a compound that causes cancer. The first carcinogens to be identified were a group of polynuclear aromatic hydrocarbons, all of which have at least four aromatic rings. Among them is benzo[a]pyrene, one of the most carcinogenic of the aromatic hydrocarbons. It forms whenever incomplete combustion of organic compounds occurs. Benzo[a]pyrene is found, for example, in cigarette smoke, automobile exhaust, and charcoal-broiled meats.

Benzo[a]pyrene causes cancer in the following way. Once it is absorbed or ingested, the body attempts to convert it into a more water-soluble compound that can be excreted easily. By a series of enzyme-catalyzed reactions, benzo[a]-pyrene is transformed into a **diol** (two —OH groups) **epoxide** (a three-membered ring, one atom of which is oxygen). This compound can bind to DNA by reacting with one of its amino groups, thereby altering the structure of DNA and producing a cancer-causing mutation.

Benzo[a]pyrene → (enzyme-catalyzed oxidation) → A diol epoxide

Test your knowledge with Problem 13-39.

13-3 What Are the Characteristic Reactions of Benzene and Its Derivatives?

By far the most characteristic reaction of aromatic compounds is substitution at a ring carbon, which we give the name **aromatic substitution**. Groups we can introduce directly on the ring include the halogens, the nitro (—NO$_2$) group, and the sulfonic acid (—SO$_3$H) group.

A. Halogenation

As noted in Section 13-1, chlorine and bromine do not react with benzene, in contrast to their instantaneous reaction with cyclohexene and other alkenes (Section 12-6C). In the presence of an iron catalyst, however, chlorine reacts rapidly with benzene to give chlorobenzene and HCl:

C$_6$H$_5$—H + Cl$_2$ $\xrightarrow{\text{FeCl}_3}$ C$_6$H$_5$—Cl + HCl

Benzene → Chlorobenzene

Treatment of benzene with bromine in the presence of FeCl$_3$ results in the formation of bromobenzene and HBr.

B. Nitration

When we heat benzene or one of its derivatives with a mixture of concentrated nitric and sulfuric acids, a nitro (—NO$_2$) group replaces one of the hydrogen atoms bonded to the ring.

C$_6$H$_5$—H + HNO$_3$ $\xrightarrow{\text{H}_2\text{SO}_4}$ C$_6$H$_5$—NO$_2$ + H$_2$O

Nitrobenzene

CHEMICAL CONNECTIONS 13C

Iodide Ion and Goiter

One hundred years ago, goiter, an enlargement of the thyroid gland caused by iodine deficiency, was common in the central United States and central Canada. This disease results from underproduction of thyroxine, a hormone synthesized in the thyroid gland. Young mammals require this hormone for normal growth and development. A deficiency of thyroxine during fetal development results in mental retardation. Low levels of thyroxine in adults result in hypothyroidism, commonly called goiter, the symptoms of which are lethargy, obesity, and dry skin.

Iodine is an element that comes primarily from the sea. Rich sources of it, therefore, are fish and other seafoods. The iodine in our diets that doesn't come from the sea most commonly is derived from food additives. Most of the iodide ion in the North American diet comes from table salt fortified with sodium iodide, commonly referred to as iodized salt. Another source is dairy products, which accumulate iodide because of the iodine-containing additives used in cattle feeds and the iodine-containing disinfectants used on milking machines and milk storage tanks.

Thyroxine

Test your knowledge with Problem 13-40.

A particular value of nitration is that we can reduce the resulting —NO_2 group to a primary amino group, —NH_2, by catalytic reduction using hydrogen in the presence of a transition-metal catalyst. In the following example, neither the benzene ring nor the carboxyl group is affected by these experimental conditions.

$$O_2N-\text{C}_6\text{H}_4-COOH + 3H_2 \xrightarrow[\text{3 atm}]{\text{Ni}} H_2N-\text{C}_6\text{H}_4-COOH + 2H_2O$$

4-Nitrobenzoic acid
(*p*-Nitrobenzoic acid)

4-Aminobenzoic acid
(*p*-Aminobenzoic acid, PABA)

Bacteria require *p*-aminobenzoic acid for the synthesis of folic acid (Section 30-4), which is in turn required for the synthesis of the heterocyclic aromatic amine bases of nucleic acids (Section 25-2). Although bacteria can synthesize folic acid from *p*-aminobenzoic acid, folic acid is a vitamin for humans and must be obtained through the diet.

C. Sulfonation

Heating an aromatic compound with concentrated sulfuric acid results in formation of an arenesulfonic acid, all of which are strong acids, comparable in strength to sulfuric acid.

$$\text{C}_6\text{H}_5-H + H_2SO_4 \longrightarrow \text{C}_6\text{H}_5-SO_3H + H_2O$$

Benzenesulfonic acid

A major use of sulfonation is in the preparation of synthetic detergents, an important example of which is sodium 4-dodecylbenzenesulfonate. To prepare this type of detergent, a linear alkylbenzene such as dodecylbenzene is treated with concentrated sulfuric acid to give an

CHEMICAL CONNECTIONS 13D

The Nitro Group in Explosives

Treatment of toluene with three moles of nitric acid in the presence of sulfuric acid as a catalyst results in nitration of toluene three times to form the explosive 2,4,6-trinitrotoluene, TNT. The presence of these three nitro groups confers explosive properties to TNT. Similarly, the presence of three nitro groups leads to the explosive properties of nitroglycerin.

In recent years, several new explosives have been discovered, all of which contain multiple nitro groups. Among them are RDX and PETN. The plastic explosive Semtex, for example, is a mixture of RDX and PETN. It was used in the destruction of Pan Am flight 103 over Lockerbie, Scotland, in December 1988.

2,4,6-Trinitrotoluene (TNT)

Trinitroglycerin (Nitroglycerin)

Cyclonite (RDX)

Pentaerythritol tetranitrate (PETN)

Test your knowledge with Problem 13-41.

---

alkylbenzenesulfonic acid. The sulfonic acid is then neutralized with sodium hydroxide.

Dodecylbenzene → (1. H_2SO_4, 2. NaOH) → Sodium 4-dodecylbenzenesulfonate, SDS (an anionic detergent)

Alkylbenzenesulfonate detergents were introduced in the late 1950s, and today they claim nearly 90% of the market once held by natural soaps. Section 18-4 discusses the chemistry and cleansing action of soaps and detergents.

13-4 What Are Phenols?

A. Structure and Nomenclature

Phenol A compound that contains an —OH group bonded to a benzene ring

The functional group of a **phenol** is an —OH group bonded to a benzene ring. Substituted phenols are named either as derivatives of phenol or by common names. ◀

Phenol

3-Methylphenol (*m*-Cresol)

1,2-Benzenediol (Catechol)

1,3-Benzenediol (Resorcinol)

1,4-Benzenediol (Hydroquinone)

▶ Phenol in crystalline form.

Phenols are widely distributed in nature. Phenol itself and the isomeric cresols (*o*-, *m*-, and *p*-cresol) are found in coal tar. Thymol and vanillin are important constituents of thyme and vanilla beans, respectively. Urushiol is the main component of the irritating oil of poison ivy. It can cause severe contact dermatitis in sensitive individuals. ◄

2-Isopropyl-5-methylphenol (Thymol)

4-Hydroxy-3-methoxybenzaldehyde (Vanillin)

Urushiol

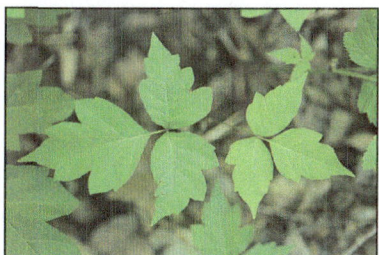

▶ Poison ivy.

B. Acidity of Phenols

Phenols are weak acids, with pK_a values of approximately 10 (Table 8-3). Most phenols are insoluble in water, but they react with strong bases such as NaOH and KOH to form water-soluble salts.

Phenol + NaOH ⟶ Sodium phenoxide + H₂O

Phenol	Sodium	Sodium phenoxide	Water
pK_a = 9.95	hydroxide	(weaker base)	pK_a = 15.7
(stronger acid)	(stronger base)		(weaker acid)

Most phenols are such weak acids that they do not react with weak bases such as sodium bicarbonate; that is, they do not dissolve in aqueous sodium bicarbonate.

C. Oxidation of Phenols

Phenols are oxidized to quinones by a variety of strong oxidizing agents. For example, they are oxidized by potassium dichromate in the presence of sulfuric acid to form quinones. Oxidation of phenol itself gives 1,4-benzenedione (*p*-quinone).

Phenol → (K₂Cr₂O₇, H₂SO₄) → 1,4-Benzenedione (*p*-Quinone)

One very important chemical property of quinones is that they are easily reduced to benzenediols, more commonly known as hydroquinones.

p-Quinone → Reduction → 1,4-Benzenediol (Hydroquinone)

A commercial process that uses this chemical property is black-and-white photography. Black-and-white film is coated with an emulsion containing silver bromide and silver iodide crystals, which become activated when they are exposed to light. The activated silver ions are reduced in the developing stage to metallic silver. In the developing process, hydoquinone is the reducing agent and is oxidized to *p*-quinone. Silver ion is the oxidizing agent and is reduced to metallic silver.

$$\text{1,4-Benzenediol (Hydroquinone)} + 2\text{Ag}^+ \longrightarrow p\text{-Quinone} + 2\text{Ag} + 2\text{H}^+$$

All silver halide not activated by light is removed in the developing process, and the result is a black image (a negative) left by deposited metallic silver where the film had been struck by light.

There are many more examples in both chemistry and biology in which the reversible oxidation/reduction of quinones or hydroquinones is important. One such example is coenzyme Q, alternatively known as ubiquinone. This name is derived from the Latin *ubique* (everywhere) + quinone. Coenzyme Q, a carrier of electrons in the respiratory chain (Section 27-5), contains a long hydrocarbon chain that serves to anchor it firmly in the nonpolar environment of the mitochondrial inner membrane. The oxidized form of coenzyme Q is a two-electron oxidizing agent. In subsequent steps of the respiratory chain, the reduced form of coenzyme Q transfers two electrons to the next link in the respiratory chain until they are eventually delivered to a molecule of oxygen, which is in turn reduced to water.

Coenzyme Q
Ubiquinone
(oxidized form)

$$+ 2\text{H}^+ + 2e^- \underset{\text{Oxidation}}{\overset{\text{Reduction}}{\rightleftarrows}}$$

Coenzyme Q
Ubiquinol
(reduced form)

Vitamin K$_2$, another important quinone, is found in animals and humans. The core of this molecule is derived from naphthalene. All members of the Vitamin K family contain a 1,4-naphthoquinone.

They differ in the length of the nonpolar side chain bonded to the naphthoquinone core.

Vitamin K$_2$

Vitamin K was discovered in 1935 as the result of a study of newly hatched chicks with a fatal disease making their blood slow to clot. It was later discovered that the delayed clotting time was caused by a deficiency of prothrombin, and that vitamin K is essential for the synthesis of prothrombin in the liver. The natural form of the vitamin in food supplements has been replaced by Menadione, a synthetic compound with hydrogen in place of the long hydrocarbon chain of natural vitamin K$_2$. Menadione is synthesized in the laboratory by chromic acid oxidation of 2-methylnaphthalene.

2-Methylnaphthalene → 2-Methyl-1,4-naphthaquinone Menadione (H$_2$CrO$_4$ Oxidation)

D. Phenols as Antioxidants

An important reaction for living systems, as well as some foods and other materials that contain carbon–carbon double bonds is **autoxidation**—that is, oxidation requiring oxygen and no other reactant. If you open a bottle of cooking oil that has stood for a long time, you may notice a hiss of air entering the bottle. This sound occurs because the consumption of oxygen by autoxidation of the oil creates a negative pressure inside the bottle.

Cooking oils contain esters of polyunsaturated fatty acids. We will discuss the structure and chemistry of esters in Chapter 19. The important point here is that all vegetable oils contain fatty acids with long hydrocarbon chains, many of which have one or more carbon–carbon double bonds (see Problems 12-70 and 12-71 for the structures of several of these fatty acids). Autoxidation takes place at a —CH$_2$— group adjacent to one or more of their double bonds.

Autoxidation The reaction of a CH$_2$ group with oxygen (O$_2$) to form a hydroperoxide R—OOH

—CH$_2$CH=CH—CH— + O$_2$ $\xrightarrow{\text{Light or heat}}$ —CH$_2$CH=CH—CH—
Section of a fatty acid hydrocarbon chain → A hydroperoxide (O—O—H)

Autoxidation is a radical-chain process that converts an R—H group to an R—O—O—H group, called a hydroperoxide. This process begins when a hydrogen atom with one of its electrons (H·) is removed from a carbon adjacent to one of the double bonds in a hydrocarbon chain. The carbon

CHEMICAL CONNECTIONS 13E

FD & C No. 6 (a.k.a. Sunset Yellow)

Did you ever wonder what gives gelatin desserts their red, green, orange, or yellow color? Or what gives margarines their yellow color? Or what gives maraschino cherries their red color? If you read the content labels, you will see code names such as FD & C Yellow No. 6 and FD & C Red No. 40.

At one time, the only colorants for foods were compounds obtained from plant or animal materials. Beginning as early as the 1890s, however, chemists discovered a series of synthetic food dyes that offered several advantages over natural dyes, such as greater brightness, better stability, and lower cost. Opinion remains divided on the safety of their use. No synthetic food colorings are allowed, for example, in Norway and Sweden. In the United States, the Food and Drug Administration (FDA) has certified seven synthetic dyes for use in foods, drugs, and cosmetics (FD & C)—two yellows, two reds, two blues, and one green. When these dyes are used alone or in combinations, they can approximate the color of almost any natural food.

Following are structural formulas for Allura Red (Red No. 40) and Sunset Yellow (Yellow No. 6). These and the other five food dyes certified in the United States have in common three or more benzene rings and two or more ionic groups, either the sodium salt of a carboxyl group, —COO$^-$Na$^+$, or the sodium salt of a sulfonic acid group, —SO$_3^-$Na$^+$. These ionic groups make the dyes soluble in water.

To return to our original questions, maraschino cherries are colored with FD & C Red No. 40 and margarines are colored with FD & C Yellow No 6. Gelatin desserts use either one or a combination of the seven certified food dyes to create their color.

Allura Red
(FD & C Red No. 40)

Sunset Yellow
(FD & C Yellow No. 6)

Test your knowledge with Problems 13-43 through 13-45.

---

losing the H· has only seven electrons in its valence shell, one of which is unpaired. An atom or a molecule with an unpaired electron is called a **radical**.

Mechanism of Autoxidation

Step 1: Chain Initiation—Formation of a Radical from a Nonradical Compound
Removal of a hydrogen atom (H·) may be initiated by light or heat. The product formed is a carbon radical; that is, it contains a carbon atom with one unpaired electron.

$$-CH_2CH=CH-\overset{H}{\underset{|}{CH}}- \xrightarrow{\text{light or heat}} -CH_2CH=CH-\overset{\cdot}{CH}-$$

Section of a fatty acid hydrocarbon chain → A carbon radical

CHEMICAL CONNECTIONS 13F

Capsaicin, for Those Who Like It Hot

Capsaicin, the pungent principal from the fruit of various species of peppers (*Capsicum* and *Solanaceae*), was isolated in 1876, and its structure was determined in 1919. Capsaicin contains both a phenol and a phenol ether.

Capsaicin
(from various types of peppers)

The inflammatory properties of capsaicin are well known; the human tongue can detect as little as one drop in 5 L of water. We all know of the burning sensation in the mouth and sudden tearing in the eyes caused by a good dose of hot chili peppers. For this reason, capsaicin-containing extracts from these flaming foods are used in sprays to ward off dogs or other animals that might nip at your heels while you are running or cycling.

Paradoxically, capsaicin is able to both cause and relieve pain. Currently, two capsaicin-containing creams, Mioton and Zostrix, are prescribed to treat the burning pain associated with postherpetic neuralgia, a complication of the disease known as shingles. They are also prescribed for diabetics to relieve persistent foot and leg pain.

Red chilies being dried in the sun.

Test your knowledge with Problems 13-46 through 13-49.

Step 2a: Chain Propagation—Reaction of a Radical and O₂ to Form a New Radical The carbon radical reacts with oxygen, itself a diradical, to form a hydroperoxy radical. The new covalent bond of the hydroperoxy radical forms by the combination of one electron from the carbon radical and one electron from the oxygen diradical.

These two unpaired electrons combine to form a C—O single bond

—CH₂CH=CH—CH— + ·O—O· ⟶ —CH₂CH=CH—CH—
                                              |
                                              O—O·

Oxygen is a diradical

A hydroperoxy radical

Step 2b: Chain Propagation—Reaction of a Radical and a Molecule to Form a Hydroperoxide and a New Radical
The hydroperoxy radical removes a hydrogen atom (H·) from a new fatty acid hydrocarbon chain to complete the formation of a hydroperoxide and at the same time produce a new carbon radical.

$$\underset{}{-CH_2CH=CH-\overset{\overset{O-O\cdot}{|}}{CH}-} \;+\; \underset{\text{Section of a new fatty acid hydrocarbon chain}}{-CH_2CH=CH-\overset{\overset{H}{|}}{CH}-} \longrightarrow$$

$$\underset{\text{A hydroperoxide}}{-CH_2CH=CH-\overset{\overset{O-O-H}{|}}{CH}-} \;+\; \underset{\text{A new carbon radical}}{-CH_2CH=CH-\overset{\cdot}{C}H-}$$

The most important point about the pair of chain propagation steps (Steps 2a and 2b) is that they form a continuous cycle of reactions, in the following way. The new radical formed in Step 2b reacts with another molecule of O_2 by Step 2a to give a new hydroperoxy radical. This new hydroperoxy radical then reacts with a new hydrocarbon chain to repeat Step 2b, and so forth. This cycle of propagation steps repeats over and over in a chain reaction. Thus, once a radical is generated in Step 1, the cycle of propagation steps repeats many thousands of times and, in so doing, generates thousands of hydroperoxide molecules. The number of times the cycle of chain propagation steps repeats is called the **chain length**.

Hydroperoxides themselves are unstable and, under biological conditions, degrade to short-chain aldehydes and carboxylic acids with unpleasant "rancid" smells. These odors may be familiar to you if you have ever smelled old cooking oil or aged foods that contain polyunsaturated fats or oils. Similar formation of hydroperoxides in the low-density lipoproteins (Section 27-4) deposited on the walls of arteries leads to cardiovascular disease in humans. Many effects of aging are thought to result from the formation and subsequent degradation of hydroperoxides.

Fortunately, nature has developed a series of defenses against the formation of these and other destructive hydroperoxides, including vitamin E (Section 30-6). This compound is one of "nature's scavengers." It inserts itself into either Step 2a or 2b, donates an H· from its —OH group to the carbon radical, and converts the carbon radical back to its original hydrocarbon chain. Because the vitamin E radical is stable, it breaks the cycle of chain propagation steps, thereby preventing further formation of destructive hydroperoxides. While some hydroperoxides may form, their numbers are very small and they are easily decomposed to harmless materials by one of several possible enzyme-catalyzed reactions.

Vitamin E

Butylated hydroxytoluene (BHT)

Butylated hydroxyanisole (BHA)

Unfortunately, vitamin E is removed in the processing of many foods and food products. To make up for this loss, phenols such as BHT and BHA are added to foods to "retard spoilage" (as they say on the packages) by autoxidation. Likewise, similar compounds are added to other materials, such as plastics and rubber, to protect them against autoxidation.

Summary of Key Questions

Section 13-1 What Is the Structure of Benzene?

- **Benzene** and its alkyl derivatives are classified as **aromatic hydrocarbons**, or **arenes**.
- The first structure for benzene was proposed by Friederich August Kekulé in 1872.
- The theory of **resonance**, developed by Linus Pauling in the 1930s, provided the first adequate structure for benzene.
- Benzene is represented as a hybrid of two equivalent Lewis contributing structures.

Section 13-2 How Do We Name Aromatic Compounds?

- Aromatic compounds are named according to the IUPAC system.
- The C_6H_5- group is named **phenyl**.
- Two substituents on a benzene ring may be located by numbering the atoms of the ring or by using the locators *ortho* (*o*), *meta* (*m*), and *para* (*p*).
- **Polynuclear aromatic hydrocarbons** contain two or more benzene rings, each sharing two adjacent carbon atoms with another ring.

Section 13-3 What Are the Characteristic Reactions of Benzene and Its Derivatives?

- A characteristic reaction of aromatic compounds is **aromatic substitution**, in which another atom or group of atoms is substituted for a hydrogen atom of the aromatic ring.
- Typical aromatic substitution reactions are halogenation, nitration, and sulfonation.

Section 13-4 What Are Phenols?

- The functional group of a phenol is an —OH group bonded to a benzene ring.
- Phenol and its derivatives are weak acids, with pK_a values of approximately 10.0.
- Vitamin E, a phenolic compound, is a natural antioxidant.
- Phenolic compounds such as BHT and BHA are synthetic antioxidants.
- Oxidation of phenols by strong oxidizing agents yields quinones. Oxidation of phenol itself gives *p*-quinone.
- *p*-Quinones can in turn be reduced to hydroquinones.
- Coenzyme Q, a quinone important in human metabolism, plays a key role in the respiratory chain as a carrier of electrons.

Summary of Key Reactions

1. **Halogenation (Section 13-3A)** Treatment of an aromatic compound with Cl_2 or Br_2 in the presence of an $FeCl_3$ catalyst substitutes a halogen for an H.

    $$C_6H_6 + Cl_2 \xrightarrow{FeCl_3} C_6H_5-Cl + HCl$$
    Chlorobenzene

2. **Nitration (Section 13-3B)** Heating an aromatic compound with a mixture of concentrated nitric and sulfuric acids substitutes a nitro group for an H.

    $$C_6H_6 + HNO_3 \xrightarrow[heat]{H_2SO_4} C_6H_5-NO_2 + H_2O$$
    Nitrobenzene

3. **Sulfonation (Section 13-3C)** Heating an aromatic compound with concentrated sulfuric acid substitutes a sulfonic acid group for an H.

    $$C_6H_6 + H_2SO_4 \xrightarrow{heat} C_6H_5-SO_3H + H_2O$$
    Benzenesulfonic acid

4. **Reaction of Phenols with Strong Bases (Section 13-4B)** Phenols are weak acids that react with strong bases to form water-soluble salts.

    $$C_6H_5-OH + NaOH \longrightarrow C_6H_5-O^-Na^+ + H_2O$$

Problems

Orange-numbered problems are applied.

Section 13-1 What Is the Structure of Benzene?

13-2 Answer true or false.

(a) Alkenes, alkynes, and arenes are unsaturated hydrocarbons.

(b) Aromatic compounds were so named because many of them have pleasant odors.

(c) According to the resonance model of bonding, benzene is best described as a hybrid of two equivalent contributing structures.

(d) Benzene is a planar molecule.

13-3 What is the difference in structure between a saturated and an unsaturated hydrocarbon?

13-4 Define *aromatic compound*.

13-5 Why are alkenes, alkynes, and aromatic compounds said to be unsaturated?

13-6 Do aromatic rings have double bonds? Are they unsaturated? Explain.

13-7 Can an aromatic compound be a saturated compound?

13-8 Draw at least two structural formulas for each of the following. (Several constitutional isomers are possible for each part.)
(a) An alkene with six carbons
(b) A cycloalkene with six carbons
(c) An alkyne with six carbons
(d) An aromatic hydrocarbon with eight carbons

13-9 Write a structural formula and the name for the simplest (a) alkane, (b) alkene, (c) alkyne, and (d) aromatic hydrocarbon.

13-10 Account for the fact that the six-membered ring in benzene is planar but the six-membered ring in cyclohexane is not.

13-11 Explain why the compound 1,4-dichlorobenzene does not show *cis-trans* isomerism.

13-12 One analogy often used to explain the concept of a resonance hybrid is to relate a rhinoceros to a unicorn and a dragon. Explain the reasoning in this analogy and how it might relate to a resonance hybrid.

Section 13-2 How Do We Name Aromatic Compounds?

13-13 Answer true or false.
(a) A phenyl group has the molecular formula C_6H_5— and is represented by the symbol Ph—.
(b) Para substituents occupy adjacent carbons on a benzene ring.
(c) 4-Bromobenzoic acid can be separated into *cis* and *trans* isomers.
(d) Naphthalene is a planar molecule.
(e) Benzene, naphthalene, and anthracene are polynuclear aromatic hydrocarbons (PAHs).
(f) Benzo[a]pyrene causes cancer by binding to DNA and producing a cancer-causing mutation.

13-14 Name these compounds.

(a) [structure: benzene with NO2 and Cl para]
(b) [structure: benzene with CH3 and Br ortho]
(c) $C_6H_5CH_2CH_2CH_2Cl$
(d) $C_6H_5\overset{Br}{\underset{CH_3}{C}}CH_2CH_3$

(e) [structure: benzene with NH2 and NO2 ortho]
(f) [structure: benzene with OH and C6H5 ortho]
(g) [structure: C6H5 and H on one carbon, H and C6H5 on the other of C=C]
(h) [structure: benzene with CH3, Cl, Cl]

13-15 Draw structural formulas for these compounds.
(a) 1-Bromo-2-chloro-4-ethylbenzene
(b) 4-Bromo-1,2-dimethylbenzene
(c) 2,4,6-Trinitrotoluene
(d) 4-Phenyl-1-pentene
(e) *p*-Cresol
(f) 2,4-Dichlorophenol

13-16 We say that naphthalene, anthracene, phenanthrene, and benzo[a]pyrene are polynuclear aromatic hydrocarbons. In this context, what does "polynuclear" mean? What does "aromatic" mean? What does "hydrocarbon" mean?

13-17 Following is the structural formula of styrene, the monomer from which the polymer polystyrene (PS) is prepared.

[structure: C6H5—CH=CH2] $\xrightarrow{\text{Polymerization}}$ Polystyrene

Styrene

Draw a section of a polystyrene chain showing at least three monomer units.

Section 13-3 What Are the Characteristic Reactions of Benzene and Its Derivatives?

13-18 Answer true or false.
(a) Benzene does not undergo the addition reactions that are characteristic of alkenes.
(b) A defining feature of aromatic compounds is that they are highly unsaturated but do not undergo characteristic alkene addition reactions.
(c) Nitration of benzene adds a —NO_2 group to one of the carbons of the aromatic ring.
(d) Halogenation of an alkene is an addition reaction; halogenation of an arene is a substitution reaction.

13-19 Suppose you have unlabeled bottles of benzene and cyclohexene. What chemical reaction could you use to tell which bottle contains which chemical? Explain what you would do, what you would expect to see, and how you would explain your observations. Write an equation for a positive test.

13-20 Three products with the molecular formula C_6H_4BrCl form when bromobenzene is treated with chlorine, Cl_2, in the presence of $FeCl_3$ as a catalyst. Name and draw a structural formula for each product.

13-21 The reaction of bromine with toluene in the presence of FeCl$_3$ gives a mixture of three products, all with the molecular formula C$_7$H$_7$Br. Name and draw a structural formula for each product.

13-22 What reagents and/or catalysts are necessary to carry out each conversion?
  (a) Benzene to nitrobenzene
  (b) 1,4-dichlorobenzene to 2-bromo-1,4-dichlorobenzene
  (c) Benzene to aniline

13-23 What reagents and/or catalysts are necessary to carry out each conversion? Each conversion requires two steps.
  (a) Benzene to 3-nitrobenzenesulfonic acid
  (b) Benzene to 1-bromo-4-chlorobenzene

13-24 Aromatic substitution can be done on naphthalene. Treatment of naphthalene with concentrated H$_2$SO$_4$ gives two (and only two) different sulfonic acids. Draw a structural formula for each.

Section 13-4 What Are Phenols?

13-25 Answer true or false.
  (a) Phenols and alcohols have in common the presence of an —OH group.
  (b) Phenols are weak acids and react with strong bases to give water-soluble salts.
  (c) The pK_a of phenol is smaller than that of acetic acid.
  (d) Autoxidation converts an R—H group to an R—OH group.
  (e) A carbon radical has only seven electrons in the valence shell of one of its carbons, and this carbon bears a positive charge.
  (f) A characteristic of a chain initiation step is conversion of a nonradical to a radical.
  (g) Autoxidation is a radical-chain reaction.
  (h) A characteristic of the chain propagation step is reaction of a radical and a molecule to form a new radical and a new molecule.
  (i) Vitamin E and other natural antioxidants function by interrupting the cycle of chain propagation steps that occurs in autoxidation.

13-26 Both phenol and cyclohexanol are only slightly soluble in water. Account for the fact that phenol dissolves in aqueous sodium hydroxide but cyclohexanol does not.

13-27 Define autoxidation.

13-28 Autoxidation is described as a *radical-chain reaction*. What is meant by the term "radical" in this context? By the term "chain"? By the term "chain length"?

13-29 Show that if you add Steps 2a and 2b of the radical-chain mechanism for the autoxidation of a fatty acid hydrocarbon chain, you arrive at the following net equation:

$$\underset{\text{Section of a fatty acid hydrocarbon chain}}{-CH_2CH=CH-\overset{H}{\underset{|}{CH}}-} + \cdot O-O \cdot \longrightarrow$$

$$\underset{\text{A hydroperoxide}}{-CH_2CH=CH-\overset{O-O-H}{\underset{|}{CH}}-}$$

13-30 How does vitamin E function as an antioxidant?

13-31 What structural features are common to vitamin E, BHT, and BHA (the three antioxidants presented in Section 13-4C)?

13-32 Black-and-white photography is a commercial process that involves a phenol. Black-and-white film is coated with an emulsion containing silver bromide or silver iodide crystals that become activated by exposure to light. The activated silver ions then react with hydroquinone in the developing stage as shown in the following balanced equation.

$$\underset{\substack{\text{1,-Benzenediol}\\ \text{(Hydroquinone)}}}{\text{OH-C}_6\text{H}_4\text{-OH}} + 2Ag^+ \longrightarrow \underset{\substack{\text{1,4-Benzoquinone}\\ \text{(p-Quinone)}}}{\text{O=C}_6\text{H}_4\text{=O}} + 2Ag^0 + 2H^+$$

All silver halide not activated by light is removed in the fixing process, and the result is a black image (a negative) left by the deposited metallic silver where the film had been struck by light. In this redox reaction state:
  (a) What is reduced and what is the reducing agent?
  (b) What is oxidized and what is the oxidizing agent?

13-33 Following is the structural formula of albuterol (Proventil), one of the most widely used inhalation bronchodilators.

Albuterol

  (a) Name the functional groups present in albuterol.
  (b) Draw a structural formula for the product formed when albuterol is treated with one equivalent of aqueous sodium hydroxide.
  (c) Draw a structural formula for the product formed when albuterol is treated with one equivalent of hydrochloric acid.

Chemical Connections

13-34 (Chemical Connections 13A) From what parts of its common name are the letters DDT derived?

13-35 (Chemical Connections 13A) What are the advantages and disadvantages of using DDT as an insecticide?

13-36 (Chemical Connections 13A) Would you expect DDT to be soluble or insoluble in water? Explain.

13-37 (Chemical Connections 13A) One of the degradation products of DDT is dichlorodiphenyldichloroethylene (DDE), which is formed by the loss of

HCl from adjacent carbons of DDT. DDE inhibits the enzyme responsible for the incorporation of calcium ion into bird eggshells. Draw a structural formula for DDE.

13-38 (Chemical Connections 13A) What is meant by the term *biodegradable*?

13-39 (Chemical Connections 13B) What is a carcinogen? What kind of carcinogen is found in cigarette smoke?

13-40 (Chemical Connections 13C) In the absence of iodine in the diet, goiter develops. Explain why goiter is a regional disease.

13-41 (Chemical Connections 13D) Calculate the molecular weight of each explosive in this Chemical Connection. In which explosive do the nitro groups contribute the largest percentage of molecular weight?

13-42 (Chemical Connections 13E) What are the differences in structure between Allura Red and Sunset Yellow?

13-43 (Chemical Connections 13E) Which features of Allura Red and Sunset Yellow make them water-soluble?

13-44 (Chemical Connections 13E) What color would you get if you mixed Allura Red and Sunset Yellow? (*Hint*: Remember the color wheel.)

13-45 (Chemical Connections 13E)
(a) Review the structural formulas for lycopene and β-carotene (Problems 12-70 and 12-71), two natural food-coloring agents, and compare their structural formulas with those of Allura Red and Sunset Yellow.
(b) What structural feature or features do these four compounds have in common that might account for the fact that each is colored?
(c) Why are these two food and cosmetic dyes soluble in water while lycopene and β-carotene are not?

13-46 (Chemical Connections 13F) From what types of plants is capsaicin isolated?

13-47 (Chemical Connections 13F) Identify the phenol group in capsaicin.

13-48 (Chemical Connections 13F) How many *cis-trans* isomers are possible for capsaicin? Is the structural formula shown in this Chemical Connection the *cis* isomer or the *trans* isomer?

13-49 (Chemical Connections 13F) In what ways is capsaicin used in medicine?

Additional Problems

13-50 The structure for naphthalene given in Section 13-2D is only one of three possible contributing structures. Draw the other two.

13-51 Draw structural formulas for these compounds.
(a) 1-Phenylcyclopropanol (b) Styrene
(c) *m*-Bromophenol (d) 4-Nitrobenzoic acid
(e) Isobutylbenzene (f) *m*-Xylene

13-52 2,6-Di-*tert*-butyl-4-methylphenol (BHT, Section 13-4C) is an antioxidant added to processed foods to "retard spoilage." How does BHT accomplish this goal?

13-53 Write the structural formula for the product of each reaction.

(a) benzene + HNO_3 $\xrightarrow{H_2SO_4}$

(b) 1,4-dimethylbenzene + Br_2 $\xrightarrow{FeCl_3}$

(c) 1,4-dibromobenzene + H_2SO_4 $\longrightarrow$

13-54 Styrene reacts with bromine to give a compound with the molecular formula $C_8H_8Br_2$. Draw a structural formula for this compound.

13-55 When toluene is treated with Br_2 in the presence of an $AlBr_3$ catalyst, a mixture of three compounds is formed, all with the molecular formula C_7H_7Br. Draw structural formulas and name each of these products.

Challenge Problem

13-56 Four alternatives to the structure of benzene proposed by Kekulé are shown in the text. In this problem, we ask you to propose additional alternatives for compounds with the molecular formula C_6H_6. As you think about possible alternatives keep in mind the building blocks available to you. Also keep in mind that carbon must be tetravalent, that is, each carbon must have no more and no fewer than four bonds.

- Three- four-, five-, and six-membered rings
- Carbon–carbon single, double, and triple bonds
- The possibility for *cis/trans* isomerism for carbon–carbon double bonds

Looking Ahead

13-57 Benzene, as we have seen in this chapter, is the simplest aromatic compound. Pyidine is an analog of benzene in which a CH group is replaced by a nitrogen atom. Pyrimidine is an analog of benzene in which two CH groups are replaced by nitrogen atoms. Each nitrogen-containing compound shows the characteristic reactions of benzene and its derivatives—each is highly unsaturated but does not undergo the characteristic addition reactions of alkenes.

Benzene   Pyridine   Pyrimidine

(a) Show by the use of curved arrows that benzene, pyridine, and pyrimidine can be represented as hybrids of two contributing structures.

(b) Show that each aromatic compound has an aromatic sextet—that is, a loop of six electrons within a cyclic system.

(c) Predict the bond angles in pyridine and pyrimidine and the shape of each molecule.

14 Alcohols, Ethers, and Thiols

KEY QUESTIONS

- 14-1 What Are the Structures, Names, and Physical Properties of Alcohols?
- 14-2 What Are the Characteristic Reactions of Alcohols?
- 14-3 What Are the Structures, Names, and Physical Properties of Ethers?
- 14-4 What Are the Structures, Names, and Physical Properties of Thiols?
- 14-5 What Are the Most Commercially Important Alcohols?

Fermentation vats of wine grapes at the Beaulieu Vineyards, California.

In this chapter, we study the physical and chemical properties of alcohols and ethers, two classes of oxygen-containing organic compounds. We also study thiols, a class of sulfur-containing organic compounds. Thiols are like alcohols in structure, except that they contain an —SH group rather than an —OH group.

$$CH_3CH_2OH \qquad CH_3CH_2OCH_2CH_3 \qquad CH_3CH_2SH$$

Ethanol   Diethyl ether   Ethanethiol
(An alcohol)   (An ether)   (A thiol)

These three compounds are certainly familiar to you. Ethanol is the fuel additive in E85 and E15, the alcohol in alcoholic beverages, and an important laboratory and industrial solvent. Solutions in which ethanol is the solvent are called tinctures. Diethyl ether was the first inhalation anesthetic used in general surgery. It is also an important laboratory and industrial solvent. Ethanethiol, like other low-molecular-weight thiols, has a stench. Traces of ethanethiol are added to natural gas so that gas leaks can be detected by the smell of the thiol.

14-1 What Are the Structures, Names, and Physical Properties of Alcohols?

A. Structure of Alcohols

The functional group of an **alcohol** is an —OH (hydroxyl) group bonded to a tetrahedral carbon atom (Section 10-4A). Figure 14-1 shows a Lewis structure and a ball-and-stick model of methanol, CH₃OH, the simplest alcohol.

B. Nomenclature

IUPAC names of alcohols are derived in the same manner as those for alkenes and alkynes, with the exception that the ending of the parent alkane is changed from *-e* to *-ol*. The ending *-ol* tells us that the compound is an alcohol.

1. Select as the parent alkane the longest carbon chain that contains the —OH group and number it from the end that gives the —OH group the lower number. In numbering the parent chain, the location of the —OH group takes precedence over alkyl, aryl, and halogen groups.
2. Change the ending of the parent alkane from *-e* to *-ol* and use a number to show the location of the —OH group. For cyclic alcohols, numbering begins at the carbon bearing the —OH group; this carbon is automatically carbon 1.
3. Name and number substituents and list them in alphabetical order.

To derive the common name for an alcohol, name the alkyl group bonded to —OH and then add the word "alcohol." Following are IUPAC names and, in parentheses, common names for eight low-molecular-weight alcohols.

FIGURE 14-1 Methanol, CH₃OH. (a) Lewis structure and (b) ball-and-stick model. The H—C—O bond angle is 108.6°, very close to the tetrahedral angle of 109.5°.

Interactive Example 14-1 — Systematic Names of Alcohols

Write the IUPAC name of each alcohol.

(a) [structure]   (b) [structure]

Strategy

Follow the steps outlined above.

Step 1: Identify the parent chain.

Step 2: Change the ending of the parent alkane from *-e* to *-ol* and use a number to locate the —OH group.

Step 3: Name and number substituents and list them in alphabetical order.

Step 4: Specify configuration if *cis-trans* isomerism exists.

Solution

(a) The parent alkane is pentane. Number the parent chain from the direction that gives the lower number to the carbon bearing the —OH group. This alcohol is 4-methyl-2-pentanol.

(b) The parent cycloalkane is cyclohexane. Number the atoms of the ring beginning with the carbon bearing the —OH group as carbon 1 and specify that the methyl and hydroxyl groups are *trans* to each other. This alcohol is *trans*-2-methylcyclohexanol.

Problem 14-1

Write the IUPAC name of each alcohol.

We classify alcohols as **primary (1°), secondary (2°), or tertiary (3°)**, depending on the number of carbon groups bonded to the carbon bearing the —OH group (Section 10-4A).

EXAMPLE 14-2 Classification of Alcohols

Classify each alcohol as primary, secondary, or tertiary.

Strategy

Locate the carbon bearing the —OH group and count the number of carbon groups bonded to that carbon.

Solution

(a) Secondary (2°)   (b) Tertiary (3°)   (c) Primary (1°)

Problem 14-2

Classify each alcohol as primary, secondary, or tertiary.

In the IUPAC system, a compound containing two hydroxyl groups is named as a **diol**, one containing three hydroxyl groups as a **triol**, and so on. In IUPAC names for diols, triols, and so on, the final *-e* in the name of the parent alkane is retained—for example, 1,2-ethanediol.

Diol A compound with —OH (hydroxyl) groups

As with many other organic compounds, common names for certain diols and triols have persisted. Compounds containing hydroxyl groups on adjacent carbons are often referred to as **glycols**. Ethylene glycol and propylene glycol are synthesized from ethylene and propylene, respectively—hence their common names. Ethylene glycol is colorless; the color of most antifreezes comes from additives. ◄ Review Section 6-8A for a discussion of freezing-point depression.

Glycols Compounds with hydroxyl (—OH) groups on adjacent carbons

CH₂—CH₂
| |
OH   OH
1,2-Ethanediol
(Ethylene glycol)

CH₃—CH—CH₂
     | |
     OH   OH
1,2-Propanediol
(Propylene glycol)

CH₂—CH—CH₂
| | |
OH   OH   OH
1,2,3-Propanetriol
(Glycerol, Glycerin)

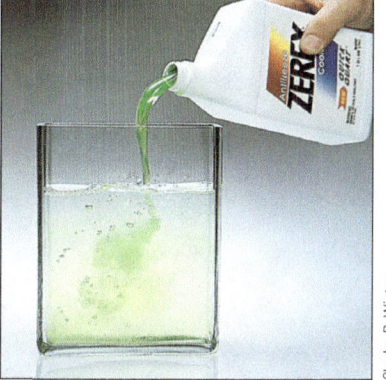

► Ethylene glycol is a polar molecule that dissolves readily in the polar solvent water.

C. Physical Properties of Alcohols

The most important physical property of alcohols is the polarity of their —OH groups. Because of the large difference in electronegativity (Table 3-5) between oxygen and carbon (3.5 − 2.5 = 1.0) and between oxygen and hydrogen (3.5 − 2.1 = 1.4), both the C—O and O—H bonds of alcohols are polar-covalent and alcohols are polar molecules, as illustrated in Figure 14-2 for methanol.

Alcohols have higher boiling points than do alkanes, alkenes, and alkynes of similar molecular weight (Table 14-1), because alcohol molecules associate with one another in the liquid state by **hydrogen bonding** (Section 5-7C). The strength of hydrogen bonding between alcohol molecules is approximately 2 to 5 kcal/mol, which means that extra energy is required to separate hydrogen-bonded alcohols from their neighbors (Figure 14-3).

Because of increased London dispersion forces (Section 5-7A) between larger molecules, the boiling points of all types of compounds, including alcohols, increase with increasing molecular weight.

(a)   (b)

FIGURE 14-2 Polarity of the C—O—H bonds in methanol. (a) There are partial positive charges on carbon and hydrogen and a partial negative charge on oxygen. (b) An electron density map showing the partial negative charge (in red) around oxygen and a partial positive charge (in blue) around the hydrogen of the OH group.

FIGURE 14-3 The association of ethanol molecules in the liquid state. Each O—H can participate in up to three hydrogen bonds (one through hydrogen and two through oxygen). Only two of these three possible hydrogen bonds are shown for the topmost molecule here.

Alcohols are much more soluble in water than are hydrocarbons of similar molecular weight (Table 14-1), because alcohol molecules interact by hydrogen bonding with water molecules. Methanol, ethanol, and 1-propanol are soluble in water in all proportions. As molecular weight increases, the water solubility of alcohols becomes more like that of hydrocarbons of similar molecular weight. Higher-molecular-weight alcohols are much less soluble in water because the size of the hydrocarbon portion of their molecules (which decreases water solubility) becomes large relative to the size of the —OH group (which increases water solubility).

Table 14-1 Boiling Points and Solubilities in Water of Four Sets of Alcohols and Alkanes of Similar Molecular Weight

Structural Formula	Name	Molecular Weight (amu)	Boiling Point (°C)	Solubility in Water
CH_3OH	methanol	32	65	infinite
CH_3CH_3	ethane	30	−89	insoluble
CH_3CH_2OH	ethanol	46	78	infinite
$CH_3CH_2CH_3$	propane	44	−42	insoluble
$CH_3CH_2CH_2OH$	1-propanol	60	97	infinite
$CH_3CH_2CH_2CH_3$	butane	58	0	insoluble
$CH_3CH_2CH_2CH_2OH$	1-butanol	74	117	8 g/100 g
$CH_3CH_2CH_2CH_2CH_3$	pentane	72	36	insoluble

14-2 What Are the Characteristic Reactions of Alcohols?

In this section, we study the acidity of alcohols, their dehydration to alkenes, and their oxidation to aldehydes, ketones, and carboxylic acids.

A. Acidity of Alcohols

Alcohols have about the same pK_a values as water (Table 8-3), which means that aqueous solutions of alcohols have approximately the same pH as that of pure water. In Section 13-4B, we studied the acidity of phenols, another class of compounds that contains an —OH group. Phenols are weak acids that react with aqueous sodium hydroxide to form water-soluble salts.

Phenol + NaOH $\xrightarrow{H_2O}$ Sodium phenoxide (a water-soluble salt) + H_2O

Alcohols are considerably weaker acids than phenols and do not react in this manner.

CHEMICAL CONNECTIONS 14A

The Importance of Hydrogen Bonding in Drug-Receptor Interactions

Hydrogen bonds have directionality in that the donor and acceptor groups must be oriented appropriately with respect to each other for hydrogen bonding to occur. Important hydrogen bond donors in biological molecules include —OH groups (proteins, carbohydrates) and —NH groups (proteins, nucleic acids). Important hydrogen bond acceptors are any N or O with a lone pair of electrons, such as C=O groups (proteins, carbohydrates, nucleic acids), —OH groups (proteins, carbohydrates), and COO⁻ groups (proteins).

With directionality comes the potential for hydrogen bonds to organize molecules at many levels ranging from the folding of biological molecules to the specific binding and recognition between a pharmaceutical and its receptor. The drug atorvastatin (Lipitor) is used to treat high cholesterol. Cholesterol is synthesized in the liver from the two-carbon acetyl group of acetyl coenzyme A (acetyl-CoA). A key intermediate in the sequence of reactions leading to the synthesis of cholesterol is a six-carbon molecule named mevalonate (Section 29-4). Atorvastatin specifically binds to and blocks the action of HMG-CoA reductase, a key enzyme in the biosynthesis of mevalonate. Atorvastatin binds to this enzyme in preference to the large number of other potential enzyme targets because (1) the drug has a shape complementary to the catalytic cavity (the active site) of HMG-CoA reductase (*a*) and (2) it can form at least nine specific hydrogen bonds with functional groups at the active site on the enzyme (*b*).

The complementary shape and pattern of hydrogen bonding ensure that atorvastatin binds to HMG-CoA reductase and inhibits its ability to catalyze the formation of mevalonate. The hallmark of this and other effective drugs is their ability to bind strongly with their intended target molecules, while at the same time not interacting with other molecules that could lead to unwanted side effects.

(*a*) A space-filling model of the cholesterol-lowering drug atorvastatin (Lipitor) bound to the active site of its enzyme target HMG-CoA reductase (shown as a yellow surface). The shape of the drug is complementary to the active site of the enzyme.

(*b*) Hydrogen bonding (shown in red) between atorvastatin and the functional groups at the active site of the enzyme HMG-CoA reductase. The nine hydrogen bonds (shown in red), many of which involve hydroxyl groups on atorvastatin or the enzyme surface, help to provide the specificity that directs the binding of the drug to its target enzyme.

Test your knowledge with Problem 14-50

B. Acid-Catalyzed Dehydration of Alcohols

We can convert an alcohol to an alkene by eliminating a molecule of water from adjacent carbon atoms in a reaction called **dehydration**. In the laboratory, dehydration of an alcohol is most often brought about by heating it with either 85% phosphoric acid or concentrated sulfuric acid. Primary alcohols—the most difficult to dehydrate—generally require heating in

Dehydration Elimination of a molecule of water from an alcohol. In the dehydration of an alcohol, OH is removed from one carbon and an H is removed from an adjacent carbon.

CHEMICAL CONNECTIONS 14B

Nitroglycerin: An Explosive and a Drug

In 1847, Ascanio Sobrero (1812–1888) discovered that 1,2,3-propanetriol, more commonly called glycerin, reacts with nitric acid in the presence of sulfuric acid to give a pale yellow, oily liquid called nitroglycerin. Sobrero also discovered the explosive properties of this compound; when he heated a small quantity of it, it exploded!

$$\begin{array}{c}CH_2\text{—}OH\\|\\CH\text{—}OH\\|\\CH_2\text{—}OH\end{array} + 3HNO_3 \xrightarrow{H_2SO_4} \begin{array}{c}CH_2\text{—}ONO_2\\|\\CH\text{—}ONO_2\\|\\CH_2\text{—}ONO_2\end{array} + 3H_2O$$

1,2,3-Propanetriol (Glycerol, Glycerin)   1,2,3-Propanetriol trinitrate (Nitroglycerin)

Nitroglycerin very soon became widely used for blasting in the construction of canals, tunnels, roads, and mines and, of course, for warfare.

One problem with the use of nitroglycerin was soon recognized: it was difficult to handle safely, and accidental explosions occurred all too frequently. This problem was solved by the Swedish chemist Alfred Nobel (1833–1896), who discovered that a clay-like substance called diatomaceous earth absorbs nitroglycerin so that it will not explode without a fuse. Nobel gave the name *dynamite* to this mixture of nitroglycerin, diatomaceous earth, and sodium carbonate.

Surprising as it may seem, nitroglycerin is used in medicine to treat angina pectoris, the symptoms of which are sharp chest pains caused by reduced flow of blood in the coronary artery. Nitroglycerin, which is available in liquid form (diluted with alcohol to render it nonexplosive), tablet form, and paste form, relaxes the smooth muscles of blood vessels, causing dilation of the coronary artery. This dilation, in turn, allows more blood to reach the heart.

When Nobel became ill with heart disease, his physicians advised him to take nitroglycerin to relieve his chest pains. He refused, saying he could not understand how the explosive could relieve chest pains. It took science more than 100 years to find the answer. We now know that nitric oxide, NO, derived from the nitro groups of nitroglycerin, relieves the pain (see Chemical Connections 24E).

(a) Nitroglycerin is more stable when absorbed onto an inert solid, a combination called dynamite. (b) The fortune of Alfred Nobel (1833–1896), built on the manufacture of dynamite, now funds the Nobel Prizes.

(a)   (b)

Test your knowledge with Problems 14-51 through 14-53.

---

concentrated sulfuric acid at temperatures as high as 180°C. Secondary alcohols undergo acid-catalyzed dehydration at somewhat lower temperatures. Tertiary alcohols generally undergo acid-catalyzed dehydration at temperatures only slightly above room temperature.

$$CH_3CH_2OH \xrightarrow[180°C]{H_2SO_4} CH_2\text{=}CH_2 + H_2O$$

Ethanol   Ethylene

$$\text{Cyclohexanol} \xrightarrow[140°C]{H_2SO_4} \text{Cyclohexene} + H_2O$$

$$\begin{array}{c}CH_3\\|\\CH_3CCH_3\\|\\OH\end{array} \xrightarrow[50°C]{H_2SO_4} \begin{array}{c}CH_3\\|\\CH_3C\text{=}CH_2\end{array} + H_2O$$

2-Methyl-2-propanol (*tert*-Butyl alcohol)   2-Methylpropene (Isobutylene)

Thus, the ease of acid-catalyzed dehydration of alcohols follows this order:

1° alcohols   2° alcohols   3° alcohols

→ Ease of dehydration of alcohols

When the acid-catalyzed dehydration of an alcohol yields isomeric alkenes, the alkene having the greater number of alkyl groups on the double bond generally predominates. In the acid-catalyzed dehydration of 2-butanol, for example, the major product is 2-butene, which has two alkyl groups (two methyl groups) on its double bond. The minor product is 1-butene, which has only one alkyl group (an ethyl group) on its double bond.

$$CH_3CH_2\underset{\underset{OH}{|}}{C}HCH_3 \xrightarrow[\text{heat}]{H_3PO_4} CH_3CH=CHCH_3 + CH_3CH_2CH=CH_2 + H_2O$$

2-Butanol   2-Butene   1-Butene
       (80%)   (20%)

EXAMPLE 14-3   Acid-Catalyzed Dehydration of Alcohols

Draw structural formulas for the alkenes formed by the acid-catalyzed dehydration of each alcohol. For each part, predict which alkene will be the major product.

(a) 3-Methyl-2-butanol   (b) 2-Methylcyclopentanol

Strategy

In the acid-catalyzed dehydration of an alcohol, H and OH are removed from adjacent carbon atoms. When dehydration yields isomeric alkenes, the alkene with the greater number of alkyl groups on the carbon atoms of the double bond generally predominates.

Solution

(a) Elimination of H_2O from carbons 2 and 3 gives 2-methyl-2-butene; elimination of H_2O from carbons 1 and 2 gives 3-methyl-1-butene. 2-Methyl-2-butene has three alkyl groups (three methyl groups) on its double bond and is the major product. 3-Methyl-1-butene has only one alkyl group (an isopropyl group) on its double bond and is the minor product.

$$\overset{4}{C}H_3\overset{3}{\underset{\underset{OH}{|}}{C}H}\overset{\overset{CH_3}{|}}{\overset{2}{C}H}\overset{1}{C}H_3 \xrightarrow[\text{dehydration}]{\underset{\text{acid-catalyzed}}{H_2SO_4}} CH_3\overset{\overset{CH_3}{|}}{C}=CHCH_3 + CH_3\overset{\overset{CH_3}{|}}{C}HCH=CH_2 + H_2O$$

3-Methyl-2-butanol   2-Methyl-2-butene   3-Methyl-1-butene
         (major product)

(b) The major product, 1-methylcyclopentene, has three alkyl groups on its double bond. The minor product, 3-methylcyclopentene, has only two alkyl groups on its double bond.

2-Methylcyclopentanol $\xrightarrow[\text{dehydration}]{\underset{\text{acid-catalyzed}}{H_2SO_4}}$ 1-Methylcyclopentene + 3-Methylcyclopentene + H_2O
          (major product)

Problem 14-3

Draw structural formulas for the alkenes formed by the acid-catalyzed dehydration of each alcohol. For each part, predict which alkene will be the major product.

(a) 2-Methyl-2-butanol (b) 1-Methylcyclopentanol

In Section 12-6B, we studied the acid-catalyzed hydration of alkenes to give alcohols. In this section, we study the acid-catalyzed dehydration of alcohols to give alkenes. In fact, hydration–dehydration reactions are reversible. Alkene hydration and alcohol dehydration are competing reactions, and the following equilibrium exists:

$$\underset{\text{An alkene}}{\diagup\!\!\!\!\diagdown\!\!C\!=\!C\!\diagup\!\!\!\!\diagdown} + H_2O \underset{\text{dehydration}}{\overset{\text{hydration}}{\rightleftarrows}} \underset{\text{An alcohol}}{-\underset{H}{\overset{|}{C}}-\underset{OH}{\overset{|}{C}}-}$$

In accordance with Le Chatelier's principle (Section 7-7), large amounts of water (in other words, using dilute aqueous acid) favor alcohol formation, whereas scarcity of water (using concentrated acid) or experimental conditions where water is removed (heating the reaction mixture above 100°C) favor alkene formation. Thus, depending on the experimental conditions, we can use the hydration–dehydration equilibrium to prepare both alcohols and alkenes, each in high yields.

EXAMPLE 14-4 Acid-Catalyzed Dehydration of Alcohols and Hydration of Alkenes

In part (a), acid-catalyzed dehydration of 2-methyl-3-pentanol gives predominantly Compound A. Treatment of Compound A with water in the presence of sulfuric acid in part (b) gives Compound B. Propose structural formulas for Compounds A and B.

(a) $CH_3\underset{OH}{\overset{\overset{\displaystyle CH_3}{|}}{CH}}CHCH_2CH_3 \xrightarrow[\text{dehydration}]{\text{acid-catalyzed}}^{H_2SO_4}$ Compound A (C_6H_{12}) + H_2O

(b) Compound A (C_6H_{12}) + $H_2O \xrightarrow{H_2SO_4}$ Compound B ($C_6H_{14}O$)

Strategy

The key to part (a) is that when acid-catalyzed dehydration of an alcohol can yield isomeric alkenes, the alkene with the greater number of alkyl groups on the carbon atoms of the double bond generally predominates. After the structural formula of A is determined, use Markovnikov's Rule to predict the structural formula of compound B.

Solution

(a) Acid-catalyzed dehydration of 2-methyl-3-pentanol gives predominantly 2-methyl-2-pentene, an alkene with three substituents on its double bond: two methyl groups and one ethyl group.

$$\underset{\text{2-Methyl-3-pentanol}}{CH_3\underset{\underset{OH}{|}}{\overset{\overset{CH_3}{|}}{C}H}CHCH_2CH_3} \xrightarrow[\text{dehydration}]{\underset{\text{acid-catalyzed}}{H_2SO_4}} \underset{\underset{\text{(major product)}}{\text{2-Methyl-2-pentene}}}{CH_3\overset{\overset{CH_3}{|}}{C}=CHCH_2CH_3} + H_2O$$

(b) Acid-catalyzed addition of water to this alkene gives 2-methyl-2-pentanol in accordance with Markovnikov's rule (Section 12-6B).

$$\underset{\text{Compound A }(C_6H_{12})}{CH_3\overset{\overset{CH_3}{|}}{C}=CHCH_2CH_3} + H_2O \xrightarrow[\text{hydration}]{\underset{\text{acid-catalyzed}}{H_2SO_4}} \underset{\text{Compound B }(C_6H_{14}O)}{CH_3\underset{\underset{OH}{|}}{\overset{\overset{CH_3}{|}}{C}}CH_2CH_2CH_3}$$

Problem 14-4

Acid-catalyzed dehydration of 2-methylcyclohexanol gives predominantly Compound C (C_7H_{12}). Treatment of Compound C with water in the presence of sulfuric acid gives Compound D ($C_7H_{14}O$). Propose structural formulas for Compounds C and D.

C. Oxidation of Primary and Secondary Alcohols

A primary alcohol can be oxidized to an aldehyde or to a carboxylic acid, depending on the experimental conditions. Following is a series of transformations in which a primary alcohol is oxidized first to an aldehyde and then to a carboxylic acid. The letter O in brackets over the reaction arrow indicates that each transformation involves oxidation.

$$\underset{\underset{\text{alcohol}}{\text{A primary}}}{CH_3-\underset{\underset{H}{|}}{\overset{\overset{OH}{|}}{C}}-H} \xrightarrow{[O]} \underset{\text{An aldehyde}}{CH_3-\overset{\overset{O}{\|}}{C}-H} \xrightarrow{[O]} \underset{\underset{\text{acid}}{\text{A carboxylic}}}{CH_3-\overset{\overset{O}{\|}}{C}-OH}$$

Recall from Section 4-4 that according to one definition, oxidation is either the loss of hydrogens or the gain of oxygens. Using this definition, conversion of a primary alcohol to an aldehyde is an oxidation reaction because the alcohol loses hydrogens. Conversion of an aldehyde to a carboxylic acid is also an oxidation reaction because the aldehyde gains an oxygen.

The reagent most commonly used in the laboratory for the oxidation of a primary alcohol to a carboxylic acid is potassium dichromate, $K_2Cr_2O_7$, dissolved in aqueous sulfuric acid. Using this reagent, oxidation of 1-octanol, for example, gives octanoic acid. This experimental condition is more than sufficient to oxidize the intermediate aldehyde to a carboxylic acid.

$$CH_3(CH_2)_6CH_2OH \xrightarrow[H_2SO_4]{K_2Cr_2O_7} CH_3(CH_2)_6\overset{\overset{O}{\|}}{C}H \xrightarrow[H_2SO_4]{K_2Cr_2O_7} CH_3(CH_2)_6\overset{\overset{O}{\|}}{C}OH$$

1-Octanol   Octanal   Octanoic acid

Secondary alcohols are oxidized to ketones by potassium dichromate. Menthol, a secondary alcohol present in peppermint and other mint oils, is used in liqueurs, cigarettes, cough drops, perfumery, and nasal inhalers. Its oxidation product, menthone, is also used in perfumes and artificial flavors.

2-Isopropyl-5-methyl-cyclohexanol (Menthol)   2-Isopropyl-5-methyl-cyclohexanone (Menthone)

Tertiary alcohols resist oxidation because the carbon bearing the —OH is bonded to three carbon atoms and, therefore, cannot form a carbon–oxygen double bond.

EXAMPLE 14-5 Oxidation of Alcohols

Draw a structural formula for the product formed by oxidation of each of the following alcohols with potassium dichromate.

(a) 1-Hexanol   (b) 2-Hexanol

Strategy

Oxidation of 1-hexanol, a primary alcohol, gives either an aldehyde or a carboxylic acid, depending on the experimental conditions. Oxidation of 2-hexanol, a secondary alcohol, gives a ketone.

Solution

(a) Hexanal or Hexanoic acid

(b) 2-Hexanone

Problem 14-5

Draw the product formed by oxidation of each of the following alcohols with potassium dichromate.

(a) Cyclohexanol   (b) 2-Pentanol

CHEMICAL CONNECTIONS 14C

Breath-Alcohol Screening

Potassium dichromate oxidation of ethanol to acetic acid is the basis for the original breath-alcohol screening test used by law enforcement agencies to determine a person's blood alcohol content (BAC). The test is based on the difference in color between the dichromate ion (reddish orange) in the reagent and the chromium(III) ion (green) in the product.

$$CH_3CH_2OH + Cr_2O_7^{2-} \xrightarrow[H_2O]{H_2SO_4}$$

Ethanol   Dichromate ion (reddish orange)

$$CH_3\overset{\overset{\displaystyle O}{\|}}{C}OH + Cr^{3+}$$

Acetic acid   Chromium(III) ion (green)

In its simplest form, breath-alcohol screening uses a sealed glass tube that contains a potassium dichromate–sulfuric acid reagent impregnated on silica gel. To administer the test, the ends of the tube are broken off, a mouthpiece is fitted to one end, and the other end is inserted into the neck of a plastic bag. The person being tested then blows into the mouthpiece to inflate the plastic bag.

As breath containing ethanol vapor passes through the tube, reddish orange dichromate ion is reduced to green chromium(III) ion. To estimate the concentration of ethanol in the breath, one measures how far the green color extends along the length of the tube. When it reaches beyond the halfway point, the person is judged as having a sufficiently high blood alcohol content to warrant further, more precise testing.

The test described here measures the alcohol content of the breath. The legal definition of being under the influence of alcohol, however, is based on alcohol content in the blood, not in the breath. The correlation between these two measurements is based on the fact that air deep within the lungs is in equilibrium with blood passing through the pulmonary arteries, and thus an equilibrium is established between blood alcohol and breath alcohol. Based on tests in persons drinking alcohol, researchers have determined that 2100 mL of breath contains the same amount of ethanol as 1.00 mL of blood.

Person forces breath through mouthpiece into the tube.

Glass tube containing potassium dichromate–sulfuric acid coated on silica gel particles

As the person blows into the tube, the plastic bag becomes inflated.

Test your knowledge with Problems 14-54 and 14-55.

14-3 What Are the Structures, Names, and Physical Properties of Ethers?

A. Structure

The functional group of an **ether** is an atom of oxygen bonded to two carbon atoms. Figure 14-4 shows a Lewis structure and a ball-and-stick model of dimethyl ether, CH_3OCH_3, the simplest ether.

Ether A compound containing an oxygen atom bonded to two carbon atoms

FIGURE 14-4 Dimethyl ether, CH_3OCH_3. (a) Lewis structure and (b) ball-and-stick model. The C—O—C bond angle is 110.3°, close to the tetrahedral angle of 109.5°.

B. Nomenclature

Although the IUPAC system can be used to name ethers, chemists almost invariably use common names for low-molecular-weight ethers. Common names are derived by listing the alkyl groups bonded to oxygen in alphabetical order and adding the word *ether*. Alternatively, one of the groups on oxygen is named as an alkoxy group. The —OCH_3 group, for example, is named "methoxy" to indicate a <u>meth</u>yl group bonded to <u>oxy</u>gen.

$CH_3CH_2OCH_2CH_3$ ⬡—OCH_3

Diethyl ether   Cyclohexyl methyl ether
                                      (Methoxycyclohexane)

EXAMPLE 14-6   Common Names for Ethers

Write the common name for each ether.

(a) $CH_3\underset{\underset{CH_3}{|}}{\overset{\overset{CH_3}{|}}{C}}OCH_2CH_3$ (b) ⬡—O—⬡

Strategy

To derive the common name of an ether, list the groups bonded to oxygen in alphabetical order.

(a) Note that the alphabetizing terms are butyl and ethyl.

Solution

(a) The groups bonded to the ether oxygen are *tert*-butyl and ethyl. The compound's common name is *tert*-butyl ethyl ether.
(b) Two cyclohexyl groups are bonded to the ether oxygen. The compound's common name is dicyclohexyl ether.

Problem 14-6

Write the common name for each ether.

(a) ⋏⋎—O—⋏   (b) ⬠—OCH_3

Cyclic ethers Ethers in which the ether oxygen is one of the atoms of a ring

In **cyclic ethers**, one of the atoms in a ring is oxygen. These ethers are also known by their common names. Ethylene oxide is an important building block for the organic chemical industry (Section 14-5). Tetrahydrofuran is a useful laboratory and industrial solvent.

△   ⬠
O   O

Ethylene oxide   Tetrahydrofuran
                   (THF)

CHEMICAL CONNECTIONS 14D

Ethylene Oxide: A Chemical Sterilant

Ethylene oxide is a colorless, flammable gas with a boiling point of 11°C. Because it is such a highly strained molecule (the normal tetrahedral bond angles of both C and O are compressed from the normal tetrahedral angle of 109.5° to approximately 60°), ethylene oxide reacts with the amino (—NH_2) and sulfhydryl (—SH) groups present in biological materials.

At sufficiently high concentrations, it reacts with enough molecules in cells to cause the death of microorganisms. This toxic property is the basis for ethylene oxide's use as a fumigant in foodstuffs and textiles and its use in hospitals to sterilize surgical instruments.

$$RNH_2 + \underset{O}{\triangle} \longrightarrow RNH-CH_2CH_2O-H$$

$$RSH + \underset{O}{\triangle} \longrightarrow RS-CH_2CH_2O-H$$

Test your knowledge with Problem 14-56.

C. Physical Properties

Ethers are polar compounds in which oxygen bears a partial negative charge and each attached carbon bears a partial positive charge (Figure 14-5). However, only weak forces of attraction exist between ether molecules in the pure liquid form. Consequently, the boiling points of ethers are close to those of hydrocarbons of similar molecular weight.

The effect of hydrogen bonding on physical properties is illustrated dramatically by comparing the boiling points of ethanol (78°C) and its constitutional isomer dimethyl ether (−24°C). The difference in boiling point between these two compounds is due to the presence in ethanol of a polar O—H group, which is capable of forming hydrogen bonds. This hydrogen

FIGURE 14-5 Ethers are polar molecules, but only weak attractive interactions exist between ether molecules in the liquid state. Shown on the right is an electron density map of diethyl ether.

CHEMICAL CONNECTIONS 14E

Ethers and Anesthesia

Before the mid-1800s, surgery was performed only when absolutely necessary, because no truly effective general anesthetic was available. More often than not, patients were drugged, hypnotized, or simply tied down.

In 1772, Joseph Priestley isolated nitrous oxide, N_2O, a colorless gas. In 1799, Sir Humphry Davy demonstrated this compound's anesthetic effect, and named it "laughing gas." In 1844, an American dentist, Horace Wells, introduced nitrous oxide into general dental practice. One patient awakened prematurely, however, screaming with pain; another died during surgery. Wells was forced to withdraw from practice, became embittered and depressed, and committed suicide at age 33. In the same period, a Boston chemist, Charles Jackson, anesthetized himself with diethyl ether; he then persuaded a dentist, William Morton, to use it. Subsequently, they persuaded a surgeon, John Warren, to give a public demonstration of surgery under anesthesia. The operation was completely successful, and soon general anesthesia by diethyl ether became routine practice for general surgery.

Diethyl ether was easy to use and caused excellent muscle relaxation. Blood pressure, pulse rate, and respiration were usually only slightly affected. Diethyl ether's chief drawbacks are its irritating effect on the respiratory passages and its aftereffect of nausea and its extreme volatility/flammability.

Among the inhalation anesthetics used today are several halogenated ethers, the most important of which are enflurane and isoflurane.

This photo shows the first use of ether as an anesthetic in 1848. Dr. Robert John Collins was removing a tumor from the patient's neck and dentist W. T. G. Morton—who discovered the anesthetic properties—administered the ether.

Enflurane (Ethrane)

Isoflurane (Forane)

Test your knowledge with Problems 14-57 through 14-59.

---

bonding increases intermolecular associations, thereby giving ethanol a higher boiling point than dimethyl ether.

CH₃CH₂OH Ethanol bp 78°C

CH₃OCH₃ Dimethyl ether bp −24°C

Ethers are more soluble in water than hydrocarbons of similar molecular weight and shape, but far less soluble than isomeric alcohols. Their greater solubility reflects the fact that the oxygen atom of an ether carries a partial negative charge and forms hydrogen bonds with water.

D. Reactions of Ethers

Ethers resemble hydrocarbons in their resistance to chemical reaction. For example, they do not react with oxidizing agents, such as potassium dichromate. Likewise, they do not react with reducing agents such as H_2 in the presence of a metal catalyst (Section 12-6D). Furthermore, most acids and bases at moderate temperatures do not affect them. Because of their general inertness to chemical reaction and their good solvent properties, ethers are excellent solvents in which to carry out many organic reactions. The most important ether solvents are diethyl ether and tetrahydrofuran.

14-4 What Are the Structures, Names, and Physical Properties of Thiols?

A. Structure

The functional group of a **thiol** is an —SH (**sulfhydryl) group** bonded to a tetrahedral carbon atom. Figure 14-6 shows a Lewis structure and a ball-and-stick model of methanethiol, CH_3SH, the simplest thiol.

Thiol A compound that contains an —SH (sulfhydryl) group bonded to a tetrahedral carbon atom

B. Nomenclature

The sulfur analog of an alcohol is called a thiol (*thi*- from the Greek: *theion*, sulfur) or, in the older literature, a **mercaptan**, which literally means "mercury capturing." Thiols react with Hg^{2+} in aqueous solution to give sulfide salts as insoluble precipitates. Thiophenol, C_6H_5SH, for example, gives $(C_6H_5S)_2Hg$.

In the IUPAC system, thiols are named by selecting the longest carbon chain that contains the —SH group as the parent alkane. To show that the compound is a thiol, we add the suffix *-thiol* to the name of the parent alkane. The parent chain is numbered in the direction that gives the —SH group the lower number.

Common names for simple thiols are derived by naming the alkyl group bonded to —SH and adding the word *mercaptan*.

FIGURE 14-6 Methanethiol, CH_3SH. (*a*) Lewis structure and (*b*) ball-and-stick model. The H—S—C bond angle is 100.3°, somewhat smaller than the tetrahedral angle of 109.5°.

Mercaptan A common name for any molecule containing an —SH group

$$CH_3CH_2SH \qquad CH_3\overset{\underset{\displaystyle CH_3}{|}}{C}HCH_2SH$$

Ethanethiol   2-Methyl-1-propanethiol
(Ethyl mercaptan)   (Isobutyl mercaptan)

EXAMPLE 14-7 Systematic Names of Thiols

Write the IUPAC name of each thiol.

(a) ~~~~SH   (b) with SH on middle carbon of a 4-carbon chain

Strategy

To derive the IUPAC name of a thiol, select as the parent alkane the longest carbon chain that contains the —SH group. Show that the compound is a thiol by adding the suffix *-thiol* to the name of the parent alkane. Number the parent chain in the direction that gives the —SH group the lower number.

Solution

(a) The parent alkane is pentane. Show the presence of the —SH group by adding "thiol" to the name of the parent alkane. The IUPAC name of this thiol is 1-pentanethiol. Its common name is pentyl mercaptan.

(b) The parent alkane is butane. The IUPAC name of this thiol is 2-butanethiol. Its common name is *sec*-butyl mercaptan.

Problem 14-7
Write the IUPAC name of each thiol.

(a) (CH₃)₂CHCH₂CH₂SH   (b) (CH₃)₂CHCH(SH)CH₃

▶ The scent of the spotted skunk, native to the Sonoran Desert, is a mixture of two thiols, 2-butene-1-thiol and 3-methyl-1-butanethiol.

The most notable property of low-molecular-weight **thiols** is their stench. They are responsible for unpleasant odors such as those from skunks, rotten eggs, and sewage. The scent of skunks is due primarily to two thiols: ◀

$$CH_3CH=CHCH_2SH \qquad CH_3\overset{\overset{\displaystyle CH_3}{|}}{C}HCH_2CH_2SH$$

2-Butene-1-thiol   3-Methyl-1-butanethiol

C. Physical Properties

Because of the small difference in electronegativity between sulfur and hydrogen ($2.5 - 2.1 = 0.4$), we classify the S—H bond as nonpolar covalent. Because of this lack of polarity, thiols show little association by hydrogen bonding. Consequently, they have lower boiling points and are less soluble in water and other polar solvents than are alcohols of similar molecular weight. Table 14-2 gives boiling points for three low-molecular-weight thiols. Shown for comparison are the boiling points of alcohols with the same number of carbon atoms.

Table 14-2 Boiling Points of Three Sets of Thiols and Alcohols with the Same Number of Carbon Atoms

Thiol	Boiling Point (°C)	Alcohol	Boiling Point (°C)
methanethiol	6	methanol	65
ethanethiol	35	ethanol	78
1-butanethiol	98	1-butanol	117

Earlier we illustrated the importance of hydrogen bonding in alcohols by comparing the boiling points of ethanol (78°C) and its constitutional isomer dimethyl ether (−24°C). By contrast, the boiling point of ethanethiol is 35°C and that of its constitutional isomer dimethyl sulfide is 37°C. Because the boiling points of these constitutional isomers are almost identical, we know that little or no association by hydrogen bonding occurs between thiol molecules.

$$CH_3CH_2SH \qquad CH_3SCH_3$$

Ethanethiol   Dimethyl sulfide
bp 35°C   bp 37°C

D. Reactions of Thiols

Thiols are weak acids ($pK_a = 10$), comparable in strength to phenols (Section 13-4B). Thiols react with strong bases such as NaOH to form thiolate salts.

$$CH_3CH_2SH + NaOH \xrightarrow{H_2O} CH_3CH_2S^-Na^+ + H_2O$$

Ethanethiol   Sodium
(pK_a 10)   ethanethiolate

The most common reaction of thiols in biological systems is their oxidation to **disulfides**, the functional group of which is a **disulfide (—S—S—) bond**. Thiols are readily oxidized to disulfides by molecular oxygen. In fact, they are so susceptible to oxidation that they must be protected from contact with air during their storage. Disulfides, in turn, are easily reduced to thiols by several reducing agents. This easy interconversion between thiols and disulfides is very important in protein chemistry, as we will see in Chapters 22 and 23.

Disulfides Compounds containing an (—S—S—) group

$$2\,HOCH_2CH_2SH \underset{\text{reduction}}{\overset{\text{oxidation}}{\rightleftharpoons}} HOCH_2CH_2S-SCH_2CH_2OH$$

A thiol A disulfide

To derive the common name of a disulfide, list the names of the groups bonded to sulfur and add the word *disulfide*.

14-5 What Are the Most Commercially Important Alcohols?

As you study the alcohols described in this section, you should pay particular attention to two key points. First, they are derived almost entirely from petroleum, natural gas, or coal—all nonrenewable resources. Second, many are themselves starting materials for the synthesis of valuable commercial products, without which our modern industrial society could not exist.

At one time, methanol was derived by heating hard woods in a limited supply of air—hence the name "wood alcohol." Today methanol is obtained entirely from the catalytic reduction of carbon monoxide. Methanol, in turn, is the starting material for the preparation of several important industrial and commercial chemicals, including acetic acid and formaldehyde. Treatment of methanol with carbon monoxide in the presence of a rhodium catalyst gives acetic acid. Partial oxidation of methanol gives formaldehyde. An important use of this one-carbon aldehyde is in the preparation of phenol-formaldehyde and urea-formaldehyde glues and resins, which are used as molding materials and as adhesives for plywood and particle board for the construction industry.

$$\text{Coal or methane} \xrightarrow{[O]} \underset{\text{Carbon monoxide}}{CO} \xrightarrow{2H_2} \underset{\text{Methanol}}{CH_3OH} \begin{array}{c} \xrightarrow[\text{catalyst}]{CO} \underset{\text{Acetic acid}}{CH_3COOH} \\ \xrightarrow[\text{oxidation}]{O_2} \underset{\text{Formaldehyde}}{CH_2O} \end{array}$$

The bulk of the ethanol produced worldwide is prepared by acid-catalyzed hydration of ethylene, itself derived from the cracking of the ethane separated from natural gas (Section 12-1). Ethanol is also produced by fermentation of the carbohydrates in plant materials, particularly corn and molasses. The majority of the fermentation-derived ethanol is used as an "oxygenate" additive to produce E85, which is a blend of up to 85% ethanol in gasoline. Combustion of E85 produces less air pollution than combustion of gasoline itself.

$$\underset{\text{Ethylene}}{CH_2=CH_2} \begin{array}{c} \xrightarrow{H_2O,\ H_2SO_4} \underset{\text{Ethanol}}{CH_3CH_2OH} \xrightarrow[180°C]{H_2SO_4} \underset{\text{Diethyl ether}}{CH_3CH_2OCH_2CH_3} + H_2O \\ \xrightarrow[\text{catalyst}]{O_2} \underset{\substack{\text{Ethylene}\\\text{oxide}}}{H_2C\overset{O}{-}CH_2} \xrightarrow{H_2O,\ H_2SO_4} \underset{\text{Ethylene glycol}}{HOCH_2CH_2OH} \end{array}$$

Acid-catalyzed dehydration of ethanol gives diethyl ether, an important laboratory and industrial solvent. Ethylene is also the starting material for the preparation of ethylene oxide. This compound itself has few direct uses. Rather, ethylene oxide's importance derives from its role as an intermediate in the production of ethylene glycol, a major component of automobile antifreeze. Ethylene glycol freezes at −12°C and boils at 199°C, which makes it ideal for this purpose. In addition, reaction of ethylene glycol with the methyl ester of terephthalic acid gives the polymer poly(ethylene terephthalate), abbreviated PET or PETE (Section 19-6B). Ethylene glycol is also used as a solvent in the paint and plastics industry and in the formulation of printer's inks, inkpads, and inks for ballpoint pens.

Isopropyl alcohol, the alcohol in rubbing alcohol, is made by acid-catalyzed hydration of propene. It is also used in hand lotions, aftershave lotions, and similar cosmetics. A multistep process also converts propene to epichlorohydrin, one of the key components in the production of epoxy glues and resins.

Glycerin, also known as glycerol, is a by-product of the manufacture of soaps by saponification of animal fats and tropical oils (Section 21-3). The bulk of the glycerin used for industrial and commercial purposes, however, is prepared from propene. Perhaps the best-known use of glycerin is for the manufacture of nitroglycerin. Glycerin is also used as an emollient in skin care products and cosmetics, in liquid soaps, and in printing inks.

$$CH_2=CHCH_3 \text{ (Propene)} \xrightarrow{\text{several steps}} ClCH_2\overset{\overset{O}{\frown}}{CH}-CH_2 \text{ (Epichlorohydrin)} \xrightarrow{\text{several steps and other reagents}} \text{Epoxy glues and resins}$$

$$CH_2=CHCH_3 \xrightarrow{H_2O, H_2SO_4} CH_3\underset{\underset{OH}{|}}{C}HCH_3 \text{ (Isopropyl alcohol)}$$

$$CH_2=CHCH_3 \xrightarrow{\text{several steps}} \underset{\underset{HO}{|}}{C}H_2\underset{\underset{HO}{|}}{C}H\underset{\underset{OH}{|}}{C}H_2 \text{ (Glycerin, Glycerol)}$$

Summary of Key Questions

Section 14-1 What Are the Structures, Names, and Physical Properties of Alcohols?

- The functional group of an **alcohol** is an **—OH (hydroxyl)** group bonded to a tetrahedral carbon atom.
- The functional group of an **ether** is an atom of oxygen bonded to two alkyl or aryl groups.
- The IUPAC name of an alcohol is derived by changing the -e of the parent alkane to -ol. The parent chain is numbered from the end that gives the carbon bearing the —OH group the lower number.
- The common name for an alcohol is derived by naming the alkyl group bonded to the —OH group and adding the word "alcohol."
- Alcohols are classified as **1°, 2°,** or **3°**, depending on the number of carbon groups bonded to the carbon bearing the —OH group.
- Compounds containing hydroxyl groups on adjacent carbons are called **glycols**.
- Alcohols are polar compounds in which oxygen bears a partial negative charge and both the carbon and hydrogen bonded to it bear partial positive charges. Alcohols associate in the liquid state by **hydrogen bonding**. As a consequence, their boiling points are higher than those of hydrocarbons of similar molecular weight.
- Because of increased London dispersion forces, the boiling points of alcohols increase with their increasing molecular weight.
- Alcohols interact with water by hydrogen bonding and are more soluble in water than are hydrocarbons of similar molecular weight.
- Alcohols have about the same pK_a values as pure water. For this reason, aqueous solutions of alcohols have the same pH as that of pure water.

Section 14-3 What Are the Structures, Names, and Physical Properties of Ethers?

- Common names for ethers are derived by naming the two groups bonded to oxygen followed by the word "ether."
- In a **cyclic ether**, oxygen is one of the atoms in a ring.
- Ethers are weakly polar compounds. Their boiling points are close to those of hydrocarbons of similar molecular weight.
- Because ethers form hydrogen bonds with water, they are more soluble in water than are hydrocarbons of similar molecular weight.

Section 14-4 What Are the Structures, Names, and Physical Properties of Thiols?

- A **thiol** contains an —SH (**sulfhydryl**) group.
- Thiols are named in the same manner as alcohols, but the suffix -e of the parent alkane is retained and -*thiol* is added.
- Common names for thiols are derived by naming the alkyl group bonded to —SH and adding the word **"mercaptan."**
- Because the S—H bond is nonpolar, the physical properties of thiols resemble those of hydrocarbons of similar molecular weight.
- Thiols are oxidized to disulfides by mild oxidizing agents.

Summary of Key Reactions

1. Acid-Catalyzed Dehydration of an Alcohol (Section 14-2B)
Where isomeric alkenes are possible, the major product is generally the more substituted alkene.

$$CH_3CH_2\underset{\underset{OH}{|}}{C}HCH_3 \xrightarrow[\text{heat}]{H_3PO_4}$$

$$CH_3CH=CHCH_3 + CH_3CH_2CH=CH_2 + H_2O$$
Major product

2. Oxidation of a Primary Alcohol (Section 14-2C)
Oxidation of a primary alcohol by potassium dichromate gives either an aldehyde or a carboxylic acid, depending on the experimental conditions.

$$CH_3(CH_2)_6CH_2OH \xrightarrow[H_2SO_4]{K_2Cr_2O_7} CH_3(CH_2)_6\overset{O}{\overset{\|}{C}}H \xrightarrow[H_2SO_4]{K_2Cr_2O_7}$$

$$CH_3(CH_2)_6\overset{O}{\overset{\|}{C}}OH$$

3. Oxidation of a Secondary Alcohol (Section 14-2C)
Oxidization of a secondary alcohol by potassium dichromate gives a ketone.

$$CH_3(CH_2)_4\underset{\underset{OH}{|}}{C}HCH_3 \xrightarrow[H_2SO_4]{K_2Cr_2O_7} CH_3(CH_2)_4\overset{O}{\overset{\|}{C}}CH_3$$

4. Acidity of Thiols (Section 14-4D) Thiols are weak acids, with pK_a values of approximately 10. They react with strong bases to form water-soluble thiolate salts.

$$CH_3CH_2SH + NaOH \xrightarrow{H_2O} CH_3CH_2S^-Na^+ + H_2O$$
Ethanethiol   Sodium
(pK_a 10)   ethanethiolate

5. Oxidation of a Thiol to a Disulfide (Section 14-4D)
Oxidation of a thiol gives a disulfide. Reduction of a disulfide gives two thiols.

$$2HOCH_2CH_2SH \underset{\text{reduction}}{\overset{\text{oxidation}}{\rightleftharpoons}} HOCH_2CH_2S-SCH_2CH_2OH$$
A thiol   A disulfide

Problems

Orange-numbered problems are applied.

Section 14-1 What Are the Structures, Names, and Physical Properties of Alcohols?

14-8 Answer true or false.
(a) The functional group of an alcohol is the —OH (hydroxyl) group.
(b) The parent name of an alcohol is the name of the longest carbon chain that contains the —OH group.
(c) A primary alcohol contains one —OH group, and a tertiary alcohol contains three —OH groups.
(d) In the IUPAC system, the presence of three —OH groups is shown by the ending –*triol*.
(e) A glycol is a compound that contains two —OH groups. The simplest glycol is ethylene glycol, $HOCH_2CH_2OH$.
(f) Because of the presence of an —OH group, all alcohols are polar compounds.
(g) The boiling points of alcohols increase with increasing molecular weight.
(h) The solubility of alcohols in water increases with increasing molecular weight.

14-9 What is the difference in structure between a primary, a secondary, and a tertiary alcohol?

14-10 Which of the following are secondary alcohols?

(a) [cyclohexane with CH₃ and OH on same carbon] (b) $(CH_3)_3COH$

(c) [CH₃CH₂CH=CHCH(OH)CH₂CH₃ type structure] (d) [cyclopentane-OH]

14-11 Which of the alcohols in Problem 14-10 are primary? Which are tertiary?

14-12 Write the IUPAC name of each compound.

(a) [CH₃CH₂CH₂CH₂CH₂OH] (b) HO–CH₂CH₂CH₂–OH

(c) [HOCH₂CH(OH)CH₂...] (d) HO–CH₂CH(CH₃)CH₃ type

(e) [trans-cyclohexane-1,2-diol] (f) [2,6-dimethylcyclohexanol]

14-13 Draw a structural formula for each alcohol.
(a) Isopropyl alcohol
(b) Propylene glycol
(c) 5-Methyl-2-hexanol
(d) 2-Methyl-2-propyl-1,3-propanediol
(e) 1-Octanol
(f) 3,3-Dimethylcyclohexanol

14-14 Draw a structural formula for each of the following alcohols.
(a) Isobutyl alcohol
(b) 1,4-Butanediol
(c) 5-Methyl-1-hexanol
(d) 1,3-Pentanediol
(e) *trans*-1,4-Cyclohexanediol
(f) 1-Chloro-2-propanol

14-15 Both alcohols and phenols contain an —OH group. What structural feature distinguishes these two classes of compounds? Illustrate your answer by drawing the structural formulas of a phenol with six carbon atoms and an alcohol with six carbon atoms.

14-16 Name the functional groups in each compound.

(a) Prednisone (a synthetic anti-inflammatory steroid)

(b) Estradiol (a female sex hormone; Section 21-10)

14-17 Explain in terms of noncovalent interactions why the low-molecular-weight alcohols are soluble in water but the low-molecular-weight alkanes and alkynes are not.

14-18 Explain in terms of noncovalent interactions why the low-molecular-weight alcohols are more soluble in water than the low-molecular-weight ethers.

14-19 Why does the water solubility of low-molecular-weight alcohols decrease as molecular weight increases?

14-20 Show hydrogen bonding between methanol and water in the following ways.
(a) Between the oxygen of methanol and a hydrogen of water
(b) Between the hydrogen of methanol's OH group and the oxygen of water

14-21 Show hydrogen bonding between the oxygen of diethyl ether and a hydrogen of water.

14-22 Arrange these compounds in order of increasing boiling point. Values in °C are −42, 78, 117, and 198.
(a) $CH_3CH_2CH_2CH_2OH$
(b) CH_3CH_2OH
(c) $HOCH_2CH_2OH$
(d) $CH_3CH_2CH_3$

14-23 Arrange these compounds in order of increasing boiling point. Values in °C are 0, 35, and 97.
(a) $CH_3CH_2CH_2OH$
(b) $CH_3CH_2OCH_2CH_3$
(c) $CH_3CH_2CH_2CH_3$

14-24 2-Propanol (isopropyl alcohol) is commonly used as rubbing alcohol to cool the skin. 2-Hexanol, also a liquid, is not suitable for this purpose. Why?

14-25 Explain why glycerol is much thicker (more viscous) than ethylene glycol, which in turn is much more viscous than ethanol.

14-26 From each pair, select the compound that is more soluble in water.
(a) CH_3OH or CH_3OCH_3
(b) CH_3CHCH_3 with OH or CH_3CCH_3 with =CH₂
(c) $CH_3CH_2CH_2SH$ or $CH_3CH_2CH_2OH$
(d) CH_3CH_2Cl or NaCl
(e) [CH₃CH(OH)CH₂CH₃] or [CH₃CH₂C(=O)CH₃]

14-27 Arrange the compounds in each set in order of decreasing solubility in water.
(a) Ethanol, butane, and diethyl ether
(b) 1-Hexanol, 1,2-hexanediol, and hexane

Synthesis of Alcohols (Review Chapter 12)

14-28 Give the structural formula of an alkene or alkenes from which each alcohol can be prepared.
(a) 2-Butanol
(b) 1-Methylcyclohexanol
(c) 3-Hexanol
(d) 2-Methyl-2-pentanol
(e) Cyclopentanol

Section 14-2 What Are the Characteristic Reactions of Alcohols?

14-29 Answer true or false.
(a) The two most important reactions of alcohols are their acid-catalyzed dehydration to give alkenes and their oxidation to aldehydes, ketones, and carboxylic acids.
(b) The acidity of alcohols is comparable to that of water.
(c) Water-insoluble alcohols and water-insoluble phenols react with strong bases to give water-soluble salts.
(d) Acid-catalyzed dehydration of cyclohexanol gives cyclohexane.
(e) When the acid-catalyzed dehydration of an alkene can yield isomeric alkenes, the alkene with the greater number of hydrogens on the carbons of the double bond generally predominates.
(f) The acid-catalyzed dehydration of 2-butanol gives predominantly 1-butene.
(g) The oxidation of a primary alcohol gives either an aldehyde or a carboxylic acid depending on experimental conditions.
(h) The oxidation of a secondary alcohol gives a carboxylic acid.
(i) Acetic acid, CH_3COOH, can be prepared from ethylene, $CH_2=CH_2$, by treatment of ethylene with H_2O/H_2SO_4, followed by treatment with $K_2Cr_2O_7/H_2SO_4$.
(j) Treatment of propene, $CH_3CH=CH_2$, with H_2O/H_2SO_4 followed by treatment with $K_2Cr_2O_7/H_2SO_4$ gives propanoic acid, CH_3CH_2COOH.

14-30 Show how to distinguish between cyclohexanol and cyclohexene by a simple chemical test. Tell what you would do, what you would expect to see, and how you would interpret your observation.

14-31 Compare the acidity of alcohols and phenols, which are both classes of organic compounds that contain an —OH group.

14-32 Both 2,6-diisopropylcyclohexanol and the intravenous anesthetic Propofol are insoluble in water. Show how these two compounds can be distinguished by their reaction with aqueous sodium hydroxide.

2,6-Diisopropylcyclohexanol   2,6-Diisopropylphenol (Propofol)

14-33 Write equations for the reaction of 1-butanol, a primary alcohol, with these reagents.
(a) H_2SO_4, heat
(b) $K_2Cr_2O_7$, H_2SO_4

14-34 Write equations for the reaction of 2-butanol with these reagents.
(a) H_2SO_4, heat
(b) $K_2Cr_2O_7$, H_2SO_4

14-35 Write equations for the reaction of each of the following compounds with $K_2Cr_2O_7/H_2SO_4$.
(a) 1-Octanol
(b) 1,4-Butanediol

14-36 Show how to convert cyclohexanol to these compounds.
(a) Cyclohexene
(b) Cyclohexane
(c) Cyclohexanone
(d) Bromocyclohexane

14-37 Show reagents and experimental conditions to synthesize each compound from 1-propanol.

14-38 Show how to convert this alcohol to compounds (a) and (b).

cyclopentyl—CH₂OH

(a) cyclopentyl—CH(=O)   (b) cyclopentyl—COOH

14-39 Name two important alcohols derived from ethylene and give two important uses of each.

14-40 Name two important alcohols derived from propene and give two important uses of each.

Section 14-3 What Are the Structures, Names, and Physical Properties of Ethers?

14-41 Answer true or false.
  (a) Ethanol and dimethyl ether are constitutional isomers.
  (b) The solubility of low-molecular-weight ethers in water is comparable to that of low-molecular-weight alcohols in water.
  (c) Ethers undergo many of the same reactions that alcohols do.

14-42 Write the common name for each ether.
  (a) cyclopentyl—O—cyclopentyl   (b) [CH₃(CH₂)₄]₂O

  (c) CH₃CH(CH₃)OCH(CH₃)CH₃

14-43 Write the common name for each of the following ethers.
  (a) ethyl isopropyl ether (structure)
  (b) dibutyl ether (structure)
  (c) diphenyl ether (structure)

Section 14-4 What Are the Structures, Names, and Physical Properties of Thiols?

14-44 Answer true or false.
  (a) The functional group of a thiol is the —SH (sulfhydryl) group.
  (b) The parent name of a thiol is the name of the longest carbon chain that contains the —SH group.
  (c) The S—H bond is nonpolar covalent.
  (d) The acidity of ethanethiol is comparable to that of phenol.
  (e) Both phenols and thiols are classified as weak acids.
  (f) The most common biological reaction of thiols is their oxidation to disulfides.
  (g) The functional group of a disulfide is the —S—S— group.
  (h) Conversion of a thiol to a disulfide is a reduction reaction.

14-45 Write the IUPAC name of each thiol.
  (a) $CH_3CH_2CH(SH)CH_3$ (b) $CH_3CH_2CH_2CH_2SH$
  (c) cyclohexyl—SH

14-46 Write the common name for each thiol in Problem 14-45.

14-47 Following are structural formulas for 1-butanol and 1-butanethiol. One of these compounds has a boiling point of 98°C and the other has a boiling point of 117°C. Which compound has which boiling point?

$CH_3CH_2CH_2CH_2OH$ $CH_3CH_2CH_2CH_2SH$
1-Butanol   1-Butanethiol

14-48 Explain why methanethiol, CH_3SH, has a lower boiling point (6°C) than methanol, CH_3OH (65°C), even though methanethiol has a higher molecular weight.

Section 14-5 What Are the Most Commercially Important Alcohols?

14-49 Answer true or false.
  (a) Today, the major carbon sources for the synthesis of methanol are coal and methane (natural gas), both nonrenewable resources.
  (b) Today the major carbon sources for the synthesis of ethanol are petroleum and natural gas, both nonrenewable resources.
  (c) Intermolecular acid-catalyzed dehydration of ethanol gives diethyl ether.
  (d) Conversion of ethylene to ethylene glycol involves oxidation to ethylene oxide, followed by acid-catalyzed hydration (addition of water) to ethylene oxide.
  (e) Ethylene glycol is soluble in water in all proportions.
  (f) A major use of ethylene glycol is as automobile antifreeze.

Chemical Connections

14-50 (Chemical Connections 14A) As stated in the CC, "Hydrogen bonds have directionality in that the donor and acceptor groups must be oriented appropriately with respect to each other for hydrogen bonding to occur." Describe the directionality of hydrogen bonding observed in figure (b) of the CC.

14-51 (Chemical Connections 14B) When was nitroglycerin discovered? Is this substance a solid, a liquid, or a gas?

14-52 (Chemical Connections 14B) What was Alfred Nobel's discovery that made nitroglycerin safer to handle?

14-53 (Chemical Connections 14B) What is the relationship between the medical use of nitroglycerin to relieve the sharp chest pains (angina) associated with heart disease and the gas nitric oxide, NO?

14-54 (Chemical Connections 14C) What is the color of dichromate ion, $Cr_2O_7^{2-}$? What is the color of chromium(III) ion, Cr^{3+}? Explain how the conversion of one to the other is used in breath-alcohol screening.

14-55 (Chemical Connections 14C) The legal definition of being under the influence of alcohol is based on blood alcohol content. What is the relationship between breath alcohol content and blood alcohol content?

14-56 (Chemical Connections 14D) What does it mean to say that ethylene oxide is a highly strained molecule?

14-57 (Chemical Connections 14E) What are the advantages and disadvantages of using diethyl ether as an anesthetic?

14-58 (Chemical Connections 14E) Show that enflurane and isoflurane are constitutional isomers.

14-59 (Chemical Connections 14E) Would you expect enflurane and isoflurane to be soluble in water? Would you expect them to be soluble in organic solvents such as hexane?

Additional Problems

14-60 Write a balanced equation for the complete combustion of ethanol, the alcohol blended with gasoline to produce E85.

14-61 Knowing what you do about electronegativity, the polarity of covalent bonds, and hydrogen bonding, would you expect an N—H—N hydrogen bond to be stronger than, the same strength as, or weaker than an O—H—O hydrogen bond?

14-62 Draw structural formulas and write IUPAC names for the eight isomeric alcohols with the molecular formula $C_5H_{12}O$.

14-63 Draw structural formulas and write common names for the six isomeric ethers with the molecular formula $C_5H_{12}O$.

14-64 Explain why the boiling point of ethylene glycol (198°C) is so much higher than that of 1-propanol (97°C), even though their molecular weights are about the same.

14-65 Following is a structural formula for Erythromycin A, a macrolactone (macrolide) antibiotic.

Erythromycin A
$C_{31}H_{57}NO_9$

(a) How many hydroxyl groups are present? Classify each as primary (1°), secondary (2°), or tertiary (3°).

(b) How many amine groups are present? Classify each as primary (1°), secondary (2°), or tertiary (3°).

(c) Locate the ester group within the large 15-member ring.

(d) Four of the hydroxyl groups within Erythromycin A are involved in intramolecular (internal) hydrogen bonds. One of these is pointed out on the structural formula. Note that this hydrogen bond creates a five-membered ring. Locate the other three intramolecular hydrogen bonds and specify the size of the ring created by each.

14-66 1,4-Butanediol, hexane, and 1-pentanol have similar molecular weights. Their boiling points, arranged from lowest to highest, are 69°C, 138°C, and 230°C. Which compound has which boiling point?

14-67 Of the three compounds given in Problem 14-66, one is insoluble in water, another has a solubility of 2.3 g/100 g water, and one is infinitely soluble in water. Which compound has which solubility?

14-68 Each of the following compounds is a common organic solvent. From each pair of compounds, select the solvent with the greater solubility in water.

(a) CH_2Cl_2 or CH_3CH_2OH

(b) $CH_3CH_2OCH_2CH_3$ or CH_3CH_2OH

14-69 Show how to prepare each compound from 2-methyl-1-propanol.
 (a) 2-Methylpropene
 (b) 2-Methyl-2-propanol
 (c) 2-Methylpropanoic acid, $(CH_3)_2CHCOOH$

14-70 Show how to prepare each compound from 2-methylcyclohexanol.

(a) 2-methylcyclohexanone
(b) 1-methylcyclohexene (methyl on double bond carbon)
(c) 1-methylcyclohexanol
(d) methylcyclohexane

14-71 The mechanism of the acid-catalyzed dehydration of an alcohol to an alkene is the reverse of the acid-catalyzed hydration of an alkene. The dehydration mechanism occurs by the following three steps.
 Step 1: Add a proton.
 Step 2: Break a bond to form stable molecules or ions.
 Step 3: Take away a proton.

These three steps are illustrated here by the dehydration of 2-butanol to give 2-butene. Use curved arrows to show the flow of electrons in each step; that is, show how each bond-making or bond-breaking step occurs.

Step 1: $CH_3-CH_2-\underset{\underset{\text{2-butanol}}{}}{\overset{\overset{\ddot{\text{:}\ddot{O}H}}{|}}{CH}}-CH_3 + H^+ \rightleftharpoons$

$CH_3-CH_2-\underset{}{\overset{\overset{H\diagdown \overset{..}{\overset{+}{O}}\diagup H}{|}}{CH}}-CH_3$

Step 2: $CH_3-CH_2-\underset{}{\overset{\overset{H\diagdown \overset{..}{\overset{+}{O}}\diagup H}{|}}{CH}}-CH_3 \rightleftharpoons$

$\underset{\text{A carbocation intermediate}}{CH_3-CH_2-\overset{+}{CH}-CH_3 + H_2O}$

Step 3: $CH_3-\overset{\overset{H}{|}}{CH}-\overset{+}{CH}-CH_3 \rightleftharpoons$

$\underset{\text{2-butene}}{CH_3-CH=CH-CH_3 + H^+}$

14-72 Use the reactions we learned in Chapter 12 and this chapter to show how to bring about the following chemical transformations. Some transformations will require only one step. Others will require two or more steps.

(a) 2-methyl-1-propanol → 2-methyl-2-propanol
(b) 1-pentene → 2-pentene
(c) 1-pentene → 2-pentanone
(d) propene → acetone (propan-2-one)
(e) cyclopentylmethanol → cyclopentanecarboxylic acid
(f) 1-methylcyclopentanol → methylcyclopentane

Looking Ahead

14-73 Lipoic acid is a growth factor for many bacteria and protozoa and an essential component of several enzymes involved in human metabolism.

Lipoic acid

(a) Name the two functional groups in lipoic acid.
(b) At one stage in its function in human metabolism, the disulfide bond of lipoic acid is reduced to two thiol groups. Draw a structural formula for this reduced form of lipoic acid.

14-74 Following is a structural formula for the amino acid cysteine:

$$HS-CH_2-\underset{\underset{NH_2}{|}}{CH}-\overset{\overset{O}{||}}{C}-OH$$

(a) Name the three functional groups in cysteine.
(b) In the human body, cysteine is oxidized to cystine, a disulfide. Draw a structural formula for cystine.

14-75 As we will see in Chapter 16, the carbonyl group of aldehydes and ketones reacts with water to form a compound called a hydrate. The hydration of a carbonyl group is catalyzed by acid. Formaldehyde, for example, reacts as shown below. When dissolved in water, formaldehyde exists almost entirely as its hydrate. This aqueous solution is given the name formalin.

$$\underset{\text{Formaldehyde}}{\overset{H}{\underset{H}{\diagdown}}C=O + H_2O} \quad \underset{\rightleftharpoons}{\overset{H_3O^+}{\longrightarrow}} \quad \underset{\text{Formaldehyde hydrate (Formalin)}}{\overset{H}{\underset{H}{\diagdown}}\overset{OH}{\underset{OH}{\diagup}}C}$$

The mechanism for this acid-catalyzed hydration is similar to the mechanism for the acid-catalyzed hydration of an alkene to give an alcohol and occurs by the following three steps.

Step 1: Add a proton.

Step 2: Make a new covalent bond between an electrophile (an electron-seeking species) and a nucleophile (an electron-donating species).

Step 3: Take a proton away.

Write an equation for each step and show by the use of curved arrows how each of these steps occurs. Show all bond-forming steps and bond-breaking steps.

14-76 Draw the structural formula for the hydrate formed by the reaction of acetone with water.

$$CH_3-\overset{\overset{O}{\|}}{C}-CH_3 + H_2O \underset{}{\overset{H_3O^+}{\rightleftharpoons}}$$

Acetone

Challenge problems

14-77 With the exception of the acid-catalyzed conversion of ethylene to ethanol, primary alcohols cannot be prepared by the acid-catalyzed hydration of alkenes. The acid-catalyzed conversion of almost all other alkenes gives either secondary or tertiary alcohols. Explain why these two generalizations are valid.

14-78 Consider alkenes A, B, and C. each of which has the same molecular formula, C_6H_{12}. Alkenes B and C can each be separated into *cis* and *trans* isomers. Upon catalytic reduction using H_2 in the presence of a transition metal catalyst (Ni, Pd, or Pt), alkenes A, B, and C all give hexane as the only product. Acid-catalyzed hydration of alkene C gives one alcohol with the molecular formula $C_6H_{14}O$. Acid catalyzed-hydration of alkene B gives an equal mixture of two alcohols, each with the molecular formula $C_6H_{14}O$. Acid-catalyzed hydration of alkene C gives only a single alcohol with the molecular formula $C_6H_{14}O$. Propose structural formulas for alkenes A, B, and C and the alcohols formed by acid-catalyzed hydration of each, consistent with these experimental results.

14-79 On acid-catalyzed hydration of alkenes A and C in the previous problem, each gives only one alcohol with the molecular formula $C_6H_{14}O$. Draw the structural formulas for five other alkenes with the molecular formula C_6H_{12}, each of which will also give only one alcohol upon acid-catalyzed hydration.

15 Chirality: The Handedness of Molecules

KEY QUESTIONS

- 15-1 What Is Enantiomerism?
  - *How To . . .* Draw Enantiomers
- 15-2 How Do We Specify the Configuration of a Stereocenter?
- 15-3 How Many Stereoisomers Are Possible for Molecules with Two or More Stereocenters?
- 15-4 What Is Optical Activity, and How Is Chirality Detected in the Laboratory?
- 15-5 What Is the Significance of Chirality in the Biological World?

Median cross section through the shell of a chambered nautilus found in the deep waters of the Pacific Ocean. The shell shows handedness; this cross section is a left-handed spiral.

15-1 What Is Enantiomerism?

In Chapters 11 through 14, we studied two types of stereoisomers, namely the *cis-trans* isomers of certain disubstituted cycloalkanes and appropriately substituted alkenes. Recall that stereoisomers have the same connectivity of their atoms but a different orientation of their atoms in space.

cis-1,2-Dimethyl-cyclohexane and *trans*-1,2-Dimethyl-cyclohexane   *cis*-2-Butene and *trans*-2-Butene

In this chapter, we study the relationship between objects and their **mirror images**; that is, we study stereoisomers called enantiomers and diastereomers. Figure 15-1 summarizes the relationship among these isomers and those we studied in Chapters 11 through 14.

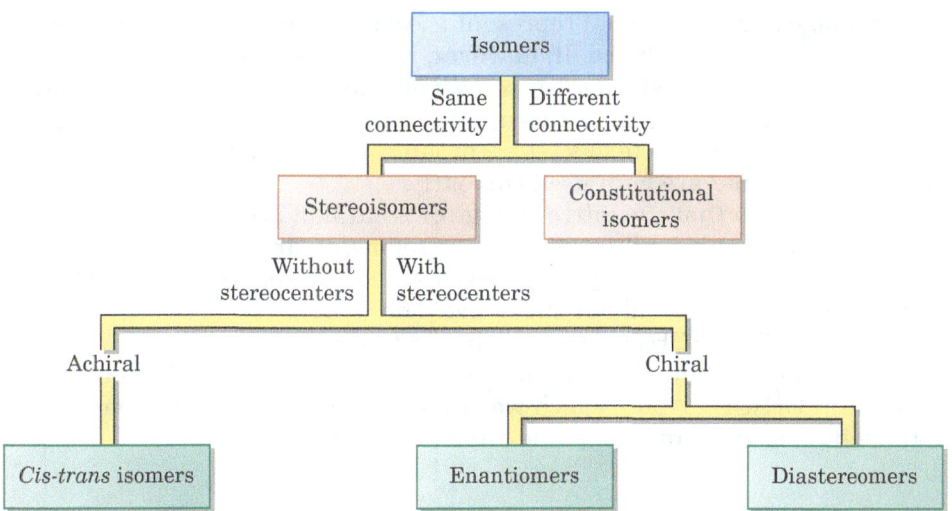

FIGURE 15-1 Relationships among isomers. In this chapter, we study enantiomers and diastereomers.

The significance of enantiomers is that, except for inorganic compounds and a few simple organic compounds, the vast majority of molecules in the biological world show this type of isomerism. Furthermore, approximately one half of all medications used to treat humans display this property.

As an example of enantiomerism, let us consider 2-butanol. As we discuss this molecule, we will focus on carbon-2, the carbon bearing the —OH group. What makes this carbon interesting is the fact that it has four different groups bonded to it: CH_3, H, OH, and CH_2CH_3.

$$\underset{\text{2-Butanol}}{CH_3\overset{\overset{\displaystyle OH}{|}}{C}HCH_2CH_3}$$

This structural formula does not show the three-dimensional shape of 2-butanol or the orientation of its atoms in space. To do so, we must consider the molecule as a three-dimensional object. On the left is what we will call the "original molecule" and a ball-and-stick model of it. In this drawing, the —OH and —CH_3 groups are in the plane of the paper, —H is behind the plane (shown as a broken wedge), and —CH_2CH_3 is in front of it (shown as a solid wedge). In the middle is a mirror. On the right is a **mirror image** of the original molecule along with a ball-and-stick model of the mirror image. Every molecule—and, in fact, every object in the world around us—has a mirror image.

The question we now need to ask is "What is the relationship between the original molecule of 2-butanol and its mirror image?" To answer this question, we need to imagine that we can pick up the mirror image and move it in 3-D space in any way we wish. If we can move the mirror image in space and find that it fits over the original molecule so that every bond, atom, and detail of the mirror image matches exactly the bonds, atoms, and details of the original, then the two are **superposable**. In other words, the mirror image and the original represent the same molecule; they are merely oriented differently in space. If, however, no matter how we turn the mirror image in space, it will not fit exactly on the original with every detail matching, then the two are **nonsuperposable**; they are different molecules (Figure 15-2). The terms "superposable" and "superimposable" mean the same thing and are both used currently.

One way to see that the mirror image of 2-butanol is not superposable on the original molecule is illustrated in the following drawings. Imagine that we hold the mirror image by the C—OH bond and rotate the bottom part of the molecule by 180° about this bond. The —OH group retains its position in space. The —CH₃ group, which was to the right and in the plane of the paper, remains in the plane of the paper but is now to the left. Similarly, the —CH₂CH₃ group, which was in front of the plane of the paper and to the left, is now behind the plane and to the right.

Now move the mirror image in space and try to fit it on the original molecule so that all bonds and atoms match.

By turning the mirror image as we did, its —OH and —CH₃ groups now fit exactly on top of the —OH and —CH₃ groups of the original molecule. However, the —H and the —CH₂CH₃ groups of the two do not match. The —H is away from us in the original but toward us in the mirror image; the —CH₂CH₃ group is toward us in the original but away from us in the mirror image. We conclude that the original molecule of 2-butanol and its mirror image are nonsuperposable and therefore are different compounds.

To summarize, we can turn and rotate the mirror image of 2-butanol in any direction in space, but as long as no bonds are broken and rearranged, we can make only two of the four groups bonded to carbon-2 of the mirror image coincide with those on its original molecule. Because 2-butanol and its mirror image are not superposable, they are isomers. Isomers such as these are called **enantiomers**. Enantiomers, like gloves, always occur in pairs.

Enantiomers Stereoisomers that are nonsuperposable mirror images; refers to a relationship between pairs of objects

FIGURE 15-2 Mirror images. (*a*) Two woodcarvings. The mirror images cannot be superposed on the actual model. The man's right arm rests on the camera in the mirror image, but in the actual statue, the man's left arm rests on the camera. (*b*) Left- and right-handed sea shells. If you cup a right-handed shell in your right hand with your thumb pointing from the narrow end to the wide end, the opening will be on your right.

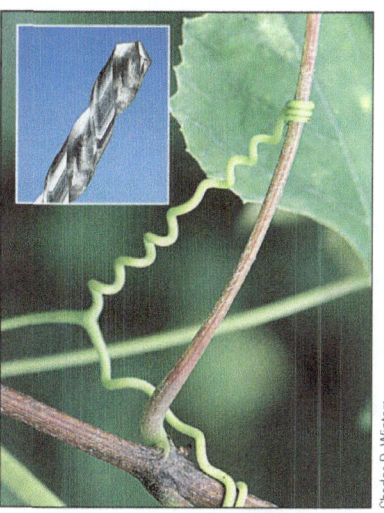

▶ The threads of a drill or screw twist along the axis of the helix, and some plants climb by sending out tendrils that twist helically. The drill bit shown here has a left-handed twist, and the tendril has a right-handed twist.

Objects that are not superposable on their mirror images are said to be **chiral** (pronounced "ki-ral," rhymes with spiral; from the Greek: *cheir*, "hand"); that is, they show handedness. We encounter chirality in three-dimensional objects of all sorts. Our left hand is chiral, and so is our right hand. Thus, our hands have an enantiomeric relationship. A spiral binding on a notebook is chiral. A machine screw with a right-handed twist is chiral. A ship's propeller is chiral. As you examine the objects in the world around you, you will undoubtedly conclude that the vast majority of them are chiral. ◀

Chiral From the Greek *cheir*, "hand"; an object that is not superposable on its mirror image

The most common cause of enantiomerism in organic molecules is the presence of a carbon with four different groups bonded to it. Let us examine this statement further by considering 2-propanol, which has no such carbon atom. In this molecule, carbon-2 is bonded to three different groups, but no carbon is bonded to four different groups.

On the left is a three-dimensional representation of 2-propanol; on the right is its mirror image. Also shown are ball-and-stick models of each molecule.

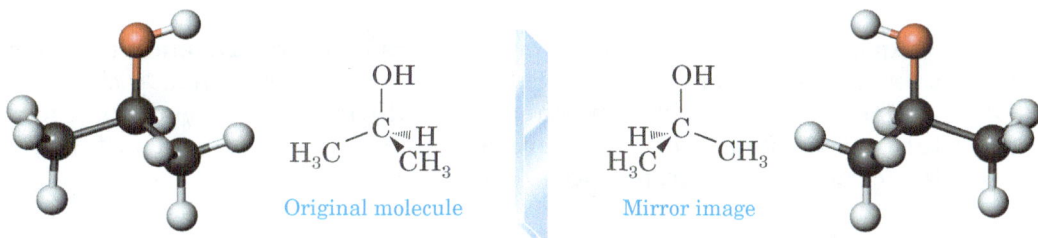

Original molecule ⎪ Mirror image

The question we now ask is "What is the relationship of the mirror image to the original?" This time, let us rotate the mirror image by 120° about the C—OH bond and then compare it to the original. After performing this rotation, we see that all atoms and bonds of the mirror image fit exactly on the original. Thus, the structures we first drew for the original molecule and its mirror image are, in fact, the same molecule—just viewed from different perspectives (Figure 15-3).

If an object and its mirror image are superposable, then they are identical and enantiomerism is not possible. We say that such an object is **achiral**

Achiral An object that lacks chirality; an object that is superposable on its mirror image

FIGURE 15-3 Rotation of the mirror image about the C—OH bond by 120° does not change the configuration of the stereocenter, but makes it easier to see that the mirror image is superposable on the original molecule.

Stereocenter A tetrahedral carbon atom that has four different groups bonded to it

(without chirality); that is, it has no handedness. Examples of achiral objects include an undecorated cup, an unmarked baseball bat, a regular tetrahedron, a cube, and a sphere.

To repeat, the most common cause of chirality in organic molecules is a tetrahedral carbon atom with four different groups bonded to it. We call such a chiral carbon atom a **stereocenter**. 2-Butanol has one stereocenter; 2-propanol has none. As another example of a molecule with a stereocenter, let us consider 2-hydroxypropanoic acid, more commonly named lactic acid. Lactic acid is a product of anaerobic glycolysis. (See Section 28-2 and Chemical Connections 28A.) It is also what gives sour cream its sour taste.

Figure 15-4 shows three-dimensional representations of lactic acid and its mirror image. In these representations, all bond angles about the central carbon atom are approximately 109.5° and the four bonds from it are directed toward the corners of a regular tetrahedron. Lactic acid displays enantiomerism or chirality; that is, the original molecule and its mirror image are not superposable but rather are different compounds.

FIGURE 15-4 Three-dimensional representations of lactic acid and its mirror image.

Racemic mixture A mixture of equal amounts of two enantiomers

An equimolar mixture of two enantiomers is called a **racemic mixture**, a term derived from the name racemic acid (Latin: *racemus*, "a cluster of grapes"). Racemic acid is the name originally given to the equimolar mixture of the enantiomers of tartaric acid that forms as a by-product during the fermentation of grape juice to produce wine.

HOW TO...

Draw Enantiomers

Now that we know what enantiomers are, we can think about how to represent their three-dimensional structures on a two-dimensional surface. Let us take one of the enantiomers of 2-butanol as an example.

Following are a molecular model of one enantiomer and four different three-dimensional representations of it.

$$
\begin{array}{cccc}
 & & & \\
\text{OH} & \text{H}\quad\text{OH} & \text{H}\quad\text{OH} & \text{OH} \\
| & \backslash\;/ & & \\
\text{H}_3\text{C}-\overset{\displaystyle|}{\text{C}}\cdots\text{H} & \text{H}_3\text{C}\;\text{C}\;\text{CH}_2\text{CH}_3 & & \\
\quad\text{CH}_2\text{CH}_3 & & & \\
(1) & (2) & (3) & (4)
\end{array}
$$

In our initial discussions of 2-butanol, we used representation (1) to show the tetrahedral geometry of the stereocenter. In this representation, two groups (OH and CH$_3$) are in the plane of the paper, one (CH$_2$CH$_3$) is coming out of the plane toward us, and one (H) is behind the plane and going away from us. We can turn representation (1) slightly in space and tip it a bit to place the carbon framework in the plane of the paper. Doing so gives us representation (2), in which there are still two groups in the plane of the paper, plus one coming toward us and one going away from us.

For a more abbreviated representation of this enantiomer of 2-butanol, we can change representation (2) into the line-angle formula (3). Although we do not normally show hydrogens in a line-angle formula, we do so here just to remind ourselves that the fourth group on the stereocenter is really there and that it is H. Finally, we can carry the abbreviation a step further and write 2-butanol as the line-angle formula (4). Here, we omit the H on the stereocenter, but we know that it must be there (carbon needs four bonds), and we know that it must be behind the plane of the paper. Clearly, the abbreviated formulas (3) and (4) are the easiest to write, and we will rely on this type of representation throughout the remainder of the text.

When you have to write three-dimensional representations of stereocenters, try to keep the carbon framework in the plane of the paper and the other two atoms or groups of atoms on the stereocenter toward and away from you, respectively. Using representation (4) as a model, we get the following alternative representations of its mirror image:

One enantiomer of 2-butanol

Alternative representations of its mirror image

EXAMPLE 15-1 Drawing Mirror Images

Each of the following molecules has one stereocenter marked by an asterisk. Draw three-dimensional representations for the enantiomers of each molecule.

(a) $CH_3\overset{*}{C}HCH_2CH_3$ with Cl on the stereocenter

(b) phenyl–$\overset{*}{C}HCH_3$ with NH_2 on the stereocenter

Strategy

First, draw the carbon stereocenter showing the tetrahedral orientation of its four bonds. One way to do this is to draw two bonds in the plane of the paper, a third bond toward you in front of the plane, and the fourth bond away from you behind the plane. Next, place the four groups bonded to the stereocenter on these positions. This completes the stereodrawing of one enantiomer. To draw the other enantiomer, interchange any two of the groups on the original stereodrawing.

Solution

To draw an original of (a), for example, place the CH_3 and CH_2CH_3 groups in the plane of the paper. Place H away from you and the Cl toward you; this orientation gives the enantiomer of (a) on the left. Its mirror image is on the right.

(a) [3D structures of the two enantiomers of 2-chlorobutane, shown as mirror images]

(b) [3D structures of the two enantiomers of 1-phenylethylamine, shown as mirror images]

Problem 15-1

Each of the following molecules has one stereocenter marked by an asterisk. Draw three-dimensional representations for the enantiomers of each molecule.

(a) cyclopentyl–$\overset{*}{C}HCH_3$ with COOH on the stereocenter

(b) $CH_3\overset{*}{C}HCHCH_3$ with OH on the starred carbon and CH_3 on the other

15-2 How Do We Specify the Configuration of a Stereocenter?

Because enantiomers are different compounds, each must have a different name. The over-the-counter drug Ibuprofen, for example, displays enantiomerism and can exist as the pair of enantiomers shown here:

The inactive enantiomer of Ibuprofen

The active enantiomer of Ibuprofen

Only one enantiomer of Ibuprofen is biologically active. It reaches therapeutic concentrations in the human body in approximately 12 minutes, whereas the racemic mixture takes approximately 30 minutes to achieve this feat. However, in this case, the inactive enantiomer is not wasted. The body converts it to the active enantiomer, but this process takes time.

What we need is a system to differentiate the enantiomers of Ibuprofen (or one of any other pair of enantiomers, for that matter) without having to draw and point to one or the other of the enantiomers. To do so, chemists have developed the **R,S system**. The first step in assigning an R or S configuration to a stereocenter is to arrange the groups bonded to it in order of priority. Priority is based on atomic number: the higher the atomic number, the higher the priority. If a priority cannot be assigned on the basis of the atoms bonded directly to the stereocenter, look at the next atom or set of atoms and continue to the first point of difference; that is, continue until you can assign a priority.

Table 15-1 shows the priorities of the most common groups we encounter in organic and biochemistry. In the R,S system, a C=O is treated as if carbon were bonded to two oxygens by single bonds; thus, CH=O has a higher priority than —CH$_2$OH, in which carbon is bonded to only one oxygen.

R,S system A set of rules for specifying the configuration about a stereocenter

Table 15-1 R,S Priorities of Some Common Groups

Atom or Group	Reason for Priority: First Point of Difference (Atomic Number)
—I	iodine (53)
—Br	bromine (35)
—Cl	chlorine (17)
—SH	sulfur (16)
—OH	oxygen (8)
—NH$_2$	nitrogen (7)
—COH (C=O)	carbon to oxygen, oxygen, then oxygen (6 ⟶ 8, 8, 8)
—CNH$_2$ (C=O)	carbon to oxygen, oxygen, then nitrogen (6 ⟶ 8, 8, 7)
—CH (C=O)	carbon to oxygen, oxygen, then hydrogen (6 ⟶ 8, 8, 1)
—CH$_2$OH	carbon to oxygen (6 ⟶ 8)
—CH$_2$NH$_2$	carbon to nitrogen (6 ⟶ 7)
—CH$_2$CH$_3$	carbon to carbon (6 ⟶ 6)
—CH$_2$H	carbon to hydrogen (6 ⟶ 1)
—H	hydrogen (1)

Increasing priority ↑

EXAMPLE 15-2 Using the R,S System

Assign priorities to the groups in each set.

(a) —CH$_2$OH and —CH$_2$CH$_2$OH
(b) —CH$_2$CH$_2$OH and —CH$_2$NH$_2$

Strategy and Solution

(a) The first point of difference is O of the —OH group compared to C of the —CH$_2$OH group.

—CH$_2$OH —CH$_2$CH$_2$OH
Higher priority Lower priority

(b) The first point of difference is C of the CH$_2$OH group compared to N of the NH$_2$ group.

—CH$_2$CH$_2$OH —CH$_2$NH$_2$
Lower priority **Higher priority**

Problem 15-2

Assign priorities to the groups in each set.

(a) —CH$_2$OH and —CH$_2$CH$_2$$\overset{\overset{\text{O}}{\|}}{\text{C}}$OH

(b) —CH$_2$NH$_2$ and —CH$_2$$\overset{\overset{\text{O}}{\|}}{\text{C}}$OH

Group of lowest priority points away from you

R Used in the R,S system to show that when the lowest-priority group is away from you, the order of priority of groups on a stereocenter is clockwise

S Used in the R,S system to show that when the lowest-priority group is away from you, the order of priority of groups on a stereocenter is counterclockwise

To assign an R or S configuration to a stereocenter:

1. Assign a priority from 1 (highest) to 4 (lowest) to each group bonded to the stereocenter.
2. Orient the molecule in space so that the lowest-priority group (4) is directed away from you, as would be, for instance, the steering column of a car. The three higher-priority groups (1–3) then project toward you, as would the spokes of a steering wheel.
3. Read the three groups projecting toward you in order from highest (1) to lowest (3) priority.
4. If reading the groups 1-2-3 proceeds in a clockwise direction (to the right), the configuration is designated as **R** (Latin: *rectus*, "straight"); if reading the groups 1-2-3 proceeds in a counterclockwise direction (to the left), the configuration is **S** (Latin: *sinister*, "left"). You can also visualize this system as follows: Turning the steering wheel to the right equals R, and turning it to the left equals S.

EXAMPLE 15-3 Assigning an *R* or *S* Configuration

Assign an *R* or *S* configuration to each stereocenter.

(a)
```
      OH
      |
      C┄┄H
     / \
   H₃C  CH₂CH₃
```
2-Butanol

(b)
```
    H₂N  H
      \ /
       C
      / \
    H₃C  COOH
```
Alanine

Strategy and Solution

View each molecule through the stereocenter and along the bond from the stereocenter to the group of lowest priority.

(a) The order of decreasing priority about the stereocenter in this enantiomer of 2-butanol is —OH > —CH₂CH₃ > —CH₃ > —H. Therefore, view the molecule along the C—H bond with the H pointing away from you. Reading the other three groups in the order 1-2-3 follows in the clockwise direction. Therefore, the configuration is *R* and this enantiomer is (*R*)-2-butanol.

(*R*)-2-Butanol

With —H, the lowest-priority group, pointing away from you, this is what you see

(b) The order of decreasing priority in this enantiomer of alanine is —NH₂ > —COOH > —CH₃ > —H. View the molecule along the C—H bond with H pointing away from you. Reading the groups in the order 1-2-3 follows in a clockwise direction; therefore, the configuration is *R* and this enantiomer is (*R*)-alanine.

(*R*)-Alanine

With —H, the lowest-priority group, pointing away from you, this is what you see

Problem 15-3

Assign an *R* or *S* configuration to the single stereocenter in glyceraldehyde, the simplest carbohydrate (Chapter 20).

```
            O
            ‖
       H    C—H
        \  /
         C
        / \
      HO   CH₂OH
```
Glyceraldehyde

Now let us return to our three-dimensional drawing of the enantiomers of Ibuprofen and assign each an R or S configuration. In order of decreasing priority, the groups bonded to the stereocenter are —COOH (1) > —C_6H_5 (2) > —CH_3 (3) > H (4). In the enantiomer on the left, reading the groups on the stereocenter in order of priority is clockwise, and therefore, this enantiomer is (R)-Ibuprofen. Its mirror image is (S)-Ibuprofen.

(R)-Ibuprofen
(the inactive enantiomer)

(S)-Ibuprofen
(the active enantiomer)

The R,S system can be used to specify the configuration of any stereocenter in any molecule. It is not, however, the only system used for this purpose. There is also a D,L system, which is used primarily to specify the configuration of carbohydrates (Chapter 20) and amino acids (Chapter 22).

In closing, note that the purpose of this section is to show you how chemists assign a configuration to a stereocenter that specifies the relative orientation of the four groups on the stereocenter. What is important is that when you see a name such as (S)-Naproxen or (R)-Plavix, you realize that the compound is chiral and that the compound is not a racemic mixture. Rather, it is a pure enantiomer. We use the symbol (R,S) to show that a compound is a racemic mixture, as for example, (R,S)-Naproxen.

15-3 How Many Stereoisomers Are Possible for Molecules with Two or More Stereocenters?

For a molecule with n stereocenters, a maximum of 2^n stereoisomers are possible. We have already verified that for a molecule with one stereocenter, $2^1 = 2$ stereoisomers (one pair of enantiomers) are possible. For a molecule with two stereocenters, a maximum of $2^2 = 4$ stereoisomers (two pairs of enantiomers) is possible; for a molecule with three stereocenters, a maximum of $2^3 = 8$ stereoisomers (four pairs of enantiomers) is possible; and so forth.

A. Molecules with Two Stereocenters

We begin our study of molecules with two stereocenters by considering 2,3,4-trihydroxybutanal, a molecule with two stereocenters.

CHO
|
*CHOH
|
*CHOH
|
CH_2OH

2,3,4-Trihydroxybutanal
(2 stereocenters;
4 stereoisomers
are possible)

The maximum number of stereoisomers possible for this molecule is $2^2 = 4$, each of which is drawn in **Figure 15-5**.

15-3 How Many Stereoisomers Are Possible for Molecules with Two or More Stereocenters? 369

$$\begin{array}{cc}
\text{CHO} & \text{CHO} \\
\text{H}-\text{C}-\text{OH} & \text{HO}-\text{C}-\text{H} \\
\text{H}-\text{C}-\text{OH} & \text{HO}-\text{C}-\text{H} \\
\text{CH}_2\text{OH} & \text{CH}_2\text{OH} \\
(a) & (b)
\end{array} \qquad \begin{array}{cc}
\text{CHO} & \text{CHO} \\
\text{H}-\text{C}-\text{OH} & \text{HO}-\text{C}-\text{H} \\
\text{HO}-\text{C}-\text{H} & \text{H}-\text{C}-\text{OH} \\
\text{CH}_2\text{OH} & \text{CH}_2\text{OH} \\
(c) & (d)
\end{array}$$

One pair of enantiomers (Erythrose)   A second pair of enantiomers (Threose)

FIGURE 15-5 The four stereoisomers of 2,3,4-trihydroxybutanal.

Stereoisomers (a) and (b) are nonsuperposable mirror images and are, therefore, a pair of enantiomers. Stereoisomers (c) and (d) are also nonsuperposable mirror images and are a second pair of enantiomers. We describe the four stereoisomers of 2,3,4-trihydroxybutanal by saying that they consist of two pairs of enantiomers. Enantiomers (a) and (b) are named **erythrose**. Erythrose is synthesized in erythrocytes (red blood cells); hence the derivation of its name. Enantiomers (c) and (d) are named **threose**. Erythrose and threose belong to the class of compounds called carbohydrates, which we will discuss in Chapter 20.

We have specified the relationship between (a) and (b) and that between (c) and (d). What is the relationship between (a) and (c), between (a) and (d), between (b) and (c), and between (b) and (d)? The answer is that they are **diastereomers**—stereoisomers that are not mirror images.

Diastereomers Stereoisomers that are not mirror images

EXAMPLE 15-4   Enantiomers and Diastereomers

1,2,3-Butanetriol has two stereocenters (carbon-2 and 3); thus, $2^2 = 4$ stereoisomers are possible for it. Following are three-dimensional representations for each.

$$\begin{array}{cccc}
\overset{1}{\text{CH}_2\text{OH}} & \text{CH}_2\text{OH} & \text{CH}_2\text{OH} & \text{CH}_2\text{OH} \\
\text{H}-\overset{2}{\text{C}}-\text{OH} & \text{H}-\text{C}-\text{OH} & \text{HO}-\text{C}-\text{H} & \text{HO}-\text{C}-\text{H} \\
\text{HO}-\overset{3}{\text{C}}-\text{H} & \text{H}-\text{C}-\text{OH} & \text{HO}-\text{C}-\text{H} & \text{H}-\text{C}-\text{OH} \\
\overset{4}{\text{CH}_3} & \text{CH}_3 & \text{CH}_3 & \text{CH}_3 \\
(1) & (2) & (3) & (4)
\end{array}$$

(a) Which stereoisomers are pairs of enantiomers?
(b) Which stereoisomers are diastereomers?

Strategy
First, identify those structures that are mirror images. These, then, are the pairs of enantiomers. All other pairs of structures are diastereomers.

Solution
(a) Enantiomers are stereoisomers that are nonsuperposable mirror images. Compounds (1) and (4) are one pair of enantiomers and compounds (2) and (3) are a second pair of enantiomers.

(b) Diastereomers are stereoisomers that are not mirror images. Compounds (1) and (2), (1) and (3), (2) and (4), and (3) and (4) are diastereomers.

The diagram shows the relationship among these four stereoisomers.

Problem 15-4

3-Amino-2-butanol has two stereocenters (carbons 2 and 3); thus, $2^2 = 4$ stereoisomers are possible for it.

$$
\begin{array}{cccc}
\overset{1}{C}H_3 & CH_3 & CH_3 & CH_3 \\
\overset{2}{H\!-\!C\!-\!OH} & H\!-\!C\!-\!OH & HO\!-\!C\!-\!H & HO\!-\!C\!-\!H \\
\overset{3}{H_2N\!-\!C\!-\!H} & H\!-\!C\!-\!NH_2 & H\!-\!C\!-\!NH_2 & H_2N\!-\!C\!-\!H \\
\overset{4}{C}H_3 & CH_3 & CH_3 & CH_3 \\
(1) & (2) & (3) & (4)
\end{array}
$$

(a) Which stereoisomers are pairs of enantiomers?
(b) Which sets of stereoisomers are diastereomers?

We can analyze chirality in cyclic molecules with two stereocenters in the same way we analyzed it in acyclic compounds.

EXAMPLE 15-5 Enantiomerism in Cyclic Compounds

How many stereoisomers are possible for 3-methylcyclopentanol?

Strategy and Solution

Carbon-1 and 3 of this compound are stereocenters. Therefore, $2^2 = 4$ stereoisomers are possible for this molecule. The *cis* isomer exists as one pair of enantiomers, the *trans* isomer exists as a second pair of enantiomers.

H$_3$C⬩⟨ ⟩⬩OH   HO⬩⟨ ⟩⬩CH$_3$

cis-3-Methylcyclopentanol
(a pair of enantiomers)

H$_3$C⫸⟨ ⟩⬩OH   HO⬩⟨ ⟩⫷CH$_3$

trans-3-Methylcyclopentanol
(a second pair of enantiomers)

Problem 15-5

How many stereoisomers are possible for 3-methylcyclohexanol?

EXAMPLE 15-6 Locating Stereocenters

Mark the stereocenters in each compound with an asterisk. How many stereoisomers are possible for each?

Strategy

A stereocenter is a carbon atom that has four different groups bonded to it. Therefore, you are being asked to identify each carbon bonded to four different groups.

Solution

Each stereocenter is marked with an asterisk, and the number of stereoisomers possible for it appears under each compound. In (a), the carbon bearing the two methyl groups is not a stereocenter; this carbon has only three different groups bonded to it.

(a) $2^1 = 2$ (b) $2^2 = 4$ (c) $2^2 = 4$

Problem 15-6

Mark all stereocenters in each compound with an asterisk. How many stereoisomers are possible for each?

B. Molecules with Three or More Stereocenters

The 2^n rule applies equally well to molecules with three or more stereocenters. The following disubstituted cyclohexanol has three stereocenters, each marked with an asterisk. A maximum of $2^3 = 8$ stereoisomers is possible for this molecule. Menthol, one of the eight, has the configuration shown in the middle and on the right. Menthol is present in peppermint and other mint oils.

2-Isopropyl-5-methyl-cyclohexanol (three stereocenters; eight stereoisomers are possible)

Menthol (one of the eight possible stereoisomers)

Menthol drawn as a chair conformation (note that the three groups on the cyclohexane ring are all equatorial)

CHEMICAL CONNECTIONS 15A

Chiral Drugs

Some common drugs used in human medicine—for example, aspirin—are achiral. Others, such as the penicillin and erythromycin classes of antibiotics and the drug Captopril, are chiral and are sold as single enantiomers. Captopril is very effective for the treatment of high blood pressure and congestive heart failure. It is manufactured and sold as the (*S,S*)-stereoisomer.

Captopril

A large number of chiral drugs, however, are sold as racemic mixtures. The popular analgesic Ibuprofen (the active ingredient in Motrin, Advil, and many other non-aspirin analgesics) is an example.

Recently, the U.S. Food and Drug Administration established new guidelines for the testing and marketing of chiral drugs. After reviewing these guidelines, many drug companies have decided to develop only single enantiomers of new chiral drugs.

In addition to regulatory pressure, pharmaceutical developers must deal with patent considerations. If a company has a patent on a racemic mixture of a drug, a new patent can often be taken out on one of its enantiomers.

Test your knowledge with Problem 15-30.

Cholesterol, a more complicated molecule, has eight stereocenters. To identify them, remember to add an appropriate number of hydrogens to complete the tetravalence of each carbon you think might be a stereocenter.

Cholesterol has eight stereocenters; 256 stereoisomers are possible

This is the stereoisomer found in human metabolism

15-4 What Is Optical Activity, and How Is Chirality Detected in the Laboratory?

A. Plane-Polarized Light

Optically active Characterized by rotation of the plane of polarized light

Plane-polarized light Light with waves vibrating in only parallel planes

As we have already established, the two members of a pair of enantiomers are different compounds, and we must expect, therefore, that some of their properties differ. One such property relates to their effect on the plane of polarized light. Each member of a pair of enantiomers rotates the plane of polarized light; for this reason, each enantiomer is said to be **optically active**. To understand how optical activity is detected in the laboratory, we must first understand what **plane-polarized light** is and how a polarimeter, the instrument used to detect optical activity, works.

Ordinary light consists of waves vibrating in all planes perpendicular to its direction of propagation. Certain materials, such as a Polaroid sheet (a plastic film like that used in polarized sunglasses), selectively transmit

FIGURE 15-6 Schematic diagram of a polarimeter with its sample tube containing a solution of an optically active compound. The analyzer has been turned clockwise by α degrees to restore the light field.

light waves vibrating only in parallel planes. Electromagnetic radiation vibrating in only parallel planes is said to be **plane polarized**.

B. A Polarimeter

A **polarimeter** consists of a light source emitting unpolarized light, a polarizer, an analyzer, and a sample tube (Figure 15-6). If the sample tube is empty, the intensity of light reaching the detector (in this case, your eye) is at its maximum when the axes of the polarizer and analyzer are parallel to each other. If the analyzer is turned either clockwise or counterclockwise, less light is transmitted. When the axis of the analyzer is at right angles to the axis of the polarizer, the field of view is dark (no light passes through).

When a solution of an optically active compound is placed in the sample tube, it rotates the plane of the polarized light. If it rotates the plane clockwise, we say it is **dextrorotatory**; if it rotates the plane counterclockwise, we say it is **levorotatory**. Each member of a pair of enantiomers rotates the plane of polarized light by the same number of degrees, but in opposite directions. If one enantiomer is dextrorotatory, the other is levorotatory. Thus, racemic mixtures (as well as achiral compounds) do not display optical activity.

The number of degrees by which an optically active compound rotates the plane of polarized light is called its **specific rotation** and is given the symbol [α]. Specific rotation is defined as the observed rotation of an optically active substance at a concentration of 1 g/mL in a sample tube that is 10 cm long. A dextrorotatory compound is indicated by a plus sign in parentheses, (+), and a levorotatory compound is indicated by a minus sign in parentheses, (−). It is common practice to report the temperature (in °C) at which the measurement is made and the wavelength of light used. The most common wavelength of light used in polarimetry is the sodium D line, the same wavelength responsible for the yellow color of sodium-vapor lamps.

Following are specific rotations for the enantiomers of lactic acid measured at 21°C and using the D line of a sodium-vapor lamp as the light source. The (+) enantiomer of lactic acid is produced by muscle tissue in humans. The (−) enantiomer is found in sour cream and sour milk.

Dextrorotatory The clockwise (to the right) rotation of the plane of polarized light in a polarimeter

Levorotatory The counterclockwise (to the left) rotation of the plane of polarized light in a polarimeter

Specific rotation The number of degrees by which an optically active compound rotates the plane of plane-polarized light in a polarimeter

(S)-(+)-Lactic acid
$[\alpha]_D^{21} = +2.6°$

(R)-(−)-Lactic acid
$[\alpha]_D^{21} = -2.6°$

15-5 What Is the Significance of Chirality in the Biological World?

Except for inorganic salts and a few low-molecular-weight organic substances, the majority of molecules in living systems—both plant and animal—are chiral. ◄ Although these molecules can exist as a number of stereoisomers, almost invariably, only one stereoisomer is found in nature. Of course, instances do occur in which more than one stereoisomer is found, but these isomers rarely exist together in the same biological system.

A. Chirality in Biomolecules

Perhaps the most conspicuous examples of chirality among biological molecules are the enzymes, all of which have many stereocenters. Consider chymotrypsin, an enzyme in the intestines of animals that catalyzes the digestion of proteins (Chapter 23). Chymotrypsin has 251 stereocenters. The maximum number of stereoisomers possible is 2^{251}—a staggeringly large number, almost beyond comprehension. Fortunately, nature does not squander its precious energy and resources unnecessarily; any given organism produces and uses only one of these stereoisomers.

B. How an Enzyme Distinguishes Between a Molecule and Its Enantiomer

An enzyme catalyzes a biological reaction of a molecule by positioning the molecule at a **binding site** on the enzyme surface. An enzyme with specific binding sites for three of the four groups on a stereocenter can distinguish between a chiral molecule and its enantiomer or one of its diastereomers. Assume, for example, that an enzyme involved in catalyzing a reaction of glyceraldehyde has three binding sites: one specific for —H, a second specific for —OH, and a third specific for —CHO. Assume further that the three sites are arranged on the enzyme surface as shown in Figure 15-7. The enzyme can distinguish (R)-glyceraldehyde (the natural or biologically active form) from its enantiomer because the natural enantiomer can be adsorbed with three groups interacting with their appropriate binding sites; for the S enantiomer, at best only two groups can interact with these three binding sites.

Because interactions between molecules in living systems take place in a chiral environment, it should come as no surprise that a molecule and its enantiomer will elicit a different physiological response when compared to the same molecule's diastereomer. As we have already seen, (S)-Ibuprofen is

▶ The horns of this African Gazelle show chirality; one horn is the mirror image of the other.

FIGURE 15-7 A schematic diagram of an enzyme surface that can interact with (R)-glyceraldehyde at three binding sites, but with (S)-glyceraldehyde at only two of these sites.

(R)-Glyceraldehyde fits the three binding sites on the surface

(S)-Glyceraldehyde fits only two of the three binding sites

active as a pain and fever reliever, while its *R* enantiomer is inactive. The *S* enantiomer of the closely related analgesic Naproxen is also the active pain reliever of this compound, but its *R* enantiomer is a liver toxin!

(*S*)-Ibuprofen

(*S*)-Naproxen

Summary of Key Questions

Section 15-1 What Is Enantiomerism?

- A **mirror image** is the reflection of an object in a mirror.
- **Enantiomers** are a pair of stereoisomers that are nonsuperposable mirror images.
- A **racemic mixture** contains equal amounts of two enantiomers and does not rotate the plane of polarized light.
- **Diastereomers** are stereoisomers that are not mirror images.
- An object that is not superposable on its mirror image is said to be **chiral**; it has handedness. An **achiral** object lacks chirality (handedness); that is, it has a superposable mirror image.
- The most common cause of chirality in organic molecules is the presence of a tetrahedral carbon atom with four different groups bonded to it. Such a carbon is called a **stereocenter**.

Section 15-2 How Do We Specify the Configuration of a Stereocenter?

- We use the ***R,S* system** to specify the configuration of a stereocenter.
- To apply this convention:
  (1) Each atom or group of atoms bonded to the chiral center is assigned a priority based on atomic number; the higher the atomic number the higher the priority. The atoms or groups of atoms bonded to the chiral center are numbered from highest priority to lowest priority.
  (2) The molecule is oriented in space so that the group of lowest priority is directed away from the observer.
  (3) The remaining three groups are read in order from highest to lowest priority. If the order of groups is clockwise, the configuration is ***R*** (Latin: *rectus*, right). If the order is counterclockwise, the configuration is ***S*** (Latin: *sinister*, left).
  (4) To invert the configuration of a chiral center (*R* to *S* or *vice versa*) switch the location of any two groups bonded to the chiral center.

Section 15-3 How Many Stereoisomers Are Possible for Molecules with Two or More Stereocenters?

- For a molecule with n stereocenters, a maximum of 2^n stereoisomers is possible.

Section 15-4 What Is Optical Activity, and How Is Chirality Detected in the Laboratory?

- **Plane-polarized** light is light with waves that vibrate in only parallel planes.
- We use a **polarimeter** to measure optical activity. A compound is said to be **optically active** if it rotates the plane of polarized light.
- If a compound rotates the plane clockwise, it is **dextrorotatory**; if it rotates the plane counterclockwise, it is **levorotatory**.
- Each member of a pair of enantiomers rotates the plane of polarized light an equal number of degrees, but in opposite directions.

Section 15-5 What Is the Significance of Chirality in the Biological World?

- An enzyme catalyzes biological reactions of molecules by positioning them at binding sites on its surface. An enzyme with binding sites specific for three of the four groups on a stereocenter can distinguish between a molecule and its enantiomer or one of its diastereomers.

Problems

Orange-numbered problems are applied.

Section 15-1 What Is Enantiomerism?

15-7 Answer true or false.
(a) The *cis* and *trans* stereoisomers of 2-butene are achiral.
(b) The carbonyl carbon of an aldehyde, ketone, carboxylic acid, or ester cannot be a stereocenter.
(c) Stereoisomers have the same connectivity of their atoms.
(d) Constitutional isomers have the same connectivity of their atoms.
(e) An unmarked cube is achiral.
(f) A human foot is chiral.
(g) Every object in nature has a mirror image.
(h) The most common cause of chirality in organic molecules is the presence of a tetrahedral carbon atom with four different groups bonded to it.
(i) If a molecule is not superposable on its mirror image, the molecule is chiral.

15-8 What does the term "chiral" mean? Give an example of a chiral molecule.

15-9 What does the term "achiral" mean? Give an example of an achiral molecule.

15-10 Define the term "stereoisomer." Name three types of stereoisomers.

15-11 In what way are constitutional isomers different from stereoisomers? In what way are they the same?

15-12 Which of the following objects are chiral (assume that there is no label or other identifying mark)?
(a) Pair of scissors (b) Tennis ball
(c) Paper clip (d) Beaker
(e) The swirl created in water as it drains out of a sink or bathtub

15-13 2-Pentanol is chiral, but 3-pentanol is not. Explain.

15-14 2-Butene exists as a pair of *cis-trans* isomers. Is *cis*-2-butene chiral? Is *trans*-2-butene chiral? Explain.

15-15 Explain why the carbon of a carbonyl group cannot be a stereocenter.

15-16 Which of the following compounds contain stereocenters?
(a) 2-Chloropentane (b) 3-Chloropentane
(c) 3-Chloro-1-butene (d) 1,2-Dichloropropane

15-17 Which of the following compounds contain stereocenters?
(a) Cyclopentanol
(b) 1-Chloro-2-propanol
(c) 2-Methylcyclopentanol
(d) 1-Phenyl-1-propanol

15-18 Using only C, H, and O, write structural formulas for the lowest-molecular-weight chiral molecule of each class.
(a) Alkane (b) Alkene
(c) Alcohol (d) Aldehyde
(e) Ketone (f) Carboxylic acid

15-19 Draw the mirror image for each molecule:

15-20 Draw the mirror image for each molecule:

Section 15-3 How Many Stereoisomers Are Possible for Molecules with Two or More Stereocenters?

15-21 Answer true or false.
(a) For a molecule with two stereocenters, $2^2 = 4$ stereoisomers are possible.
(b) For a molecule with three stereocenters, $3^2 = 9$ stereoisomers are possible.
(c) Enantiomers, like gloves, occur in pairs.
(d) 2-Pentanol and 3-pentanol are both chiral and show enantiomerism.
(e) 1-Methylcyclohexanol is achiral and does not show enantiomerism.
(f) Diastereomers are stereoisomers that are not mirror images.

15-22 Mark each stereocenter in these molecules with an asterisk. Note that not all contain stereocenters.

(a) PhCH(OH)CH₂N(CH₃)₂

(b) CH₃CH₂C(=O)CH(OH)CH₂CH₃

(c) (CH₃)₂CHCH(NH₂)COOH

(d) CH₃CH₂C(=O)CH₂CH₃

15-23 Mark each stereocenter in these molecules with an asterisk. Note that not all contain stereocenters.

(a) HOCH₂CH(OH)CH₂OH

(b) HOCH₂CH(OH)CH₂CH₃

(c) CH₂=CHCH₂CH(OH)CH₃ (with OH)

(d) CH₃CH₂CH(OH)CH₂CH₃

15-24 Label all stereocenters in each molecule with an asterisk. How many stereoisomers are possible for each molecule?

(a) CH₃CH(OH)CH(OH)COOH

(b) HOOCCH₂CH(COOH)CH(OH)COOH — (CH₂COOH / CHCOOH / HO—CHCOOH)

(c) 4-isopropyl-2-hydroxycyclohexane (cyclohexane ring with OH and isopropyl substituents)

(d) 4-isopropyl-1-methylcyclohex-1-ene

15-25 Label all stereocenters in each molecule with an asterisk. How many stereoisomers are possible for each molecule?

(a) 2-methylcyclopentanol (cyclopentane with OH and CH₃)

(b) acyclic terpene alcohol with internal C=C and terminal OH group

(c) 2-hydroxytetrahydrofuran

(d) decalin-2-one with angular methyl group

15-26 For centuries, Chinese herbal medicine has used extracts of *Ephedra sinica* to treat asthma. The asthma-relieving component of this plant is ephedrine, a very potent dilator of the air passages of the lungs. The naturally occurring stereoisomer is levorotatory and has the following structure.

Ephedrine $[\alpha]_D^{21} = -41°$

(a) Mark each stereocenter in epinephrine with an asterisk.

(b) How many stereoisomers are possible for this compound?

15-27 The specific rotation of naturally occurring ephedrine, shown in Problem 15-26, is −41°. What is the specific rotation of its enantiomer?

15-28 What is a racemic mixture? Is a racemic mixture optically active? That is, will it rotate the plane of polarized light?

Section 15-4 What Is Optical Activity, and How Is Chirality Detected in the Laboratory?

15-29 Answer true or false.

(a) If a chiral compound is dextrorotatory, its enantiomer is levorotatory by the same number of degrees.

(b) A racemic mixture is optically inactive.

(c) All stereoisomers are optically active.

(d) Plane-polarized light consists of light waves vibrating in parallel planes.

Chemical Connections

15-30 (Chemical Connections 15A) What does it mean to say that a drug is *chiral*? If a drug is chiral, will it be optically active? That is, will it rotate the plane of polarized light?

15-31 (Chemical Connections 15A) When a successful drug is discovered, there often follows an effort to synthesize compounds closely related in structure to the original, in hopes of discovering new drugs that are even more effective. Following are structural formulas for Captopril and three related angiotensin converting enzyme (ACE) inhibitors, which are used to treat hypertension. Which of these three are chiral, and how many stereoisomers are possible for each? Compare these three structures with that of Captopril and determine the similarities in structure among the four drugs.

Chapter 15 Chirality: The Handedness of Molecules

Captopril

Quinapril (Accupril)

Enalopril (Altace)

Ramipril (Vasotec)

Additional Problems

15-32 Which of the eight alcohols with a molecular formula of $C_5H_{12}O$ are chiral?

15-33 Write the structural formula of an alcohol with the molecular formula $C_6H_{14}O$ that contains two stereocenters.

15-34 Which carboxylic acids with a molecular formula of $C_6H_{12}O_2$ are chiral?

15-35 Following are structural formulas for three of the drugs most widely prescribed to treat depression. Label all stereocenters in each and state the number of stereoisomers possible for each.

(a) Fluoxetine (Prozac)

(b) Sertraline (Zoloft)

(c) Paroxetine (Paxil)

15-36 Label the four stereocenters in amoxicillin, which belongs to the family of semisynthetic penicillins.

Amoxicillin

15-37 Consider a cyclohexane ring substituted with one hydroxyl group and one methyl group. Draw a structural formula for a compound of this composition that:
  (a) Does not show *cis-trans* isomerism and has no stereocenters.
  (b) Shows *cis-trans* isomerism but has no stereocenters.
  (c) Shows *cis-trans* isomerism and has two stereocenters.

15-38 The next time you have the opportunity to view a collection of seashells that have a helical twist, study the chirality (handedness) of their twists. For each kind of shell, do you find an equal number of left-handed and right-handed twists or, for example, do they all have the same handedness?

15-39 The next time you have an opportunity to examine any of the seemingly endless varieties of blond spiral pasta (rotini, fusilli, radiatori, tortiglione, and so forth), examine their twists. Do the twists of any one kind all have a right-handed twist or a left-handed twist, or are they a racemic mixture?

15-40 Think about the helical coil of a telephone cord or the spiral binding on a notebook. Suppose that you view the spiral from one end and find that it has a left-handed twist. If you view the same spiral from the other end, does it have a left-handed twist from that end as well or does it have a right-handed twist?

15-41 Compound **A**(C_5H_8) is not optically active and cannot be separated into enantiomers. It reacts with bromine in carbon tetrachloride to discharge the purple color of bromine and form Compound **B**($C_5H_8Br_2$). When Compound **A** is treated with H_2 in the presence of a transition metal catalyst, it is converted to compound **C**(C_5H_{10}). When treated with HCl, compound **A** is converted to compound **D**(C_5H_9Cl). Given this information, propose structural formulas for compounds **A**, **B**, **C**, and **D**. *Hint:* There are at least three possibilities for Compound **A** and, in turn, three possibilities for Compounds **B**, **C**, and **D**.

Looking Ahead

15-42 Following is a chair conformation of glucose, the most prevalent carbohydrate in the biological world (Chapter 20).

  (a) Identify the five stereocenters in this molecule.
  (b) How many stereoisomers are possible?
  (c) How many pairs of enantiomers are possible?

15-43 Triamcinolone acetonide, the active ingredient in Azmacort Inhalation Aerosol, is a steroid used to treat bronchial asthma.

Triamcinolone acetonide

  (a) Label the eight stereocenters in this molecule.
  (b) How many stereoisomers are possible for it? (Of these, the stereoisomer with the configuration shown here is the active ingredient in Azmacort.)

Challenge Problems

15-44 Consider the structure of the immunosuppressant *FK-506*, a molecule shown to disrupt calcineurin-mediated signal transduction in *T*-lymphocytes.

(a) What is the molecular formula of this immunosuppressant?
(b) How many stereocenters are present in *FK-506*? Determine the maximum number of stereoisomers possible.
(c) Identify and label the various functional groups present.
(d) Consider the two stereocenters in this structure labeled with asterisks (*). Determine the absolute configuration of each stereocenter.
(e) *FK-506* has been shown to exhibit moderate solubility in various organic solvents. Is this immunosuppressant expected to be soluble in ethanol (CH_3CH_2OH)?
(f) Consider the carbon atom labeled "**1**." Describe the geometry and approximate bond angles about this carbon atom.
(g) Draw the alternative chair conformations of the cyclohexane ring at the lower right of *FK-506* and label the more stable conformation.
(h) Are there any aromatic components present in *FK-506*?
(i) Patients taking *FK-506* have reported several side effects from this medication, including headaches, nausea or diarrhea, and slight shaking. Would you expect the enantiomer of this drug to result in the same side effects?

15-45 *Oseltamivir phosphate* (sold under the trade name Tamiflu®) is a prescription antiviral drug that is used in the treatment of both *Influenza virus Type A* and *Influenza virus Type B*.

Oseltamivir phosphate
(Tamiflu)

(a) Name all functional groups present in *Oseltamivir*. Is the amine group primary, secondary, or tertiary?
(b) What is the molecular formula of *Oseltamivir*? Note that the structural formula above is that of the phosphate salt. This question refers to the molecular formula of the free base.
(c) Place a check mark in the box next to all the words that describe *Oseltamivir*:

☐ Chiral
☐ Achiral
☐ Optically active
☐ Optically inactive
☐ Racemic

(d) Label all stereocenters (if any) present in *Oseltamivir* as either *R* or *S*.
(e) Draw an enantiomer of *Oseltamivir*.
(f) Draw a diastereomer of *Oseltamivir*.
(g) *Oseltamivir* is synthesized starting from shikimic acid, which can be partially hydrogenated to form the compound shown below on the right. Draw alternative chair conformations of this molecule and label the more stable and less stable conformations.

Shikimic acid

(h) What is the maximum number of stereoisomers possible for *Oseltamivir*?

15-46 Consider *Lunesta*, a nonbenzodiazepine hypnotic agent (i.e., sleep-inducing drug) that is frequently advertised on TV commercials. Answer the following questions with respect to the given structure:

Lunesta

(a) Determine the molecular formula for *Lunesta*.
(b) Identify the functional groups present in *Lunesta*.
(c) How many of *Lunesta's* rings are aromatic?
(d) Fill in the blanks as shown: *Lunesta* has _____ stereocenter(s) and therefore _____ possible stereoisomer(s). Of the possible stereocenter(s), _____ is/are *R* and _____ is/are *S*.
(e) Does *Lunesta* have an enantiomer? Does it have a diastereomer?
(f) Which of the following is true about an enantiomer of *Lunesta*? Identify all that apply:
 (1) The enantiomer rotates plane-polarized light in the opposite direction as *Lunesta*.
 (2) The enantiomer is a mirror image of *Lunesta*.
 (3) The enantiomer has the opposite biological effects as *Lunesta* (i.e., it keeps you awake).
 (4) *Lunesta* does not have an enantiomer.
(g) Draw an enantiomer of *Lunesta*.
(h) Examine the derivative of the representation of the six-membered ring found in *Lunesta*. Draw the alternative chair conformations of this ring and label the more stable chair conformation. (Chapter 11)

15-47 Following are structural formulas for the enantiomers of carvone.
(a) Assign an *R* or *S* configuration to the single stereocenter in each.
(b) How do you account for the fact that each has such a different smell when they are so similar a structure?

(−) Carvone
Spearmint oil

(+) Carvone
Caraway and dill seed oil

16 Amines

KEY QUESTIONS

16-1 What Are Amines?

16-2 How Do We Name Amines?

16-3 What Are the Physical Properties of Amines?

16-4 How Do We Describe the Basicity of Amines?

16-5 What Are the Characteristic Reactions of Amines?

This inhaler delivers a puff of albuterol (Proventil), a potent synthetic bronchodilator whose structure is patterned after that of epinephrine. See Chemical Connections 16E.

16-1 What Are Amines?

Carbon, hydrogen, and oxygen are the three most common elements in organic compounds. Because of the wide distribution of amines in the biological world, nitrogen is the fourth most common element of organic compounds. Amines are derivatives of ammonia in which one, two or three hydrogens are replaced by alkyl or aryl groups. The most important chemical property of amines is their basicity.

Amines (Section 10-4B) are classified as **primary (1°)**, **secondary (2°)**, or **tertiary (3°)**, depending on the number of carbon groups bonded to nitrogen.

$$CH_3-NH_2 \qquad CH_3-\underset{H}{\overset{|}{N}}-CH_3 \qquad CH_3-\underset{CH_3}{\overset{|}{N}}-CH_3$$

Methylamine (a 1° amine)   Dimethylamine (a 2° amine)   Trimethylamine (a 3° amine)

Amines are further classified as aliphatic or aromatic. An **aliphatic amine** is one in which all the carbons bonded to nitrogen are derived

Aliphatic amine An amine in which nitrogen is bonded only to alkyl groups or hydrogens

CHEMICAL CONNECTIONS 16A

Amphetamines (Pep Pills)

Amphetamine, methamphetamine, and phentermine—all synthetic amines—are powerful stimulants of the central nervous system. Like most other amines, they are stored and administered as their salts. The sulfate salt of amphetamine is named Benzedrine, the hydrochloride salt of the S enantiomer of methamphetamine is named Methedrine, and the hydrochloride salt of phentermine is named Fastin.

These three amines have similar physiological effects and are referred to by the general name **amphetamines**. Structurally, they have in common a benzene ring with a three-carbon side chain and an amine nitrogen on the second carbon of the side chain. Physiologically, they share an ability to reduce fatigue and diminish hunger by raising the glucose level of the blood. Because of these properties, amphetamines are widely prescribed to counter mild depression, reduce hyperactivity in children, and suppress appetite in people who are trying to lose weight. These drugs are also used illegally to reduce fatigue and elevate mood.

Abuse of amphetamines can have severe effects on both body and mind. They are addictive, concentrate in the brain and nervous system, and can lead to long periods of sleeplessness, loss of weight, and paranoia.

The action of amphetamines is similar to that of epinephrine (Chemical Connections 16E), the hydrochloride salt of which is named adrenaline.

Amphetamine (Benzedrine)   (s)-Methamphetamine (Methedrine)   Phentermine (Fastin)

Test your knowledge with Problems 16-33 and 16-34.

---

from alkyl groups. An **aromatic amine** is one in which one or more of the groups bonded to nitrogen are aryl groups.

Aniline (a 1° aromatic amine)   N-Methylaniline (a 2° aromatic amine)   Benzyldimethylamine (a 3° aliphatic amine)

Aromatic amine An amine in which nitrogen is bonded to one or more aromatic rings

An amine in which the nitrogen atom is part of a ring is classified as a **heterocyclic amine**. When the ring is saturated, the amine is classified as a **heterocyclic aliphatic amine**. When the nitrogen is part of an aromatic ring (Section 13-1), the amine is classified as a **heterocyclic aromatic amine**. Two of the most important heterocyclic aromatic amines are pyridine and pyrimidine, in which nitrogen atoms replace first one and then two CH groups of a benzene ring. Pyrimidine and purine serve as the building blocks for the amine bases of DNA and RNA (Chapter 25).

Heterocyclic amine An amine in which nitrogen is one of the atoms of a ring

Heterocyclic aromatic amine An amine in which nitrogen is one of the atoms of an aromatic ring

Pyridine   Pyrimidine   Imidazole   Purine   Pyrrole
(heterocyclic aromatic amines)

CHEMICAL CONNECTIONS 16B

Alkaloids

Alkaloids are basic nitrogen-containing compounds found in the roots, bark, leaves, berries, or fruits of plants. In almost all alkaloids, the nitrogen atom is part of a ring. The name "alkaloid" was chosen because these compounds are alkali-like (*alkali* is an older term for a basic substance) and react with strong acids to give water-soluble salts. Thousands of different alkaloids, many of which are used in modern medicine, have been extracted from plant sources.

When administered to animals, including humans, alkaloids have pronounced physiological effects. Whatever their individual effects, most alkaloids are toxic in large enough doses. For some, the toxic dose is very small!

(S)-Coniine is the toxic principle of water hemlock (a member of the carrot family). Its ingestion can cause weakness, labored respiration, paralysis, and eventually death. It was the toxic substance in the "poison hemlock" used in the death of Socrates. Water hemlock is easily confused with Queen Anne's lace, a type of wild carrot—a mistake that has killed numerous people.

(S)-Nicotine occurs in the tobacco plant. In small doses, it is an addictive stimulant. In larger doses, this substance causes depression, nausea, and vomiting. In still larger doses, it is a deadly poison. Solutions of nicotine in water are used as insecticides.

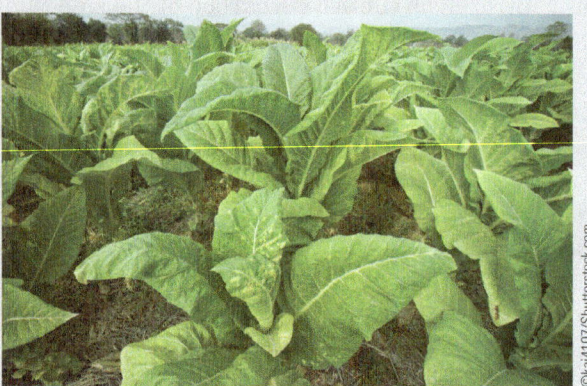

Tobacco plants.

Cocaine is a central nervous system stimulant obtained from the leaves of the coca plant. In small doses, it decreases fatigue and gives a sense of well-being. Prolonged use of cocaine leads to physical addiction and depression.

Test your knowledge with Problems 16-35 through 16-38.

Interactive Example 16-1 — Structure of Amines

How many hydrogen atoms does piperidine have? How many hydrogen atoms does pyrrolidine have? Write the molecular formula of each amine.

Pyrrolidine   Piperidine
(heterocyclic aliphatic amines)

Strategy

Remember that hydrogen atoms bonded to carbon are not shown in line-angle formulas. To determine the number of hydrogens present, add a sufficient number to give four bonds to each carbon and three bonds to each nitrogen.

Solution

Piperidine has 11 hydrogen atoms, and its molecular formula is $C_5H_{11}N$. Pyridine has 5 hydrogen atoms, and its molecular formula is C_5H_5N.

Problem 16-1

How many hydrogen atoms does pyrrolidine have? How many does purine have? Write the molecular formula of each amine.

16-2 How Do We Name Amines?

A. IUPAC Names

IUPAC names for aliphatic amines are derived just as they are for alcohols. The final -e of the parent alkane is dropped and replaced with -**amine**. Indicate the location of the amino group on the parent chain by a number.

2-Propanamine   Cyclohexanamine   1,6-Hexanediamine

IUPAC nomenclature retains the common name **aniline** for $C_6H_5NH_2$, the simplest aromatic amine. Its simple derivatives are named using numbers to locate substituents or, alternatively, using the locators *ortho* (*o*), *meta* (*m*), and *para* (*p*). Several derivatives of aniline have common names that remain in use. Among them is **toluidine** for a methyl-substituted aniline.

Aniline   4-Nitroaniline   3-Methylaniline
             (*p*-Nitroaniline)   (*m*-Toluidine)

Unsymmetrical secondary and tertiary amines are commonly named as *N*-substituted primary amines. The largest group bonded to nitrogen is taken as the parent amine; the smaller groups bonded to nitrogen are named, and their locations are indicated by the prefix *N* (indicating that they are bonded to nitrogen).

N-Methylaniline   *N,N*-Dimethylcyclopentanamine

EXAMPLE 16-2   IUPAC Names of Amines

Write the IUPAC name for each amine. Assign an R or S configuration to the stereocenter in (c).

(a) [structure: butane with NH₂ on C2] (b) H₂N(CH₂)₅NH₂ (c)

Strategy

The parent chain is the longest chain that contains the amino group. Number the parent chain from the end that gives the amino group the lowest possible number.

Solution

(a) The parent alkane has four carbon atoms and is butane. The amino group is on carbon-2, giving the IUPAC name 2-butanamine.

(b) The parent chain has five carbon atoms and is pentane. There are amino groups on carbons 1 and 5, giving the IUPAC name 1,5-pentanediamine. The common name of this diamine is cadaverine, which should give you a hint of where it occurs in nature and its odor. Cadaverine, one of the end products of decaying flesh, is quite poisonous.

(c) The parent chain has three carbon atoms and is propane. To have the lowest numbers possible, number the chain from the end that places the phenyl group on carbon-1 and the amino group on carbon-2. The priorities for determining R or S configuration are $NH_2 > C_6H_5CH_2 > CH_3 > H$. This amine's systematic name is (R)-1-phenyl-2-propanamine. It is the (R)-enantiomer of the stimulant amphetamine.

Problem 16-2

Write a structural formula for each amine.

(a) 2-Methyl-1-propanamine   (b) Cyclopentanamine
(c) 1,4-Butanediamine

B. Common Names

Common names for most aliphatic amines list the groups bonded to nitrogen in alphabetical order in one word ending in the suffix **-amine**.

Propylamine   sec-Butylamine   Diethylmethylamine   Cyclohexylamine

EXAMPLE 16-3   Common Names of Amines

Write a structural formula for each amine.

(a) Isopropylamine   (b) Cyclohexylmethylamine
(c) Triethylamine

Strategy and Solution

In these common names, the names of the groups bonded to carbon are listed in alphabetical order followed by the suffix **-amine**.

(a) $(CH_3)_2CHNH_2$ (b) ⬡—$NHCH_3$ (c) $(CH_3CH_2)_3N$

or

Problem 16-3

Write a structural formula for each amine.

(a) 2-Aminoethanol   (b) Diphenylamine

(c) Diisopropylamine

When four atoms or groups of atoms are bonded to a nitrogen atom—as, for example, in NH_4^+ and $CH_3NH_3^+$—nitrogen bears a positive charge and is associated with an anion as a salt. The compound is named as a salt of the corresponding amine. The ending **-amine** (or aniline, pyridine, or the like) is replaced with **-ammonium** (or *anilinium*, *pyridinium*, or the like), and the name of the anion (chloride, acetate, and so on) is added. ◂

$$(CH_3CH_2)_3NH^+Cl^-$$

Triethylammonium chloride

16-3 What Are the Physical Properties of Amines?

Like ammonia, low-molecular-weight amines have very sharp, penetrating odors. Trimethylamine, for example, is the pungent principle in the smell of rotting fish. Two other particularly pungent amines are 1,4-butanediamine (putrescine) and 1,5-pentanediamine (cadaverine).

Amines are polar compounds because of the difference in electronegativity between nitrogen and hydrogen (3.0 − 2.1 ≈ 0.9). Both primary and secondary amines have N—H bonds and can form hydrogen bonds with one another (Figure 16-1). Tertiary amines do not have a hydrogen bonded to nitrogen and, therefore, do not form hydrogen bonds with one another.

An N—H----N hydrogen bond is weaker than an O—H----O hydrogen bond, because the difference in electronegativity between nitrogen and hydrogen (3.0 − 2.1 ≈ 0.9) is less than that between oxygen and hydrogen (3.5 − 2.1 ≈ 1.4). To see the effect of hydrogen bonding between alcohols and amines of similar molecular weight, compare the boiling points of ethane, methanamine, and methanol. Ethane is a nonpolar hydrocarbon, and the only attractive forces between its molecules are weak London dispersion forces (Section 5-7A). Both methanamine and methanol are polar molecules that interact in the pure liquid by hydrogen bonding. Methanol has the highest boiling point of the three compounds, because the hydrogen bonding between its molecules is stronger than that between methanamine molecules.

▸ Several over-the-counter mouthwashes contain an *N*-alkylpyridinium chloride as an antibacterial agent.

Hydrogen bonding

FIGURE 16-1 Hydrogen bonding between two molecules of a secondary amine.

	CH_3CH_3	CH_3NH_2	CH_3OH
Molecular weight (amu)	30.1	31.1	32.0
Boiling point (°C)	−88.6	−6.3	65.0

CHEMICAL CONNECTIONS 16C

Tranquilizers

Most people face anxiety and stress at some time in their lives, and each person develops various ways to cope with these factors. Perhaps this strategy involves meditation, exercise, psychotherapy, or drugs. One modern coping technique is to use tranquilizers, drugs that provide relief from the symptoms of anxiety or tension.

The first modern tranquilizers were derivatives of a compound called benzodiazepine. The first of these compounds, chlorodiazepoxide, better known as Librium, was introduced in 1960 and was soon followed by more than two dozen related compounds. Diazepam, better known as Valium, became one of the most widely used of these drugs.

Librium, Valium, and other benzodiazepines are central nervous system sedatives/hypnotics. As sedatives, they diminish activity and excitement, thereby exerting a calming effect. As hypnotics, they produce drowsiness and sleep.

Benzodiazepine

Chlorodiazepoxide
(Librium)

Diazepam
(Valium)

Test your knowledge with Problems 16-39 through 16-41.

All classes of amines form hydrogen bonds with water and are more soluble in water than are hydrocarbons of comparable molecular weight. Most low-molecular-weight amines are completely soluble in water, but higher-molecular-weight amines are only moderately soluble in water or are insoluble.

16-4 How Do We Describe the Basicity of Amines?

Like ammonia, amines are weak bases, and aqueous solutions of amines are basic. The following acid–base reaction between an amine and water is written using curved arrows to emphasize that in this proton-transfer reaction (Section 8-1), the unshared pair of electrons on nitrogen forms a new covalent bond with hydrogen and displaces a hydroxide ion.

$$CH_3-\underset{H}{\overset{H}{N}}: + H-\overset{..}{\underset{..}{O}}-H \rightleftharpoons CH_3-\underset{H}{\overset{H}{N^+}}-H + :\overset{..}{\underset{..}{O}}-H$$

Methylamine
(a base)

Methylammonium
hydroxide

The base dissociation constant, K_b, for the reaction of an amine with water, has the following form, illustrated here for the reaction of methylamine with water to give methylammonium hydroxide. pK_b is defined as the negative logarithm of K_b.

$$K_b = \frac{[CH_3NH_3^+][OH^-]}{[CH_3NH_2]} = 4.37 \times 10^{-4}$$

$$pK_b = -\log 4.37 \times 10^{-4} = 3.360$$

All aliphatic amines have approximately the same base strength, pK_b 3.0 – 4.0, and are slightly stronger bases than ammonia (Table 16-1). Aromatic amines and heterocyclic aromatic amines (pK_b 8.5 – 9.5) are considerably weaker bases than aliphatic amines. One additional point about the basicities of amines: while aliphatic amines are weak bases by comparison with inorganic bases such as NaOH, they are strong bases among organic compounds.

Table 16-1 Approximate Base Strengths of Amines

Class	pK_b	Example	Name	
Aliphatic	3.0–4.0	$CH_3CH_2NH_2$	Ethanamine	Stronger base
Ammonia	4.74			↑
Aromatic	8.5–9.5	⟨phenyl⟩—NH_2	Aniline	Weaker base

Given the basicities of amines, we can determine which form of an amine exists in body fluids—say, blood. In a normal healthy person, the pH of blood is approximately 7.40, which is slightly basic. If an aliphatic amine is dissolved in blood, it is present predominantly as its protonated or conjugate acid form.

Dopamine

Conjugate acid of dopamine
(the major form present
in blood plasma)

We can show that an aliphatic amine such as dopamine dissolved in blood is present largely as its protonated or conjugate acid form in the following way. Assume that the amine RNH_2 has a pK_b of 3.50 and that it is dissolved in blood, pH 7.40. We first write the base dissociation constant for the amine and then solve for the ratio of RNH_3^+ to RNH_2.

$$RNH_2 + H_2O \rightleftharpoons RNH_3^+ + OH^-$$

$$K_b = \frac{[RNH_3^+][OH^-]}{[RNH_2]}$$

$$\frac{K_b}{[OH^-]} = \frac{[RNH_3^+]}{[RNH_2]}$$

We now substitute the appropriate values for K_b and [OH$^-$] in this equation. Taking the antilog of 3.50 gives a K_b of 3.2×10^{-4}. Calculating the concentration of hydroxide requires two steps. First, recall from Section 8-8 that pH + pOH = 14. If the pH of blood is 7.40, then its pOH is 6.60 and its

[OH⁻] is 2.5×10^{-7}. Substituting these values in the appropriate equation gives a ratio of 1300 parts RNH_3^+ to 1 part RNH_2.

$$\frac{3.2 \times 10^{-4}}{2.5 \times 10^{-7}} = \frac{[RNH_3^+]}{[RNH_2]} = 1300$$

As this calculation demonstrates, more than 99.9% of an aliphatic amine present in blood is in the protonated (conjugate acid) form. Thus, even though we may write the structural formula for dopamine as the free amine, it is present in blood as the protonated form. It is important to realize, however, that the free amine and conjugate acid forms are always in equilibrium, so some of the free base form is nevertheless present in solution.

Aromatic amines, by contrast, are considerably weaker bases than aliphatic amines and are present in blood largely in the unprotonated form. Performing the same type of calculation for an aromatic amine, $ArNH_2$, with a pK_b of approximately 10, we find that the aromatic amine is more than 99.0% in its unprotonated ($ArNH_2$) form.

EXAMPLE 16-4 Basicity of Amines

Select the stronger base in each pair of amines.

Strategy

Determine whether the amine is an aromatic or an aliphatic amine. Aliphatic amines are stronger bases than aromatic amines.

Solution

(a) Morpholine (B), a 2° aliphatic amine, is the stronger base. Pyridine (A), a heterocyclic aromatic amine, is the weaker base.

(b) Benzylamine (D), a 1° aliphatic amine, is the stronger base. Even though it contains an aromatic ring, it is not an aromatic amine because the amine nitrogen is not bonded to the aromatic ring. o-Toluidine (C), a 1° aromatic amine, is the weaker base.

Problem 16-4

Select the stronger base from each pair of amines.

16-5 What Are the Characteristic Reactions of Amines?

The most important chemical property of amines is their basicity. Amines, whether soluble or insoluble in water, react quantitatively with strong acids to form water-soluble salts, as illustrated by the reaction of (R)-norepinephrine (noradrenaline) with aqueous HCl to form a hydrochloride salt.

CHEMICAL CONNECTIONS 16D

The Solubility of Drugs in Body Fluids

Many drugs have "•HCl" or some other acid as part of their chemical formula and occasionally as part of their generic name. Invariably, these drugs are amines that are insoluble in aqueous body fluids such as blood plasma and cerebrospinal fluid. For the administered drug to be absorbed and carried by body fluids, it must be treated with an acid to form a water-soluble salt. Methadone, a narcotic analgesic, is marketed as its water-soluble hydrochloride salt. Novocain, one of the first local anesthetics, is the hydrochloride salt of procaine.

Methadone · HCl

Procaine · HCl
(Novocain, a local anesthetic)

These two drugs are amino salts and are labeled as hydrochlorides.

There is another reason besides increased water solubility for preparing these and other amine drugs as salts. Amines are very susceptible to oxidation and decomposition by atmospheric oxygen, with a corresponding loss of biological activity. By comparison, their amine salts are far less susceptible to oxidation; they retain their effectiveness for a much longer time.

Test your knowledge with Problem 16-42

(R)-Norepinephrine
(only slightly soluble in water)
+ HCl $\xrightarrow{H_2O}$
(R)-Norepinephrine hydrochloride
(a water-soluble salt)

EXAMPLE 16-5 Basicity of Amines

Complete the equation for each acid–base reaction and name the salt formed.

(a) $(CH_3CH_2)_2NH + HCl \longrightarrow$

(b) [pyridine] $+ CH_3COOH \longrightarrow$

Strategy

Each acid–base reaction involves a proton transfer from the acid to the amino group (a base). The product is named as an ammonium salt.

Solution

Problem 16-5
Complete the equation for each acid–base reaction and name the salt formed.

(a) $(CH_3CH_2)_3N + HCl \longrightarrow$ (b) ⬡NH + $CH_3COOH \longrightarrow$

The basicity of amines and the solubility of amine salts in water gives us a way to separate water-insoluble amines from water-insoluble nonbasic compounds. Figure 16-2 is a flowchart for the separation of aniline from cyclohexanol, a neutral compound.

FIGURE 16-2 Separation and purification of an amine and a neutral compound.

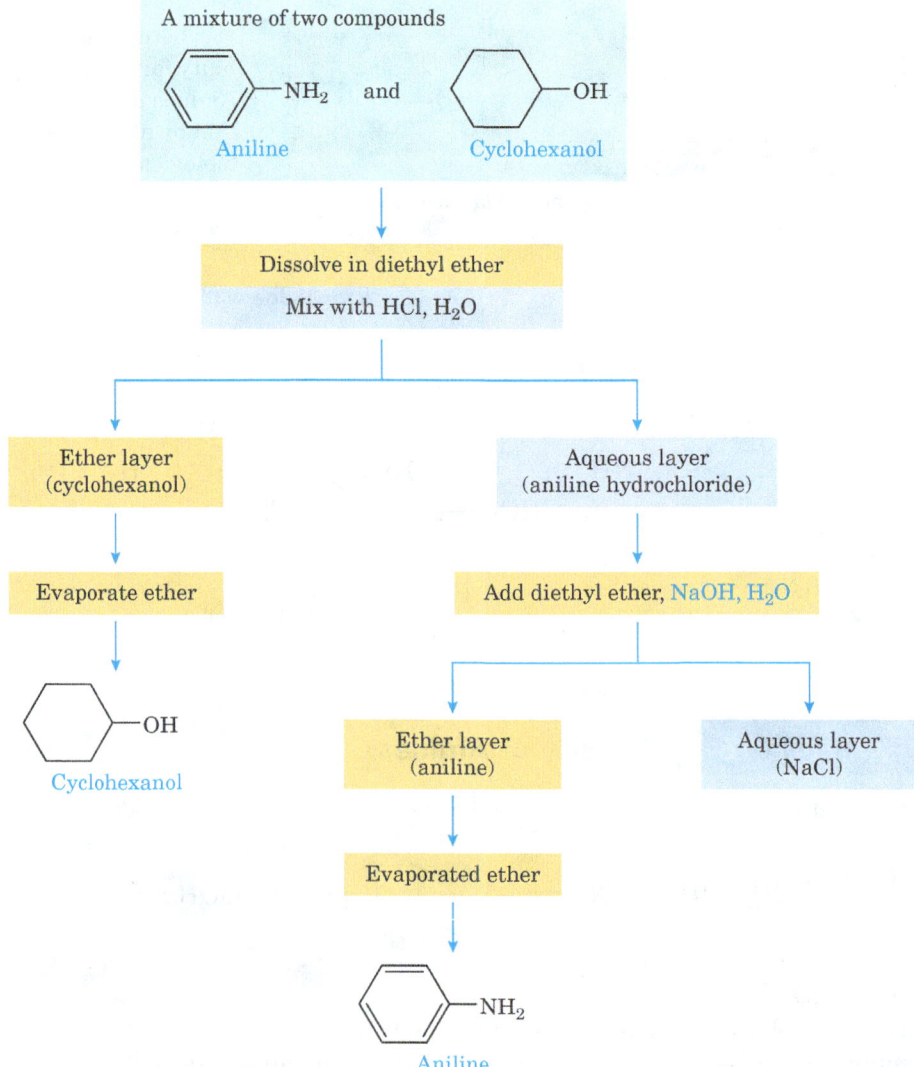

CHEMICAL CONNECTIONS 16E

Epinephrine: A Prototype for the Development of New Bronchodilators

Epinephrine was first isolated in pure form in 1897 and its structure determined in 1901. It occurs in the adrenal gland (hence the common name adrenalin) as a single enantiomer with the R configuration at its stereocenter. Epinephrine is commonly referred to as a catecholamine: the common name of 1,2-dihydroxybenzene is catechol (Section 13-4A), and amines containing a benzene ring with *ortho*-hydroxyl groups are called catecholamines.

Early on, it was recognized that epinephrine is a vasoconstrictor, a bronchodilator, and a cardiac stimulant. The fact that it has these three major effects stimulated much research, one line of which sought to develop compounds that are even more effective bronchodilators than epinephrine but, at the same time, are free from epinephrine's cardiac-stimulating and vasoconstricting effects.

Soon after epinephrine became commercially available, it emerged as an important treatment of asthma and hay fever. It has been marketed for the relief of bronchospasms under several trade names, including Bronkaid Mist and Primatene Mist.

enzyme-catalyzed reaction that converts one of the two —OH groups on the catechol unit to an —OCH$_3$ group. A strategy to circumvent this enzyme-catalyzed inactivation was to replace the catechol unit with one that would allow the drug to bind to the catecholamine receptors in the bronchi but would not be inactivated by this enzyme.

In terbutaline (Brethaire), inactivation is prevented by placing the —OH groups *meta* to each other on the aromatic ring. In addition, the isopropyl group of isoproterenol is replaced with a *tert*-butyl group. In albuterol (Proventil), the commercially most successful of the antiasthma medications, one —OH group of the catechol unit is replaced with a —CH$_2$OH group and the isopropyl group is replaced with a *tert*-butyl group. When terbutaline and albuterol were introduced into clinical medicine in the 1960s, they almost immediately replaced isoproterenol as the drugs of choice for the treatment of asthmatic attacks. The R enantiomer of albuterol is 68 times more effective in the treatment of asthma than the S enantiomer.

Epinephrine

Terbutaline

(R)-Isoproterenol

(R)-Albuterol

One of the most important of the first synthetic catecholamines was isoproterenol, the levorotatory enantiomer of which retains the bronchodilating effects of epinephrine but is free from its cardiac-stimulating effects. (R)-Isoproterenol was introduced into clinical medicine in 1951; for the next two decades, it was the drug of choice for the treatment of asthmatic attacks. Interestingly, the hydrochloride salt of (S)-isoproterenol is a nasal decongestant and was marketed under several trade names, including Sudafed.

A problem with the first synthetic catecholamines (and with epinephrine itself) is that they are inactivated by an

In their search for a longer-acting bronchodilator, scientists reasoned that extending the side chain on nitrogen might strengthen the binding of the drug to the adrenoreceptors in the lungs, thereby increasing the duration of the drug's action. This line of reasoning led to the synthesis and introduction of salmeterol (Serevent), a bronchodilator that is approximately ten times more potent than albuterol and remains active for much longer.

Salmeterol

Test your knowledge with Problem 16-44.

Summary of Key Questions

Section 16-1 What Are Amines?
- **Amines** are derivatives of ammonia in which one, two, or three hydrogens are replaced by alkyl or aryl groups.
- Amines are classified as **primary**, **secondary**, or **tertiary**, depending on the number of carbon atoms bonded to nitrogen.
- In an **aliphatic amine**, all carbon atoms bonded to nitrogen are derived from alkyl groups.
- In an **aromatic amine**, one or more of the groups bonded to nitrogen are aryl groups.
- In a **heterocyclic amine**, the nitrogen atom is part of a ring.

Section 16-2 How Do We Name Amines?
- In IUPAC nomenclature, aliphatic amines are named by changing the final -e of the parent alkane to -**amine** and using a number to locate the amino group on the parent chain.
- In the common system of nomenclature, aliphatic amines are named by listing the carbon groups bonded to nitrogen in alphabetical order as one word ending in the suffix -**amine**.

Section 16-3 What Are the Physical Properties of Amines?
- Amines are polar compounds. Primary and secondary amines associate by intermolecular hydrogen bonding.
- All classes of amines form hydrogen bonds with water and are more soluble in water than are hydrocarbons of comparable molecular weight.

Section 16-4 How Do We Describe the Basicity of Amines?
- Amines are weak bases, and aqueous solutions of amines are basic.
- The base ionization constant for an amine in water is denoted by the symbol K_b.
- Aliphatic amines are stronger bases than aromatic amines.

Section 16-5 What Are the Characteristic Reactions of Amines?
- All amines, whether soluble or insoluble in water, react with strong acids to form water-soluble salts.
- We can use this property to separate water-insoluble amines from water-insoluble nonbasic compounds.

Summary of Key Reactions

1. **Basicity of Aliphatic Amines (Section 16-4)** Most aliphatic amines have about the same basicity (pK_b 3.0 – 4.0) and are slightly stronger bases than ammonia (pK_b 4.74).

    $CH_3NH_2 + H_2O \rightleftharpoons CH_3NH_3^+ + OH^-$ $pK_b = 3.36$

2. **Basicity of Aromatic Amines (Section 16-4)** Most aromatic amines (pK_b 9.0 – 10.0) are considerably weaker bases than ammonia and aliphatic amines.

    $C_6H_5-NH_2 + H_2O \rightleftharpoons C_6H_5-NH_3^+ + OH^-$ $pK_b = 9.36$

3. **Reaction with Acids (Section 16-5)** All amines, whether water-soluble or water-insoluble, react quantitatively with strong acids to form water-soluble salts.

    PhN(CH$_3$)$_2$ + HCl ⟶ PhN$^+$H(CH$_3$)$_2$ Cl$^-$

    Insoluble in water   A water-soluble salt

Problems

Orange-numbered problems are applied.

Section 16-1 What Are Amines?

16-6 Answer true or false.
  (a) *tert*-Butylamine is a 3° amine.
  (b) In an aromatic amine, one or more of the groups bonded to nitrogen is an aromatic ring.
  (c) In a heterocyclic amine, the amine nitrogen is one of the atoms of a ring.
  (d) The Lewis structures of both NH_4^+ and CH_4 show the same number (eight) of valence electrons, and the VSEPR model predicts tetrahedral geometry for each.
  (e) There are four constitutional isomers with the molecular formula C_3H_9N.

16-7   We use the terms *primary, secondary*, and *tertiary* to classify both alcohols and amines. Cyclohexanol, for example, is classified as a secondary alcohol, and cyclohexanamine is classified as a primary amine. In each compound, the functional group is bonded to a carbon of the cyclohexane ring. Explain why one compound is classified as secondary, whereas the other is classified as primary.

16-8   What is the difference in structure between an aliphatic amine and an aromatic amine?

16-9   In what way are pyridine and pyrimidine related to benzene?

Section 16-2 How Do We Name Amines?

16-10   Answer true or false.
   (a) In the IUPAC system, primary aliphatic amines are named as alkanamines.
   (b) The IUPAC name of $CH_3CH_2CH_2CH_2CH_2NH_2$ is 1-pentylamine.
   (c) 2-Butanamine is chiral and shows enantiomerism.
   (d) *N,N*-Dimethylaniline is a 3° aromatic amine.

16-11   Draw a structural formula for each amine.
   (a) 2-Butanamine
   (b) 1-Octanamine
   (c) 2,2-Dimethyl-1-propanamine
   (d) 1,5-Pentanediamine
   (e) 2-Bromoaniline
   (f) Tributylamine

16-12   Draw a structural formula for each amine.
   (a) 4-Methyl-2-pentanamine
   (b) *trans*-2-Aminocyclohexanol
   (c) *N,N*-Dimethylaniline
   (d) Dicyclohexylamine
   (e) *sec*-Butylamine
   (f) 2,4-Dimethylaniline

16-13   Classify each amino group as primary, secondary, or tertiary and as aliphatic or aromatic.

   (a) Serotonin (a neurotransmitter)

   (b) Benzocaine (a topical anesthetic)

   (c) Diphenhydramine (the hydrochloride salt is the antihistamine Benadryl)

   (d) Lysine (an amino acid)

   (e) Chloroquine (an antimalaria drug)

   (f) 4-Aminobutanoic acid (a neurotransmitter)

16-14   There are eight constitutional isomers with the molecular formula $C_4H_{11}N$.
   (a) Name and draw a structural formula for each amine.
   (b) Classify each amine as primary, secondary, or tertiary.
   (c) Which of these amines are chiral?

16-15   There are eight primary amines with the molecular formula $C_5H_{13}N$.
   (a) Name and draw a structural formula for each amine.
   (b) Which of these amines are chiral?

Section 16-3 What Are the Physical Properties of Amines?

16-16   Answer true or false.
   (a) Hydrogen bonding between 2° amines is stronger than that between 2° alcohols.
   (b) Primary and secondary amines generally have higher boiling points than hydrocarbons with comparable carbon skeletons.
   (c) The boiling points of amines increase as the molecular weight of the amine increases.

16-17 Propylamine (bp 48°C), ethylmethylamine (bp 37°C), and trimethylamine (bp 3°C) are constitutional isomers with the molecular formula C_3H_9N. Account for the fact that trimethylamine has the lowest boiling point of the three and propylamine has the highest.

16-18 Account for the fact that 1-butanamine (bp 78°C) has a lower boiling point than 1-butanol (bp 117°C).

16-19 2-Methylpropane (bp −12°C), 2-propanol (bp 82°C), and 2-propanamine (bp 32°C) all have approximately the same molecular weight, yet their boiling points are quite different. Explain the reason for these differences.

16-20 Account for the fact that most low-molecular-weight amines are very soluble in water, whereas low-molecular-weight hydrocarbons are not.

Section 16-4 How Do We Describe the Basicity of Amines?

16-21 Answer true or false.
(a) Aqueous solutions of amines are basic.
(b) Aromatic amines such as aniline in general are weaker bases than aliphatic amines such as cyclohexanamine.
(c) Aliphatic amines are stronger bases than inorganic bases such as NaOH and KOH.
(d) Water-insoluble amines react with strong aqueous acids such as HCl to form water-soluble salts.
(e) If the pH of an aqueous solution of a 1° aliphatic amine, RNH_2, is adjusted to pH 2.0 by the addition of concentrated HCl, the amine will be present in solution almost entirely as its conjugate acid, RNH_3^+.
(f) If the pH of an aqueous solution of a 1° aliphatic amine, RNH_2, is adjusted to pH 10.0 by the addition of NaOH, the amine will be present in solution almost entirely as the free base, RNH_2.
(g) For a 1° aliphatic amine, the concentrations of RNH_3^+ and RNH_2 will be equal when the pH of the solution is equal to the pK_b of the amine.

16-22 Compare the base strengths of amines with those of alcohols.

16-23 Write a structural formula for each amine salt.
(a) Ethyltrimethylammonium hydroxide
(b) Dimethylammonium iodide
(c) Tetramethylammonium chloride
(d) Anilinium bromide

16-24 Name these amine salts.
(a) $CH_3CH_2NH_3^+Cl^-$
(b) $(CH_3CH_2)_2NH_2^+Cl^-$
(c) C₆H₅—$NH_3^+HSO_4^-$

16-25 From each pair of compounds, select the stronger base.

(a) piperidine or pyridine

(b) cyclohexyl—$N(CH_3)_2$ or phenyl—$N(CH_3)_2$

(c) phenyl—$NHCH_3$ or phenyl—CH_2NH_2

16-26 The pK_b of amphetamine is approximately 3.2.

Amphetamine

(a) Which form of amphetamine (the free base or its conjugate acid) would you expect to be present at pH 1.0, the pH of stomach acid?
(b) Which form of amphetamine would you expect to be present at pH 7.40, the pH of blood plasma?

16-27 Guanidine, pK_a 13.6, is a very strong base, almost as basic as hydroxide ion.

$$\underset{\text{Guanidine}}{H_2N-\overset{\overset{NH}{\|}}{C}-NH_2} + H_2O \rightleftharpoons \underset{\text{Guanidinium ion}}{H_2N-\overset{\overset{NH_2^+}{\|}}{C}-NH_2} + OH^-$$

$$pK_b = 13.6$$

(a) Complete the Lewis structure for guanidine, showing all valence electrons.
(b) The remarkable basicity of guanidine is attributed to the fact that the positive charge on the guanidinium ion is delocalized by resonance over the three nitrogen atoms. This delocalization increases the stability of the guanidinium ion relative to the ammonium ion or substituted ammonium ions.
Draw three equivalent contributing structures for the guanidinium ion and show by the use of curved arrows how these three contributors are related.
(c) Propose an explanation for the fact that protonation occurs on the C=NH nitrogen rather than on one of the —NH_2 nitrogens. (*Hint*: Consider the resonance stabilization of the structure formed by protonation on a —NH_2 nitrogen compared with the resonance stabilization of the structure formed by protonation on the =NH nitrogen.)
(d) Predict the N—C—N bond angles in the hybrid.
(e) Which is the stronger acid, the ammonium ion or the guanidinium ion?

16-28 Following is the structural formula of metformin, the hydrochloride salt of which is marketed as the antidiabetic medication Glucophage. Metformin was introduced into clinical practice in the United States in 1995 for the treatment of type 2 diabetes. More than 25 million prescriptions for this drug were written in 2000, making it the most commonly prescribed brand-name diabetes medication in the nation.

Metformin

(a) Complete the Lewis structure for metformin, showing all valence electrons.
(b) Which nitrogen is the most likely site of protonation?
(c) Draw the structural formula of Glucophage.

Section 16-5 What Are the Characteristic Reactions of Amines?

16-29 Suppose you have two test tubes, one containing 2-methylcyclohexanol and the other containing 2-methylcyclohexanamine (both of which are insoluble in water) and that you do not know which test tube contains which compound. Describe a simple chemical test by which you could tell which compound is the alcohol and which is the amine.

16-30 Complete the equations for the following acid–base reactions.

(a) CH_3COH (Acetic acid) + Pyridine →

(b) 1-Phenyl-2-propanamine (Amphetamine) + HCl →

(c) Methamphetamine + H_2SO_4 →

16-31 Pyridoxamine is one form of vitamin B_6.

Pyridoxamine (Vitamin B_6)

(a) Which nitrogen atom of pyridoxamine is the stronger base?
(b) Draw a structural formula for the salt formed when pyridoxamine is treated with one mole of HCl.

16-32 Many tumors of the breast are correlated with estrogen levels in the body. Drugs that interfere with estrogen binding have antitumor activity and may even help prevent tumor occurrence. A widely used antiestrogen drug is tamoxifen.

Tamoxifen

(a) Name the functional groups in tamoxifen.
(b) Classify the amino group in tamoxifen as primary, secondary, or tertiary.
(c) How many stereoisomers are possible for tamoxifen?
(d) Would you expect tamoxifen to be soluble or insoluble in water? In blood?

Chemical Connections

16-33 (Chemical Connections 16A) What are the differences in structure between the natural hormone epinephrine (Chemical Connections 16E) and the synthetic pep pill amphetamine? Between amphetamine and methamphetamine?

16-34 (Chemical Connections 16A) What are the possible negative effects of illegal use of amphetamines such as methamphetamine?

16-35 (Chemical Connections 16B) What is an alkaloid? Are all alkaloids basic to litmus?

16-36 (Chemical Connections 16B) Identify all stereocenters in coniine and nicotine. How many stereoisomers are possible for each?

16-37 (Chemical Connections 16B) Which of the two nitrogen atoms in nicotine is converted to its salt by reaction with one mole of HCl? Draw a structural formula for this salt.

16-38 (Chemical Connections 16B) Cocaine has four stereocenters. Identify each. Draw a structural formula for the salt formed by treatment of cocaine with one mole of HCl.

16-39 (Chemical Connections 16C) What structural feature is common to all benzodiazepines?

16-40 (Chemical Connections 16C) Is Librium chiral? Is Valium chiral?

16-41 (Chemical Connections 16C) Benzodiazepines affect neural pathways in the central nervous system that are mediated by GABA, whose IUPAC name is 4-aminobutanoic acid. Draw a structural formula for GABA.

16-42 (Chemical Connections 16D) Suppose you saw this label on a decongestant: phenylephrine · HCl. Should you worry about being exposed to a strong acid such as HCl? Explain.

16-43 (Chemical Connections 16D) Give two reasons why amine-containing drugs are most commonly administered as their salts.

16-44 (Chemical Connections 16E) Classify each amino group in epinephrine and albuterol as primary, secondary, or tertiary. In addition, list the similarities and differences between the structural formulas of these two compounds.

Additional Problems

16-45 Draw a structural formula for a compound with the given molecular formula that is:
(a) A 2° aromatic amine, C_7H_9N
(b) A 3° aromatic amine, $C_8H_{11}N$
(c) A 1° aliphatic amine, C_7H_9N
(d) A chiral 1° amine, $C_4H_{11}N$
(e) A 3° heterocyclic amine, $C_5H_{11}N$
(f) A trisubstituted 1° aromatic amine, $C_9H_{13}N$
(g) A chiral quaternary ammonium salt, $C_9H_{22}NCl$

16-46 Arrange these three compounds in order of decreasing ability to form intermolecular hydrogen bonds: CH_3OH, CH_3SH, and $(CH_3)_2NH$.

16-47 Consider these three compounds: CH_3OH, CH_3SH, and $(CH_3)_2NH$.
(a) Which is the strongest acid?
(b) Which is the strongest base?
(c) Which has the highest boiling point?
(d) Which forms the strongest intermolecular hydrogen bonds in the pure state?

16-48 Arrange these compounds in order of increasing boiling point: $CH_3CH_2CH_2CH_3$, $CH_3CH_2CH_2OH$, and $CH_3CH_2CH_2NH_2$. Boiling point values from lowest to highest are $-0.5°C$, $7.2°C$, and $77.8°C$.

16-49 Account for the fact that amines have about the same solubility in water as alcohols of similar molecular weight.

16-50 If you dissolve $CH_3CH_2CH_2OH$ and $CH_3CH_2CH_2NH_2$ in the same container of water and lower the pH of the solution to 2.0 by adding HCl, would anything happen to the structures of these compounds? Write the formula of the species present in solution at pH 2.0.

16-51 The compound phenylpropanolamine hydrochloride is used as both a decongestant and an anorexic. The IUPAC name of this compound is 1-phenyl-2-amino-1-propanol.
(a) Draw a structural formula for 1-phenyl-2-amino-1-propanol.
(b) How many stereocenters are present in this molecule? How many stereoisomers are possible for it?

16-52 Procaine was one of the first local anesthetics. Its hydrochloride salt is marketed as Novocain.

(a) Is procaine chiral? Does it contain a stereocenter?
(b) Which nitrogen atom of procaine is the stronger base?
(c) Draw a structural formula for the salt formed by treating procaine with one mole of HCl, showing which nitrogen is protonated and bears the positive charge.

16-53 Following are two structural formulas for gabapentin, a drug used to treat epilepsy. Is gabapentin better represented by structure (A) or (B)? Explain.

16-54 Several poisonous plants, including *Atropa belladonna*, contain the alkaloid atropine. The name "belladonna" (which means "beautiful lady") probably comes from the fact that Roman women used extracts from this plant to make themselves more attractive. Atropine is widely used by ophthalmologists and optometrists to dilate the pupils for eye examination.

(a) Classify the amino group in atropine as primary, secondary, or tertiary.
(b) Locate all stereocenters in atropine.
(c) Account for the fact that atropine is almost insoluble in water (1 g in 455 mL of cold water) but atropine hydrogen sulfate is very soluble (1 g in 5 mL of cold water).
(d) Account for the fact that a dilute aqueous solution of atropine is basic (pH approximately 10.0).

16-55 Epibatadine, a colorless oil isolated from the skin of the Equadorian poison arrow frog *Epipedobates tricolor*, has several times the analgesic potency of morphine. It is the first chlorine-containing, non-opioid (nonmorphine-like in structure) analgesic ever isolated from a natural source.
(a) Which of the two nitrogen atoms in epibatadine is the stronger base?
(b) Mark the three stereocenters in this molecule.

Epibatadine

16-56 Following are two structural formulas for 4-aminobutanoic acid, a neurotransmitter. Is this compound better represented by structural formula (A) or (B)? Explain.

$$H_2N\text{-}CH_2CH_2CH_2\text{-}COOH \quad \text{or} \quad H_3\overset{+}{N}\text{-}CH_2CH_2CH_2\text{-}COO^-$$

(A)   (B)

16-57 Alanine, $C_3H_7O_2N$, is one of the 20 amino acid building blocks of proteins (Chapter 22). Alanine contains a primary amino group (—NH_2) and a carboxyl group (—COOH) and has one stereocenter. Given this information, draw a structural formula for alanine.

Challenge Problems

16-58 Following is a structural formula of desosamine, a sugar component of several macrolide antibiotics, including the erythromycins. The configuration shown here is that of the natural product. Erythromycin is produced by a strain of *Streptomyces erythreus* originally found in a soil sample from the Philippine Archipelago.

Desosamine

(a) Name all the functional groups in desosamine. (Chapter 10)
(b) How many stereocenters are present in desosamine? How many stereoisomers are possible for it? How many pairs of enantiomers are possible for it?
(c) Draw the alternative chair conformations for desosamine and label which groups are equatorial and which are axial.
(d) Which of the alternative chair conformations for desosamine is more stable?

16-59 Morphine and its O-methylated derivative, codeine, are among the most effective painkillers known. However, they possess two serious drawbacks: they are addictive, and repeated use induces a tolerance to the drug.

R = H; Morphine
R = CH_3; Codeine

Many morphine analogs have been prepared in an effort to find drugs that are equally effective as painkillers but that have less risk of physical dependence and potential abuse. Following are three of these.

Meperidine (Demerol)

Methadone

Propoxyphene (Darvon)

(a) List the structural features common to each of these molecules.
(b) The Beckett-Casey rules are a set of empirical rules to predict the structure of molecules that will bind to morphine receptors and act as analgesics. According to these rules, to provide an effective morphine-like analgesic, a molecule must have (1) an aromatic ring bonded to (2) a quaternary carbon and (3) a nitrogen at a distance equal to two carbon–carbon single bond lengths from the quaternary center. Show that these structural requirements are present in the four molecules shown in this problem.

Aldehydes and Ketones

17

Benzaldehyde is found in the kernels of bitter almonds, and cinnamaldehyde is found in Ceylonese and Chinese cinnamon oils.

KEY QUESTIONS

- 17-1 What Are Aldehydes and Ketones?
- 17-2 How Do We Name Aldehydes and Ketones?
- 17-3 What Are the Physical Properties of Aldehydes and Ketones?
- 17-4 What Are the Characteristic Reactions of Aldehydes and Ketones?
- 17-5 What Is Keto-Enol Tautomerism?

17-1 What Are Aldehydes and Ketones?

In this and the three following chapters, we study the physical and chemical properties of compounds containing the **carbonyl group**, C=O. Because the carbonyl group is present in aldehydes, ketones, and carboxylic acids and their derivatives as well as in carbohydrates, it is one of the most important functional groups in organic chemistry. Its chemical properties are straightforward, and an understanding of its characteristic reaction patterns leads very quickly to an understanding of a wide variety of organic and biochemical reactions.

The functional group of an **aldehyde** is a carbonyl group bonded to a hydrogen atom (Section 10-4C). In methanal (formaldehyde), the simplest aldehyde, the carbonyl group is bonded to two hydrogen atoms. In other aldehydes, it is bonded to one hydrogen atom and one carbon group. The functional group of a **ketone** is a carbonyl group bonded to two carbon groups (Section 10-4C). Acetone is the simplest ketone.

Methanal (Formaldehyde) — HCH with =O
Ethanal (Acetaldehyde) — CH₃CH with =O
Propanone (Acetone) — CH₃CCH₃ with =O

Because aldehydes always contain at least one hydrogen bonded to the C=O group, they are often written RCH=O or RCHO. Similarly, ketones are often written RCOR'.

17-2 How Do We Name Aldehydes and Ketones?

A. IUPAC Names

The IUPAC names for aldehydes and ketones follow the familiar pattern of selecting as the parent alkane the longest chain of carbon atoms that contains the functional group (Section 11-4A). To name an aldehyde, we change the suffix *-e* of the parent alkane to *-al*. Because the carbonyl group of an aldehyde can appear only at the end of a parent chain and numbering must start with it as carbon 1, there is no need to use a number to locate the aldehyde group.

For **unsaturated aldehydes**, we show the presence of the carbon–carbon double bond and the aldehyde by changing the ending of the parent alkane from *-ane* to *-enal*: "-en-" to show the carbon–carbon double bond and "-al" to show the aldehyde. We show the location of the carbon–carbon double bond by the number of its first carbon.

Hexanal   3-Methylbutanal   2-Propenal (Acrolein)

In the IUPAC system, we name ketones by selecting as the parent alkane the longest chain that contains the carbonyl group and then indicating the presence of this group by changing the *-e* of the parent alkane to *-one*. The parent chain is numbered from the direction that gives the smaller number to the carbonyl carbon. While the systematic name of the simplest ketone is 2-propanone, the IUPAC retains its common name, acetone.

Acetone   5-Methyl-3-hexanone   2-Methylcyclohexanone

Interactive Example 17-1   IUPAC Names for Aldehydes and Ketones

Write the IUPAC name for each compound:

Strategy and Solution

(a) The longest chain has six carbons, but the longest chain that contains the carbonyl carbon has only five carbons. Its IUPAC name is 2-ethyl-3-methylpentanal.

2-Ethyl-3-methylpentanal

(b) Number the six-membered ring beginning with the carbonyl group. Its IUPAC name is 3,3-dimethylcyclohexanone.

(c) This molecule is derived from benzaldehyde. Its IUPAC name is 2-ethylbenzaldehyde.

Problem 17-1
Write the IUPAC name for each compound.

(a) (b) (c)

EXAMPLE 17-2 Structural Formulas for Ketones

Draw structural formulas for all ketones with the molecular formula $C_6H_{12}O$ and write the IUPAC name of each. Which of these ketones are chiral?

Strategy and Solution

There are six ketones with this molecular formula: two with a six-carbon chain, three with a five-carbon chain and a methyl branch, and one with a four-carbon chain and two methyl branches. Only 3-methyl-2-pentanone has a stereocenter and is chiral.

2-Hexanone   3-Hexanone   4-Methyl-2-pentanone

3-Methyl-2-pentanone   2-Methyl-3-pentanone   3,3-Dimethyl-2-butanone

Problem 17-2
Draw structural formulas for all aldehydes with the molecular formula $C_6H_{12}O$ and write the IUPAC name of each. Which of these aldehydes are chiral?

In naming aldehydes or ketones that also contain an —OH or —NH₂ group elsewhere in the molecule, the parent chain is numbered to give the carbonyl group the lower number. An —OH substituent is indicated by *hydroxy*, and an —NH₂ substituent is indicated by *amino-*. Hydroxy and amino substituents are numbered and alphabetized along with any other substituents that might be present.

EXAMPLE 17-3 Naming Difunctional Aldehydes and Ketones

Write the IUPAC name for each compound.

Strategy and Solution

(a) We number the parent chain beginning with CHO as carbon 1. There is a hydroxyl group on carbon 3 and a methyl group on carbon 4. The IUPAC name of this compound is 3-hydroxy-4-methylpentanal. Note that this hydroxyaldehyde is chiral and can exist as a pair of enantiomers.

(b) The longest chain that contains the carbonyl is six carbons; the carbonyl group is on carbon-2, and the amino group is on carbon 3. The IUPAC name of this compound is 3-amino-4-ethyl-2-hexanone. Note that this ketoamine is also chiral and can exist as a pair of enantiomers.

Problem 17-3

Write the IUPAC name for each compound.

B. Common Names

We derive the common name for an aldehyde from the common or IUPAC names of the corresponding carboxylic acid. The word "acid" is dropped and the suffix *-ic* or *-oic* is changed to *-aldehyde*. Because we have not yet studied names for carboxylic acids, we are not in a position to discuss common names for aldehydes. We can, however, illustrate how they are derived by reference to two common names with which you are familiar. The name formaldehyde is derived from formic acid, and the name acetaldehyde is derived from acetic acid.

| HCH | HCOH | CH₃CH | CH₃COH |
| Formaldehyde | Formic acid | Acetaldehyde | Acetic acid |

We derive common names for ketones by naming each alkyl or aryl group bonded to the carbonyl group as a separate word, followed by the word

CHEMICAL CONNECTIONS 17A

From Moldy Clover to a Blood Thinner

In 1933, a disgruntled farmer delivered a pail of unclotted blood to the laboratory of Dr. Karl Link at the University of Wisconsin, along with tales of cows bleeding to death from minor cuts. Over the next couple of years, Link and his collaborators discovered that when cows are fed moldy clover, their blood clotting is inhibited and they can bleed to death from minor cuts and scratches. From the moldy clover, they isolated the anticoagulant dicoumarol, a substance that delays or prevents blood clotting. Dicoumarol exerts its anticoagulation effect by interfering with vitamin K activity. Within a few years after its discovery, dicoumarol became widely used to treat victims of heart attack and others at risk for developing blood clots.

Dicoumarol is a derivative of coumarin, the compound that gives sweet clover its pleasant smell. Coumarin, which does not interfere with blood clotting, is converted to dicoumarol as sweet clover becomes moldy.

In a search for even more potent anticoagulants, Link developed warfarin (named for the Wisconsin Alumni Research Foundation), now used primarily as a rat poison. When rats consume it, their blood fails to clot and they bleed to death. Warfarin is also used as a blood anticoagulant in humans. The S enantiomer shown here is more active than the R enantiomer. The commercial product is sold as a racemic mixture.

Coumarin (from sweet clover) → as sweet clover becomes moldy → Dicoumarol (an anticoagulant)

(S)-Warfarin (a synthetic anticoagulant)

Test your knowledge with Problem 17-46.

"ketone." The alkyl or aryl groups are generally listed in order of increasing molecular weight. 2-Butanone, more commonly called methyl ethyl ketone (MEK), is used as a solvent for paints and varnishes.

Ethyl isopropyl ketone   Methyl ethyl ketone   Dicyclohexyl ketone

17-3 What Are the Physical Properties of Aldehydes and Ketones?

Oxygen is more electronegative than carbon (3.5 compared with 2.5; see Table 3-5). Therefore, a carbon–oxygen double bond is polar, with oxygen bearing a partial negative charge and carbon bearing a partial positive charge (Figure 17-1).

In liquid aldehydes and ketones, intermolecular attractions occur between the partial positive charge on the carbonyl carbon of one molecule and the partial negative charge on the carbonyl oxygen of another. There is no possibility for hydrogen bonding between aldehyde or ketone molecules, which explains why these compounds have lower boiling points than alcohols (Section 14-1C) and carboxylic acids (Section 18-3), compounds in which hydrogen bonding between molecules does occur.

Polarity of a carbonyl group

FIGURE 17-1 The polarity of a carbonyl group. The carbonyl oxygen bears a partial negative charge and the carbonyl carbon bears a partial positive charge.

Table 17-1 Boiling Points of Six Compounds of Comparable Molecular Weight

Name	Structural Formula	Molecular Weight	Boiling Point (°C)
diethyl ether	$CH_3CH_2OCH_2CH_3$	74	34
pentane	$CH_3CH_2CH_2CH_2CH_3$	72	36
butanal	$CH_3CH_2CH_2CHO$	72	76
2-butanone	$CH_3CH_2COCH_3$	72	80
1-butanol	$CH_3CH_2CH_2CH_2OH$	74	117
propanoic acid	CH_3CH_2COOH	74	141

Table 17-1 lists structural formulas and boiling points of six compounds of similar molecular weight. Of the six, pentane and diethyl ether have the lowest boiling points. The boiling point of 1-butanol, which can associate by intermolecular hydrogen bonding, is higher than that of either butanal or 2-butanone. Propanoic acid, in which intermolecular association by hydrogen bonding is the strongest, has the highest boiling point.

Because the oxygen atom of each carbonyl group is a hydrogen bond acceptor, the low-molecular-weight aldehydes and ketones are more soluble in water than are nonpolar compounds of comparable molecular weight. Formaldehyde, acetaldehyde, and acetone are infinitely soluble in water. As the hydrocarbon portion of the molecule increases in size, aldehydes and ketones become less soluble in water.

Most aldehydes and ketones have strong odors. The odors of ketones are generally pleasant, and many are used in perfumes and as flavoring agents. The odors of aldehydes vary. You may be familiar with the smell of formaldehyde; if so, you know that it is not pleasant. Many higher aldehydes, however, have pleasant odors and are used in perfumes.

17-4 What Are the Characteristic Reactions of Aldehydes and Ketones?

A. Oxidation

Aldehydes are oxidized to carboxylic acids by a variety of oxidizing agents, including potassium dichromate (Section 14-2C).

Hexanal $\xrightarrow{K_2Cr_2O_7 / H_2SO_4}$ Hexanoic acid

Aldehydes are also oxidized to carboxylic acids by the oxygen in the air. In fact, aldehydes that are liquid at room temperature are so sensitive to oxidation that they must be protected from contact with air during storage. Often this is done by sealing the aldehyde in a container under an atmosphere of nitrogen.

$$\underset{\text{Benzaldehyde}}{\text{C}_6\text{H}_5\text{-CHO}} + \text{O}_2 \longrightarrow \underset{\text{Benzoic acid}}{\text{C}_6\text{H}_5\text{-COOH}}$$

Ketones, in contrast, resist oxidation by most oxidizing agents, including potassium dichromate and molecular oxygen.

The fact that aldehydes are so easy to oxidize and ketones are not allows us to use simple chemical tests to distinguish between these types of compounds. Suppose that we have a compound we know is either an aldehyde or a ketone. To determine which it is, we can treat the compound with a mild oxidizing agent. If it can be oxidized, it is an aldehyde; otherwise, it is a ketone. One reagent that has been used for this purpose is Tollens' reagent.

Tollens' reagent is prepared from silver nitrate and ammonia in water. When these two compounds are mixed, silver ion combines with NH_3 to form the complex ion $Ag(NH_3)_2^+$. When this solution is added to an aldehyde, the aldehyde acts as a reducing agent and reduces the complexed silver ion to silver metal. If this reaction is carried out properly, the silver metal precipitates as a smooth, mirror-like deposit on the inner surface of the reaction vessel, leading to the name **silver-mirror test**. ◄ If the remaining solution is then acidified with HCl, the carboxylic anion $RCOO^-$ formed during the aldehyde's oxidation is converted to the carboxylic acid $RCOOH$.

$$\underset{\substack{\text{Aldehyde} \\ }}{\text{R-CHO}} + \underset{\text{Tollens' reagent}}{2Ag(NH_3)_2^+} + 3OH^- \longrightarrow \underset{\text{Carboxylate anion}}{\text{R-COO}^-} + \underset{\text{Silver mirror}}{2Ag} + 4NH_3 + 2H_2O$$

Today, silver(I) is rarely used for the oxidation of aldehydes because of its high cost and because of the availability of other more convenient methods for this oxidation. This reaction, however, is still used for making (silvering) mirrors.

► A silver mirror has been deposited on the inside of this flask by the reaction between an aldehyde and Tollens' reagent.

EXAMPLE 17-4 Oxidation of Aldehydes and Ketones

Draw a structural formula for the product formed by treating each compound with Tollens' reagent followed by acidification with aqueous HCl.

(a) Pentanal
(b) 4-Hydroxybenzaldehyde

Strategy and Solution

The aldehyde group in each compound is oxidized to a carboxylate anion, $R-COO^-$. Acidification with HCl converts the carboxylate anion to a carboxylic acid, $R-CO_2H$.

(a) Pentanoic acid: $CH_3CH_2CH_2CH_2COOH$

(b) 4-Hydroxybenzoic acid: $HO-C_6H_4-COOH$

Problem 17-4

Complete equations for these oxidations.
(a) Hexanedial + $O_2 \longrightarrow$
(b) 3-Phenylpropanal + $Ag(NH_3)_2^+ \longrightarrow$

B. Reduction

In Section 12-5A, we saw that the C=C double bond of an alkene is reduced by hydrogen in the presence of a transition metal catalyst to a C—C single bond. The same is true of the C=O double bond of an aldehyde or ketone. Aldehydes are reduced to primary alcohols and ketones are reduced to secondary alcohols.

Pentanal + H$_2$ → (Transition metal catalyst) → 1-Pentanol

Cyclopentanone + H$_2$ → (Transition metal catalyst) → Cyclopentanol

The reduction of a C=O double bond under these conditions is slower than the reduction of a C=C double bond. Thus, if the same molecule contains both C=O and C=C double bonds, the C=C double bond is reduced selectively.

The reagent most commonly used in the laboratory for the reduction of an aldehyde or a ketone is sodium borohydride, NaBH$_4$. This reagent behaves as a source of hydride ions, H:$^-$. In the hydride ion, hydrogen has two valence electrons and bears a negative charge. In a reduction by sodium borohydride, hydride ion is attracted to and then adds to the partially positive carbonyl carbon, which leaves a negative charge on the carbonyl oxygen. Reaction of this alkoxide intermediate with aqueous acid gives the alcohol. Note that this reaction fits a pattern we have seen already in discussing reaction mechanisms, namely the reaction of an electrophile and a nucleophile to form a new covalent bond.

H:$^-$ + C=Ö: ⟶ H—C—Ö:$^-$ →(H$_3$O$^+$)→ H—C—Ö—H

Hydride ion — Alkoxide ion

Of the two hydrogens added to the carbonyl group in this reduction, one comes from the reducing agent and the other comes from aqueous acid. Reduction of cyclohexanone, for example, with this reagent gives cyclohexanol:

cyclohexanone →(NaBH$_4$)→ alkoxide intermediate →(H$_3$O$^+$)→ cyclohexanol

An advantage of using NaBH$_4$ over the H$_2$/metal reduction is that NaBH$_4$ does not reduce carbon–carbon double bonds. The reason for this selectivity is quite straightforward. There is no polarity (no partial positive or negative charges) on a carbon–carbon double bond. Therefore, a C=C double bond has no partially positive site to attract the negatively charged hydride ion. In the following example, NaBH$_4$ selectively reduces the aldehyde to a primary alcohol:

Cinnamaldehyde →(1. NaBH$_4$; 2. H$_2$O)→ Cinnamyl alcohol

In biological systems, the agent for the reduction of aldehydes and ketones is the reduced form of the coenzyme nicotinamide adenine dinucleotide, abbreviated NADH (Section 27-3). This reducing agent, like $NaBH_4$, delivers a hydride ion to the carbonyl carbon of the aldehyde or ketone. Reduction of pyruvate, for example, by NADH gives lactate:

$$CH_3-\underset{Pyruvate}{\overset{O}{\underset{\|}{C}}}-COO^- \xrightarrow{NADH} CH_3-\underset{H}{\overset{O^-}{\underset{|}{C}}}-COO^- \xrightarrow{H_3O^+} CH_3-\underset{H}{\overset{O-H}{\underset{|}{C}}}-COO^-$$
Lactate

Pyruvate is the end product of glycolysis, a series of enzyme-catalyzed reactions that converts glucose to two molecules of this ketoacid (Section 28-1). Under anaerobic conditions, NADH reduces pyruvate to lactate. The build-up of lactate in the bloodstream leads to acidosis, and in muscle tissue, it is associated with muscle fatigue. When blood lactate reaches a concentration of about 0.4 mg/100 mL, muscle tissue becomes almost completely exhausted.

C. Addition of Alcohols and Formation of Hemiacetals and Acetals

Addition of a molecule of alcohol to the carbonyl group of an aldehyde or a ketone forms a **hemiacetal** (a half-acetal). The functional group of a hemiacetal is a carbon bonded to one —OH group and one —OR group. In

Hemiacetal A molecule containing a carbon bonded to one —OH group and one —OR group; the product of adding one molecule of alcohol to the carbonyl group of an aldehyde or ketone forming a hemiacetal, the H of the alcohol adds to the carbonyl oxygen and the OR group of the alcohol adds to the carbonyl carbon. Shown here are the hemiacetals formed by addition of one molecule of ethanol to benzaldehyde and to cyclohexanone:

Hemiacetals are generally unstable and are only minor components of an equilibrium mixture, except in one very important type of molecule. When a hydroxyl group is part of the same molecule that contains the carbonyl group and a five- or six-membered ring can form, the compound exists almost entirely in a cyclic hemiacetal form. In this case, the —OH group adds to the C=O group of the same molecule. We will have much more to say about cyclic hemiacetals when we consider the chemistry of carbohydrates in Chapter 20.

Acetals Molecules containing two —OR groups bonded to the same carbon

Hemiacetals can react further with alcohols to form **acetals** plus water. This reaction is acid-catalyzed. The functional group of an acetal is a carbon bonded to two —OR groups.

Mechanism for Acid-Catalyzed Formation of an Acetal

Step 1: Add a proton to the carbonyl oxygen, giving a resonance-stabilized cation.

The result of this step is to form a resonance-stabilized cation that renders the carbonyl carbon a stronger electrophile and makes it more susceptible to attack by a nucleophile.

[Step 1 reaction scheme showing protonation of acetone]

Step 2: Reaction of an electrophile and a nucleophile to form a new covalent bond.
This step adds the first of the —OCH$_3$ groups required for formation of the acetal and results in a tetrahedral carbonyl addition intermediate.

[Structure: protonated acetone + HÖ-CH$_3$ ⇌ tetrahedral intermediate with OH and O(+)(H)-CH$_3$]

Step 3: Proton transfer from one oxygen to another oxygen.
To make the transfer clearer, the transferred proton is shown in color.

[Structure showing proton transfer between the two oxygens]

Step 4: Break a bond to form stable molecules or ions.
In this case, the stable molecules and ions are a water molecule and a 3° carbocation.

[Structure: protonated intermediate ⇌ carbocation with O-CH$_3$ + H$_2$Ö:]

Step 5: Reaction of an electrophile and a nucleophile to form a new covalent bond.
The result of this step is to add the second —OCH$_3$ group to what was the carbonyl carbon.

[Structure: carbocation + HÖ-CH$_3$ ⇌ protonated acetal]

Step 6: Take a proton away.
Transfer of a proton to solvent gives the acetal and regenerates the acid catalyst.

[Structure: protonated acetal ⇌ acetal + H$^+$]

All steps in hemiacetal and acetal formation are reversible. As with any other equilibrium, we can make this one go in either direction by using Le Chatelier's principle (Section 7-7). If we want to drive it to the right (formation of the acetal), we either use a large excess of alcohol or remove water from the equilibrium mixture. If we want to drive it to the left (hydrolysis of the acetal to the original aldehyde or ketone and alcohol), we use a large excess of water.

EXAMPLE 17-6 Formation of Hemiacetals and Acetals

Show the reaction of 2-butanone with one molecule of ethanol to form a hemiacetal and then with a second molecule of ethanol to form an acetal.

Strategy and Solution

Given are structural formulas for the hemiacetal and then the acetal.

2-Butanone ⇌ A hemiacetal ⇌ An acetal + H₂O

Problem 17-6

Show the reaction of benzaldehyde with one molecule of methanol to form a hemiacetal and then with a second molecule of methanol to form an acetal.

EXAMPLE 17-7 Recognizing the Presence of a Hemiacetal and an Acetal

Identify all hemiacetals and acetals in the following structures and tell whether each is formed from an aldehyde or a ketone.

(a) $CH_3CH_2C(OCH_3)_2CH_2CH_3$ (b) $CH_3CH_2OCH_2CH_2OH$ (c) [cyclic structure with OH and O]

Strategy

An acetal contains a carbon atom bonded to two —OR groups; a hemiacetal contains a carbon atom bonded to one —OH group and one —OR group.

Solution

Compound (a) is an acetal derived from a ketone. Compound (b) is neither a hemiacetal nor an acetal because it does not have a carbon bonded to two oxygens; its functional groups are an ether and a primary alcohol. Compound (c) is a cyclic hemiacetal derived from an aldehyde.

(a) $CH_3CH_2C(OCH_3)_2CH_2CH_3 + H_2O \xrightarrow{H^+} CH_3CH_2COCH_2CH_3 + 2CH_3OH$

An acetal → 2-Pentanone

(c) A hemiacetal $\xrightarrow[H_2O]{H^+}$ [open chain with OH and CHO] Redraw the carbon chain → 5-Hydroxypentanal

Problem 17-7

Identify all hemiacetals and acetals in the following structures and tell whether each is formed from an aldehyde or a ketone.

(a) CH₃CH₂C(OH)(OCH₂CH₃)CH₂CH₃   (b) CH₃OCH₂CH₂OCH₃   (c) [tetrahydropyran with OCH₃]

17-5 What Is Keto-Enol Tautomerism?

A carbon atom adjacent to a carbonyl group is called an α-carbon, and a hydrogen atom bonded to it is called an α-hydrogen.

CH₃—C(=O)—CH₂—CH₃

(α-hydrogens on the α-carbons)

A carbonyl compound that has a hydrogen on an α-carbon is in equilibrium with a constitutional isomer called an **enol**. The name "enol" is derived from the IUPAC designation of it as both an alkene (*-en-*) and an alcohol (*-ol*).

Enol A molecule containing an —OH group bonded to a carbon of a carbon–carbon double bond

CH₃—C(=O)—CH₃ ⇌ CH₃—C(OH)=CH₂

Acetone (keto form) ⇌ Acetone (enol form)

Keto and enol forms are examples of **tautomers**, constitutional isomers in equilibrium with each other that differ in the location of a hydrogen atom and a double bond. This type of isomerism is called **keto-enol tautomerism**. For any pair of keto-enol tautomers, the keto form generally predominates at equilibrium.

Tautomers Constitutional isomers that differ in the location of a hydrogen atom and a double bond

EXAMPLE 17-8 Keto-Enol Tautomerism

Draw structural formulas for the two enol forms for each ketone.

(a) [2-methylcyclohexanone]   (b) [2-methyl-3-pentanone type structure]

Strategy and Solution

Any aldehyde or ketone with one hydrogen on its α-carbon can show keto-enol tautomerism.

(a) [structure: 2-methylcyclohexanone enol ⇌ 2-methylcyclohex-1-en-1-ol tautomer]

(b) [structure: hex-2-en-2-ol ⇌ hex-1-en-2-ol] Cis & trans isomers

Problem 17-8

Draw a structural formula for the keto form of each enol.

(a) cyclohexenol   (b) cyclohexene-1,2-diol   (c) cyclohexane=CHOH

Summary of Key Questions

Section 17-1 What Are Aldehydes and Ketones?
- An **aldehyde** contains a carbonyl group bonded to at least one hydrogen atom.
- A **ketone** contains a carbonyl group bonded to two carbon groups.

Section 17-2 How Do We Name Aldehydes and Ketones?
- We derive the IUPAC name for an aldehyde by changing the *-e* of the parent alkane to *-al*.
- We derive the IUPAC name for a ketone by changing the *-e* of the parent alkane to *-one* and using a number to locate the carbonyl carbon.

Section 17-3 What Are the Physical Properties of Aldehydes and Ketones?
- Aldehydes and ketones are polar compounds. They have higher boiling points and are more soluble in water than nonpolar compounds of comparable molecular weight.

Section 17-4 What Are the Characteristic Reactions of Aldehydes and Ketones?
- Aldehydes are oxidized to carboxylic acids, but ketones resist oxidation.
- **Tollens' reagent** is used to test for the presence of aldehydes.
- Aldehydes can be reduced to primary alcohols and ketones to secondary alcohols.
- Addition of a molecule of alcohol to an aldehyde or a ketone produces a **hemiacetal**. A hemiacetal can react with another molecule of alcohol to produce an **acetal**.

Section 17-5 What Is Keto-Enol Tautomerism?
- A molecule containing an —OH group bonded to a carbon of a carbon–carbon double bond is called an **enol**.
- Constitutional isomers that differ in the location of a hydrogen atom and a double bond are called **tautomers**.

Summary of Key Reactions

1. **Oxidation of an Aldehyde to a Carboxylic Acid (Section 17-4A)**

   The aldehyde group is among the most easily oxidized organic functional groups. Oxidizing agents include $K_2Cr_2O_7$, Tollens' reagent, and O_2.

   $$R-\underset{Aldehyde}{\overset{O}{\underset{\|}{C}}}-H + 2Ag(NH_3)_2^+ + 3OH^- \longrightarrow$$

   $$R-\underset{\underset{anion}{Carboxylate}}{\overset{O}{\underset{\|}{C}}}-O^- + 2\underset{\underset{mirror}{Silver}}{Ag} + 4NH_3 + 2H_2O$$

2. **Reduction (Section 17-4B)**
Aldehydes are reduced to primary alcohols and ketones to secondary alcohols by H_2 in the presence of a transition metal catalyst such as Pt or Ni. They are also reduced to alcohols by sodium borohydride, $NaBH_4$, followed by protonation.

cyclohexanone + H_2 $\xrightarrow{\text{Transition metal catalyst}}$ cyclohexanol

cinnamaldehyde $\xrightarrow[\text{2. }H_2O]{\text{1. }NaBH_4}$ cinnamyl alcohol

3. **Addition of Alcohols to Form Hemiacetals (Section 17-4C)**
Hemiacetals are only minor components of an equilibrium mixture of an aldehyde or a ketone and an alcohol, except when the —OH and C=O groups are parts of the same molecule and a five- or six-membered ring can form.

4. **Addition of Alcohols to Form Acetals (Section 17-4C)**
Formation of acetals is catalyzed by acid. Acetals are hydrolyzed in aqueous acid to an aldehyde or a ketone and two molecules of an alcohol.

cyclohexanone =O + 2CH_3OH $\xrightleftharpoons{H^+}$ acetal with OCH_3, OCH_3 + H_2O

Cyclohexanone   Methanol   An acetal

5. **Keto-Enol Tautomerism (Section 17-5)**
The keto form generally predominates at equilibrium.

$$CH_3\overset{O}{\overset{\|}{C}}CH_3 \rightleftharpoons CH_3\overset{OH}{\overset{|}{C}}=CH_2$$

Keto form   Enol form
(approximately 99.9%)

Problems

Orange-numbered problems are applied.

Section 17-1 What Are Aldehydes and Ketones?

17-9 Answer true or false.
   (a) The one aldehyde and the one ketone with a molecular formula of C_3H_6O are constitutional isomers.
   (b) Aldehydes and ketones both contain a carbonyl group.
   (c) The VSEPR model predicts bond angles of 120° about the carbonyl carbon of aldehydes and ketones.
   (d) The carbonyl carbon of a ketone is a stereocenter.

17-10 What is the difference in structure between an aldehyde and a ketone?

17-11 What is the difference in structure between an aromatic aldehyde and an aliphatic aldehyde?

17-12 Is it possible for the carbon atom of a carbonyl group to be a stereocenter? Explain.

17-13 Which compounds contain carbonyl groups?

(a) $CH_3\overset{OH}{\overset{|}{C}H}CH_3$

(b) $CH_3CH_2\overset{O}{\overset{\|}{C}H}$

(c) Ph–$\overset{O}{\overset{\|}{C}}OCH_2CH_3$

(d) cyclopentanone

(e) tetrahydrofuran-2-ol (cyclic with —OH)

(f) $CH_3CH_2CH_2\overset{O}{\overset{\|}{C}}OH$

17-14 Following are structural formulas for two steroid hormones.

Cortisone

Aldosterone

(a) Name the functional groups in each.

(b) Mark all stereocenters in each hormone and state how many stereoisomers are possible for each.

17-15 Draw structural formulas for the four aldehydes with the molecular formula $C_5H_{10}O$. Which of these aldehydes are chiral?

Section 17-2 How Do We Name Aldehydes and Ketones?

17-16 Answer true or false.

(a) An aldehyde is named as an alkanal, and a ketone is named as an alkanone.

(b) The names for aldehydes and ketones are derived from the name of the longest carbon chain that contains the carbonyl group.

(c) In an aromatic aldehyde, the carbonyl carbon is bonded to an aromatic ring.

17-17 Draw structural formulas for these aldehydes.

(a) Formaldehyde (b) Propanal
(c) 3,7-Dimethyloctanal (d) Decanal
(e) 4-Hydroxybenzaldehyde
(f) 2,3-Dihydroxypropanal

17-18 Draw structural formulas for these ketones.

(a) Ethyl isopropyl ketone
(b) 2-Chlorocyclohexanone
(c) 2,4-Dimethyl-3-pentanone
(d) Diisopropyl ketone
(e) Acetone
(f) 2,5-Dimethylcyclohexanone

17-19 Write the IUPAC names for these compounds.

17-20 Write the IUPAC names for these compounds.

17-21 Explain why each name is incorrect. Write the correct IUPAC name for the intended compound.

(a) 3-Butanone (b) 1-Butanone
(c) 4-Methylbutanal (d) 2,2-Dimethyl-3-butanone

17-22 Explain why each name is incorrect. Write the correct IUPAC name for the intended compound.

(a) 2-Pentanal (b) Cyclopentanal
(c) 3-Ethyl-2-butanone (d) 5-Aminobenzaldehyde

Section 17-3 What Are the Physical Properties of Aldehydes and Ketones?

17-23 Answer true or false.

(a) Aldehydes and ketones are polar compounds.

(b) Aldehydes have lower boiling points than alcohols with comparable carbon skeletons.

(c) Low-molecular-weight aldehydes and ketones are very soluble in water.

(d) There is no possibility for hydrogen bonding between molecules of aldehydes and ketones.

17-24 In each pair of compounds, select the one with the higher boiling point.

(a) Acetaldehyde or ethanol
(b) Acetone or 3-pentanone
(c) Butanal or butane
(d) Butanone or 2-butanol

17-25 Acetone is completely soluble in water, but 4-heptanone is completely insoluble in water. Explain.

17-26 Account for the fact that acetone has a higher boiling point (56°C) than ethyl methyl ether (11°C) even though their molecular weights are almost the same.

17-27 Pentane, 1-butanol, and butanal all have approximately the same molecular weights but different boiling points. Arrange them in order of increasing boiling point. Explain the basis for your ranking.

17-28 Show how acetaldehyde can form hydrogen bonds with water.

17-29 Why can't two molecules of acetone form a hydrogen bond with each other?

Section 17-4 What Are the Characteristic Reactions of Aldehydes and Ketones?

17-30 Answer true or false.
(a) The reduction of an aldehyde always gives a primary alcohol.
(b) The reduction of a ketone always gives a secondary alcohol.
(c) The oxidation of an aldehyde gives a carboxylic acid.
(d) The oxidation of a primary alcohol gives a ketone.
(e) Tollens' reagent can be used to distinguish between an aldehyde and a ketone.
(f) Sodium borohydride, $NaBH_4$, reduces an aldehyde to a primary alcohol.
(g) The addition of one molecule of alcohol to the carbonyl group of a ketone gives a hemiacetal.
(h) The reaction of an aldehyde with two molecules of alcohol gives an acetal, plus a molecule of water.
(i) The formation of hemiacetals and acetals is reversible.
(j) The cyclic hemiacetal formed from 4-hydroxypentanal has two stereocenters and can exist as a mixture of $2^2 = 4$ stereoisomers.

17-31 Draw a structural formula for the principal organic product formed when each compound is treated with $K_2Cr_2O_7/H_2SO_4$. If there is no reaction, say so.
(a) Butanal (b) Benzaldehyde
(c) Cyclohexanone (d) Cyclohexanol

17-32 Draw a structural formula for the principal organic product formed when each compound in Problem 17-31 is treated with Tollens' reagent. If there is no reaction, explain why.

17-33 What simple chemical test could you use to distinguish between the members of each pair of compounds? Tell what you would do, what you would expect to observe, and how you would interpret your experimental observation.
(a) Pentanal and 2-pentanone
(b) 2-Pentanone and 2-pentanol

17-34 Explain why liquid aldehydes are often stored under an atmosphere of nitrogen rather than in air.

17-35 Suppose that you take a bottle of benzaldehyde (a liquid, bp 179°C) from a shelf and find a white solid in the bottom of the bottle. The solid turns litmus red; that is, it is acidic. Yet aldehydes are neutral compounds. How can you explain these observations?

17-36 Explain why the reduction of an aldehyde always gives a primary alcohol and the reduction of a ketone always gives a secondary alcohol.

17-37 Write a structural formula for the principal organic product formed by treating each compound with H_2/transition metal catalyst. Which products are chiral?

(a) $CH_3CCH_2CH_3$ with C=O
(b) $CH_3(CH_2)_4CH$ with C=O
(c) 2-methylcyclopentanone
(d) 2-hydroxybenzaldehyde (salicylaldehyde)

17-38 Draw a structural formula for the principal organic product formed by treating each compound in Problem 17-37 with $NaBH_4$ followed by H_2O.

17-39 1,3-Dihydroxy-2-propanone, more commonly known as dihydroxyacetone, is the active ingredient in artificial tanning agents such as Man-Tan and Magic Tan.
(a) Write a structural formula for this compound.
(b) Would you expect it to be soluble or insoluble in water?
(c) Write a structural formula for the product formed by its reduction with $NaBH_4$.

17-40 Draw a structural formula for the product formed by treatment of butanal with each set of reagents.
(a) H_2/transition metal catalyst
(b) $NaBH_4$, then H_2O
(c) $Ag(NH_3)_2^+$ (Tollens' reagent)
(d) $K_2Cr_2O_7/H_2SO_4$

17-41 Draw a structural formula for the product formed by treatment of acetophenone, $C_6H_5COCH_3$, with each set of reagents given in Problem 17-40.

Section 17-5 What Is Keto-Enol Tautomerism?

17-42 Mark each statement true or false.
(a) Keto and enol tautomers are constitutional isomers.
(b) For a pair of keto-enol tautomers, the keto form generally predominates.

17-43 Which of these compounds undergo keto-enol tautomerism?
(a) CH_3CH with C=O
(b) CH_3CCH_3 with C=O
(c) benzaldehyde (Ph–CHO)
(d) acetophenone (Ph–COCH$_3$)
(e) benzophenone (Ph–CO–Ph)
(f) cyclohexanone

17-44 Draw all enol forms of each aldehyde and ketone.

(a) CH₃CH₂CHO (b) CH₃COCH₂CH₃

(c) 2-methylcyclopentanone

17-45 Draw a structural formula for the keto form of each enol.

(a) cyclopent-1-en-1-ol (b) CH₃C(OH)=CHCH₂CH₂CH₃

(c) C₆H₅–CH=C(OH)CH₃

Chemical Connections

17-46 Warfarin is widely used in medicine as a blood anticoagulant.
(a) Identify all the functional groups in warfarin.
(b) The *S* enantiomer is more active than the *R* enantiomer. Draw a stereorepresentation of the *R* enantiomer.
(c) Warfarin, as drawn in the Chemical Connections box, contains an enol group. Draw the structural formula of the keto form of warfarin.
(d) Draw the product formed by treating warfarin with NaBH₄.
(e) Warfarin exerts its anticoagulation property in many individuals at a dose of 4 mg/day. How many molecules of warfarin are present in a 4 mg tablet?

Addition of Alcohols

17-47 What is the characteristic structural feature of a hemiacetal? Of an acetal?

17-48 Which compounds are hemiacetals, which are acetals, and which are neither?

(a) C₆H₅–CH(OCH₃)OCH₃ (b) CH₃CH₂CH(OH)OCH₃

(c) CH₃OCH₂OCH₃ (d) 2-ethyl-1,3-dioxolane-like (tetrahydrofuran with CH₂CH₃)

(e) 2-ethoxytetrahydrofuran (f) 2-hydroxytetrahydrofuran

17-49 Which compounds are hemiacetals, which are acetals, and which are neither?

(a) CH₃CH₂C(OH)(OCH₃)CH₃ (b) C₆H₅C(OCH₃)(OCH₃)CH₂CH₃

(c) γ-butyrolactone (d) 1-methoxycyclopentan-1-ol (OCH₃, OH)

(e) 2-methoxytetrahydrofuran (f) 1,1-dimethoxycyclopentane

17-50 Draw the hemiacetal and then the acetal formed in each reaction. In each case, assume an excess of the alcohol.
(a) Propanal + methanol →
(b) Cyclopentanone + methanol →

17-51 Draw the structures of the aldehydes or ketones and alcohols formed when these acetals are treated with aqueous acid and hydrolyzed.

(a) 2-ethyl-2-methyl-1,3-dioxolane (b) ortho-substituted benzene with CH(OCH₃)₂ and OCH₃

(c) 1,1-dimethoxycyclohexane (d) 2-methoxytetrahydro-2H-pyran

17-52 The following compound is a component of the fragrance of jasmine:

(2-phenyl-1,3-dioxolane)

From which carbonyl-containing compound and alcohol is this compound derived?

17-53 Consider the formation of an acetal. If the carbonyl oxygen of the starting aldehyde or ketone were labeled with oxygen-18 (Chapter 9) as a chemical tag, would you predict that the oxygen-18 tag would be found as one of the two oxygen atoms of the acetal, or as the oxygen atom of a water molecule? (Hint: think of the steps by which the reaction occurs; that is, think about the mechanism of acid-catalyzed acetal formation.)

17-54 Following is the structure of immunosuppressant FK-506, a molecule shown to disrupt calcineurin-mediated signal transduction in T-lymphocytes.

Immunosuppressant FK-506

(a) There are three carbon–carbon double bonds present in this molecule. Which of the three has the potential for *cis/trans* isomerism? Assign a *cis* or *trans* configuration to each carbon–carbon double bond that has this possibility.

(b) How many stereocenters are present in this molecule? How many stereoisomers are possible for it?

(c) Are there any aromatic components in this molecule?

(d) Consider the two carbon atoms marked with asterisks. Assign an *R* or *S* configuration of each stereocenter.

(e) Because of the presence of a 21-member ring, this molecule is described as a macrocycle. This ring is fashioned by three types of bonds, several carbon–carbon bonds, one ester, one hemiacetal, and one amide. Locate the ester and the hemiacetal.

(f) Draw the structural formula of the long chain compound that would result if the hemiacetal were to be cleaved to an alcohol and a carbonyl group.

17-55 What is the difference in meaning between the terms "hydration" and "hydrolysis"? Give an example of each.

17-56 What is the difference in meaning between the terms "hydration" and "dehydration"? Give an example of each.

17-57 List the reagents and experimental conditions that might be used to convert cyclohexanone to each of the following compounds.

17-58 Draw a structural formula for an aldehyde or a ketone that can be reduced to produce each alcohol. If none exists, say so.

(a) $CH_3\overset{OH}{\underset{|}{C}}HCH_3$ (b) phenyl-CH_2OH

(c) CH_3OH (d) cyclohexyl with OH and CH_3

17-59 Draw a structural formula for an aldehyde or a ketone that can be reduced to produce each alcohol. If none exists, say so.

(a) cyclopentyl-OH (b) cyclohexyl-CH_2OH

(c) $CH_3\overset{CH_3}{\underset{\underset{CH_3}{|}}{C}}OH$ (d) $HO\sim\sim\sim OH$

17-60 1-Propanol can be prepared by the reduction of an aldehyde, but it cannot be prepared by the acid-catalyzed hydration of an alkene. Explain why it cannot be prepared from an alkene.

17-61 Show how to bring about these conversions. In addition to the given starting material, use any other organic or inorganic reagents as necessary.

(a) $C_6H_5\overset{O}{\overset{\|}{C}}CH_2CH_3 \longrightarrow C_6H_5\overset{OH}{\underset{|}{C}}HCH_2CH_3 \longrightarrow$
$C_6H_5CH=CHCH_3$

(b) cyclopentanone → cyclopentanol → cyclopentene → cyclopentyl chloride

17-62 Show how to bring about these conversions. In addition to the given starting material, use any other organic or inorganic reagents as necessary.

(a) 1-Pentene to 2-pentanone
(b) Cyclohexene to cyclohexanone

17-63 Describe a simple chemical test by which you could distinguish between the members of each pair of compounds.

(a) Cyclohexanone and aniline
(b) Cyclohexene and cyclohexanol
(c) Benzaldehyde and cinnamaldehyde

Additional Problems

17-64 Mark the aldehyde or ketone group in these compounds.

(a) $HCCH_2CH_2CH_2CCH_3$ (with two C=O groups)

(b) cyclohexanone with a –CHO substituent

(c) $HOCH_2CHCH$ (with HO– and =O groups)

(d) 2-tetralone (bicyclic ketone)

(e) C_6H_5–CO–CH_2CH_3 (phenyl propyl ketone)

(f) 3-methoxy-4-hydroxybenzaldehyde

17-65 Draw a structural formula for the product formed by treating each compound in Problem 17-64 with sodium borohydride, $NaBH_4$.

17-66 Draw structural formulas for the (a) one ketone and (b) two aldehydes with the molecular formula C_4H_8O.

17-67 Draw structural formulas for these compounds.
  (a) 1-Chloro-2-propanone
  (b) 3-Hydroxybutanal
  (c) 4-Hydroxy-4-methyl-2-pentanone
  (d) 3-Methyl-3-phenylbutanal
  (e) 1,3-Cyclohexanedione
  (f) 5-Hydroxyhexanal

17-68 Why does acetone have a lower boiling point (56°C) than 2-propanol (82°C) even though their molecular weights are almost the same?

17-69 Propanal (bp 49°C) and 1-propanol (bp 97°C) have about the same molecular weight, yet their boiling points differ by almost 50°C. Explain this fact.

17-70 What simple chemical test could you use to distinguish between the members of each pair of compounds? Tell what you would do, what you would expect to observe, and how you would interpret your experimental observation.
  (a) Benzaldehyde and cyclohexanone
  (b) Acetaldehyde and acetone

17-71 5-Hydroxyhexanal forms a six-membered cyclic hemiacetal, which predominates at equilibrium in aqueous solution.
  (a) Draw a structural formula for this cyclic hemiacetal.
  (b) How many stereoisomers are possible for 5-hydroxyhexanal? (Chapter 15)
  (c) How many stereoisomers are possible for its cyclic hemiacetal? (Chapter 15)

17-72 The following molecule is an enediol; each carbon of the double bond carries an —OH group. Draw structural formulas for the α-hydroxyketone and the α-hydroxyaldehyde with which this enediol is in equilibrium.

$$\alpha\text{-hydroxyaldehyde} \rightleftharpoons \begin{array}{c} HC-OH \\ \| \\ C-OH \\ | \\ CH_3 \end{array} \rightleftharpoons \alpha\text{-hydroxyketone}$$

An enediol

17-73 Alcohols can be prepared by the acid-catalyzed hydration of alkenes (Section 12-6B) and by the reduction of aldehydes and ketones (Section 17-4B). Show how you might prepare each of the following alcohols by (1) acid-catalyzed hydration of an alkene and (2) reduction of an aldehyde or a ketone.
  (a) Ethanol   (b) Cyclohexanol
  (c) 2-Propanol   (d) 1-Phenylethanol

Looking Ahead

17-74 Glucose, $C_6H_{12}O_6$, contains an aldehyde group but exists predominantly in the form of the cyclic hemiacetal shown here. We will discuss this cyclic form of glucose in Chapter 20.

β-D-Glucose (cyclic structure shown with numbered carbons 1–6)

A cyclic hemiacetal is formed when the —OH group of one carbon bonds to the carbonyl group of another carbon.
  (a) Which carbon in glucose provides the —OH group and which provides the —CHO group?
  (b) Draw the alternative chair confirmations of β-D-glucose and state which of the two is the more stable.

17-75 Ribose, $C_5H_{10}O_5$, contains an aldehyde group but exists predominantly in the form of the cyclic hemiacetal shown here. We will discuss this cyclic form of ribose in Chapter 20.

β-D-Ribose (cyclic structure shown with numbered carbons 1–5)

Which carbon of ribose provides the —OH group and which provides the CHO group for formation of this cyclic hemiacetal?

17-76 Name the functional groups in each steroid hormone. Methandrostenolone is a synthetic anabolic steroid; that is, it promotes tissue and muscle growth and development.
  (a) What are the differences in the structural formula between testosterone and methandrostenolone?
  (b) Mark all stereocenters in each molecule and state the number of stereoisomers possible for each. The stereochemistry shown on the structural formulas is that of the biologically active stereoisomer.

Testosterone (a)

Methandrostenolone (b)

17-77 Sodium borohydride is a laboratory reducing agent. NADH is a biological reducing agent. In what way is the chemistry by which each reduces aldehydes and ketones similar?

Tying It Together

17-78 Complete the following equations for these reactions.

(a) cycloheptanol-OH $\xrightarrow{K_2Cr_2O_7}{H_2SO_4}$

(b) cyclopentyl-CH$_2$OH $\xrightarrow{K_2Cr_2O_7}{H_2SO_4}$

17-79 Write an equation for each conversion.
  (a) 1-Pentanol to pentanal
  (b) 1-Pentanol to pentanoic acid
  (c) 2-Pentanol to 2-pentanone
  (d) 2-Propanol to acetone
  (e) Cyclohexanol to cyclohexanone

18 Carboxylic Acids

KEY QUESTIONS

18-1 What Are Carboxylic Acids?

18-2 How Do We Name Carboxylic Acids?

18-3 What Are the Physical Properties of Carboxylic Acids?

18-4 What Are Soaps and Detergents?

18-5 What Are the Characteristic Reactions of Carboxylic Acids?

Citrus fruits are sources of citric acid, a tricarboxylic acid.

18-1 What Are Carboxylic Acids?

In this chapter, we study carboxylic acids, another class of organic compounds containing the carbonyl group. The functional group of a **carboxylic acid** is a **carboxyl group** (Section 10-4D), which can be represented in any one of three ways:

$$-\overset{\overset{\displaystyle O}{\|}}{C}-OH \quad -COOH \quad -CO_2H$$

18-2 How Do We Name Carboxylic Acids?

A. IUPAC Names

We derive the IUPAC name of an acyclic carboxylic acid from the name of the longest carbon chain that contains the carboxyl group. Drop the final *-e* from the name of the parent alkane and replace it with *-oic acid*. Number the chain beginning with the carbon of the carboxyl group. Because the carboxyl carbon is understood to be carbon 1, there is no need to give it a number. In the following examples, the common name is given in parentheses.

Hexanoic acid
(Caproic acid)

3-Methylbutanoic acid
(Isovaleric acid)

When a carboxylic acid also contains an —OH (hydroxyl) group, we indicate its presence by adding the prefix *hydroxy-*. When it contains a primary (1°) amine, we indicate the presence of the —NH₂ group with *amino-*. When a carboxylic acid also contains a ketone or aldehyde group, we indicate the presence of that group by the prefix *-oxo*.

5-Hydroxyhexanoic acid

4-Aminobenzoic acid
(*p*-Aminobenzoic acid)

CH₃CCH₂CO₂H
3-Oxobutanoic acid
(β-Ketobutyric acid
Acetoacetic acid)

CH₃CCO₂H
2-Oxopropanoic acid
(Pyruvic acid)

To name dicarboxylic acids, we add the suffix *-dioic acid* to the name of the parent alkane that contains both carboxyl groups. The numbers of the carboxyl carbons are not indicated because they can be only at the ends of the parent chain.

Ethanedioic acid
(Oxalic acid)

Propanedioic acid
(Malonic acid)

Butanedioic acid
(Succinic acid)

Pentanedioic acid
(Glutaric acid)

Hexanedioic acid
(Adipic acid)

The name *oxalic acid* is derived from one of its sources in the biological world—plants of the genus *Oxalis*, one of which is rhubarb. Oxalic acid also occurs in human and animal urine, and calcium oxalate is a major component of kidney stones. Succinic acid is an intermediate in the citric acid cycle (Section 27-4). Adipic acid is one of the two monomers required for the synthesis of the polymer nylon-66 (Section 19-6B).

B. Common Names

Common names for aliphatic carboxylic acids, many of which were known long before the development of IUPAC nomenclature, are often derived from the name of a natural source from which the acid could be isolated. Table 18-1 lists several of the unbranched aliphatic carboxylic acids found in the biological world along with the common name of each. Those with 16, 18, and 20 carbon atoms are particularly abundant in animal fats and vegetable oils (Section 21-2) and the phospholipid components of biological membranes (Section 21-5).

Table 18-1 Several Aliphatic Carboxylic Acids and Their Common Names

Structure	IUPAC Name	Common Name	Derivation
HCOOH	methanoic acid	formic acid	Latin: *formica*, ant
CH_3COOH	ethanoic acid	acetic acid	Latin: *acetum*, vinegar
CH_3CH_2COOH	propanoic acid	propionic acid	Greek: *propion*, first fat
$CH_3(CH_2)_2COOH$	butanoic acid	butyric acid	Latin: *butyrum*, butter
$CH_3(CH_2)_3COOH$	pentanoic acid	valeric acid	Latin: *valere*, to be strong
$CH_3(CH_2)_4COOH$	hexanoic acid	caproic acid	Latin: *caper*, goat
$CH_3(CH_2)_6COOH$	octanoic acid	caprylic acid	Latin: *caper*, goat
$CH_3(CH_2)_8COOH$	decanoic acid	capric acid	Latin: *caper*, goat
$CH_3(CH_2)_{10}COOH$	dodecanoic acid	lauric acid	Latin: *laurus*, laurel
$CH_3(CH_2)_{12}COOH$	tetradecanoic acid	myristic acid	Greek: *myristikos*, fragrant
$CH_3(CH_2)_{14}COOH$	hexadecanoic acid	palmitic acid	Latin: *palma*, palm tree
$CH_3(CH_2)_{16}COOH$	octadecanoic acid	stearic acid	Greek: *stear*, solid fat
$CH_3(CH_2)_{18}COOH$	eicosanoic acid	arachidic acid	Greek: *arachis*, peanut

▶ Formic acid was first obtained in 1670 from the destructive distillation of ants, whose Latin genus is *Formica*. It is one of the components of the venom injected by stinging ants.

When common names are used, the Greek letters alpha (α), beta (β), gamma (γ), and so forth, are often added as a prefix to locate substituents. In the example below, we give the IUPAC name, followed by a common name and four-letter abbreviation. GABA is a neurotransmitter in the central nervous system.

4-Aminobutanoic acid
(γ-Aminobutyric acid; GABA)

Interactive Example 18-1 IUPAC Names for Carboxylic Acids

Write the IUPAC name for each carboxylic acid:

(a) [structure]

(b) HO—⟨benzene⟩—COOH

(c) [structure]

Strategy and Solution

(a) The longest carbon chain that contains the carboxyl group has five carbons; therefore, the parent alkane is pentane. The IUPAC name is 2-ethylpentanoic acid.
(b) 4-Hydroxybenzoic acid
(c) *trans*-3-Phenyl-2-propenoic acid (cinnamic acid)

Problem 18-1

Each of the following compounds has a well-recognized and widely used common name. A derivative of glyceric acid is an intermediate in glycolysis (Section 28-2). β-Alanine is a building block of pantothenic acid

(Section 27-5). Mevalonic acid is an intermediate in the biosynthesis of steroids (Section 27-4). Write the IUPAC name for each compound.

(a) COOH
     |
     CHOH
     |
     CH₂OH

     Glyceric acid

(b) H₂NCH₂CH₂COOH

    β-Alanine

(c) Mevalonic acid structure: HO–CH₂–C(CH₃)(OH)–CH₂–COOH

    Mevalonic acid

18-3 What Are the Physical Properties of Carboxylic Acids?

A major feature of carboxylic acids is the polarity of the carboxyl group (Figure 18-1). This group contains three polar covalent bonds: C=O, C—O, and O—H. The polarity of these bonds determines the major physical properties of carboxylic acids.

Carboxylic acids have significantly higher boiling points than other types of organic compounds of comparable molecular weight (Table 18-2). Their higher boiling points result from their polarity and the fact that hydrogen bonding between two carboxyl groups creates a dimer that behaves as a higher-molecular-weight compound.

Hydrogen bonding between two molecules

A hydrogen-bonded dimer of acetic acid

Carboxylic acids are more soluble in water than are alcohols, ethers, aldehydes, and ketones of comparable molecular weight. This increased solubility is due to their strong association with water molecules by hydrogen bonding through both their carbonyl and hydroxyl groups. The first four aliphatic carboxylic acids (formic, acetic, propanoic, and butanoic) are infinitely soluble in water. As the size of the hydrocarbon chain increases relative to that of the carboxyl group, water solubility decreases. The solubility of hexanoic acid (six carbons) in water is 1.0 g/100 mL water.

We must mention two other properties of carboxylic acids. First, the liquid carboxylic acids from propanoic acid to decanoic acid have sharp, often disagreeable odors. Butanoic acid is found in stale perspiration and is a

FIGURE 18-1 Polarity of a carboxyl group.

Table 18-2 Boiling Points and Solubilities in Water of Two Groups of Compounds of Comparable Molecular Weight

Structure	Name	Molecular Weight	Boiling Point (°C)	Solubility (g/100 mL H$_2$O)
CH$_3$COOH	acetic acid	60.1	118	infinite
CH$_3$CH$_2$CH$_2$OH	1-propanol	60.1	97	infinite
CH$_3$CH$_2$CHO	propanal	58.1	48	16
CH$_3$(CH$_2$)$_2$COOH	butanoic acid	88.1	163	infinite
CH$_3$(CH$_2$)$_3$CH$_2$OH	1-pentanol	88.1	137	2.3
CH$_3$(CH$_2$)$_3$CHO	pentanal	86.1	103	slight

major component of "locker room odor." Pentanoic acid smells even worse, and goats, which secrete C$_6$, C$_8$, and C$_{10}$ carboxylic acids (Table 18-1), are not famous for their pleasant odors. Second, carboxylic acids have a characteristic sour taste. The sour taste of pickles and sauerkraut, for example, is due to the presence of lactic acid. The sour tastes of limes (pH 1.9), lemons (pH 2.3), and grapefruit (pH 3.2) are due to the presence of citric and other acids.

18-4 What Are Soaps and Detergents?

A. Fatty Acids

Fatty acids Long, unbranched-chain carboxylic acids, most commonly with 12 to 20 carbons. They are derived from the hydrolysis of animal fats, vegetable oils, or the phospholipids of biological membranes (Chapter 21). The hydrocarbon chain may be saturated or unsaturated. In most unsaturated fatty acids, the *cis* isomer predominates. *Trans* isomers are rare.

More than 500 different **fatty acids** have been isolated from various cells and tissues. Given in Table 18-3 are common names and structural formulas for the most abundant fatty acids. The number of carbons in a fatty acid and the number of carbon–carbon double bonds in its hydrocarbon chain are shown by two numbers separated by a colon. In this notation, linoleic acid, for example, is designated as an 18:2 fatty acid; its 18-carbon chain contains two carbon–carbon double bonds.

Table 18-3 The Most Abundant Fatty Acids in Animal Fats, Vegetable Oils, and Biological Membranes

Carbon Atoms: Double Bonds*	Structure	Common Name	Melting Point (°C)
Saturated Fatty Acids			
12: 0	CH$_3$(CH$_2$)$_{10}$COOH	lauric acid	44
14: 0	CH$_3$(CH$_2$)$_{12}$COOH	myristic acid	58
16: 0	CH$_3$(CH$_2$)$_{14}$COOH	palmitic acid	63
18: 0	CH$_3$(CH$_2$)$_{16}$COOH	stearic acid	70
20: 0	CH$_3$(CH$_2$)$_{18}$COOH	arachidic acid	77
Unsaturated Fatty Acids			
16: 1	CH$_3$(CH$_2$)$_5$CH=CH(CH$_2$)$_7$COOH	palmitoleic acid	1
18: 1	CH$_3$(CH$_2$)$_7$CH=CH(CH$_2$)$_7$COOH	oleic acid	16
18: 2	CH$_3$(CH$_2$)$_4$(CH=CHCH$_2$)$_2$(CH$_2$)$_6$COOH	linoleic acid	−5
18: 3	CH$_3$CH$_2$(CH=CHCH$_2$)$_3$(CH$_2$)$_6$COOH	linolenic acid	−11
20: 4	CH$_3$(CH$_2$)$_4$(CH=CHCH$_2$)$_4$(CH$_2$)$_2$COOH	arachidonic acid	−49

*The first number is the number of carbons in the fatty acid; the second number is the number of carbon–carbon double bonds in its hydrocarbon chain.

Following are several characteristics of the most abundant fatty acids in higher plants and animals:

1. Nearly all fatty acids have an even number of carbon atoms, most between 12 and 20, in an unbranched chain.
2. The three most abundant fatty acids in nature are palmitic acid (16:0), stearic acid (18:0), and oleic acid (18:1).
3. In most unsaturated fatty acids, the *cis* isomer predominates; the *trans* isomer is rare.
4. Unsaturated fatty acids have lower melting points than their saturated counterparts. The greater the degree of unsaturation, the lower the melting point. Compare, for example, the melting points of the following 18-carbon fatty acids: Linolenic acid, with three carbon–carbon double bonds, has the lowest melting point of these four fatty acids.

Stearic acid (18:0) (mp 70°C)

Oleic acid (18:1) (mp 16°C)

Linoleic acid (18:2) (mp –5°C)

Linolenic acid (18:3) (mp –11°C)

Fatty acids can be divided into two groups: saturated and unsaturated. Saturated fatty acids have only carbon–carbon single bonds in their hydrocarbon chains. Unsaturated fatty acids have one or more C=C double bonds in the chain. All unsaturated fatty acids listed in Table 18-3 are the *cis* isomer.

Saturated fatty acids are solids at room temperature, because the regular nature of their hydrocarbon chains allows their molecules to pack together in close parallel alignment. When packed in this manner, the attractive interactions between adjacent hydrocarbon chains (London dispersion forces, Section 5-7A) are maximized. Although London dispersion forces are weak interactions, the regular packing of hydrocarbon chains allows these forces to operate over a large portion of their chains, ensuring that a considerable amount of energy is needed to separate and melt them.

CHEMICAL CONNECTIONS 18A

Trans Fatty Acids: What Are They and How Do You Avoid Them?

Animal fats are rich in saturated fatty acids, whereas plant oils (for example, corn, soybean, canola, olive, and palm oils) are rich in unsaturated fatty acids. Fats are added to processed foods to provide a desirable firmness along with a moist texture and pleasant taste. To supply the demand for dietary fats of the appropriate consistency, the *cis* double bonds of vegetable oils are partially hydrogenated. The greater the extent of hydrogenation, the higher the melting point of the triglyceride. The extent of hydrogenation is carefully controlled, usually by employing a Ni catalyst and a calculated amount of H_2 as a limiting reagent. Under these conditions, the H_2 is used up before all double bonds are reduced, so that only partial hydrogenation and the desired overall consistency is achieved. For example, by controlling the degree of hydrogenation, an oil with a melting point below room temperature can be converted to a semisolid or even a solid product.

The mechanism of catalytic hydrogenation of alkenes was discussed in Section 12-6D. Recall that a key step in this mechanism involves interaction of the carbon–carbon double bond of the alkene with the metal catalyst to form a carbon–metal bond. Because the interaction of a carbon–carbon double bond with the Ni catalyst is reversible, many of the double bonds remaining in the oil may be isomerized from the less stable *cis* configuration to the more stable *trans* configuration. Thus, the equilibration between the *cis* and *trans* configurations may occur when H_2 is the limiting reagent. For example, elaidic acid is the C_{18} *trans* fatty acid isomer of oleic acid, a common C_{18} *cis* fatty acid.

The oils used for frying in fast-food restaurants are usually partially hydrogenated plant oils, and therefore, they contain substantial amounts of *trans* fatty acids that are transferred to the foods cooked in them. Other major sources of *trans* fatty acids in the diet include stick margarine, certain commercial bakery products, creme-filled cookies, potato and corn chips, frozen breakfast foods, and cake mixes.

Recent studies have shown that consuming significant amounts of *trans* fatty acids can lead to serious health problems related to serum cholesterol levels. Low overall serum cholesterol and a decreased ratio of low-density lipoprotein (LDL) cholesterol to high-density lipoprotein (HDL) cholesterol are associated with good cardiovascular health. High serum cholesterol levels and an elevated ratio of LDL cholesterol to HDL cholesterol are linked to a high incidence of cardiovascular disease, especially atherosclerosis. Research has indicated that a diet high in either saturated fatty acids or *trans* fatty acids substantially increases the risk of cardiovascular disease.

The FDA has announced that processed foods must list the amount of *trans* fatty acids they contain, so that consumers can make better choices about the foods they eat. A diet low in saturated and *trans* fatty acids is recommended, along with consumption of more fish, whole grains, fruits, and vegetables, and especially daily exercise, which is tremendously beneficial regardless of diet.

Monounsaturated and polyunsaturated *cis* fatty acids have not produced similar health risks in most studies, although too much fat of any kind in the diet can lead to obesity, a major health problem that is associated with several diseases, one of which is diabetes. Some polyunsaturated (*cis*) fatty acids, such as those found in certain types of fish, have even been shown to have beneficial effects in some studies. These are the so-called omega-3 fatty acids.

In omega-3 fatty acids, the last carbon of the last double bond of the hydrocarbon chain ends three carbons in from the methyl terminal end of the chain. The last carbon of the hydrocarbon chain is called the omega (the last letter of the Greek alphabet) carbon—hence the designation of omega-3. The two most commonly found in health food supplements are eicosapentaenoic acid and docosahexaenoic acid.

Eicosapentaenoic acid, $C_{20}H_{30}O_2$, is an important fatty acid in the marine food chain and serves as a precursor in humans of several members of the prostacyclin and thromboxane families (Chapter 21). Note how the name of this fatty acid is derived. *Eicosa-* is the prefix indicating 20 carbons in the chain, *-pentaene-* indicates five carbon–carbon double bonds, and *–oic acid* shows the carboxyl functional group.

Docosahexaenoic acid, $C_{22}H_{32}O_2$, is found in fish oils and many phospholipids. It is a major structural component of excitable membranes in the retina and brain and is synthesized in the liver from linoleic acid.

Elaidic acid
mp 46°C
(a *trans* C_{18} fatty acid)

CHEMICAL CONNECTIONS 18A

Trans Fatty Acids: What Are They and How Do You Avoid Them? (continued)

All common *cis* unsaturated fatty acids are liquids at room temperature because the *cis* double bonds interrupt the regular packing of the chains and the London dispersion forces only act over shorter segments of the chains, so that less energy is required to melt them. The greater the degree of unsaturation, the lower the melting point, because each double bond interrupts the regular packing of the fatty acid molecules.

B. Structure and Preparation of Soaps

Natural **soaps** are most commonly prepared from a blend of tallow and coconut oils. In the preparation from tallow, the solid fats of cattle are melted with steam, and the tallow layer that forms on the top is removed. The preparation of soaps begins by boiling these triglycerides with sodium hydroxide. The reaction that takes place is called **saponification** (Latin: *saponem*, "soap"):

Saponification The hydrolysis of an ester in aqueous NaOH or KOH to an alcohol and the sodium or potassium salt of a carboxylic acid (Section 19-4A)

$$\underset{\text{A triglyceride}}{\begin{array}{c} \text{O} \\ \parallel \\ \text{CH}_2\text{OCR} \\ | \\ \text{RCOCH} \quad \text{O} \\ | \quad \parallel \\ \text{CH}_2\text{OCR} \end{array}} + 3\text{NaOH} \xrightarrow{\text{saponification}} \underset{\substack{\text{1,2,3-Propanetriol} \\ \text{(Glycerol; Glycerin)}}}{\begin{array}{c} \text{CH}_2\text{OH} \\ | \\ \text{CHOH} \\ | \\ \text{CH}_2\text{OH} \end{array}} + \underset{\text{Sodium soaps}}{3\text{RCO}^-\text{Na}^+}$$

FIGURE 18-2 Soap micelles. Nonpolar (hydrophobic) hydrocarbon chains cluster in the interior of the micelle, and polar (hydrophilic) carboxylate groups are on the surface of the micelle. Soap micelles repel each other because of their negative surface charges.

Micelles Spherical arrangements of molecules in aqueous solution such that their hydrophobic (water-hating) parts are shielded from the aqueous environment and their hydrophilic (water-loving) parts are on the surface of the sphere and in contact with the aqueous environment

FIGURE 18-3 A soap micelle with a "dissolved" oil or grease droplet.

At the molecular level, saponification corresponds to base-promoted hydrolysis (Section 19-4A) of the ester groups in triglycerides. A triglyceride is a triester of glycerol. The resulting soaps contain mainly the sodium salts of palmitic, stearic, and oleic acids from tallow and the sodium salts of lauric and myristic acids from coconut oil.

After hydrolysis is complete, sodium chloride is added to precipitate the sodium salts as thick curds of soap. The water layer is then drawn off, and glycerol is recovered by vacuum distillation. The crude soap contains sodium chloride, sodium hydroxide, and other impurities that are removed by boiling the curd in water and reprecipitating with more sodium chloride. After several purifications, the soap can be used as an inexpensive industrial soap without further processing. Other treatments transform the crude soap into pH-controlled cosmetic soaps, medicated soaps, and the like.

C. How Soap Cleans

Soap owes its remarkable cleansing properties to its ability to act as an emulsifying agent. Because the long hydrocarbon chains of natural soaps are insoluble in water, soap molecules tend to cluster in such a way as to minimize contact of their hydrocarbon chains with surrounding water molecules. The polar carboxylate groups, by contrast, tend to remain in contact with the surrounding water molecules. Thus, in water, soap molecules spontaneously cluster into **micelles** (Figure 18-2).

Many of the things we commonly think of as dirt (such as grease, oil, and fat stains) are nonpolar and insoluble in water. When soap and this type of dirt are mixed together, as in a washing machine, the nonpolar hydrocarbon inner parts of the soap micelles "dissolve" the nonpolar substances. In effect, new soap micelles form, with the nonpolar dirt molecules in the center (Figure 18-3). In this way, nonpolar organic grease, oil, and fat are "dissolved" and washed away in the polar wash water.

Soaps, however, have their disadvantages, foremost among which is the fact that they form water-insoluble salts when used in water containing Ca(II), Mg(II), or Fe(III) ions (hard water):

$$2CH_3(CH_2)_{14}COO^-Na^+ + Ca^{2+} \longrightarrow [CH_3(CH_2)_{14}COO^-]_2Ca^{2+} + 2Na^+$$

A sodium soap   Calcium salt of a fatty acid
(soluble in water as micelles)   (insoluble in water)

These water-insoluble calcium, magnesium, and iron salts of fatty acids create problems, including rings around the bathtub, films that spoil the luster of hair, and grayness and roughness that build up on textiles after repeated washings.

D. Synthetic Detergents

After the cleansing action of soaps was understood, chemists were in a position to design a synthetic detergent. Molecules of a good detergent, they reasoned, must have a long hydrocarbon chain—preferably 12 to 20 carbon atoms long—and a polar group at one end of the molecule that does not form insoluble salts with the Ca(II), Mg(II), or Fe(III) ions that are present in hard water. These essential characteristics of a soap, they recognized, could be produced in a molecule containing a sulfonate ($-SO_3^-$) group instead of a carboxylate ($-COO^-$) group. Calcium, magnesium, and iron salts of alkylsulfonic acids ($R-SO_3H$) are much more soluble in water than comparable salts of fatty acids.

The most widely used synthetic detergents today are the linear alkylbenzenesulfonates (LAS). One of the most common of these is sodium 4-dodecylbenzenesulfonate. To prepare this type of detergent, a linear alkylbenzene is treated with sulfuric acid to form an alkylbenzenesulfonic acid (Section 13-3C), followed by neutralization of the sulfonic acid with sodium hydroxide:

$$CH_3(CH_2)_{10}CH_2-\underset{\text{Dodecylbenzene}}{\bigcirc} \xrightarrow[\text{2. NaOH}]{\text{1. }H_2SO_4} CH_3(CH_2)_{10}CH_2-\underset{\substack{\text{Sodium 4-dodecylbenzenesulfonate}\\\text{(an anionic detergent)}}}{\bigcirc}-SO_3^-\ Na^+$$

The product is mixed with builders and then spray-dried to give a smooth-flowing powder. The most common builder is sodium silicate. Alkylbenzenesulfonate detergents were introduced in the late 1950s, and today they account for close to 90% of the market that was once held by natural soaps.

Among the most common additives to detergent preparations are foam stabilizers, bleaches, and optical brighteners. A common foam stabilizer added to liquid soaps, but not to laundry detergents (for obvious reasons: think of a top-loading washing machine with foam spewing out of the lid!), is the amide prepared from dodecanoic acid (lauric acid) and 2-aminoethanol (ethanolamine). The most common bleach is sodium perborate tetrahydrate, which decomposes at temperatures higher than 50°C to give hydrogen peroxide, the actual bleaching agent.

$$\underset{\substack{N\text{-(2-Hydroxyethyl)dodecanamide}\\\text{(a foam stabilizer)}}}{CH_3(CH_2)_{10}\overset{\overset{\displaystyle O}{\|}}{C}NHCH_2CH_2OH} \qquad \underset{\substack{\text{Sodium perborate tetrahydrate}\\\text{(a bleach)}}}{O=B-O-O^-\ Na^+\cdot 4H_2O}$$

Also added to laundry detergents are optical brighteners (optical bleaches). These substances are absorbed into fabrics and, after absorbing ambient light, fluoresce with a blue color, offsetting the yellow color that develops in fabric as it ages. Optical brighteners produce a "whiter-than-white" appearance. You most certainly have observed their effects if you have seen the glow of white T-shirts or blouses when they are exposed to black light (UV radiation).

18-5 What Are the Characteristic Reactions of Carboxylic Acids?

A. Acidity

Carboxylic acids are weak acids. Values of K_a for most unsubstituted aliphatic and aromatic carboxylic acids fall within the range of 10^{-4} to 10^{-5} (pK_a = 4.0–5.0). The value of K_a for acetic acid, for example, is 1.74×10^{-5}; its pK_a is 4.75 (Section 8-5).

$$CH_3COOH + H_2O \rightleftharpoons CH_3COO^- + H_3O^+ \qquad K_a = \frac{[CH_3COO^-][H_3O^+]}{[CH_3COOH]} = 1.74 \times 10^{-5}$$

$$pK_a = 4.75$$

Substituents with high electronegativity (especially —OH, —Cl, and —NH_3^+) near the carboxyl group increase the acidity of carboxylic acids, often by several orders of magnitude. Compare, for example, the acidities of acetic acid and the chlorine-substituted acetic acids. Both dichloroacetic acid and trichloroacetic acid are stronger acids than acetic acid (pK_a 4.75) and H_3PO_4 (pK_a 2.12). Dentists use a 50% aqueous solution of trichloroacetic acid to cauterize gums. This strong acid stops the bleeding, kills diseased tissue, and allows the growth of healthy gum tissue.

Formula:	CH_3COOH	$ClCH_2COOH$	$Cl_2CHCOOH$	Cl_3CCOOH
Name:	Acetic acid	Chloroacetic acid	Dichloroacetic acid	Trichloroacetic acid
pK_a:	4.75	2.86	1.48	0.70

Increasing acid strength ⟶

Electronegative atoms on the carbon adjacent to a carboxyl group increase acidity because they pull electron density away from the O—H bond, thereby facilitating ionization of the carboxyl group and making it a stronger acid.

One final point about carboxylic acids: when a carboxylic acid is dissolved in an aqueous solution, the form of the carboxylic acid present depends on the pH of the solution in which it is dissolved. Consider typical aliphatic carboxylic acids, which have pK_a values of 4.0 to 5.0. When the pH of the solution is equal to the pK_a of the carboxylic acid (that is, when the pH of the solution is in the range 4.0–5.0), the acid, RCOOH, and its conjugate base, RCOO$^-$, are present in equal concentrations, which we can demonstrate by using the Henderson–Hasselbalch equation (Section 8-11).

$$pH = pK_a + \log\frac{[A^-]}{[HA]} \qquad \text{Henderson–Hasselbalch Equation}$$

Consider the ionization of a weak acid, HA, in aqueous solution. When the pH of the solution is equal to the pK_a of the carboxylic acid, the Henderson–Hasselbalch equation reduces to:

$$\log\frac{[A^-]}{[HA]} = 0$$

Taking the antilog gives us the ratio of [A$^-$] to [HA] and tells us that the concentrations of the two are equal.

$$\frac{[A^-]}{[HA]} = 1$$

If the pH of the solution is adjusted to 2.0 or lower by the addition of a strong acid, the carboxylic acid then is present in solution almost entirely as RCOOH. If the pH of the solution is adjusted to 7.0 or higher, the carboxylic acid is present almost entirely as its anion. Thus, even in a neutral solution (pH 7.0), a carboxylic acid is present predominantly as its anion.

$$R-\overset{O}{\underset{\|}{C}}-OH \underset{H^+}{\overset{OH^-}{\rightleftharpoons}} R-\overset{O}{\underset{\|}{C}}-OH + R-\overset{O}{\underset{\|}{C}}-O^- \underset{H^+}{\overset{OH^-}{\rightleftharpoons}} R-\overset{O}{\underset{\|}{C}}-O^-$$

Predominant species when pH of the solution is ≤2.0 | Present in equal concentrations when pH of the solution = pK_a of the acid | Predominant species when pH of the solution is ≥7.0

B. Reaction with Bases

All carboxylic acids, whether soluble or insoluble in water, react with NaOH, KOH, and other strong bases to form water-soluble salts.

$$\text{C}_6\text{H}_5\text{-COOH} + \text{NaOH} \xrightarrow{\text{H}_2\text{O}} \text{C}_6\text{H}_5\text{-COO}^-\text{Na}^+ + \text{H}_2\text{O}$$

Benzoic acid (slightly soluble in water) → Sodium benzoate (60 g/100 mL water)

Sodium benzoate, a fungal growth inhibitor, is often added to baked goods "to retard spoilage." ◀ Calcium propanoate is used for the same purpose. Carboxylic acids also form water-soluble salts with ammonia and amines.

$$\text{C}_6\text{H}_5\text{-COOH} + \text{NH}_3 \xrightarrow{\text{H}_2\text{O}} \text{C}_6\text{H}_5\text{-COO}^-\text{NH}_4^+$$

Benzoic acid (slightly soluble in water) → Ammonium benzoate (20 g/100 mL water)

▶ Sodium benzoate and calcium propanoate are fungal growth inhibitors that are added to baked goods "to retard spoilage."

Carboxylic acids react with sodium bicarbonate and sodium carbonate to form water-soluble sodium salts and carbonic acid (a weaker acid). Carbonic acid, in turn, decomposes to give water and carbon dioxide, which evolves as a gas (Section 8-6E).

$$\text{CH}_3\text{COOH}(aq) + \text{NaHCO}_3(aq) \longrightarrow \text{CH}_3\text{COO}^-\text{Na}^+(aq) + \text{CO}_2(g) + \text{H}_2\text{O}(l)$$

Salts of carboxylic acids are named in the same manner as the salts of inorganic acids: the cation is named first and then the anion. The name of the anion is derived from the name of the carboxylic acid by dropping the suffix *-ic acid* from the IUPAC or common name of the acid and adding the suffix *-ate*.

EXAMPLE 18-2 Acidity of Carboxylic Acids

Complete each acid–base reaction and name the carboxylate salt formed.

(a) CH₃CH₂CH₂COOH + NaOH ⟶

(b) CH₃CH(OH)COOH + NaHCO₃ ⟶
   (s)-Lactic acid

> **Strategy and Solution**
>
> Each carboxylic acid is converted to its sodium salt. In (b), carbonic acid is formed and decomposes to carbon dioxide and water.
>
> (a) $\text{CH}_3\text{CH}_2\text{CH}_2\text{COOH}$ + NaOH ⟶ $\text{CH}_3\text{CH}_2\text{CH}_2\text{COO}^-\text{Na}^+$ + H_2O
> Butanoic acid   Sodium butanoate
>
> (b) $\text{CH}_3\text{CH(OH)COOH}$ + NaHCO$_3$ ⟶ $\text{CH}_3\text{CH(OH)COO}^-\text{Na}^+$ + H_2O + CO_2
> (S)-Lactic acid   Sodium (S)-lactate
>
> **Problem 18-2**
>
> Write equations for the reaction of each acid in Example 18-2 with ammonia and name the carboxylate salt formed.

A consequence of the water solubility of carboxylic acid salts is that water-insoluble carboxylic acids can be converted to water-soluble ammonium or alkali metal salts and extracted into aqueous solution. The salt, in turn, can be transformed back to the free carboxylic acid by treatment with HCl, H$_2$SO$_4$, or another strong acid. These reactions allow an easy separation of water-insoluble carboxylic acids from water-insoluble nonacidic compounds.

Shown in Figure 18-4 is a flowchart for the separation of benzoic acid, a water-insoluble carboxylic acid, from benzyl alcohol, a nonacidic

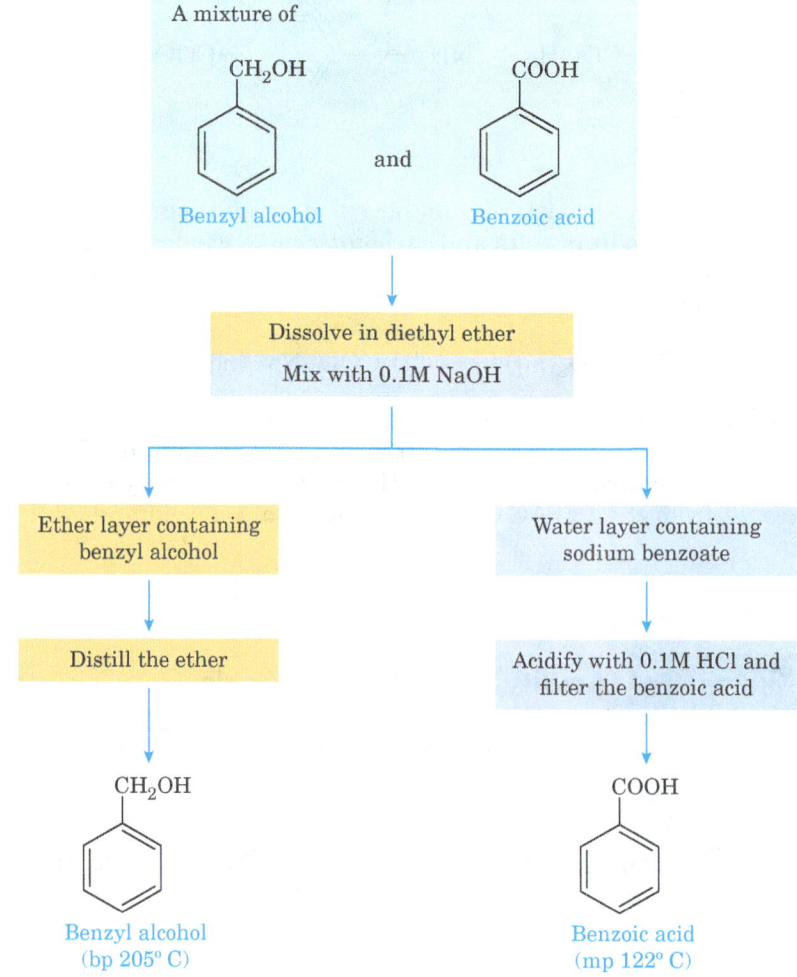

FIGURE 18-4 Flowchart for separation of benzoic acid from benzyl alcohol.

compound. First, the mixture of benzoic acid and benzyl alcohol is dissolved in diethyl ether. When the ether solution is shaken with aqueous NaOH or another strong base, benzoic acid is converted to its water-soluble sodium salt. Then the ether and aqueous phases are separated. The ether solution is distilled, yielding first diethyl ether (bp 35°C) and then benzyl alcohol (bp 205°C). The aqueous solution is acidified with HCl, and benzoic acid precipitates as a white crystalline solid (mp 122°C), which is recovered by filtration.

C. Reduction

The carboxyl group is among the organic functional groups most resistant to reduction. It is not affected by catalytic reduction under conditions that readily reduce alkenes to alkanes (Section 12-5D) or by sodium borohydride, ($NaBH_4$), which reduces aldehydes to 1° alcohols and ketones to 2° alcohols (Section 17-4B).

The most common reagent for the reduction of a carboxylic acid to a 1° alcohol is the very powerful reducing agent lithium aluminum hydride, $LiAlH_4$. Reduction of a carboxyl group with this reagent is commonly carried out in diethyl ether. The initial product is an aluminum alkoxide, which is then treated with water to give the primary alcohol and lithium and aluminum hydroxides. These two hydroxides are insoluble in diethyl ether and are removed by filtration. Evaporation of the ether solvent yields the primary alcohol.

3-Cyclopentene-carboxylic acid →(1. $LiAlH_4$, ether; 2. H_2O)→ 4-Hydroxymethylcyclopentene + LiOH + $Al(OH)_3$

D. Fischer Esterification

Treatment of a carboxylic acid with an alcohol in the presence of an acid catalyst—most commonly, concentrated sulfuric acid—gives an **ester**. This method of forming an ester is given the special name **Fischer esterification**, after the German chemist Emil Fischer (1852–1919). As an example of Fischer esterification, treating acetic acid with ethanol in the presence of concentrated sulfuric acid gives ethyl acetate and water:

Removal of OH and H gives the ester

$CH_3C(=O)$—OH + H—OCH_2CH_3 ⇌(H_2SO_4) $CH_3COCH_2CH_3$ + H_2O

Ethanoic acid (Acetic acid) + Ethanol (Ethyl alcohol) → Ethyl ethanoate (Ethyl acetate)

Ester A compound in which the —OH of a carboxyl group, RCOOH, is replaced with an alkoxy (—OR) or aryloxy (—OAr) group

Fischer esterification The process of forming an ester by refluxing a carboxylic acid and an alcohol in the presence of an acid catalyst, commonly sulfuric acid

We study the structure, nomenclature, and reactions of esters in detail in Chapter 19. In this chapter, we discuss only their preparation from carboxylic acids.

In the process of Fischer esterification, the alcohol adds to the carbonyl group of the carboxylic acid to form a tetrahedral carbonyl addition intermediate. Note how closely this step resembles the addition of an alcohol to the carbonyl group of an aldehyde or ketone to form a hemiacetal (Section 17-4C).

In the case of Fischer esterification, the intermediate collapses by loss of a water molecule to give the ester.

$$\text{CH}_3\overset{\overset{\displaystyle O}{\|}}{\text{C}}\text{—OH} + \overset{\overset{\displaystyle H}{|}}{\text{O}}\text{CH}_2\text{CH}_3 \underset{}{\overset{H_2SO_4}{\rightleftarrows}} \left[\text{CH}_3\overset{\overset{\displaystyle O-H}{|}}{\underset{\underset{\displaystyle OH}{|}}{C}}\text{—OCH}_2\text{CH}_3 \right] \underset{}{\overset{H_2SO_4}{\rightleftarrows}} \text{CH}_3\overset{\overset{\displaystyle O}{\|}}{\text{C}}\text{OCH}_2\text{CH}_3 + \text{H}_2\text{O}$$

A tetrahedral carbonyl addition intermediate

Acid-catalyzed esterification is reversible, and at equilibrium, the quantities of remaining carboxylic acid and alcohol are generally appreciable. By controlling the experimental conditions, however, we can use Fischer esterification to prepare esters in high yields. If the alcohol is inexpensive compared with the carboxylic acid, we can use a large excess of the alcohol (one of the starting reagents) to drive the equilibrium to the right and achieve a high conversion of carboxylic acid to its ester. Alternatively, we can remove water (one of the products of the reaction) as it forms and drive the equilibrium to the right (review Le Chatelier's principle, Section 7-7).

CHEMICAL CONNECTIONS 18B

Esters as Flavoring Agents

Flavoring agents are the largest class of food additives. At present, more than 1000 synthetic and natural flavors are available. The majority of these are concentrates or extracts from the material whose flavor is desired; they are often complex mixtures of tens to hundreds of compounds. A number of ester flavoring agents are synthesized industrially. Many have flavors very close to the target flavor, and so adding only one or a few of them is sufficient to make ice cream, soft drinks, or candies taste natural. The table shows the structures of a few of the esters used as flavoring agents.

Structure	Name	Flavor
H–C(=O)–OEt	Ethyl formate	Rum
CH₃–C(=O)–O–CH₂CH₂CH(CH₃)₂	Isopentyl acetate	Banana
CH₃–C(=O)–O–(CH₂)₇CH₃	Octyl acetate	Orange
CH₃CH₂CH₂–C(=O)–OMe	Methyl butanoate	Apple
CH₃CH₂CH₂–C(=O)–OEt	Ethyl butanoate	Pineapple
2-NH₂–C₆H₄–C(=O)–OMe	Methyl 2-aminobenzoate (Methyl anthranilate)	Grape

EXAMPLE 18-3 Fischer Esterification

Complete these Fischer esterification reactions (assume an excess of the alcohol). The stoichiometry of each reaction is indicated in the problem.

(a) Ph−COOH + CH$_3$OH $\xrightleftharpoons{H^+}$

Benzoic acid

(b) HOOC−CH$_2$−CH$_2$−COOH + 2 EtOH $\xrightleftharpoons{H^+}$

Butanedioic acid (Succinic acid)   (excess)

Strategy and Solution
Substitution of the —OR group of the alcohol for the —OH group of the carboxyl group gives an ester. Here are the structural formulas and names for the ester produced in each reaction.

(a) Ph−C(=O)−OCH$_3$

Methyl benzoate

(b) EtO−C(=O)−CH$_2$−CH$_2$−C(=O)−OEt

Diethyl butanedioate (Diethyl succinate)

Problem 18-3
Complete these Fischer esterification reactions:

(a) (CH$_3$)$_2$CHCOOH + HO−C$_6$H$_{11}$ $\xrightleftharpoons{H^+}$

(b) HO−CH$_2$CH$_2$CH$_2$−COOH $\xrightleftharpoons{H^+}$ (a cyclic ester)

E. Decarboxylation

Decarboxylation is the loss of CO$_2$ from a carboxyl group. Almost any carboxylic acid, when heated to a very high temperature, undergoes thermal decarboxylation:

$$\text{RCOOH} \xrightarrow[\text{high temperature}]{\text{decarboxylation}} \text{RH} + \text{CO}_2$$

Most carboxylic acids, however, are quite resistant to moderate heat and melt or even boil without decarboxylation. Exceptions are carboxylic acids that have a carbonyl group β to the carboxyl group (a β-ketoacid). This type of carboxylic acid undergoes decarboxylation quite readily upon mild heating. For example, when 3-oxobutanoic acid (acetoacetic acid) is heated moderately, it undergoes decarboxylation to give acetone and carbon dioxide:

$$\underset{\substack{\text{3-Oxobutanoic acid}\\\text{(Acetoacetic acid)}\\\text{(a }\beta\text{-ketoacid)}}}{\text{CH}_3\text{COCH}_2\text{COOH}} \xrightarrow{\text{warm}} \underset{\text{Acetone}}{\text{CH}_3\text{COCH}_3} + \text{CO}_2$$

Decarboxylation by moderate heating is a unique property of β-ketoacids and is not observed with other classes of ketoacids.

Mechanism: Decarboxylation of a β-Ketocarboxylic Acid

Step 1: Redistribution of six electrons in a cyclic six-membered transition state gives carbon dioxide and an enol:

[Cyclic six-membered transition state] $\xrightarrow{(1)}$ Enol of a ketone $\xrightleftharpoons{(2)}$ acetone + CO_2

Step 2: Keto–enol tautomerism (Section 17-5) of the enol gives the more stable keto form of the product:

$$\text{H}_3\text{C–C(OH)=CH}_2 \rightleftharpoons \text{CH}_3\text{–C(=O)–CH}_3$$

An important example of decarboxylation of a β-ketoacid in the biological world occurs during the oxidation of foodstuffs in the tricarboxylic acid (TCA) cycle (Chapter 28). Oxalosuccinic acid, one of the intermediates in this cycle, undergoes spontaneous decarboxylation to produce α-ketoglutaric acid. Only one of the three carboxyl groups of oxalosuccinic acid has a carbonyl group in the position β to it, and this carboxyl group is lost as CO_2:

Only this carboxyl has a C=O beta to it

$$\underset{\text{Oxalosuccinic acid}}{\text{HOOC–CH(COOH)–CH}_\alpha\text{–CO}_\beta\text{–COOH}} \longrightarrow \underset{\alpha\text{-Ketoglutaric acid}}{\text{HOOC–CH}_2\text{–CO–COOH}} + \text{CO}_2$$

Note that thermal decarboxylation is a reaction unique to β-ketoacids—it does not occur with α-ketoacids. In the biochemistry chapters that follow, however, we will see examples of decarboxylation of α-ketoacids—for example, the decarboxylation of α-ketoglutarate. Because the decarboxylation of α-ketoacids requires an oxidizing agent (NAD^+), this reaction is called oxidative decarboxylation.

CHEMICAL CONNECTIONS 18C

Ketone Bodies and Diabetes

3-Oxobutanoic acid (acetoacetic acid) and its reduction product, 3-hydroxybutanoic acid, are synthesized in the liver from acetyl-CoA (Section 28-5), a product of the metabolism of fatty acids and certain amino acids.

3-Oxobutanoic acid (Acetoacetic acid)

3-Hydroxybutanoic acid (β-Hydroxybutyric acid)

3-Oxobutanoic acid and 3-hydroxybutanoic acid are known collectively as ketone bodies.

The concentration of ketone bodies in the blood of healthy, well-fed humans is approximately 0.01 mM/L. However, in persons suffering from starvation or diabetes mellitus, the concentration of ketone bodies may increase to as much as 500 times the normal level. Under these conditions, the concentration of acetoacetic acid increases to the point where it undergoes spontaneous decarboxylation to form acetone and carbon dioxide. Acetone is not metabolized by humans and is excreted through the kidneys and the lungs. The odor of acetone is responsible for the characteristic "sweet smell" on the breath of severely diabetic patients.

Summary of Key Questions

Section 18-1 What Are Carboxylic Acids?
- The functional group of a **carboxylic acid** is the **carboxyl group**, —COOH.

Section 18-2 How Do We Name Carboxylic Acids?
- IUPAC names of carboxylic acids are derived from the name of the parent alkane by dropping the suffix *-e* and adding *-oic acid*. Because the carboxyl group can only occur at the end of a carbon chain, there is no need to give it a number because it is automatically carbon-1.
- Dicarboxylic acids are named as *-dioic acids*.
- Common names for many carboxylic and dicarboxylic acids are still widely used.

Section 18-3 What Are the Physical Properties of Carboxylic Acids?
- Carboxylic acids are polar compounds and can participate in hydrogen bonding as both a hydrogen-bond donor and a hydrogen-bond acceptor. Consequently, they have higher boiling points and are more soluble in water than alcohols, aldehydes, ketones, and ethers of comparable molecular weight.

Section 18-4 What Are Soaps and Detergents?
- Fatty acids are long, unbranched-chain carboxylic acids. They can be saturated or unsaturated.
- A **triglyceride** is a triester of glycerol.
- A **micelle** is a spherical arrangement of molecules in an aqueous environment in which the hydrocarbon parts are on the inside and the hydrophilic parts are on the surface.

Section 18-5 What Are the Characteristic Reactions of Carboxylic Acids?
- Carboxylic acids are weak acids that react with strong bases to form water-soluble salts.
- Treatment of a carboxylic acid with an alcohol in the presence of an acid catalyst gives an ester.
- When exposed to a very high temperature, carboxylic acids can undergo decarboxylation.

Summary of Key Reactions

1. **Acidity of Carboxylic Acids (Section 18-5A)** Values of pK_a for most unsubstituted carboxylic acids are within the range of 4 to 5.

$$CH_3COH + H_2O \rightleftharpoons CH_3CO^- + H_3O^+$$

$$K_a = \frac{[CH_3COO^-][H_3O^+]}{[CH_3COOH]} = 1.74 \times 10^{-5}$$

$$pK_a = 4.75$$

2. Reaction of Carboxylic Acids with Bases (Section 18-5B)
Carboxylic acids, whether water-soluble or insoluble, react with alkali metal hydroxides, carbonates and bicarbonates, and ammonia and amines to form water-soluble salts.

$$\text{Ph-COOH} + \text{NaOH} \xrightarrow{H_2O}$$

Benzoic acid
(slightly soluble in water)

$$\text{Ph-COO}^-\text{Na}^+ + H_2O$$

Sodium benzoate
(60 g/100 mL water)

$$CH_3COOH + NaHCO_3 \longrightarrow$$

$$CH_3COO^-Na^+ + CO_2 + H_2O$$

3. Reduction by Lithium Aluminum Hydride (Section 18-5C)
Lithium aluminum hydride reduces a carboxyl group to a primary alcohol. This reagent does not normally reduce carbon–carbon double bonds, but it does reduce aldehydes to 1° alcohols and ketones to 2° alcohols.

3-Cyclopentene-carboxylic acid $\xrightarrow[\text{2. H}_2\text{O}]{\text{1. LiAlH}_4,\text{ ether}}$

4-Hydroxymethyl-cyclopentene + LiOH + Al(OH)$_3$

4. Fischer Esterification (Section 18-5D)
Fischer esterification is reversible.

In **Fischer esterification**, a carboxylic acid and an alcohol are heated in the presence of a strong acid, most commonly sulfuric acid, to produce an ester.

$$CH_3COH + CH_3CH_2OH \xrightleftharpoons{H_2SO_4}$$

Ethanoic acid   Ethanol
(Acetic acid) (Ethyl alcohol)

$$CH_3COCH_2CH_3 + H_2O$$

Ethyl ethanoate
(Ethyl acetate)

One way to force the equilibrium to the right is to use an excess of the alcohol. Alternatively, water can be removed from the reaction mixture as it is formed.

5. Decarboxylation (Section 18-5E)
Thermal decarboxylation is a unique property of β-ketoacids. The immediate products of thermal decarboxylation of a β-ketoacid are carbon dioxide and an enol. Loss of CO_2 is followed immediately by keto-enol tautomerism.

3-Oxobutanoic acid $\xrightarrow{\text{warm}}$ Acetone + CO_2
(Acetoacetic acid)
(a β-ketoacid)

Problems

Orange-numbered problems are applied.

For preparation of carboxylic acids, review Chapters 14, Alcohols, Esters, and Thiols, and 17, Aldehydes and Ketones

Section 18-1 What Are Carboxylic Acids?

18-4 Answer true or false.
   (a) The functional groups of a carboxylic acid are a carbonyl group bonded to a hydroxyl group.
   (b) The VSEPR model predicts bond angles of 180° about the carbonyl carbon of a carboxyl group.
   (c) The VSEPR model predicts bond angles of 109.5° about the oxygen of the OH group of a carboxyl group.
   (d) The carbonyl carbon of a carboxyl group can be a stereocenter, depending on its location within a molecule.
   (e) Carboxylic acids can be prepared by chromic acid oxidation of primary alcohols and of aldehydes.
   (f) The product of chromic acid oxidation of hexanoic acid is 1-hexanol.

Section 18-2 How Do We Name Carboxylic Acids?

18-5 Answer true or false.
   (a) The general name of an aliphatic carboxylic acid is alkanoic acid.
   (b) A molecule containing two COOH groups is called a dicarboxylic acid.
   (c) Ethanedioic acid (oxalic acid) is the simplest dicarboxylic acid.
   (d) 3-Methylbutanoic acid is chiral.
   (e) The simplest carboxylic acid is methanoic acid (common name: formic acid), HCO_2H.

(f) Benzoic acid is an aromatic carboxylic acid.

(g) Formic acid, which is the common name for HCO_2H, is derived from the word *formica*, which is the Latin name for ants.

(h) (S)-Lactic acid, CH_3—CHOH—COOH, contains two functional groups: a 2° alcohol and a carboxyl group.

18-6 Name and draw structural formulas for the four carboxylic acids with the molecular formula $C_5H_{10}O_2$. Which of these carboxylic acids are chiral?

18-7 Write the IUPAC name for each carboxylic acid.

(a) [structure: isobutyl-CH2-COOH] (b) [structure: CH3-CH2-CH(NH2)-COOH]

(c) [structure: CH3-CH2-CH2-CH2-CH2-COOH, pentyl chain ending in COOH]

18-8 Write the IUPAC name for each carboxylic acid.

(a) HOOC—CH(OH)—COOH   (b) salicylic acid (o-hydroxybenzoic acid)

(c) CCl_3COOH

18-9 Draw a structural formula for each carboxylic acid.
(a) 4-Nitrophenylacetic acid
(b) 4-Aminobutanoic acid
(c) 4-Phenylbutanoic acid
(d) *cis*-3-Hexenedioic acid

18-10 Draw a structural formula for each carboxylic acid.
(a) 2-Aminopropanoic acid
(b) 3,5-Dinitrobenzoic acid
(c) Dichloroacetic acid
(d) *o*-Aminobenzoic acid

18-11 Draw a structural formula for each carboxylate salt.
(a) Sodium benzoate   (b) Lithium acetate
(c) Ammonium acetate   (d) Disodium adipate
(e) Sodium salicylate   (f) Calcium butanoate

18-12 Calcium oxalate is a major component of kidney stones. Draw a structural formula for this compound.

18-13 The monopotassium salt of oxalic acid is present in certain leafy vegetables, including rhubarb. Both oxalic acid and its salts are poisonous in high concentrations. Draw a structural formula for monopotassium oxalate.

Section 18-3 What Are the Physical Properties of Carboxylic Acids?

18-14 Answer true or false.

(a) Carboxylic acids are polar compounds.

(b) The most polar bond of a carboxyl group is the C—O single bond.

(c) Carboxylic acids have significantly higher boiling points than aldehydes, ketones, and alcohols of comparable molecular weight.

(d) The low-molecular-weight carboxylic acids (formic, acetic, propanoic, and butanoic acids) are infinitely soluble in water.

(e) The following compounds are arranged in order of increasing boiling point:

$$\underset{I}{HOCCH_2CH_2COH} < \underset{II}{HOCCH_2CH_2CH_3} <$$

$$\underset{III}{HOCCH_2CH_2CH_2CH_3} < \underset{IV}{HOCCH_2CH_2CH_2COH}$$

18-15 Draw a structural formula for the dimer formed when two molecules of formic acid interact by hydrogen bonding.

18-16 Propanedioic (malonic) acid forms an intramolecular hydrogen bond in which the H of one COOH group forms a hydrogen bond with an O of the other COOH group. Draw a structural formula to show this internal hydrogen bonding. (There are two possible answers.)

18-17 Hexanoic (caproic) acid has a solubility in water of about 1 g/100 mL water. Which part of the molecule contributes to water solubility, and which part prevents solubility?

18-18 Propanoic acid and methyl acetate are constitutional isomers, and both are liquids at room temperature. One of these compounds has a boiling point of 141°C; the other has a boiling point of 57°C. Which compound has which boiling point? Explain.

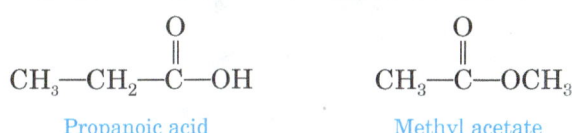

Propanoic acid   Methyl acetate

18-19 The following compounds have approximately the same molecular weight: hexanoic acid, heptanal, and 1-heptanol. Arrange them in order of increasing boiling point.

18-20 The following compounds have approximately the same molecular weight: propanoic acid, 1-butanol, and diethyl ether. Arrange them in order of increasing boiling point.

18-21 Arrange these compounds in order of increasing solubility in water: acetic acid, pentanoic acid, decanoic acid.

Section 18-4 What Are Soaps and Detergents?

18-22 Answer true or false.

(a) Fatty acids are long-chain carboxylic acids, with most consisting of between 12 to 20 carbons in an unbranched chain.

(b) An unsaturated fatty acid contains one or more carbon–carbon double bonds in its hydrocarbon chain.

(c) In most unsaturated fatty acids found in animal fats, vegetable oils, and biological membranes, the *cis* isomer predominates.

(d) In general, unsaturated fatty acids have lower melting points than saturated fatty acids with the same number of carbon atoms.

(e) Natural soaps are sodium or potassium salts of fatty acids.

(f) Soaps remove grease, oil, and fat stains by incorporating these substances into the nonpolar interior of soap micelles.

(g) "Hard water," by definition, is water that contains Ca^{2+}, Mg^{2+}, or Fe^{3+} ions, all of which react with soap molecules to form water-insoluble salts.

(h) The structure of synthetic detergents is patterned after that of natural soaps.

(i) The most widely used synthetic detergents are the linear alkylbenzenesulfonates (LAS).

(j) Present-day synthetic detergents do not form water-insoluble salts with hard water.

(k) Most detergent preparations contain foam stabilizers, a bleach, and optical brighteners (optical bleaches).

18-23 Characterize the structural features necessary to make a good synthetic detergent.

18-24 The detergents illustrated in this chapter are classified as anionic detergents. Following are structural formulas for two other classifications of synthetic detergents: cationic detergents and neutral detergents.

$$CH_3(CH_2)_6CH_2\overset{+}{N}(CH_3)(CH_2C_6H_5)CH_3 \;\; Cl^-$$

Benzyldimethyloctylammonium chloride
(a cationic detergent)

$$\underset{\underset{HOCH_2}{|}}{\overset{\overset{HOCH_2}{|}}{HOCH_2CCH_2OC(CH_2)_{14}CH_3}}$$

Pentaerythrityl palmitate
(a neutral detergent)

Explain how each compound is able to function as a detergent.

18-25 Following are structural formulas for two more cationic detergents. Each is a mild surface-acting detergent and fungicide and is used as a topical antiseptic and disinfectant. They are examples of quaternary ammonium detergents, commonly called "quats." Account for the detergent properties of each.

Cetylpyridinium chloride   Benzylcetyldimethylammonium chloride

Section 18-5 What Are the Characteristic Reactions of Carboxylic Acids?

18-26 Answer true or false.

(a) Carboxylic acids are weak acids compared to mineral acids such as HCl, H_2SO_4, and HNO_3.

(b) Phenols, alcohols, and carboxylic acids have in common the presence of an —OH group.

(c) Carboxylic acids are stronger acids than alcohols but weaker acids than phenols.

(d) The order of acidity of the following carboxylic acids is:

A > B > C > D

(e) The order of acidity of the following carboxylic acids is:

A > B > C > D

(f) The reaction of benzoic acid with aqueous sodium hydroxide gives sodium benzoate.

(g) A mixture of the following compounds is extracted in order with (1) 1 M HCl, (2) 1 M NaOH, and (3) diethyl ether. Only compound II is extracted into the basic layer.

I (aniline, NH_2)   II (benzoic acid, COOH)   III (acetophenone)

(h) Conversion of compound I to compound II is best accomplished by reduction with $NaBH_4$.

I → II

(i) The following ester can be prepared by treating benzoic acid with 1-butanol in the presence of a catalytic amount of H_2SO_4:

[structure: benzene ring with –C(=O)OCH₂CH₂CH₂CH₃ and –OH ortho substituents]

(j) Thermal decarboxylation of this β-ketoacid gives benzoic acid and carbon dioxide:

[structure: C₆H₅–C(=O)–CH₂–C(=O)OH]

(k) Thermal decarboxylation of this β-ketoacid gives 2-pentanone and carbon dioxide.

$$CH_3CH_2CH_2\overset{O}{\underset{\|}{C}}CH_2\overset{O}{\underset{\|}{C}}OH$$

18-27 Alcohols, phenols, and carboxylic acids all contain an —OH group. Which are the strongest acids? Which are the weakest acids?

18-28 Arrange these compounds in order of increasing acidity: benzoic acid, benzyl alcohol, phenol.

18-29 Complete the equations for these acid–base reactions.

(a) [C₆H₅]–CH₂COOH + NaOH ⟶

(b) CH₂=CH–CH₂–COOH + NaHCO₃ ⟶

(c) [benzene ring with –COOH and –OCH₃ ortho] + NaHCO₃ ⟶

(d) CH₃CH(OH)COOH + H₂NCH₂CH₂OH ⟶

(e) CH₂=CH–CH₂–COO⁻Na⁺ + HCl ⟶

18-30 Complete the equations for these acid–base reactions.

(a) [benzene ring with –OH and –CH₃ ortho] + NaOH ⟶

(b) [benzene ring with –COO⁻Na⁺ and –OH ortho] + HCl ⟶

(c) [benzene ring with –COOH and –OCH₃ ortho] + H₂NCH₂CH₂OH ⟶

(d) [C₆H₁₁]–COOH + NaHCO₃ ⟶

18-31 Formic acid is one of the components responsible for the sting of biting ants and is injected under the skin by bee and wasp stings. The pain can be relieved by rubbing the area of the sting with a paste of baking soda ($NaHCO_3$) and water, which neutralizes the acid. Write an equation for this reaction.

18-32 Starting with the definition of K_a of a weak acid, HA, as

$$HA + H_2O \rightleftharpoons A^- + H_3O^+ \qquad K_a = \frac{[A^-][H_3O^+]}{[HA]}$$

show that:

$$\frac{[A^-]}{[HA]} = \frac{K_a}{[H_3O^+]}$$

18-33 Using the equation from Problem 18-32 that shows the relationship between K_a, $[H_3O^+]$, $[A^-]$, and $[HA]$, calculate the ratio of $[A^-]$ to $[HA]$ in a solution whose pH is:
(a) 2.0 (b) 5.0
(c) 7.0 (d) 9.0
(e) 11.0
Assume that the pK_a of the weak acid is 5.0.

18-34 The pK_a of acetic acid is 4.75. What form(s) of acetic acid are present at pH 2.0 or lower? At pH 4.75? At pH 8.0 or higher?

18-35 The normal pH range for blood plasma is 7.35 to 7.45. Under these conditions, would you expect the carboxyl group of lactic acid (pK_a 4.07) to exist primarily as a carboxyl group or as a carboxylate anion? Explain.

18-36 The pK_a of ascorbic acid (Vitamin C, see Chemical Connections 20B) is 4.10. Would you expect ascorbic acid dissolved in blood plasma, pH 7.35–7.45, to exist primarily as ascorbic acid or as ascorbate anion? Explain.

18-37 Complete the equations for the following acid–base reactions. Assume one mole of NaOH per mole of amino acid. (*Hint*: Review Section 8-4.)

(a) $CH_3\underset{NH_3^+}{\underset{|}{CH}}COOH + NaOH \xrightarrow{H_2O}$

(b) $CH_3\underset{NH_3^+}{\underset{|}{CH}}COO^-Na^+ + NaOH \xrightarrow{H_2O}$

18-38 Which is the stronger base: $CH_3CH_2NH_2$ or $CH_3CH_2COO^-$? Explain.

18-39 Complete the equations for the following acid–base reactions. Assume one mole of HCl per mole of amino acid.

(a) $CH_3\underset{NH_2}{\underset{|}{CH}}COO^-Na^+ + HCl \xrightarrow{H_2O}$

(b) $CH_3\underset{NH_3^+}{\underset{|}{CH}}COO^-Na^+ + HCl \xrightarrow{H_2O}$

Section 18-5D Fischer Esterification

18-40 Define and give an example of Fischer esterification.

18-41 Complete these examples of Fischer esterification. In each case, assume an excess of the alcohol.

(a) $CH_3COOH + HO-CH_2CH(CH_3)_2 \xrightleftharpoons{H^+}$

(b) $CH_3COOH + HO-\text{(cyclohexyl)} \xrightleftharpoons{H^+}$

(c) benzene-1,2-dicarboxylic acid (phthalic acid) $+ CH_3CH_2OH \xrightleftharpoons{H^+}$

18-42 From what carboxylic acid and alcohol is each ester derived?

(a) $CH_3CO-\text{(1,4-cyclohexylene)}-OCCH_3$ (diacetate)

(b) cyclohexyl–$COCH_3$

(c) $CH_3OCCH_2CH_2COCH_3$ (dimethyl succinate)

(d) isobutyl ester of pent-2-enoic acid

18-43 Methyl 2-hydroxybenzoate (methyl salicylate) has the odor of oil of wintergreen. This compound is prepared by Fischer esterification of 2-hydroxybenzoic acid (salicylic acid) with methanol. Draw a structural formula for methyl 2-hydroxybenzoate.

18-44 Show how you could convert cinnamic acid to each compound.

trans-3-Phenyl-2-propenoic acid
(Cinnamic acid)

→ PhCH$_2$CH$_2$COOH

→ PhCH=CHCH$_2$OH

→ PhCH=CHCOOCH$_2$CH$_3$

Additional Problems

18-45 Give the expected organic product formed when phenylacetic acid, $C_6H_5CH_2COOH$, is treated with each of the following reagents:

(a) $NaHCO_3, H_2O$
(b) $NaOH, H_2O$
(c) NH_3, H_2O
(d) $LiAlH_4$ then H_2O
(e) $NaBH_4$ then H_2O
(f) $CH_3OH + H_2SO_4$ (catalyst)
(g) H_2/Ni

18-46 Procaine (its hydrochloride salt is marketed as Novocaine) was one of the first local anesthetics developed for infiltration and regional anesthesia. It is synthesized by the following Fischer esterification:

4-Aminobenzoic acid + 2-Diethylaminoethanol $\xrightarrow{\text{Fischer esterification}}$ Procaine

(a) Draw the structural formula of Procaine.
(b) Which of the two nitrogen atoms in Procaine is the stronger base?
(c) Draw the structural formula of Novocaine (the hydrochloride salt of Procaine).
(d) Would you predict Procaine or Novocaine to be more soluble in blood?

18-47 Methylparaben and propylparaben are used as preservatives in foods, beverages, and cosmetics.

4-Aminobenzoic acid (*p*-Aminobenzoic acid)

→ Methyl 4-aminobenzoate (Methylparaben)

→ Propyl 4-aminobenzoate (Propylparaben)

Show how each of these preservatives can be prepared from 4-aminobenzoic acid.

18-48 4-Aminobenzoic acid is prepared from benzoic acid by the following two steps.

Benzoic acid —(1)→ 4-Nitrobenzoic acid

—(2)→ 4-Aminobenzoic acid

Show reagents and experimental conditions to bring about each step.

Looking Ahead

18-49 When 5-hydroxypentanoic acid is treated with an acid catalyst, it forms a lactone (a cyclic ester). Draw a structural formula for this lactone.

18-50 We have seen that esters can be prepared by treatment of a carboxylic acid and an alcohol in the presence of an acid catalyst. Suppose you start instead with a dicarboxylic acid such as 1,6-hexanedioic acid (adipic acid) and a diol such as 1,2-ethanediol (ethylene glycol).

HOCCH$_2$CH$_2$CH$_2$CH$_2$COH +

1,6-Hexanedioic acid
(Adipic acid)

HOCH$_2$CH$_2$OH ⟶ A polyester

1,2-Ethanediol
(Ethylene glycol)

Show how Fischer esterification in this case can produce a polymer (a macromolecule with a molecular weight several thousand times that of the starting materials).

Tying It Together

18-51 Draw the structural formula of a compound with the given molecular formula that, on oxidation by potassium dichromate in aqueous sulfuric acid, gives the carboxylic acid or dicarboxylic acid shown.

(a) C$_6$H$_{14}$O oxidation ⟶

(b) C$_6$H$_{12}$O oxidation ⟶

(c) C$_6$H$_{14}$O$_2$ oxidation ⟶

18-52 Complete the equations for these oxidations:

(a) CH$_3$(CH$_2$)$_4$CH$_2$OH $\xrightarrow{K_2Cr_2O_7 / H_2SO_4}$

(b) (aromatic ring with CHO, OCH$_3$, OH) + Ag(NH$_3$)$_2^+$ ⟶

(c) HO—(cyclohexane)—CH$_2$OH $\xrightarrow{K_2Cr_2O_7 / H_2SO_4}$

18-53 In Chapter 27, we will study a metabolic pathway called the tricarboxylic acid cycle (TCA), also known as the citric acid or Krebs cycle. We have already seen examples of most of the reactions in this pathway. Following is an outline of the pathway beginning with the molecule for which the pathway is named. A particular enzyme that is highly specific catalyzes each of these reactions. Each enzyme-catalyzed reaction gives a high yield of the target molecule.

Citric acid —(1)→ Aconitic acid —(2)→ Isocitric acid —(3)→ Oxalosuccinic acid —(4)→ α-Ketoglutaric acid —(5)→ Succinic acid —(6)→ Fumaric acid —(7)→ Malic acid —(8)→ Oxaloacetic acid

(a) Which of these TCA-cycle intermediates are chiral? Which intermediate has the greatest number of chiral centers? Which intermediates show *cis-trans* isomerism?

(b) Name the type of reaction that takes place in Steps 1–3.

(c) Notice that the hydration of aconitic acid to give isocitric acid does not follow Markovnikov's rule. If the hydration of aconitic acid were to follow Markovnikov's rule, what product would be formed? Offer an explanation for the formation of this non-Markovnikov product.

(d) What type of reactions take place in Steps 4–8?

(e) The only reaction we have not studied is Step 5. Because it involves a loss of CO_2, it is classified as a decarboxylation. Show that it also involves oxidation. Because it involves both oxidation and decarboxylation, it is classified as an oxidative decarboxylation.

(f) Reaction 4 is also classified as a decarboxylation. Does this decarboxylation also involve an oxidation?

18-54 Azithromycin is a broad-spectrum antibiotic derived from erythromycin. It is one of a large number of what are called macrocyclic antibiotics, so named because they contain a large ring as part of their structure. Like many of the macrocyclic antibiotics, azithromycin has an incredibly complex structure, and it was an enormous challenge for chemists to determine its structural formula.

Azithromycin
$C_{38}H_{72}N_2O_{12}$

(a) What is the size of the large ring in this antibiotic?

(b) Name the functional groups present in azithromycin.

(c) Is azithromycin chiral? If so, how many stereoisomers are possible for this structural formula?

(d) Azithromycin contains one twenty-one membered ring and two five-membered rings. The larger ring is closed by an intramolecular ester bond. Show that each smaller ring is attached to the larger ring by one arm of an acetal group. If each acetal were to be cleaved by hydrolysis to an alcohol and a carbonyl-containing compound, draw the structural formula of the two carbonyl-containing compounds that would be formed. (Hint: It may be difficult to identify the acetal group but remember that the characteristic structural feature of an acetal is a carbon bonded to two —OR groups.

18-55 Below is the structural formula of a compound known as the *queen substance*, which is secreted in the mandibular gland of queen honey bees (*Apis mellifera*). The *queen substance* inhibits the development of ovaries in worker bees, prevents queen cell formation, and attracts male bees (drones) to virgin queens for the purpose of mating. Write the IUPAC name of this compound. Note that in the IUPAC system, when the carboxylic acid also contains an aldehyde or ketone group in the molecule, the presence of the aldehyde or ketone group is indicated by the prefix *–oxo*.

Queen substance

APPENDIX 1
Exponential Notation

The **exponential notation** system is based on powers of 10 (see table). For example, if we multiply $10 \times 10 \times 10 = 1000$, we express this as 10^3. The 3 in this expression is called the **exponent** or the **power**, and it indicates how many times we multiplied 10 by itself and how many zeros follow the 1.

There are also negative powers of 10. For example, 10^{-3} means 1 divided by 10^3:

$$10^{-3} = \frac{1}{10^3} = \frac{1}{1000} = 0.001$$

Numbers are frequently expressed like this: 6.4×10^3. In a number of this type, 6.4 is the **coefficient** and 3 is the exponent, or power of 10. This number means exactly what it says:

$$6.4 \times 10^3 = 6.4 \times 1000 = 6400$$

Similarly, we can have coefficients with negative exponents:

$$2.7 \times 10^{-5} = 2.7 \times \frac{1}{10^5} = 2.7 \times 0.00001 = 0.000027$$

For numbers greater than 10 in exponential notation, we proceed as follows: *Move the decimal point to the left,* to just after the first digit. The (positive) exponent is equal to the number of places we moved the decimal point.

Exponential notation is also called scientific notation.

For example, 10^6 means a one followed by six zeros, or 1,000,000, and 10^2 means 100.

APP. 1-1 Examples of Exponential Notation

10,000	$= 10^4$
1000	$= 10^3$
100	$= 10^2$
10	$= 10^1$
1	$= 10^0$
0.1	$= 10^{-1}$
0.01	$= 10^{-2}$
0.001	$= 10^{-3}$

EXAMPLE

$37500 = 3.75 \times 10^4$ — 4 because we went four places to the left

Four places to the left — Coefficient

$628 = 6.28 \times 10^2$

Two places to the left — Coefficient

$859{,}600{,}000{,}000 = 8.596 \times 10^{11}$

Eleven places to the left — Coefficient

We don't really have to place the decimal point after the first digit, but by doing so we get a coefficient between 1 and 10, and that is the custom.

Using exponential notation, we can say that there are 2.95×10^{22} copper atoms in a copper penny. For large numbers, the exponent is always *positive*. Note that we do not usually write out the zeros at the end of the number.

For small numbers (less than 1), we move the decimal point *to the right*, to just after the first nonzero digit, and use a *negative exponent*.

> **EXAMPLE**
>
> $$0.00346 = 3.46 \times 10^{-3}$$
>
> Three places to the right
>
> $$0.000004213 = 4.213 \times 10^{-6}$$
>
> Six places to the right

In exponential notation, a copper atom weighs 2.3×10^{-25} pounds.

To convert exponential notation into fully written-out numbers, we do the same thing backward.

> **EXAMPLE**
>
> Write out in full: (a) 8.16×10^7 (b) 3.44×10^{-4}
>
> **Solution**
>
> (a) $8.16 \times 10^7 = 81{,}600{,}000$ (b) $3.44 \times 10^{-4} = 0.000344$
>
> Seven places to the right (add enough zeros)   Four places to the left

When scientists add, subtract, multiply, and divide, they are always careful to express their answers with the proper number of digits, called significant figures. This method is described in Appendix II.

A. Adding and Subtracting Numbers in Exponential Notation

We are allowed to add or subtract numbers expressed in exponential notation *only if they have the same exponent*. All we do is add or subtract the coefficients and leave the exponent as it is.

> **EXAMPLE**
>
> Add 3.6×10^{-3} and 9.1×10^{-3}.
>
> **Solution**
>
> $$\begin{array}{r} 3.6 \times 10^{-3} \\ +\ 9.1 \times 10^{-3} \\ \hline 12.7 \times 10^{-3} \end{array}$$
>
> The answer could also be written in other, equally valid ways:
>
> $$12.7 \times 10^{-3} = 0.0127 = 1.27 \times 10^{-2}$$

When it is necessary to add or subtract two numbers that have different exponents, we first must change them so that the exponents are the same.

A calculator with exponential notation changes the exponent automatically.

EXAMPLE

Add 1.95×10^{-2} and 2.8×10^{-3}.

Solution

To add these two numbers, we make both exponents -2. Thus, $2.8 \times 10^{-3} = 0.28 \times 10^{-2}$. Now we can add:

$$\begin{array}{r} 1.95 \times 10^{-2} \\ + \; 0.28 \times 10^{-2} \\ \hline 2.23 \times 10^{-2} \end{array}$$

B. Multiplying and Dividing Numbers in Exponential Notation

To multiply numbers in exponential notation, we first multiply the coefficients in the usual way and then algebraically *add* the exponents.

EXAMPLE

Multiply 7.40×10^5 by 3.12×10^9.

Solution

$$7.40 \times 3.12 = 23.1$$

Add exponents:

$$10^5 \times 10^9 = 10^{5+9} = 10^{14}$$

Answer:

$$23.1 \times 10^{14} = 2.31 \times 10^{15}$$

EXAMPLE

Multiply 4.6×10^{-7} by 9.2×10^4.

Solution

$$4.6 \times 9.2 = 42$$

Add exponents:

$$10^{-7} \times 10^4 = 10^{-7+4} = 10^{-3}$$

Answer:

$$42 \times 10^{-3} = 4.2 \times 10^{-2}$$

To divide numbers expressed in exponential notation, the process is reversed. We first divide the coefficients and then algebraically *subtract* the exponents.

> **EXAMPLE**
>
> Divide: $\dfrac{6.4 \times 10^8}{2.57 \times 10^{10}}$
>
> **Solution**
>
> $$6.4 \div 2.57 = 2.5$$
>
> Subtract exponents:
>
> $$10^8 \div 10^{10} = 10^{8-10} = 10^{-2}$$
>
> Answer:
>
> $$2.5 \times 10^{-2}$$

> **EXAMPLE**
>
> Divide: $\dfrac{1.62 \times 10^{-4}}{7.94 \times 10^7}$
>
> **Solution**
>
> $$1.62 \div 7.94 = 0.204$$
>
> Subtract exponents:
>
> $$10^{-4} \div 10^7 = 10^{-4-7} = 10^{-11}$$
>
> Answer:
>
> $$0.204 \times 10^{-11} = 2.04 \times 10^{-12}$$

Scientific calculators do these calculations automatically. All that is necessary is to enter the first number, press $+$, $-$, $\times$, or $\div$, enter the second number, and press $=$. (The method for entering numbers of this form varies; consult the instructions that come with the calculator.) Many scientific calculators also have a key that will automatically convert a number such as 0.00047 to its scientific notation form (4.7×10^{-4}), and vice versa. For problems relating to exponential notation, see Chapter 1, Problems 1-17 through 1-24.

APPENDIX II
Significant Figures

If you measure the volume of a liquid in a graduated cylinder, you might find that it is 36 mL, to the nearest milliliter, but you cannot tell if it is 36.2, or 35.6, or 36.0 mL because this measuring instrument does not give the last digit with any certainty. A buret gives more digits, and if you use one you should be able to say, for instance, that the volume is 36.3 mL and not 36.4 mL. But even with a buret, you could not say whether the volume is 36.32 or 36.33 mL. For that, you would need an instrument that gives still more digits. This example should show you that *no measured number can ever be known exactly.* No matter how good the measuring instrument, there is always a limit to the number of digits it can measure with certainty.

We define the number of **significant figures** as the number of digits of a measured number that have uncertainty only in the last digit.

What do we mean by this definition? Assume that you are weighing a small object on a laboratory balance that can weigh to the nearest 0.1 g, and you find that the object weighs 16 g. Because the balance weighs to the nearest 0.1 g, you can be sure that the object does not weigh 16.1 g or 15.9 g. In this case, you would write the weight as 16.0 g. To a scientist, there is a difference between 16 g and 16.0 g. Writing 16 g says that you don't know the digit after the 6. Writing 16.0 g says that you do know it: It is 0. However, you don't know the digit after that. Several rules govern the use of significant figures in reporting measured numbers.

A. Determining the Number of Significant Figures

In Section 1.3, we saw how to determine the number of significant figures in a reported number. We summarize those guidelines here:

1. Nonzero digits are always significant.
2. Zeros at the beginning of a number are never significant.
3. Zeros between nonzero digits are always significant.
4. Zeros at the end of a number that contains a decimal point are always significant.
5. Zeros at the end of a number that contains no decimal point may or may not be significant.

We use periods as decimal points throughout this text to indicate the significant figures in numbers with trailing zeros. For example, 1000. mL has four significant figures; 20. m has two significant figures.

B. Multiplying and Dividing

The rule in multiplication and division is that the final answer should have the *same* number of significant figures as there are in the number with the *fewest* significant figures.

> **EXAMPLE**
>
> Do the following multiplications and divisions:
> (a) 3.6×4.27
> (b) 0.004×217.38
> (c) $\dfrac{42.1}{3.695}$
> (d) $\dfrac{0.30652 \times 138}{2.1}$
>
> **Solution**
> (a) 15 (3.6 has two significant figures)
> (b) 0.9 (0.004 has one significant figure)
> (c) 11.4 (42.1 has three significant figures)
> (d) 2.0×10^1 (2.1 has two significant figures)

C. Adding and Subtracting

In addition and subtraction, the rule is completely different. The number of significant figures in each number doesn't matter. The answer is given to the *same number of decimal places* as the term with the fewest decimal places.

> **EXAMPLE**
>
> Add or subtract:
>
> (a) 320.0|84
>   80.4|7
>   200.2|3
>   20.0|
>   620.8|
>
> (b) 61|.4532
>   13|.7
>   22|
>   0|.003
>   97|
>
> (c) 14.26|
>   −1.05|041
>   13.21|
>
> **Solution**
>
> In each case, we add or subtract in the normal way but then round off so that the only digits that appear in the answer are those in the columns in which every digit is significant.

D. Rounding Off

When we have too many significant figures in our answer, it is necessary to round off. In this book we have used the rule that if *the first digit dropped* is 5, 6, 7, 8, or 9, we raise *the last digit kept* to the next number; otherwise, we do not.

> **EXAMPLE**
>
> In each case, drop the last two digits:
> (a) 33.679   (b) 2.4715   (c) 1.1145   (d) 0.001309   (e) 3.52
>
> **Solution**
> (a)   33.679 = 33.7
> (b)   2.4715 = 2.47
> (c)   1.1145 = 1.11
> (d) 0.001309 = 0.0013
> (e)   3.52 = 4

E. Counted or Defined Numbers

All of the preceding rules apply to *measured* numbers and **not** to any numbers that are *counted* or *defined*. Counted and defined numbers are known exactly. For example, a triangle is defined as having 3 sides, not 3.1 or 2.9. Here, we treat the number 3 as if it has an infinite number of zeros following the decimal point.

> **EXAMPLE**
>
> Multiply 53.692 (a measured number) × 6 (a counted number).
>
> **Solution**
>
> $$322.15$$
>
> Because 6 is a counted number, we know it exactly, and 53.692 is the number with the fewest significant figures. All we really are doing is adding 53.692 six times.

For problems relating to significant figures, see Chapter 1, Problems 1-25 to 1-30.

Answers

Chapter 1 Matter, Energy, and Measurement

1-1 multiplication (a) 4.69×10^5 (b) 2.8×10^{-15}; division (a) 2.00×10^{18} (b) 1.37×10^5
1-2 (a) 147°F (b) 8.3°C
1-3 13.8 km
1-4 743 mi/h
1-5 1.3 mg/min
1-6 78.5 g
1-7 2.43 g/mL
1-8 1.016 g/mL
1-9 4.8×10^4 cal = 48 kcal
1-10 46°C
1-11 0.0430 cal/g · deg
1-13 (a) Matter is anything that has mass and takes up space. (b) Chemistry is the science that studies matter.
1-15 Dr. X's claim that the extract cured diabetes would be classified as (c) a hypothesis. No evidence had been provided to prove or disprove the claim.
1-17 (a) 3.51×10^{-1} (b) 6.021×10^2 (c) 1.28×10^{-4} (d) 6.28122×10^5
1-19 (a) 6.65×10^{17} (b) 1.2×10^1 (c) 3.9×10^{-16} (d) 3.5×10^{-23}
1-21 (a) 1.3×10^5 (b) 9.40×10^4 (c) 5.139×10^{-3}
1-23 4.45×10^6
1-25 (a) 2 (b) 5 (c) 5 (d) 5 (e) ambiguous, better to write as 3.21×10^4 (three significant figures) or 32100. (five significant figures) (f) 3 (g) 2
1-27 (a) 92 (b) 7.3 (c) 0.68 (d) 0.0032 (e) 5.9
1-29 (a) 1.53 (b) 2.2 (c) 0.00048
1-31 330 min = 5.6 h
1-33 (a) 20 mm (b) 1 inch (c) 1 mile
1-35 Weight would change slightly. Mass is independent of location, but weight is a force exerted on a body influenced by gravity. The influence of the Earth's gravity decreases with increasing distance from sea level.
1-37 (a) 77°F, 298 K (b) 104°F, 313 K (c) 482°F, 523 K, (d) −459°F, 0 K
1-39 (a) 0.0964 L (b) 27.5 cm (c) 4.57×10^4 g (d) 4.75 m (e) 21.64 mL (f) 3.29×10^3 cc (g) 44 mL (h) 0.711 kg (i) 63.7 cc (j) 7.3×10^4 mg (k) 8.34×10^4 mm (l) 0.361 g
1-41 512 fl oz.
1-43 50 mi/h
1-45 4 tablets
1-47 16 mg
1-49 42 cc/h
1-51 420 min
1-53 solids and liquids
1-55 No, melting is a physical change.
1-57 bottom: manganese; top: sodium acetate; middle: calcium chloride
1-59 0.8 mL
1-61 water
1-63 One should raise the temperature of water to 4°C. During this temperature change, the density of the crystals decreases, while the density of water increases. This brings the less dense crystals to the surface of the more dense water.
1-65 The motion of the wheels of the car generates kinetic energy, which is stored in your battery as potential energy.
1-67 0.34 cal/g · °C
1-69 334 mg
1-71 The body shivers. Further temperature lowering results in unconsciousness and then death.
1-73 Methanol, because its higher specific heat allows it to retain the heat longer.
1-75 0.732
1-77 kinetic: (b), (d), (e); potential: (a), (c)
1-79 the European car
1-81 kinetic energy
1-83 The largest is 41 g. The smallest is 4.1310×10^{-8} kg.
1-85 10.9 h
1-87 The heavy water. When converting the specific heat given in J/g · °C to cal/g · °C, one finds that the specific heat of heavy water is 1.008 cal/g · °C, which is somewhat greater than that of ordinary water.
1-89 (a) 1.57 g/mL (b) 1.25 g/mL
1-91 two
1-93 60 J would raise the temperature by 4.5°C; thus, the final temperature will be 24.5°C.
1-95 0.80 mL
1-97 To do this calculation, you need a conversion factor from kilometers to miles. Table 1-3 gives 1 mile = 1.609 km.

$$95 \text{ km} \times \frac{1 \text{ mile}}{1.609 \text{ km}} = 59 \text{ km}$$

If you use the other possible conversion factor:

$$95 \text{ km} \times \frac{1.609 \text{ km}}{\text{mi}} = \frac{153 \text{ km}^2}{\text{mi}}$$

Both the numbers and the units are incorrect.
1-99 In photosynthesis, the radiant energy of sunlight is converted to chemical energy in the sugars produced.
1-101 Converting 30°C from the Celsius to Fahrenheit temperature scale gives 86°F. You are most likely to be wearing a T-shirt and shorts.
1-103 Cells that have been exposed to several cycles of freezing and thawing will have expanded quite a bit. The expansion process tends to break open the cells to make their contents available for fractionation and further study.
1-105 We use the specific heat of water and the information that a liter of water weighs 1000. grams.

$$\text{Amount of heat} = SH \times m \times (T_2 - T_1)$$

$$\text{Amount of heat} = \frac{1.00 \text{ cal}}{\text{g} \cdot °C} \times 2.000 \text{ L} \times \frac{1000. \text{ grams}}{\text{L}} \times 4.85°C$$

$$\text{Amount of heat} = 9.70 \times 10^3 \text{ calories}$$

1-107 Determining the amount of substance and its effectiveness can be done together. You separate the components of the original material and, in the process, determine its amount. One possible way is to weigh the amounts of recovered material. You would then test the substance to see whether the individual compound yields the predicted results.
1-109 4.85×10^3 calories
1-111 (a) 20. mL   (b) No; 5 gtts/min

Chapter 2 Atoms

2-1 (a) $NaClO_3$ (b) AlF_3
2-2 (a) The mass number is $15 + 16 = 31$.
(b) The mass number is $86 + 136 = 222$.
2-3 (a) The element is phosphorus (P); its symbol is $^{31}_{15}P$.
(b) The element is radon (Rn); its symbol is $^{222}_{86}Rn$.
2-4 (a) The atomic number of mercury (Hg) is 80; that of lead (Pb) is 82.
(b) An atom of Hg has 80 protons; an atom of Pb has 82 protons.
(c) The mass number of this isotope of Hg is 200; the mass number of this isotope of Pb is 202.
(d) The symbols of these isotopes are $^{200}_{80}Hg$ and $^{202}_{82}Pb$.
2-5 The atomic number of iodine (I) is 53. The number of neutrons in each isotope is 72 for iodine-125 and 78 for iodine-131. The symbols for these two isotopes are $^{125}_{53}I$ and $^{131}_{53}I$, respectively.
2-6 Lithium-7 is the more abundant isotope (92.50%). The natural abundance of lithium-6 is 7.50%.
2-7 The element is aluminum (Al). Its Lewis dot structure is:

$$\cdot Al:$$

2-9 (a) F (b) T (c) T (d) T (e) F (f) T
(g) T (h) T (i) F (j) F (k) T (l) F
(a) False: Matter is divided into pure substances and mixtures.
(e) False: Mixtures can be separated into their component pure substances.
(i) False: Technetium, promethium, and all of the elements beyond uranium are man-made.
(j) False: H, O, C, N, Ca, and P are the six most important elements in the human body.
(l) False: The combining ratio is based on the ratio of atoms, not the ratio of masses.
2-11 (a) Oxygen (b) Lead (c) Calcium
(d) Sodium (e) Carbon (f) Titanium (g) Sulfur
(h) Iron (i) Hydrogen (j) Potassium (k) Silver (l) Gold
2-13 (a) Americium (b) Berkelium (c) Californium
(d) Dubnium (e) Europium (f) Francium
(g) Gallium (h) Germanium (i) Hafnium
(j) Hassium (k) Holmium (l) Lutetium
(m) Magnesium (n) Polonium (o) Rhenium
(p) Ruthenium (q) Scandium (r) Strontium
(s) Ytterbium, Yttrium, Terbium (t) Thulium
2-15 (a) K_2O (b) Na_3PO_4 (c) $LiNO_3$
2-17 (a) The law of conservation of mass states that matter can be neither created nor destroyed. Dalton's theory explains this because if all matter is made up of indestructible atoms, then any chemical reaction just changes the attachments between atoms and does not destroy the atoms themselves.
(b) The law of constant composition states that any compound is always made up of elements in the same proportion by mass.

Dalton's theory explains this because molecules consist of tightly bound groups of atoms, each of which has a particular mass. Therefore, each element in a compound always constitutes a fixed proportion of the total mass.
2-19 No. CO and CO_2 are different compounds, and each obeys the law of constant composition for that compound.
2-21 (a) F (b) T (c) T (d) F (e) T (f) T (g) T (h) T
(i) F (j) F (k) T (l) F (m) T (n) F (o) T (p) T (q) T
(r) F (s) T (t) F
(a) False: Electrons and protons have equal but opposite charges. Electrons have a much lighter mass than protons.
(d) False: 1 amu has a mass of 1.6605×10^{-24} grams.
(i) False: Electrons and protons have opposite charges and attract each other.
(j) False: The size of an atom includes the space occupied by its electrons. The nucleus is a small fraction of the size of an atom.
(l) False: The mass number is the number of protons and neutrons.
(n) False: ^1H has no neutrons. ^2H has one neutron, and ^3H has two neutrons.
(r) False: Atomic weights are averages of the known isotopes.
(t) False: Density is mass/volume.
2-23 The statement is true in the sense that the number of protons (the atomic number) determines the identity of the element.
2-25 (a) The element with 22 protons is titanium (Ti).
(b) The element with 76 protons is osmium (Os).
(c) The element with 34 protons is selenium (Se).
(d) The element with 94 protons is plutonium (Pu).
2-27 Each would still be the same element, because the number of protons has not changed.
2-29 Radon (Rn) has an atomic number of 86, so each isotope has 86 protons. The number of neutrons is mass number − atomic number.
(a) Radon-210 has $210 − 86 = 124$ neutrons
(b) Radon-218 has $218 − 86 = 132$ neutrons
(c) Radon-222 has $222 − 86 = 136$ neutrons
2-31 Two more neutrons: tin-120
Three more neutrons: tin-121
Six more neutrons: tin-124
2-33 (a) An ion is an atom or a group of bonded atoms with an unequal number of protons and electrons.
(b) Isotopes are atoms of the same element with the same number of protons in their nuclei but a different number of neutrons.
2-35 Rounded to four significant figures, the calculated value is 12.01 amu; the value given in the Periodic Table is 12.011 amu.

$$\frac{98.90}{100} \times 12.00 + \frac{1.10}{100} \times 13.003 = 12.01 \text{ amu}$$

2-37 Carbon-11 has 6 protons, 6 electrons, and 5 neutrons.
2-39 Americium-241 (Am) has an atomic number of 95. This isotope has 95 protons, 95 electrons, and $241 − 95 = 146$ neutrons.
2-41 (a) T (b) F (c) F (d) F (e) F (f) T (g) T
(h) T (i) T
(b) False: The main group elements go from group 1A through 8A.
(c) False: Very roughly, the nonmetals exist in a diagonal starting in the lower right corner, moving up to the middle of the Periodic Table. Nonmetals exist above the diagonal and metals below it.
(d) False: There are more metals than nonmetals.
(e) False: Horizontal rows are called periods.

2-43 (a) Groups 2A, 3B, 4B, 5B, 6B, 7B, 8B, 1B, and 2B contain only metals. Note that Group 1A contains one nonmetal, hydrogen.
(b) No groups contain metalloids exclusively.
(c) Only Groups 7A and 8A contain nonmetals exclusively.
2-45 Elements in the same group of the Periodic Table should have similar properties.
N, P, and As; I and F; Ne and He; Mg; Ca, and Ba; K and Li
2-47 (a) Aluminum > silicon (b) Arsenic > phosphorus
(c) Gallium > germanium (d) Gallium > aluminum
2-49 (a) T (b) T (c) T (d) F (e) T (f) F (g) T
(h) T (i) F (j) T (k) T (l) T (m) T (n) T (o) F
(p) F (q) T (r) T (s) T (t) F
(d) False: Principal energy level 1 can contain a maximum of 2 electrons, principal energy level 2 can contain a maximum of 8 electrons, principal energy level 3 can contain a maximum of 18 electrons, and principal energy level 4 can contain a maximum of 32 electrons.
(f) False: A $2s$ electron is easier to remove than a $1s$ electron because it is further from the influence of the positively charged nucleus.
(i) False: The three $2p$ orbitals are at right angles to each other.
(o) False: Paired electrons have spins in opposite directions.
(p) False: Each box represents an orbital, and each orbital can accommodate two electrons. When an orbital is completely filled, one electron in the pair must spin in the opposite direction from the other.
(t) False: Group 6A elements have six electrons in their valence shells, only two of which are unpaired.
2-51 The group number describes the number of electrons in the valence shell of an element in the group.
2-53 (a) Li(3): $1s^2 2s^1$ (b) Ne(10): $1s^2 2s^2 2p^6$ (c) Be(4): $1s^2 2s^2$
(d) C(6): $1s^2 2s^2 2p^2$ (d) Mg(12): $1s^2 2s^2 2p^6 3s^2$
2-55 (a) He(2): $1s^2$ (b) Na(11): $1s^2 2s^2 2p^6 3s^1$
(c) Cl(17): $1s^2 2s^2 2p^6 3s^2 3p^5$ (d) P(15): $1s^2 2s^2 2p^6 3s^2 3p^3$
(e) H(1): $1s^1$
2-57 In (a), (b), and (c), the outer shell electron configurations are the same. The only difference is the number of the valence shell being filled.
2-59 The element might be in Group 2A, all of which have two valence electrons. It might also be helium, in Group 8A.
2-61 (a) T (b) T (c) T (d) F (e) F (f) T (g) F (h) T
(d) False: Helium, a group 8A element, has only two valence electrons.
(e) False: The period number has nothing to do with the number of valence electrons.
(g) False: Period 3 has eight elements.
2-63 (a) T (b) T (c) T (d) T (e) F (f) T
(e) False: Ionization energy decreases going from top to bottom within a column of the Periodic Table.
2-65 (a) Fact: The atomic radius of an anion is always larger than that of the atom from which it is derived. For anions, the nuclear charge is unchanged, but an electron added to the valence shell introduces new repulsions and the electron cloud swells because of the increased electron-to-electron repulsions.
(b) Fact: The atomic radius of a cation is always smaller than that of the atom from which it is derived. When an electron is removed from an atom, the nuclear charge remains the same but there are fewer electrons repelling each other.
Consequently, the positive nucleus attracts the remaining electrons more strongly, causing the electron cloud to contract.
2-67 Here are ground state electron configurations for each O, O^+, and N, N^+ pair.

One of these electrons is lost

O $1s^2\, 2s^2\, 2p_x^2\, 2p_y^1\, 2p_z^1$ ⟶ $O^+\, 1s^2\, 2s^2\, 2p_x^1\, 2p_y^1\, 2p_z^1 + e^-$

This electron is lost

N $1s^2\, 2s^2\, 2p_x^1\, 2p_y^1\, 2p_z^1$ ⟶ $N^+\, 1s^2\, 2s^2\, 2p_x^1\, 2p_y^1\ \ \ \ \ \ + e^-$

The electron removed from O is one of the paired electrons in the doubly occupied $2p_x$ orbital, whereas the electron removed from N is an electron from the singly occupied $2p_z$ orbital. There is some repulsion between the two paired electrons in the case of oxygen, which means that it is easier to remove an electron from O than it is to remove an electron from the singly occupied $2p_z$ orbital for nitrogen.
2-69 Sulfur and iron are essential components of proteins, and calcium is a major component of bones and teeth.
2-71 Calcium is an essential element in human bones and teeth. Because strontium behaves chemically much like calcium, strontium-90 gets into our bones and teeth and gives off radioactivity directly into our bodies for many years.
2-73 Copper can be hardened by hammering.
2-75 (a) Metals (b) Nonmetals (c) Metals
(d) Nonmetals (e) Metals (f) Metals
2-77 (a) The largest atomic radius in Group 2A is radium (Ra).
(b) The smallest atomic radius in Group 2A is beryllium (Be).
(c) The largest atomic radius in the second Period is Neon (Ne).
(d) The smallest atomic radius in the second Period is Lithium (Li).
(e) The largest ionization energy in Group 7A is fluorine (F).
(f) The lowest ionization energy in Group 7A is astatine (At).
2-79

	Atomic Number	Protons	Neutrons	Electrons
(a) Phosphorus-32	15	15	17	15
(b) Molybdenum-98	42	42	56	42
(c) Calcium-44	20	20	24	20
(d) Hydrogen-3	1	1	2	1
(e) Gadolinium-158	64	64	94	64
(f) Bismuth-212	83	83	129	83

2-81 Isotopes of elements 37 to 53 contain more neutrons than protons. Here are some elements and the percent of their mass attributed to neutrons:
(a) Carbon-12 has 6 protons and 6 neutrons. Neutrons make up 50% of its mass.
(b) Calcium-40 has 20 protons and 20 neutrons. Neutrons make up 50% of its mass.
(c) Iron-55 has 26 protons and 29 neutrons. Neutrons make up 53% of its mass.
(d) Bromine-79 has 35 protons and 44 neutrons. Neutrons make up 56% of its mass.

(e) Platinum-195 has 78 protons and 117 neutrons. Neutrons make up 60% of its mass.
(f) Uranium-238 has 92 protons and 146 neutrons. Neutrons make up 61% of its mass.

2-83 Rounded to three significant figures, the calculated value from this information is 10.8 amu. The value given in the Periodic Table is 10.81.

$$\frac{19.9}{100} \times 10.013 + \frac{80.1}{100} \times 11.009 = 10.8 \text{ amu}$$

$$1.992 \qquad 8.818$$

2-85 The dimensions of the solution are protons per grain. Using the concept of dimensional analysis developed in Chapter 1, arrange the calculation so that these dimensions result.

$$1.0 \times 10^{-2} \frac{\text{grams}}{\text{grain}} \times \frac{1 \text{ proton}}{1.67 \times 10^{-24} \text{ grams}}$$

$$= 6.0 \times 10^{21} \text{ protons/grain}$$

2-87 The atomic number of this element is 54, which means that the element is xenon (Xe). This isotope of xenon has 54 protons, 54 electrons, and $131 - 54 = 77$ neutrons.

2-89 (a) Ionization energy generally decreases down a column in the Periodic Table, so the ionization energy of element 117 should be less than that of At (85).
(b) Ionization energy generally increases from left to right across a row in the Periodic Table, so the ionization energy of element 117 should be greater than that of radium Ra (88).

2-91 The key is to consider the electron configuration of a lithium atom and a lithium ion. Lithium has an atomic number of 3.

$$\text{Li}(3) \ 1s^2 2s^1 \longrightarrow \text{Li}^+ + e^-$$

$$\text{Li}^+ \ 1s^2 \longrightarrow \text{Li}^{2+} + e^-$$

In forming lithium ion, Li^+, a lithium atom loses one electron from its valence shell, that is, from its $2s$ orbital. In forming Li^{2+} ion, a lithium ion now loses an electron from its $1s$ orbital. A $1s$ orbital is smaller than a $2s$ orbital, and electrons in a $1s$ orbital are held more tightly than those in a $2s$ orbital. $1s$ electrons are held more tightly because they are closer to the positively charged nucleus. Therefore, the second ionization energy is considerably larger than the first ionization energy.

2-93 Atomic size decreases in going across a period and increases in going down a column. In order of increasing atomic size, these elements are:

$$\text{C}(77) < \text{B}(83) < \text{Al}(143) < \text{Na}(186)$$

2-95 The group 3A elements are boron, aluminum, gallium, indium, and thallium. The characteristic that is common for all is that each has three valence electrons.

2-97 Ca^{3+} would be formed by the loss of three electrons from Ca; the third ionization energy is prohibitively large (see Problem 2-91).

2-99 6.85 g of unreacted Mg will remain.

2-101 See answers in RED below.

Symbol	Atomic number	Atomic weight	Mass number	# of protons	# of neutrons	# of electrons
H	1	1.0079	1	1	0	1
Li	3	6.941	7	3	4	3
Al	13	26.9815	27	13	14	13
Fe	26	55.845	58	26	32	26
Pt	78	195.084	195	78	117	78
Ca	20	40.078	37	20	17	20
S	16	32.066	32	16	16	16

2-103 108.904 amu

2-105 Element 118 will fall in Group 8A. It will have the chemical properties of the other elements in this family, and it will have a melting point higher than –71°C and a boiling point above –62°C; in other words, it will be a gas at room temperature.

Chapter 3 Chemical Bonds

3-1 By losing two electrons, a Mg atom becomes Mg^{2+} and acquires a complete octet. By gaining two electrons, a sulfur atom becomes a sulfide ion, S^{2-}, with an eight-valence electron configuration the same as that of argon.
(a) Mg (12 electrons): $1s^2 2s^2 2p^6 3s^2 \longrightarrow \text{Mg}^{2+}$ (10 electrons): $1s^2 2s^2 2p^6 + 2e^-$
(b) S (16 electrons): $1s^2 2s^2 2p^6 3s^2 3p^4 + 2e^- \longrightarrow \text{S}^{2-}$ (18 electrons): $1s^2 2s^2 2p^6 3s^2 3p^6$

3-2 Each pair of elements is in the same column of the Periodic Table, and electronegativity increases from bottom to top within a column. Therefore:
(a) Li > K   (b) N > P   (c) C > Si

3-3 (a) KCl   (b) CaF_2 (c) Fe_2O_3

3-4 (a) Magnesium oxide   (b) Barium iodide
(c) Potassium chloride

3-5 (a) MgCl_2 (b) Al_2O_3 (c) LiI

3-6 (a) Iron(II) oxide, ferrous oxide   (b) Iron(III) oxide, ferric oxide

3-7 (a) Potassium hydrogen phosphate   (b) Aluminum sulfate
(c) Iron(II) carbonate, ferrous carbonate

3-8 (a) S—H (2.5 – 2.1 = 0.4); nonpolar covalent
(b) P—H (2.1 – 2.1 = 0.0); nonpolar covalent
(c) C—F (4.0 – 2.5 = 1.5); polar covalent
(d) C—Cl (3.0 – 2.5 = 0.5); polar covalent

3-9 (a) $\overset{\delta+}{\text{C}}-\overset{\delta-}{\text{N}}$ (b) $\overset{\delta+}{\text{N}}-\overset{\delta-}{\text{O}}$ (c) $\overset{\delta+}{\text{C}}-\overset{\delta-}{\text{Cl}}$

3-10 (a) H—C(H)(H)—C(H)(H)—H (b) H—C(H)(H)—Cl̈: (c) H—C≡N:

3-11 (a) H—C(H)(H)—C(H)(H)—H **4 single bonds**

(b) H₂C=CH₂ (with H's on each carbon) **2 single bonds and 1 double bond**

(c) H₂C=C=CH₂ **2 double bonds**

(d) H—C≡C—H **1 single bond and 1 triple bond**

3-12 (a) Nitrogen dioxide (b) Phosphorus tribromide (c) Sulfur dichloride (d) Boron trifluoride

3-13 (a) H—C(=Ö:⁻)(Ö:⁻) with + on C
(b) H—C(Ö:⁻)=Ö̈
(c) CH₃—C(=Ö:⁻ above)=Ö⁺—CH₃

3-14 (a) A valid pair of contributing structures.
(b) Not a valid pair. The contributing structure on the right has 10 electrons in the valence shell of carbon and thus violates the octet rule. The valence shell of carbon consists of one 2s orbital and three 2p orbitals, which can hold a maximum of 8 valence electrons, hence the octet rule.

3-15 Given are three-dimensional structures showing all bonds and unshared electron pairs.

(a) H₃C—Ö:H with 109.5° angles
(b) H₃C—Cl̈: with 109.5° angles
(c) H—C(=Ö)—O—H with 109.5° and 120° angles

3-16 (a) H₂S; the difference in electronegativity between H and S is 2.5 − 2.1 = 0.4. Therefore, H—S bonds are nonpolar and the molecule is nonpolar.

H—S̈—H

Nonpolar

(b) HCN contains a polar C—N bond and is a polar molecule.

H—C≡N:

Polar

(c) C₂H₆ contains no polar bonds and is not a polar molecule.

H₃C—CH₃

Nonpolar

3-17 (a) F (b) T (c) F (d) T (e) T (f) F (g) T (h) F (i) F
(a) False: It helps us understand the bonding patterns of the Group 1A-7A elements.
(c) False: Atoms that gain electrons become anions.
(f) False: Sodium typically forms a positive ion, a cation, by losing its single 3s electron.
(h) False: Phosphorus, sulfur, and chlorine can expand their valence shells to accommodate more than eight electrons.
(i) False: They couldn't be more different, starting with the most obvious difference—that of their charges.

$$\text{Li: } 1s^2 2s^1 \longrightarrow \text{Li}^+ \, 1s^2 + e^-$$

3-19 (a) A lithium (3) atom has the electron configuration $1s^2 2s^1$. When a Li atom loses its single 2s electron, it forms lithium ion, Li⁺, which has the electron configuration $1s^2$. This configuration is the same as that of helium, the noble gas nearest Li in atomic number.

$$\text{Li: } 1s^2 2s^1 \longrightarrow \text{Li}^+ \, 1s^2 + e^-$$

(b) An oxygen (8) atom has the electron configuration $1s^2 2s^2 2p^4$. When an O atom gains two electrons, it forms O^{2-}, which has the electron configuration $1s^2 2s^2 2p^6$. This configuration is the same as that of neon, the noble gas nearest oxygen in atomic number.

$$\text{O: } 1s^2 2s^2 2p^4 + 2e^- \longrightarrow O^{2-}: 1s^2 2s^2 2p^6 \text{ (complete octet)}$$

3-21 (a) Mg^{2+} (b) F^- (c) Al^{3+} (d) S^{2-} (e) K^+ (f) Br^-
3-23 The stable ions are: (a) I^-, (c) Na^+, and (d) S^{2-}.
3-25 Being intermediate in electronegativity, carbon and silicon are reluctant to accept electrons from a metal or lose electrons to a halogen to form ionic bonds. Instead, carbon and silicon share electrons in nonpolar covalent and polar covalent bonds.
3-27 (a) T (b) F (c) F (d) T (e) F (f) T (g) F (h) T (i) T (j) F (k) F (l) F (m) T (n) T
(b) False: H⁺ is named hydrogen ion, or more commonly, a proton, because it is a nucleus consisting of a single proton. The hydronium ion is H_3O^+.
(c) False: H⁺ has one proton and no neutrons.
(e) False: –ous refers to the ion with the lower charge; –ic refers to the ion with the higher charge.
(g) False: The anion derived from a bromine atom is named bromide ion.
(j) False: The prefix "bi" indicates the presence of a single hydrogen in this polyatomic ion.
(k) False: The hydrogen phosphate ion has a charge of −2, and the dihydrogen phosphate ion has a charge of −1.
(l) False: The numbers in the superscripts and subscripts are reversed. The phosphate ion is PO_4^{3-}.
3-29 (a) T (b) T (c) T (d) T (e) F (f) T (g) T (h) T (i) F (j) F (k) T (l) F (m) T (n) T (o) F
(e) False: Ionic bonds usually form between elements on the far left and the far right of the Periodic Table.
(i) False: Electronegativity is a periodic property.
(j) False: Electronegativity increases going from left to right in a period and decreases going down a group in the Periodic Table.
(l) False: Fluorine is the most electronegative element, and francium is the least electronegative element.
(o) False: The opposite is true.
3-31 Electronegativity generally increases going from left to right across a row of the Periodic Table because the number of positive charges in the nucleus of each successive element

in the row increases going from left to right. The increasing nuclear charge exerts a stronger and stronger pull on the valence electrons.

3-33 Electrons are shifted toward the more electronegative atom.
(a) Cl (b) O (c) O (d) Cl (e) Negligible
(f) Negligible (g) O

3-35 (a) C—Cl polar covalent (b) C—Li polar covalent
(c) C—N polar covalent

3-37 (a) T (b) F (c) T (d) T (e) T (f) F
(g) F (h) F (i) F
(b) False: Ionic bonds form by the transfer of one or more electrons from the atom of lower electronegativity to the atom of higher electronegativity.
(f) False: The formula of calcium hydroxide is $Ca(OH)_2$.
(g) False: The formula of aluminum sulfide is Al_2S_3.
(h) False: The formula of iron(III) oxide is Fe_2O_3.
(i) The formula of barium oxide is BaO.

3-39 (a) NaBr (b) Na_2O (c) $AlCl_3$ (d) $BaCl_2$ (e) MgO

3-41 Sodium chloride in the solid state forms a lattice in which each Na^+ ion is surrounded by six Cl^- ions, and each Cl^- ion is surrounded by six Na^+ ions.

3-43 (a) $Fe(OH)_3$ (b) $BaCl_2$ (c) $Ca_3(PO_4)_2$ (d) $NaMnO_4$

3-45 (a) The formula $(NH_4)_2PO_4$ is not correct. The correct formula is $(NH_4)_3PO_4$.
(b) The formula Ba_2CO_3 is not correct. The correct formula is $BaCO_3$.
(c) The formula Al_2S_3 is correct.
(d) The formula MgS is correct.

3-47 (a) T (b) F (c) T (d) T (e) F (f) F (g) T (h) F
(i) T (j) F (k) F
(b) False: The name includes no indication of the number of ions present.
(e) False: The systematic name of Fe_2O is iron(III) oxide.
(f) False: The systematic name of $FeCO_3$ is iron(II) carbonate.
(h) False: The systematic name of K_2HPO_4 is potassium hydrogen phosphate.
(j) False: The name of PCl_3 is phosphorus trichloride.
(k) False: The correct formula is $(NH_4)_2CO_3$.

3-49 The formula for potassium nitrite is KNO_2.

3-51 (a) Na^+, Br^- (b) Fe^{2+}, SO_4^{2-} (c) Mg^{2+}, PO_4^{3-}
(d) K^+, $H_2PO_4^-$ (e) Na^+, HCO_3^- (f) Ba^{2+}, NO_3^-

3-53 (a) KBr (b) CaO (c) HgO (d) $Cu_3(PO_4)_2$
(e) Li_2SO_4 (f) Fe_2S_3

3-55 (a) T (b) F (c) F (d) T (e) T (f) T (g) T (h) F
(i) T (j) T (k) F (l) T (m) F (n) T
(b) False: They will form a nonpolar covalent bond.
(c) False: A bond formed by sharing two electrons is a single bond. A double bond is a bond formed by sharing two pairs of electrons.
(h) False: The order given here is reversed.
(k) Ethane, C_2H_6, must show 14 valence electrons.
(m) False: The Lewis structure for the ammonium ion, NH_4^+, must show eight valence electrons.

3-57 (a) A single bond results when one electron pair is shared between two atoms.
(b) A double bond results when two electron pairs are shared between two atoms.
(c) A triple bond results when three electron pairs are shared between two atoms.

3-59 (a) H—C(H)(H)—H (b) H—C≡C—H
(c) $H_2C=CH_2$ (d) BF_3
(e) $H_2C=O$ (f) $Cl_2C—CCl_2$ with Cl's

3-61 The total number of valence electrons for each compound:
(a) NH_3 has 8 (b) C_3H_6 has 18 (c) $C_2H_4O_2$ has 24
(d) C_2H_6O has 20 (e) CCl_4 has 32 (f) HNO_2 has 18
(g) CCl_2F_2 has 32 (h) O_2 has 12

3-63 (a) A bromine atom has seven electrons in its valence shell. (b) A bromine molecule has two bromine atoms bonded by a single covalent bond. (c) A bromide ion is an anion; it has a complete octet of eight valence electrons and a charge of -1.
(a) :Br· (b) :Br—Br: (c) :Br:$^-$

3-65 Hydrogen has the electron configuration $1s^1$. Hydrogen's valence shell has only a $1s$ orbital, which can hold only two electrons.

3-67 Nitrogen has five valence electrons. By sharing three more electrons with other atoms or another atom, nitrogen can achieve the outer shell electron configuration of neon, the noble gas nearest to it in atomic number. The three shared pairs of electrons may be in the form of three single bonds, one double bond and one single bond, or one triple bond. With these combinations, there is one unshared pair of electrons on nitrogen.

3-69 Oxygen has six valence electrons. By sharing two electrons with another atom or atoms, oxygen can achieve the outer shell electron configuration of neon, the noble gas nearest it in atomic number. The two shared pairs of electrons may be in the form of one double bond or two single bonds. With either of these configurations, there are two unshared pairs of electrons on oxygen.

3-71 O^{6+} has a charge too concentrated and too large for a small ion.

3-73 (a) BF_3 does not obey the octet rule because in this compound, boron has only six electrons in its valence shell.
(b) CF_2 does not obey the octet rule because in this compound, carbon has only four electrons in its valence shell.
(c) BeF_2 does not obey the octet rule because in this compound, beryllium has only four electrons in its valence shell.
(d) C_2H_4, ethylene, obeys the octet rule. In this compound, each carbon has a double bond to the other carbon and single bonds to two hydrogen atoms, giving each carbon a complete octet.
(e) CH_3 does not obey the octet rule. In this compound, carbon has single bonds to three hydrogens, which gives carbon only six electrons in its valence shell.
(f) N_2 obeys the octet rule. Each nitrogen has one triple bond and one unshared pair of electrons and, therefore, eight electrons in its valence shell.

(g) NO does not obey the octet rule. This compound has 11 valence electrons, and any Lewis structure drawn for it will show either oxygen or nitrogen with only seven electrons in its valence shell.

3-75 (a) Sulfur dioxide (b) Sulfur trioxide (c) Phosphorus trichloride (d) Carbon disulfide

3-77 (a) An acceptable structure for the ozone molecule must show 18 valence electrons.
(b) Two equivalent contributing structures for ozone are:

(c) The curved arrows in (b) show the redistribution of electron pairs between the two contributing structures.
(d) The two contributing structures for ozone are equivalent and, therefore, make equal contributors to the hybrid. In each contributing structure, three regions of electron density surround the central oxygen, and therefore, the O—O—O bond angles are predicted to be 120°.
(e) This structure is not acceptable because the central oxygen atom has 10 electrons in its valence shell, which violates the octet rule. The valence shell of oxygen has one $2s$ orbital and three $2p$ orbitals, which between them can hold no more than eight electrons, hence the octet rule.

3-79 (a) T (b) F (c) T (d) F (e) T (f) T (g) F (h) T (i) T (j) T (k) F (l) T (m) T
(b) False: A prediction of bond angles must also consider nonbonding pairs of electrons.
(d) False: In CO_2, carbon is surrounded by two regions of electron density, and the VSEPR model predicts a bond angle of 180°.
(g) False: Four regions of electron density around a central atom result in bond angles of approximately 109.5°.
(k) False: The oxygen atom in H_2O is surrounded by four regions of electron density, and therefore, the VSEPR model predicts the H—O—H bond angle of approximately 109.5°.

3-81 (a) H_2O has 8 valence electrons, and H_2O_2 has 14 valence electrons.
(b) Each Lewis structure must show two bonds to oxygen and two unshared pairs on each oxygen. Lewis structures are:

H—Ö—H   H—Ö—Ö—H
Water   Hydrogen peroxide

(c) Predict bond angles of 109.5° about each oxygen atom.

3-83 (a) methane Tetrahedral (109.5°) (b) PH_3 Pyramidal (109.5°) (c) CF_4 Tetrahedral (109.5°)
(d) SO_2 Bent (120°) (e) SO_3 Trigonal planar (120°) (f) CCl_3F Tetrahedral (109.5°)

(g) NH_3 Pyramidal (109.5°) (h) PCl_3 Pyramidal (109.5°)

3-85 (a) T (b) T (c) F (d) T (e) T (f) T (g) T (h) T
(c) False: If the dipole moments of polar bonds cancel each other by acting in equal but opposite directions, then the molecule will be nonpolar.

3-87 (a) The Lewis structure of BF_3 is:

(b) The predicted F—B—F bond angles are 120°.
(c) BF_3 has three polar bonds, but the three bond dipole moments cancel each other by acting in opposite directions, and therefore, the molecule is nonpolar.

3-89 No, it is not possible to have a polar molecule with all nonpolar bonds.

3-91 The individual C—Cl bond dipoles in CCl_4 act in equal but opposite directions, canceling each other's effect on the molecular dipole.

3-93 Sodium iodide, NaI, and potassium iodide, KI, are used as iodide sources in table salt. Iodide is necessary for proper thyroid function and for the formation of thyroid hormones.

3-95 Potassium permanganate is used as an external antiseptic.

3-97 Nitric oxide, NO, quickly oxidizes in air to nitrogen dioxide, NO_2, which then dissolves in rainwater to form nitric acid, HNO_3.

3-99 (a) $SiCl_4$ (b) PH_3 (c) H_2S

3-101 The predicted shape is created by putting together the bases of two square-based pyramids. This shape is called octahedral, because it has eight faces.

3-103 To arrive at the atom-atom distance in H_2O and H_2S, add the atomic radii:
H—O = 103 pm   H—S = 141 pm

3-105 (a) The following types of geometries are present in vitamin E: tetrahedral and trigonal planar.
(b) Bond angles about the carbon atom participating in four single bonds are 109.5°, and that about the single oxygen atom is also 109.5°.
(c) Vitamin E has only one polar bond, the —OH group, and large regions containing only nonpolar covalent bonds. Because the nonpolar regions are so large compared with the size of the one polar covalent region, predict that vitamin E is a nonpolar molecule.

3-107 (a) The most polar bond in ephedrine is the O—H bond.
(b) Predict that ephedrine is a polar molecule because it has polar covalent O—H, C—O, N—H, and C—N bonds.

3-109 Both are polar molecules, with the negative end of the dipole determined by the more electronegative fluorine atoms.

Freon-11   Freon-12

3-111 The compound is white zinc oxide, ZnO.
3-113 The lead-containing compound is primarily lead(IV) oxide, PbO_2.
3-115 Fe^{2+} is utilized in over-the-counter iron supplements.
3-117 (a) $CaSO_3$ (b) $Ca(HSO_3)_2$ (c) $Ca(OH)_2$
(d) $CaHPO_4$
3-119 Perchloroethylene has four polar C—Cl bonds, but given its geometry, the molecule is nonpolar.
3-121 (a) The Lewis structure for tetrafluoroethylene is:

(b) Predict 120° for each F—C—F bond angle.
(c) No, it does not have a dipole moment.
3-123 (a) The borohydride ion, BH_4^-, has $(3 + 4 + 1) = 8$ valence electrons.
(b) The Lewis structure of the borohydride ion shows boron surrounded by four regions of electron density.

(c) Predict each H—B—H bond angle to be 109.5°.
3-125 (a)

(b)

(c) O—H bond
(d) Polar
3-127 (a)

(b)

(c)

(d) O—H bond
(e) Polar
(f) Yes; resonance is expected for this antibiotic as shown.

Chapter 4 Chemical Reactions

4-1 The balanced equation is

$$6CO_2(g) + 6H_2O(\ell) \xrightarrow{\text{photosynthesis}} C_6H_{12}O_6(aq) + 6O_2(g)$$

4-2 The balanced equation is

$$2C_6H_{14}(g) + 19O_2(g) \longrightarrow 12CO_2(g) + 14H_2O(g)$$

4-3 The balanced equation is

$$3K_2C_2O_4(aq) + Ca_3(AsO_4)_2(s) \longrightarrow 2K_3AsO_4(aq) + 3CaC_2O_4(s)$$

4-4 The net ionic equation is

$$Cu^{2+}(aq) + S^{2-}(aq) \longrightarrow CuS(s)$$

4-5 (a) Ni^{2+} gained two electrons so is reduced. Cr lost two electrons so is oxidized. Ni^{2+} is the oxidizing agent, and Cr is the reducing agent.
(b) CH_2O gained hydrogens so is reduced. H_2 gains oxygens in being converted to CH_3OH and so is oxidized. CH_2O is the oxidizing agent, and H_2 is the reducing agent.

4-6 (a) ibuprofen, $C_{13}H_{18}O_2 = 206.1$ amu
(b) $Ba_3(PO_4)_2 = 601$ amu

4-7 1500. g H_2O is 83.3 mol H_2O.

4-8 2.84 mol Na_2S is 222 g Na_2S.

4-9 In 2.5 mol of glucose, there are 15 mol of C atoms, 30 mol of H atoms, and 15 mol of O atoms.

4-10 0.062 g $CuNO_3$ contains 4.9×10^{-4} mol Cu^+.

4-11 235 g H_2O contains 7.86×10^{24} molecules H_2O.

4-12 (a) The balanced equation is

$$2Al_2O_3(s) \xrightarrow{\text{electrolysis}} 4Al(s) + 3O_2(g)$$

(b) It requires 51 g of aluminum oxide to prepare 27 g of aluminum.

4-13 From the balanced equation, we see that the molar ratio of CO required to produce CH_3COOH is 1:1. Therefore, it requires 16.6 moles of CO to produce 16.6 moles of CH_3COOH.

4-14 From the balanced equation, we see that the molar ratio of ethylene to ethanol is 1:1. Therefore, 7.24 mol of ethylene gives 7.24 mole of ethanol, which is 334 g of ethanol.

4-15 (a) H_2 (1.1 mole) is in excess, and C (0.50 mole) is the limiting reagent.
(b) 8.0 g CH_4 is produced.

4-16 The percent yield is 80.87%.

4-17 Following are the balanced equations.
(a) $HI + NaOH \longrightarrow NaI + H_2O$
(b) $Ba(NO_3)_2 + H_2S \longrightarrow BaS + 2HNO_3$
(c) $CH_4 + 2O_2 \longrightarrow CO_2 + 2H_2O$
(d) $2C_4H_{10} + 13O_2 \longrightarrow 8CO_2 + 10H_2O$
(e) $2Fe + 3CO_2 \longrightarrow Fe_2O_3 + 3CO$

4-19 $CO_2(g) + Ca(OH)_2(aq) \longrightarrow CaCO_3(s) + H_2O(\ell)$

4-21 $2Mg(s) + O_2(g) \longrightarrow 2MgO(s)$

4-23 $2C(s) + O_2(g) \longrightarrow 2CO(g)$

4-25 $2AsH_3(g) \xrightarrow{\text{heat}} 2As(s) + 3H_2(g)$

4-27 $2NaCl(aq) + 2H_2O(\ell) \xrightarrow{\text{electrolysis}} Cl_2(g) + 2NaOH(aq) + H_2(g)$

4-29 The following chemical reactions are balanced net ionic equations.
(a) $Ag^+(aq) + Br^-(aq) \longrightarrow AgBr(s)$
(b) $Cd^{2+}(aq) + S^{2-}(aq) \longrightarrow CdS(s)$
(c) $2Sc^{3+}(aq) + 3SO_4^{2-}(aq) \longrightarrow Sc_2(SO_4)_3(s)$
(d) $Sn^{2+}(aq) + 3Fe^{2+}(aq) \longrightarrow Sn(s) + 2Fe^{3+}(aq)$
(e) $2K(s) + 2H_2O(\ell) \longrightarrow 2K^+(aq) + 2OH^-(aq) + H_2(g)$

4-31 (a) $Ca_3(PO_4)_2$ will precipitate.

$$3Ca^{2+}(aq) + 2PO_4^{3-}(aq) \longrightarrow Ca_3(PO_4)_2(s)$$

(b) No precipitate will form (Group 1 chlorides and sulfates are soluble).
(c) $BaCO_3$ will precipitate.

$$Ba^{2+}(aq) + CO_3^{2-}(aq) \longrightarrow BaCO_3(s)$$

(d) $Fe(OH)_2$ will precipitate.

$$Fe^{2+}(aq) + 2OH^-(aq) \longrightarrow Fe(OH)_2(s)$$

(e) $Ba(OH)_2$ will precipitate.

$$Ba^{2+}(aq) + 2OH^-(aq) \longrightarrow Ba(OH)_2(s)$$

(f) Sb_2S_3 will precipitate.

$$2Sb^{2+}(aq) + 3S^{2-}(aq) \longrightarrow Sb_2S_3(s)$$

(g) $PbSO_4$ will precipitate.

$$Pb^{2+}(aq) + SO_4^{2-}(aq) \longrightarrow PbSO_4(s)$$

4-33 The net ionic equation is

$$SO_3^{2-}(aq) + 2H^+(aq) \longrightarrow SO_2(g) + H_2O(\ell)$$

4-35 (a) KCl (soluble: all Group 1 chlorides are soluble).
(b) NaOH (soluble: all sodium salts are soluble).
(c) $BaSO_4$ (insoluble: most sulfates are insoluble).
(d) Na_2SO_4 (soluble: all sodium salts are soluble).
(e) Na_2CO_3 (soluble: all sodium salts are soluble).
(f) $Fe(OH)_2$ (insoluble: most hydroxides are insoluble).

4-37 (a) T (b) T (c) T (d) T (e) T (f) F (g) F (h) T (i) T (j) T (k) T (l) T (m) T (n) T

4-39 (a) C_7H_{12} is oxidized (the carbons gain oxygens in going to CO_2), and O_2 is reduced.
(b) O_2 is the oxidizing agent and C_7H_{12} is the reducing agent.

4-41 (a) F (b) F (c) T (d) F (e) T

4-43 (a) sucrose, $C_{12}H_{22}O_{11}$ 342.3 amu
(b) glycine, $C_2H_5NO_2$ 75.07 amu
(c) DDT, $C_{14}H_9Cl_5$ 354.5 amu

4-45 (a) 32 g CH_4 = 2.0 mol CH_4
(b) 345.6 g NO = 11.52 mol NO
(c) 184.4 g ClO_2 = 2.734 mol ClO_2
(d) 720. g glycerine = 7.82 mol glycerine

4-47 (a) 18.1 mol CH_2O = 18.1 mol O atoms
(b) 0.41 mol $CHBr_3$ = 1.2 mol Br atoms
(c) 3.5×10^3 mol $Al_2(SO_4)_3$ = 4.2×10^4 mol O atoms
(d) 87 g HgO = 0.40 mol Hg atoms

4-49 (a) 25.0 g TNT (MW = 227 g/mol) contains 1.99×10^{23} N atoms.
(b) 40 g ethanol (MW = 46 g/mol) = 1.0×10^{24} mol C atoms
(c) 500. mg aspirin (MW 180.2 g/mol) = 6.68×10^{21} O atoms
(d) 2.40 g NaH_2PO_4 (MW 120 g/mol) = 1.20×10^{22} Na atoms

4-51 (a) 100. molecules CH_2O (MW 30 g/mol) = 4.98×10^{-21} g CH_2O.
(b) 3000. molecules CH_2O (MW 30 g/mol) = 1.495×10^{-19} g CH_2O.
(c) 5.0×10^6 molecules CH_2O = 2.5×10^{16} grams CH_2O molecules.
(d) 2.0×10^{24} molecules CH_2O = 100 g CH_2O.

4-53 3.9 mg cholesterol (MW 386.7 g/mol) = 6.1×10^{18} molecules cholesterol.

4-55 (a) 1 mol O_2 requires 0.67 mol of N_2.
(b) 0.67 mol of N_2O_3 are produced from 1 mol of O_2.
(c) To produce 8 mol N_2O_3 requires 12 mol O_2.
4-57 1.50 mol $CHCl_3$ requires 319 g of Cl_2.
4-59 (a) $2NaClO_2(aq) + Cl_2(g) \longrightarrow 2ClO_2(g) + 2NaCl(aq)$
(b) 5.5 kg of $NaClO_2$ will yield 4.10 kg of ClO_2.
4-61 To produce 5.1 g of glucose requires 7.5 g of CO_2.
4-63 To completely react with 0.58 g of Fe_2O_3, we need 0.13 g C.
4-65 51.1 g of salicyclic acid.
4-67 The theoretical yield from 5.6 g of ethane is 12 g of chloroethane. The percentage yield is 68%.
4-69 (a) T (b) F (c) T (d) T (e) T (f) T
4-71 (a) endothermic (22.0 kcal appears as a reactant).
(b) exothermic (124 kcal appears as a product).
(c) exothermic (94.0 kcal appears as a product).
(d) endothermic (9.80 kcal appears as a reactant).
(e) exothermic (531 kcal appears as a product).
4-73 1.6×10^2 kcal of heat is evolved in burning 0.37 mol of acetone.
4-75 Ethanol has a greater heat of combustion per gram (7.09 kcal/g) than glucose (3.72 kcal/g).
4-77 156.0 kcal will produce 88.68 g of Fe metal.
4-79 Hydroxyapatite is composed of calcium ions, phosphate ions, and hydroxide ions.
4-81 Li is oxidized, and I_2 is reduced. I_2 is the oxidizing agent, and Li is the reducing agent.
4-83 Cu^+ is oxidized. The species that is oxidized during the course of the reaction gives up an electron and is the reducing agent. Therefore, Cu^+ is the reducing agent.
4-85 $2C_5H_{12}O(\ell) + 15O_2(g) \longrightarrow 10CO_2(g) + 12H_2O(g)$
4-87 488 mg of aspirin (MW 180.2 g/mol) is equal to 2.71×10^{-3} mol aspirin.
4-89 N_2 is the limiting reagent, and H_2 is in excess.
4-91 4×10^{10} molecules of hemoglobin are present in a red blood cell.
4-93 29.7 kg N_2 = 1061 mol N_2 and 3.31 kg H_2 = 1655 mol H_2
(a) From the balanced chemical equation, we see that the two gases react in the ratio $3H_2/N_2$. Complete reaction of 1061 mol N_2 requires 3183 mol H_2, but less than this number of moles of H_2 is present. Therefore, H_2 is the limiting reagent.
(b) Under the balanced equation are moles of each present before reaction, moles reacting, and moles present after complete reaction.

	N_2	+	$3H_2$	$\longrightarrow$	$2NH_3$
Before rexn	1061		1655		0
Reacting	551		1655		0
After rexn	510		0		1102

551 mol N_2 = 14.3 kg of N_2 remains after the reaction.
(c) 1102 moles of NH_3 = 18.7 kg of NH_3 formed.
4-95 (a) 441 mg of furan = 6.48×10^{-3} mol of furan
(b) 0.060 L of furan = 2.0×10^{24} atoms of C
(c) 9.86×10^{25} molecules of furan = 1.11×10^4 g of furan
4-97 2.84×10^5 g $KClO_4(aq)$
4-99 (a) Following are balanced equations for each oxidation.
$C_{16}H_{32}O_2(s) + 23O_2(g) \longrightarrow 16CO_2 + 16H_2O(\ell) + 238.5$ kcal/mol
$C_6H_{12}O_6(s) + 6O_2(g) \longrightarrow 6CO_2 + 6H_2O(\ell) + 670$ kcal/mol
(b) The heat of combustion of palmitic acid is 9.302 kcal/gram. The heat of combustion of glucose is 3.72 kcal/gram.

(c) Palmitic acid has the greater heat of combustion per mole.
(d) Palmitic acid also has the greater heat of combustion per gram.
4-101 64 kg CO_2
4-103 (a) 103 mol caffeine
(b) 3.82×10^{23} molecules caffeine
(c) 4.3×10^{19} atoms N
(d)

(e)

All carbon atoms contained in both rings are trigonal planar

(f)

All carbon atoms contained in both rings have bond angles of 120°

(g) C—O bond
(h) Polar
(i) $2C_8H_{10}N_4O_2(\ell) + 27O_2(g) \longrightarrow 16CO_2(g) + 10H_2O(g) + 8NO_2(g)$
(j) 9.2 kcal
(k) When 8.00 g of caffeine is mixed with 20.3 g of oxygen gas, 3.71 g of $H_2O(g)$ will be produced.

Chapter 5 Gases, Liquids, and Solids

5-1 0.41 atm
5-2 16.4 atm
5-3 0.053 atm
5-4 4.84 atm
5-5 0.422 mol Ne
5-6 9.91 g He
5-7 0.107 atm of H_2O vapor
5-8 (a) Yes, there can be hydrogen bonding between water and methanol, because a hydrogen atom on each molecule is bonded to an electronegative oxygen atom. The O—H hydrogen can form a hydrogen bond to an oxygen lone pair on another molecule.
(b) No, there is no polarity to a C—H bond, and therefore, it cannot participate in hydrogen bonding.
5-9 The heat of vaporization of water is 540 cal/g. 45.0 kcal is sufficient to vaporize 83.3 g H_2O.
5-10 The heat required to heat 1.0 g of iron to melting = 2.3×10^2 cal.
Heat (up to melting) = 166 cal
Heat to melt = 63.7 cal
5-11 According to the phase diagram of water (Figure 5-20), the vapor will first condense to liquid water and then freeze to give ice.
5-13 As the volume of a gas decreases, the concentration of gas molecules per unit of volume increases and the number of gas molecules colliding with the walls of the container increases. Because gas pressure results from the collisions of gas molecules with the walls of the container, as volume decreases, pressure increases.
5-15 The volume of a gas can be decreased by (1) increasing the pressure on the gas or (2) lowering the temperature (cooling) of the gas. (3) The volume of the gas can be decreased by removing some of the gas.
5-17 7.37 L
5-19 2.0 atm of CO_2 gas
5-21 615 K
5-23 6.2 L of SO_2 gas upon heating
5-25 The pressure read by the manometer is the difference between the gas in the bulb and the atmospheric pressure: 833 mm Hg − 760 mm Hg = 73 mm Hg
5-27 2.6 atm of halothane
5-29

V_1	T_1	P_1	V_2	T_2	P_2
6.35 L	10°C	0.75 atm	**4.6 L**	0°C	1.0 atm
75.6 L	0°C	1.0 atm	**88 L**	35°C	735 torr
1.06 L	75°C	0.55 atm	3.2 L	0°C	**0.14 atm**

5-31 The volume of the balloon will be 3×10^6 L.
5-33 The new temperature is 300 K.
5-35 1.87 atm
5-37 (a) 2.33 mol of gas are present.
(b) No, the only information you need to know about the gas is that it is an ideal gas.
5-39 Using the ideal gas law $PV = nRT$ and n(moles) = mass/MW, the following equation can be derived and solved for the molecular weight of the gas.

$$MW = \frac{(\text{mass})RT}{PV} = \frac{(8.00\text{g})(0.0821 \text{ L} \cdot \text{atm} \cdot \text{mol}^{-1} \cdot \text{K}^{-1})(273\text{K})}{(2.00 \text{ atm})(22.4\text{L})} = 4.00 \text{ g/mol}$$

5-41 At constant temperature, gas density increases as pressure increases. At constant pressure, gas density decreases as temperature increases.
5-43 (a) 24.7 mol O_2 are needed to fill the chamber.
(b) 790 g of O_2 are needed to fill the chamber.
5-45 5.5 L of air contains 1.16 L of O_2, which, under these conditions, is 0.050 mol of O_2.
0.050 mol of O_2 contains 3.0×10^{22} molecules of O_2.
5-47 (a) The mass of one mol of air is 28.95 grams.
(b) The density of air is 1.29 g/L.
5-49 The density of each gas is
(a) SO_2 = 2.86 g/L (b) CH_4 = 0.714 g/L (c) H_2 = 0.0892 g/L
(d) He = 0.179 g/L (e) CO_2 = 1.96 g/L
Gas comparison: SO_2 and CO_2 are more dense than air; He, H_2, and CH_4 are less dense than air.
5-51 The density of octane is 0.7025 g/mL.
The mass of 1.00 mL of octane is 0.07025 g.
Using the ideal gas equation, this mass of octane vapor occupies 0.197 L.
5-53 When 3.50 g of Na(s) is reacted at a temperature of 18°C and a pressure of 0.995 atm, 1.83 L of H_2(g) is produced.
5-55 (a) T (b) F (c) T (d) F
(d) False: $P_T = P_{N_2} + P_{O_2} + P_{CO_2} + P_{H_2O}$
$P_{N_2} = (0.740)(1.0 \text{ atm}) = 0.740 \text{ atm}$ (562.4 mm Hg)
$P_{O_2} = (0.194)(1.0 \text{ atm}) = 0.194 \text{ atm}$ (147.5 mm Hg)
$P_{H_2O} = (0.062)(1.0 \text{ atm}) = 0.062 \text{ atm}$ (47.1 mm Hg)
$P_{CO_2} = (0.004)(1.0 \text{ atm}) = 0.004 \text{ atm}$ (3.0 mm Hg)
$P_T = 1.00 \text{ atm}$ (760.0 mm Hg)

5-57 0.194 atm oxygen, 0.004 atm carbon dioxide, 0.062 atm water vapor, and 0.740 atm nitrogen
5-59 (a) T (b) F (c) T (d) F (e) T (f) T (g) T (h) F (i) T (j) T
5-61 (a) F (b) F (c) T (d) T (e) F (f) T (g) T (h) T (i) F (j) F (k) F (l) F (m) T (n) T (o) F
5-63 Gases behave most ideally under low pressure and high temperature to minimize non-ideal intermolecular interactions. Therefore, choice (c) best suits these conditions.
5-65 (a) CCl_4 is nonpolar; London dispersion forces
(b) CO is polar; dipole−dipole interactions
The most polar molecule (CO) will have the largest surface tension.
5-67 Yes, London dispersion forces range from 0.001 to 0.2 kcal/mol, whereas the lower end of dipole−dipole attractive forces can be as low as 0.1 kcal/mol.
5-69 (a) T (b) F (c) F (d) T (e) F (f) T (g) T (h) T (i) T (j) T (k) F (l) F (m) T (n) F (o) T (p) F
5-71 (a) T (b) T (c) T (d) F (e) T (f) F (g) T (h) F (i) F (j) F (k) T
5-73 (a) T (b) T (c) F (d) T (e) T (f) F (g) F (h) F (i) F (j) T (k) T
5-75 1.53 kcal is required to vaporize one mol of CF_2Cl_2.
5-77 The vapor pressures are approximately:
(a) 90 mm Hg (b) 120 mm Hg (c) 490 mm Hg
5-79 (a) HI > HBr > HCl The increasing size in this series increases London dispersion forces.
(b) H_2O_2 > HCl > O_2. O_2 has only weak London dispersion intermolecular forces to overcome for boiling to occur, whereas HCl is a polar molecule with stronger dipole−dipole attractions to overcome for boiling to occur. H_2O_2 has the strongest intermolecular forces (hydrogen bonding) to overcome for boiling to occur.

5-81 The difference between heating water from 0°C to 37°C and heating ice from 0°C to 37°C is the heat of fusion.
The energy required to heat 100. g of ice from 0°C to 37°C is 11700. cal.
The energy required to heat 100. g of water from 0°C to 37°C is 3700 cal.
5-83 The name of the phase change is sublimation, which is the conversion of a solid to a gas, bypassing the liquid phase.
5-85 1.00 mL of Freon-11 is 1.08×10^{-2} mol of Freon-11. Vaporizing this volume of Freon-11 from the skin will remove 6.96×10^{-2} kcal.
5-87 When a person lowers the diaphragm, the volume of the chest cavity increases, thus lowering the pressure in the lungs relative to atmospheric pressure. Air at atmospheric pressure is then drawn into the lungs, beginning the breathing process.
5-89 The first tapping sound one hears is the systolic pressure, which occurs when the sphygmomanometer pressure matches the blood pressure and the ventricle contracts, pushing blood into the arm.
5-91 When water freezes, it expands (water is one of the few substances that expands on freezing) and will break the bottle when the expansion exceeds the volume of the bottle.
5-93 Compressing a liquid or a solid is difficult because their molecules or atoms are already close together and there is very little empty space between them.
5-95 34 psi = 2.3 atm
5-97 Aerosol cans already contain gases under pressure. Gay-Lussac's law predicts that the pressure inside the can will increase as it is heated, with the potential of explosive rupturing of the can causing injury.
5-99 112 mL
5-101 Water, which forms strong intermolecular hydrogen bonds, has the highest boiling point. Boiling points of these three compounds are:
(a) pentane, C_5H_{12} (36°C) (b) chloroform, $CHCl_3$ (61°C)
(c) water, H_2O (100°C)
5-103 (a) As a gas is compressed under pressure, the molecules are forced closer together and the intermolecular forces pull molecules together, forming a liquid.
(b) 9.1 kg of propane (c) 2.1×10^2 moles of propane
(d) 4.6×10^3 L of propane
5-105 The density of the gas is 3.00 g/L.
Using the ideal gas law,

$$MW = \frac{massRT}{PV}$$

and the molecular weight of the gas is 91.9 g/mol.
5-107 313K (40°C)
5-109 The temperature of a liquid drops during evaporation because as the molecules with higher kinetic energy leave the liquid and enter the gas phase, the average kinetic energy of molecules remaining in the liquid decreases. The temperature of the liquid is directly proportional to the average kinetic energy of molecules in the liquid phase and as the average kinetic energy decreases, the temperature decreases.
5-111 (a) The pressure on the body at 100 feet is 3.0 atm.
(b) At 1.00 atm, P_{N_2} = 593 mm Hg (0.780 atm) and thus makes up 78.0% of the gas mixture, which does not change at a depth of 100 feet. At this depth, the total pressure on the lungs, which is equalized by pressure of air delivered by the SCUBA tank, is 3.0 atm and the partial pressure of N_2 is 2.34 atm.

(c) At 2 atm, P_{O_2} = 158 mm Hg (0.208 atm) and thus makes up 20.8% of the gas mixture at 2 atm, which does not change at a depth of 100 feet. At this depth, the total pressure on the lungs, which is equalized by pressure of air delivered by the scuba tank, is 3.0 atm. Thus, at 100 feet, the partial pressure of O_2 = 0.63 atm.
(d) As a diver ascends from 100 ft, the external pressure on the lungs decreases, and therefore, the volume of gases in the lungs increases. If the diver does not exhale during a rapid ascent, the diver's lungs could overinflate due to the expansion of gases in the lungs, causing injury.
5-113 9.33 g NH_4Cl
5-115 1.26 g dry NH_4NO_2
5-117 4250 cal
5-119 (a) $NH_4NO_3(s) \longrightarrow N_2O(g) + 2H_2O(g)$
(b) 0.737 atm $N_2O(g)$ and 1.47 atm $H_2O(g)$
(c) 2.21 atm
(d)

$$\left[:\ddot{N}=N=\ddot{O}: \longleftrightarrow :\ddot{N}-N\equiv O: \longleftrightarrow :N\equiv N-\ddot{O}: \right]$$

Chapter 6 Solutions and Colloids

6-1 To 11 g of KBr, add a quantity of water sufficient to dissolve the KBr. Following dissolution of the KBr, add water to the 250 mL mark, stopper, and mix.
6-2 1.7% w/v
6-3 First, calculate the number of moles and mass of KCl needed, which is 2.12 mol and 158 g of KCl. To prepare the solution, place 158 g of KCl in a 2 L volumetric flask, add some water until the solid has dissolved, and then fill the flask with water to the 2.0 L mark.
6-4 Because the units of molarity are moles of solute/L of solution, grams of KSCN must be converted to moles of KSCN and mL of solution must be converted to L of solution. When these conversions are complete, you should find that the concentration of the solution is 0.0133M.
6-5 First, convert grams of glucose into moles of glucose; then convert moles of glucose into mL of solution. 10.0 g of glucose is 0.0556 mol of glucose. This mass of glucose is contained in 185 mL of the given solution.
6-6 First, convert 100 gallons to liters of solution. 3.9×10^2 g $NaHSO_3$ must be added to the 100-gallon barrel.
6-7 Place 15.0 mL of 12.0 M HCl solution into a 300 mL volumetric flask, add some water, swirl to mix completely, and then fill the flask with water to the 300 mL mark.
6-8 Place 0.13 mL of the 15% KOH solution into a 20 mL volumetric flask, add some water, swirl until completely dissolved, and then fill the flask with water to the 20 mL mark.
6-9 The Na^+ concentration is 0.24 ppm Na^+.
6-10 215 g of CH_3OH (molecular weight 32.0 g/mol) is 6.72 mol of CH_3OH.
$\Delta T = (1.86°C/mol)(6.72 mol) = 12.5°C$. The freezing point is lowered by 12.5°C. The new freezing point is $-12.5°C$.
6-11 Compare the number of moles of ions or molecules in each solution. The solution with the most ions or molecules in solution will have the lowest freezing point.

Solution	Particles in solution
(a) 6.2 M NaCl	$2 \times 6.2\,M = 12.4\,M$ ions
(b) 2.1 M $Al(NO_3)_3$	$4 \times 2.1\,M = 8.4\,M$ ions
(c) 4.3 M K_2SO_3	$3 \times 4.3\,M = 12.9\,M$ ions

Solution (c) has the highest concentration of solute particles (ions); therefore, it will have the lowest freezing point.

6-12 The boiling point is raised by 3.50°C. The new boiling point is 103.5°C.

6-13 The molarity of the solution prepared by dissolving 3.3 g Na_3PO_4 in 100 mL of water is 0.20M Na_3PO_4. Each formula unit of Na_3PO_4 dissolved in water gives 3 Na^+ ions and 1 PO_4^{3-} ion, for a total of 4 particles. The osmolarity of the solution is (0.20 M)(4 ions) = 0.80 osmol.

6-14 The osmolarity of red blood cells is 0.30 osmol.

Solution	Mol particles/L
(a) 0.1 M Na_2SO_4	$3 \times 0.1\,M$ = 0.30 osmol
(b) 1.0 M Na_2SO_4	$3 \times 1.0\,M$ = 3.0 osmol
(c) 0.2 M Na_2SO_4	$3 \times 0.2\,M$ = 0.6 osmol

Solution (a) has the same osmolarity as red blood cells and, therefore, is isotonic with red blood cells.

6-15 (a) T (b) T (c) T (d) T

6-17 The solvent is water.

6-19 (a) both tin and copper are solids
(b) solid solute (caffeine, flavorings) and liquid solvent (water).
(c) both CO_2 and H_2O (steam) are gases.
(d) gas (CO_2) and liquid (ethanol) solutes in a liquid solvent (water).

6-21 Mixtures of gases are true solutions because they mix in all proportions, molecules are distributed uniformly, and the component gases do not separate upon standing.

6-23 The prepared aspartic acid solution was unsaturated. Over the two days, some of the solvent (water) evaporated and the solution had become saturated. When water continued evaporating, the remaining water could not hold all the dissolved solute, so the excess aspartic acid precipitated as a white solid.

6-25 (a) NaCl is an ionic solid and will be dissolved in the water layer.
(b) Camphor is a nonpolar molecular compound and will dissolve in the nonpolar diethyl ether layer.
(c) KOH is an ionic solid and will be dissolved in the water layer.

6-27 Isopropyl alcohol would be a good first choice. The oil base in the paint is nonpolar. Both benzene and hexane are nonpolar solvents and may dissolve the oil-based paint, thus destroying the painting.

6-29 The solubility of aspartic acid in water at 25°C is 0.250 g in 50.0 mL of water. The cooled solution of 0.251 g of aspartic acid in 50.0 mL water will be supersaturated by 0.001 g of aspartic acid.

6-31 According to Henry's law, the solubility of a gas in a liquid is directly proportional to pressure. A closed bottle of a carbonated beverage is under pressure. After the bottle is opened, the pressure is released and the carbon dioxide becomes less soluble and escapes, leaving the contents "flat."

6-33 (a) $\dfrac{1\text{ min}}{1.0 \times 10^6 \text{ min}} \times 10^6 = 1 \text{ ppm}$

$\dfrac{1\text{ p}}{1.05 \times 10^6 \text{ p}} \times 10^6 = 1 \text{ ppm}$

(b) $\dfrac{1\text{ min}}{1.05 \times 10^9 \text{ min}} \times 10^9 = 1 \text{ ppb}$

$\dfrac{1\text{ p}}{1.05 \times 10^9 \text{ p}} \times 10^9 = 1 \text{ ppm}$

6-35 (a) Dissolve 76 mL of ethanol in 204 mL of water (to give 280 mL of solution).

(b) Dissolve 8.0 mL of ethyl acetate in 427 mL of water (to give 435 mL to solution).
(c) Dissolve 0.13 L of benzene in 1.52 L chloroform (to give 1.65 L of solution).

6-37 (a) 4.15% w/v casein (b) 0.030% w/v vitamin C
(c) 1.75% w/v sucrose

6-39 (a) Place 19.5 g NH_4Br in a 175 mL volumetric flask, add some water, swirl until completely dissolved, and then fill the flask with water to the 175 mL mark.
(b) Place 167 g of NaI in a 1.35 L volumetric flask, add some water, swirl until completely dissolved, and then fill the flask with water to the 1.35 L mark.
(c) Place 2.4 g of ethanol in a 330 mL volumetric flask, add some water, swirl until completely dissolved, and then fill the flask with water to the 330 mL mark.

6-41 0.2 M NaCl

6-43 0.509 M glucose
0.0202 M K^+
$7.25 \times 10^{-4}\,M$ Na^+

6-45 2.5 M sucrose

6-47 The total volume of the dilution is 30.0 mL. Starting with 5.00 mL of the stock solution, add 25.0 mL of water to reach a final volume of 30.0 mL. Note that this is a dilution by a factor of 6.

6-49 Place 2.1 mL of the 30.0% H_2O_2 into a 250 mL volumetric flask, add some water, swirl until completely mixed, and then fill the flask with water to the 250 mL mark.

6-51 (a) 3.85×10^4 ppm Captopril (b) 6.8×10^4 ppm Mg^{2+}
(c) 8.3×10^2 ppm Ca^{2+}

6-53 Assume the density of the lake water to be 1.00 g/mL. The dioxin concentration is 0.01 ppb dioxin.
No, the dioxin level in the lake did not reach a dangerous level.

6-55 (a) 10 ppm Fe or 1×10^1 ppm
(b) 3×10^3 ppm Ca
(c) 2 ppm vitamin A

6-57 (a) KCl; An ionic compound, very soluble in water: a strong electrolyte
(b) Ethanol; A covalent compound: a nonelectrolyte
(c) NaOH; An ionic compound, very soluble in water: a strong electrolyte
(d) HCl; A strong acid that dissociates completely in water: a strong electrolyte
(e) Glucose; A covalent compound, very soluble in water: a nonelectrolyte

6-59 Water dissolves ethanol by forming hydrogen bonds. The O—H group of ethanol is both a hydrogen bond acceptor and a hydrogen bond donor.

6-61 (a) T (b) T

6-63 (a) homogeneous (b) heterogeneous (c) colloidal
(d) heterogeneous (e) colloidal (f) colloidal

6-65 As the temperature of the solution decreased, the protein molecules must have aggregated and formed a colloidal mixture. The turbid appearance is the result of the Tyndall effect.

6-67 (a) 1.0 mol NaCl, freezing point −3.72°C
(b) 1.0 mol $MgCl_2$ freezing point −5.58°C
(c) 1 mol $(NH_4)_2CO_3$ freezing point −5.58°C
(d) 1 mol $Al(HCO_3)_3$ freezing point −7.44°C

6-69 Methanol dissolves in water but does not dissociate; it is a nonelectrolyte. It would require 344 g of CH_3OH in 1000. g of water to lower the freezing point to −20°C.

6-71 Acetic acid, a weak acid, is only weakly dissociated in water. KF is a strong electrolyte, completely dissociating in water and nearly doubling the effect on freezing-point depression compared with that of acetic acid.

6-73 In each case, side with greater osmolarity rises.
(a) B (b) B (c) A (d) B (e) neither (f) neither

6-75 (a) 0.39 M Na_2CO_3 = 0.39 M × 3 particles/formula unit = 1.2 osmol
(b) 0.62 M $Al(NO_3)_3$ = 0.62 × 4 particles/formula unit = 2.5 osmol
(c) 4.2 M LiBr = 4.2 × 2 particles/formula unit = 8.4 osmol
(d) 0.009 M K_3PO_4 = 0.009M × 4 particles/formula unit = 0.04 osmol

6-77 Cells in hypertonic solutions undergo crenation (shrink).
(a) 0.3% NaCl = 0.3 osmol NaCl
(b) 0.9 M glucose = 0.9 osmol glucose
(c) 0.9% glucose = 0.05 osmol glucose
Solution (b) has a concentration greater than the isotonic solution, so it will crenate red blood cells.

6-79 Carbon dioxide (CO_2) dissolves in rainwater to form a dilute solution of carbonic acid (H_2CO_3), which is a weak acid.

6-81 Nitrogen narcosis is the intoxication caused by the increased solubility of nitrogen in the blood as a result of high pressures as divers descend.

6-83 23 mg Ca^{2+} ion

6-85
$$CaCO_3(s) + H_2SO_4(aq) \longrightarrow CaSO_4(s) + CO_2(g) + H_2O(\ell)$$
$$CaSO_4 + 2H_2O \longrightarrow CaSO_4 \cdot 2H_2O$$
Gypsum dihydrate

6-87 The minimum pressure required for reverse osmosis in the desalinization of seawater exceeds 100 atm (the osmotic pressure of seawater).

6-89 Yes, the change altered the tonicity. A 0.2% $NaHCO_3$ solution is 0.05 osmol. A 0.2% solution of $KHCO_3$ is 0.04 osmol. This difference arises because of the difference in formula weight of $NaHCO_3$ (84 g/mol) compared with that of $KHCO_3$ (100.1 g/mol). The error in replacing $NaHCO_3$ with $KHCO_3$ results in a hypotonic solution and an electrolyte imbalance by reducing the number of ions (osmolarity) in solution.

6-91 When a cucumber is placed in a saline solution, the osmolarity of the saline is greater than the water in the cucumber; so water moves from the cucumber to the saline solution. When a prune (a partially dehydrated plum) is placed in the same solution, it expands because the osmolarity in the prune is greater than the saline solution; so water moves from the saline solution to inside the prune.

6-93 The solubility of a gas is directly proportional to the pressure (Henry's law) and inversely proportional to the temperature. The dissolved carbon dioxide formed a saturated solution in water when bottled under 2 atm pressure. When the bottles are opened at atmospheric pressure, the gas becomes less soluble in water. The excess carbon dioxide escapes through bubbles and frothing. In the other bottle, the solution of carbon dioxide in water is unsaturated at lower temperature and does not lose carbon dioxide.

6-95 Methanol is more efficient at lowering the freezing point of water. A given mass of methanol (32 g/mol) contains a greater number of moles than the same mass of ethylene glycol (62 g/mol).

6-97
$$CO_2(g) + H_2O(\ell) \longrightarrow H_2CO_3(aq)$$
Carbonic acid
$$SO_2(g) + H_2O(\ell) \longrightarrow H_2SO_3(aq)$$
Sulfurous acid

6-99 Place 39 mL of 35% HNO_3 into a 300 mL volumetric flask, add some water, swirl until completely mixed, and then fill the flask with water to the 300 mL mark.

6-101 6 × 10^{-3} g of pollutant

6-103 Assume that the density of the pool water is 1.00 g/mL. The Cl_2 concentration in the pool is 355 ppm. 7.09 kg of Cl_2 must be added to reach this concentration.

6-105 78.2 mL H_2SO_4 solution

6-107 (a) One mole of $MgCl_2$ dissociates to produce three moles of ions: one mole of Mg^{2+} and two moles of Cl^-. The freezing point will be lowered by $3 \times \left(\frac{1.86°C}{mol}\right) \times 0.263$ mol = 1.47°C, and the solution will freeze at −1.47°C.
(b) One mole of $MgCl_2$ dissociates to produce three moles of ions: one mole of Mg^{2+} and two moles of Cl^-. The boiling point will be raised by $3 \times \left(\frac{0.512°C}{mol}\right) \times 0.263$ mol = 0.404°C, and the solution will boil at 100.404°C.

6-109 (a) From the balanced chemical equation, $Ca(s) + 2HBr(aq) \longrightarrow CaBr_2(aq) + H_2(g)$, we see that the two reactants react in the ratio of 2HBr/Ca. The complete reaction of 0.0364 mol Ca requires 0.0728 mol of HBr, but less than this number of moles of HBr is present. Therefore, HBr is the limiting reagent, producing 0.0187 mol $H_2(g)$.
(b) The volume of dry H_2 produced is 0.469 L $H_2(g)$.
(c) 0.709 g of Ca(s) remains after the reaction.

6-111 (a) 0.049 atm
(b) 2.8 atm
(c) 0.31 M
(d) 1.39 × 10^4 g/mol
(e) 8.45 × 10^3 g/mol

6-113 0.040 M glycerin ($C_3H_8O_3$) > 0.025 M NaBr > 0.015 M $Al(NO_3)_3$

Chapter 7 Reaction Rates and Chemical Equilibrium

7-1 rate of O_2 formation = 0.022 L O_2/min

7-2 rate = 4 × 10^{-2} mol H_2O_2/L · min for disappearance of H_2O_2

7-3 $K = \dfrac{[H_2SO_4]}{[SO_3][H_2O]}$

7-4 $K = \dfrac{[N_2][H_2]^3}{[NH_3]^2}$

7-5 $K = 0.602$

7-6 Le Chatelier's principle predicts that adding Br_2 (a product) will shift the equilibrium to the left—that is, toward the formation of more NOBr(g).

7-7 Because oxygen's solubility in water is exceeded, oxygen bubbles out of the solution, driving the equilibrium toward the right.

7-8 If the equilibrium shifts to the right with the addition of heat, heat must have been a reactant, and the reaction is endothermic.

7-9 The equilibrium in a reaction where there is an increase in pressure favors the side with fewer moles of gas. Therefore, this equilibrium shifts to the right.

7-11 rate of formation of $CH_3I = 7.3 \times 10^{-3}$ M CH_3I/min

7-13 Reactions involving ions in aqueous solution are faster because they do not require bond breaking and have low activation energies. In addition, the attractive force between positive and negative ions provides energy to drive the reaction. Reactions between covalent compounds require the breaking of covalent bonds and have higher activation energies and, therefore, slower reaction rates.

7-15
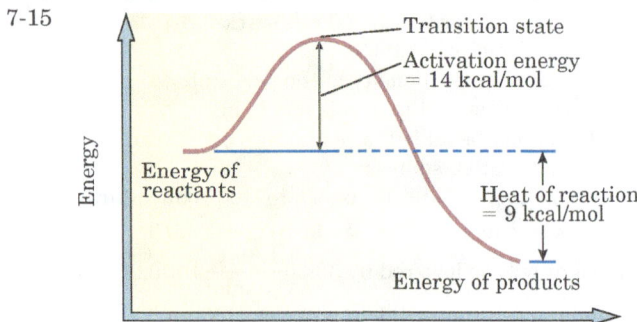

7-17 A general rule for the effect of temperature on the rate of reaction states that for every temperature increase of 10°C, the reaction rate doubles. In this case, a reaction temperature of 50°C would predict completion of the reaction in 1 h.

7-19 You might (a) increase the temperature, (b) increase the concentration of reactants, or (c) add a catalyst.

7-21 A catalyst increases the rate of a reaction by providing an alternative reaction pathway with lower activation energy.

7-23 Other examples of irreversible reactions include digesting a piece of candy, rusting of iron, exploding TNT, and the reaction of Na or K metal with water.

7-25 (a) $K = [H_2O]^2[O_2]/[H_2O_2]^2$
(b) $K = [N_2O_4]^2[O_2]/[N_2O_5]^2$
(c) $K = [O_2]^6/[H_2O]^6[CO_2]^6$

7-27 $K = 0.667$

7-29 $K = 0.099$

7-31 Products are favored in (b) and (c). Reactants are favored in (a), (d), and (e).

7-33 No, the rate of reaction is independent of the energy difference between products and reactants—that is, it is independent of the heat of reaction.

7-35 The reaction reaches equilibrium quickly, but the position of equilibrium favors the reactants. It would not be a very good industrial process unless products are constantly drawn off to shift the equilibrium to the right.

7-37 (a) right (b) right (c) left (d) left (e) no shift

7-39 (a) Adding Br_2 (a reactant) will shift the equilibrium to the right.
(b) The equilibrium constant will remain the same.

7-41 (a) no change (b) no change (c) smaller

7-43 As temperatures increase, the rates of most chemical processes increase. A high body temperature is dangerous because metabolic processes (including digestion, respiration, and biosynthesis of essential compounds) take place at a faster rate than is safe for the body. As temperatures decrease, so do the rates of most chemical reactions. As body temperature decreases below normal, the vital chemical reactions will slow to rates slower than is safe for the body.

7-45 The capsule with the tiny beads will act faster than the solid coated-pill form. The small bead size increases the drug's surface area, allowing the drug to react faster and deliver its therapeutic effects more quickly.

7-47 Assuming that there is an excess of AgCl from the previous recipe, the recipe does not need to be changed. The desert conditions add nothing that would affect the coating process.

7-49 At $-5°C$, the rate is 0.70 moles per liter per second. At 45°C, the rate is 22 moles per liter per second.

7-51

Profile 2 represents the addition of a catalyst.

7-53 0.14 M

7-55 The rate of a gaseous reaction could be increased by decreasing the volume of the container. This would increase the number of collisions between molecules.

7-57 $4NH_3 + 7O_2 \longrightarrow 4NO_2 + 6H_2O$

7-59 The reaction with spherical molecules will proceed more rapidly since some collisions between the rod-like molecules will not interact with the proper orientations.

7-61 Initial rate = 0.030 moles I_2 per liter per second

7-63 (a) If the initial concentration of acetic acid is 0.10 M and the final concentration is 0.098 M, then a change of 0.0020 M has occurred. Therefore, the equilibrium concentration of each product is 0.0020 M.
(b) $K = 4.1 \times 10^{-5}$

7-65 Monitoring the disappearance of a reactant is a possible way to determine the rate of a reaction. It will do just as well as monitoring the formation of product because the stoichiometry of the reaction relates the concentrations of products and reactants to one another.

7-67 Some reactions are so fast that they are over before you can turn on a stopwatch or timer. You need specialized instruments with sophisticated electronics to follow the rates of very fast reactions.

7-69 The rate of conversion of diamond to graphite is so slow that it does not take place in any measurable length of time.

7-71 When you add sodium chloride, the presence of more chloride ions increases the concentration of one of the products of the reaction. The equilibrium shifts to the left, increasing the amount of solid silver chloride.

7-73 $[HF] = 1.0 \times 10^{-4} M$

7-75

Change	Quantity	Effect
Increase the pressure	Concentration of NO_3^-	**Increase**
Add some Zn	Concentration of NO_2	**No change**
Decrease the H^+	Concentration of Zn^{2+}	**Decrease**
Add Pt catalyst	Equilibrium constant, K	**No change**
Add some Ar	Concentration of H^+	**No change**
Decrease the Zn^{2+}	Equilibrium constant, K	**No change**
Increase the temperature	Concentration of Zn	**Increase**

7-77 Activation energy for the reverse reaction = 10 kJ.

7-79 $K = \dfrac{[H_2]^2[O_2]}{[H_2O]^2}$

$[H_2O] = 5.8 \times 10^{-5} M$

7-81 (a) No; the concentration of reactants is favored over the concentration of products.
(b) The reaction must shift to the right (to the products side) in order to establish equilibrium.

7-83 $K = 210$

Chapter 8 Acids and Bases

8-1 acid reaction: $HPO_4^{2-} + H_2O \rightleftharpoons PO_4^{3-} + H_3O^+$;
base reaction: $HPO_4^{2-} + H_2O \rightleftharpoons H_2PO_4^- + OH^-$

8-2 (a) toward the left;

$H_3O^+ + I^- \rightleftharpoons H_2O + HI$
Weaker acid, Weaker base, Stronger base, Stronger acid

(b) toward the left;

$CH_3COO^- + H_2S \rightleftharpoons CH_3COOH + HS^-$
Weaker base, Weaker acid, Stronger acid, Stronger base

8-3 pK_a is 9.31
8-4 (a) ascorbic acid (b) aspirin
8-5 $Pb(s) + 2I^-(aq) \longrightarrow PbI_2(s)$
8-6 1.0×10^{-2}
8-7 (a) 2.46 (b) 7.9×10^{-5}, acidic
8-8 pOH = 4, pH = 10
8-9 0.0960 M
8-10 (a) 9.25 (b) 4.74
8-11 9.44
8-12 7.7
8-13 (a) An Arrhenius acid produces H_3O^+ ions in aqueous solution.
(b) An Arrhenius base produces OH^- ions in aqueous solution.
8-15 (a) $LiOH(s) \xrightarrow{H_2O} Li^+(aq) + OH^-(aq)$
(b) $(CH_3)_2NH(aq) + H_2O(\ell) \rightleftharpoons (CH_3)_2NH_2^+(aq) + OH^-(aq)$
(c) $Sr(OH)_2(s) \xrightarrow{H_2O} Sr^{2+}(aq) + 2OH^-(aq)$
(d) $CH_3CH_2NH_2(aq) + H_2O(\ell) \rightleftharpoons CH_3CH_2NH_3^+(aq) + OH^-(aq)$

8-17 (a) strong (b) weak (c) strong (d) weak (e) weak
8-19 (a) A Brønsted-Lowry acid is a proton donor.
(b) A Brønsted-Lowry base is a proton acceptor.
8-21 (a) HPO_4^{2-} (b) HS^- (c) CO_3^{2-}
(d) $CH_3CH_2O^-$ (e) OH^-
8-23 (a) H_3O^+ (b) $H_2PO_4^-$ (c) $CH_3NH_3^+$ (d) HPO_4^{2-}
8-25 acid reaction: $HPO_3^{2-} + H_2O \rightleftharpoons PO_3^{3-} + H_3O^+$
base reaction: $HPO_3^{2-} + H_2O \rightleftharpoons H_2PO_3^- + OH^-$
8-27 The equilibrium favors the side with the weaker acid–weaker base combination. Equilibria (b) and (c) lie to the left; equilibrium (a) lies to the right.

(a) $C_6H_5OH + C_2H_5O^- \rightleftharpoons C_6H_5O^- + C_2H_5OH$
Stronger acid, Stronger base, Weaker base, Weaker acid

(b) $HCO_3^- + H_2O \rightleftharpoons H_2CO_3 + OH^-$
Weaker base, Weaker acid, Stronger acid, Stronger base

(c) $CH_3COOH + H_2PO_4^- \rightleftharpoons CH_3COO^- + H_3PO_4$
Weaker acid, Weaker base, Stronger base, Stronger acid

8-29 (a) the pK_a of a weak acid (b) the K_a of a strong acid
8-31 (a) 0.10 M HCl (b) 0.10 M H_3PO_4 (c) 0.010 M H_2CO_3
(d) 0.10 M NaH_2PO_4 (e) 0.10 M aspirin
8-33 Only (b) and (h) are redox reactions. The others are acid–base reactions.
(a) $Na_2CO_3 + 2HCl \longrightarrow 2NaCl + CO_2 + H_2O$
(b) $Mg + 2HCl \longrightarrow MgCl_2 + H_2$
(c) $NaOH + HCl \longrightarrow NaCl + H_2O$
(d) $Fe_2O_3 + 6HCl \longrightarrow 2FeCl_3 + 3H_2O$
(e) $NH_3 + HCl \longrightarrow NH_4Cl$
(f) $CH_3NH_2 + HCl \longrightarrow CH_3NH_3Cl$
(g) $NaHCO_3 + HCl \longrightarrow NaCl + H_2O + CO_2$
(h) $2Al + 6HCl \longrightarrow 2AlCl_3 + 3H_2$
8-35 (a) $10^{-3} M$ (b) $10^{-10} M$ (c) $10^{-7} M$ (d) $10^{-15} M$
8-37 (a) pH = 8 (basic) (b) pH = 10 (basic) (c) pH = 2 (acidic) (d) pH = 0 (acidic) (e) pH = 7 (neutral)
8-39 (a) pH = 8.5 (basic) (b) pH = 1.2 (acidic)
(c) pH = 11.1 (basic) (d) pH = 6.3 (acidic)
8-41 (a) pOH = 1.0, $[OH^-] = 0.10 M$
(b) pOH = 2.4, $[OH^-] = 4.0 \times 10^{-3} M$
(c) pOH = 2.0, $[OH^-] = 1.0 \times 10^{-2} M$
(d) pOH = 5.6, $[OH^-] = 2.5 \times 10^{-6} M$
8-43 0.348 M
8-45 (a) 12 g of NaOH diluted to 400 mL of solution:

$400 \text{ mL sol} \left(\dfrac{1 \text{ L sol}}{1000 \text{ mL sol}}\right)\left(\dfrac{0.75 \text{ mol NaOH}}{1 \text{ L sol}}\right) \times \left(\dfrac{40.0 \text{ g NaOH}}{1 \text{ mol NaOH}}\right) = 12 \text{ g NaOH}$

(b) 12 g of $Ba(OH)_2$ diluted to 1.0 L of solution:

$\left(\dfrac{0.071 \text{ mol Ba(OH)}_2}{1 \text{ L sol}}\right)\left(\dfrac{171.4 \text{ Ba(OH)}_2}{1 \text{ mol Ba(OH)}_2}\right) = 12 \text{ g Ba(OH)}_2$

(c) 2.81 g KOH diluted to 500 mL
(d) 49.22 g sodium acetate diluted to 2 liters
8-47 5.66 mL
8-49 3.30×10^{-3} mol

8-51 The point at which the observed change occurs during a titration. It is usually so close to the equivalence point that the difference between the two becomes insignificant.

8-53 (a)
$H_3O^+ + CH_3COO^- \rightleftharpoons CH_3COOH + H_2O$ (removal of H_3O^+)
(b)
$HO^- + CH_3COOH \rightleftharpoons CH_3COO^- + H_2O$ (removal of OH^-)

8-55 Yes, the conjugate acid becomes the weak acid and the weak base becomes the conjugate base.

8-57 The pH of a buffer can be changed by altering the weak acid/conjugate base ratio, according to the Henderson–Hasselbalch equation. The buffer capacity can be changed without a change in pH by increasing or decreasing the amount of weak acid/conjugate base mixture while keeping the ratio of the two constant.

8-59 This would occur in a couple of cases. One is very common: You are using a buffer, such as Tris with a pK_a of 8.3, but you do not want the solution to have a pH of 8.3. If you wanted a pH of 8.0, for example, you would need unequal amounts of the conjugate acid and base, with there being more conjugate acid. Another case might be a situation where you are performing a reaction that you know will generate H^+ but you want the pH to be stable. In that situation, you might start with a buffer that was initially set to have more of the conjugate base so that it could absorb more of the H^+ that you know will be produced.

8-61 No, 100 mL of 0.1 M phosphate at pH 7.2 has a total of 0.01 mole of weak acid and conjugate base with equimolar amounts of each. 20 mL of 1 M NaOH has 0.02 mole of base, so there is more total base than there is buffer to neutralize it. This buffer would be ineffective.

8-63 (a) According to the Henderson–Hasselbalch equation, no change in pH will be observed as long as the weak acid/conjugate base ratio remains the same.
(b) The buffer capacity increases with increasing amounts of weak acid/conjugate base concentrations; therefore, 1.0 mol amounts of each, diluted to 1 L, would have a greater buffer capacity than 0.1 mol of each diluted to 1 L.

8-65 From the Henderson–Hasselbach equation,
$pH = pK_a + \log(A^-/HA)$
$A^-/HA = 10$, $\log(A^-/HA) = 1$ since $10^1 = 10$
$pH = pK_a + 1$

8-67 When 0.10 mol of sodium acetate is added to 0.10 M HCl, the sodium acetate completely neutralizes the HCl to acetic acid and sodium chloride. The pH of the solution is determined by the incomplete ionization of acetic acid.
$K_a = \dfrac{[CH_3COO^-][H_3O^+]}{[CH_3COOH]}$ $[H_3O^+]=[CH_3COO^-] = x$
$\sqrt{x^2} = \sqrt{K_a[CH_3COOH]} = \sqrt{(1.8 \times 10^{-5})(0.10)}$
$x = [H_3O^+] = 1.34 \times 10^{-3} M$
$pH = -\log[H_3O^+] = 2.9$

8-69 $TRIS\text{-}H^+ + NaOH \longrightarrow TRIS + H_2O + Na^+$

8-71 The only parameter you need to know about a buffer is its pK_a. Choosing a buffer involves identifying the acid form that has a pK_a within one unit of the desired pH.

8-73 Choosing a buffer involves identifying the acid form that has a pK_a within one unit of the desired pH (a pH of 8.15). The TRIS buffer with a $pK_a = 8.3$ best fits this criteria.

8-75 $Mg(OH)_2$ is a weak base used in flame-retardant plastics.

8-77 (a) Respiratory acidosis is caused by hypoventilation, which occurs due to a variety of breathing difficulties such as a windpipe obstruction, asthma, or pneumonia. (b) Metabolic acidosis is caused by starvation or heavy exercise.

8-79 Sodium bicarbonate is the weak base form of one of the blood buffers. It tends to raise the pH of blood, which is the purpose of the sprinter's trick, so that the person can absorb more H^+ during the event. By putting $NaHCO_3$ into the system, the following reaction will occur:
$HCO_3^- + H^+ \rightleftharpoons H_2CO_3$. The loss of H^+ means that the blood pH will rise.

8-81 (a) Benzoic acid is soluble in aqueous NaOH.
$C_6H_5COOH + NaOH \rightleftharpoons C_6H_5COO^- + H_2O$
$pK_a = 4.19$　　　　　　　　$pK_a = 15.56$
(b) Benzoic acid is soluble in aqueous $NaHCO_3$.
$C_6H_5COOH + NaHCO_3 \rightleftharpoons CH_3C_6H_4O^- + H_2CO_3$
$pK_a = 4.19$　　　　　　　　$pK_a = 6.37$
(c) Benzoic acid is soluble in aqueous Na_2CO_3.
$C_6H_5COOH + CO_3^{2-} \rightleftharpoons CH_3C_6H_4O^- + HCO_3^-$
$pK_a = 4.19$　　　　　　　　$pK_a = 10.25$

8-83 The strength of an acid is not important to the amount of NaOH that would be required to hit a phenolphthalein end point. Therefore, the more concentrated acid, the acetic acid, would require more NaOH.

8-85 $3.70 \times 10^{-3} M$

8-87 0.9 M

8-89 Yes, a pH of 0 is possible. A 1.0 M solution of HCl has $[H_3O^+] = 1.0 M$. $pH = -\log[H_3O^+] = -\log[1.0 M] = 0$

8-91 The qualitative relationship between acids and their conjugate bases states that the stronger the acid, the weaker its conjugate base. This can be quantified in the equation $K_b \times K_a = K_w$ or $K_b = 1.0 \times 10^{-14}/K_a$, where K_b is the base dissociation equilibrium constant for the conjugate base, K_a is the acid dissociation equilibruim constant for the acid, and K_w is the ionization equilibrium constant for water.

8-93 Yes, the strength of the acid is irrelevant. Both acetic acid and HCl have one H^+ to give up, so equal moles of either will require equal moles of NaOH to titrate to an end point.

8-95 You would need a ratio of 0.182 parts of the conjugate base to 1 part of the conjugate acid.

8-97 An equilibrium will favor the side of the weaker acid/weaker base. The larger the pK_a value, the weaker the acid.

8-99 (a) $HCOO^- + H_3O^+ \rightleftharpoons HCOOH + H_2O$
(b) $HCOOH + OH^- \rightleftharpoons HCOO^- + H_2O$

8-101 (a) 0.050 mol (b) 0.0050 mol (c) 0.50 mol

8-103 According to the Henderson–Hasselbalch equation,
$$pH = 7.21 + \log \dfrac{[HPO_4^{2-}]}{[H_2PO_4^-]}$$
As the concentration of $H_2PO_4^-$ increases, the $\log \dfrac{[HPO_4^{2-}]}{[H_2PO_4^-]}$ becomes negative, lowering the pH and becoming more acidic.

8-105 No, a buffer will have a pH equal to its pK_a only if equimolar amounts of the conjugate acid and base forms are present. If this is the basic form of Tris, then just putting any amount of it into water will give a pH much higher than the pK_a value.

8-107 (a) $pH = 7.1$, $[H_3O^+] = 7.9 \times 10^{-8} M$, basic
(b) $pH = 2.0$, $[H_3O^+] = 1.0 \times 10^{-2} M$, acidic
(c) $pH = 7.4$, $[H_3O^+] = 4.0 \times 10^{-8} M$, basic

(d) pH = 7.0, $[H_3O^+]$ = 1.0×10^{-7} M, neutral
(e) pH = 6.6, $[H_3O^+]$ = 2.5×10^{-7} M, acidic
(f) pH = 7.4, $[H_3O^+]$ = 4.0×10^{-8} M, basic
(g) pH = 6.5, $[H_3O^+]$ = 3.2×10^{-7} M, acidic
(h) pH = 6.9, $[H_3O^+]$ = 1.3×10^{-7} M, acidic

8-109 4.9 : 1, or 5 : 1 to one significant figure

8-111 (a) The concentration of H^+ is 1.0×10^{-2} M. Therefore, the number of moles of H^+ present is equal to 6.0×10^{-3} mol. (b) From the balanced chemical equation, $HCl + NaHCO_3 \longrightarrow NaCl + CO_2 + H_2O$, the amount of sodium hydrogen carbonate needed to completely neutralize the stomach acid, is equal to 0.504 g $NaHCO_3$.

8-113 (a) 0.500 M HF(aq)
(b) $K_a = 3.57 \times 10^{-4}$

8-115 From the balanced chemical equation, 2 moles of NaOH are needed for every 1 mole of the unknown diprotic acid. Therefore, the molarity of the unknown acid is 0.120 M, and its molar mass is 41.7 g/mol.

8-117 (a) When pH = 8.2, $[H_3O^+] = 6 \times 10^{-9}$ M and $[OH^-] = 2 \times 10^{-6}$ M
When pH = 8.1, $[H_3O^+] = 8 \times 10^{-9}$ M and $[OH^-] = 1 \times 10^{-6}$ M
(b) When the pH decreases from 8.2 to 8.1, $[H_3O^+]$ increases by 2×10^{-9} M. This increase in concentration is one-third of the original concentration, or an approximately 30% increase in acidity.
(c) When CO_2 dissolves in seawater, it forms carbonic acid (H_2CO_3), which releases hydrogen ions into solution. These hydrogen ions then combine with carbonate ions in the water to form bicarbonate (HCO_3^-). The formation of bicarbonate through this chemical reaction removes carbonate ions from seawater, making them less useful for organisms.

Chapter 9 Nuclear Chemistry

9-1 $^{139}_{53}I \longrightarrow {}^{139}_{54}Xe + {}^{0}_{-1}e$

9-2 $^{223}_{90}Th \longrightarrow {}^{4}_{2}He + {}^{219}_{88}Ra$

9-3 $^{74}_{33}As \longrightarrow {}^{0}_{+1}e + {}^{74}_{32}Ge$

9-4 $^{201}_{81}Tl + {}^{0}_{-1}e \longrightarrow {}^{201}_{80}Hg + \gamma$

9-5 Barium-122 has decayed through five half-lives, leaving 0.31 g. 10 g $\longrightarrow$ 5.0 g $\longrightarrow$ 2.5 g $\longrightarrow$ 1.25 g $\longrightarrow$ 0.625 g $\longrightarrow$ 0.31 g

9-6 The dose is 1.5 mL.

9-7 The intensity at 3.0 m is 3.3×10^{-3} mCi.

9-9 Alpha rays are He^{2+} ions (${}^{4}_{2}He$) whereas protons are positively charged H^+ ions (${}^{1}_{1}H$).

9-11 (a) 4.0×10^{-5} cm, which is visible light (blue).
(b) 3.0 cm (microwave radiation)
(c) 2.7×10^{-5} cm (ultraviolet light)
(d) 2.0×10^{-8} cm (X-ray)

9-13 (a) Infrared has the longest wavelength.
(b) X-rays have the highest energy.

9-15 (a) nitrogen-13 (b) phosphorus-33 (c) lithium-9
(d) calcium-39

9-17 oxygen-16

9-19 $^{151}_{62}Sm \longrightarrow {}^{0}_{-1}e + {}^{151}_{63}Eu$

9-21 $^{51}_{24}Cr + {}^{0}_{-1}e \longrightarrow {}^{51}_{23}V$

9-23 $^{248}_{96}Cm + {}^{28}_{10}X \longrightarrow {}^{116}_{51}Sb + {}^{160}_{55}Cs$
The bombarding nucleus was neon $^{28}_{10}Ne$.

9-25 (a) beta emission (b) gamma emission (c) positron emission (d) alpha emission

9-27 Gamma emission does not result in transmutation.

9-29 $^{239}_{94}Pu + {}^{4}_{2}He \longrightarrow {}^{240}_{95}Am + {}^{1}_{1}H + 2{}^{1}_{0}n$

9-31 Iodine-125 decayed through approximately six half-lives, with 0.31 mg remaining: 20 mg $\longrightarrow$ 10 mg $\longrightarrow$ 5.0 mg $\longrightarrow$ 2.5 mg $\longrightarrow$ 1.25 mg $\longrightarrow$ 0.625 mg $\longrightarrow$ 0.31 mg

9-33 The plutonium underwent four half-lives since the glacier deposited it. There were 16 mg of plutonium/kg at the time of deposition.
16 mg $\longrightarrow$ 8 mg $\longrightarrow$ 4 mg $\longrightarrow$ 2 mg $\longrightarrow$ 1 mg

9-35 The rate of radioactive decay is independent of all conditions and is a property of each specific isotope. There is no way we can increase or decrease the rate.

9-37 (a) The iodine-131 remaining after two hours will be 8.88×10^8 counts/s. (b) After 24 days, three half-lives have passed: 1/2 × 1/2 × 1/2 = 1/8, or 12.5% of the original amount remains. 24.0 mCi × 0.125 = 3.0 mCi.

9-39 Gamma radiation has the greatest penetrating power; therefore, protection from it requires the largest amount of shielding.

9-41 30 m

9-43 The curie (Ci) measures radiation intensity.

9-45 0.63 cc

9-47 At 20 cm, the intensity would be 3×10^3 Bq (8×10^{-2} µCi).

9-49 Person A was exposed to the larger dose of radiation and injured more seriously.

9-51 Iodine-131 is concentrated in the thyroid and would be expected to induce the cancer.

9-53 (a) Cobalt-60 is used for (4) cancer therapy.
(b) Thallium-201 is used in (1) heart scans and exercise stress tests. (c) Tritium is used for (2) measuring water content of the body. (d) Mercury-197 is used for (3) kidney scans.

9-55 The product of fusion of hydrogen-2 and hydrogen-3 nuclei is helium-4 plus a neutron and energy.

9-57 $^{209}_{83}Bi + {}^{58}_{26}Fe \longrightarrow {}^{1}_{0}n + {}^{266}_{109}Mt$

9-59 $^{10}_{5}B + {}^{1}_{0}n \longrightarrow {}^{11}_{5}B$
$^{11}_{5}B \longrightarrow {}^{7}_{3}Li + {}^{4}_{2}He$

9-61 The assumption of a constant carbon-14 to carbon-12 ratio rests on two assumptions: (1) that carbon-14 is continuously generated in the upper atmosphere by the production and decay of nitrogen-15 and (2) that carbon-14 is incorporated into carbon dioxide, CO_2, and other carbon compounds and then distributed worldwide as part of the carbon cycle. The continual formation of carbon-14; transfer of the isotope within the oceans, atmosphere, and biosphere; and decay of living matter keep the supply of carbon-14 constant.

9-63 2003 − 1350 = 653 years (if the experiment was run in 2003). 653 years/5730 years = 0.111 half-lives.

9-65 Radon-222 decays by alpha emission to polonium-218.

$^{222}_{86}Rn \longrightarrow {}^{218}_{84}Po + {}^{4}_{2}He$

9-67 Hydrogen is a major constituent of aqueous body fluids and fatty tissue and is thus the most convenient marker for MRI studies. Hydrogen atoms in the body are in different chemical environments, absorbing frequencies of different energies in the presence of an external magnetic field that can be imaged at specific depths.

9-69

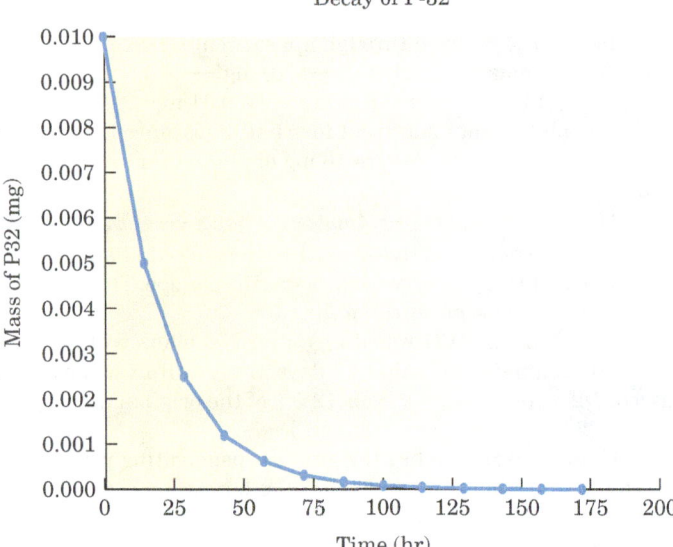

Decay of P-32

9-71 Neon-19 decays to fluorine-19, and sodium-20 decays to neon-20.

$$^{19}_{10}\text{Ne} \longrightarrow {}^{0}_{+1}e + {}^{19}_{9}\text{F}$$

$$^{20}_{11}\text{Na} \longrightarrow {}^{0}_{+1}e + {}^{20}_{10}\text{Ne}$$

9-73 Both the curie and the becquerel have units of disintegrations/second, a measure of radiation intensity.
9-75 (a) Natural sources = 82%
(b) Diagnostic medical sources = 11%
(c) Nuclear power plants = 0.1%
9-77 X-rays will cause more ionization than radar waves. X-rays have higher energy.
9-79 The decay product is neptunium-237. 1000/432 = 2.3 half-lives, so somewhat less than 25% of the original americium will remain after 1000 years.
9-81 One sievert is equal to 100 rem. This is sufficient to cause radiation sickness but not certain death.
9-83 (a) Radioactive elements are constantly decaying to other elements or isotopes, and these decay products are mixed with the original sample.
(b) Beta emission results from the decay of a neutron in the nucleus to a proton (the increase in atomic number) and an electron (the beta particle).
9-85 Oxygen-16 is stable because it has an equal number of protons and neutrons. The others are unstable because the numbers of protons and neutrons are unequal. In this case, the greater the difference in numbers of protons and neutrons, the faster the isotope decays.
9-87 The new element is darmstadtium-266.

$$^{208}_{82}\text{Pb} + {}^{64}_{28}\text{Ni} \longrightarrow {}^{266}_{110}\text{Ds} + 6\,{}^{1}_{0}n$$

9-89 The intermediate nucleus is boron-11.

$$^{10}_{5}\text{B} + {}^{1}_{0}n \longrightarrow {}^{11}_{5}\text{B}$$

$$^{11}_{5}\text{B} \longrightarrow {}^{7}_{3}\text{Li} + {}^{4}_{2}\text{He}$$

9-91 $^{206}_{80}\text{Hg}$

9-93 6 alpha and 4 beta particles are emitted.

Chapter 10 Organic Chemistry

10-1 Following are Lewis structures showing all valence electrons and all bond angles.

10-2 Of the four alcohols with the molecular formula $C_4H_{10}O$, two are 1°, one is 2°, and one is 3°. For the Lewis structures of the 3° alcohol and one of the 1° alcohols, some C—CH_3 bonds are drawn longer to avoid crowding in the formulas.

$CH_3CH_2CH_2CH_2OH$ — Primary (1°)

$CH_3CH_2CHCH_3$ — Secondary (2°)

CH_3CHCH_2OH with CH_3 — Primary (1°)

CH_3COH with CH_3 groups — Tertiary (3°)

10-3 The three secondary (2°) amines with the molecular formula $C_4H_{11}N$ are:

$CH_3CH_2CH_2NHCH_3$ $CH_3CHNHCH_3$ (with CH_3) $CH_3CH_2NHCH_2CH_3$

10-4 The three ketones with the molecular formula $C_5H_{10}O$ are:

$CH_3CH_2CH_2CCH_3$ (O=) $CH_3CH_2CCH_2CH_3$ (O=) CH_3CCHCH_3 (O=) with CH_3

10-5 The two carboxylic acids with the molecular formula $C_4H_8O_2$ are:

$$CH_3CH_2CH_2\overset{\overset{O}{\|}}{C}OH \quad \text{and} \quad CH_3\overset{\overset{O}{\|}}{C}HCOH$$
$$\phantom{CH_3CH_2CH_2COH \quad \text{and} \quad CH_3CHCOH}|$$
$$\phantom{CH_3CH_2CH_2COH \quad \text{and} \quad CH_3CH}CH_3$$

10-6 The four esters with the molecular formula $C_4H_8O_2$ are:

$$\overset{\overset{O}{\|}}{H}COCH_2CH_2CH_3 \qquad \overset{\overset{O}{\|}}{H}COCHCH_3$$
$$|$$
$$CH_3$$

(1)   (2)

$$CH_3\overset{\overset{O}{\|}}{C}OCH_2CH_3 \qquad CH_3CH_2\overset{\overset{O}{\|}}{C}OCH_3$$

(3)   (4)

10-7 (a) T (b) T (c) F (d) F
(c) False: Carbon isn't even close. Silicon and oxygen are the two most abundant elements in the Earth's crust.
(d) False: Most organic compounds are insoluble in water.
10-9 Assuming that each is pure, there are no differences in their chemical or physical properties.
10-11 Wöhler heated ammonium chloride and silver cyanate, both inorganic compounds, and obtained urea, an organic compound.
10-13 The four principal elements that make up organic compounds and the number of bonds each typically forms are:
H forms one bond.
C forms four bonds.
O forms two bonds.
N forms three bonds.
10-15 Following are Lewis dot structures for each element:

(a) ·C̈· (b) ·Ö· (c) ·N̈· (d) :F̈·

  (4)   (6)   (5)   (7)

10-17 (a) H—Ö—Ö—H   (b) H—N̈—N̈—H
                                           H H
       Hydrogen peroxide

                                            Hydrazine

        H   H
        | |
(c) H—C—Ö—H   (d) H—C—S̈—H
        | |
        H   H
     Methanol   Methanethiol

        H   H
        | |
(e) H—C—N̈—H   (f) H—C—C̈l:
        | | |
        H H   H
    Methylamine   Chloromethane

10-19 Following is a Lewis structure for each ion.

(a) H—Ö—C(=Ö)—Ö:⁻   (b) ⁻:Ö—C(=Ö)—Ö:⁻

(c) CH_3—C(=Ö)—Ö:⁻   (d) :C̈l:⁻

10-21 To use the VSEPR model to predict bond angles and the geometry about atoms of carbon, nitrogen, and oxygen: (1) Write the Lewis structure for the target molecule showing all valence electrons. (2) Determine the number of regions of electron density around an atom of C, O, or N. (3) If you find four regions of electron density, predict bond angles of 109.5°. If you find three regions, predict bond angles of 120°. If you find two regions, predict bond angles of 180°.
10-23 You would find two regions of electron density around oxygen and, therefore, predict 180° for the C—O—H bond angle. The actual bond angle is approximately 109.5°.
10-25 (a) 120° about C and 109.5° about O
(b) 109.5° about N
(c) 120° about N
10-27 A functional group is a group of atoms that undergoes a predictable set of chemical reactions.

10-29 (a) —C(=Ö)—   (b) —C(=Ö)—Ö—H   (c) —Ö—H

(d) —N̈—H   (e) —C(=Ö)—Ö—   (f) —C(=Ö)—N̈H₂
     |
     H

10-31 When applied to alcohols, tertiary (3°) means that the carbon bearing the –OH group is bonded to three other carbon atoms.
10-33 When applied to amines, tertiary (3°) means that the amine nitrogen is bonded to three other carbon groups.
10-35 (a) The four primary (1°) alcohols with the molecular formula $C_5H_{12}O$ are:

$$CH_3CH_2CH_2CH_2CH_2OH \qquad CH_3CH_2CHCH_2OH$$
$$|$$
$$CH_3$$

$$CH_3\overset{\overset{CH_3}{|}}{\underset{\underset{CH_3}{|}}{C}}CH_2OH \qquad CH_3\overset{\overset{}{}}{\underset{\underset{CH_3}{|}}{C}HCH_2CH_2OH}$$

(b) The three secondary (2°) alcohols with the molecular formula $C_5H_{12}O$ are:

    OH   OH   OH
    | | |
$CH_3CHCH_2CH_2CH_3$ $CH_3CH_2CHCH_2CH_3$ $CH_3CHCHCH_3$
                                                                             |
                                                                             CH_3

(c) The one tertiary (3°) alcohol with the molecular formula $C_5H_{12}O$ is:

$$CH_3CH_2\overset{\overset{CH_3}{|}}{\underset{\underset{CH_3}{|}}{C}}-OH$$

10-37 The eight carboxylic acids with the molecular formula $C_6H_{12}O_2$ are:

a six-carbon chain	a five-carbon chain with a one-carbon branch	a four-carbon chain with two carbons as branches
$CH_3CH_2CH_2CH_2CH_2CO_2H$	$CH_3CHCH_2CH_2CO_2H$ | CH_3	$CH_3CHCHCO_2H$ | | CH_3 CH_3
	$CH_3CH_2CHCH_2CO_2H$ | CH_3	$CH_3CH_2CHCO_2H$ | CH_2CH_3
	$CH_3CH_2CH_2CHCO_2H$ | CH_3	CH_3 | $CH_3CH_2CCO_2H$ | CH_3
		CH_3 | $CH_3CCH_2CO_2H$ | CH_3

10-39 Taxol was discovered during a survey of indigenous plants for those containing phytochemicals that exhibited anti-tumor activity. It was sponsored by the National Cancer Institute with the goal of discovering new chemicals for fighting cancer.

10-41 The arrows point to atoms and show bond angles about each atom.

(a) CH_3—CH_2—CH_2—$\ddot{O}H$ (109.5°)

(b) CH_3—CH_2—$\overset{O}{\underset{\|}{C}}$—H (109.5°, 120°)

(c) CH_3—CH=CH_2 (109.5°, 120°)

(d) CH_3—C≡C—CH_3 (109.5°, 180°)

(e) CH_3—$\overset{O}{\underset{\|}{C}}$—$\ddot{O}$—$CH_3$ (109.5°, 120°, 109.5°)

(f) CH_3—$\underset{\ddot{}}{N}$—CH_3 with CH_3 (109.5°)

10-43 Predict 109.5° for C—P—C bond angles.

CH_3—$\ddot{P}$—CH_3
       \|
      CH_3 (109.5°)

10-45 The eight aldehydes with the molecular formula $C_6H_{12}O$ are below. The aldehyde functional group is written CHO.

a six-carbon chain	a five-carbon chain with a one-carbon branch	a four-carbon chain with two carbons as branches
$CH_3CH_2CH_2CH_2CH_2CHO$	$CH_3CHCH_2CH_2CHO$ | CH_3	CH_3 | $CH_3CHCHCHO$ | CH_3
	$CH_3CH_2CHCH_2CHO$ | CH_3	CH_3CH_2CHCHO | CH_2CH_3
	$CH_3CH_2CH_2CHCHO$ | CH_3	CH_3 | CH_3CH_2CCHO | CH_3
		CH_3 | CH_3CCH_2CHO | CH_3

10-47 (a) nonpolar covalent (b) nonpolar covalent (c) nonpolar covalent (d) polar covalent (e) polar covalent (f) polar covalent (g) polar covalent (h) polar covalent

10-49 Under each formula is given the difference in electronegativity between the atoms of the most polar bond.

(a) H—C—O—H with H's; O—H ($3.5-2.1 = 1.4$), δ^- on O, δ^+ on H

(b) H—C—N—H with H's; N—H ($3.0-2.1 = 0.9$)

(c) H—S—C—C—N—H; N—H ($3.0-2.1 = 0.9$)

(d) H—C—C—C—H with =O; C=O ($3.5-2.5 = 1.0$)

(e) H₂C=O; C=O ($3.5-2.5 = 1.0$)

(f) H—C—C—O—H with =O; O—H ($3.5-2.1 = 1.4$)

10-51 (a) All bond angles are approximately 120°, and the molecule is planar.
(b) Naphthalene is nonpolar.

10-53 The following all have the molecular formula $C_4H_8O_2$.

(a) Two carboxylic acids: $CH_3CH_2CH_2\overset{O}{\underset{\|}{C}}$—OH $CH_3CH\overset{O}{\underset{\|}{C}}$—OH
                                                                                \|
                                                                               CH_3

(b) Four esters: CH₃CH₂C(=O)—OCH₃   CH₃C(=O)—OCH₂CH₃
HC(=O)—OCH₂CH₂CH₃   HC(=O)—OCHCH₃ | CH₃

(c) One ketone + 2° alcohol: CH₃CHCCH₃ with OH and C=O

(d) One aldehyde + 3° alcohol: CH₃C(OH)(CH₃)—C(=O)—H

10-55 (a) The molecular formula of lactic acid is C₃H₆O₃.
(b) The two functional groups are a 2° hydroxyl group and a carboxyl group.
(c) You should predict bond angles of 120° about the carbonyl carbon and bond angles of 109.5° about the two other carbons and about the oxygen of each hydroxyl group.
(d) The C=O, O—H, and O—H bonds are polar covalent.
(e) Lactic acid is a polar molecule.

10-57 Convert grams of salicylic acid (138 g/mol) to moles of salicylic acid. From the balanced equation, see that one mole of salicylic acid gives one mole of aspirin (180 g/mol). Finally, convert moles of aspirin to grams of aspirin. Doing this math gives 157 grams of aspirin.

$$120 \times \frac{180}{138} = 157 \text{ grams of aspirin}$$

10-59 (a) Curved arrows show the repositioning of two pairs of electrons.

(A) H—C(H)(H)—C(=O)—N(H)—H   ⟷   (B) H—C(H)(H)—C(—Ö:⁻)=N⁺(H)—H

(b) Hydroxide ion shows 6 + 1 + 1 = 8 valence electrons. In both the hydroxide ion and the oxygen atom of contributing structure B, the oxygen in question has a complete octet. That is, each has a single bond and three nonbonding electron pairs and bears a negative charge.

H—Ö:⁻

(c) The ammonium ion contains 5 + 4 − 1 = 8 valence electrons. In both the ammonium ion and one nitrogen atom of contributing structure B, the nitrogen atom has a complete octet; it has four bonds and bears a positive charge.

H—N⁺(H)(H)—H

(d) If the hybrid is best represented by contributing structure A, then the H—N—H bond angles will be approximately 109.5°.
(e) If the hybrid is best represented by contributing structure B, then the H—N—H bond angles will be 120°.
(f) The discovery that the actual H—N—H bond angles about each amide nitrogen atom in a protein are 120° suggests that contributing structure B makes a greater contribution to the hybrid than contributing structure A.

Chapter 11 Alkanes

11-1 This alkane is octane, and its molecular formula is C₈H₁₈.
11-2 (a) Constitutional isomers   (b) Same compound
11-3 Line-angle formulas for the three constitutional isomers with the molecular formula C₅H₁₂ are:

11-4 (a) 5-Isopropyl-2-methyloctane; its molecular formula is C₁₂H₂₆.
(b) 4-Isopropyl-4-propyloctane; its molecular formula is C₁₄H₃₀.
11-5 (a) Isobutylcyclopentane, C₉H₁₈
(b) sec-Butylcycloheptane, C₁₁H₂₂
(c) 1-Ethyl-1-methylcyclopropane, C₆H₁₂
11-6 The chair conformation with the three methyl groups equatorial is:

11-7 Cycloalkanes (a) and (c) show cis-trans isomerism:

(a) cis-1,3-Dimethylcyclopentane
    trans-1,3-Dimethylcyclopentane

(c) cis-1,3-Dimethylcyclohexane
    trans-1,3-Dimethylcyclohexane

11-8 In order of increasing boiling point, they are:
(a) 2,2-Dimethylpropane (9.5°C), 2-Methylbutane (27.8°C), Pentane (36.1°C)
(b) 2,2,4-Trimethylhexane (127°C), 3,3-Dimethylheptane (137°C), Nonane (151°C)

11-9 The two chloroalkanes with their IUPAC and common names are:

1-Chloropropane (Propyl chloride)   2-Chloropropane (Isopropyl chloride)

11-11 (a) A *hydrocarbon* is a compound that contains only hydrogen and carbon atoms.
(b) An *alkane* is a saturated hydrocarbon.
(c) A *saturated hydrocarbon* contains only carbon–carbon single bonds.

11-13 In a *line-angle formula*, each line terminus and vertex represents a carbon atom. Single, double, and triple carbon–carbon bonds are represented by one, two, and three lines, respectively.

11-15 (a) $C_{10}H_{22}$ (b) C_8H_{18} (c) $C_{11}H_{24}$

11-17 (a) T (b) T (c) F (d) F
(c) False: Constitutional isomers have the same molecular formula but a different connectivity of their atoms.
(d) False: Constitutional isomers are different compounds and, therefore, have different physical properties.

11-19 Structures (a) and (g) represent the same compound. Compounds (a,g), (c), (d), (e), and (f) each have the molecular formula $C_4H_{11}N$ and represent constitutional isomers.

11-21 Sets (b), (c), (e), and (f) represent pairs of constitutional isomers.

11-23 (a) T (b) T (c) T

11-25 (a) IUPAC names and boiling points are:

CH$_3$ CH$_3$
| |
CH$_2$CHCH$_3$ CH$_2$CHCH$_2$CH$_3$

2-Methylpropane   2-Methylbutane
(Isobutane)   (Isopentane)
bp −11.7°C   bp 27.7°C

(b) Both hydrocarbons are gases at room temperature and are added to shaving gels as propellants and foaming agents.

11-27 (a) T (b) F (c) F
(b) False: They are not constitutional isomers because each has a different molecular formula: hexane is C_6H_{14}, and cyclohexane is C_6H_{12}.
(c) False: The parent name of a cycloalkane is the name of the unbranched alkane with the same number of carbon atoms that are in the ring with the prefix *cyclo-*.

11-29

11-31 A condensed structural formula shows only the order of bonding of the atoms in the compound. It does not show bond angles or the molecular shape.

11-33 (a) F (b) F (c) T (d) T (e) F
(a) False: *Cis-trans* isomers have the same atom connectivity but a different three-dimensional arrangement of their atoms in space.
(b) False: *Cis-trans* isomerism exists where there is hindered rotation about carbon–carbon bonds, such as in rings and carbon–carbon double bonds.
(e) False: They are stereoisomers.

11-35 No. *Cis-trans* isomerism results when there is restricted rotation about carbon–carbon bonds. Alkanes have free rotation about all of their carbon–carbon bonds.

11-37 Structural formulas for the six cycloalkanes with the molecular formula C_5H_{10} are:

Cyclopentane   Methylcyclobutane   1,1-Dimethylcyclopropane

cis-1,2-Dimethyl-cyclopropane   trans-1,2-Dimethyl-cyclopropane   Ethylcyclopropane

11-39 (a) The three six-membered rings are in chair conformations. The one five-membered ring is nearly planar, approximating an envelope conformation.
(b) The hydroxyl group on ring **A** is equatorial.
(c) The methyl group between rings **A** and **B** is axial to each ring.

11-41 (a) T (b) F (c) F (d) F (e) T (f) T
(b) False: Alkanes are all less dense than water.
(c) False: *Cis-trans* isomers have different physical properties.
(d) False: Usually, the least branched alkane isomer has the highest boiling point.

11-43 Heptane, C_7H_{16}, has a boiling point of 98°C and a molecular weight of 100. Its molecular weight is approximately 5.5 times that of water. Although considerably smaller, water molecules associate in the liquid by relatively strong hydrogen bonding, whereas the much larger heptane molecules associate only by relatively weak London dispersion forces.

11-45 Alkanes are nonpolar and insoluble in very polar water.

11-47 Boiling points of unbranched alkanes are related to their surface areas. The larger the surface area, the greater the strength of dispersion forces, and the higher the boiling point. The relative increase in size per CH_2 group is greatest between CH_4 and CH_3CH_3 and becomes progressively smaller as the molecular weight increases. Therefore, the increase in boiling point per added CH_2 group is greatest between CH_4 and CH_3CH_3 and becomes progressively smaller for higher alkanes.

11-49 (a) F (b) T (c) T
(a) False: Alkane combustions are exothermic.

Hydro-carbon	Component of	Heat of combustion (kcal/mol)	Molar mass (g/mol)	Heat of combustion (kcal/g)
CH_4	Natural gas	212	16.0	13.3
C_3H_8	LPG	530	44.1	12.0

11-51 On a gram-per-gram basis, methane is the better source of heat energy.

11-53 The three chloropentanes are:

1-Chloropentane   2-Chloropentane   3-Chloropentane

11-55
(a) Chlorocyclohexane

(b) 1-Chlorohexane   2-Chlorohexane   3-Chlorohexane

(c) 1-Chloro-2,3-dimethyl-butane   2-Chloro-2,3-dimethyl-butane

11-57 (a) T (b) T (c) T (d) F
(d) False: Chloroform is the common name for $CHCl_3$.
11-59 Octane rating indicates the relative smoothness with which a gasoline blend burns in an automobile engine. The higher the octane rating, the less the engine knocks. The reference hydrocarbons are 2,2,4-trimethylpentane (isooctane), which is assigned an octane rating of 100, and heptane, which is assigned an octane rating of 0.
11-61 2,2,4-Trimethylpentane has a higher heat of combustion per mole and per gram.

Hydrocarbon	Molar mass (g/mol)	Heat of combustion (kcal/mol)	Heat of combustion (kcal/g)
2,2,4-Trimethyl-pentane	114.2	1304	22.4
Ethanol	46.0	327	7.11

11-63 The presence of Freons in the stratosphere results in destruction of the stratospheric ozone layer.
11-65 (a) Constitutional isomers (b) Constitutional isomers
(c) Constitutional isomers (d) Not isomers (e) Not isomers
(f) Constitutional isomers

11-67 Compounds (a) and (c) show *cis-trans* isomerism.

(a) *cis* and *trans* isomers of 3-methylcyclohexanol

(c) *cis* and *trans* isomers of 4-methylcyclohexanol

11-69 Dodecane (a) does not dissolve in water, (b) dissolves in hexane, (c) burns when ignited, (d) is a liquid at room temperature, and (e) is less dense than water.
11-71 In the following drawings, all hydrogens are shown.

11-73 Water, a polar molecule, cannot penetrate the surface layer created by this nonpolar hydrocarbon.

Chapter 12 Alkenes and Alkynes

12-1 (a) 3,3-Dimethyl-1-pentene (b) 2,3-Dimethyl-2-butene
(c) 3,3-Dimethyl-1-butyne
12-2 (a) *trans*-3,4-Dimethyl-2-pentene
(b) *cis*-4-Ethyl-3-heptene
12-3 (a) 1-Isopropyl-4-methylcyclohexene (b) Cyclooctene
(c) 4-*tert*-Butylcyclohexene
12-4 Line-angle formulas for the other two heptadienes are:

cis,trans-2,4-Heptadiene   *cis,cis*-2,4-Heptadiene

12-5 Four stereoisomers are possible (two pair of *cis-trans* isomers).

12-6 (a)

12-7 Propose a two-step mechanism similar to that which describes addition of HCl to propene.
Step 1: Add a proton. Reaction of H^+ with carbon-2 of the carbon–carbon double bond gives a 3° carbocation intermediate.

A 3° carbocation intermediate

Step 2: Reaction of an electrophile and a nucleophile to form a new covalent bond. Reaction of the 3° carbocation intermediate with bromide ion completes the valence shell of carbon and gives the product.

12-8 The product from each acid-catalyzed hydration is the same alcohol.

$$CH_3CCH_2CH_3$$ with CH$_3$ and OH substituents

2-Methyl-2-butanol

12-9 Propose a three-step mechanism similar to that which describes the acid-catalyzed hydration of propene.
Step 1: Add a proton. Reaction of the carbon-carbon double bond with H⁺ gives a 3° carbocation intermediate.

[cyclohexene-CH₃ + H⁺ → cyclohexyl⁺-CH₃]

A 3° carbocation intermediate

Step 2: Reaction of an electrophile and a nucleophile to form a new covalent bond. Reaction of the 3° carbocation intermediate with water completes the valence shell of carbon and gives an oxonium ion.

[cyclohexyl⁺-CH₃ + :Ö–H → cyclohexyl(CH₃)–O⁺H₂]

An oxonium ion

Step 3: Take a proton away. Loss of H⁺ from the oxonium ion completes the reaction and generates a new H⁺ catalyst.

[oxonium ion → alcohol + H⁺]

12-10 (a) CH₃–C(CH₃)(H₃C)–CH(Br)–CH₂Br   (b) cyclohexane with Cl and CH₂Cl

12-11 (a) F (b) F (c) F (d) T
(a) False: There are three classes of unsaturated hydrocarbons: alkenes, alkynes, and arenes. Arenes are discussed in Chapter 13.
(b) False: In the United States and other areas of the world that have vast supplies of natural gas, ethylene is derived from cracking the ethane that is separated from natural gas. In areas of the world that do not have vast reserves of natural gas, ethylene is derived from the thermal cracking of petroleum.
(c) False: They are not constitutional isomers because they have different molecular formulas. Ethylene is C₂H₄, and acetylene is C₂H₂.
12-13 A saturated hydrocarbon contains only carbon–carbon single bonds. An unsaturated hydrocarbon contains one or more carbon–carbon double or triple bonds, or benzene-like rings.

12-15 There are 6 alkenes with the molecular formula C_5H_{10}.

1-Pentene   trans-2-Pentene   cis-2-Pentene

2-Methyl-1-butene   3-Methyl-1-butene   2-Methyl-2-butene

12-17 There are several possibilities for each part. Here is one for each.

(a) trans-4-Bromo-2-pentene   (b) 4-Bromo-1-pentene

12-19 (a) 109.5°, 120° (cyclopentane)   (b) 120° (cyclohexene with CH₂OH)
(c) HC≡C–CH=CH₂ 180°, 120°   (d) 120°

12-21 (a) T (b) F (c) F (d) F (e) T (f) F
(b) False: Its name is 2-butene.
(c) False: To show cis-trans isomerism, each carbon of a double bond must have two different groups bonded to it.
(d) False: The four atoms bonded to the carbons atoms of the double bond all lie in the same plane.
(f) False: It has no stereocenters and no possibility for cis-trans isomerism.

12-23 (a) CH₂=CHCH₂Cl   (b) cyclohexene-CH₃
(c) 1-methylcyclohexene   (d) substituted alkene
(e) methylenecyclopropane   (f) alkyne structure

12-25 (a) 2,5-Dimethyl-1-hexene
(b) 1,3-Dimethylcyclopentene
(c) 2-Methyl-1-butene (d) 2-Propyl-1-pentene
12-27 (a) The longest chain is four carbon atoms. The correct name is 2-methyl-1-butene.
(b) The ring is numbered incorrectly. The correct name is 4-isopropylcyclohexene.
(c) The parent chain is not numbered correctly. The correct name is 3-methyl-2-hexene.
(d) The chain is numbered incorrectly. The correct name is 2-ethyl-3-methyl-1-pentene
(e) The ring is numbered incorrectly. The correct name is 3,3-dimethylcyclohexene.

(f) The longest chain is seven carbon atoms. The correct name is 3-methyl-3-heptene.

12-29 Only (b) 2-hexene, (c) 3-hexene, and (e) 3-methyl-2-hexene will show *cis-trans* isomerism.

(b)

    trans-2-Hexene   *cis*-2-Hexene

(c)

    trans-3-Hexene   *cis*-3-Hexene

(e)

  trans-3-Methyl-2-hexene   *cis*-3-Methyl-2-hexene

12-31 Here are drawings of the *cis* and *trans* isomers of cyclodecene.

    cis-Cyclodecene   *trans*-Cyclodecene

12-33 Here are the *cis* and *trans* isomers:

Oleic acid

Elaidic acid

12-35 Only parts (b) and (d) show *cis-trans* isomerism.

(b) *cis*-1,2-Dimethylcyclohexane   (d) *cis*-4,5-Dimethylcyclohexene

12-37 β-Ocimene has the molecular formula $C_{10}H_{16}$. It is drawn here in alternative line-angle drawings.

β-Ocimene

12-39 (a) T (b) T (c) T (d) T (e) F (f) T (g) T (h) T (i) T (j) T (k) F (l) F (m) F (n) F
(e) False: A carbocation carbon has three bonds and a positive charge.
(k) False: Catalytic reduction of cyclohexene gives cyclohexane.

(l) False: H^+ comes from the acid catalyst, and —OH comes from H_2O.
(m) False: It is neither oxidation nor reduction. It is a hydration (addition).
(n) False: Both hydrations give 2-butanol.

12-41 (a) HBr (b) H_2O/H_2SO_4 (c) HI (d) Br_2

12-43 (a) $CH_3-CH_2-\overset{CH_3}{\underset{+}{C}}-CH_2-CH_3$ (3°) and

$CH_3-CH_2-\overset{CH_3}{\underset{+}{CH}}-CH-CH_3$ (2°)

(b) $CH_3-CH_2-CH_2-\overset{+}{CH}-CH_3$ (2°) and

$CH_3CH_2-\overset{+}{CH}-CH_2-CH_3$ (2°)

(c) [cyclopentyl cation] (3°) and [cyclopentyl cation] (2°)

(d) [cyclohexyl]—CH_3^+ (3°) and [cyclohexyl]—CH_2^+ (1°)

12-45 (a) [cyclohexane with CH_3, Br, Br] (b) [cyclopentane with CH_3, Cl, CH_3, Cl]

12-47 (a) $CH_3-\overset{CH_3}{\underset{|}{C}}=CH-CH_3$

(b) $CH_3-\overset{CH_3}{\underset{|}{C}}-CH=CH_2$

(c) $CH_2=CH-CH_2-CH_2-CH_3$

12-49 The only possibility for the formation of a 1° alcohol is if the molecule contains two $=CH_2$ groups ($CH_2=CH_2$). In ethylene, both carbons are $=CH_2$ and protonation of either gives a $CH_3-CH_2^+$ cation intermediate, which converts to ethanol, a 1° alcohol. For any other terminal alkene ($RCH_2CH=CH_2$ or $R_2C=CH_2$), H^+ will always add to the terminal carbon because it is the carbon of the double bond with the greater number of H atoms bonded to it. The result will be either a 2° or a 3° carbocation intermediate, which will lead to a 2° alcohol or a 3° alcohol.

12-51 Acid-catalyzed hydration of 2-pentene gives a mixture of 2-pentanol and 3-pentanol.

$$CH_2=CH-CH=CH-CH_2-CH_3 + H_2O \xrightarrow{H_2SO_4}$$

$$CH_3-\underset{\underset{OH}{|}}{CH}-CH_2-CH_2-CH_3 \;+$$

2-Pentanol

$$CH_3-CH_2-\underset{\underset{OH}{|}}{CH}-CH_2-CH_3$$

3-Pentanol

Addition of a proton to 2-pentene gives two different 2° carbocation intermediates, one of which reacts with water to give 2-pentanol. The other reacts with water to give 3-pentanol.

Addition of a proton to either carbon of the double bond of 3-hexene gives the same 2° carbocation, which then reacts with a water molecule to give 3-hexanol.

12-53

1,5,9-Cyclododecatriene + 3Br₂ → Hexabromocyclododecane

12-55 Six of these alkenes give only a single alcohol. One of them gives two constitutional isomers.

4-Methyl-1-pentene + H₂O $\xrightarrow{H_2SO_4}$

4-Methyl-2-pentene + H₂O $\xrightarrow{H_2SO_4}$

2-Methyl-2-pentene + H₂O $\xrightarrow{H_2SO_4}$

2-Methyl-1-pentene + H₂O $\xrightarrow{H_2SO_4}$

3,3-Dimethyl-1-butene + H₂O $\xrightarrow{H_2SO_4}$

2,3-Dimethyl-1-butene + H₂O $\xrightarrow{H_2SO_4}$

2,3-Dimethyl-2-butene + H₂O $\xrightarrow{H_2SO_4}$

12-57 The unsaturated hydrocarbon is 2-methyl-1,3-butadiene.

12-59

(a) + H₂ $\xrightarrow{Pd}$

(b) + H₂O $\xrightarrow{H_2SO_4}$

(c) + HBr →

(d) + Br₂ →

12-61 One function of ethylene as a plant growth regulator is to promote ripening of fruit.

12-63 A *pheromone* is a chemical produced by an animal and serves as a stimulus to another individual of the same species for one or more behavioral responses.

12-65 The molecular weight of 11-tetradecenyl acetate is 254.4 amu, and its molar mass is 254.4 g/mol. 1.0×10^{-12} g of 11-tetradecenyl acetate is equal to 2.0×10^9 molecules.

$$1.0 \times 10^{-12} \text{g} \times \frac{\text{mol}}{254.4 \text{ g}} \times \frac{6.02 \times 10^{23} \text{ molecules}}{\text{mol}}$$

$$= 2 \times 10^9 \text{ molecules}$$

12-67 It is longer for the all-*trans* isomer.

12-69 V is a code letter for poly(vinyl chloride), also known as PVC. PP is the code for polypropylene, and PS is the code for polystyrene.

12-71 There are 13 carbon-carbon double bonds in lycopene, only 11 of which have the possibility for *cis-trans* isomerism. For the double bonds on either end of the long hydrocarbon chain, no *cis-trans* isomerism is possible.

12-73 (a) [cyclohexene] (b) [cyclopentyl-CH₃] (c) [1-methylcyclopentene] (d) [methylenecyclopentane]

12-75 Following is the equation for the acid-catalyzed hydration of 3-hexene. Because of the symmetrical structure of 3-hexene, the *cis* and *trans* isomers will both give 3-hexanol. Because the same reactive intermediate is formed from either alkene, the same alcohol is formed from either alkene.

$$CH_3CH_2CH{=}CHCH_2CH_3 + H_2O \xrightarrow{H_2SO_4}$$
3-Hexene

$$CH_3CH_2\overset{OH}{\underset{|}{C}}HCH_2CH_2CH_3$$
3-Hexanol

12-77 Reagents are shown over the reaction arrows:
[cyclopentane diagram with (a) Br₂ → dibromide, (b) H₂O/H₂SO₄ → OH, (c) HI → I, (d) H₂/Pt → cyclopentane]

12-79 Two *cis-trans* isomers are possible for oleic acid, four for linoleic acid, and eight for linolenic acid. Note that each is a C_{18} fatty acid. The all-*cis* isomer of each is drawn:

Oleic acid
(A monounsaturated fatty acid)

Linolenic acid
(A polyunsaturated fatty acid)

Linolenic acid
(A polyunsaturated fatty acid)

12-81 **Step 1:** An oxidation that converts a carbon-carbon single bond of an alkane to a carbon–carbon double bond of an alkene.
Step 2: A hydration of a carbon–carbon double bond to give a 2° alcohol.
Step 3: An oxidation of a 2° alcohol to a ketone.

It is common in organic and biochemistry to call a carbon adjacent to a functional group an α-carbon and the next one a β-carbon. In this biochemical sequence, the β-carbon from the carbonyl group of the ester is converted from a —CH_2— group to a ketone group by an oxidation reaction, hence the term β-oxidation.

Chapter 13 Benzene and Its Derivatives

13-1 (a) 2,4,6-Tri-*tert*-butylphenol (b) 2,4-Dichloroaniline (c) 3-Nitrobenzoic acid

13-3 A saturated compound contains only single covalent bonds. An unsaturated compound contains one or more double or triple bonds, or benzene-like rings. The most common types of double and triple bonds are:

$$\text{C}={\text{C}} \quad \text{C}={\ddot{\text{O}}} \quad \text{C}={\ddot{\text{N}}}$$
$$-\text{C}\equiv\text{C}- \quad -\text{C}\equiv\text{N}{:}$$

13-5 The members of each class of hydrocarbons contain fewer hydrogens than an alkane or cycloalkane with the same number of carbon atoms. Alternatively, each class of hydrocarbons contains one or more carbon–carbon double or triple bonds.

13-7 No

13-9 (a) CH_4 (b) $CH_2{=}CH_2$
         Methane   Ethene
                               (Ethylene)

(c) HC≡CH
Ethyne (Acetylene)

(d) Benzene (C₆H₆ with all H shown)

13-11 Benzene consists of six carbons, each surrounded by three regions of electron density, which gives 120° for all bond angles. Bond angles of 120° in benzene can be maintained only if the molecule is planar. All C, H, and Cl atoms in 1,4-dichlorobenzene lie in the same plane.

13-13 (a) T (b) F (c) F (d) T (e) F (f) F
(b) False: *para* refers to a 1,4-disubstituted benzene ring.
(c) False: Benzene rings are planar, and therefore, *cis-trans* isomers are not possible.
(e) False: Benzene contains only one ring and is not a PAH (polynuclear aromatic hydrocarbon). A PAH must have two or more adjacent aromatic rings.
(f) False: Benzo[a]pyrene does not directly bind with DNA, but instead, its metabolic product, a diol epoxide, binds with DNA and causes mutations.

13-15 (a) 1-bromo-2-chloro-4-ethylbenzene (b) 1,2-dimethyl-4-bromobenzene (c) 2-methyl-1,3,5-trinitrobenzene (d) (3-phenyl-but-1-ene or similar) (e) 4-methylphenol (f) 2,4-dichlorophenol

13-17 The repeating unit is circled and shown in bold.
C₆H₅ **C₆H₅** C₆H₅ C₆H₅ C₆H₅ C₆H₅

13-19 Only cyclohexene will react with a solution of bromine in dichloromethane. A solution of Br₂/CH₂Cl₂ is reddish-purple, whereas benzene and 1,2-dibromocyclohexane are both colorless. To tell which bottle contains which compound, place a small quantity of each compound in separate test tubes and add a few drops of Br₂/CH₂Cl₂ solution. If the red color disappears, the compound is cyclohexene, which is converted to 1,2-dibromocyclohexane. If the reddish-purple color remains, the compound is benzene.

Cyclohexene (colorless) + Br₂ (red) $\xrightarrow{CH_2Cl_2}$ 1,2-dibromocyclohexane (colorless)

Benzene + Br₂ $\xrightarrow{CH_2Cl_2}$ No reaction

13-21 2-Bromotoluene (*o*-Bromotoluene), 3-Bromotoluene (*m*-Bromotoluene), 4-Bromotoluene (*p*-Bromotoluene)

13-23 (a) Nitration using HNO₃/H₂SO₄ followed by sulfonation using H₂SO₄. The order of the steps may be reversed.
(b) Bromination using Br₂/FeCl₃ followed by chlorination using Cl₂/FeCl₃. The order of the steps may be reversed.

13-25 (a) T (b) T (c) F (d) F (e) F (f) T (g) T (h) T (i) T
(c) False: The pK_a of phenol (10) is larger than that of acetic acid (4.7).
(d) False: In autoxidation, a C—H is converted to a C—O—O—H (hydroperoxide) group.
(e) False: A carbon radical has no charge.

13-27 Autoxidation is the reaction of a C—H group with oxygen, O₂, to form a hydroperoxide group, C—O—O—H.

13-29 Adding the equations for Step 2a and 2b and then canceling species that appear on both sides of the reaction arrow does, in fact, give the net equation for hydroperoxide formation.

13-31 The common group in vitamin E, BHT, and BHA is a substituted phenol.

13-33 (a) The functional groups in albuterol are a primary (1°) alcohol, a secondary (2°) alcohol, a secondary (2°) amine, and a phenolic —OH group.
(b) When albuterol is treated with one equivalent of NaOH, the phenolic —OH group is the only group acidic enough to react and form a sodium salt.

Albuterol + NaOH ⟶ (sodium phenoxide salt) + H₂O

(c) When albuterol is treated with one equivalent of HCl, the 2° amine is converted to an ammonium salt.

Albuterol + HCl ⟶ [ammonium salt product]

13-35 The advantages of DDT are that it is extremely effective in killing the insect hosts that transmit diseases such as malaria and typhus. It is also effective in killing crop-destroying insect pests. Disadvantages are that it persists in the environment and is toxic to many higher organisms.

13-37 Loss of hydrogen and chlorine from adjacent carbons gives a carbon–carbon double bond.

DDE

13-39 By definition, a carcinogen is a cancer-causing substance. The carcinogens present in cigarette smoke belong to the class of compounds called polynuclear aromatic hydrocarbons (PAHs).

13-41 Cyclonite (RDX) has the largest percentage of nitrogen.

Explosive	Mol Wt	% N
TNT	227.1	18.50
Nitroglycerin	227.1	18.50
Cyclonite	222.1	37.84
PETN	316.1	17.71

13-43 The functional groups most responsible for the water solubility of these dyes are the ionic —$SO_3^-Na^+$ groups in each.

13-45 (a) Both lycopene and β-carotene have an extended system of single-double-single bonds.
(b) The extended system of single-double-single bonds.
(c) Both lycopene and β-carotene are nonpolar covalent compounds and, therefore, are insoluble in water. The food dyes have some polar covalent bonds, but their water solubility is due primarily to the presence of ionic groups and the fact that they are ionic compounds.

13-47 The phenolic group is the —OH group bonded to the benzene ring, shown here in boldface.

[Capsaicin structure]

13-49 Capsaicin is used in medicine as a topical anesthetic.

13-51
(a) 1-Phenylcyclopropanol
(b) Styrene
(c) m-Bromophenol
(d) 4-Nitrobenzoic acid
(e) Isobutylbenzene
(f) m-Xylene

13-53
(a) benzene + HNO_3 $\xrightarrow{H_2SO_4}$ nitrobenzene + H_2O

(b) p-xylene + Br_2 $\xrightarrow{FeCl_3}$ 2-bromo-1,4-dimethylbenzene + HBr

(c) 1,4-dibromobenzene + H_2SO_4 ⟶ 2,5-dibromobenzenesulfonic acid + H_2O

13-55

+ Br$_2$ $\xrightarrow{AlBr_3}$ o-Bromotoluene

+ m-Bromotoluene + p-Bromotoluene

13-57 (a) Each is represented as a hybrid of two contributing structures.

Benzene

Pyridine

Pyrimidine

(b) Each has a loop of six electrons (two from the second bond of each double bond). It is only these six electrons that become redistributed from one contributing structure to the next. Note that the unshared pairs of electrons on nitrogen in pyridine and pyrimidine are not a part of the aromatic sextet.
(c) As predicted by the VSEPR model, all bond angles about each carbon and nitrogen are 120°. Because of the required planarity of a double bond and the four atoms bonded to it, predict that just as is the case with benzene, pyridine and pyrimidine are also planar molecules.

Chapter 14 Alcohols, Ethers, and Thiols

14-1 (a) 2-Heptanol (b) 2,2-Dimethyl-1-propanol
(c) cis-3-Isopropylcyclohexanol
14-2 (a) Primary (1°) (b) Secondary (2°) (c) Primary (1°)
(d) Tertiary (3°)
14-3 The structure of the major alkene product from each reaction is enclosed in a box. In each case, the major product contains the more substituted double bond.

(a) $\boxed{CH_3C=CHCH_3}$ + $CH_2=CCH_2CH_3$ (with CH$_3$ substituents)

(b) [cyclopentene with CH$_3$] + [methylenecyclopentane]

14-4

2-Methyl-cyclohexanol $\xrightarrow{H_2SO_4}$ 1-Methyl-cyclohexene (C) $\xrightarrow[H_2SO_4]{H_2O}$ 1-Methyl-cyclohexanol (D)

14-5 Each secondary alcohol is oxidized to a ketone.

(a) cyclohexanone (b) $CH_3\overset{O}{\underset{\|}{C}}CH_2CH_2CH_3$

14-6 (a) Ethyl isobutyl ether (b) Cyclopentyl methyl ether
14-7 (a) 3-Methyl-1-butanethiol (b) 3-Methyl-2-butanethiol
14-9 The difference is in the number of carbon atoms bonded directly to the carbon bearing the —OH group. For primary alcohols, there is zero or a single carbon group; for a secondary alcohol, there are two, and for a tertiary alcohol; there are three.
14-11 None of the alcohols in problem 14-10 is primary. The alcohols in parts (a) and (b) are tertiary. The alcohols in parts (c) and (d) are secondary.
14-13

(a) isopropanol (OH) (b) 1,2-propanediol structure
(c) 4-methyl-2-pentanol (d) 2-ethyl-2-propyl-1,3-propanediol
(e) heptanol (f) 3,3-dimethylcyclohexanol

14-15 Phenol contains an —OH group bonded to a benzene ring. An alcohol contains an —OH group bonded to a tetravalent carbon. Following are structural formulas for phenol and cyclohexanol.

Phenol   Cyclohexanol

14-17 Low-molecular-weight alcohols form hydrogen bonds with water molecules through both the oxygen and hydrogen atoms of their —OH groups, and because of this association, they dissolve in water. Low-molecular-weight alkanes, alkenes, and alkynes are nonpolar compounds and do not associate in any way with water molecules. The large amount of hydrogen bonding between an alcohol molecule and water molecules makes the low-molecular-weight alcohols soluble in water. Low-molecular-weight alkanes, alkenes, and alkynes are insoluble in water.
14-19 As the molecular weight of an alcohol increases, the hydrocarbon part of the molecule becomes a larger and larger

part of its surface area. The physical properties of alcohols, including their solubility in water, then become more like those of hydrocarbons of similar carbon skeletons.

14-21 Hydrogen bonding between diethyl ether and water is shown in the following illustration.

$$CH_3CH_2 \overset{\delta-}{-} \underset{\underset{\overset{\delta+}{H}}{\|}}{O} \overset{CH_2CH_3}{\underset{}{}}$$
$$H - O^{\delta-}$$
$$\quad\quad\quad\quad | $$
$$\quad\quad\quad\quad H$$

14-23 In order of increasing boiling point, they are:

$CH_3CH_2CH_2CH_3$ $CH_3CH_2OCH_2CH_3$
0°C   35°C

$CH_3CH_2CH_2OH$
97°C

14-25 The thickness (viscosity) of these three liquids is related to the degree of hydrogen bonding between their molecules in the liquid state. Hydrogen bonding is strongest between molecules of glycerol, weaker between molecules of ethylene glycol, and weakest between molecules of ethanol.

14-27 In order of decreasing solubility in water, they are:
(a) Ethanol > Diethyl ether > Butane
(b) 1,2-Hexanediol > 1-Hexanol > Hexane

14-29 (a) T (b) T (c) F (d) F (e) F (f) F (g) T (h) F (i) T (j) F
(c) False: Only phenols react with strong bases to form water-soluble salts.
(d) False: Acid-catalyzed dehydration of cyclohexanol gives cyclohexene.
(e) False: The carbon–carbon double bond with the greater number of substituents predominates in the acid-catalyzed dehydration of alcohols.
(f) False: Acid-catalyzed dehydration of 2-butanol gives trans-2-butene as the major product.
(h) False: The oxidation of a 2° alcohol gives a ketone.
(j) False: The product will be 2-propanone (acetone).

14-31 Phenols (pK_a 10) are weak acids. Alcohols (pK_a 16-18) are considerably weaker acids than phenols, and are even weaker acids than water.

14-33

(a) $CH_3CH_2CH_2CH_2OH \xrightarrow[\text{heat}]{H_2SO_4} CH_3CH_2CH=CH_2 + H_2O$

(b) $CH_3CH_2CH_2CH_2OH \xrightarrow[H_2SO_4]{K_2Cr_2O_7} CH_3CH_2CH_2CO_2H$

14-35

(a) ~~~~~~OH $\xrightarrow[H_2SO_4]{K_2Cr_2O_7}$ ~~~~~~C(=O)OH

(b) $HOCH_2CH_2CH_2CH_2OH \xrightarrow[H_2SO_4]{K_2Cr_2O_7} HOCCH_2CH_2COH$ (with two C=O)

14-37 (a) H_2SO_4, heat
(b) H_2O/H_2SO_4
(c) $K_2Cr_2O_7/H_2SO_4$
(d) HBr
(e) Br_2
(f) H_2/Pd
(g) $K_2Cr_2O_7/H_2SO_4$
(h) $K_2Cr_2O_7/H_2SO_4$
(i) $K_2Cr_2O_7/H_2SO_4$

14-39 Ethanol and ethylene glycol are important alcohols derived from ethylene. Ethanol is used as a solvent; is the alcohol in alcoholic beverages; is the starting material for the synthesis of diethyl ether, also an important solvent; and is blended with gasoline to improve octane number and reduce air pollution. Ethylene glycol is widely used as an automotive antifreeze and is one of the two starting materials for the synthesis of the polymer poly(ethylene terephthalate), better known as PET (Chapter 18).

14-41 (a) T (b) F (c) F
(b) False: Low-molecular-weight alcohols are more soluble in water than low-molecular-weight ethers.
(c) False: Ethers are unreactive to most of the reaction conditions that transform alcohols into other compounds.

14-43 (a) Ethyl isopropyl ether (b) Dibutyl ether
(c) Diphenyl ether

14-45 (a) 2-Butanethiol (b) 1-Butanethiol
(c) Cyclohexanethiol

14-47 1-Butanol has the higher boiling point because its molecules can associate in the liquid state by hydrogen bonding. The polarity of an —SH group is so low that there is little association among thiol molecules in the liquid state.

$CH_3CH_2CH_2CH_2OH$ $CH_3CH_2CH_2CH_2SH$
1-Butanol   1-Butanethiol
(bp 117°C)   (bp 98°C)

14-49 (a) T (b) T (c) T (d) T (e) T (f) T
All statements are true. There are no false statements.

14-51 Nitroglycerin was discovered in 1847 by Ascanio Sobrero. It is a pale yellow, oily liquid.

14-53 One of the products that the body derives from the metabolism of nitroglycerin is nitric oxide, NO, which causes the coronary artery to dilate, thus relieving angina.

14-55 The correlation is that "air deep within the lungs is in equilibrium with blood passing through the pulmonary arteries, and thus an equilibrium is established between blood alcohol and breath alcohol."

14-57 Diethyl ether is easy to use and causes excellent muscle relaxation. Blood pressure, pulse rate, and respiration are usually only slightly affected. Its chief disadvantages are its irritating effect on the respiratory passages and its aftereffect of nausea. Furthermore, it is highly flammable and must not be used whenever there is danger of electrical sparking, which may ignite it.

14-59 Yes, expect them to be soluble in nonpolar organic solvents such as hexane.

14-61 Because the electronegativity of oxygen (3.5) is greater than that of nitrogen (3.0), an O—H----O hydrogen bond is stronger than an N—H----N hydrogen bond.

14-63 The six ethers with molecular formulas $C_5H_{12}O$ are here grouped first by the four with a methyl group bonded to oxygen and then the two with an ethyl group bonded to oxygen.

$CH_3-O-CH_2CH_2CH_2CH_3$
Butyl methyl ether

$CH_3CH_2-O-CH_2CH_2CH_3$
Ethyl propyl ether

$CH_3-O-\underset{CH_3}{\underset{|}{CH}}CH_2CH_3$
sec-Butyl methyl ether

$CH_3-CH_2-O-\underset{CH_3}{\underset{|}{CH}}CH_3$
Ethyl isopropyl ether

$CH_3-O-CH_2\underset{CH_3}{\underset{|}{CH}}CH_3$
Isobutyl methyl ether

$CH_3-O-\underset{CH_3}{\overset{CH_3}{\underset{|}{\overset{|}{C}}}}CH_3$
tert-Butyl methyl ether

14-65 (a) There are three 3° alcohols and two 2° alcohols.
(b) There is only one 3° amine.
(c) The ester group is at the bottom of the large ring on the left. Note that because the ester group is part of a ring, the large ring can be called either a cyclic ester or a lactone.
(d) The four internal hydrogen bonds are marked. Three involve O—H hydrogen bonding and one involves N—H hydrogen bonding.

14-67 Predict that 1,4-butanediol, which has two —OH groups is infinitely soluble in water. 1-Pentanol, which has only one —OH group, has the solubility of 2.3 g/100 mL of water, and hexane, a nonpolar hydrocarbon, is insoluble in water.

14-69 Acid-catalyzed dehydration of 2-methyl-1-propanol gives 2-methylpropene. Acid-catalyzed hydration of 2-methylpropene gives 2-methyl-2-propanol. Oxidation of 2-methyl-1-propanol by $K_2Cr_2O_7/H_2SO_4$ gives 2-methylpropanoic acid.

$CH_3-\underset{CH_3}{\underset{|}{CH}}-CH_2OH \xrightarrow[\text{heat}]{H_2SO_4} CH_3-\underset{CH_3}{\underset{|}{C}}=CH_2 \xrightarrow[H_2SO_4]{H_2O} CH_3-\underset{OH}{\overset{CH_3}{\underset{|}{\overset{|}{C}}}}-CH_3$

2-Methyl-1-propanol   2-Methylpropene   2-Methyl-2-propanol

$\xrightarrow[H_2SO_4]{K_2Cr_2O_7} CH_3-\underset{CH_3}{\underset{|}{CH}}-COOH$
Propanoic acid

14-71 Curved arrows are used to show the redistribution of electron pairs in bond making and bond breaking in this three-step reaction mechanism.

Step 1: $CH_3-CH_2-\overset{:\ddot{O}H}{\underset{|}{CH}}-CH_3 + H^+ \rightleftharpoons CH_3-CH_2-\overset{H\;\;H}{\underset{|}{\overset{\overset{+}{O}}{CH}}}-CH_3$

Step 2: $CH_3-CH_2-\overset{H\;\;H}{\underset{|}{\overset{\overset{+}{O}}{CH}}}-CH_3 \rightleftharpoons CH_3-CH_2-\overset{+}{CH}-CH_3 + H_2O$
A carbocation intermediate

Step 3: $CH_3-\overset{H}{\underset{|}{CH}}-\overset{+}{CH}-CH_3 \rightleftharpoons CH_3-CH=CH-CH_3 + H^+$

14-73 The two functional groups in lipoic acid are a disulfide group and a carboxyl group. Reduction of the disulfide group in lipoic acid produces two sulfhydryl groups.

Dihydrolipoic acid

14-75 The three-step mechanism results in the addition of a molecule of water to a carbonyl group.
Step 1: **Add a proton** to oxygen. The effect of this step is to increase the positive charge on the carbonyl carbon and make it a stronger electrophile.

Step 2: **Reaction of an electrophile and a nucleophile to form a new covalent bond**.

Step 3: **Take a proton away**. This step completes the formation of formaldehyde hydrate and regenerates the acid catalyst.

14-77 Review your answer to Problem 12-49.

14-79 The five other alkenes with the molecular formula C_5H_{10} and the alcohol each produces on acid-catalyzed hydration are:

4-Methyl-1-pentene + H_2O $\xrightarrow{H_2SO_4}$

2-Methyl-2-pentene + H_2O $\xrightarrow{H_2SO_4}$

2-Methyl-1-pentene + H_2O $\xrightarrow{H_2SO_4}$

3,3-Dimethyl-1-pentene + H_2O $\xrightarrow{H_2SO_4}$

3-Methyl-2-pentene + H_2O $\xrightarrow{H_2SO_4}$

Chapter 15 Chirality: The Handedness of Molecules

15-1 The enantiomers of each part are drawn with two groups in the plane of the paper, a third group toward you in front of the plane, and the fourth group away from you behind the plane.

15-2 The group of higher priority in each set is circled.

(a) ⟨—CH$_2$OH⟩ and —CH$_2$CH$_2$COH (with C=O)

(b) ⟨—CH$_2$NH$_2$⟩ and —CH$_2$COH (with C=O)

15-3 The configuration is R, and the compound is (R)-glyceraldehyde.

15-4 (a) Compounds 1 and 3 are one pair of enantiomers. Compounds 2 and 4 are a second pair of enantiomers.
(b) Diastereomers are stereoisomers that are not mirror images. Compounds 1 and 2, 1 and 4, 2 and 3, and 3 and 4 are diastereomers.

15-5 Carbons 1 and 3 of 3-methylcyclohexanol are stereocenters. Therefore, $2^2 = 4$ stereoisomers are possible for this molecule. The *cis* isomer exists as one pair of enantiomers and the *trans* isomer exists as a second pair of enantiomers.

3-Methylcyclohexanol

15-6 Each stereocenter is marked by an asterisk, and the number of stereoisomers possible is shown under the structural formula.

(a) HO-C$_6$H$_3$(OH)-CH$_2$CH(NH$_2$)COOH $2^1 = 2$

(b) CH$_2$=CHCHCH$_2$CH$_3$ (with OH) $2^1 = 2$

(c) cyclohexane with OH and NH$_2$ $2^2 = 4$

15-7 (a) T (b) T (c) T (d) F (e) T (f) T (g) T (h) T (i) T
(d) False: Constitutional isomers have a different connectivity of their atoms.

15-9 An achiral object has no handedness; it is an object whose mirror image is superposable on the original. Examples are methane, CH$_4$, and benzene, C$_6$H$_6$.

15-11 Both constitutional isomers and stereoisomers have the same molecular formula. Whereas stereoisomers have the same connectivity, constitutional isomers have a different connectivity of their atoms.

15-13 2-Pentanol has one stereocenter (carbon 2). 3-Pentanol has no stereocenter.

2-Pentanol   3-Pentanol

15-15 The carbon of a carbonyl group has only three groups bonded to it. To be a stereocenter, a carbon must have four different groups bonded to it.

15-17 Compounds (b), (c), and (d) contain stereocenters, here marked by asterisks, and are chiral.

15-19 Following are the mirror images of each molecule.

15-21 (a) T (b) F (c) T (d) F (e) T (f) T
(b) False: For a molecule with three stereocenters, $2^3 = 8$ stereoisomers are possible.
(d) False: 2-Pentanol is chiral, but 3-pentanol is achiral. Only 2-pentanol shows enantiomerism.

15-23 Parts (b) and (c) contain stereocenters.

15-25 Stereocenters are marked with an asterisk. Under each is the number of stereoisomers possible.

(a) [cyclopentane with OH and CH3, one stereocenter marked]
$2^2 = 4$
(Two pair of enantiomers)

(b) [acyclic structure with OH]
2
(One pair of cis + trans isomers)

(c) [tetrahydrofuran with OH]
$2^1 = 2$
(One pair of enantiomers)

(d) [decalone structure]
$2^2 = 4$
(Two pair of enantiomers)

15-27 The specific rotation of its enantiomer is +41°.
15-29 (a) T (b) T (c) F (d) T
(c) False: *Cis-trans* stereoisomers do not need to be chiral; therefore, there is no requirement that they need to be optically active.
15-31 All three structures are chiral. The stereocenters in each are marked with an asterisk. Under the name of each is the number of possible stereoisomers.

Captopril
($2^2 = 4$ stereoisomers)

Enalopril (Altace)
($2^3 = 8$ stereoisomers)

Quinapril (Accupril)
($2^3 = 8$ stereoisomers)

Ramipril (Vasotec)
($2^5 = 32$ stereoisomers)

The partial skeletons of the three (left) compared to that of Captopril (right).

all of the molecules share structural features similar to captopril

In addition, the second, third, and fourth molecules have a larger skeletal unit in common:

15-33 Only one alcohol with the molecular formula $C_6H_{14}O$ contains two stereocenters.

3-Methyl-2-pentanol

15-35 All three drugs are chiral. Stereocenters are marked with asterisks.

(a) Fluoxetine (Prozac)
($2^1 = 2$)

(b) Sertraline (Zoloft) ($2^2 = 4$)

(c) Paroxetine (Paxil) ($2^2 = 4$)

15-37
(a) 1-methylcyclohexanol with OH and CH₃
(b) 4-methylcyclohexanol
(c) 2-methylcyclohexanol or 3-methylcyclohexanol

15-39 The majority have a right-handed twist because the machines that make them all impart the same twist.

15-41 Compound A could contain a carbon–carbon triple bond, but if it did, treatment with H_2 in the presence of a transition metal catalyst would add two moles of hydrogen to give C_5H_{12}. So from the evidence of catalytic hydrogenation, we can eliminate the presence of a carbon–carbon triple bond, and because catalytic reduction only added two hydrogens, we know that Compound A contains one carbon–carbon double bond. Any alkene with five carbons in a chain (for example, 2-pentene) has the molecular formula C_5H_{10}. So if Compound A is an alkene, it cannot be an open-chain alkene. One possibility for Compound A is cyclopentene, which has the correct molecular formula.

D (chlorocyclopentane) ←HCl— A (cyclopentene) —H₂/Pt→ C (cyclopentane); A —Br₂→ B (1,2-dibromocyclopentane)

Two other possibilities for Compound A are: methylenecyclobutane and vinylcyclopropane

15-43 This molecule has eight stereocenters, here marked with asterisks. There are $2^8 = 256$ possible stereoisomers (128 pairs of enantiomers).

Triamcinolone acetonide

15-45 (a) Tamiflu contains five functional groups.
- A carbon-carbon double bond (an alkene)
- An ether
- An ester
- An amide
- 1° amine

Oseltamivir (Tamiflu)

(b) Its molecular formula is $C_{16}H_{28}N_2O_4$.
(c) It is chiral and contains three stereocenters marked with asterisks and has $2^3 = 8$ possible stereoisomers. If it were isolated from a natural source, it would be optically active and would rotate the plane of polarized light. If, however, it were synthesized in the laboratory, it would be a mixture of stereoisomers—racemic and optically inactive.
(d) There are three stereocenters in Tamiflu. Their configuration is $3R, 4R, 5S$.

Oseltamivir (Tamiflu)

(e) The enantiomer of Tamiflu is its nonsuperimposable mirror image.

Oseltamivir (Tamiflu)

Mirror

Mirror Image of Tamiflu

(f) A diastereomer is a stereoisomer that is not a mirror image. For a diastereomer of Tamiflu, invert the configuration at any one of the stereocenters. The diastereomers shown here differ in the configuration at the carbon atom bearing the 1° amine group.

Oseltamivir (Tamiflu)

Diastereomer of Tamiflu (a stereoisomer that is not the mirror image)

(g) The orientation of each substituent on the six-membered ring is marked (e) for equatorial or (a) for axial. The more stable chair conformation is drawn on the left. In it, more groups on the ring are equatorial.

More stable chair (e, e, e, a)

Less stable chair (a, a, a, e)

15-47 (a) The configuration of each stereocenter is marked and given in the complete name.

(R)-(−)-Carvone
Spearmint oil

(S)-(+)-Carvone
Caraway and dill seed

(b) Each enantiomer is a different compound, and we can expect them to interact differently with the highly stereoselective olfactory receptors in the nose and nasal passages.

Chapter 16 Amines

16-1 Pyrrolidine has nine hydrogens; its molecular formula is C_4H_9N. Purine has four hydrogens; its molecular formula is $C_5H_4N_4$.

16-2

(a) isobutyl-NH$_2$

(b) cyclopentyl-NH$_2$

(c) H$_2$N—(CH$_2$)$_3$—NH$_2$

16-3

(a) HO—CH$_2$CH$_2$—NH$_2$

(b) Ph—NH—Ph

(c) (iPr)$_2$NH

16-4 The stronger base is circled.

(a) pyridine or [cyclohexyl-NH$_2$]

(b) [CH$_3$NH$_2$] or Ph—NH$_2$

16-5 The product of each reaction is an ammonium salt.

(a) $(CH_3CH_2)_3\overset{+}{N}HCl^-$
Triethylammonium chloride

(b) piperidinium$^+$ $CH_3CO_2^-$
Piperidinium acetate

16-7 In the classification of alcohols, the terms 1°, 2°, and 3° refer to the number of carbons bonded to the carbon bearing the —OH group. In the classification of amines, the terms 1°, 2°, and 3° refer to the number of carbons bonded to the nitrogen atom of the amine.

16-9 Each compound has a six-membered ring with three double bonds, that is, each has an aromatic sextet.

16-11 Following is a structural formula for each amine:

(a) sec-butylamine (CH₃CH₂CH(NH₂)CH₃ structure shown)
(b) $CH_3(CH_2)_6CH_2NH_2$
(c) neopentylamine ((CH₃)₃CCH₂NH₂ structure shown)
(d) $H_2N(CH_2)_5NH_2$
(e) 2-bromoaniline (benzene ring with NH₂ and Br)
(f) $(CH_3CH_2CH_2CH_2)_3N$

16-13 Each amine is classified by type.

(a) Serotonin-like structure: HO-indole-CH₂CH₂NH₂
   - NH₂ group: 1° aliphatic amine
   - ring N–H: heterocyclic aromatic amine

(b) Ethyl 4-aminobenzoate: H₂N–C₆H₄–COOCH₂CH₃
   - 1° aromatic amine

(c) Diphenhydramine-type: (C₆H₅)₂CH–O–CH₂CH₂–N(CH₃)₂
   - 3° aliphatic amine

(d) Lysine: H₂N–(CH₂)₄–CH(NH₂)–COOH
   - 1° aliphatic amines

(e) Chloroquine-type structure
   - 2° aromatic amine (H–N)
   - 3° aliphatic amine (N with ethyl)
   - heterocyclic aromatic amine (quinoline N)

(f) $H_2N{\diagup}{\diagdown}COOH$
   1° aliphatic amine

16-15 For the molecular formula $C_5H_{13}N$, there are eight constitutional isomers. Three of these amines are chiral. The stereocenters in each are marked with asterisks.

$CH_3CH_2CH_2CH_2CH_2NH_2$ $CH_3CH_2CH_2\overset{*}{C}H(NH_2)CH_3$
1-Pentanamine   2-Pentanamine

$CH_3CH_2CH(NH_2)CH_2CH_3$ $CH_3CH_2\overset{*}{C}H(CH_3)CH_2NH_2$
3-Pentanamine   2-Methyl-1-butanamine

$CH_3CH_2C(NH_2)(CH_3)CH_3$ $CH_3\overset{*}{C}H(NH_2)CH(CH_3)CH_3$
2-Methyl-2-butanamine   3-Methyl-2-butanamine

$H_2NCH_2CH_2CH(CH_3)CH_3$ $(CH_3)_3CCH_2NH_2$
3-Methyl-1-butanamine   2,2-Dimethyl-1-propanamine

16-17 Both propylamine (a 1° amine) and ethylmethylamine (a 2° amine) have an N—H group, and hydrogen bonding occurs between their molecules in the liquid state. Because of this intermolecular force of attraction, these two amines have higher boiling points than trimethylamine (a 3° amine), which has no N—H group and, therefore, cannot participate in intermolecular hydrogen bonding.

16-19 2-Methylpropane is a nonpolar hydrocarbon and the only attractive forces between its molecules in the liquid state are the very weak London dispersion forces. Both 2-propanol and 2-propanamine are polar molecules and associate in the liquid state by hydrogen bonding. Hydrogen bonding is stronger between alcohol molecules than between amine molecules because of the greater strength of an O—H—O hydrogen bond compared with an H—N—H hydrogen bond. It takes more energy (a higher temperature) to separate an alcohol molecule in the liquid state from its neighbors than to separate an amine molecule in the liquid state from its neighbors, and therefore, the alcohol has the higher boiling point.

16-21 (a) T (b) T (c) F (d) T (e) T (f) T (g) T
(c) False: Aliphatic amines are weaker bases than inorganic bases such as NaOH and KOH.

16-23

(a) $(CH_3)_3\overset{+}{N}CH_2CH_3\,OH^-$ (b) $(CH_3)_2NH_2^+\,I^-$

(c) $(CH_3)_4N^+Cl^-$ (d) $C_6H_5{-}NH_3^+\,Br^-$

A46 Answers

16-25 The stronger base is denoted with a rectangle. The determining factor is that aliphatic amines are stronger bases than aromatic amines.

(a) [piperidine (boxed)] or [pyridine]

(b) [C₆H₁₁—N(CH₃)₂ (boxed)] or [C₆H₅—NHCH₃]

(c) [C₆H₅—N(CH₃)₂] or [C₆H₅—CH₂NH₂ (boxed)]

16-27 (a) Here is the completed Lewis structure for guanidine:

$$H_2\ddot{N}-\underset{\underset{Guanidine}{}}{\overset{\overset{\ddot{N}H}{\|}}{C}}-\ddot{N}H_2$$

(b) Here are the three equivalent contributing structures. Each has an identical pattern of bonding:

$$H_2\ddot{N}-\overset{\overset{+}{NH_2}}{\underset{\|}{C}}-\ddot{N}H_2 \longleftrightarrow H_2\overset{+}{N}=\underset{\underset{}{}}{\overset{\overset{\ddot{N}H_2}{|}}{C}}-\ddot{N}H_2 \longleftrightarrow H_2\ddot{N}-\overset{\overset{\ddot{N}H_2}{|}}{\underset{}{C}}=\overset{+}{N}H_2$$

(c) If protonation occurs on the =NH nitrogen, then all three nitrogens become equivalent; that is, they have identical patterns of chemical bonding.
(d) Predict all N—C—N bond angles in the hybrid to be 120°.
(e) Guanidine is a considerably stronger base than ammonia.

16-29 Both compounds are insoluble in water. Treat each with aqueous HCl. The amine will react with HCl to form a water-soluble salt. The alcohol does not react with aqueous HCl and is insoluble in it. Thus, the amine dissolves in aqueous HCl, whereas the alcohol does not.

16-31 The primary aliphatic amine is the stronger base and forms a water-soluble salt in HCl. The salt is named pyridoxamine hydrochloride, which is a form of vitamin B_6.

[Structure: Pyridoxamine hydrochloride — pyridine ring with $CH_2NH_3^+$ Cl^-, HO, CH_2OH, HO substituents]

Pyridoxamine hydrochloride

16-33 Epinephrine has two phenolic —OH groups, whereas [amphetamine] has none. Epinephrine has a 2° alcohol group [in the si]de chain, whereas amphetamine has none. [Epinephrine] is a 2° amine, whereas amphetamine has a 1° [amine;] both are chiral, but their stereocenters are at [different positi]ons within the molecules.

16-35 Alkaloids are basic nitrogen-containing compounds found in the roots, bark, leaves, berries, and fruits of plants. In almost all alkaloids, the nitrogen atom is present as a member of a ring; that is, it is present as a heterocyclic amine. By definition, alkaloids are nitrogen-containing bases, and therefore, they turn red litmus blue.

16-37 The tertiary aliphatic amine in the five-membered ring is the stronger base and is protonated by aqueous HCl.

[Structure of (S)-Nicotine hydrochloride]

(S)-Nicotine hydrochloride

16-39 The common structural feature is a benzene ring fused to a seven-membered ring containing two nitrogen atoms. This parental compound, from which a number of psychoactive compounds are synthesized, is named benzodiazepine.

16-41 The structural formula of 4-aminobutanoic acid is drawn on the left, showing the 1° amino group (a base) and the carboxyl group (an acid). It is drawn on the right as an internal salt called a zwitterion.

a 1° amine   a carboxyl group

$$H_2NCH_2CH_2CH_2\overset{\overset{O}{\|}}{C}OH \qquad H_3\overset{+}{N}CH_2CH_2CH_2\overset{\overset{O}{\|}}{C}O^-$$

un-ionized amino and carboxyl groups   an internal salt

16-43 Amine salts are more soluble in water and body fluids and are less reactive to oxidation by atmospheric oxygen.

16-45 Following is one possible structural formula for each part:

(a) [C₆H₅—NHCH₃]   (b) [C₆H₅—N(CH₃)₂ with CH₃ on N]

(c) [C₆H₅—CH₂NH₂]   (d) $CH_3CHCH_2CH_3$ with NH_2

(e) [N-methylpyrrolidine, CH₃ on N]   (f) [2,3-dimethyl-... trimethylaniline with NH₂]

(g) $CH_3—\overset{\overset{CH_2CH_3}{|}}{\underset{\underset{CH_3CHCH_3}{|}}{\overset{+}{N}}}—CH_2CH_2CH_3 \quad Cl^-$

16-47 (a) CH_3SH is the strongest acid.
(b) $(CH_3)_2NH$ is the strongest base.
(c) CH_3OH has the highest boiling point.
(d) Molecules of CH_3OH form the strongest hydrogen bonds.

16-49 Both alcohols and amines can interact with water by hydrogen bonding. Because amines and alcohols with the same molecular weight have roughly the same solubility in water, the strength of the hydrogen bonds between their molecules and water must be comparable.

16-51 (a) Following is a structural formula for 1-phenyl-2-amino-1-propanol.

Ph—*CH(OH)—*CH(CH₃)—NH₂ + HCl ⟶ Ph—*CH(OH)—*CH(CH₃)—NH₃⁺ Cl⁻

1-Phenyl-2-amino-1-propanol

(b) The molecule has two stereocenters, and $2^2 = 4$ stereoisomers are possible; two pair of enantiomers.

16-53 Structural formula A is the better representation. The molecule has within its structure an —NH₂ group (a base) and a —COOH group (an acid). They react and form the internal salt seen in A.

16-55 The 2° aliphatic amine is more basic than the heterocyclic aromatic amine. The three stereocenters are marked with asterisks.

Epibatadine

16-57 Alanine is classified as an amino acid. Note that as drawn on the right, the 1° amino group and the carboxyl group have reacted to form an internal salt.

CH₃—CH(NH₂)—C(=O)—OH   CH₃—CH(NH₃⁺)—C(=O)—O⁻

Alanine   Alanine (internal salt)

16-59 (a) The following skeleton is the structural feature common to meperidine, methadone, and propoxyphene:

(b) The features that are consistent with the Beckett-Casey rules are outlined as follows:

nitrogen two C—C single bonds away (Morphine)

nitrogen two C—C single bonds away (Meperidine (Demerol))

nitrogen two C—C single bonds away (Methadone), 4° center

nitrogen two C—C single bonds away (Propoxyphene (Darvon)), 4° center

Chapter 17 Aldehydes and Ketones

17-1 (a) 3,3-Dimethylbutanal (b) Cyclopentanone (c) 1-Phenyl-1-propanone

17-2 Following are line-angle formulas for the eight aldehydes with the molecular formula $C_6H_{12}O$. In the three that are chiral, the stereocenter is marked with an asterisk.

Hexanal

4-Methylpentanal

3-Methylpentanal

2-Methylpentanal

2,3-Dimethylbutanal

3,3-Dimethylbutanal

2,2-Dimethylbutanal

2-Ethylbutanal

17-3 (a) 2,3-Dihydroxypropanal (b) 2-Aminobenzaldehyde (c) 5-Amino-2-pentanone

17-4 Each aldehyde is oxidized to a carboxylic acid.

(a) Hexanedioic acid (Adipic acid)

(b) 3-Phenylpropanoic acid

17-5 Each primary alcohol comes from the reduction of an aldehyde. Each secondary alcohol comes from the reduction of a ketone.

17-6 Shown first is the hemiacetal and then the acetal.

C₆H₅–CHO →(CH₃OH) C₆H₅–CH(OH)(OCH₃) →(CH₃OH)
Benzaldehyde   A hemiacetal

→ C₆H₅–CH(OCH₃)₂ + H₂O
An acetal

17-7 (a) A hemiacetal formed from 3-pentanone (a ketone) and ethanol.
(b) Neither a hemiacetal nor an acetal. This compound is the dimethyl ether of ethylene glycol.
(c) A cyclic acetal derived from 5-hydroxypentanal and methanol.

17-8 Following is the keto form of each enol.

(a) cyclohexanone (b) 2-hydroxycyclohexanone (c) cyclohexanecarbaldehyde

17-9 (a) T (b) T (c) T (d) F
(d) False: A carbonyl carbon of an aldehyde or a ketone has only three groups bonded to it. Therefore, it cannot be a stereocenter.

17-11 In an aromatic aldehyde, the —CHO group is bonded to an aromatic ring. In an aliphatic aldehyde, it is bonded to a tetrahedral carbon atom.

17-13 Compounds (b), (c), (d), and (f) contain carbonyl groups.

17-15 Of the four aldehydes with this molecular formula, only 2-methylbutanal is chiral. Its stereocenter is marked with an asterisk.

Pentanal   3-Methylbutanal
2-Methylbutanal   2,2-Dimethylpropanal

17-17
(a) H–CO–H (b) CH₃CH₂CHO
(c) (branched chain with CHO)
(d) CH₃(CH₂)₈CHO
(e) p-hydroxybenzaldehyde
(f) HOCH₂CH(OH)CHO

17-19 (a) 4-Heptanone (b) 2-Methylcyclopentanone
(c) cis-2-Methyl-2-butenal (d) 2-Hydroxypropanal
(e) 1-Phenyl-2-propanone (f) Hexanedial

17-21 (a) The carbon chain is numbered incorrectly. The correct name is 2-butanone.
(b) The carbonyl group is on carbon-1, and therefore, the compound is an aldehyde. Its name is butanal.
(c) The longest (parent) carbon chain is five carbons long, and the correct name is pentanal.
(d) The carbon chain must be numbered to give the carbonyl group the lowest possible number. The correct name is 3,3-dimethyl-2-butanone.

17-23 (a) T (b) T (c) T (d) T
All statements are true.

17-25 The carbonyl group of acetone forms hydrogen bonds with water. These hydrogen bonds are sufficient to make any proportion of acetone soluble in water. 4-Heptanone contains a carbonyl group, which, through its hydrogen bonding with water molecules, promotes water solubility. It also contains two 3-carbon hydrocarbon groups bonded to the carbonyl carbon, which inhibit water solubility. In 4-heptanone, the combined hydrophobic effect of the two hydrocarbon groups is greater than the hydrophilic effect of the single carbonyl group, making 4-heptanone insoluble in water.

17-27 Pentane is a nonpolar hydrocarbon, and the only attractive forces between its molecules in the liquid state are the very weak London dispersion forces. Pentane, therefore, has the lowest boiling point. Butanal and 1-butanol are both polar molecules. Because 1-butanol has a polar —OH group, its molecules can associate by hydrogen bonding. The intermolecular attraction between molecules of 1-butanol are greater than those between molecules of butanal. 1-Butanol, therefore, has a higher boiling point than butanal.

Pentane (bp 36°C)   Butanal (bp 76°C)   1-Butanol (bp 117°C)

17-29 Acetone has no —OH group through which to form intermolecular hydrogen bonds.

17-31 Aldehydes are oxidized by this reagent to a carboxylic acid, ketones are not affected, and 2° alcohols are oxidized to ketones.

(a) CH₃CH₂CH₂COOH   (b) C₆H₅–COOH
(c) NR   (d) cyclohexanone

17-33 (a) Treat each with Tollens' reagent. Only pentanal gives a silver mirror.
(b) Treat each with K₂Cr₂O₇/H₂SO₄. Only 2-pentanol is oxidized (to 2-pentanone) by this reagent, which will cause the red color of Cr₂O₇²⁻ ion to disappear and be replaced with the green color of Cr³⁺ ion.

17-35 The white solid is benzoic acid, C₆H₅COOH, formed by the oxidation of benzaldehyde in the air.

17-37 These experimental conditions reduce an aldehyde to a primary alcohol and a ketone to a secondary alcohol. Products (a) and (c) are chiral and will be formed as racemic mixtures.

(a) CH₃CH(OH)CH₂CH₃   (b) CH₃(CH₂)₄CH₂OH
(c) 2-methylcyclopentanol   (d) 2-(hydroxymethyl)phenol

17-39 (a) Following is a structural formula for dihydroxyacetone:

$$HOCH_2-\overset{\overset{O}{\|}}{C}-CH_2OH$$

1,3-Dihydroxy-2-propanone
(Dihydroxyacetone)

(b) Because dihydroxyacetone has two hydroxyl groups and one carbonyl group, all of which can interact with water molecules by hydrogen bonding, predict that it is soluble in water.
(c) Treatment of dihydroxyacetone with NaBH$_4$ followed by the addition of H$_2$O reduces the carbonyl group to a hydroxyl group.

$$HOCH_2-\overset{\overset{OH}{|}}{CH}-CH_2OH$$

1,2,3-Propanetriol
(Glycerol, glycerin)

17-41

(a) C$_6$H$_5$–CH(OH)CH$_3$ (b) C$_6$H$_5$–CH(OH)CH$_3$

(c) No reaction (d) No reaction

17-43 Only parts (a), (b), (d), and (f) have an alpha hydrogen, and only these four can undergo keto-enol tautomerism.

17-45 Following are the keto forms of each enol:

(a) cyclopentanone

(b) CH$_3$COCH$_2$CH$_2$CH$_2$CH$_3$

(c) C$_6$H$_5$–CH$_2$COCH$_3$

17-47 The characteristic structural feature of a hemiacetal is a carbon atom bonded to an —OH group and either an —OR or an —OAr group. The characteristic structural feature of an acetal is a carbon atom bonded to two —OR or —OAr groups.

17-49 (a) Hemiacetal (b) Acetal (c) Neither
(d) Hemiacetal (e) Cyclic acetal (f) Acetal

17-51 Following are structural formulas for the products of each hydrolysis.

(a) CH$_3$CH$_2$COCH$_2$CH$_3$ + HOCH$_2$CH$_2$OH

(b) CH$_3$O–C$_6$H$_4$–CHO + 2CH$_3$OH

(c) cyclohexanone =O + 2CH$_3$OH

(d) 2-hydroxycyclohexyl-CHO + CH$_3$OH

17-53 If you follow the steps in the formation of an acetal, you will find that oxygen-18 appears in the water molecules formed along with the acetal.

17-55 *Hydration* refers to the addition of one or more molecules of water to a substance. An example of hydration is the acid-catalyzed addition of water to propene to give 2-propanol. *Hydrolysis* refers to the reaction of a substance with water, generally with the breaking (lysis) of one or more bonds in the substance. An example of hydrolysis is the acid-catalyzed hydrolysis of an acetal with a molecule of water to give an aldehyde or a ketone and two molecules of alcohol. Another example of hydrolysis is the reverse of Fischer esterification (Chapter 18), namely the hydrolysis of an ester to give a carboxylic acid and an alcohol.

17-57 (a) NaBH$_4$
(b) H$_2$SO$_4$, heat
(c) HBr
(d) H$_2$/Pd
(e) Br$_2$

17-59

(a) cyclopentanone (b) cyclohexane-CHO

(c) None exist (d) H–CO–CH$_2$CH$_2$–CO–H

17-61 Each conversion can be brought about in two steps.

(a) C$_6$H$_5$COCH$_2$CH$_3$ $\xrightarrow{NaBH_4}$ C$_6$H$_5$CH(OH)CH$_2$CH$_3$ $\xrightarrow[heat]{H_2SO_4}$ C$_6$H$_5$CH=CHCH$_3$

(b) cyclopentanone =O $\xrightarrow{NaBH_4}$ cyclopentanol–OH $\xrightarrow[heat]{H_2SO_4}$ cyclopentene $\xrightarrow{HCl}$ cyclopentyl–Cl

17-63 (a) Both cyclohexanone and aniline are insoluble in water. Distinguish between this pair by a test based on the fact that cyclohexanone is neutral to both acids and bases. Aniline, however, is a weak organic base and reacts with hydrochloric acid to form a water-soluble salt. Place a small amount of each in a test tube and to each test tube add some dilute hydrochloric acid. Observe that aniline dissolves in aqueous HCl and forms a homogeneous solution. Cyclohexanone does not react with dilute aqueous HCl and forms a heterogeneous mixture.
(b) Cyclohexene and cyclohexanone: Distinguish between the members of this pair based on the fact that cyclohexene is an alkene and reacts instantaneously with a solution of bromine (Br$_2$) in dichloromethane to discharge the red-purple color of bromine to form colorless 1,2-dibromocyclohexane. Cyclohexanone does not react with bromine under these conditions.
(c) Benzaldehyde and cinnamaldehyde: Distinguish between the members of this set by the fact that cinnamaldehyde contains a carbon–carbon double bond and will react with bromine to discharge the red-purple color of bromine and give a colorless product.

17-65 The —OH group formed by each NaBH$_4$ reduction is shown in red.

(a) HO—CH$_2$CH$_2$CH$_2$CH$_2$CH(OH)CH$_3$ (b) cyclohexyl(OH)–CH$_2$OH

(c) HOCH₂CHCH₂OH (with OH on middle C)

(d) 2-hydroxy-tetralin (tetrahydronaphthalen-2-ol)

(e) C₆H₅—CH(OH)CH₂CH₃

(f) 2-methoxy-4-(hydroxymethyl)phenol (H₃CO, HO on benzene ring with CH₂OH)

17-67

(a) ClCH₂C(=O)CH₃
1-Chloro-2-propanone

(b) CH₃CH(OH)CH₂CHO
3-Hydroxybutanal

(c) (CH₃)₂C(OH)CH₂C(=O)CH₃
4-Hydroxy-4-methyl-2-pentanone

(d) C₆H₅C(CH₃)₂CH₂CHO
3-Methyl-3-phenylbutanal

(e) 1,3-Cyclohexanedione

(f) CH₃CH(OH)CH₂CH₂CH₂CHO
5-Hydroxyhexanal

17-69 2-Propanol has a higher boiling point than propanal because of the greater attraction between molecules 2-propanol due to hydrogen bonding through its —OH group.

17-71 (a) Reaction of the aldehyde group and the hydroxyl group forms a six-membered cyclic hemiacetal.

5-Hydroxyhexanal (open chain, carbons numbered 6,5,4,3,2,1 with OH on C5 and CHO on C1) → Redraw to show —OH close to CHO → cyclic hemiacetal (six-membered ring with OH)

(b) 5-Hydroxyhexanal has one stereocenter and two stereoisomers (one pair of enantiomers) are possible.
(c) The cyclic hemiacetal has two stereocenters and four stereoisomers (two pair of enantiomers) are possible.

17-73 Each type of preparation is shown:

(a) CH₂=CH₂ + H₂O —H₂SO₄→ CH₃CH₂OH ←— NaBH₄ + CH₃CHO

(b) cyclohexene + H₂O —H₂SO₄→ cyclohexanol ←— NaBH₄ + cyclohexanone

(c) C₆H₅CH=CH₂ + H₂O —H₂SO₄→ C₆H₅CH(OH)CH₃ ←— NaBH₄ + C₆H₅C(=O)CH₃

(d) C₆H₅CH=CH₂ + H₂O —H₂SO₄→ C₆H₅CH(OH)CH₃ ←— NaBH₄ + C₆H₅CCH₃(=O)

17-75 (a) Carbon 4 of ribose provides the —OH group and carbon 1 provides the —CHO group.
(b) Following is a structural formula for the free aldehyde form of ribose. Note that ribose has three stereocenters. Carbon 1 provides the —CHO group and carbon 4 provides the —OH group. No effort is made in the following structure to show the configuration of any of the stereocenters. We will take up how to show the configuration in Chapter 20 (Carbohydrates).

$$\begin{array}{c} \text{CHO} \\ | \\ \text{CHOH} \\ | \\ \text{CHOH} \\ | \\ \text{CHOH} \\ | \\ \text{CH}_2\text{OH} \end{array}$$

Ribose

17-77 Sodium borohydride and NADH have in common that each is a source of hydride ion, (H⁻). Each reduction occurs by transfer of a hydride ion from the reagent to the carbonyl carbon. This transfer is an example of the step we have used in several reaction mechanisms, namely reaction of an electrophile (the carbonyl carbon) and a nucleophile (the hydride ion) to form a new covalent bond.

17-79 Each transformation is an oxidation using the same reagents.

(a) CH₃CH₂CH₂CH₂CH₂OH + K₂Cr₂O₇ —H₂SO₄→
1-Pentanol
CH₃CH₂CH₂CH₂CHO
Pentanal

(b) CH₃CH₂CH₂CH₂CH₂OH + K₂Cr₂O₇ —H₂SO₄→
1-Pentanol
CH₃CH₂CH₂CH₂CO₂H
Pentanoic acid

(c) 2-Pentanol + K₂Cr₂O₇ —H₂SO₄→ 2-Pentanone

(d) 2-Propanol + K₂Cr₂O₇ —H₂SO₄→ Propanone

(e) Cyclohexanol + K₂Cr₂O₇ —H₂SO₄→ Cyclohexanone

Chapter 18 Carboxylic Acids

18-1 (a) 2,3-Dihydroxypropanoic acid
(b) 3-Aminopropanoic acid
(c) 3,5-Dihydroxy-3-methylpentanoic acid

18-2

(a) Butanoic acid + NH_3 → Ammonium butanoate (propyl-COOH + NH_3 → propyl-COO$^-$ NH_4^+)

(b) (s)-Lactic acid (CH$_3$CH(OH)COOH) + NH_3 → Ammonium (s)-lactate (CH$_3$CH(OH)COO$^-$ NH_4^+)

18-3

(a) (CH$_3$)$_2$CHCOOH + HO–cyclohexyl $\underset{}{\overset{H^+}{\rightleftharpoons}}$ (CH$_3$)$_2$CHCOO–cyclohexyl + H_2O

(b) HO–CH$_2$CH$_2$CH$_2$–COOH $\underset{}{\overset{H^+}{\rightleftharpoons}}$ γ-butyrolactone + H_2O

18-5 (a) T (b) T (c) T (d) F (e) T (f) T (g) T (h) T
(d) False: 3-methylbutanoic acid has no stereocenter and is therefore achiral.

18-7 (a) 3,4-Dimethylpentanoic acid
(b) 2-Aminobutanoic acid (c) Hexanoic acid

18-9

(a) 4-NO$_2$-C$_6$H$_4$-CH$_2$COOH

(b) H$_2$NCH$_2$CH$_2$CH$_2$COOH

(c) C$_6$H$_5$-CH$_2$CH$_2$CH$_2$COOH

(d) HO–CO–CH=CH–CO–OH (cis-butenedioic acid)

18-11

(a) C$_6$H$_5$–CO–O$^-$ Na$^+$
(b) CH$_3$–CO–O$^-$ Li$^+$
(c) CH$_3$–CO–O$^-$ NH$_4^+$

(d) Na$^+$ $^-$O–CO–(CH$_2$)$_3$–CO–O$^-$ Na$^+$

(e) 2-hydroxybenzoate sodium salt (salicylate Na$^+$)

(f) (CH$_3$CH$_2$CH$_2$COO$^-$)$_2$Ca^{2+}

18-13

HOC–CO$^-$K$^+$ (with two C=O)
Monopotassium oxalate

18-15 The dimer drawn here shows two hydrogen bonds:

(formic acid dimer with two hydrogen bonds between O–H···O)

18-17 The carboxyl group contributes to water solubility; the hydrocarbon chain prevents water solubility.

18-19 In order of increasing boiling point, they are:

CH$_3$CH$_2$CH$_2$CH$_2$CH$_2$CH$_2$CHO
Heptanal
(bp 153°C)

CH$_3$CH$_2$CH$_2$CH$_2$CH$_2$CH$_2$CH$_2$OH
1-Heptanol
(bp 176°C)

CH$_3$CH$_2$CH$_2$CH$_2$CH$_2$COH
Hexanoic acid
(bp 205°C)

18-21 In order of increasing solubility in water, they are decanoic acid, pentanoic acid, and acetic acid.

18-23 The structural features necessary to make a good detergent are (1) a hydrocarbon chain of 12–20 carbon atoms and (2) a polar end group that will not form insoluble salts with Ca^{2+}, Mg^{2+}, or Fe^{3+} ions.

18-25 Each has a long, hydrophobic, 16-carbon hydrocarbon chain and a polar end group that will not form insoluble salts in hard water.

18-27 Carboxylic acids (pK_a 4.0–5.0) are the strongest acids, and alcohols (pK_a 16–18) are the weakest acids. Phenols (pK_a 10) are intermediate in acidity.

18-29 Following are completed equations for these acid-base reactions:

(a) C$_6$H$_5$—CH$_2$COOH + NaOH ⟶ C$_6$H$_5$—CH$_2$COO$^-$Na$^+$ + H$_2$O

(b) CH$_3$CH=CHCH$_2$COOH + NaHCO$_3$ ⟶ CH$_3$CH=CHCH$_2$COO$^-$Na$^+$ + CO$_2$ + H$_2$O

(c) 2-methoxybenzoic acid + NaHCO$_3$ ⟶ sodium 2-methoxybenzoate + CO$_2$ + H$_2$O

(d) CH$_3$CH(OH)COOH + H$_2$NCH$_2$CH$_2$OH ⟶ CH$_3$CH(OH)COO$^-$ + H$_3$$^+NCH_2CH_2$OH

(e) CH$_3$CH=CHCH$_2$COO$^-$Na$^+$ + HCl ⟶ CH$_3$CH=CHCH$_2$COOH + NaCl

18-31 Formic acid reacts with NaHCO$_3$ in an acid-base reaction to convert formic acid to nonirritating sodium formate.

H—C(=O)—OH + NaHCO$_3$ ⟶ H—C(=O)—O$^-$Na$^+$ + CO$_2$ + H$_2$O
Formic acid   Sodium formate

18-33 (a) 1×10^{-3}
(b) 1
(c) 1×10^2
(d) 1×10^4
(e) 1×10^6

18-35 The pK_a of lactic acid is 4.07. At this pH, lactic acid would be present as 50% lactic acid, CH$_3$CH(OH)COOH, and 50% as lactate anion, CH$_3$CH(OH)COO$^-$. At pH 7.35 to 7.45, which is more basic than pH 4.07, lactic acid would be present as lactate anion, CH$_3$CH(OH)COO$^-$.

18-37 Recall from Chapter 8 that carboxylic acids are stronger acids than ammonium ion. Therefore, in part (a), the most acidic hydrogen is that of the COOH group.

(a) CH$_3$CH(NH$_3$$^+$)COOH + NaOH ⟶ CH$_3$CH(NH$_3$$^+$)COO$^-Na^+$ + H$_2$O

(b) CH$_3$CH(NH$_3$$^+$)COO$^-Na^+$ + NaOH ⟶ CH$_3$CH(NH$_2$)COO$^-$Na$^+$ + H$_2$O

18-39 In part (a), the amino (—NH$_2$) group is the stronger base than the carboxylate (—COO$^-$) group.

(a) CH$_3$CH(NH$_2$)COO$^-$Na$^+$ + HCl ⟶ CH$_3$CH(NH$_3$$^+Cl^-$)COO$^-Na^+$

(b) CH$_3$CH(NH$_3$$^+$)COO$^-Na^+$ + HCl ⟶ CH$_3$CH(NH$_3$$^+$)COOH + NaCl

18-41 Following are the structural formulas for the ester formed in each reaction:

(a) isobutyl acetate

(b) CH$_3$CO—O—cyclohexyl

(c) diethyl benzene-1,2-dicarboxylate (COOCH$_2$CH$_3$ ortho-substituted)

18-43 Following is a structural formula for methyl 2-hydroxybenzoate:

(2-hydroxybenzoic acid methyl ester, with COOCH$_3$ and OH groups ortho on benzene)

Methyl 2-hydroxybenzoate
(Methyl salicylate)

18-45 Following are the expected organic products for each reaction:

(a) C$_6$H$_5$CH$_2$CO$_2$$^-Na^+$

(b) C$_6$H$_5$CH$_2$CO$_2$$^-Na^+$

(c) [benzyl-CO₂⁻ NH₄⁺]   (d) [benzyl-CH₂CH₂OH]

(e) No reaction

(f) [benzyl-CO₂CH₃]

(g) No reaction

18-47 Each product can be formed by Fischer esterification with the appropriate alcohol. Note that 4-aminobenzoic acid contains an aromatic amine group, which will react with the acid catalyst to form a salt. Therefore, after Fischer esterification is completed, sufficient base (NaOH) must be added to convert the amine salt to the free amine. Thus, formation of the ester and isolation of the final product requires two steps: (1) formation of the ester and (2) neutralization of the amine salt/acid catalyst.

[Reaction scheme: 4-Aminobenzoic acid (p-Aminobenzoic acid) → 1) CH₃OH/H₂SO₄, 2) NaOH/H₂O → Methyl 4-aminobenzoate (Methylparaben); and → 1) CH₃CH₂CH₂OH/H₂SO₄, 2) NaOH/H₂O → Propyl 4-aminobenzoate (Propylparaben)]

18-49 The product is a 6-membered lactone.

CH₂CH₂CH₂CH₂C(=O)—OH (with OH on first carbon) →(acid catalyst)→ [6-membered lactone] + H₂O

5-Hydroxypentanoic acid   a lactone

18-51 Carboxylic acids can be formed by the oxidation of primary alcohols (—CH₂OH) or of aldehydes (CHO).

(a) [pentanol chain ending in OH]   (b) [pentanal chain ending in CHO]

(c) HO—[chain]—OH

18-53 (a) Only three of these intermediates are chiral: isocitric acid, oxalosuccinic acid, and malic acid. Isocitric acid has two stereocenters and may potentially form $2^2 = 4$ stereoisomers (two pairs of enantiomers). Two of these intermediates are capable of *cis-trans* isomerism: aconitic acid and fumaric acid.
(b) **Step 1**: Dehydration of a 3° alcohol to give an alkene.
    Step 2: Hydration of an alkene to give an alcohol.
    Step 3: Oxidation of a 2° alcohol to a ketone.
(c) If the hydration were to follow Markovnikov's rule, citric acid would be formed. The fact that the hydration proceeds as it does is due to the presence of the enzyme. This enzyme positions aconitic acid at the reactive site on the enzyme surface in such a way that only one carbocation intermediate can be formed, and water is then added to it in a regiospecific way.
(d) **Step 4**: Decarboxylation of a β-ketoacid. Only one of the three carboxyl groups in oxalosuccinic acid includes a carbonyl group beta to it.
    Step 5: Oxidative decarboxylation.
    Step 6: Oxidation of an alkane to an alkene.
    Step 7: Hydration of a carbon–carbon double bond to give a 2° alcohol.
    Step 8: Oxidation of a 2° alcohol to a ketone.
(e) This is a decarboxylation because CO_2 is lost, and it is an oxidation because an oxygen is added to the ketone carbonyl carbon to make it a new carboxylic acid.
(f) Yes, it is a decarboxylation, but it is not also an oxidation because no new oxygen is added.

18-55 The parent chain is a decanoic acid. Modify the ending of the acid name to show the presence of the carbon–carbon double bond between carbons 2 and 3 and specify its configuration, which gives *trans*-2-decenoic acid. Finally show the presence of a ketone carbonyl group at carbon 9, which completes the name and gives *trans*-9-oxo-2-decenoic acid.

[Structure of *trans*-9-oxo-2-decenoic acid with carbons numbered 1–10]

trans-9-oxo-2-decenoic acid

Glossary

A site (*Section 26-5*) The site on the large ribosomal subunit where the incoming tRNA molecule binds.

Absolute zero (*Section 1-4*) The lowest possible temperature; the zero point of the Kelvin temperature scale.

Abzymes (*Section 23-8*) Immunoglobins generated by using a transition state analog as an antigen.

Acetal (*Section 17-4C*) A molecule containing two —OR groups bonded to the same carbon.

Acetyl group (*Section 27-3*) The group CH_3CO—.

Achiral (*Section 15-1*) An object that lacks chirality; an object that is superposable on its mirror image.

Acid (*Section 8-1*) A substance that produces H^+ ions in aqueous solution.

Acid–base reaction (*Section 8-3*) A proton-transfer reaction.

Acid ionization constant (K_a) (*Section 8-5*) An equilibrium constant for the ionization of an acid in aqueous solution to H_3O^+ and its conjugate base. K_a is also called an **acid dissociation constant**.

Acid rain (*Chemical Connections 6A*) Rain with acids other than carbonic acid dissolved in it.

Acidic polysaccharide (*Section 20-6*) A polysaccharide that is important in connective tissue and contains carboxyl groups or sulfuric ester groups.

Acidosis (*Chemical Connections 8C*) A condition in which the pH of blood is lower than 7.35.

Acquired immunity (*Section 31-1*) The second line of defense that vertebrates have against invading organisms.

Actinide series (*Section 22-10*) The 14 elements (90–103) immediately following actinium in period 7 in which the 5f shell is being filled.

Activating receptor (*Section 31-8*) A receptor on a cell of the innate immune system that triggers activation of the immune cell in response to a foreign antigen.

Activation (*Section 23-3*) In the context of enzymology, activation refers to any process that initiates or increases the action of an enzyme.

Activation energy (*Section 7-2*) The minimum energy necessary to cause a chemical reaction.

Activation of an amino acid (*Section 26-5*) The process by which an amino acid is bonded to an AMP molecule and then to the 3′—OH of a tRNA molecule.

Activation of an enzyme (*Section 23-3*) Any process by which an inactive enzyme is transformed into an active enzyme.

Active site (*Section 23-3 and Chemical Connections 23B*) A three-dimensional cavity of an enzyme with specific chemical properties to accommodate the substrate.

Active transport (*Section 21-5B*) The energy-requiring process of moving substances into a cell against a concentration gradient.

Activity series (*Section 8-6B*) The ranking of elements in order of their reducing abilities in aqueous solution.

Actual yield (*Section 4-7C*) The mass of product actually formed or isolated in a chemical reaction.

Acyl group (*Section 19-1A*) An R—CO or Ar—CO group.

Adaptive immunity (*Section 31-1*) Acquired immunity with specificity and memory.

Adenovirus (*Section 26-9*) A common vector used in gene therapy.

Adenosine deaminase (ADA) (*Section 26-9*) An enzyme involved in purine catabolism, the lack of which leads to the disease Severe Combined Immune Deficiency (SCID).

Adhesion molecules (*Section 31-5*) Various protein molecules that help to bind an antigen to the T-cell receptor and dock the T cell to another cell via an MHC.

Adrenergic neurotransmitter (*Section 24-4*) A monoamine neurotransmitter or hormone, the most common of which are epinephrine (adrenaline), serotonin, histamine, and dopamine.

Advanced glycation end-products (AGEs) (*Section 22-7*) Chemical products of sugars and proteins linking together to produce an imine.

Affinity maturation (*Section 31-4*) The process of mutation of T cells and B cells in response to an antigen.

Agonist (*Section 24-1*) A molecule that mimics the structure of a natural neurotransmitter or hormone, binds to the same receptor, and elicits the same response.

α-helix (*Section 22-9*) A type of repeating secondary structure of a protein in which the chain adopts a helical conformation stabilized by hydrogen bonding between a peptide backbone N—H and the backbone C=O four amino acids farther up the chain.

AIDS (*Section 31-9*) Acquired immune deficiency syndrome. The disease caused by the human immunodeficiency virus, which attacks and depletes T cells.

Alcohol (*Section 10-4A*) A compound containing an —OH (hydroxyl) group bonded to a tetrahedral carbon atom.

Aldehyde (*Sections 10-4C and 17-1*) A compound containing a carbonyl group bonded to a hydrogen; a —CHO group.

Alditols (*Section 20-3*) The products formed when the CHO group of a monosaccharide is reduced to CH_2OH.

Aldoses (*Section 20-1*) Monosaccharides containing an aldehyde group.

Aliphatic amine (*Section 16-1*) An amine in which nitrogen is bonded only to alkyl groups.

Aliphatic hydrocarbons (*Section 11-1*) Alkanes.

Alkali metals (*Section 2-5C*) Elements, except hydrogen, in Group 1A of the Periodic Table.

Alkaloids (*Chemical Connections 16B*) Basic nitrogen-containing compounds of plant origin, many of which have physiological activity when administered to humans.

Alkalosis (*Chemical Connections 8D*) A condition in which the pH of blood is greater than 7.45.

Alkanes (*Section 11-1*) Saturated hydrocarbons whose carbon atoms are arranged in an open chain—that is, not arranged in a ring.

Alkenes (*Section 12-1*) Unsaturated hydrocarbons that contain a carbon–carbon double bond.

Alkyl group (*Section 11-4A*) A group derived by removing a hydrogen atom from an alkane; is given the symbol —R.

Alkynes (*Section 12-1*) Unsaturated hydrocarbons that contain a carbon–carbon triple bond.

Allosteric protein (*Chemical Connections 22G*) A protein that exhibits a behavior where binding of one molecule at one site changes the ability of the protein to bind another molecule at a different site.

Allosterism (Allosteric enzyme) (*Section 23-6*) An enzyme regulation in which the binding of a regulator on one site of the enzyme modifies the ability of the enzyme to bind the substrate at the active site. Allosteric enzymes often have multiple polypeptide chains with the possibility of chemical communication between the chains.

Allotropes (*Section 5-9*) Elements that exist in different forms in the same physical state with different physical and chemical properties.

Alloys (*Section 6-2*) Homogeneous mixtures of metals.

Alpha (α-) amino acid (*Section 22-2*) An amino acid in which the amino group is bonded to the carbon atom next to the —COOH carbon.

Alpha particles (α) (*Section 9-2*) Helium nuclei, He^{2+}, ^4_2He.

Amide (*Section 19-1C*) A derivative of a carboxylic acid in which the —OH of the carboxyl group is replaced by an amine.

Amine (*Section 11-4B*) A functional group in which a nitrogen atom is bonded to one, two, or three carbon groups: RNH_2, R_2NH, or R_3N.

Amino acid (*Section 22-2*) An organic compound containing an amino group and a carboxyl group.

Amino acid neurotransmitter (*Section 24-4*) A neurotransmitter or hormone that is an amino acid.

Amino acid pool (*Section 28-1*) The free amino acids found both inside and outside cells throughout the body.

Amino group (*Section 10-4B*) An —NH_2 group.

Amino sugars (*Section 20-1*) Monosaccharides in which an —OH group is replaced by an —NH_2 group.

Aminoacyl-tRNA synthetases (*Section 26-3*) Enzymes that link the correct amino acid to a tRNA molecule.

-ammonium (*Section 16-2*) A functional group in which a nitrogen atom is bonded to four groups and bears a positive charge; NH_4^+, RNH_3^+, $R_2NH_2^+$, R_3NH^+, R_4N^+.

Amorphous solids (*Section 5-9*) Solids whose atoms, molecules, or ions are not in an orderly arrangement.

Amphetamines (*Chemical Connections 16A*) Amphetamine is 1-phenyl-2-propanamine. Amphetamines are a class of compounds that have within their structure the same atomic skeleton as that of amphetamine namely a three-carbon chain with a benzene ring on the first carbon and an amine nitrogen on the second carbon.

Amphiprotic (*Section 8-3*) A substance that can act as either an acid or a base.

Amylase (*Section 30-3*) An enzyme that catalyzes the hydrolysis of α-1,4-glycosidic bonds in dietary starches.

Amylopectin (*Section 20-5A*) A polysaccharide used to store energy in plants made of glucose residues linked α 1→4 with branches linked α 1→6.

Anabolism (*Section 27-1*) The biochemical process of building up larger molecules from smaller ones.

Anaerobic pathway (*Section 28-2*) One in which the reactions take place in the absence of O_2.

Aneuploid cell (*Chemical Connections 31C*) A cell with the wrong number of chromosomes.

Anhydride (*Section 19-1A*) A compound derived from another or others by the loss of the elements of water. A carboxylic acid anhydride is formally derived by loss of the elements of water from two carboxyl groups. The characteristic structural of a carboxylic acid anhydride is an oxygen atom bonded to two acyl groups bonded to the same oxygen.

Anhydrous (*Section 6-6B*) A crystal without its water of hydration that was heated at a high temperature.

Aniline (*Section 16-2A*) The simplest aromatic amine, with the molecular formula C_6H_5—NH_2.

Anion (*Section 3-2*) An ion with a negative electric charge.

Anode (*Section 6-6C*) The negatively charged electrode.

Anomeric carbon (*Section 20-2*) The hemiacetal carbon of the cyclic form of a monosaccharide.

Anomers (*Section 20-2*) Monosaccharides that differ in configuration only at their anomeric carbons.

Antagonist (*Section 24-8*) A molecule that binds to a neurotransmitter receptor but does not elicit the natural response.

Antibody (*Section 31-1*) A defense glycoprotein synthesized by the immune system of vertebrates that interacts with an antigen; also called an immunoglobulin.

Anticodon (*Section 26-3*) A sequence of three nucleotides on tRNA, also called a codon recognition site, complementary to the codon in mRNA.

Antigen (*Sections 31-1 and 31-3*) A substance foreign to the body that triggers an immune response.

Antigen-presenting cells (*APCs*) (*Section 31-2*) Cells that cleave foreign molecules and present them on their surfaces for binding to T cells or B cells.

Antisense strand (*Section 26-2*) The strand of DNA that acts as the template for transcription. Also called the template strand and the (−) strand.

AP site (*Section 25-7*) The ribose and phosphate that are left after a glycosylase removes a purine or pyrimidine base during DNA repair.

Apoenzyme (*Section 23-2*) The protein portion of an enzyme that has cofactors or prosthetic groups.

Aqueous solution (*Section 4-3*) A solution in which the solvent is water.

Ar— (*Section 13-1*) The symbol used for an aryl group.

Arene (*Section 13-1*) A compound containing one or more benzene rings.

Aromatic amine (*Section 16-1*) An amine in which nitrogen is bonded to one or more aromatic rings.

Aromatic compounds (*Section 13-1*) Benzenes or one of its derivatives.

Aromatic sextet (*Section 13-1B*) The closed loop of six electrons (two from the second bond of each double bond) characteristic of a benzene ring.

Aromatic substitution (*Section 13-3*) A characteristic reaction of aromatic compounds in which a hydrogen of the compound is replaced by another atom or group or atoms.

Aryl group (*Section 13-1*) A group derived from an arene by removal of an H atom and given the symbol Ar—.

Atmosphere (*Section 5-2*) A unit of pressure equal to 760 mm Hg at sea level.

Atom (*Section 2-3*) The smallest particle of an element that retains the chemical properties of the element.

Atomic energy (*Section 9-9*) The energy produced from a nuclear fission reaction, resulting in products that have less mass than the starting materials.

Atomic mass unit (*Section 2-4A*) A unit of the scale of relative masses of atoms: 1 amu = 1.6605×10^{-24} g. By definition, 1 amu is 1/12 the mass of a carbon atom containing 6 protons and 6 neutrons.

Atomic number (*Section 2-4C*) The number of protons in the nucleus of at atom.

Atomic weight (*Section 2-4E*) The weighted average of the masses, in atomic mass units, of the naturally occurring isotopes of the element.

Attenuated vaccine (*Section 31-6A*) A vaccine made from a weakened virus or bacterium.

Autoxidation (*Section 13-4C*) The reaction of a C—H group with oxygen, O_2, to form a hydroperoxide, R—OOH.

Avogadro's law (*Section 5-4*) Equal volumes of gases at the same temperature and pressure contain the same number of molecules.

Avogadro's number (*Section 4-6*) 6.02×10^{23} formula units per mole; the amount of any substance that contains the same number of formula units as the number of atoms in 12 g of carbon-12.

Axial bonds (*Section 11-7B*) A bond from a carbon atom of a six-membered ring that extends from the ring roughly parallel to the imaginary axis of the ring.

Axial position (*Section 11-7B*) A position on a chair conformation of a cyclohexane ring that extends from the ring parallel to the imaginary axis of the ring.

Axon (*Section 24-2*) The long part of a nerve cell that comes out of the main cell body and eventually connects with another nerve cell or tissue cell.

B cell (*Section 31-1*) A type of lymphocyte that is produced in and matures in the bone marrow. B cells produce antibody molecules.

Baking soda (*Section 8-6E*) A common household product that consists of $NaHCO_3$ (sodium bicarbonate).

Barometer (*Section 5-2*) An instrument used to measure atmospheric pressure.

Basal caloric requirement (*Section 30-2*) The caloric requirement for an individual at rest, usually given in Cal/day.

Base (*Section 8-1*) An Arrhenius base is a substance that ionizes in aqueous solution to give hydroxide (OH^-) ions.

Bases (*Section 25-2*) Purines and pyrimidines, which are components of nucleosides in DNA and RNA.

Batteries (*Section 4-4*) A voltaic cell where electricity is generated from a chemical reaction.

Becquerel (Bq) (*Section 9-5A*) A measure of radioactive decay equal to one disintegration per second.

Bent (*Section 3-10*) A shape where a central atom is surrounded by two regions of electron density to atoms, and two regions of unshared pairs of electrons on the central atom.

Beta particles (β) (*Section 9-2*) Electrons, $_{-1}^{0}\beta$.

Binary compound (*Section 3-6A*) A compound containing only two elements.

Binary covalent compound (*Section 3-8A*) A compound containing two elements.

Binary ionic compound (*Section 3-6A*) A compound containing two elements present as ions.

Binding protein (*Section 26-2*) A protein that binds to nucleosomes, making DNA more accessible for transcription.

Binding site (*Section 15-5B*) A site on the surface of an enzyme that binds a molecule or molecules whose reaction the enzyme is designated to catalyze.

Biochemical pathway (*Section 27-1*) Series of consecutive biochemical reactions.

Bleaching (*Section 4-4*) The process where colored compounds become colorless in the presence of bleaches.

Boiling point (*Section 5-8C*) The temperature at which the vapor pressure of a liquid is equal to the atmospheric pressure.

Boiling-point elevation (*Section 6-8B*) The increase in the boiling point of a liquid caused by adding a solute.

Bond angle (*Section 3-10*) The angle between two atoms bonded to a central atom.

Bonding electrons (*Section 3-7C*) Valence electrons involved in forming a covalent bond—that is, shared electrons.

β-oxidation (*Section 28-5*) The biochemical pathway that degrades fatty acids to acetyl CoA by removing two carbons at a time and yielding energy.

Boyle's law (*Section 5-3A*) The volume of a gas at constant temperature is inversely proportional to the pressure applied to the gas.

β-pleated sheet (*Section 22-9*) A type of secondary protein structure in which the backbone of two protein chains in the same or different molecules is held together by hydrogen bonds.

Brønsted–Lowry acid (*Section 8-3*) A proton donor.

Brønsted–Lowry base (*Section 8-3*) A proton acceptor.

Brownian motion (*Section 6-8*) The random motion of colloidal-size particles.

Buffer (*Section 8-10*) A solution that resists change in pH when limited amounts of an acid or a base are added to it; the most common example is an aqueous solution containing a weak acid and its conjugate base.

Buffer capacity (*Section 8-10*) The extent to which a buffer solution can prevent a significant change in the pH of a solution upon addition of a strong acid or strong base.

Calorie (*Section 1-9*) The amount of heat necessary to raise the temperature of 1 g of liquid water by 1°C.

Cannabinoid receptor (*Chemical Connections 30G*) A class of cell membrane receptors that are activated by natural and synthetic cannabinoids.

Carbocation (*Section 12-5A*) A species containing a carbon atom with only three bonds to it and bearing a positive charge.

Carbohydrates (*Section 20-1*) Polyhydroxyaldehydes or polyhydroxyketones or substances that give these compounds on hydrolysis.

Carbonyl group (*Section 10-4C*) A C=O group.

Carboxyl group (*Sections 10-4D and 18-1*) A —COOH group.

Carboxylic acid (*Section 10-4D*) A compound containing a —COOH group.

Carboxylic ester (*Section 10-4*) A derivative of a carboxylic acid in which a carbon replaces the H of the carboxyl group.

Carcinogen (*Section 26-7*) A chemical mutagen that can cause cancer.

Catabolism (*Section 27-1*) The biochemical process of breaking down molecules to supply energy.

Catalyst (*Section 7-4D*) A substance that increases the rate of a chemical reaction by providing an alternative pathway with a lower activation energy.

Catalytic hydrogenation (*Section 12-5D*) An addition reaction in which hydrogen, H_2, is used to convert a carbon-carbon or carbon oxygen double bond to a carbon-carbon single bond or carbon oxygen single bond and for which a catalyst is required, most commonly a transition metal such as Pd, Pt, or Ni.

Catalytic reduction (*Section 12-5D*) A reduction in which a catalyst is required. A specific example is a catalytic hydrogenation.

Cathode (*Section 6-6C*) The positively charged electrode.

Cation (*Section 3-2*) An ion with a positive electric charge.

Cell reprogramming (*Chemical Connections 31C*) A technique used in whole-mammal cloning in which a somatic cell is reprogrammed to behave like a fertilized egg.

Cellulose (*Section 20-5C*) A linear polysaccharide found in plant cell walls made of glucose residues linked β 1→4.

Celsius scale (°C) (*Section 1-4*) A temperature scale based on 0° as the freezing point of water and 100° as the normal boiling point of water.

Central dogma of molecular biology (*Section 26-1*) A doctrine stating the basic directionality of heredity when DNA leads to RNA, which leads to protein. This doctrine is true in almost all life forms except certain viruses.

Ceramide (*Section 21-7*) A feature of lipid structure in which a fatty acid is bonded to sphingosine by an amide bond.

Cerebrosides (*Section 21-8*) Glycolipids in which a ceramide is bonded to a sugar moiety.

Chain-growth polymer (*Section 12-6*) A polymer formed by the stepwise addition of monomers to a growing polymer chain.

Chain length (*Section 13-4D*) The number of times that a cycle of chain propagation steps repeats.

Chain reaction (*Section 9-9*) A nuclear reaction that results from fusion of a nucleus with another particle (most commonly a neutron) followed by decay of the fused nucleus to smaller nuclei and more neutrons. The newly formed neutrons continue the process, which results in a chain reaction.

Chair conformation (*Section 11-7B*) The most stable conformation of a cyclohexane ring; all bond angles are approximately 109.5°.

Chaperone (*Section 22-11*) A protein that helps other proteins to fold into the biologically active conformation and enables partially denatured proteins to regain their biologically active conformation.

Charles's law (*Section 5-3B*) The volume of a gas at constant pressure is inversely proportional to the temperature in Kelvin.

Chemical change (*Section 1-1*) Matter can change, or be made to change, from one form to another.

Chemical equation (*Section 4-2*) A representation using chemical formulas of the process that occurs when reactants are converted to products.

Chemical equilibrium (*Section 7-5*) A state in which the rate of the forward reaction equals the rate of the reverse reaction.

Chemical kinetics (*Section 7-1*) The study of the rates of chemical reactions.

Chemical messengers (*Section 24-1*) Any chemical that is released from one location and travels to another location before acting. They may be hormones, neurotransmitters, or ions.

Chemical properties (*Section 1-1*) Chemical reactions that a substance undergoes.

Chemical reaction (*Section 1-1*) Substances are used up (disappear) and others are formed to take their place.

Chemiosmotic theory (*Section 27-6*) Mitchell's proposal that electron transport is accompanied by an accumulation of protons in the intermembrane space of the mitochondrion, which in turn creates osmotic pressure; as protons flow from an area of high concentration to an area of low concentration, they are driven back to the mitochondrion under this pressure and generate ATP.

Chemistry (*Section 1-1*) The science that deals with matter.

Chemokine (*Section 31-7*) A chemotactic cytokine that facilitates the migration of leukocytes from the blood vessels to the site of injury or inflammation.

Chiral (*Section 15-1*) From the Greek *cheir*, meaning "hand"; an object that is not superposable on its mirror image.

Chlorofluorocarbons (CFCs) (*Section 11-11A*) A type of hydrocarbon in which atoms of chlorine and/or fluorine are substituted for hydrogen atoms.

Cholesterol (*Section 21-9A*) The most abundant steroid in the body; occurs in cell membranes.

Cholinergic neurotransmitter (*Section 24-1*) A neurotransmitter or hormone based on acetylcholine.

Chromatin (*Section 25-6*) A complex of DNA with histones and nonhistone proteins that exists in eukaryotic cells between cell divisions.

Chromatin remodelers (CR) (*Chemical Connections 26G*) Chemicals that modify chromatin structure.

Chromosomes (*Section 25-1*) Structures within the nucleus of eukaryotes that contain DNA and protein and that are replicated as units during mitosis. Each chromosome is made up of one long DNA molecule that contains many heritable genes.

Circular RNA (*Section 25-4*) An RNA molecule created by alternative splicing of introns during eukaryotic mRNA transcription. It has many binding sites for miRNA.

Cis (*Section 11-8*) A prefix meaning "on the same side."

***Cis-trans* isomerism** (*Section 11-8*) Isomers that have the same connectivity of their atoms but a different arrangement of their atoms in space due to the presence of either a ring or a carbon–carbon double bond.

***Cis-trans* isomers** (*Sections 1-8 and 12-7*) Isomers that have the same (1) molecular formula, (2) connectivity of their atoms, but (3) a different arrangement of their atoms in space due to the presence of either a ring or a carbon–carbon double bond.

Citric acid cycle (*Section 27-3*) A central biochemical pathway.

Cloning (*Section 25-8*) A process whereby DNA is amplified by inserting it into a host and having the host replicate it along with the host's own DNA.

Cluster determinant (*Section 31-5*) A set of membrane proteins on T cells that helps the binding of antigens to the T-cell receptors.

Cobalamin (*Chemical Connections 30G*) A term used to refer to compounds having vitamin B_{12} activity.

Coding strand (*Section 26-2*) The DNA strand that is not used as a template for transcription, but which has a sequence that is the same as the RNA produced. Also called the (+) strand and the sense strand.

Codon (*Section 26-3*) A three-nucleotide sequence on mRNA that specifies a particular amino acid.

Codon recognition site (*Section 26-3*) A sequence of three bases on tRNA that recognizes the codon on mRNA.

Coenzymes (*Section 23-3*) Organic molecules, frequently B vitamins, that acts as cofactors.

Cofactors (*Section 23-3*) Nonprotein parts of enzymes necessary for its catalytic function.

Colligative property (*Section 6-8*) A property of a solution that depends only on the number of solute particles and not on the chemical identity of the solute particles.

Colloid (*Section 6-7*) A two-part mixture in which suspended solute particles range from 1 to 1000 nm in size.

Combined gas law (*Section 5-3C*) The pressure, volume, and temperature in Kelvin of two samples of the same gas are related by the equation $P_1V_1/T_1 = P_2V_2/T_2$.

Combustion (*Section 4-4*) Burning in air.

Common nomenclature (*Section 11-4B*) Refers to names in wide use before the IUPAC system of nomenclature was devised. Many of these names are still used today.

Competitive inhibitors (*Section 23-3*) Compounds that decrease the activity of an enzyme by competing with the substrate for the active site.

Complementary base pairs (*Section 25-3*) The combination of a purine and a pyrimidine base that hydrogen bond together in DNA.

Complete protein (*Section 30-5*) A protein source that contains sufficient quantities of all amino acids required for normal growth and development.

Compound (*Section 2-2B*) A pure substance made up of two or more elements in a fixed ratio by mass; properties are different from those of a mixture of its constituent elements.

Concentration (*Section 6-5*) The amount of solute dissolved in a given quantity of solvent.

Condensation (*Section 5-7*) The change of a substance from the vapor or gaseous state to the liquid state.

Condensed structural formula (*Section 10-4A*) A structural formula that shows all carbon and hydrogen atoms, as for example $CH_3CH_2CH_2CH_2CH_3$.

Configuration (*Section 11-8*) The arrangement of atoms about a stereocenter—that is, the relative arrangements of the parts of a molecule in space.

Conformations (*Section 11-7*) Any three-dimensional arrangements of atoms in a molecule that result from rotation about a single bond.

Conjugate acid (*Section 8-3*) According to the Brønsted–Lowry theory, a substance formed when a base accepts a proton.

Conjugate acid–base pair (*Section 8-3*) A pair of molecules or ions that are related to one another by the gain or loss of a proton.

Conjugate base (*Section 8-3*) According to the Brønsted–Lowry theory, a substance formed when an acid donates a proton to another molecule or ion.

Conjugated proteins (*Section 22-12*) Proteins that contain a nonprotein part, such as the heme part of hemoglobin.

Consensus sequence (*Section 26-2*) A sequence of DNA in the promoter region that is relatively conserved from species to species.

Constitutional isomers (*Section 11-3*) Compounds with the same molecular formula but a different order of attachment (connectivity) of their atoms. Constitutional isomers have also been called structural isomers, an older term that is still in use.

Contributing structure (*Sections 3-9B and 13-1B*) Representations of a molecule or ion that differ only in the distribution of valence electrons.

Control site (*Section 26-6*) A DNA sequence that is part of a prokaryotic operon. This sequence is upstream of the structural gene DNA and plays a role in controlling whether the structural gene is transcribed.

Conversion factors (*Section 1-5*) Ratios of two different units.

Coordinate covalent bond (*Section 22-10*) A covalent bond between two elements, typically between a metal and a ligand, in which one of the bonded atoms provides both of the electrons for the bond.

Coordination compounds (*Section 22-10*) A kind of transition metal compound in which a central transition metal is bonded to a group of other ions or neutral molecules and may be electrically neutral or charged.

Cosmic rays (*Section 9-6*) High-energy particles, mainly protons, from outer space bombarding the Earth.

Covalent bond (*Section 3-4A*) A bond resulting from the sharing of electrons between two atoms.

Crenation (*Section 6-8C*) An osmotic process in which water flows out of red blood cells and into a solution through a semipermeable membrane, causing the cells to shrivel.

Crystalline solids (*Section 5-9*) Solids whose atoms, molecules, or ions are in an orderly arrangement.

Crystallization (*Section 5-7*) The formation of a solid from a liquid.

C-terminus (*Section 22-6*) The amino acid at the end of a peptide chain that has a free carboxyl group.

Curie (Ci) (*Section 9-5A*) A measure of radioactive decay equal to 3.7×10^{10} disintegrations per second.

Curved arrow (*Section 3-9A*) A representation that indicates where a pair of electrons originates (the tail of the arrow) and where it is repositioned in an alternative contributing structure (the head of the arrow).

Cyclic ethers (*Section 14-3B*) Ethers in which oxygen is one of the atoms of a ring.

Cyclic hydrocarbon (*Section 11-6*) A hydrocarbon that contains carbon atoms joined to form a ring.

Cycloalkane (*Section 11-6*) A saturated hydrocarbon that contains carbon atoms bonded to form a ring.

Cycloalkene (*Section 12-3D*) An alkene that contains carbon atoms joined to form a ring.

Cyclooxygenase (COX) (*Section 21-12*) An enzyme that catalyzes the first step in the synthesis of prostaglandins from arachidonic acid.

Cystine (*Section 22-11*) A dimer of cysteine in which the two amino acids are covalently bonded by a disulfide bond between their side chain —SH groups.

Cytokines (*Section 31-6*) Glycoproteins that traffic between cells and alter the function of a target cell.

D-monosaccharide (*Section 20-1*) A monosaccharide that, when written as a Fischer projection, has the —OH group on its penultimate carbon to the right.

Dalton's law (*Section 5-5*) The pressure of a mixture of gases is equal to the sum of the partial pressure of each gas in the mixture.

Debranching enzyme (*Section 30-3*) The enzyme that catalyzes the hydrolysis of the 1,6-glycosidic bonds in starch and glycogen.

Decarboxylation (*Section 18-5E*) The process that leads to loss of CO_2 from a carboxyl (—COOH) group.

Decay, nuclear (*Section 9-3F*) The change of a radioactive nucleus of one element into the nucleus of another element.

Dehydration (*Section 14-2B*) The elimination of a molecule of water from an alcohol. An OH is removed from one carbon, and an H is removed from an adjacent carbon.

Dehydrogenase (*Section 23-2*) A class of enzymes that catalyze oxidation–reduction reactions, often using NAD^+ as the oxidizing agent.

Denaturation (*Section 22-13*) The loss of the secondary, tertiary, and quaternary structure of a protein by a chemical or physical agent that leaves the primary structure intact.

Dendrites (*Section 24-2*) Hair-like projections that extend from the cell body of a nerve cell on the opposite side from the axon.

Dendritic cells (*Sections 31-1 and 31-2*) Important cells in the innate immune system that are often the first cells to defend against invaders.

Density (*Section 1-7*) The ratio of mass to volume for a substance.

Deoxyribonucleic acid (DNA) (*Section 25-2*) The macromolecule of heredity in eukaryotes and prokaryotes. It is composed of chains of nucleotide monomers of a nitrogenous base, 2-deoxy-D-ribose, and phosphate.

Detergent (*Section 18-4D*) A synthetic soap. The most common are the linear alkylbenzene sulfonic acids (LAS).

Dextrorotatory (*Section 15-4B*) The clockwise (to the right) rotation of the plane of polarized light in a polarimeter.

Dialysis (*Section 6-8*) A process in which a solution containing particles of different sizes is placed in a bag made of a semipermeable membrane. The bag is placed in a solvent or solution containing only small molecules. The solution in the bag reaches equilibrium with the solvent outside, allowing the small molecules to diffuse across the membrane but retaining the large molecules.

Diastereomers (*Section 15-3A*) Stereoisomers that are not mirror images of each other.

Diatomic elements (*Section 2-3B*) Substances that consist of two atoms of the same element per molecule.

Dietary Reference Intakes (DRI) (*Section 30-1*) The current numerical system for reporting nutrient requirements; an average daily requirement for nutrients published by the U.S. Food and Drug Administration.

Diet faddism (*Section 30-1*) An exaggerated belief in the effects of nutrition upon health and disease.

Digestion (*Section 30-1*) The process in which the body breaks down large molecules into smaller ones that can then be absorbed and metabolized.

Diol (*Section 14-1B*) A compound containing two —OH (hydroxyl) groups.

Dipeptide (*Section 22-6*) A peptide made up of two amino acids.

Dipole (*Section 3-7B*) A chemical species in which there is a separation of charge; there is a positive pole in one part of the species and a negative pole in another part.

Dipole–dipole interaction (*Section 5-7B*) The attraction between the positive end of one dipole and the negative end of another dipole in the same or different molecule.

Diprotic acids (*Section 8-3*) Acids that can give up two protons.

Disaccharides (*Section 20-4*) Carbohydrates containing two monosaccharide units joined by a glycosidic bond.

Discriminatory curtailment diets (*Section 30-1*) Diets that avoid certain food ingredients that are considered harmful to the health of an individual—for example, low-sodium diets for people with high blood pressure.

Dissociation (*Section 4-3*) An ionic compound that dissolves in water and separates into positive and negative ions.

Disulfide (*Section 14-4D*) A compound containing an —S—S— group.

Disulfide bond (*Section 14.4D*) A sulfur–sulfur bond in a disulfide group (—S—S—).

DNA (*Section 25-2*) Deoxyribonucleic acid.

DNA fingerprint (*Chemical Connections 25C*) A pattern of DNA fragments generated by electrophoresis that is used in forensic science.

Double bond (*Section 3-7C*) A bond formed by sharing two pairs of electrons; represented by two lines between the two bonded atoms.

Double-headed arrows (*Section 3-9A*) Symbols used to show that the structures on either side of them are resonance-contributing structures.

Double helix (*Section 25-3*) The arrangement in which two strands of DNA are coiled around each other in a screw-like fashion.

Dynamic equilibrium (*Section 7-5*) A state in which the rate of the forward reaction equals the rate of the reverse reaction.

Effective collision (*Section 7-2*) A collision between two molecules or ions that results in a chemical reaction.

EGF (*Section 31-6*) Epidermal growth factor; a cytokine that stimulates epidermal cells during healing of wounds.

Electrolyte (*Section 6-6C*) A substance that, when dissolved in water, produces a solution that conducts electricity.

Electromagnetic spectrum (*Section 9-3*) The array of electromagnetic phenomena by wavelength.

Electron (*Section 2-4*) A subatomic particle with a mass of approximately 1/1837 amu and a charge of -1; it is found outside the nucleus.

Electron capture (*Section 9-3F*) A reaction in which a nucleus captures an extranuclear electron and then undergoes a nuclear decay.

Electron configuration (*Section 2-6C*) A description of the orbitals that the electrons of an atom occupy.

Electron pushing (*Section 3-9A*) A representation that shows how electron pairs are redistributed from one contributing structure to the next.

18-Electron rule (*Section 22-10*) When transition metals form compounds, maximum stability is achieved with 18 electrons in bonds and unshared pairs, analogous to the octet rule with second row elements.

Electron transport chain (*Section 27-3*) The pathway in which electrons are passed to oxygen in the central metabolic pathway.

Electronegativity (*Section 3-4B*) A measure of an atom's attraction for the electrons it shares in a chemical bond with another atom.

Electrophile (*Section 12-5A*) An electron-poor species that can accept a pair of electrons to form a new covalent bond.

Electrophoresis (*Chemical Connections 25C*) A laboratory technique involving the separation of molecules in an electric field.

Element (*Section 2-4A*) A substance that consists of identical atoms.

Elongation (*Section 26-2*) The phase of protein synthesis during which activated tRNA molecules deliver new amino acids to ribosomes where they are joined by peptide bonds to form a polypeptide.

Elongation factor (*Section 26-5*) A small protein molecule that is involved in the process of tRNA binding and movement of the ribosome on the mRNA during elongation.

Embryonal carcinoma cell (*Chemical Connections 31C*) A cell that is multipotent and is derived from carcinomas.

Embryonic stem cell (*Chemical Connections 31C*) Stem cells derived from embryonic tissue. Embryonic tissue is the richest source of stem cells.

Emulsion (*Section 6-7*) A system, such as fat in milk, consisting of a liquid with or without an emulsifying agent in an immiscible liquid, usually as droplets larger than colloidal size.

Enantiomers (*Section 15-1*) Stereoisomers that are nonsuperposable mirror images; refers to a relationship between pairs of objects.

End point (*Section 8-9*) The point in a titration where a visible change occurs.

Endocannabinoid (*Chemical Connections 30G*) A group of lipids and their receptors that are involved in mood, pain sensations, and other processes.

Endocrine gland (*Section 24-2*) A gland such as the pancreas, pituitary, and hypothalamus that produces hormones involved in the control of chemical reactions and metabolism.

Endothermic (*Section 4-8*) A chemical reaction that absorbs heat.

Energy (*Section 1-8*) The capacity to do work. The SI base unit is the joule (J).

Enhancer (*Section 26-6*) A DNA sequence that is not part of the promoter region that binds a transcription factor, enhancing transcription and speeding up protein production.

Enkephalins (*Section 24-6*) Pentapeptides found in nerve cells of the brain that act to control the perception of pain.

Enol (*Section 17-5*) A molecule containing an —OH group bonded to a carbon of a carbon–carbon double bond.

Envelope conformation (*Section 11-7B*) A puckered conformation of cyclopentane in which four carbons of the ring lie in a plane and the fifth carbon is bent out of the plane, like an envelope with its flap bent upward.

Enzyme activity (*Section 23-4*) The rate at which an enzyme-catalyzed reaction proceeds, commonly measured as the amount of product produced per minute.

Enzyme specificity (*Chemical Connections 23A*) The limitation of an enzyme to catalyze one specific reaction with one specific substrate.

Enzyme-substrate complex (*Section 23-5*) A part of an enzyme reaction mechanism where the enzyme is bound to the substrate.

Enzymes (*Section 23-1*) Biological catalysts that increase the rate of a chemical reaction by providing an alternative pathway with a lower activation energy.

Epigenetics (*Section 26-10*) The study of heritable processes that alter gene expression without altering the actual DNA.

Epigenome (*Chemical Connections 26G*) The totality of changes to DNA and chromatin in epigenetics.

Epimutation (*Chemical Connections 26G*) Mutation in the DNA scaffolding that does not affect the DNA sequence.

Epitope (*Section 31-3*) The smallest number of amino acids on an antigen that elicits an immune response.

Epoxide (*Chemical Connections 13B*) A three-membered ring, in which two atoms of the ring are carbons and the third is oxygen.

Equatorial orientation (*Section 11-7B*) A position on a chair conformation of a cyclohexane ring that extends from the ring roughly perpendicular to the imaginary axis of the ring.

Equilibrium (*Section 5-8B*) A condition in which two opposing physical forces are equal.

Equilibrium constant (*Section 7-6*) The ratio of product concentrations to reactant concentrations (with exponents that depend on the coefficients of the balanced equation).

Equilibrium expression (*Section 7-6*) The ratio of the multiplied product concentrations divided by multiplied reactant concentrations (with exponents that depend on the coefficients of the balanced equation).

Equivalence point (*Section 8-9*) The point in an acid–base titration at which there is a stoichiometric amount of acid and base.

Ergogenic aid (*Chemical Connections 30E*) A substance that can be consumed to enhance athletic performance.

Essential amino acids (*Sections 29-5 and 30-5*) Amino acids that the body cannot synthesize in the required amounts and so must be obtained in the diet.

Essential fatty acid (*Section 30-4*) A fatty acid required in the diet.

Ester (*Sections 10-4E and 18-5D*) A compound in which the —OH of a carboxyl group, RCOOH, is replaced by an alkoxy (—OR) group or aryloxy (—OAr) group.

Ether (*Section 14-3A*) A compound containing an oxygen atom bonded to two carbon atoms.

Ex vivo (*Section 26-9*) A type of gene therapy where somatic cells are removed from the patient, altered with the gene therapy, and then returned to the patient.

Excitatory neurotransmitters (*Section 24-4*) Neurotransmitters that increase the transmission of nerve impulses.

Exons (*Section 25-5*) Nucleotide sequences in mRNA that code for a protein.

Exothermic (*Section 4-8*) A chemical reaction that gives off heat.

Exponential notation (*Section 1-3*) An easy way to express both large and small numbers based on powers of 10.

Expression cassette (*Section 26-9*) A gene sequence containing a gene that was incorporated into a vector and introduced via gene therapy, replacing some of the vector's own DNA.

Extended helix (*Section 22-9*) A type of helix found in collagen, caused by a repeating sequence.

External innate immunity (*Section 31-1*) The innate protection against foreign invaders characteristic of the skin barrier, tears, and mucus.

Fact (*Section 1-2*) A statement based on direct experience.

Factor-Label method (*Section 1-5*) A procedure in which the equations are set up so that all the unwanted units cancel and only the desired units remain.

Familial DNA searches (*Chemical Connections 25C*) Techniques where police can use DNA samples already collected to not only pinpoint exact DNA matches, but also use the DNA database to match family members with DNA left at a crime scene.

Families (*Section 2-5*) The elements in a vertical column in the Periodic Table.

Fats (*Section 21-3*) Mixtures of triglycerides containing a high proportion of long-chain, saturated fatty acids.

Fatty acid (*Section 18-4A*) A long, unbranched chain carboxylic acid, most commonly containing 12 to 30 carbon atoms. They are derived from animal fats, vegetable oils, or the phospholipids of biological membranes. The hydrocarbon chain may be saturated or unsaturated. In most unsaturated fatty acids, the *cis* isomer predominates. *Trans* isomers are rare.

Feedback control (*Section 23-6*) A type of enzyme regulation where the product of a series of reactions inhibits the enzyme that catalyzes the first reaction in the series.

Fiber (*Section 30-1*) The cellulosic, non-nutrient component in our food.

Fiberscope (*Chemical Connections 22H*) A medical device used to aim lasers accurately for various surgeries, such as laser vision correction.

Fibrous proteins (*Section 22-1*) Proteins used for structural purposes. Fibrous proteins are insoluble in water and have a high percentage of secondary structures, such as alpha helices and/or beta-pleated sheets.

Fischer esterification (*Section 18-5D*) The process of forming an ester by refluxing a carboxylic acid and an alcohol in the presence of an acid catalyst, commonly sulfuric acid.

Fischer projections (*Section 20-1B*) Two-dimensional representations of showing the configuration of a stereocenter; horizontal lines represent bonds projecting forward from the stereocenter, and vertical lines represent bonds projecting toward the rear.

Fission, nuclear (*Section 9-9*) The fragmentation of a heavier nucleus into two or more smaller nuclei.

Fluid mosaic model (*Section 21-5A*) The model for membrane structure in which proteins and lipid molecules move freely with respect to each other.

Formin 2 (*Chemical Connections 26G*) A gene required for memory in mice.

Formula weight (FW) (*Section 4-5*) The sum of the atomic weights of all atoms is a compound's formula expressed in atomic mass units (amu). Formula weight can be used for both ionic and molecular compounds.

Free radicals (*Chemical Connections 9C*) Compounds that have unpaired electrons.

Freezing-point depression (*Section 6-8A*) The decrease in the freezing point of a liquid caused by adding a solute.

Frequency (ν) (*Section 9-2*) The number of wave crests that pass a given point per unit of time.

Functional group (*Section 10-4*) An atom or group of atoms within a molecule that shows a characteristic set of physical and chemical properties.

Furanose (*Section 20-2*) A five-membered cyclic hemiacetal form of a monosaccharide.

Fusion, nuclear (*Section 9-8*) The combining of two or more nuclei to form a heavier nucleus.

Gamma rays (γ) (*Section 9-2*) Forms of electromagnetic radiation characterized by very short wavelength and very high energy.

Gangliosides (*Section 21-8*) Glycolipids in which a ceramide is bonded to an oligosaccharide.

Gaseous state (*Section 5-1*) The state of matter where molecules at high temperatures possess a high kinetic energy that the intermolecular attractive forces between them are too weak to hold together.

Gases (*Section 1-6*) The forms of matter that have no definite shape or volume, expands and are highly compressible.

Gay-Lussac's law (*Section 5-3C*) The pressure of a gas at constant volume is directly proportional to its temperature in Kelvin.

Geiger-Müller counter (*Section 9-5*) An instrument for measuring ionizing radiation.

Gene (*Section 25-1*) The unit of heredity; a DNA segment that codes for a protein.

Gene expression (*Section 26-1*) The activation of a gene to produce a specific protein. It involves both transcription and translation.

Gene regulation (*Section 26-6*) The various methods used by organisms to control which genes will be expressed and when.

Gene therapy (*Section 26-9*) The process of treating a disease by introducing a functional copy of a gene to an organism that was lacking it.

General transcription factor (**GTF**) (*Section 26-6*) Proteins that make a complex with the DNA being transcribed and the RNA polymerase.

Genetic code (*Section 26-4*) The sequence of triplets of nucleotides (codons) that determines the sequence of amino acids in a protein.

Genetic engineering (*Section 26-8*) The process by which genes are inserted into cells.

Genome (*Chemical Connections 25D*) The complete DNA sequence of an organism.

Globular protein (*Section 22-1*) Protein that is used mainly for nonstructural purposes and is largely soluble in water.

Glucogenic (*Section 28-9*) Refers to amino acids whose carbon skeletons can lead to production of sugars.

Gluconeogenesis (*Section 29-2*) The process by which glucose is synthesized in the body.

Glycerophospholipids (*Section 21-4*) Lipids that contain the alcohol glycerol, two fatty acids, and a phosphate group.

Glycogen (*Section 20-5B*) A polysaccharide used to store energy in animals made of glucose residues linked α-1$\rightarrow$4 with branches linked α-1$\rightarrow$6 glycosidic bonds.

Glycogenesis (*Section 29-2*) The conversion of glucose to glycogen.

Glycolipids (*Section 21-4*) Complex lipids that contain carbohydrates.

Glycols (*Section 14-1B*) Compounds with hydroxyl (—OH) groups on adjacent carbons.

Glycolysis (*Section 28-2*) The biochemical pathway that breaks down glucose to pyruvate, which yields chemical energy in the form of ATP and reduced coenzymes.

Glycoside (*Section 20-3*) A carbohydrate in which the —OH group on its anomeric carbon is replaced by an —OR group.

Glycosidic bond (*Section 20-3*) The bond from the anomeric carbon of a glycoside to an —OR group.

Gp120 (*Section 31-5*) A 120,000-molecular-weight glycoprotein on the surface of the human immunodeficiency virus that binds strongly to the CD4 molecules on T cells.

G-Protein (*Section 24-5*) A protein that is either stimulated or inhibited when a hormone binds to a receptor and subsequently alters the activity of another protein such as adenyl cyclase.

Gram (*Section 1-4C*) The SI base unit of mass.

Gray (Gy) (*Section 9-6*) The SI unit of the amount of radiation absorbed from a source; 1 Gy = 100 rad.

Ground state (*Section 2-6*) The electron configuration of the lowest energy of an atom.

Guanosine (*Section 25-2*) A nucleoside made of D-ribose and guanine.

Half-life (*Section 9-4*) The time it takes for one half of a sample of radioactive material to decay.

Halogens (*Section 2-5*) Elements in Group 7A of the Periodic Table.

Haworth projection (*Section 20-2*) A way to view furanose and pyranose forms of monosaccharides; the ring is drawn flat and viewed through its edge, with the anomeric carbon on the right and the oxygen atom to the rear.

HDL (*Section 21-9*) High-density lipoprotein; "good cholesterol."

HDPE (*Section 12-7C*) *H*igh-*d*ensity *p*oly*e*thylene.

Heat (*Section 1-9A*) The form of energy that most frequently accompanies chemical reactions.

Heat of combustion (*Section 4-8*) The heat given off in a combustion reaction.

Heat of fusion (*Section 5-10A*) The heat necessary to melt a solid.

Heat of reaction (*Section 4-8*) The heat given off or absorbed in a chemical reaction.

Heat of vaporization (*Section 5-10A*) The heat necessary to vaporize a liquid.

Heating curve (*Section 5-10A*) A graphical representation that shows how a substance changes phases while heat is applied at various temperatures.

Helicases (*Section 25-6*) Unwinding proteins that act at a replication fork to unwind DNA so that DNA polymerase can synthesize a new DNA strand.

Helix-Turn-Helix (*Section 26-6*) A common motif for a transcription factor.

Helper T cells (*Section 31-2*) A type of T cell that helps in the response of the acquired immune system against invaders but does not kill infected cells directly.

Hemiacetals (*Section 17-4C*) Molecules containing a carbon bonded to one —OH and one —OR group; the product of adding one molecule of alcohol to the carbonyl group of an aldehyde or ketone.

Hemodialysis (*Chemical Connections 6G*) The procedure to remove toxic waste products from the blood using a machine and dialyzer.

Hemolysis (*Section 6-8C*) An osmotic process in which water flows into red blood cells through the cell's semipermeable membrane, causing the cells to burst.

Henderson–Hasselbalch equation (*Section 8-11*) A mathematical relationship between pH, the pK_a of a weak acid (represented by the general formula HA), and the concentrations of the weak acid and its conjugate base.

Henry's law (*Section 6-4C*) The solubility of a gas in a liquid is directly proportional to the pressure of the gas above the liquid.

Heterocyclic aliphatic amine (*Section 16-1*) A heterocyclic amine in which nitrogen is bonded only to alkyl groups.

Heterocyclic amine (*Section 16-1*) An amine in which nitrogen is one of the atoms of a ring.

Heterocyclic aromatic amine (*Section 16-1*) An amine in which nitrogen is one of the atoms of an aromatic ring.

Heterogeneous catalysts (*Section 7-4D*) Catalysts in a separate phase from the reactants—for example, the solid platinum, Pt(s), in the reaction between CO(g) and H_2(g) to produce $CH_3OH(\ell)$.

Highly active antiretroviral therapy (HAART) (*Section 31-9*) An aggressive treatment against AIDS involving the use of several different drugs.

Histone (*Section 25-6*) A basic (pH > 7) protein that is found in complexes with DNA in eukaryotes.

HIV (*Sections 31-4 and 31-9*) Human immunodeficiency virus.

Homogeneous catalysts (*Section 8-4D*) Catalysts in the same phase as the reactants—for example, enzymes in body tissues.

Hormone (*Section 24-2*) A chemical messenger released by an endocrine gland into the bloodstream and transported there to reach its target cell.

Hybridization (*Section 25-8*) A process whereby two strands of nucleic acids or segments thereof form a double-stranded structure through hydrogen bonding of complementary base pairs.

Hybridoma (*Section 31-4*) A combination of a myeloma cell with a B cell to produce monoclonal antibodies.

Hydrated (*Section 6-6A*) When a solid ionic compound is dissolved in water, the water molecules surround the ions when the combined force of attraction of the water molecules is greater than the force of attraction of the ionic bonds.

Hydrates (*Section 6-6B*) Substances that contain water in their crystals.

Hydration (*Section 12-6B*) The addition of water.

Hydrocarbon (*Section 11-1*) A compound that contains only carbon and hydrogen atoms.

Hydrogen bonding (*Section 5-7C*) A noncovalent force of attraction between the partial positive charge on a hydrogen atom bonded to an atom of high electronegativity, most commonly oxygen or nitrogen, and the partial negative charge on a nearby oxygen or nitrogen.

Hydrogenation (*Section 12-5D*) Addition of hydrogen atoms to a double or triple bond using H_2 in the presence of a transition metal catalyst, most commonly Ni, Pd, or Pt. Also called catalytic reduction or catalytic hydrogenation.

Hydrolase (*Section 23-2*) An enzyme that catalyzes a hydrolysis reaction.

Hydrolysis (*Section 19-4A*) A chemical reaction of decomposition characterized by splitting of a bond and addition of the elements of water.

Hydronium ion (*Section 8-1*) The H_3O^+ ion.

Hydrophobic interaction (*Section 22-11*) Interaction by London dispersion forces between hydrophobic groups.

Hydroxyl group (*Section 10-4A*) An —OH group bonded to a tetrahedral carbon atom.

Hygroscopic (*Section 6-6B*) A compound able to absorb water vapor from the air.

Hyperbolic (*Chemical Connections 22G*) Refers to a graph in which a curve rises quickly and then levels off.

Hyperthermia (*Chemical Connections 1B*) Having a body temperature higher than normal.

Hyperthermophile (*Section 23-4*) An organism that lives at extremely high temperatures.

Hypertonic solutions (*Section 6-8C*) Solutions in which the osmolarity (and hence osmotic pressure) is greater than red blood cells.

Hypothermia (*Chemical Connections 1B*) Having a body temperature lower than normal.

Hypothesis (*Section 1-2*) A statement that is proposed, without actual proof, to explain certain facts and their relationship.

Hypotonic solutions (*Section 6-8C*) Solutions in which the osmolarity (and hence osmotic pressure) is lower than red blood cells.

Ideal gas (*Section 5-4*) A gas whose physical properties are described accurately by the ideal gas law.

Ideal gas constant (R) (*Section 5-4*) $0.0821 \cdot L \cdot atm \cdot mol^{-1} \cdot K^{-1}$.

Ideal gas law (*Section 5-4*) $PV = nRT$.

Immune system (*Section 31-1B*) The cells and molecules involved in the vertebrate system that fight against diseases attacking the body.

Immunization (*Section 31-6*) The process of stimulating the immune system to fight a particular disease.

Immunogen (*Section 31-3*) Another term for antigen.

Immunoglobulin superfamily (*Section 31-1*) A family of molecules based on a similar structure that includes the immunoglobulins, T cell receptors, and other membrane proteins that are involved in cell communications. All molecules in this class have a certain portion that can react with antigens.

Immunoglobulins (*Section 31-4*) Antibody proteins generated against and capable of binding specifically to an antigen.

In vivo (*Section 26-9*) A type of gene therapy where a virus is used to directly infect the patient's cells.

Inactivated vaccine (*Section 31-6A*) A vaccine made from a killed disease agent that is no longer capable of reproducing.

Indicator, acid–base (*Section 8-8*) A substance that changes color within a given pH range.

Induced-fit model (*Section 23-5*) A model explaining the specificity of enzyme action by comparing the active site to a glove and the substrate to a hand.

Inhibition (*Section 23-3*) The process by which a compound binds to an enzyme and lowers its activity.

Inhibition of enzymatic activity (*Section 23-3*) Any reversible or irreversible process that makes an enzyme less active.

Inhibitor (*Section 23-3*) A compound that binds to an enzyme and lowers its activity.

Inhibitory neurotransmitters (*Section 24-4*) Neurotransmitters that decreases the transmission of nerve impulses.

Inhibitory receptor (*Section 31-7*) A receptor on the surface of a cell of the innate immune system that recognizes antigens on healthy cells and prevents activation of the immune system.

Initial rate (*Section 7-1*) The initial change in concentration of a substance with respect to time.

Initial rates of reaction (*Section 7-4B*) The rate at the beginning of reactions, when the change in concentration is directly proportional to time.

Initiation factor (*Section 26-5B*) Protein that aids in the initiation of transcription or translation.

Initiation of protein synthesis (*Section 26-5*) The first step in the process whereby the base sequence of a mRNA is translated into the primary structure of a polypeptide.

Initiation signal (*Section 26-2*) A sequence on DNA that identifies the location where transcription is to begin.

Innate immunity (*Section 31-1*) The first line of defense against foreign invaders, which includes skin resistance to penetration, tears, mucus, and nonspecific macrophages that engulf bacteria.

Inner transition elements (*Section 2-5A*) Elements 58 to 71 and 90 to 103 of the Periodic Table.

Interleukin (*Section 31-7*) A cytokine that controls and coordinates the action of leukocytes.

Internal innate immunity (*Section 31-1*) The type of innate immunity that is used once a pathogen has already penetrated a tissue.

International System of Units (SI) (*Section 1-4*) A system of units of measurement based in part on the metric system.

International Union of Pure and Applied Chemistry (IUPAC) (*Section 11-4*) An international organization representing chemical societies throughout the world. Among other duties, this body is charged with establishing rules for chemical nomenclature, including establishing the rules and conventions for the designation of configuration.

Introns (*Section 25-5*) Nucleotide sequences in mRNA that do not code for a protein.

Ion (*Section 2-8B*) An atom with an unequal number of protons and electrons.

Ion product of water (K_w) (*Section 8-7*) The concentration of H_3O^+ multiplied by the concentration of OH^-; $[H_3O^+][OH^-] = 1 \times 10^{-14}$.

Ionic bond (*Section 3-4A*) A chemical bond resulting from the attraction between a positive ion and a negative ion.

Ionic compound (*Section 3-5A*) A compound formed by the combination of positive and negative ions.

Ionization energy (*Section 2-8B*) The energy required to remove the most loosely held electron from an atom in the gas phase.

Ionizing radiation (*Section 9-5*) Radiation that causes one or more electrons to be ejected from an atom or a molecule, thereby producing positive ions.

Isoelectric point (pI) (*Section 22-3*) The pH at which a molecule has no net charge.

Isoenzymes (*Section 23-6*) Enzymes that perform the same function but have different combinations of subunits and thus different quaternary structures; also called isozymes.

Isomerase (*Section 23-2*) An enzyme that catalyzes an isomerization reaction.

Isotonic (*Section 6-8C*) Solutions that have the same osmolarity.

Isotonic solution (*Section 6-8C*) A solution that has the same salt concentration as cells and the blood.

Isotopes (*Section 2-4D*) Atoms with the same number of protons but different number of neutrons.

Isozymes (*Section 23-6*) Two or more enzymes that perform the same functions but have different combinations of subunits and thus different quaternary structures.

Joule (J) (*Section 1-9*) The SI base unit for heat; 4.184 J is 1 cal.

Kelvin (*Section 1-4E*) The SI base unit of temperature; also called the absolute scale.

Keto-enol tautomerism (*Section 17-5*) A type of isomerism involving keto (from ketone) and enol tautomers. Tautomers are constitutional isomers that differ in the location of a hydrogen atom and a double bond.

Ketoacidosis (*Chemical Connections 28C*) A physiological condition often found in diabetes, marked by low blood pH and high levels of blood ketones.

Ketone (*Sections 10-4C and 17-1*) A compound containing a carbonyl group bonded to two carbons.

Ketone bodies (*Section 28-7*) A collective name for acetone, acetoacetate, and β-hydroxybutyrate; compounds produced from acetyl CoA in the liver that are used as a fuel for energy production by muscle cells and neurons.

Ketoses (*Section 20-1*) Monosaccharides containing a ketone group.

Killer T cells (*Section 31-2*) T cells that kill invading foreign cells by cell-to-cell contact. Also called cytotoxic T cells.

Kinases (*Chemical Connections 23B*) Classes of enzymes that covalently modify a protein with a phosphate group, usually through the —OH group on the side chain of a serine, threonine, or tyrosine.

Kinetic energy (*Section 1-8*) The energy of motion; energy that is in the process of doing work.

Kinetic molecular theory (*Section 5-6*) A set of assumptions about the molecules of a gas, which gives an idealized picture of the molecules of a gas and their interactions with one another.

Kwashiorkor (*Section 30-5*) A disease caused by insufficient protein intake and characterized by a swollen stomach, skin discoloration, and retarded growth.

L-Monosaccharide (*Section 20-1*) A monosaccharide that, when written as a Fischer projection, has the —OH group on its penultimate carbon to the left.

Lactam (*Section 19-1B*) A cyclic amide.

Lactone (*Section 19-1B*) A cyclic ester.

Lactose (*Section 20-4B*) A disaccharide made of glucose and galactose linked β-1→4 glycosidic bonds.

Lagging strand (*Section 25-6*) A discontinuously synthesized DNA that elongates in a direction away from the replication fork.

Lanthanide series (*Section 22-10*) The 14 elements (58–71) immediately following lanthanum in period 6 in which the $4f$ shell is being filled.

Law of conservation of energy (*Section 1-8*) Energy can be neither created nor destroyed.

Law of conservation of mass (*Section 2-3A*) Matter can neither be created nor destroyed.

Law of constant composition (*Section 2-3A*) Any compound is always made up of elements in the same proportion by mass.

LDL (*Section 21-9*) Low-density lipoprotein; "bad cholesterol."

LDPE (*Section 12-7B*) Low-density polyethylene.

Le Chatelier's principle (*Section 7-7*) When a stress is applied to a system in chemical equilibrium, the position of the equilibrium shifts in the direction that will relieve the applied stress.

Leading strand (*Section 25-6*) The continuously synthesized DNA strand that elongates toward the replication fork.

Leucine zipper (*Section 26-6*) A common motif for a transcription factor.

Leukocytes (*Section 31-2*) White blood cells, which are the principal parts of the acquired immunity system and act via phagocytosis or antibody production.

Leukotrienes (*Section 21-12*) Substances derived from white blood cells that have three double bonds and are of pharmaceutical importance.

Levorotatory (*Section 15-4B*) The counterclockwise rotation of the plane of polarized light in a polarimeter.

Lewis dot structure (*Section 2-6F*) The symbol of the element surrounded by a number of dots equal to the number of electrons in the valence shell of an atom of that element.

Lewis structures (*Section 3-7C*) Formulas for a molecule or an ion showing all pairs of bonding electrons as single, double, or triple lines and all nonbonding (unshared) electrons as pairs of Lewis dots.

Ligands (*Section 22-10*) Ions or neutral molecules bonded to a central transition metal in a coordination compound.

Ligase (*Section 23-2*) A class of enzymes that catalyzes a reaction joining two molecules. They are often called synthetases or synthases.

Limiting reagent (*Section 4-7*) The reactant that is consumed, leaving an excess of another reagent or reagents unreacted.

Line-angle formula (*Section 11-1*) An abbreviated way to draw structural formulas in which each vertex and line terminus represents a carbon atom and each line represents a bond.

Lipase (*Section 30-4*) An enzyme that catalyzes the hydrolysis of an ester bond between a fatty acid and glycerol.

Lipid bilayers (*Section 21-5A*) Aggregates of lipid molecules in which the polar head groups are in contact with water and the hydrophobic parts are not.

Lipids (*Section 21-1*) A family of substances that are insoluble in water but soluble in nonpolar solvents and solvents of low polarity.

Lipoproteins (*Section 21-9B*) Spherically shaped clusters containing both lipid molecules and protein molecules.

Liquid state (*Section 5-1*) The state of matter where molecules at lower temperatures move more slowly, to the point where the forces of attraction become more important.

Liquids (*Section 1-6*) The forms of matter that have no definite shape, have a definite volume that remains the same when poured, and are slightly compressible.

Liter (*Section 1-4B*) The SI base unit of volume.

Lock-and-key model (*Section 23-5*) A model explaining the specificity of enzyme action by comparing the active site to a lock and the substrate to a key.

London dispersion forces (*Section 5-7A*) Extremely weak attractive forces between atoms or molecules caused by the electrostatic attraction between temporary induced dipoles.

Long non-coding RNA (lncRNA) (*Section 25-4*) An RNA molecule greater than 100 nucleotides in length that does not fit into any of the other categories of RNA. Some estimates identify more than 200,000 types in mammalian cells.

Low-density polyethylene (LDPE) (*Section 12-6B*) Polyethylene with a density between 0.91 and 0.94 g/cm³.

Lyase (*Section 23-2*) A class of enzymes that catalyzes the addition of two atoms or groups of atoms to a double bond or their removal to form a double bond.

Lymph (*Section 31-2*) The fluid that bathes vertebrate cells and travels through lymphatic vessels.

Lymphocytes (*Sections 31-1 and 31-2*) White blood cells that spend most of their time in the lymphatic tissues. Those that mature in the bone marrow are B cells. Those that mature in the thymus are T cells.

Lymphoid organs (*Section 31-2*) The main organs of the immune system, such as the lymph nodes, spleen, and thymus, that are connected by lymphatic capillary vessels.

Macrophages (*Sections 31-1 and 31-2*) Ameboid white blood cells that move through tissue fibers, engulfing dead cells and bacteria by phagocytosis, and then display some of the engulfed antigens on its surface.

Main-group elements (*Section 2-5A*) Elements in the A groups (Groups 1A, 2A, and 3A–8A) of the Periodic Table.

Major groove (*Section 25-3*) The side of a DNA double helix that is narrower.

Major histocompatibility complex (MHC) (*Sections 31-2 and 31-3*) A transmembrane protein complex that brings the epitope of an antigen to the surface of the infected cell to be presented to the T cells.

Maloney murine leukemia virus (MMLV) (*Section 26-9*) A common vector used for gene therapy.

Maltose (*Section 20-4C*) A disaccharide made of two glucose residues linked by α-1$\rightarrow$4 glycoside bonds.

Manometer (*Section 5-2*) An instrument used to measure the pressure of a gas in a container.

Marasmus (*Section 30-2*) Another term for chronic starvation, whereby the individual does not have adequate caloric intake. It is characterized by arrested growth, muscle wasting, anemia, and general weakness.

Markovnikov's rule (*Section 12-6A*) In the addition of HX or H_2O to an alkene, hydrogen adds to the carbon of the double bond having the greater number of hydrogens.

Mass (*Section 1-4*) The quantity of matter in an object; the SI base unit is the kilogram; often referred to as weight.

Mass number (*Section 2-4B*) The sum of the number of protons and neutrons in the nucleus of an atom.

Matter (*Section 1-1*) Anything that has mass and takes up space.

Memory cell (*Section 31-2*) A type of T cell that stays in the blood after an infection is over and acts as a quick line of defense if the same antigen is encountered again.

Memory Molecule (*Chemical Connections 23B*) A colloquialism for the protein PKMz that stabilizes memories in the cerebral cortex.

Mercaptan (*Section 14-4B*) A common name for any molecule containing an —SH group.

Messenger RNA (mRNA) (*Section 25-4*) The RNA that carries genetic information from DNA to the ribosome and acts as a template for protein synthesis.

Meta (*m*) (*Section 13-2B*) Refers to groups occupying the 1 and 3 positions on a benzene ring.

Metabolic acidosis (*Chemical Connections 8C*) The lowering of the blood pH due to metabolic effects such as starvation or intense exercise.

Metabolism (*Section 27-1*) The sum of all chemical reactions in a cell.

Metal-binding finger (*Section 26-6*) A type of transcription factor containing heavy metal ions, such as Zn^{2+}, that is involved in helping RNA polymerase bind to the DNA being transcribed.

Metalloids (*Section 2-5B*) Elements that display some of the properties of metals and some of the properties of nonmetals. Six elements are classified as metalloids.

Metals (*Section 2-5B*) Elements that are solid at room temperature (except for mercury which is a liquid), shiny, conduct electricity, ductile (they can be drawn into wires), malleable, and form alloys. In their reactions, metals tend to give up electrons.

Meter (*Section 1-4*) The SI base unit of length.

Methylene group (*Section 11-2*) A —CH_2— group.

Metric system (*Section 1-4*) A system in which measurements of parameter are related by powers of 10.

Micelle (*Section 18-4*) A spherical arrangement of molecules in aqueous solution such that their hydrophobic (water-hating) parts are shielded from the aqueous environment in the interior and their hydrophilic (water-loving) parts are on the surface of the sphere and in contact with the aqueous environment.

Micro RNA (miRNA) (*Section 25-4*) A small RNA of 22 nucleosides that is involved in the regulation of genes and the development of an organism.

Millimeters of mercury (mm Hg) (*Section 5-2*) The most commonly used units to measure pressure.

Millirem (*Section 9-6*) A measure of the effect of radiation when a person absorbs one thousandth of a rem or roentgen equivalent for man.

Mini-satellite (*Section 25-5*) A small repetitive DNA sequence that is sometimes associated with cancer when it mutates.

Minor groove (*Section 25-3*) The side of a DNA double helix that is wider.

Mirror image (*Section 15-1*) The reflection of an object in a mirror.

Miscible (*Section 6-4*) Liquids that mix in all proportions.

Mixture (*Section 2-2C*) A combination of two or more pure substances.

Molar mass (*Section 4-6*) The mass of one mole of a substance expressed in grams; the formula weight of a compound expressed in grams.

Molarity (*Section 6-5*) The number of moles of solute dissolved in 1 L of solution.

Mole (mol) (*Section 4-6*) The formula weight of a substance expressed in grams.

Molecular weight (MW) (*Section 4-5*) The sum of the atomic weights of all atoms in a molecular compound expressed in atomic mass units (amu).

Molecule (*Section 2-3*) A tightly bound combination of two or more atoms that act as a single unit.

Monatomic elements (*Section 2-3B*) Substances that consist of single atoms that are not connected to each other.

Monoamine Oxidase Inhibitors (MAOIs) (*Chemical Connections 24G*) An early class of antidepressant that works by inhibiting monoamine oxidase, which breaks down dopamine, serotonin, and norepinephrine.

Monoclonal antibodies (*Section 31-4*) Antibodies produced by clones of a single B cell specific to a single epitope.

Monomer (*Section 12-6A*) From the Greek *mono*, "single," and *meros*, "part"; the simplest nonredundant unit from which a polymer is synthesized.

Monoprotic acids (*Section 8-3*) Acids that can give up only one proton.

Monosaccharides (*Section 20-1*) Carbohydrates that cannot be hydrolyzed to a simpler compound.

Multiclonal antibodies (*Chemical Connections 31B*) The type of antibodies found in its serum after a vertebrate is exposed to an antigen.

Multipotent stem cell (*Chemical Connections 31C*) A stem cell capable of differentiating into many, but not all, cell types.

Mutagen (*Section 26-7*) A chemical substance that induces a base change or mutation in DNA.

Mutarotation (*Section 20-2*) The change in a specific rotation at a given wavelength that occurs when an α or β form of a carbohydrate is converted to an equilibrium mixture of the two forms.

Mutation (*Section 26-7*) An error in the copying of a sequence of bases in DNA replication.

Natural gas (*Section 11-5*) A biofuel, consisting of a mixture of approximately 90 to 95% methane, 5 to 10% ethane, and several other low-boiling alkanes—chiefly propane, butane, and 2-methylpropane.

Natural killer cells (*Sections 31-1 and 31-2*) Cells of the innate immune system that attack infected or cancerous cells.

Negative modulation (*Section 23-6*) The process whereby an allosteric regulator inhibits enzymatic action.

Net ionic equation (*Section 4-3*) A chemical equation that does not contain spectator ions, where both atoms and charges are balanced.

Neuron (*Section 24-1*) Another name for a nerve cell.

Neuropeptide Y (*Section 24-6*) A brain peptide that affects the hypothalamus and is an appetite-stimulating agent.

Neurotransmitter (*Section 24-2*) A chemical messenger between a neuron and another cell, which may be a neuron, a muscle cell, or the cell of a gland.

Neutralizing antibody (*Section 31-8*) A type of antibody that completely destroys its target antigen.

Neutron (*Section 2-4*) A subatomic particle with a mass of approximately 1 amu and a charge of zero; it is found in the nucleus.

Noble gases (*Section 2-5C*) Elements in Group 8A of the Periodic Table, which are gases under normal temperature and pressure, and form either no compounds or very few compounds.

Nonbonding electrons (*Section 3-7C*) Valence electrons not involved in forming covalent bonds—that is, unshared electrons.

Noncompetitive inhibitors (*Section 23-3*) Compounds that bind to an enzyme and change the shape of the active site so that substrate cannot bind.

Nonelectrolyte (*Section 6-6C*) A substance that does not conduct electricity.

Nonmetals (*Section 2-5B*) Elements that do not have the characteristic properties of a metal and, in their reactions, tends to accept electrons. Eighteen elements are classified as nonmetals.

Nonpolar covalent (*Section 3-7B*) A covalent bond between two atoms whose difference in electronegativity is less than 0.5.

Nonsuperposable (*Section 15-1*) In the context of molecular structure, to lay one structure on another and find that all like parts do not coincide.

Norepinephrine-Dopamine Reuptake Inhibitors (NDRIs) (*Chemical Connections 24G*) A type of antidepressant that works by blocking the reuptake of norepinephrine and dopamine.

Norepinephrine Reuptake Inhibitors (NRIs) (*Chemical Connections 24G*) A type of antidepressant that works by blocking the reuptake of the neurotransmitter norepinephrine.

Normal boiling point (*Section 5-8C*) The temperature at which a liquid boils under a pressure of 1 atm.

N-terminus (*Section 22-4*) The amino acid at the end of a peptide chain that has a free amino group.

Nuclear fission (*Section 9-9*) The process of splitting a nucleus into smaller nuclei.

Nuclear fusion (*Section 9-8*) Joining together atomic nuclei to form a heavier nucleus than the starting nuclei.

Nuclear reaction (*Section 9-3A*) A reaction that changes an atomic nucleus (usually to the nucleus of another element).

Nucleic acids (*Section 25-3*) A polymer composed of nucleotides.

Nucleophile (*Section 12-5A*) An electron-rich species that can donate a pair of electrons to form a new covalent bond.

Nucleophilic attack (*Section 23-5*) A chemical reaction where an electron-rich atom such as oxygen or sulfur bonds to an electron-deficient atom such as a carbonyl carbon.

Nucleoside (*Section 25-2*) The combination of a heterocyclic aromatic amine bonded by a glycosidic bond to either D-ribose or 2-deoxy-D-ribose.

Nucleosome (*Section 25-3*) Combinations of DNA and proteins.

Nucleotide (*Section 25-2*) A phosphoric ester of a nucleoside.

Nucleus (*Section 2-4A*) The center of the atom which hosts the protons and neutrons.

Nutrients (*Section 30-1*) Components of food and drink that provide energy, replacement, and growth.

Octane rating (*Chemical Connections 11B*) The percent of 2,2,4-trimethylpentane in a mixture of 2,2,4-trimethylpentane and heptane that has the antiknock properties of a test gasoline.

Octet rule (*Section 3-2*) When undergoing chemical reactions, atoms of Group 1A–7A elements tend to gain, lose, or share electrons to achieve an election configuration having eight valence electrons.

—OH (hydroxyl) group (*Section 10-4*) An –OH group bonded to a tetrahedral carbon atom.

Oils (*Section 21-3*) Mixtures of triglycerides containing a high proportion of long-chain, unsaturated fatty acids.

Okazaki fragments (*Section 25-6*) Short segments of DNA made up of about 200 nucleotides in higher organisms and 2000 nucleosides in prokaryotes.

Oligosaccharide (*Section 20-4*) A carbohydrate containing from six to ten monosaccharide units, each joined to the next by a glycosidic bond.

Open complex (*Section 26-6*) The complex of DNA, RNA polymerase, and general transcription factors that must be formed before transcription can take place. In this complex, the DNA is being separated so that it can be transcribed.

Optically active (*Section 15-4B*) Characterized by rotation of the plane of polarized light.

Orbital box diagrams (*Section 2-6D*) Diagrams that consist of a box to represent an orbital, an arrow with its head up to represent a single electron, and a pair of arrows with heads in opposite directions to represent two electrons with paired spins.

Orbitals (*Section 2-6A*) Regions of space around a nucleus that can hold a maximum of two electrons.

Organic chemistry (*Section 10-1*) The study of the compounds of carbon.

Origin of replication (*Section 25-6*) The point in a DNA molecule where replication starts.

***Ortho* (*o*)** (*Section 13-2B*) Refers to groups occupying the 1 and 2 positions on a benzene ring.

Osmolarity (*Section 6-8C*) Molarity multiplied by the number of particles in solution in each formula unit of solute.

Osmosis (*Section 6-8*) The passage of solvent molecules from a less concentrated solution across a semipermeable membrane into a more concentrated solution.

Osmotic pressure (Π) (*Section 6-8C*) The amount of external pressure that must be applied to a more concentrated solution to stop the passage of solvent molecules into it from across a semipermeable membrane.

Oxidation (*Section 4-4*) The loss of electrons; the gain of oxygen atoms or the loss of hydrogen atoms.

Oxidation-reduction reaction (*Section 4-4*) This reaction involves the transfer of electrons from one species to another.

Oxidative deamination (*Section 28-8*) The reaction in which the amino group of an amino acid is removed and an α-ketoacid is formed.

Oxidative phosphorylation pathway (*Section 27-3*) The pathway in which transfer of electrons to oxygen is coupled to production of ATP.

Oxidizing agent (*Section 4-4*) An entity that accepts electrons in an oxidation–reduction reaction.

Oxidoreductase (*Section 23-2*) A class of enzymes that catalyzes an oxidation–reduction reaction.

Oxonium ion (*Section 12-6B*) An ion in which oxygen is bonded to three other atoms and bears a positive charge.

p53 (*Chemical Connections 26E*) A common and important tumor suppressor protein with a molecular weight of 53,000 that is found to be mutated in a large number of cancer types.

P site (*Section 26-5*) The site on the large ribosomal subunit where the current peptide is bound before peptidyl transferase links it to the amino acid attached to the A site during elongation.

***Para* (p)** (*Section 13-2B*) Refers to groups occupying the 1 and 4 positions on a benzene ring.

Parenteral nutrition (*Chemical Connections 29A*) The technical term for intravenous feeding.

Partial pressure (*Section 5-5*) The pressure that a gas in a mixture of gases would exert if it were alone in a container.

Parts per billion (ppb) (*Section 6-5D*) The concentration of a solution in grams of solute per 10^9 (billion) grams of solution.

Parts per million (ppm) (*Section 6-5D*) The concentration of a solution in grams of solute per 10^6 (million) grams of solution.

Passive transport (*Section 21-5B*) The process by which a substance enters a cell without input of energy by the cell.

Pentose phosphate pathway (*Section 28-2*) The biochemical pathway that produces ribose and NADPH from glucose-6-phosphate or, alternatively, releases energy.

Peptide backbone (*Section 22-7*) The repeating pattern of peptide bonds in a polypeptide or protein.

Peptide bond (*Section 22-6*) An amide bond that links two amino acids.

Peptides (*Section 22-6*) Short chains of amino acids linked via peptide bonds.

Peptidergic neurotransmitter (*Section 24-6*) A type of neurotransmitter or hormone that is based on a peptide, such as glucagon, insulin, and the enkephalins.

Peptidyl transferase (*Section 26-5*) The enzymatic activity of the ribosomal complex that is responsible for the formation of peptide bonds between the amino acids of the growing peptide.

Percent concentration (% w/v) (*Section 6-5A*) The number of grams of solute in 100 mL of solution.

Percent yield (*Section 4-7C*) The actual yield divided by the theoretical yield times 100.

Perforin (*Section 31-2*) A protein produced by killer T cells that punches holes in the membrane of target cells.

Periods (*Section 2-5*) Horizontal rows of the Periodic Table.

Peroxide (*Section 12-7B*) A compound that contains an —O—O— bond; for example, hydrogen peroxide, H—O—O—H.

Petroleum (*Section 11-5*) A thick viscous liquid mixture of thousands of compounds, most of them hydrocarbons, formed from the decomposition of marine plants and animals.

pH (*Section 8-8*) The negative logarithm of the hydronium ion concentration; $pH = -\log[H_3O^+]$.

Ph— (*Section 13-2A*) The symbol for a phenyl group, C_6H_5—.

Phagocytosis (*Section 31-4*) The process by which large particulates, including bacteria, are pulled inside a white cell called a phagocyte.

Phase (*Section 5-10A*) Any part of a system that looks uniform or homogenous throughout.

Phase changes (*Section 5-10*) Changes from one physical state (gas, liquid, or solid) to another.

Phase diagram (*Section 5-10A*) A graphical representation that shows the phases of a substance as a function of temperature and pressure.

Phenol (*Section 13-4*) A compound that contains an —OH group bonded to a benzene ring.

Phenyl group (*Section 13-2A*) C_6H_5—, the aryl group derived by removing a hydrogen atom from benzene. The name is derived from *phene*, an earlier name for benzene.

Pheromone (*Chemical Connections 12C*) A chemical secreted by an organism to influence the behavior of another member of the same species.

Phospholipids (*Section 21-4*) Lipids that contain an alcohol, two fatty acids, and a phosphate group.

Phosphoric anhydride (*Section 19-5A*) A compound derived by loss of the elements of water from two molecules of phosphoric acid. The characteristic structural feature of a phosphoric anhydride is two phosphoryl groups bonded to the same oxygen.

Phosphoric esters (*Section 19-5B*) Phosphoric acid has three —OH groups and can form mono-, di-, and triesters, in which one, two, or three of the —OH groups are replaced by —OR or —OAr groups.

Phosphorylation (*Section 27-3*) The bonding of a phosphate group to a molecule, particularly to ATP

Photons (*Section 9-2*) The smallest unit of electromagnetic radiation.

Photosynthesis (*Section 29-2*) The process in which plants synthesize carbohydrates from CO_2 and H_2O with the help of sunlight and chlorophyll.

Physical changes (*Section 1-1*) Changes in matter in which it does not lose its identity.

Physical properties (*Section 1-1*) Characteristics of a substance that are not chemical properties; those properties that are not a result of a chemical change.

Piwi-Associated RNA (piRNA) (*Section 25-4*) An RNA molecule that is 21–31 nucleotides long, found only in animal cells, and distinct from miRNA and siRNA. Its role is to bind to and lead to the destruction of transposons.

Plane-polarized light (*Section 15-4A*) Light vibrating in only parallel planes.

Plasma cell (*Section 31-2*) A cell derived from a B cell that has been exposed to an antigen.

Plasmids (*Section 26-8*) Small circular DNAs of bacterial origin often used to construct recombinant DNA.

Pluripotent stem cell (*Chemical Connections 31F*) A stem cell that is capable of developing into every cell type.

pOH (*Section 8-8*) The negative logarithm of the hydroxide ion concentration; $pOH = -\log[OH^-]$.

Polar covalent (*Section 3-7C*) A covalent bond between two atoms whose difference in electronegativity is between 0.5 and 1.9.

Polarimeter (*Section 15-4B*) An instrument for measuring the ability of a compound to rotate the plane of polarized light.

Polyamides (*Section 19-6A*) Polymers in which each monomer unit is joined to the next by an amide bond, as for, example, Nylon 66.

Polyatomic elements (*Section 2-3B*) Substances that consist of multiple atoms of the element per molecule.

Polyatomic ion (*Section 3-3C*) An ion that contains more than one atom.

Polycarbonate (*Section 19-6C*) A polyester in which the carboxyl groups are derived from carbonic acid.

Polyester (*Section 19-6B*) A polymer in which each monomer unit is joined to the next by an ester bond, as, for example, poly(ethylene terephthalate).

Polymer (*Section 12-6A*) From the Greek *poly*, "many", and *meros*, "parts"; any long-chain molecule synthesized by bonding together many single parts called monomers.

Polymerase (*Section 26-2*) An enzyme that synthesizes DNA and RNA from its nucleotide subunits.

Polymerase chain reaction (PCR) (*Section 25-8*) An automated technique for amplifying DNA using a heat-stable DNA polymerase from thermophilic bacteria.

Polynuclear aromatic hydrocarbons (PAHs) (*Section 13-2D*) Hydrocarbons containing two or more benzene rings, each of which shares two carbon atoms with another benzene ring.

Polypeptides (*Section 22-6*) Long chains of amino acids bonded via peptide bonds.

Polysaccharides (*Section 20-4*) Carbohydrates containing a large number of monosaccharide units, each joined to the next by one or more glycosidic bonds.

Positive cooperativity (*Chemical Connections 22G*) A type of allosterism where the binding of one molecule of a protein makes it easier to bind another of the same molecule.

Positive modulation (*Section 23-6*) The process whereby an allosteric regulator increases enzymatic action.

Positron (β^+) (*Section 9-3D*) A particle with the mass of an electron but a charge of +1, $_{+1}^{0}\beta$.

Positron emission tomography (PET) (*Section 9-7A*) The detection of positron-emitting isotopes in different tissues and organs; a medical imaging technique.

Post-transcription process (*Section 26-2*) A process such as splicing or capping that alters RNA after it is initially made during transcription.

Postsynaptic (*Section 24-2*) The membrane on the side of the synapse nearest the dendrite of the neuron receiving the transmission.

Potential energy (*Section 1-8*) Energy that is being stored; energy that is available for later use.

Pre-initiation complex (*Section 26-6A*) In translation, the complex containing the 30S ribosomal subunit and the initial tRNA molecule as well as initiation factors.

Precipitation reaction (*Section 4-3*) Positive and negative ions combine to form a water-insoluble compound.

Pressure (*Section 5-2*) The force per unit area exerted against a surface.

Presynaptic (*Section 24-2*) The membrane on the side of the synapse nearest the dendrite of the axon of the neuron transmitting the signal.

Primary (1°) alcohol (*Section 10-4A*) An alcohol in which the carbon atom bearing the —OH group is bonded to only one other carbon group, a —CH_2OH group.

Primary (1°) amine (*Section 10-4B*) An amine in which nitrogen is bonded to one carbon group and two hydrogens.

Primary structure, of DNA (*Section 25-3*) The order of the bases in DNA.

Primary structure, of proteins (*Section 22-8*) The order of amino acids in a protein.

Primer (*Section 25-6*) Short pieces of DNA or RNA that initiate DNA replication.

Principal energy levels (*Section 2-6A*) Energy level containing orbitals of the same number (1, 2, 3, 4, and so forth).

Proenzymes (*Section 23-6*) Enzymes in an inactive form that become active after undergoing a chemical change; also called zymogens.

Progenitor cells (*Chemical Connections 31D*) Another term for stem cells.

Prokaryote (*Section 25-5*) An organism that has no true nucleus or organelles.

Promoter (*Section 26-2*) An upstream DNA sequence that is used for RNA polymerase recognition and binding to DNA.

Proportional counter (*Section 9-5A*) An instrument which contains a gas such as helium or argon that can detect when a radioactive nucleus emits alpha or beta particles or gamma rays.

Prostaglandin (*Section 21-12*) A derivative of 20-carbon arachidonic acid that contains a five-membered ring and are of pharmaceutical importance.

Prosthetic group (*Section 22-12*) The non-amino-acid part of a conjugated protein.

Proteasomes (*Section 26-6*) Large protein complexes that are involved in the degradation of other proteins.

Protein complementation (*Section 30-5*) A diet that combines proteins of varied sources to arrive at a complete protein.

Protein microarray (*Chemical Connections 22F*) An automated technique used to study proteomics that is based on having thousands of protein samples imprinted on a chip.

Protein modification (*Section 23-6*) The process of affecting enzymatic activity by covalently modifying the enzyme, such as phosphorylating a particular amino acid.

Proteins (*Section 22-1*) Long chains of amino acids linked via peptide bonds. There must usually be a minimum of 30 to 50 amino acids in a chain before it is considered a protein (instead of a peptide).

Proteome (*Chemical Connections 22F*) The complement of proteins expressed by a genome.

Proton (*Section 2-4A*) A subatomic particle with a charge of +1 and a mass of approximately 1 amu; found in a nucleus.

Proton channel (*Section 27-6*) The part of the proton-translocation ATPase that allows the protons to cross the membrane.

Proton gradient (*Section 27-6*) A continuous variation in the H^+ concentration along a given region.

Proton-translocation ATPase (*Section 27-6*) The protein on the inner mitochondrial membrane that produces ATP.

Pyramidal (*Section 3-10*) A shape where a central atom is surrounded by a triangular-based pyramid with three regions of electron density to atoms, and a fourth region which contains an unshared pair of electrons on the central atom.

Pyranose (*Section 20-2*) A six-membered cyclic hemiacetal form of a monosaccharide.

Quaternary structure (*Section 22-12*) The spatial relationship and interactions between subunits in a protein that has more than one polypeptide chain.

R (*Section 15-2*) From the Latin *rectus*, meaning "straight, correct"; used in the R,S system to show that when the lowest-priority group is away from you, the order of priority of groups on a stereocenter is clockwise.

R— (*Section 11-4A*) A symbol used to represent an alkyl group.

R form (*Section 23-6*) The more active form of an allosteric enzyme.

R,S system (*Section 15-2*) A set of rules for specifying configuration about a stereocenter.

Racemic mixture (*Section 15-1*) A mixture of equal amounts of two enantiomers.

Radiation, nuclear (*Section 9-3*) Radiation emitted from a nucleus during nuclear decay. Includes alpha particles, beta particles, gamma rays, and positrons.

Radical (*Section 13-4*) An atom or a molecule with one or more unpaired electrons.

Radioactive (*Section 9-2*) Refers to a substance that emits radiation during nuclear decay.

Radioactive dating (*Chemical Connections 9A*) The process of establishing the age of a substance by analyzing radioisotope abundance as compared with a current relative abundance.

Radioactive isotopes (*Section 9-3*) Radiation-emitting isotopes of an element.

Radioactivity (*Section 9-2*) Another name for nuclear radiation. Includes alpha particles, beta particles, gamma rays, and positrons.

Rads (*Section 9-5*) Radiation absorbed doses. The SI unit is the gray (Gy).

Random coil (*Section 22-9*) Protein that does not exhibit any repeated pattern.

Rate constant (*Section 7-4B*) A proportionality constant, k, between the molar concentrations of reactants and the rate of reaction; rate = k[compound].

Rate of a reaction (*Section 7-1*) The change in concentration of a reactant (or product) per unit time.

Reaction mechanism (*Section 12-6A*) A step-by-step description of how a chemical reaction occurs.

Receptor (*Section 24–1*) A membrane protein that can bind a chemical messenger and then perform a function such as synthesizing a second messenger or opening an ion channel.

Recognition site (*Section 26-3*) The area of the tRNA molecule that recognizes the mRNA codon.

Recombinant DNA techniques (*Section 26-8*) DNAs from two sources that have been combined into one molecule.

Recommended Daily Allowances (RDA) (*Section 30-1*) *also* **Recommended Dietary Allowances**; average daily requirements for nutrients published by the U.S. Food and Drug Administration.

Redox reaction (*Section 4-4*) An oxidation–reduction reaction.

Reducing agent (*Section 4-4*) An entity that donates electrons in an oxidation–reduction reaction.

Reducing sugar (*Section 20-3*) A carbohydrate that reacts with a mild oxidizing agent under basic conditions to give an aldonic acid; the carbohydrate reduces the oxidizing agent.

Reduction (*Section 4-4*) The gain of electrons; the loss of oxygen atoms or the gain of hydrogen atoms.

Regioselective (*Section 12-6A*) A reaction in which one direction of bond forming or bond breaking occurs in preference to all other directions.

Regulator (*Section 23-6*) A molecule that binds to an allosteric enzyme and changes its activity. This change could be positive or negative.

Regulatory site (*Section 23-6*) A site other than the active site where a regulator binds to an allosteric site and affects the rate of reaction.

Relative humidity (*Section 5-8B*) The ratio of the actual partial pressure of water vapor in the air to the equilibrium vapor pressure of water at a relevant temperature.

Rems (*Section 9-6*) Roentgen equivalent for man; a biological measure of radiation.

Replication (*Section 25-6*) The process whereby DNA is duplicated to form two exact replicas of an original DNA molecule.

Replication fork (*Section 25-6*) The point on a DNA molecule where replication is proceeding.

Residues (*Section 22-6*) Another term for amino acids in a peptide chain.

Resonance (*Section 3-9*) A theory that many molecules and ions are best represented as hybrids of two or more Lewis contributing structures.

Resonance contributors (*Section 3-9A*) Representations of a molecule or ion that differ only in the distribution of valence electrons.

Resonance hybrid (*Section 3-9A*) A molecule best described as a hybrid of two or more Lewis contributing structures.

Resonance structures (*Section 3-9A*) Theories that many molecules and ions are best described as a hybrid of two or more Lewis contributing structures.

Respiration (*Section 4-4*) The process where humans and animals obtain their energy through the oxidation of carbon-containing compounds in the presence of oxygen.

Respiratory acidosis (*Chemical Connections 8C*) The lowering of the blood pH due to difficulty breathing.

Response element (*Section 26-6*) A sequence of DNA upstream from a promoter that interacts with a transcription factor to stimulate transcription in eukaryotes. Response elements may control several similar genes based on a single stimulus.

Restriction endonuclease (*Section 26-8*) An enzyme, usually purified from bacteria, that cuts DNA at a specific base sequence.

Retrovaccination (*Section 31-8*) A process whereby scientists have an antibody they want to use and try to develop molecules to elicit it.

Retrovirus (*Section 26-1*) A virus such as HIV that has an RNA genome.

Reuptake (*Section 24-4*) The transport of a neurotransmitter from its receptor back through the presynaptic membrane into the neuron.

Reverse osmosis (*Chemical Connections 6F*) When pressure greater than the osmotic pressure is applied to a more concentrated solution, solvent molecules flow to the more dilute solution.

Reversible reaction (*Section 7-5*) A process that can go back and forth between states along exactly the same path.

Ribonucleic acid (RNA) (*Section 25-2*) A type of nucleic acid consisting of nucleotide monomers, a nitrogenous base, D-ribose, and phosphate.

Ribosomal RNA (rRNA) (*Section 25-5*) The type of RNA that is complexed with proteins and makes up the ribosomes used in the translation of mRNA into protein.

Ribosome (*Section 25-4*) Small spherical bodies in the cell made of protein and RNA, the site of protein synthesis.

Ribozymes (*Section 23-1*) Enzymes that are made up of ribonucleic acid. The currently recognized ribozymes catalyze cleavage of part of their own sequences in mRNA and tRNA.

RNA (*Section 25-2*) Ribonucleic acid.

Roentgen (R) (*Section 9-6*) The amount of radiation that produces ions having 2.58×10^{-4} coulomb per kilogram.

Rusting (*Section 4-4*) The process where iron is oxidized to a mixture of iron oxides.

S (*Section 15-2*) From the Latin *sinister*, meaning "left"; used in the *R,S* system to show that when the lowest-priority group is away from you, the order of priority of groups on a stereocenter is counterclockwise.

Saponification (*Section 18-4B*) The hydrolysis of an ester in aqueous NaOH or KOH to give an alcohol and the sodium or potassium salt of a carboxylic acid.

Satellites (*Section 25-5*) Short sequences of DNA that are repeated hundreds of thousands of times but do not code for any protein in RNA.

Saturated (*Section 6-4*) A solution in which the solvent contains all the solute it can hold at a given temperature.

Saturated hydrocarbons (*Section 11-1*) Hydrocarbons that contain only carbon–carbon single bonds.

Saturation curve (*Section 23-4*) A graph of enzyme rate versus substrate concentration. At high levels of substrate, the enzyme becomes saturated and the velocity does not increase linearly with increasing substrate.

Scientific method (*Section 1-2*) A method of acquiring knowledge by testing theories.

Scintillation counter (*Section 9-5A*) An instrument containing a phosphor that emits light on exposure to ionizing radiation.

Second genetic code (*Section 26-5A*) The specific recognition by an enzyme, aminoacyl-tRNA synthetase, of its proper tRNA and amino acid.

Secondary (2°) alcohol (*Section 10-4A*) An alcohol in which the carbon atom bearing the —OH group is bonded to two other carbon groups.

Secondary (2°) amine (*Section 10-4B*) An amine in which nitrogen is bonded to two carbon groups and one hydrogen.

Secondary messengers (*Section 24-1*) Molecules that are created or released due to the binding of a hormone or neurotransmitter, which then proceed to carry and amplify the signal inside the cell.

Secondary structure of DNA (*Section 25-3*) Specific forms of DNA due to pairing of complementary bases.

Secondary structure of proteins (*Section 22-9*) Repeating structures within polypeptides that are based solely on interactions of the peptide backbone. Examples are the alpha helix and the beta-pleated sheet.

Secondary structures of proteins (*Section 22-9*) Repetitive conformations of the protein backbone.

Seeding (*Section 6-4B*) A process used to crystallize excess solute of a supersaturated solution by adding a crystal of the solute.

Selective Serotonin Reuptake Inhibitors (SSRIs) (*Chemical Connections 24G*) A common type of antidepressant that blocks reabsorption of the neurotransmitter serotonin.

Semiconservative (*Section 25-6*) Replication of DNA strands whereby each daughter molecule has one parental strand and one newly synthesized strand.

Semipermeable membrane (*Section 6-8C*) A substance that contains very tiny pores that only allow solvent molecules to pass through the pores while still retaining the solvated solute particles.

Sense strand (*Section 26-2*) The DNA strand that is not used as a template for transcription but has a sequence that is the same as the RNA produced. Also called the coding strand and the (+) strand.

Serotonin-Norepinephrine Reuptake Inhibitors (SNRIs) (*Chemical Connections 24G*) A type of antidepressant the blocks the reuptake of both serotonin and norepinephrine.

Severe Combined Immune Deficiency (SCID) (*Section 26-9*) A disease caused by several possible missing enzymes that leads to the organism having no immune system.

—SH (*Section 14-4A*) A sulfhydryl group.

Shells (*Section 2-6A*) All orbitals of a principal energy level of an atom.

Shine–Dalgarno Sequence (*Section 26-5*) A sequence on the mRNA that attracts the ribosome for translation.

SI (*Section 1-4*) International System of Units.

Sickle cell anemia (*Chemical Connections 22D*) A disease caused by a single amino acid substitution in normal hemoglobin that causes the red blood cells to form a sickle shape.

Side chains (*Section 22-7*) The unique part of an amino acid; the side chain is attached to the alpha carbon, and the nature of the side chain determines the characteristics of the amino acid.

Sievert (Sv) (*Section 9-6*) A biological measure of radiation. One sievert is the value of 100 rem.

Sieving portions (*Section 26-6*) Parts of a ribosome that allow only certain tRNA molecules to enter.

Sigmoidal (*Chemical Connections 22G*) Refers to an S-shaped curve on a graph.

Signal transduction (*Section 24-5*) A cascade of events through which the signal of a neurotransmitter or hormone delivered to its receptor is carried inside the target cell and amplified into many signals that can cause protein modification, enzyme activation, or the opening of membrane channels.

Significant figures (*Section 1-3*) Numbers that are known with certainty.

Silencer (*Section 26-6*) A DNA sequence that is not part of the promoter that binds a transcription factor suppressing transcription.

Single bond (*Section 3-7C*) A bond formed by sharing one pair of electrons; represented by a single line between two bonded atoms.

Small interfering RNA (siRNA) (*Section 25-4*) Small RNA molecules that are involved in the degradation of specific mRNA molecules.

Small nuclear ribonucleoprotein particles (snRNPs) (*Section 25-4*) Combinations of RNA and protein that are used in RNA splicing reactions.

Small nuclear RNA (snRNA) (*Section 25-4*) Small RNA molecules (100–200 nucleotides) located in the nucleus that are distinct from tRNA and rRNA.

Soap (*Section 18-4B*) A sodium or potassium salt of a fatty acid.

Solenoid (*Section 25-3*) A coil wound in the form of a helix.

Solidification (*Section 5-7*) The change of a substance from the liquid state to the solid state.

Solid state (*Section 5-1*) The state of matter where molecules at extremely low temperatures no longer have enough energy to move past each other.

Solids (*Section 1-6*) The forms of matter that have a definite shape and definite volume, and are essentially incompressible.

Solubility (*Section 6-4*) The maximum amount of solute that can be dissolved in a solvent at a specific temperature and pressure.

Solute (*Section 6-2*) The substance or substances that are dissolved in a solvent to produce a solution.

Solvated (*Section 6-6A*) When a solid ionic compound is dissolved in a solvent, the solvent molecules surround the ions when the combined force of attraction to the solvent molecules is greater than the force of attraction of the ionic bonds.

Solvent (*Section 6-2*) A liquid in which a solute is dissolved to form a solution.

Specific gravity (*Section 1-7*) The density of a substance compared to water as a standard.

Specific heat (*Section 1-9*) The amount of heat (calories) necessary to raise the temperature of 1 g of a substance by 1°C.

Specific rotation (*Section 15-4B*) The number of degrees [α] by which a chiral compound at a concentration of 1g/L in sample tube 10 cm long rotates the plane of plane-polarized light.

Specificity (*Section 31-1*) A characteristic of acquired immunity based on the fact that cells make specific antibodies to a wide range of specific pathogens.

Spectator ions (*Section 4-3*) Ions that appear unchanged on both sides of a chemical equation.

Sphingolipids (*Section 21-4*) Lipids that contain the alcohol sphingosine, two fatty acids, and a phosphate group.

Splicing (*Section 25-4*) The removal of an internal RNA segment and the joining of the remaining ends of the RNA molecule.

Standard temperature and pressure (STP) (*Section 5-4*) The pressure of one atmosphere and 0°C (273 K).

Step-growth polymerization (*Section 19-6*) A polymerization in which chain growth occurs in a stepwise manner between difunctional monomers—as, for example, between adipic acid and hexamethylenediamine to form nylon-66.

Step-growth polymers (*Section 19-6*) Polymers in which chain growth occurs in a stepwise manner between difunctional monomers, as for example between hexanedioic acid (adipic acid) and 1,6-hexanediamine to form Nylon 66.

Stereocenter (*Section 15-1*) An atom, most commonly a tetrahedral carbon atom, at which exchange of two groups produces a stereoisomer.

Stereoisomers (*Section 11-8*) Isomers that have the same connectivity (the same order of attachment of their atoms) but different orientations of their atoms in space.

Steroid hormones (*Section 24-7*) A class of hormone based on the steroid backbone of four fused rings.

Steroids (*Section 21-9*) Lipids with a characteristic fused-ring structure.

Stoichiometry (*Section 4-7*) The quantitative relationship between reactants and products in a chemical reaction as expressed by a balanced chemical equation.

(−) Strand (*Section 26-2*) The strand of DNA used as a template for transcription. Also called the template strand and the antisense strand.

(+) Strand (*Section 26-2*) The DNA strand that is not used as a template for transcription but has a sequence that is the same as the RNA produced. Also called the coding and the sense strand.

Strong acid (*Section 8-2*) An acid that ionizes completely in aqueous solution.

Strong base (*Section 8-2*) A base that ionizes completely in aqueous solution.

Strong electrolytes (*Section 6-6C*) Compounds that dissociate completely.

Structural formula (*Section 3-7C*) A formula showing how atoms in a molecule or ion are bonded to each other. Similar to a Lewis structure except that a structural formula shows only bonding pairs of electrons.

Structural genes (*Section 26-2*) Genes that code for the product proteins.

Sublimation (*Section 5-10*) A phase change from the solid state directly to the vapor state.

Subshells (*Section 2-6*) All the orbitals of an atom having the same principal energy level and the same letter designation ($s, p, d,$ or f).

Substance P (*Section 24-6*) An 11-amino acid peptidergic neurotransmitter involved in the transmission of pain signals.

Substrate (*Section 23-3*) The compound or compounds whose reactions an enzyme catalyzes.

Substrate specificity (*Section 23-1*) The limitation of an enzyme to catalyze specific reactions with specific substrates.

Subunit (*Section 23-6*) An individual polypeptide chain of an enzyme that has multiple chains.

Subunit vaccine (*Section 31-6A*) A vaccine made by injecting a host with pieces of a pathogen.

Sulfhydryl group (*Section 14-4A*) An —SH group.

Superposable (*Section 15-1*) In the context of molecular structure, to lay one structure on another and find that all like parts coincide.

Supersaturated solution (*Section 6-4*) A solution in which the solvent has dissolved an amount of solute beyond the maximum amount at a specific temperature and pressure.

Surface presentation (*Section 31-1*) The process whereby a portion of an antigen from a foreign pathogen that infected a cell is brought to the surface of the cell.

Surface tension (*Section 5-8A*) The layer on the surface of a liquid produced by the strength of the intermolecular attractions between the molecules of liquid at the surface layer.

Suspensions (*Section 6-7*) Systems where the size of colloidal particles is larger than 1000 nm, although the systems are unstable and separates into phases.

Synapse (*Section 24-2*) A small aqueous space between the tip of a neuron and its target cell.

T cell (*Section 31-1*) A type of lymphoid cell that matures in the thymus and that reacts with antigens via bound receptors on its cell surface. T cells can differentiate into memory T cells or killer T cells.

T-cell receptor (*Section 31-1*) A glycoprotein of the immunoglobulin superfamily on the surface of T cells that interacts with the epitope presented by the MHC (major histocompatibility complex).

T-cell receptor complex (*Section 31-5*) The combination of T-cell receptors, antigen, and cluster determinants (CD) that are all involved in the T cell's ability to bind antigen.

T form (*Section 23-6*) The form of an allosteric enzyme that is less active.

Tautomers (*Section 17-5*) Constitutional isomers that differ in the location of an H atom.

Template strand (*Section 26-2*) The strand of DNA used as a template for transcription. Also called the (−) strand and the antisense strand.

Termination (*Sections 26-2 and 26-5*) The final stage of translation during which a termination sequence on mRNA tells the ribosomes to dissociate and release the newly synthesized peptide.

Termination sequence (*Section 26-2*) A sequence of DNA that tells RNA polymerase to terminate synthesis.

Tertiary (3°) alcohol (*Section 10-4A*) An alcohol in which the carbon atom bearing the —OH group is bonded to three other carbon groups.

Tertiary (3°) amine (*Section 10-4B*) An amine in which nitrogen is bonded to three carbon groups.

Tertiary structure (*Section 22-11*) The overall 3-D conformation of a polypeptide chain, including the interactions of the side chains and the position of every atom in the polypeptide.

Tetrahedral (*Section 3-10*) Shape where a central atom is surrounded by four regions of electron density to atoms, and the maximum angle between any two regions of electron density is 109.5°.

Theoretical yield (*Section 4-7*) The mass of product that should be formed in a chemical reaction according to the stoichiometry of the balanced equation.

Theory (*Section 1-2*) The formulation of an apparent relationship among certain observed phenomena, which has been verified.

Thermal cracking (*Section 12-1*) A process in which a molecule or molecules are heated at a high temperature, which causes covalent bonds to break and smaller molecules to form.

Thiol (*Section 14-4A*) A compound containing an —SH (sulfhydryl) group bonded to a tetrahedral carbon atom.

Threose (*Section 15-3A*) A four-carbon-carbohydrate.

Thromboxanes (*Section 21-12*) Derivatives of 20-carbon arachidonic acid that contain a cyclic ether as part of their structure and are of pharmaceutical importance.

Tissue necrosis factor (**TNF**) (*Section 31-6*) A type of cytokine produced by T cells and macrophages that has the ability to lyse susceptible tumor cells.

Titration (*Section 8-9*) An analytical procedure whereby we react a known volume of a solution of known concentration with a known volume of a solution of unknown concentration.

TNF (*Section 31-6*) Tumor necrosis factor; a type of cytokine produced by T cells and macrophages that has the ability to lyse tumor cells.

Toluidine (*Section 16-2A*) A methyl-substituted aniline. Three constitutional isomers are possible: 2-methylaniline, 3-methylaniline, and 4-methylaniline, alternatively named *o*-toluidine, *m*-toluidine, and *p*-toluidine.

Trans (*Section 11-8*) A prefix meaning "across from."

Transamination (*Section 28-8*) The exchange of the amino group of an amino acid and a keto group of an α-ketoacid.

Transcription (*Section 25-4*) The process whereby DNA is used as a template for the synthesis of RNA.

Transcription factor (*Section 26-2*) Binding proteins that facilitate the binding of RNA polymerase to the DNA to be transcribed or that bind to a remote location and stimulate transcription.

Transesterification (*Section 19-4B*) Exchange of the —OR or —OAr group of an ester for another —OR of —OAr group.

Transfer RNA (tRNA) (*Section 25-4*) The RNA that transports amino acids to the site of protein synthesis on ribosomes.

Transferase (*Section 23-2*) A class of enzymes that catalyzes a reaction where a group of atoms such as an acyl group or amino group is transferred from one molecule to another.

Transition analogs (*Section 23-8*) Molecules that mimic the transition state of a chemical reaction and are used as enzyme inhibitors.

Transition elements (*Section 2-5A*) The elements in the B columns (Groups 3–12 in the new numbering system).

Transition state (*Sections 7-3 and 23-5*) An unstable species formed during the highest energy of a chemical reaction; a maximum on an energy diagram.

Translation (*Section 26-1*) The process in which information encoded in a mRNA is used to assemble a specific protein.

Translocated (*Section 26-5*) The part of translation where the ribosome moves down the mRNA a distance of three bases so that the new codon is on the A site.

Transmutation (*Section 9-3B*) Changing one element into another element.

Transporter (*Section 24-4C*) A protein molecule carrying small molecules such as glucose or glutamic acid across a membrane.

Transuranium elements (*Section 9-8*) Elements with atomic numbers greater than 92, are artificial, and have been prepared by a fusion process.

Triacylglycerol (*Section 21-2*) A kind of lipid formed by bonding glycerol to three fatty acids with ester bonds.

Triglycerides (*Section 21-2*) Kinds of lipid formed by bonding glycerol to three fatty acids with ester bonds.

Trigonal planar (*Section 3-10*) A shape where a central atom is surrounded by three regions of electron density to atoms, and the maximum angle between any two regions of electron density is 120°.

Triol (*Section 14-1B*) A compound containing three —OH (hydroxyl) groups.

Tripeptide (*Section 22-6*) A peptide made up of three amino acids.

Triple bond (*Section 3-7C*) A bond formed by sharing three pairs of electrons; represented by three lines between the two bonded atoms.

Triple helix (*Section 22-12*) The collagen triple helix is composed of three peptide chains. Each chain is itself a left-handed helix. These chains are twisted around each other in a right-handed helix.

Triple point (*Section 5-10A*) The temperature at which a substance exhibits all three phases of matter at one time.

Triprotic acid (*Section 8-3*) An acid that can give up three protons.

Trisaccharides (*Section 20-4*) Carbohydrates containing three monosaccharide units, each joined to the next by a glycosidic bond.

tRNA^{fmet} (*Section 26-5*) The special tRNA molecule that initiates translation.

Tumor suppression factor (*Chemical Connections 26F*) A protein that controls replication of DNA so that cells do not divide constantly. Many cancers are caused by mutated tumor suppression factors.

Tumor suppressor genes (*Chemical Connections 26E*) Genes that make proteins that control cell growth.

Tyndall effect (*Section 6-7*) Light passing through and scattered by a colloid viewed at a right angle.

Unsaturated (*Section 6-4*) A solution in which the solvent can dissolve additional solute at a given temperature.

Unsaturated aldehydes (*Section 17-2A*) Aldehydes that have within their structural formula a carbon–carbon double or triple bond.

Unsaturated hydrocarbon (*Section 11-1*) A hydrocarbon that contains one or more carbon–carbon double or triple bond or benzene ring.

Unwinding proteins (*Section 25-6*) Special proteins that help unwind DNA so that it can be replicated.

Urea cycle (*Section 28-8*) A cyclic pathway that produces urea from ammonia and carbon dioxide.

Vaccination (*Section 31-6*) The treatment of people with vaccines that stimulate their immune systems.

Valence electrons (*Section 2-6F*) Electron in the outermost occupied (valence) shell of an atom.

Valence shell (*Section 2-6F*) The outermost occupied shell of an atom.

Vapor pressure (*Section 5-8B*) The pressure of gas in equilibrium with its liquid form in a closed container.

Vesicle (*Section 24-2*) A compartment containing a neurotransmitter that fuses with a presynaptic membrane and releases its contents when a nerve impulse arrives.

Vitamins (*Section 30-6*) Organic substances required in small quantities in the diet of most species, which generally function as cofactors in important metabolic reactions.

VLDL (*Section 21-9*) Very-low-density lipoprotein.

Voltaic cell (*Chemical Connections 4B*) A device that uses a redox reaction to generate an electric current.

Volume (*Section 1-4*) The space that a substance occupies; the base SI unit is the cubic meter (m^3).

VSEPR model (*Section 10-3*) Valence-shell electron-pair repulsion model.

Water of hydration (*Section 6-6B*) The result of the attraction between ions and water molecules, where the water molecules are an integral part of the crystal structure.

Wavelength (λ) (*Section 9-2*) The distance from the crest of one wave to the crest of the next.

Weak acid (*Section 8-2*) An acid that is only partially ionized in aqueous solution.

Weak base (*Section 8-2*) A base that is only partially ionized in aqueous solution.

Weak electrolytes (*Section 6-6C*) Compounds that only partially dissociate.

Weight (*Section 1-4*) The result of a mass acted upon by gravity; the base unit of measure is a gram (g).

X-ray (*Section 9-2*) A type of electromagnetic radiation with a wavelength shorter than ultraviolet light but longer than gamma rays.

Zwitterions (*Section 22-3*) Molecules that have equal numbers of positive and negative charges, giving it a net charge of zero.

Zymogens (*Section 23-6*) Enzymes in an inactive form that become active after undergoing a chemical change; also called proenzymes.

The tendency of atoms to react in ways that achieve an outer shell of eight valence electrons is particularly common among Group 1A–7A elements and is given the special name of the **octet rule**. An atom with almost eight valence electrons tends to gain the needed electrons to have eight electrons in its valence shell and an electron configuration like that of the noble gas nearest to it in atomic number. In gaining electrons, the atom becomes a negatively charged ion called an **anion**. An atom with only one or two valence electrons tends to lose the number of electrons required to have an electron configuration like the noble gas nearest it in atomic number. In losing electrons, the atom becomes a positively charged ion called a **cation**. It is important to note that when an atom gains or loses electrons to achieve an outer shell of eight valence electrons, it still retains its same atomic number (Z) as noted in Chapter 2. That is, the resulting ion, which achieves the same electron configuration as a noble gas, is still uniquely characterized by its original atomic number, which identifies the number of protons in the nucleus. For example, although both Cl^- (Example 3-1) and Ar have the same number of 18 electrons, Cl^- has 17 protons in its nucleus while Ar has 18 protons in its nucleus, thus distinguishing between these two species. Therefore, when an ion forms, the number of protons (and neutrons) in the nucleus remains unchanged; only the number of electrons in the valence shell changes.

Noble Gas	Electron Configuration
He	$1s^2$
Ne	$[He]2s^2\,2p^6$
Ar	$[Ne]3s^2\,3p^6$
Kr	$[Ar]4s^2\,4p^6\,3d^{10}$
Xe	$[Kr]5s^2\,5p^6\,4d^{10}$

Octet rule When undergoing a chemical reaction, atoms of Group 1A–7A elements tend to gain, lose, or share sufficient electrons to achieve an electron configuration having eight valence electrons

Anion An ion with a negative electric charge

Cation An ion with a positive electric charge

EXAMPLE 3-1 The Octet Rule

Show how the following chemical changes obey the octet rule:

(a) A sodium atom loses an electron to form a sodium ion, Na^+.

$$Na \longrightarrow Na^+ + e^-$$
A sodium atom   A sodium ion   An electron

(b) A chorine atom gains an electron to form a chloride ion, Cl^-.

$$Cl + e^- \longrightarrow Cl^-$$
A chlorine atom   An electron   A chloride ion

Strategy

To see how each chemical change follows the octet rule, first write the condensed ground-state electron configuration (Section 2-6C) of the atom involved in the chemical change and of the ion it forms and then compare them.

Solution

(a) The condensed ground-state electron configurations for Na and Na^+ are:

$$Na\ (11\ electrons): 1s^2 2s^2 2p^6 3s^1$$

$$Na^+\ (10\ electrons): 1s^2 2s^2 2p^6$$

A Na atom has one electron ($3s^1$) in its valence shell. The loss of this one valence electron changes the Na atom to a Na ion, Na^+, which has a complete octet of electrons in its valence shell ($2s^2 2p^6$) and the same electron configuration as Ne, the noble gas nearest to it in atomic number. We can write this chemical change using Lewis dot structures (Section 2-6F):

$$Na\cdot \longrightarrow Na^+ + e^-$$
A sodium atom   A sodium ion   An electron

Chemical Bonds

Sodium chloride crystal

3-1 What Do We Need to Know Before We Begin?

In Chapter 2, we stated that compounds are tightly bound groups of atoms. In this chapter, we will see that the atoms in compounds are held together by powerful forces of attraction called chemical bonds. There are two main types: ionic bonds and covalent bonds. We begin by examining ionic bonds. To talk about ionic bonds, however, we must first discuss why atoms form the ions they do.

3-2 What Is the Octet Rule?

In 1916, Gilbert N. Lewis (Section 2-6) devised a beautifully simple model that unified many of the observations about chemical bonding and chemical reactions. He pointed out that the lack of chemical reactivity of the noble gases (Group 8A) indicates a high degree of stability of their electron configurations: helium with a filled valence shell of two electrons ($1s^2$), neon with a filled valence shell of eight electrons ($2s^2 2p^6$), argon with a valence shell of eight electrons ($3s^2 3p^6$), and so forth.

The tendency of atoms to react in valence electrons is particularly co and is given the special name of the valence electrons tends to gain the r in its valence shell and an electron nearest to it in atomic number. In ga atively charged ion called an **anion** electrons tends to lose the number c configuration like the noble gas nea trons, the atom becomes a positively tant to note that when an atom gai shell of eight valence electrons, it s as noted in Chapter 2. That is, the electron configuration as a noble ga original atomic number, which iden cleus. For example, although both C number of 18 electrons, Cl⁻ has 17 p tons in its nucleus, thus distinguishi when an ion forms, the number of p mains unchanged; only the number

EXAMPLE 3-1   The Octet

Show how the following chemical c
(a) A sodium atom loses an electro

$$Na \longrightarrow$$
A sodium   A
atom

(b) A chorine atom gains an electr

$$Cl \;\; +$$
A chlorine
atom   el

Strategy

To see how each chemical change f condensed ground-state electron c involved in the chemical change a them.

Solution

(a) The condensed ground-state e Na⁺ are:

Na (11 elect

Na⁺ (10 elect

A Na atom has one electron ($ one valence electron changes has a complete octet of electro the same electron configuratic atomic number. We can write structures (Section 2-6F):

$$Na\cdot \longrightarrow$$
A sodium   A
atom